Adaptive Radiations of
Neotropical Primates

Adaptive Radiations of Neotropical Primates

Edited by

Marilyn A. Norconk
Kent State University
Kent, Ohio

Alfred L. Rosenberger
Smithsonian Institution
Washington, D.C.

and

Paul A. Garber
University of Illinois
Urbana, Illinois

Plenum Press • New York and London

Library of Congress Cataloging-in-Publication Data

Adaptive radiations of neotropical primates / edited by Marilyn A.
  Norconk, Alfred L. Rosenberger, and Paul A. Garber.
      p.   cm.
    "Proceedings of a Conference on Neotropical Primates: Setting the
  Future Research Agenda, held February 26-27, 1995, in Washington,
  D.C."--T.p. verso.
    Includes bibliographical references and index.
    ISBN 0-306-45399-1
    1. Primates--Adaptation--Latin America--Congresses.   I. Norconk,
  Marilyn A. II. Rosenberger, Alfred L. III. Garber, Paul Alan.
  IV. Conference on Neotropical Primates: Setting the Future Research
  Agenda (1995 : Washington, D.C.)
  QL737.P9A36 1996
  599.8'045'098--dc20   96-41929
                                                         CIP

Proceedings of a conference on Neotropical Primates: Setting the Future Research Agenda,
held February 26 – 27, 1995, in Washington, D.C.

ISBN 0-306-45399-1

© 1996 Plenum Press, New York
A Division of Plenum Publishing Corporation
233 Spring Street, New York, N. Y. 10013

Printed in the United States of America

Warren G. Kinzey
1935 – 1994

PREFACE

This collection of 29 papers grew out of a symposium entitled "Setting the Future Agenda for Neotropical Primates." The symposium was held at the Department of Zoological Research, National Zoological Park, Washington D.C., on February 26–27, 1994, and was sponsored by the Wenner-Gren Foundation for Anthropological Research, Smithsonian Institution, and Friends of the National Zoo. We put the symposium together with two objectives: to honor Warren G. Kinzey for his contributions to the growing field of platyrrhine studies and to provide researchers who work in the Neotropics with the opportunity to discuss recent developments, to identify areas of research that require additional study, and especially to help guide the next generation of researchers.

The symposium provided the opportunity to recognize Warren as a mentor and collaborator to the contribution of the study of platyrrhines. Contributions to the book were expanded in order to provide a more comprehensive view of platyrrhine evolution and ecology, to emphasize the interdisciplinary nature of many of these studies, and to highlight the central role that New World monkeys play in advancing primatology. If this volume were to require major revisions after just one more decade of research, that would be a fitting testament to Warren's enthusiasm and his drive to continually update the field with new ideas and methods. Tributes to Warren and a list of his publications have been published elsewhere (Norconk, 1994, 1996; Rosenberger 1994, 1995). Warren's last book is due to be published very soon (Kinzey, in press).

Published studies on wild platyrrhines more than doubled from a total of 793 during the time period of 1973 to 1984, to 1,511 studies published from 1985 to 1993 (J. Prichard, Seattle Regional Primate Research Center, personal communication). By comparison, studies on wild cercopithecines declined by half during these same time periods. Enthusiasm for Neotropical research, in large part, has been fueled by the discovery of new extant species and fossil taxa, as well as an increased appreciation of the range of adaptive variation found in living platyrrhines. Indeed, long-term studies have yet to be completed on some genera (e.g. *Callimico goeldii*), and differences among populations of several species are yet to be explored. With habitat destruction continuing at a faster rate than new studies can be completed, there is a real danger that the ecology and evolutionary adaptations of some species will be mere footnotes in our records.

The first section of the book introduces the platyrrhines, their systematics (Schneider and Rosenberger) and their geographic distribution in far eastern Amazonia (Ferrari and Lopes), southern coastal Brazil (Rylands, da Fonseca, Leite, and Mittermeier), and northern South America (Norconk, Sussman, Phillips-Conroy). Readers are referred to Terborgh (1985) and chapters in Gentry (1990) for discussions of primate distributions in southern Amazonia and Middle America. Introductory essays on each of the four subfamilies ("Marmoset Misconceptions," "Critical Issues in Cebine Evolution and Behavior," "New Perspectives on the Pitheciines," and "On Ateline Evolution") were prepared by the editors and invited contributors, and precede the chapters in Sections II through V. These introductions were written with the intention of raising controversial or problematic issues relevant to each of the subfamilies. They provided us with the opportunity to interject our personal views and redirect scientific discourse into new, underdeveloped, or provocative areas of inquiry.

The range of research involving callitrichines is the broadest of any of the platyrrhine groups. Interest continues to grow as more and more distinct populations are identified and studied in the wild (Section II). Field studies of four species by Ferrari, Corrêa and Coutinho (*Callithrix aurita* and *Callithrix flaviceps*), Digby and Barreto (*Callithrix jacchus*), and Savage, Snowdon, and Giraldo (*Saguinus oedipus*) test traditional paradigms of monogamy, territorial behavior, and parental care that emerged from studies of captive animals. The work of these authors and others is essential if we are to understand the full range of behavioral variability of this very successful taxonomic group. Stafford and colleagues combine field and laboratory analysis to examine the locomotion of *Leontopithecus rosalia*, particularly in light of the animals' feeding specialization as extractive foragers. Garber and Dolins brought the "lab" to the field to test hypotheses derived from foraging theory. Power's laboratory experiments highlight distinctions among the larger and smaller callitrichines in the evolution of digestive physiology. Davis's work on locomotor anatomy places *Callimico goeldii* within the marmoset/tamarin clade, thus strengthening the view expressed by Schneider and Rosenberger (Section I) based on molecular and other morphological criteria.

Study of the cebines (Section III) has sufficiently matured so that there now exists a set of detailed long-term studies from which to examine differences between species as well as between populations. Fedigan, Rose, and Avila compare new data on demographics of *Cebus capucinus* after 10 years of study in Costa Rica with data collected on long-term studies of *Cebus olivaceus* in Venezuela. Boinski also takes a comparative approach to the vocalizations and travel patterns of two species of *Cebus* and two populations of *Saimiri*. Miller, using naturally occurring differences in seasonal food supply of Venezuelan *Cebus olivaceus*, evaluates individual feeding success rates in both large and small groups. The intense seasonality of forests in Argentina provided Janson with the opportunity to manipulate food supply and to better understand how foraging rules are used by *Cebus apella*. The morphological basis for the extractive foraging that characterizes *C. apella* is analyzed by Ford and Hobbs and compared to the "gracile" capuchins.

The authors of the chapters in Section IV focus their interests on seasonal influences on food abundance and feeding adaptations of the pitheciines. Wright examines the nocturnal adaptation of *Aotus* in the context of a feeding niche, comparing night monkeys with the very similar, but diurnal, titi monkeys. Müller describes the first long-term study of *Callicebus personatus* living in the most seasonal habitat of any titi, in southern Brazil. Norconk reports on the seasonal diets and the advantages of seed predation by white-faced and bearded sakis. Locomotor behavior of all three of the larger pitheciines are compared within the context of feeding and moving between feeding trees by Walker.

The chapters in the ateline section (Section V) are thematically diverse, but comparisons of ateline feeding ecology suggest that differences within and between taxa are closely tied to food availability and dispersion. This section contains both updated accounts of long-term studies (Glander and Teaford, Crockett, and Strier) and studies by Peres and Castellanos and Chanin of species that are less well known (*Lagothrix lagotricha* and *Ateles belzebuth*). Teaford and Glander report on the inter-relationships between pronounced seasonal changes in the Costa Rican dry forest, diet, and dental microwear in their long-term study of *Alouatta palliata*.   Castellanos and Chanin examine the influence of the more subtle seasonal changes in food availability in a wet forest in Venezuela in their feeding study of *Ateles belzebuth*, and Peres adds new information on the ranging patterns of woolly monkeys and compares them to other fission–fusion atelines.   Like the long-term studies of *Cebus capucinus* by Fedigan and her colleagues, the studies by Crockett (*Alouatta seniculus*) and Strier (*Brachyteles arachnoides*) present fine-grained analyses of reproductive opportunities and competition among adult females.

These papers were subjected to both internal and external reviews. We are very grateful for the care and thoroughness provided by both the contributors and the following external reviewers: Colin Chapman, David Chivers, Marian Dagosto, John Fleagle, Jeffrey French, Mary Lou Harrison, Richard Meindl, John Oates, Carlos Ruiz, Susan Shideler, Elizabeth Strasser, Fred Szalay, Suzette Tardif, and Hans Thewissen.   We thank Eileen Bermingham and Lisa Tuvalo at Plenum for their gentle prodding and careful editorial assistance.   Laura Cancino and Anthony Rylands translated the chapter summaries into Spanish or Portuguese, respectively, depending on the origin of the species in the morphological studies or location of the field study and home institution of principle authors.   We hope that the translations will be the first step in making these studies more accessible to our colleagues and particularly students in "habitat countries."

M. A. Norconk, A. L. Rosenberger, and P. A. Garber
Kent, Washington, D.C., and Urbana-Champaign

REFERENCES

Gentry, A. H. 1990. *Four Neotropical Forests.* Yale University Press, New Haven.

Kinzey, W.G. in press. *New World Primates: Ecology, Evolution, and Behavior.* Aldine de Gruyter, Hawthorne, New York.

Norconk, M.A. 1994. A personal remembrance. *Neotropical Primates* 2:19–20.

Norconk, M.A. 1996. Remembrance of Warren G. Kinzey (1935-1994). *Am. J. Primatol.* 38:281–284.

Rosenberger, A.L. 1994.   Warren G. Kinzey—A founding father of platyrrhinology. *Neotropical Primates* 2(4):18–19.

Rosenberger, A.L. 1995. Obituary: Warren Glenford Kinzey (1935-1994). *Am. J. Phys. Anthropol.* 97:207–211.

Terborgh, J. 1985.   The ecology of Amazonian primates, in G.T. Prance and T.E. Lovejoy (eds.), *Key Environments Amazonia*, pp. 284–304, Pergamon Press, Oxford.

CONTENTS

Section III. Critical Issues in Cebine Evolution and Behavior

Section IV. New Perspectives on the Pitheciines

Section V. On Atelines

SECTION I

Problems of Platyrrhine Evolution

MOLECULES, MORPHOLOGY, AND PLATYRRHINE SYSTEMATICS

H. Schneider[1]and A. L. Rosenberger[2]

[1]University Federal of Para
Center of Biological Sciences
Department of Genetics
Belem, Para 66075-900, Brazil
[2]Smithsonian Institution
Washington, DC 20008

INTRODUCTION

Phylogenetic perspective is gradually penetrating fields not always accustomed to the language of systematics, thanks, in part, to extended discussions of methodology (e.g., Harvey and Pagel, 1991; Brooks and McLennan, 1991). Two recent examples are studies of the evolution of social organization in primates: Garber's (1994) analysis of callitrichines and Di Fiore and Rendell's (1994) review of the primate order. As this welcome trend continues, the importance of classification, a reference system of ideas regarding evolution, phylogeny and adaptation, will also grow.

After decades of extensive debate, it is widely recognized that the Linnaean system of classification is imperfect in many respects. It is also no secret that taxonomic discontent is an intractable, permanent feature of evolutionary biology. The reasons for this involve analytical difficulties in choosing the best answers among several logically possible historical hypotheses, as well as issues of scholarship, and the use of a cumbersome system of linguistically and procedural rules associated with classification (that many consider arcane). In such an environment it is perhaps best for professionals to agreeably disagree. For practical purposes, however, it is also desirable to delineate classifications that are true to the better ideas, that are consistent with the broadest range of available evidence - behavioral, ecological, anatomical, paleontological, molecular, etc. This, we think, is very feasible for the platyrrhines, in spite of the fact that the co-authors of this chapter each advocate slightly different systematic schemes. The point is that classification is an organizing tool that plays a primary role in the understanding of adaptation and evolution. Given the exponential increase of information in many fields pertinent to platyrrhines, the importance of this tool will grow and its shape will be adjusted to accommodate new facts, types of analyses and interpretations.

Adaptive Radiations of Neotropical Primates
edited by Norconk *et al*. Plenum Press, New York, 1996

Central to the framework of a classification is a hypothesis of phylogeny. For the New World primates, a considerable research investment has been made in recent years to reconstruct the cladistic linkages of the modern genera. This is a critical step toward a full phylogeny, which would also specify the nature of the genealogical ties between living and extinct forms, i.e., both ancestral-descendant and collateral (sister group) relationships. The cladistic approach is a suitable way to begin. It offers a rich foundation for developing a heuristically sound classification. More, it is presently the only approach that can be applied to the certain types of data that cannot yet be extracted from fossils.

Our intention is to summarize views on the higher-level systematics and classification of the platyrrhines, emphasizing a synthesis of data developed by morphologists and new studies conducted in H. Schenider's lab employing DNA sequences. We particularly seek to accomplish the following: 1) demonstrate the growing concordance of molecular and morphological evidence; and 2) dispel the notion that platyrrhine higher-level systematics is in a state of disarray, which frequently finds its way into the literature, as in the following passage:

> "...the relationships of neotropical platyrrhine monkeys to other groups of primates and to each other remain perhaps the most poorly known for any major primate taxon" (Flynn et al., 1995).

The discussion emphasizes modern forms but with the conviction that the framework we establish accommodates the fossils as well.

Our preferred classifications of the extant groups are also presented (Table 1). Occasionally, these differ. We emphasize, however, that most of these differences reflect one of the most arbitrary features of the classification process, the selection of taxonomic ranks attributed to groups based on phylogenetic (non-arbitrary) criteria. This is a fact of life in the world of taxonomy. One of us may prefer to list a particular monophyletic group as a family but the other, while retaining the same etymology for that taxon, may consider it best classified as a subfamily. In such cases, when the taxonomic contents are precisely the same, we introduce taxa under both ranks (e.g., atelids/atelines; pitheciids/pitheciines; callitrichids/callitrichines) and, where necessary, employ more general common names (see Rosenberger, 1981) to simplify discussion. In the one place where our phylogenetic views are significantly incompatible, we use versions of the family-level term set off in quotation marks; "Cebidae" and "cebids". It appears likely to us that the rules of nomenclature will perpetually saddle us with a term deriving from "Cebidae", based on the genus *Cebus*. However, the composition of this taxonomic group (to include the nearest relatives of *Cebus*) has been a matter of dispute for more than a decade, and it will likely continue thusly well into the future.

TURNING POINT

Historical reviews (Hershkovitz, 1977; Rosenberger, 1981) mark the middle 1970s as the turning point when contemporary platyrrhine classification, a doctrine unchanged for generations (e.g., Pocock, 1925; Simpson, 1945; Napier and Napier, 1967), began to falter. Its heuristic value was undermined by revolutions in theory, method, and new scientific programs. The pressures of cladistic-based analyses, molecular systematics and new data from behavioral ecology, especially, proved that the familiar classification, which emphasized discontinuities among taxa (Table 2), was unworkable. Key to this re-

Table 1. Abbreviated classifications of the modern platyrrhine genera according to the authors

Schneider	Rosenberger
Family Atelidae	**Family Atelidae**
Subfamily Atelinae	Subfamily Atelinae
Tribe Atelini	Tribe Atelini
Ateles, Brachyteles, Lagothrix	*Ateles, Brachyteles, Lagothrix*
Tribe Alouattini	Tribe Alouattini
Alouatta	*Alouatta*
Family Pitheciidae	Subfamily Pitheciinae
Subfamily Pitheciinae	Tribe Pitheciini
Tribe Pitheciini	*Pithecia, Chiropotes, Cacajao*
Pithecia, Chiropotes, Cacajao	Tribe Homunculini
Tribe Callicebini	*Callicebus, Aotus*
Callicebus	
Family Cebidae	**Family Cebidae**
Subfamily Cebinae	Subfamily Cebinae
Cebus, Saimiri	*Cebus, Saimiri*
Subfamily Aotinae	Subfamily Callitrichinae
Aotus	*Callimico, Saguinus, Leontopithecus,*
	Callithrix, Cebuella
Subfamily Callitrichinae	
Callimico, Saguinus, Leontopithecus,	
Callithrix, Cebuella	

Table 2. Comparison of selected classifications of the modern genera

Martin, 1990	Fleagle, 1988	Hershkovitz, 1977
Family Cebidea	**Family Cebidea**	**Family Cebidea**
Subfamily Cebinae	Subfamily Cebinae	Subfamily Cebinae
Cebus, Saimiri	*Cebus, Saimiri*	*Cebus*
Subfamily Aotinae	Subfamily Aotinae	Subfamily Aotinae
Aotus, Callicebus	*Aotus, Callicebus*	*Aotus*
Subfamily Pitheciinae		Subfamily Pitheciinae
Pithecia, Chiropotes, Cacajao		*Pithecia, Chiropotes, Cacajao*
Subfamily Atelinae	Family Atelidae	Subfamily Atelinae
Ateles, Brachyteles, Lagothrix	Subfamily Atelinae	*Ateles, Brachyteles, Lagothrix*
	Ateles Brachyteles,	
	Lagothrix, Alouatta	
Subfamily Alouattinae	Subfamily Pitheciinae	Subfamily Saimiriinae
Alouatta	*Pithecia, Chiropotes, Cacajao*	*Saimiri*
Subfamily Callimiconinae		Subfamily Callicebinae
Callimico		*Callicebus*
		Subfamily Alouattinae
		Alouatta
Family Callitrichidae	**Family Callitrichidae**	**Family Callitrichidae**
Callithrix, Cebuella,	*Callithrix, Cebuella,*	*Callithrix, Cebuella,*
Leontopithecus, Saguinis	*Leontopithecus, Saguinis,*	*Leontopithecus, Saguinus*
	Callimico	
		Family Callimiconidae
		Callimico

vision was the recognition that marmosets and tamarins, often called Callitrichidae (formerly Hapalidae), were not an isolated, primitive line. Rather, it seemed likely to many investigators that callitrichines were a group sharing a complex of derived anatomical and behavioral features. Furthermore, it appeared they could be linked genealogically with other platyrrhine groups, based on a reinterpretation of the evolution of their unique blend of traits. The implications forced a paradigm shift, a reorganization of platyrrhine classification based on a phylogenetic perspective. Single or pairs of genera could no longer be treated routinely as subfamilies because of their anatomical differences. Rather, the ranks of family, subfamily, tribe and subtribe would be invoked to reflect genealogical coherence and adaptive continuity.

Not surprisingly, as the old structure toppled the higher-level systematics of the New World monkeys became the subject of rich debate. A variety of classifications were proposed, leading some workers to abstain altogether from selecting any one in particular. Today, there are camps supporting the division of platyrrhines into two or three families, with differing content (Table 2). However, in contrast to the prevailing views of prior generations, which split families into many subfamilies along the taxonomic lines of genera (see Hershkovitz, Table 2), there is a growing consensus that clusters of genera should be assigned to a smaller number of coherent subfamilies.

This new paradigm is a distinctly phylogenetic approach. There is a broad consensus which recognizes three monophyletic groups within Platyrrhini, with a relatively stable albeit "new" nomenclature. 1) The smallest-bodied, clawed monkeys, callitrichines (*Cebuella, Callithrix, Leontopithecus, Saguinus, Callimico*). 2) The specialized seed-predators, pitheciines (*Pithecia, Chiropotes, Cacajao*). 3) The largest-bodied, prehensile-tailed monkeys, atelines (*Alouatta, Lagothrix, Ateles, Brachyteles*). Within these groups there remains only a few stubborn cladistic problems, but none really affect the bigger picture of platyrrhine relationships and classification.

That there is universal consensus about the essential phylogenetics of these three groups is a significant shift and advance. For example, Hershkovitz (1977) divided callitrichines into two families, one alone for *Callimico*. He argued they were two lineages that may have arisen independently from non-platyrrhine stock. Similarly, *Alouatta* was long maintained, by almost unanimous consent, as a subfamily separate from *Lagothrix*, *Ateles*, and *Brachyteles*. Only the morphologically bizarre pitheciines have been consistently united as a group. St. George Mivart set them aside as a subfamily in 1865 and there they remained, until new ideas began to broaden the concept.

Concerning the overarching cladistics, the picture is indeed muddled by significant disagreements about the inter-relationships of the three clades (Fig. 1). Among morphologists focusing on the 16 modern genera, Rosenberger (1981 *et seq.*) and Ford (1986; Ford and Davis, 1992) place pitheciines close to the atelines, while Kay (1990) argues that atelines and callitrichines are a monophyletic group trichotomized by the unresolved position of *Aotus*. The specific causes of this disagreement are unclear. Rosenberger's data principally comes from the dentition and skull. Ford's data are derived from several long bones and tarsal elements, but also involve a reanalysis of Rosenberger's original dental characters. These dental features form the basis of Kay's study, too.

An important source of the discrepancies may be the methodological differences involved in these studies. Rosenberger employed a synthetic approach to character analysis, using distributional information such as in-group and out-group commonality, time, ontogeny in the few instances where the data exits, functional morphology and behavioral ecology to assess the evolution of each character or complex individually, and thus build

Figure 1. Hypotheses of the cladistic relationships of modern platyrrhines based on morphology, after Rosenberger (a), Ford (b) and Kay (c).

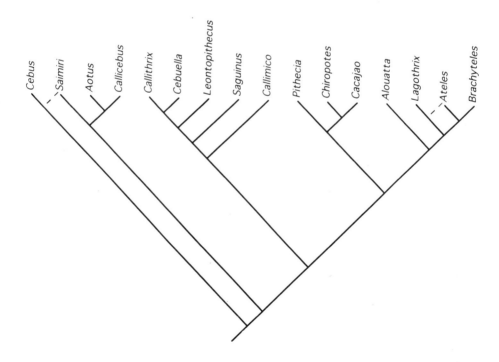

Figure 1b.

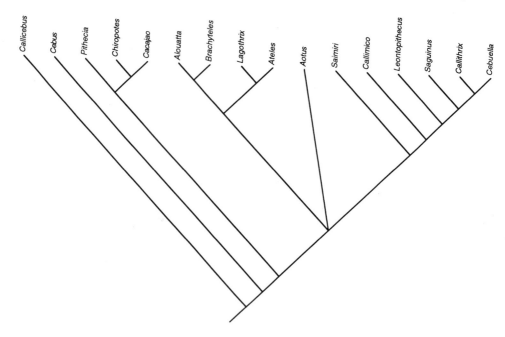

Figure 1c.

the tree. Both Ford and Kay, in contrast, employed numerical cladistic, parsimony algorithms to reconstruct relationships based on a matrix of traits.

The underlying systematics problem concerns the interpretation of four genera, *Cebus*, *Saimiri*, *Aotus*, and *Callicebus* (Fig. 1). Rosenberger (1981, 1984) proposed that *Cebus* and *Saimiri* are most closely related to the callitrichines, and *Aotus* and *Callicebus* to the saki-uakaris. Kay (1990) considered *Callicebus* and *Cebus* as two independent lineages that emerged one after the other at the beginning of the modern radiation, outside the more conventional groupings of callitrichines, atelines and pitheciines. Like Rosenberger, Kay also linked *Saimiri* with callitrichines but the affinities of *Aotus* were to remain indeterminate: Owl monkeys were rooted in a single trichotomous node shared also with atelines and callitrichines. Ford (1986; Ford and Davis, 1992) proposed two solutions for *Cebus* and *Saimiri*. One had them as independent clades, with *Cebus* a stem group of the radiation and *Saimiri* a relative of the *Aotus-Callicebus* clade. The alternative had *Cebus* and *Saimiri* united as the first-branching clade.

MOLECULAR EVIDENCE

Recently, Schneider's team analyzed the cladistic relationships of the 16 living New World monkey genera using 3.7 kilobases of DNA encompassing the whole Epsilon-globin gene (1.9 kilobases [kb]; Schneider *et al.*, 1993), and the intron 1 of the Interphotoreceptor Retinoid Binding Protein (IRBP) gene (1.8 kb; Schneider et al., in preparation). Procedures for extracting, amplifying and aligning DNA to prepare a data matrix are explained elsewhere (Sambrook *et al.*, 1989; Schneider *et al.*, 1993; Cabot and Beckenbach, 1989). Cladograms were constructed using a variety of the maximum parsimony algorithms (e.g., DNAPARS of Felsenstein, 1989; PTRALL, SURF and CONSEN of J. Czelusniak; see Schneider *et al.,* 1993 for descriptions). Distance matrix methods were also used (Kimura, 1980) to examine the data. Bootstrap analyses were performed with 2000 replications for both neighbor-joining (Saitou and Nei, 1987) and parsimony trees using the programs MEGA (Kumar *et al,* 1993) and PHYLIP 3.5 (Felsenstein, 1989). This is a re-sampling procedure that tests for consistency by iteratively deleting and replacing characters randomly and comparing the results. As shown in Figures 2 and 3, consensus trees (maximum parsimony and neighbor-joining trees, respectively) generated for the Epsilon and IRBP genes provide strong evidence for three major clades.

THREE MAJOR CLADES

Clade I - Atelids/Atelines

Cladistic arrangements obtained by both maximum parsimony and neighbor-joining distances unite *Alouatta* with *Ateles, Brachyteles,* and *Lagothrix.* Seventeen sites were identified at which *Alouatta* shares derived characters with the others, grouping this genus with the three other atelines in all bootstrap iterations. The monophyly of this group is also strongly supported by morphology (Rosenberger, 1981, 1984; Ford, 1986; Kay et al. 1987), as well as behavior and ecology (Rosenberger and Strier, 1989). The connection between *Alouatta* and the others has a long history (Rosenberger, 1981), although it has been uncommon until recently to assemble the four genera into one family or subfamily.

The first-branch position of *Alouatta* is corroborated cladistically by molecular and morphological data, with few exceptions. Dunlap *et al.* (1985) placed *Alouatta* as the sister group of *Lagothrix* on the basis of forelimb musculature, while Kay (1990; but see Kay et al., 1987) grouped *Alouatta* with *Brachyteles* on the basis of dental characteristics. Another discrepancy concerns the relationships of *Brachyteles.* Our molecular evidence suggests that *Brachyteles* and *Lagothrix* share a unique common ancestor, to the exclusion of *Ateles.* This *Brachyteles/Lagothrix* clade is supported by eleven base-pair synapomorphies, as opposed to three possibly derived sites that would link *Lagothrix* with *Ateles,* and four favoring a *Brachyteles/Ateles* clade. There are also cytogenetic similarities, recently found by determining and verifying the correct karyotype for *Brachyteles* (Koiffman and Saldhana, 1978) which was incorrectly reported in the literature since the early 1960s. *Brachyteles* and *Lagothrix* share a karyotype with 2n= 62 chromosomes, while the diploid number in *Ateles* varies from 2n=32 to 2n=34 (Dutrillaux, 1988; Pieczarka *et al.,* 1989).

Thus within atelids/atelines, these molecular and chromosomal data disagree with the morphology. Zingeser's (1973) view of an *Ateles-Lagothrix* sister-group has never been independently corroborated. The recent review by Rosenberger and Strier (1989) supported an *Ateles/Brachyteles* link based on an extensive series of derived postcranial traits and complexes, balanced by the demonstration of convergence in the dental morphology of *Alouatta* and *Brachyteles.* Ford's (1986) early analysis of the craniodental data produced an undefined *Ateles/Brachyteles/Lagothrix* trichotomy, which was later (Ford and Davis, 1992) revised to the *Ateles/Brachyteles* concept. Kay (Kay et al.. 1987; Kay, 1990) also discussed two alternative solutions to the internal relationships of this group. The *Ateles/Brachyteles* linkage proved most compelling, based on non-dental data.

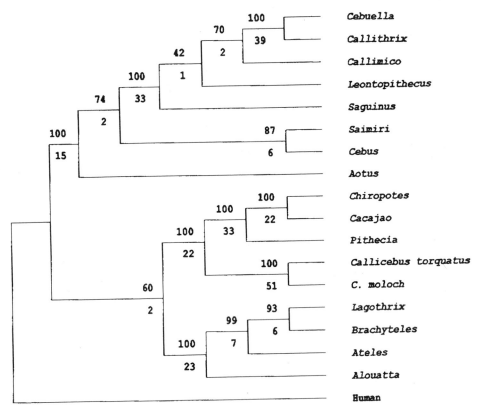

Figure 2. Phylogenetic tree produced by maximum parsimony method using DNAPARS from PHYLIP 3.5 (Felsenstein, 1989) and J. Czelusniak's programs (PTRALL, SURF, CONSEN). Numbers above nodes represent bootstrap values of 2,000 replications (SEQBOOT program). Numbers below the nodes are strength of grouping values (number of extramutations required to break the respective clade) estimated by J. Czelusniak's programs.

A potential source of the conflicting molecular and chromosomal information is the apparently very short span of time separating the emergence of the three lineages. This explanation is indicated by the very small differences in nucleotide diversity that separates the three genera. By comparison with the strongest generic dyads in the tree (Fig. 2), *Cacajao/Chiropotes* and *Cebuella/Callithrix*, the *Brachyteles/Lagothrix* linkage is less secure. Although the pair emerged as sister taxa in 93% of bootstrap iterations, it requires only 6 alternative substitutions to break the implied monophyly, compared with values of 100% & 39 and 100% & 22, respectively, for the other pairs. The *Saimiri/Cebus* link appeared in fewer iterations, 87%, and is supported at the same strength, requiring 6 "hits" to be broken.

Clade II - Pitheciids/Pitheciines

The cladistic arrangement obtained with molecular data clusters *Chiropotes* with *Cacajao* and *Pithecia* as the collateral relatives of the atelid/ateline clade, but with a twist.

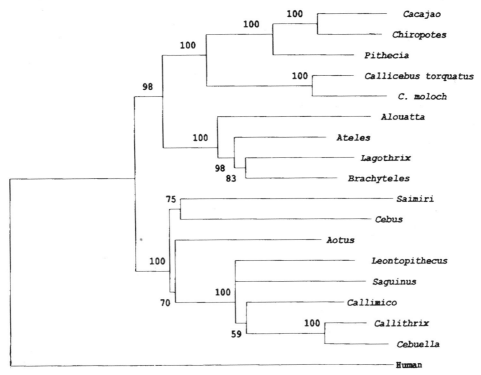

Figure 3. Phylogenetic tree obtained by the neighbor-joining method. Numbers at nodes represent bootstrap values pertaining to 2,000 replications.

The monophyly of the saki-uakaris seems unequivocal, based on morphological (Rosenberger, 1981; Ford, 1986; Kay, 1990; Kinzey, 1992; Hershkovitz, 1987) and biochemical markers (Schneider et al., 1995). Studies of immunological distances and karyotypes have reached similar conclusions using only two of the three genera; *Pithecia* and *Cacajao* (Cronin and Sarich, 1975, Dutrillaux, 1988) or *Cacajao* and *Chiropotes* (Baba *et al.*, 1979). Most previous arrangements have identified *Chiropotes* and *Cacajao* as the most closely related internal lineages, but de Boer (1974) proposed an alternative arrangement based on chromosomal data, grouping *Pithecia* (2n=46) with *Cacajao* (2n= 44 and 46), and identifying *Chiropotes* (2n=54) as the oldest member of the clade. It seems more likely that the unusual diploid number in *Chiropotes* is autapomorphic, having no immediate cladistic relevance.

Our molecular studies also group *Callicebus*, in all bootstrap iterations, as the sister-group to the saki-uakaris. This connection is as strong as that binding *Chiropotes* and *Cacajao*, involving 22 sites. Rosenberger (1981) came to the same conclusion based on dental morphology. He later modified the hypothesis by recognizing *Aotus* as a stem of the *Callicebus* lineage (Rosenberger, 1984; Rosenberger *et al.* 1990), a linkage endorsed by Ford and Davis (1992) and also acknowledged in many early classifications. As discussed above, the morphological evidence generated mixed views of the affinities of both

Callicebus and *Aotus* whereas the molecular evidence, while strongly supporting a *Callicebus*/saki-uakari clade, also suggests *Aotus* is related to the another group. We return to this matter below.

Clade III - "Cebids"

Callitrichines are the third group long recognized as a coherent unit, even before cladistic analysis deepened this understanding. Although our molecular data do not resolve well the internal cladistics of callitrichines, it confirms the morphological consensus that *Callimico* is genealogically integral to this group, a point no longer questioned seriously (but see Hershkovitz, 1977). The DNA data robustly supports the monophyly of the five callitrichine genera, at the same level of assurance found among the three monophyletic saki-uakaris (Fig. 2).

How callitrichines are related to other platyrrhines, on the other hand, is a matter of deep debate. Rosenberger argued that callitrichines are the sister-group of cebines; Ford regarded them as relatives of a pitheciine/ateline clade; Kay linked them to atelines. At one level, the DNA sequence evidence supports Rosenberger's interpretation (1981, 1990) that callitrichines and a *Cebus/Saimiri* clade form a monophyletic group. However, this node, which occurs in 74% of bootstrap trees, would collapse into an unresolved link if only two base-pair substitutions were altered.

A second matter relates to our finding that *Aotus* is universally linked with the callitrichine/cebine clade by the DNA sequences. This grouping has not been suggested previously by any of the morphological work. While it is not highly supported in comparison to the other higher taxa discussed previously (15 sites versus 33 in callitrichines, 33 in saki-uakaris, 22 in *Callicebus*/sakis-uakaris, 23 in atelines), it merits further consideration and testing. Alternative interpretations, such as aligning *Aotus* as the sister group of *Callicebus* using either Epsilon or IRBP data sets, would require a considerable increase in the number of nucleotide substitutions above those required by the most parsimonious trees (Harada et al., 1995). There are thus no potential synapomorphies in either gene that would specifically group *Aotus* with *Callicebus*, as some of the morphological data indicate (Rosenberger, 1984; Ford, 1986; Ford and Davis, 1992).

The internal bootstrap values within the "cebid" clade are also variable and, in three cases, they are relatively low. They average 79% over six nodes. In contrast, values among the six atelid internal nodes average 99%. So, although the basal nodes uniting these groups seem very different - 100% bootstraps & 15 substitutions at the *Aotus* linkage, and 60% bootstraps & 2 substitutions at the atelid dichotomy - our confidence is tempered by the weaker internal structure of the "cebid" branching sequence.

RELATIONSHIPS WITHIN THE MAJOR CLADES

Cebus and *Saimiri*

In our previous analysis of Epsilon-globin gene sequences (Schneider *et al.*, 1993), the most parsimonious tree grouped *Cebus, Saimiri, Aotus*, and the callitrichine clades as an unresolved tetrachotomy. However, after enlarging the data set with the IRBP sequences, we discovered that *Cebus* and *Saimiri* link strongly with one another (Harada et al., 1995). This corroborates one of the traditional taxonomic schemes that was confirmed cladistically by Rosenberger (1981) using craniodental and other evidence. Often overlooked is a body of neurological evidence which strongly corroborates the linkage of *Cebus* and *Saimiri* (Armstrong and Shea, in press) as well. Postcranial information (Ford, 1986) also supports a *Cebus/Saimiri* clade as an option (Ford and Davis, 1992), although Kay's (1990) numerical cladistic analysis of dental characters posits these genera as distant relatives: *Saimiri* as the sister-group to callitrichines deep within Platyrrhini, and *Cebus* as a mono-generic branch rooted at the second internal node of the platyrrhine radiation, outside 14 of the 16 modern genera. Ford's (1986) postcranial data, as well as her re-analysis of Rosenberger's morphological data set using numerical cladistics, allowed for a similar interpretation of *Cebus* as an early branch, as did the forearm muscle-based study of Dunlap et al., (1985).

In our view, the "outlier" hypothesis is untenable, for several reasons: 1) It is thoroughly inconsistent with the positive evidence of fine-grained analyses that seem to require a *Cebus/Saimiri* clade, based on a diversity of independent data sets (i.e., DNA, morphology, ecology, and behavior (see Janson and Boinski, 1989; Rosenberger, 1992). 2) A *Cebus/Saimiri* clade accords well with the broader picture of platyrrhine relationships and adaptations, which points to a close genealogical connection between (at least) *Saimiri* and the callitrichines. 3) Other evolutionary models can explain the anatomical and life-history differences that now distinguish these genera from their shared ancestral pattern (e.g., Hartwig, in press), differences which heretofore have led to doubt about their potentially close linkage. 4) The proposal that *Cebus* is an isolated outlier seems explicable by a likely methodological bias appearing in the three contravening studies where this was indicated. All used numerical cladistic procedures in which a key part of the analysis polarized traits by out-group comparison to catarrhines. In the studies of Ford and Kay this led to an over-emphasis on molar and long bone similarities of *Cebus* and parapithecids. In the case of Dunlap et al. (1985), the resemblances were between *Cebus* and living terrestrial cercopithecines. This comparative framework probably resulted in a failure to identify nonhomologies between in- and out-group, making some specialized features of *Cebus* appear more primitive than they are.

Callitrichines

The monophyly of the callitrichine clade (*Callithrix, Cebuella, Leontopithecus, Saguinus* and *Callimico)* is hardly disputed (Ford, 1986; Seuánez *et al.,* 1988, 1989), although the phylogenetic relationships between genera are still controversial. Martin (1990; Table 2) is the rare exception, classifying *Callimico* among the non-clawed monkeys while hinting it "...might in due course prove to be justifiable" (pg. 715) to reassign the species to the callitrichine group.

Ford (1986) and Rosenberger et al. (1990) considered *Leontopithecus* as the lineage most closely related to the *Callithrix-Cebuella* clade, followed by *Saguinus* and *Callimico.* Snowden (1994) corroborated this branching sequences in his analysis of long call vocalizations. On the other hand, Kay (1990) suggested *Saguinus* as the sister group of *Callithrix-Cebuella*, followed by *Leontopithecus* and *Callimico*. The latter view is also consistent with Garber's (1994) character analysis concerning the evolution of callitrichin mating and social systems.

The evidence obtained from Epsilon and IRBP sequences, while not very strong (70% of trees, 2 sites), places *Callimico* as the sister group of the *Cebuella-Callithrix* clade (Schneider, *et al.,* 1993), followed by *Leontopithecus* (42% of trees, 1 site) and

Saguinus (100%, 33 sites). As noted above, it is important to emphasize that this internal cladistic structure is not a robust outcome. There are no bootstrap values above 70% and two nodes are held together by only 1 or 2 sites (Figure 2), meaning only one or two mutations would be enough to change the position of *Leontopithecus* and *Saguinus*. In fact, since the base of the callitrichine tree is bounded by three taxa with relatively low rooting robusticity (33 for the *Saguinus* root, 1 for *Leontopithecus*, 2 for *Callimico*), and this particular topology has not appeared before in any of the morphological trees, it is likely that alternative solutions will be found with additional DNA data.

Callithrix and *Cebuella*

The sister-group relationship of *Callithrix* and *Cebuella* is unanimously agreed upon by nearly all the aforementioned researchers (Ford, Garber, Kay, Rosenberger, Snowdon), but not by Hershkovitz (1977). This linkage is the most robust of any based on our DNA sequences. Actually, the divergence values we obtained from IRBP nucleotide sequences for *Cebuella pygmaea* and *Callithrix jacchus* are within the range of values expected for species of the same genus, corroborating our previous findings with the Epsilon gene (Schneider et al., 1993). This re-opens discussion of the monophyletic status of the genus *Callithrix.*, for it is possible that *Callithrix jacchus* (and close allies) are more closely related to *Cebuella pygmaea* than to other species now classified in the genus *Callithrix*. Analyzing morphological data, both Moynihan (1976) and Rosenberger (1981) suggested that *pygmaea* should appear as a congener, a practice advocated by some systematists for years.

Recently, Nagamachi et al. (1992), observed that the karyotypes of *Cebuella pygmaea* (2n=44) and *Callithrix emiliae* (2n=44) differed from that of *Callithrix jacchus* (2n=46) by a pericentric inversion of chromosome 19 and a Robertsonian translocation (20/16 in *Cebuella* and 22/16 in *Callithrix emiliae).* *Cebuella* and the Amazonian *Callithrix emiliae* differ from each other by only a reciprocal translocation between an acrocentric autosome and the short arm of the submetacentric chromosome, which distinguishes their karyotype from that of *Callithrix jacchus*. Seuánez et al., (1989) also observed a close karyotypic resemblance between *Callithrix argentata* and *Cebuella*. In our opinion, the mounting evidence (morphological, cytogenetic and molecular) requires a detailed analysis of the interrelationships of these species, with the possibility that some will have to be shifted taxonomically from one genus to the other.

FOSSILS AND PHYLOGENY

Although our emphasis has been the modern forms, absent input from the fossil record, our picture of platyrrhine relationships and the material basis for a classification remains incomplete. Fossils hold their own clues to phylogenetic relationships. Their analysis serves to validate morphotype reconstructions by identifying intermediate morphological linkages between anatomically disparate modern forms, and by expanding our knowledge of morphological combinations. Below we discuss briefly how platyrrhine fossils contribute specifically to our knowledge of phylogeny. In this discussion, the temporal dimension provided by geological context is also highly pertinent. It may, for example, contribute explanations to straighten likely cladistic linkages, or clarify the ambiguities of others.

The earliest fossil platyrrhines (Rosenberger et al., 1991) shed little light on the relationships of later forms. The most that can be said of them now is they reinforce the likelihood that callitrichines are not the ancestral stock. Although negative evidence, this lends confidence to the prevailing interpretations of the major outlines of character evolution among New World monkeys, and their contingent cladistic hypotheses.

A striking picture that emerges from the fossil record beginning in the early Miocene is the number of the lineages and modern genera which appear to be long lived (Delson and Rosenberger, 1984). Cebines have been traced back to the early Miocene, when they are represented by *Dolichocebus* and possibly *Chilecebus* (see Flynn et al., 1995), and by *Leventiana* (Rosenberger et al., 1991b) from the middle Miocene. *Saimiri* is related to a middle Miocene species that we place in the same genus, *S. fieldsi* (Rosenberger et al., 1991a); others (e.g., Takai et al., 1994) agree with the phylogeny but maintain the original genus name for the species (*Neosaimiri*). We interpret *Callimico* as phylogenetically linked with *Mohanamico hershkovitzi* (Rosenberger et al. 1990) from the middle Miocene, although this point is disputed by Kay (1990). If correct, *Mohanamico* is the best preserved evidence demonstrating the existence of callitrichines in the fossil record. Saki-uakaris are easily recognized in the middle Miocene in *Cebupithecia sarmientoi,* whose precise affinities are still unclear. Earlier, less saki-like pitheciines are also known from Argentina (see Rosenberger et al., 1990). The genus *Aotus* is known by the species *A. dindensis*, although this interpretation is also debated (cf. Rosenberger et al., 1990; Kay ,1990). *Alouatta* relatives are well recognized in the middle Miocene by two species of *Stirtonia* (see Kay et al., 1987). It is also possible that one or both of these species could just as well be classified in the same genus as the living howler monkeys.

Is there a pattern in the fossil record that we can relate to DNA sequence data, or to other questions regarding platyrrhine cladogeny? Realistically, we understand too little to draw firm inferences. Nor would we expect simple explanations to fit complex historical puzzles. However, it would appear that the most confounding higher taxa, the close relatives of the *Callicebus, Aotus, Cebus* and *Saimiri* lines, are drawn from ancient lineages. They may be difficult to place not only because their phylogenetic signals have been obscured by the passage of time, but also because of the probability that a variety of intervening, intermediate clades have since become extinct and have not yet been resurrected in the form of a fossil. As noted below, a succession of early and rapid branching events may be difficult to tease apart with molecular data at this time.

PLATYRRHINE CLASSIFICATION: TWO OR THREE FAMILIES?

How to classify three distinct clades when their interrelationship are still murky? Our work with Epsilon sequences supports the view that the New World monkeys should be divided into two families; the Atelidae, with two subfamilies Atelinae (*Ateles, Brachyteles, Lagothrix* and *Alouatta*) and Pitheciinae (*Pithecia, Chiropotes, Cacajao* and *Callicebus*); and the "Cebidae", with *Aotus, Cebus, Saimiri*, and the callitrichines. Analyzed independently, the IRBP data places pitheciines closer to the "cebids", but the IRBP and Epsilon, when analyzed together, groups pitheciines with atelines, as suggested by the Epsilon gene. However, in this case the bootstrap values are low.

These results cannot dichotomously resolve the relationships among the three groups. Perhaps there is a message here regarding evolutionary tempo. For example, Nei (1986) proposed a method for inferring the number of cladistically informative nucleotides necessary to resolve three taxa into a dichotomous branching model. According to this method, the probability of finding the correct topology depends on the number of nucleotides examined and the trees' summed branch length. In our previous paper (Schneider *et al.*, 1993), we estimated the split between Cebidae and Atelidae occurred at 20 MYA and the Atelinae-Pitheciinae split at 17 MYA. If the temporal difference separating the origins of these two divisions were smaller than 3 MYA, we would need far more than 6,000 nucleotides to find the correct topology with a probability of 90%.

Still, there are no simple solutions to the question of classification. H. Schneider favors a three-family system while A.L. Rosenberger favors a two-family system, as summarized in Table 1. In either case, we agree on the composition of most taxonomic units below the family-level. These, we propose, have significant utility for a wide range of applications. Of great theoretical interest to us both is the alignment of *Aotus* with callitrichines and cebines based on the molecules. We are less worried about the inconsistencies of morphology- and molecule-based branching patterns within callitrichines.

CONCLUDING REMARKS

Relative to the morphological cladistic studies that have a longer history, we conclude that, despite several unanswered questions, cladistic analysis of Epsilon-globin and IRBP gene sequences provides important complementary information on the major genealogical outlines in the phylogeny of New World monkeys. These are quite consistent with evidence from the fossil record, which means that the modern forms provide a good basis for developing a platyrrhine classification, and also that understanding the relationships of fossils may best proceed by including the living genera in the analysis of fossils. The molecular and morphological studies strengthen the idea of three major modern groups, possibly diverging closely spaced in time.

Concerning what has been the muddle in the middle of platyrrhine systematics for decades - *Cebus, Saimiri, Aotus, Callicebus* - the combined evidence definitely places *Callicebus* as a relative of pitheciines. They reinforce the connection between *Saimiri* and callitrichines, the linkage of *Cebus* with *Saimiri*, and their association with callitrichines as a monophyletic "cebid" group. However, the data diverge in that the DNA adds *Aotus* as the stem group of this cluster, an interpretation that is inconsistent with morphology.

The DNA also points to the need for a reconsideration of the taxonomy of genus *Callithrix*, which may not be monophyletic. It partially confirms the branching patterns of the atelid clade, placing *Alouatta* as the oldest lineage. Problems remaining within callitrichines and atelines include: 1) precise affinities among the atelins, *Lagothrix, Ateles* and *Brachyteles* ; 2) the branching sequence within callitrichines, i.e., *Callithrix/Cebuella, Leontopithecus, Saguinus*, and *Callimico*.

ACKNOWLEDGMENTS

We would like to thank Dr. Eric Cabot for the sequence editor (ESEE200c), Dr. José Carneiro Muniz (Centro Nacional de Primatas - Belém, Pará, Brazil), Dr. Adelmar Coimbra-Filho and Dr. Alcides Pissinatti (Centro de Primatologia do Rio de Janeiro, Rio de Janeiro, RJ, Brazil), and Dr. Filomeno Encarnación, Projeto Peruano de Primatologia, Iquitos, Peru) for the samples used in this work. We also thank Dr. John Czelusniak for the parsimony programs PTRALL, CONSEN, SURF, and Steve Ferrari for his reading of the manuscript. This research was made possible thanks to grants from CNPq-Brazil (910043/91.4; 201142/91.0; 201596/92.0), NSF-USA(DEB9116098), and FINEP-Brazil (6.6.94.0034.00). A.L. Rosenberger's work on platyrrhine systematics has been supported by NSF, University of Illinois, the L.S.B. Leakey Foundation, the National Zoological Park and the Smithsonian Institution. He thanks the many curators and museums that made his research possible.

REFERENCES

Armstrong, E. and Shea, M.A. (in press) Brains of New and Old World Monkeys. In New World Primates: Ecology, Evolution: Behavior. (W.G. Kinsey, Ed.), Aldine, NY.

Baba, M.L., Darga, L., and Goodman, M. (1979). Biochemical evidence on the phylogeny of Anthropoidea. In "Evolutionary Biology of the New World Monkeys and Continental Drift" (R. Ciochon and A.B. Chiarelli, Eds.), pp. 423–443, Plenum Press, New York.

Brooks, D.R. and McLennan, D.H. (1991) *Phylogeny, Ecology, and Behavior: A Research Program in Comparative Biology*. University of Chicago Press, Chicago.

Beckenbach, A.T. (1989). Simultaneous editing of multiple nucleic acid and protein sequences with ESEE. *Comp. Applic. Biosc.* 5: 233–234.

Cabrera, A. (1958). Catalogo de los mamiferos de America del Sur. *Rev. Mus. Argentino Cienc. Nat. Bernardino Rivadavia* 4: 1–307.

Cronin, J. E., Sarich, V. S. (1975). Molecular systematics of the New World monkeys. *J. Hum. Evol.*: 4:357–375.

De Boer, L. E. M. (1974). Cytotaxonomy of the Platyrrhini (Primates). *Genen Phaenen,* 17:1–115.

Delson, E. & A.L. Rosenberger. (1984) Are there any anthropoid primate "living fossils"? In: *Casebook on Living Fossils*. N. Eldredge & S. Stanley, Eds. Fischer Publishers, New York, pp. 50–61.

Di Fiore, A. and Rendell, D. (1994) Evolution of social organization: A reappraisal for primates by using phylogenetic methods. *Proc. Nat. Acad. Sci.*, 91: 9941–9945.

Dunlap, S. S., Thorington, R. W. Jr., and Aziz, M. A. (1985). Forelimb anatomy of New World monkeys: myology, and the interpretation of primitive anthropoid models. *Am. J. Phys. Anthropol.* 6: 499–517.

Dutrillaux, B. (1988). Chromosome evolution in primates. *Folia Primatol.* 50: 134–135.

Felsenstein, J. (1989). PHYLIP - Phylogeny inference package (version 3.2). *Cladistics* 5:164–166.

Fleagle, J.G. (1988) Primate Adaptation and Evolution. Academic Press, New York., pp. xix + 486.

Flynn, J.J., Wyss, A.R., Charrier, R. and Swisher, C.C. (1995) An early Miocene anthropoid skull from the Chilean Andes. *Nature* 373: 603–607.

Ford, S.M. (1986). Systematics of the New World monkeys. In "Comparative Primate Biology, Volume I: Systematics, Evolution and Anatomy" (D.R. Swindler and J. Erwin, Eds.), pp. 73–135, Alan R. Liss, New York.

Ford, S.M. and Davis, L.C. (1992) Systematics and body size: implication for feeding adaptations in New World monkeys. *Am. J. Phys. Anthrop.* 88: 415–568.

Garber, P.A. (1994) Phylogenetic approach to the study of tamarin and marmoset social behavior. *Am. J. Primatol.* 34: 199–219.

Harada, M. L., Schneider, H., Schneider, M.P.C., Sampaio, I., Czelusniak, J., and Goodman, M. (1995). DNA Evidence on the Phylogenetic Systematics of the New World Monkeys: Support for the Sister-Grouping of *Cebus* and *Saimiri* from two Unlinked Nuclear Genes. *Molecular Phylogenetics and Evolution*. 4:331–349.

Hartwig, W.C. (in press) The effect of life history on the squirrel (Platyrrhine, *Saimiri*) monkey cranium. *Am. J. Phys. Anthropol.*

Harvey, P. and Pagel, M. (1991) *The Comparative Method in Evolutionary Biology*. Oxford University Press, Oxford.

Hershkovitz, P. *Living New World Monkeys*. University of Chicago Press, Chicago, 1977.

Hershkovitz, P. (1987) The taxonomy of South American sakis, genus *Pithecia* (Cebidae, Platyrrhini): a preliminary report and critical review with the description of a new species and a new subspecies. *Am. J. Primatol.* 12: 387–468.

Hill, W. C. O. (1962). *Primates, comparative anatomy and taxonomy. Vol.V. Cebidae, Part B*. Univ. Press, Edinburgh.

Janson, C.H. and Boinski, S. (1992) Morphological and behavioral adaptations for foraging in generalist primates. The case of the ebines. Am J. Phys. Anthrop. 88: 483–498.

Kay, R.F., Madden, R.H., Plaven, J.M., Cifelli, R.L. and Guerro-D, J. (1987) *Stirtonia victoriae*, a new species of Miocene Colombian primate. *J. Hum. Evolu.* 16: 173–196.

Kay, R.F. (1990). The phyletic relationships of extant and fossil Pitheciinae (Platyrrhini, Anthropoidea). *J. Human. Evol.* 19: 175–208.

Koiffman and Saldhana (1978) Cytogenetics of Brazilian monkeys. *J. Hum. Evol.* 3: 275–282.

Kimura, M. (1980). A simple method for estimating evolutionary rate of base substitution through comparative studies of nucleotide sequences. *J. Mol. Evol.,* 16:111–120.

Kinzey, W.G. (1992) Dietary and dental adaptations in the Pitheciinae. *Am. J. Phys. Anthrop.* 88: 499–514.

Kumar, S., Tamura, K., and Nei, M. (1993). MEGA: Molecular Evolutionary Analysis, version 1.2. The Pennsylvania State University, University Park, PA.

Martin, R.D. (1990) Primate Origins and Evolution. A Phylogenetic Reconstruction. Princeton University Press, Princeton, pp. xiv + 804.

Moynihan, M. (1976). The New World primates: adaptive radiation and the evolution of social behavior, languages, and intelligence. Princeton University Press, New Jersey.

Nagamachi, C.Y.; Pieczarka, J.C., and Barros, R.M.S. (1992). Karyotypic comparison among *Cebuella pygmaea, Callithrix jacchus* and *emiliae* (Callitrichidae, Primates) and its taxonomic implications. *Genetica*, 85: 249–257.

Napier, J.R., Napier, P.H. (1967). *A handbook of living primates*. Academic Press, New York.

Nei, M. (1986). *Stochastic errors in DNA evolution and molecular phylogeny.* Evolutionary perspectives and the New genetics. Alan R. Liss, pp. 133–147.

Pieczarka, J., Nagamachi, C.Y., and Barros, R.M.S. (1989). The karyotype of *Ateles paniscus paniscus* (Cebidae, Primates): 2n=32. *Rev. Brasil. Genet.* 12: 543–551.

Pocock, R.I. (1925) Additional notes on the external characters of some platyrrhine monkeys. Proc. Zoo. Soc. London, 1925: 27–47.

Rosenberger, A.L (1981). *Systematics: the higher taxa.* In "Ecology and behavior of Neotropical Primates, Vol. 1, (A.F. Coimbra-Filho and R. Mittermeier, Eds.), pp. 9–27, Academia Brasileira de Ciencias, Rio de Janeiro.

Rosenberger, A.L. (1983) Aspects of the systematics and evolution of marmosets. In: *A Primatologia no Brasil.* M.T. de Mello, Ed. Universidad Federal Districto Federal, Brasilia, pp. 159–180.

Rosenberger, A.L. (1984) Fossil New World monkeys dispute the molecular clock. *Journal of Human Evolution,* 13:737–742.

Rosenberger, A.L. (1992) The evolution of feeding niches in New World monkeys. American Journal of Physical Anthropology. 88:525–562.

Rosenberger, A.L., Hartwig, W.C., Takai, M., Setoguchi, T. and Shigehara, N. (1991a) Dental variability in *Saimiri* and the taxonoic status of *Neosaimri filedsi*, as early squirrel monkey from La Venta, Colombia. *Int. J. Primatol.* 12: 291–301.

Rosenberger, A.L., Setoguchi, T. and Hartwig, W.C. (1991b) *Laventiana annectens*, new genus and species: fossil evidence for the origins of callitrichine monkeys. *Proc. Nat. Acad. Sci.* 28: 315–356.

Rosenberger, A.L., W.C. Hartwig & R. Wolff. (1991) *Szalatavus attricuspis*, an early platyrrhine primate from Salla, Bolivia. *Folia Primatologica*, 56:221–233.

Rosenberger, A. L., Strier, K. B. (1989). Adaptive radiation of the ateline primates. *J. Hum. Evol.,* 18:717–750.

Rosenberger, A.L., Setoguchi, T., and Shigehara, N. (1990). The fossil record of callitrichine primates. *J. Human Evol.* 19: 209–236.

Saitou, N., and Nei, M. (1987). The Neighbor-joining method; a new method for reconstructing phylogenetic trees. *Mol. Biol. Evol.* 4:406–425.

Sambrook, J., Fritsch, E.F., and Maniatis, T. (1989). *Molecular Cloning*: A Laboratory Manual," 2nd ed., Cold Spring Harbor Laboratory Press, Cold Spring Harbor, New York.

Simons, E.L. *Primate evolution: an introduction to man's place in nature.* Macmillan, New York, 1972.

Simpson, G. G. (1945). The principles of classification and classification of mammals. *Amer. Mus. Nat. Hist.* 85: 1–350.

Schneider, H., Schneider, M.P.C., Sampaio, M.I.C.; Harada, M.L., Stanhope, M. and M. Goodman. (1993). Molecular phylogeny of the New World monkeys (Platyrrhini, Primates). *Molecular Phylogenetic and Evolution,* 2: 225–242.

Schneider, M.P.C., Schneider, H., Sampaio, M.I.C., Carvalho-Filho, N.M., Encarnación, F., Montoya, E. and Salzano, F.M. 1995 Biochemical Diversity and Genetic Distances in the Pitheciinae Subfamily (Primates, Platyrrhini). *Primates* 36:129–134.

Seuanez, H.N., Forman, L., and Alves, G. (1988). Comparative chromosome morphology in three callitrichid genera: *Cebuella, Callithrix*, and *Leontopithecus. J. Hered.* 79: 418–424.

Seuanez, H. N., Forman, L., Matayoshi, T., Fanning, T. G. (1989). The *Callimico goeldii* (Primates, Platyrrhini) genome: karyology and middle repetitive (LINE-1) DNA sequences. *Chromosoma,* 98:389–395.

Snowden, C.T. (1993) A vocal taxonomy of callitrichids. In *Marmosets and Tamarins. Systematics, Behaviour, and Ecology*. (A.B. Rylands, ed.), pp. 78–94, Oxford University Press, Oxford.

Takai, M. (1994) New specimens of *Neosaimri fieldsi* from La Venta, Colombia: a middle Miocene ancestor of living squirrel monkeys. *J. Hum. Evolu. 27*: 329–360.

Zingeser, M.R. (1973). Dentition of *Brachyteles arachnoides* with reference to alouattine and ateline affinities. *Folia Primatol.* 20: 351–390.

2

PRIMATES OF THE ATLANTIC FOREST

Origin, Distributions, Endemism, and Communities

Anthony B. Rylands,[1] Gustavo A. B. da Fonseca,[1] Yuri L. R. Leite,[2] and
Russell A. Mittermeier[3]

[1]Departamento de Zoologia, Instituto de Ciências Biológicas
Universidade Federal de Minas Gerais
31270–901 Belo Horizonte, Minas Gerais
and Conservation International
Avenida Abrahão Caram 820/302
31275–000 Belo Horizonte, Minas Gerais, Brazil
[2]Fundação Biodiversitas
Avenida do Contorno 9155
30110–130 Belo Horizonte, Minas Gerais, Brazil
[3]Conservation International
1015 Eighteenth Street, N. W.
Washington, D.C. 20036

THE ATLANTIC FOREST

The Atlantic forest extends from the north-east of Brazil, state of Rio Grande do Norte, and including inland forests (*brejos*) in the state of Ceará, south along eastern Brazil, through the southernmost state, Rio Grande do Sul, into the northeastern tip of Argentina in the province of Misiones and between the Rios Paraná, Uruguai and Iguaçu. In contrast to the Amazon, the Atlantic forest (*sensu strictu*) is typically upland, stretching along the coastal mountain chain, known as the Serra do Mar in the south between the states of Santa Catarina and northern Rio de Janeiro where it is very close to the sea, and the Serra da Mantiqueira and eastern slopes of the Serra do Espinhaço inland. It comprises a complex of vegetation types, which could be referred to as the "Atlantic forest complex", and includes principally: 1) evergreen humid tropical forest, divided into a) cool, humid, montane forest (altitudes 800 to 1,500–1,700 m), and b) lower-montane forest (altitudes 300–800 m) with deeper soils, a marked dry season, and lower humidity (except in the valleys and near to the coast where they receive orographic rainfall); and 2) inland semideciduous or dry forests (Rizzini, 1979; Rizzini *et al.*, 1988; Joly *et al.*, 1991). In terms of the primate communities, however, some other forest formations are also included in a broad definition of the Atlantic forest biome

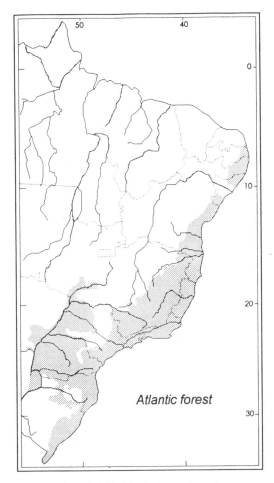

Figure 1. The Atlantic forest of Brazil.

(Fig. 1). These include: 1) coastal lowland forest on Quaternary sediments (*floresta dos tabuleiros*), floristically (genera and families) more similar to inland forest than to the montane forests, extending from the state of Pernambuco in the north to the northern part of the state of Rio de Janeiro (known as the *Hiléia Bahiana,* including as it does many otherwise Amazonian plant species), but also being found along part of the coast of the state of Paraná, some isolated areas along the coast of the state of Santa Catarina, and the mouth of the Rio Ribeira in São Paulo; 2) the *Araucaria* pine forest and Lauraceae forests of inland southern Brazil; and 3) the gallery forests extending way inland to the Rio São Francisco. These are considered distinct forest formations or subprovinces by Rizzini (1979; see also Rizzini *et al.*, 1988; Joly *et al.*, 1991). In the north-east of Brazil, the Atlantic forest is restricted to a narrow strip of tablelands along the coast. Inland the vegetation is primarily semiarid thorn scrub, although there are forest patches, known as *brejos*, occurring on massifs and isolated mountain ranges due to orographic rainfall (Andrade and Lins, 1964). These *brejos* have many Amazonian elements in their fauna and flora, and comprise the remnants of a widespread forest connecting the eastern Amazon with the Atlantic forest during the Holocene up to about 5,000 to 6,000 years ago (Bigarella, 1991).

When compared to the Amazon forests, there are certain important differences which need to be considered in terms of the ecological and physiological adaptations of the Atlantic forest primate communities. Whereas in Amazonia temperatures are high year-round, approximately 26–27°C (maximum 38.8°C and minimum 22°C), average temperatures in the Atlantic forest vary between 14°C and 21°C (maximum 35°C and minimum normally 1°C, but in the far south in mountainous regions may go below freezing). Soils also differ, with those of the Atlantic forest being generally derived from crystalline rock (granite or gneiss), being shallow and sandy with low fertility, whereas those of the Amazon basin are Tertiary sediments, intrinsically more fertile but generally heavily leached. The vegetation types and floristic composition (plant distributions) of the Atlantic forests are determined by topography affecting local temperatures and humidity. Depending on these factors, most especially humidity, lower-montane rain forests may be very similar to montane forest (Rizzini, 1979). The drier, more seasonal, semideciduous forests are less dense formations, less rich in species, and with lower canopies (18–20 m compared to 20–30 m and emergents up to 40 m in montane forest). These forests are typically found further inland bordering, or as isolated patches within, the central bush savanna (*cerrado*) of Brazil.

THE ATLANTIC FOREST PRIMATES

The Atlantic forest, covering approximately 1,272,539 km^2, is the third largest (after Amazonia and the Cerrado) and the second most speciose Brazilian biome (following Amazonia) in terms of mammal diversity, with approximately 229 species (Fonseca *et al.*, in press). Endemism of these mammalian species is very high, estimated at 32%, and a large part of this is due to the primates, which together with rodents (39 species) and marsupials (11 species), account for 84% of the Atlantic forest endemics (Fonseca *et al.*, in press). There are 24 primate species and subspecies currently recognized for the Atlantic forest. Following the classification of Rosenberger (1981), they include: two families and four subfamilies (Cebidae, Cebinae and Callitrichinae, and Atelidae, Atelinae and Pitheciinae); six genera, two of which are endemic, *Leontopithecus* and *Brachyteles*; 15 species, 11 of which are endemic; and 20 of the species and subspecies are endemic (Table 1). Endemism is, therefore, very high: one-third of the genera, 73% of the species, and 80% of the species and subspecies.

Species which also occur outside of the Atlantic forest include two Callitrichinae, *Callithrix penicillata* and *C. jacchus*. *C. penicillata* is essentially an inhabitant of the *cerrado* (bush savanna) of Central Brazil; including gallery forest, savanna forest (*cerradão*) and semideciduous forest patches. *C. jacchus* ranges throughout the north-east of Brazil, also occurring in gallery forest and forest patches in the semi-arid thorn scrub (*caatinga*). The capacity of these two species to occupy these relatively hostile habitats (high seasonality, long-periods of fruit shortage) is related to their tree-gouging and gum-feeding, which although common to all members of the genus is most highly developed, morphologically and behaviorally, in these two species (Rylands, 1984; Natori and Shigehara, 1992; Ferrari, 1993a). The tufted capuchins, *Cebus apella*, are highly adaptable and wide-ranging, occurring throughout a large part of tropical and subtropical South America. One of the endemic subspecies listed here, *C. a. xanthosternos*, may, however, warrant species status (Seuánez *et al.*, 1986). *Alouatta caraya and C. a. libidinosus* are wide ranging and typical of the Cerrado of Central Brazil, but may occur in some westernmost parts of the Atlantic forest complex. Finally, *Alouatta belzebul* is an Amazonian species, with the subspecies *belzebul* otherwise occurring east from the lower Tocantins basin to approximately 45° in the region of the basin of the Rio Mearim in Maranhão. Very few and small rem-

Table 1. Species and subspecies of Atlantic forest primates and their habitats. Taxonomy at family and subfamily level following Rosenberger (1981, 1992a; Rosenberger *et al.*, 1990). * Endemic

Species	Habitat
Cebidae - Callitrichinae	
*Callithrix aurita**	Montane forest, lower-montane forest
*Callithrix flaviceps**	Montane forest, lower-montane forest
*Callithrix geoffroyi**	Lower-montane forest, coastal lowland forest
*Callithrix kuhli**	Lower-montane forest, coastal lowland forest
Callithrix penicillata[1]	Semideciduous forest, gallery forest
Callithrix jacchus[1]	Semideciduous forest, coastal lowland, forest, gallery forest
*Leontopithecus rosalia**	Coastal lowland forest
*Leontopithecus chrysomelas**	Semideciduous forest, coastal lowland forest
*Leontopithecus chrysopygus**	Semideciduous forest
*Leontopithecus caissara**	Coastal lowland forest
Cebidae - Cebinae	
Cebus apella	
*C.a.nigritus**	Coastal lowland forest, lower-montane forest, montane forest, semideciduous forest, Lauraceae forest
*C.a.robustus**	Coastal lowland forest, lower montane forest, semideciduous forest
*C.a.xanthosternos**	Lower montane forest, semideciduous forest, gallery forest
C.a.libidinosus	Semideciduous forest, gallery forest
Atelidae - Pitheciinae	
*Callicebus personatus**	
*C.p.personatus**	Coastal lowland forest, lower montane forest, semideciduous forest, gallery forest
*C.p.melanochir**	Coastal lowland forest, lower-montane forest, semideciduous forest
*C.p.nigrifrons**	Lower montane forest, semideciduous forest, gallery forest
*C.p.barbarabrownae**	Semideciduous forest
C.personatus ssp. *[2]	Coastal lowland forest
Atelidae - Atelinae	
*Alouatta fusca**	
*A.f.fusca**	Coastal lowland forest, montane forest, lower-montane forest
*A.f.clamitans**	Coastal lowland forest, montane forest, lower montane forest, semideciduous forest, Araucaria forest, Lauraceae forest.
Alouatta belzebul belzebul	Coastal lowland forest
*Brachyteles arachnoides**	
*B.a.arachnoides**	Montane forest, lower-montane forest
*B.a.hypoxanthus**	Montane forest, lower-montane forest

[1]These species have been introduced into areas outside of their natural geographic range and in some cases have become established in other vegetation types, including montane forest, and lower montane forest.

[2]A subspecies described but not named by Kobayashi and Langguth (1994).

nant populations occur in north-east Brazil, north of the Rio São Francisco, and result undoubtedly from a relatively recent (Quaternary) connection of the Atlantic forest with eastern Amazonia (Bonvicino *et al.*, 1989; see below).

CURRENT PRIMATE DISTRIBUTIONS IN THE ATLANTIC FOREST

The analysis of the distributions of the Atlantic forest primates suffers from the widespread destruction of their natural habitats, along with, and particularly in the case of

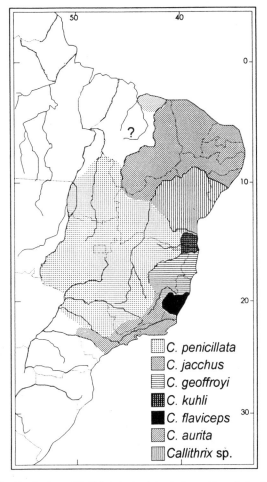

Figure 2. The distribution of *Callithrix* in the Atlantic forest (from Rylands *et al.*, 1993).

the Callitrichinae, introductions outside their natural range. Approximate distributions are shown in Figures 2–7.

Callithrix

All the *Callithrix* species are allopatric, although narrow hybrid zones have been re-corded for a number of localities at their range boundaries (Coimbra-Filho and Mitter-meier, 1973; Coimbra-Filho *et al.*, 1993a). *C. aurita* occurs in the montane forests of the south-east of the Atlantic forest, in the southern part of Minas Gerais, the state of Rio de Janeiro, and the east and north-east of the state of São Paulo (Rylands *et al.*, 1993). Hershkovitz (1977) marks the northern limit in Minas Gerais as the Rio Muriaé, but it has also been recorded in the Rio Doce State Park further north (Mittermeier *et al.*, 1982) and in the Serra do Brigadeiro further east (Ferrari and Mendes, 1991; Coimbra-Filho *et al.*, 1993a). The southeasternmost locality is the Rio Ribeira do Iguape in São Paulo, from which it extends west between the upper reaches of the Rios Tietê/Piracicaba and Paranapanema. The westernmost locality given by Hershkovitz (1977) is Boracéia, north-east of Bauru, on the upper Rio Tietê. *C. flaviceps*, is also a montane forest species, occurring to the north of *C. aurita* in the highlands of Espírito Santo, north as far as the Rio Doce, extending west into Minas Gerais as far as the municipality of Manhuaçu. It probably also occurred in northern Rio de Janeiro in the past. The southern part of the range of *C. geoffroyi* overlaps with *C. flaviceps*, and hybrid zones have been located between these species in Espírito Santo (Coimbra-Filho *et al.*, 1993a). It also occurs north of the Rio Doce, as far as the Rio Jequitinhonha and Rio Araçuaí in southern Bahia and northern Minas Gerais (Rylands *et al.*, 1988), although it is believed that the populations in the Rio Jequitinhonha valley arose from the release of animals at Belmonte at its mouth in 1975 (A.F.Coimbra-Filho, pers. comm.). Vivo (1991) limits it to the west of the Serra do Espinhaço (marking the western limit to the Atlantic forest) in Minas Gerais, where hybrids with *C. penicillata* have been recorded (Coimbra-Filho *et al.*, 1993a). *C. kuhli* occurs north of the Rio Jequitinhonha in the extreme north of Minas Gerais, and southern Bahia at least as far north as the Rio de Contas. Marmosets occurring north of the Rio de Contas, north to Salvador, have yet to be studied, showing characteristics of both *C. penicillata* and *C. kuhli* (Oliver and Santos, 1991). Coimbra-Filho *et al.* (1991) argued that true *C. kuhli* once occurred throughout the Atlantic forest of Bahia, but this is now difficult to ascertain due to deforestation (invasion by *C. penicillata* from the west) and a long history of casual introduction of *C. jacchus* from the north of the Rio São Francisco. *C. penicillata* is not, strictly speaking, an Atlantic forest marmoset. The bulk of its distribution is in the *cerrado* (bush savanna) of central Brazil. It is, however, taking a hold and probably replacing other species in numerous localities east and south (states of Minas Gerais, Espírito Santo and São Paulo) of its original range. *C. jacchus* is the species occurring throughout the Atlantic forest of north-east Brazil, north of the Rio São Francisco, and in forest patches remaining in the dry thorn scrub (caatinga). As with *C. penicillata*, a long history of introductions means that it is also now established in numerous areas throughout the south-east, including Rio de Janeiro.

Leontopithecus

The four species have very small and disjunct distributions. *L. rosalia* is now restricted to lowland forest patches in just four municipalities in the state of Rio de Janeiro (Kierulff, 1993). The black lion tamarin is known from five localities in the interior (west) of the state of São Paulo, south of the Rio Tietê, and north of the Rio Paranapanema (Valladares-Padua *et al.*, 1994). More populations remain of the golden-headed lion tamarin, *L. chrysomelas*, with records of over 100 localities through the region bounded by the Rio de Contas in the north and the Rio Jequitinhonha in the south. No evidence is available, however, for its occurrence between the lower Rios Padre and Jequitinhonha, and the south of the lower Rio de Contas (Pinto and Tavares, 1994). The distribution of the recently described black-faced lion tamarin, *L. caissara*, is restricted to part of the Island of Superagui, off the coast of the state of Paraná, and a small area of lowland forest on the continent, in the extreme north-east of Paraná and neighboring São Paulo (Martuscelli and Rodrigues, 1992; Lorini and Persson, 1994).

Callicebus

The northern subspecies, *C. p. barbarabrownae*, occurs south of the Rio Itapicuru, in the state of Bahia, north to the Rio São Francisco (Hershkovitz, 1990; Rylands 1994c).

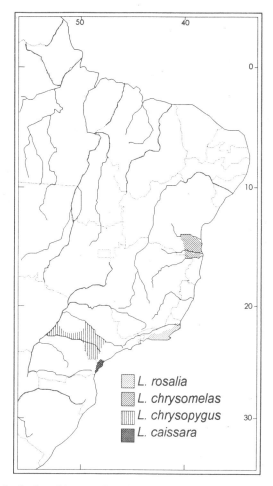

Figure 3. The distribution of *Leontopithecus* in the Atlantic forest (from Rylands *et al.*, 1993).

Kobayashi and Langguth (1994), however, have recorded a new subspecific form with markedly different pelage coloration in the state of Sergipe, near the Rio São Francisco, and as such the northernmost record for the species. *C. p. melanochir* ranges south of the Rio Paraguaçu, extending through southern Bahia, to near to the limits with Espírito Santo, along the lower reaches of the Rio Mucuri (Oliver and Santos, 1991). *C. p. personatus* occurs in Espírito Santo from the region of the lower Rio Mucuri, south into northern Rio de Janeiro (Hershkovitz, 1990). It also extends north to the middle Rio Jequitinhonha in Minas Gerais. Rylands *et al.* (1988) presented evidence that *C. p. personatus* occupies a larger part of the north of Minas Gerais than was indicated by Hershkovitz (1990), extending west along both margins of the middle and upper Jequitinhonha, at least as far as the locality of Buenopolis. *C. p. nigrifrons* is the widest ranging of the subspecies, occurring throughout a large part of the state of Minas Gerais and São Paulo, north of the Rio Tietê, and east of the Rios Paraná and Parnaíba (Kinzey, 1982; Hershkovitz, 1990).

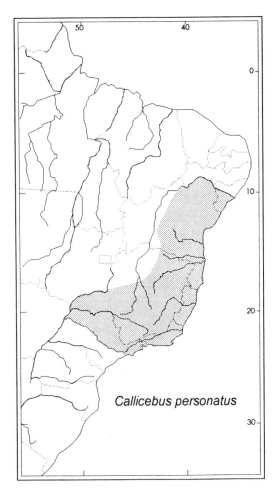

Figure 4. The distribution of *Callicebus personatus* in the Atlantic forest (from Hershkovitz, 1990; Rylands, 1994c).

Cebus

The recent literature has recognized only four subspecies of *Cebus apella* for the Atlantic forest region of Brazil: *C. a. libidinosus*, *C. a. nigritus*, *C. a. robustus*, and *C. a. xanthosternos* (see, for example, Kinzey, 1982; Mittermeier *et al.*, 1988). The black-horned capuchin, *C. a. nigritus,* is the form recognized south of the Rio Doce, in the states of Minas Gerais and Espírito Santo, Brazil. The type locality is Rio de Janeiro (= Serra dos Órgãos), although no type is preserved (Hill, 1960). Kinzey (1982), indicated a distribution which extended south from the Rio Doce in Minas Gerais and Espírito Santo, and south from the Rio Grande in Minas Gerais and São Paulo, extending throughout the states of Rio de Janeiro, São Paulo, Paraná, Santa Catarina, and Rio Grande do Sul, east of the Rio Paraná, and extending into the Misiones Province in Argentina. The robust tufted capuchin, *C. a. robustus*, occurs between the Rio Jequitinhona in the north and the Rio Doce in the south.

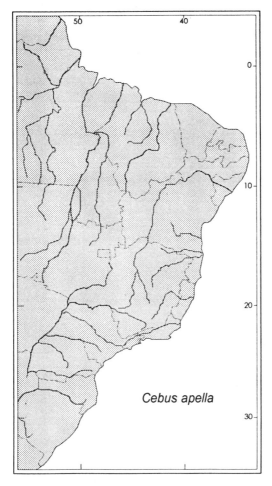

Figure 5. The distribution of *Cebus apella* in the Atlantic forest (from Hill, 1960; Kinzey 1982).

Hill (1960), however, described its distribution as eastern Brazil, from southern Bahia (Rio Jucurucú), through Espírito Santo (Colatina, Rio Piracicaba, Rio Sussuí, Rio Doce) to Rio de Janeiro and westwards into Minas Gerais (Rio Matipo). However, Kinzey's (1982) analysis placed the form *C. a. nigritus* to the south of the Rio Doce in Minas Gerais and Espírito Santo, extending into Rio de Janciro and São Paulo; a scheme followed by Oliver and Santos (1991). Torres (1988) also identified clearly different forms either side of the Rio Doce. The Rio Matipó locality cited by Hill (1960) is south of the Rio Doce (above the confluence with the Rio Piracicaba), and Kinzey (1982) regarded skins collected from there by P.Fonseca in 1919 as hybrids between *C. a. nigritus* and *C. a. robustus*. It would seem that the range of *C. a. robustus* extends north of the Rio Jucurucú in Bahia, as far as the Rio Jequitinhonha (Rylands, *et al.*, 1988; Oliver and Santos, 1991). The westernmost locality in the state of Minas Gerais is given by Pinto (1941), who obtained specimens from the headwaters of the Rio Pissarão, a mountainous region north of the Rio Piracicaba, not far from the town of Presidente Vargas. It is possible that the Serra do Espinhaço of Minas Gerais, running north-south and defining the transition from forest to bush savanna (*cerrado*) in the west, marks the western limits of the distribution of this form. Further west (and west of the Serra do Espinhaço), Kinzey (1982) recorded a specimen from Tomás Gonzaga, near Corinto, which he listed as *C. a. robustus* (= *C. a. robustus* x *C. a. libidinosus*).

The distribution of the yellow-breasted capuchin, *C. a. xanthosternos*, probably extends throughout the region north of the Rio Jequitinhonha north and west to the Rio São Francisco in Bahia, southern Sergipe, and parts of northern Minas Gerais, wherever suitable habitat is available (Coimbra-Filho *et al.*, 1991; Oliver and Santos, 1991; Rylands and Pinto, 1994). A record from the Rio Jucurucú, to the south of the Rio Jequitinhonha (C.A.Camargo in 1932; see Kinzey, 1982) is not *C. a. xanthosternos* (R.A. Mittermeier, unpubl. data), being the domain of *C. a. robustus*, also collected from this locality.

Hill (1960) described the distribution of *C. a. libidinosus* as east central Brazil along the left bank of the Rio São Francisco, which separates it from *C. a. xanthosternos*. This range includes western Minas Gerais and part of western Bahia, and extends north through the north-east states of Sergipe, Piauí, Pernambuco, Natal, and Ceará to Maranhão.

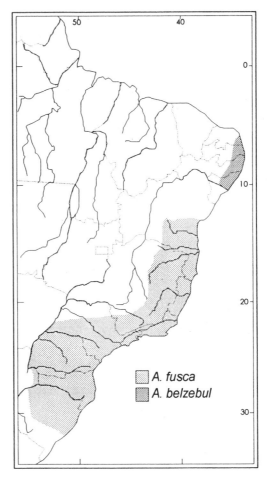

Figure 6. The distribution of *Alouatta fusca* and *Alouatta belzebul* in the Atlantic forest (from Langguth *et al.*, 1987; Bonvicino *et al.*, 1989; Rylands, 1994a, 1994b).

Alouatta

Alouatta fusca occurs throughout a large part of the Atlantic forest, south from southern Bahia, probably as far north as the Rio Paraguaçu in the past, although it is now practically extinct in the state, some few isolated populations remaining in the far south near the border with Minas Gerais. It extends through the states of Espírito Santo, eastern Minas Gerais, Rio de Janeiro, São Paulo, Paraná, Santa Catarina and Rio Grande do Sul into Argentina in the province of Misiones (Kinzey, 1982; Rylands, 1994b; Bitteti *et al.*, 1994). Kinzey (1982) indicated that the northern brown howler, *A. f. fusca*, occurred north of the Rio Doce in Minas Gerais and Espírito Santo, but recent evidence indicates that the southern form, *A. f. clamitans,* occurs north as far as the Rio Jequitinhonha, and if the sub-species are valid, *A. f. fusca* may be restricted to southern Bahia, and indeed the most endangered of all Atlantic forest primates and on the verge of extinction (Rylands *et al.*, 1988).

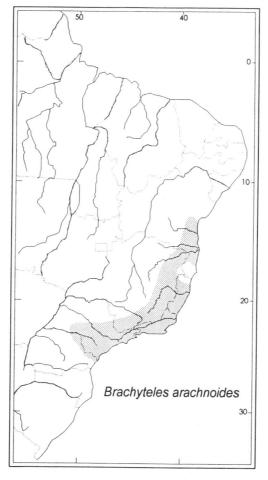

Figure 7. The distribution of *Brachyteles arachnoides* in the Atlantic forest (from Aguirre, 1971; Fonseca, 1994; Martuscelli *et al.*, 1994).

The Atlantic forest populations of the otherwise Amazonian red-handed howling monkey, *A. belzebul belzebul*, are critically endangered. A few, very small and isolated populations are known from the northeastern states of Ceará, Paraíba, Alagoas and Pernambuco, all north of the Rio São Francisco (Langguth *et al.*, 1987; Bonvicino *et al.*, 1989). *Alouatta caraya*, occurring in central Brazil, Paraguay and northern Argentina, is known to occur in some of the westernmost parts of the Atlantic forest in, for example, the state of São Paulo.

Brachyteles

The original range of *Brachyteles arachnoides* extended from southern Bahia, perhaps as far north as the Rio Paraguaçu, south in montane forests (600–1800 m altitude) through the states of Espírito Santo, eastern Minas Gerais, Rio de Janeiro and São Paulo, and the northern part of Paraná (Aguirre, 1971; Mittermeier *et al.*, 1987; Fonseca, 1994, Martuscelli *et al.*, 1994). This excludes a hiatus, extending between the north bank of the Rio Doce and the extreme south of Bahia, where it would seem that it has never existed (Aguirre, 1971). The southernmost locality recorded by Martuscelli *et al.* (1994) is the Morro Três Pontões, Guaraqueçaba, in the extreme north-east of Paraná. The northern subspecies, *B. a. hypoxanthus* occurs north of the Serra da Mantiqueira, in southern Minas Gerais, whereas the southern subspecies, *B. a. arachnoides* occurs south of this mountain range in São Paulo, Rio de Janeiro and north Paraná (Lemos de Sá *et al.*,1993; Coimbra-Filho *et al.*, 1993b). Morphological and genetic differences between the two forms of *Brachyteles* indicate that they should be considered distinct species (Lemos de Sá *et al.*, 1990, 1993).

COMPOSITION AND STRUCTURE OF PRIMATE COMMUNITIES

The maximum number of species comprising an Atlantic forest primate community is six, which could be found in the past in southern Bahia: *Callithrix kuhli, Leontopithecus chrysomelas, Callicebus personatus, Cebus apella, Alouatta fusca* and *Brachyteles arachnoides*, although both of the latter two species are now practically extinct in the region. This number may be reduced to five when considering that *Brachyteles* is an upland forest species and *Leontopithecus* occurs only in lowland forests, and therefore may never have actually occurred together (Table 1). Five is the norm for the richest communities in the south-east, for example, the Fazenda Barreiro Rico in São Paulo, the Rio Doce State Park in Minas Gerais, and the Augusto Ruschi Biological Reserve (formerly called Nova Lombardia) in Espírito Santo, all of which have a full complement of *Callithrix, Callicebus, Cebus, Alouatta* and *Brachyteles* (Mittermeier *et al.* 1982; Pinto *et al.*, 1993). In comparison to communities in the upper Amazon, for example, that of the Atlantic forest is, therefore, consistently depauperate (Rylands, 1987; see Fig. 8).

Rosenberger (1992a) provided a detailed classification of the adaptive zones of Neotropical primates in terms of the functional morphology of the dentition, diet, body size, and locomotion. His classification is followed here (Table 1), and gives the richest possible primate community (e.g., southern Bahia) as the following: 1) *Callithrix* - small, canopy-subcanopy feeders, clingers and leapers, partially scansorial, with an essentially faunivorous-frugivorous diet, but with exudates (gums) supplementing or replacing fruits at times of scarcity (notably in *C. jacchus* and *C. penicillata*); 2) *Leontopithecus* - small, canopy-subcanopy feeder, clingers and leapers, partially scansorial, with an essentially

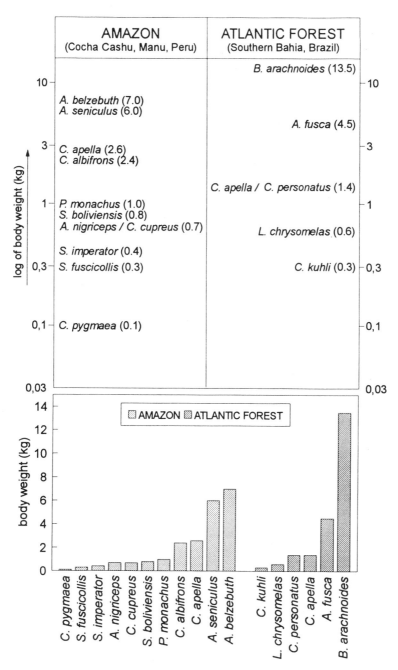

Figure 8. A comparison of the sizes of primates occurring in an Amazonian community (Cocha Cashu, Manu National Park, Peru; from Terborgh, 1983) and in southern Bahia, Brazil. Notes: 1) Weight of *Callithrix kuhli* as for *C. penicillata* in Rosenberger (1992a); 2) Approximate weight of *Callicebus p. personatus* is that of an adult female given by Hershkovitz (1990); 3) Weight of *Cebus apella xanthosternos* is that of two adult males given in Hill (1960); 4) Taxonomy of *Pithecia* according Hershkovitz (1983); 5) Taxonomy of *Saimiri* according to Hershkovitz (1984); Taxonomy of *Ateles* according to Froehlich *et al.* (1991).

faunivorous-frugivorous diet; *Callicebus* - small quadrupedal, canopy-subcanopy feeders, sclerocarpic frugivores including leaves and insects; *Cebus* - middle-sized, quadrupedal, canopy-feeding, omnivores and specialists on a diet of hard objects (durophagy, see Daegling, 1992); *Alouatta* - large, suspensory climbing, showing deliberate quadrupedalism, canopy-feeding folivores; and *Brachyteles* - large, suspensory climbing, brachiating, canopy-feeding, frugivore-folivores. Syntopy between *Callithrix* and *Leontopithecus* occurs only in southern Bahia, and is possible through different home range sizes (larger in *Leontopithecus*) and consequently differing diet diversity, and the exploitation of different small animal prey faunas, through contrasting foraging methods (Rylands, 1989).

The vegetation types occupied by the Atlantic forest primates are listed in Table 1. Whereas many of the species are apparently eclectic in their preferences (*Callithrix penicillata, C. jacchus, Cebus apella, Callicebus personatus* and *Alouatta fusca*), some are essentially lowland forest species (*Leontopithecus* spp., *Callithrix geoffroyi, Callithrix kuhli*) and others upland, montane forest species (*Callithrix aurita, C. flaviceps*, and *Brachyteles arachnoides*).

The size range of these primates is illustrated by the community in southern Bahia, and compared to that of Cocha Cashu in the Manu National Park (Terborgh, 1983). As pointed out by Rosenberger (1992b), there is a lack of middle-sized primates in the Atlantic forest, with the body weight range of 1.5–3.5 kg undifferentiated. This is due largely to the lack of *Chiropotes, Cacajao* and *Pithecia* and *Cebus albifrons/nigrivittatus*. Atlantic forest *Callicebus* are larger than *Callicebus cupreus* (although similar in size to *Callicebus torquatus*).

THE ORIGIN OF THE ATLANTIC FOREST PRIMATES

The origin of the highly endemic Atlantic forest primate communities can be understood when considering geomorphologic, climate and vegetation changes during the Cenozoic, from the middle Tertiary and during the Quaternary periods involving 1) connections and separations of the Amazonian forest with the Atlantic forest, 2) the fragmentation of the Atlantic forest through climatic fluctuations, the "refuge theory" (Haffer, 1969, 1993; Simpson and Haffer, 1978; Kinzey, 1982), and 3) habitat and feeding specialization of the primates, and changes in the forest vegetation types (not necessarily involving fragmentation) and their distributions during short climatic fluctuations due to small changes in mean annual temperatures and regional rainfall regimes during the Holocene. The first of these is considered to be relevant in terms of providing the ancestry of the Atlantic forest primates, the second and third in terms of speciation processes.

1) Amazonian Connections

Evidence from the distributions of plant genera and species, and a number of vertebrates argue clearly for past connections between the northern Atlantic forest and the eastern Amazonian forests (Bigarella *et al.*, 1975; Bigarella and Andrade-Lima, 1982). To give just one example, Cracraft and Prum (1988) demonstrated that south-east Brazil (Serra do Mar) is a composite, or biogeographic "hybrid", having component taxa with different biogeographic histories, this based on the geographical patterns of speciation of four clades of parrots and toucans.

During a Holocene post-glacial period, *c.* 5,000–6,000 B.P., the Amazon was connected to the Atlantic forest along the north-east of Brazil (Bigarella *et al.*, 1975;

Bigarella and Andrade-Lima, 1982). Cracraft and Prum (1988) provided evidence for a recent separation between the Amazonian Belém-Pará area of endemism and the Atlantic forest (see also, Bush, 1994). Only a connection of this sort could explain the now disjunct distribution of *Alouatta belzebul*, with minute, remnant and isolated populations surviving in the states of Pernambuco, Paraíba, Alagoas and Rio Grande do Norte (Bonvicino *et al.*, 1989). This connection resulted in the addition of just this one species to the Atlantic forest primate fauna. Other primates which today occur in southern Pará and eastern Maranhão (east of the Rio Tocantins) include *Saguinus midas, Cebus kaapori, Chiropotes satanas satanas, Saimiri sciureus,* and *Cebus apella apella,* but none of these extend into the northeastern part of the Atlantic forest.

There were also connection(s) during the Tertiary and Quaternary between the southern Amazon and southeastern Brazil (Bigarella *et al.*, 1975; Bigarella and Andrade-Lima, 1982; Cracraft and Prum, 1988; Bush, 1994). According to Bigarella (1991; see also Bigarella *et al.*, 1975), tropical humid conditions prevailed throughout northern South America, east of the Andes, during interglacial times in the Quaternary. There is "striking structural and floristic similarity" between Amazonian *terra firme* forest and the lowland coastal forests of northern Rio de Janeiro, Espírito Santo and southern Bahia, resulting from connections during the late Tertiary (Mori *et al.*, 1981; Silva and Shepherd, 1986; Joly *et al.*, 1991).

When considering the ancestry of the Atlantic forest primates, middle to late Tertiary connections are undoubtedly the most significant feature. It would appear that this broad belt of forest, non-existent today, sheltered primates ancestral to the genera *Brachyteles, Callithrix* and *Leontopithecus*, to the species *Callicebus personatus* and *Alouatta fusca*, and to the subspecies of *Cebus apella*. The differentiation of these forms, at least (a minimum of) one genus, species or subspecies, in the various taxa is explained as such by vicariance. The implications for the phylogeny of these groups are discussed in the next section, but, based on this conclusion, it is interesting to speculate on possible scenarios which gave rise to the Atlantic forest mix. If it is taken that the possibilities and rates of dispersion of the ancestral forms were similar and that the occupation of the Atlantic forest by primates during the Miocene and Pliocene reflects the rates of their differentiation, it would be necessary to conclude that the genera *Callicebus* and *Alouatta* and the species *Cebus apella* are older than the genera *Brachyteles, Callithrix* and *Leontopithecus*. Alternatively, one could argue that occupation was staggered, with a form ancestral to *Brachyteles* and *Leontopithecus* arriving first, followed by the, by then, differentiated genera, *Callithrix, Callicebus* and *Alouatta*, and subsequently, probably in the early Quaternary, by an already differentiated species, *Cebus apella*.

This, however, presumes that isolation events were few and absolute, undoubtedly a simplification. Repeated changes in the vegetation formations of the interior of Brazil will have been characterized by partial isolation (connections along gallery forests) and varying habitats from perhaps humid forests to highly seasonal dry transition forests. These vegetation types may have acted as filters, with differing effects according to the habitat specialization or ecological flexibility of the species determining whether they are barriers or not. *Cebus apella* (differentiated in the Atlantic forest only to subspecies level) is certainly a highly adaptable species, whereas *Brachyteles* and *Leontopithecus* are more specific in their habitat requirements. The systematic differentiation of the Atlantic forest primates may, as such, be explained, by the intrinsic characteristics of the species and genera: their ecological specialization and capacity to disperse through transition forests and along the gallery forest corridors: *Brachyteles* and *Leontopithecus* are highly limited in this respect, *Alouatta* and *Callicebus* intermediate, and *Cebus apella* the least limited. *Cal-*

lithrix, however, varies in this respect. Whereas *C. aurita* and *C. flaviceps* are limited, the evolutionary trend for the group, characterized by the sequence *C.geoffroyi, C. kuhli, C.penicillata* and *C. jacchus*, has been for their increasing capacity to occupy highly seasonal forests, and as such overcoming the finest of filters. The ancestral *Callithrix*, evidently an incipient tree-gouger and gum feeder, was probably less adaptable to marginal or highly seasonal habitats than, such as, *C. jacchus* and *C. penicillata*, and therefore more susceptible to isolation than would be the case today.

Given that *Leontopithecus, Brachyteles*, and, to a lesser extent, *Callithrix* exhibit a higher degree of habitat specialization, climatic fluctuations resulting in dynamic contraction and expansion cycles, affecting more seriously the interior of Brazil where *cerrado* is found today, would effectively result in a higher degree of divergence of the ancestral stocks of these three genera than would be the case for the more adaptable genera, *Cebus, Alouatta* and *Callicebus*. The fact that all three still tolerate *cerrado* conditions supports this contention. In other words, the first set of cycles would have provoked evolutionary cleavages of the ancestral stock, resulting in *Brachyteles, Leontopithecus* and *Callithrix* in the Atlantic forest, and *Ateles, Lagothrix* and *Saguinus* in the Amazon, and leaving *Cebus, Alouatta* and *Callicebus* unchanged: the exact genealogy is less important for our case here. As noted above, *Callithrix penicillata* and *Callithrix jacchus*, with their extreme adaptation for tree-gouging and gum-feeding, evolved out of the Atlantic forest to conquer forests in the cerrado and caatinga. Only with further deterioration of the contact zone in latter times would evolutionary events be expected to occur, through vicariance, for *Callicebus* and *Alouatta*. If this interpretation is correct, *Brachyteles* and *Leontopithecus* are older than, for instance, the species of *Callicebus, Alouatta* and *Callithrix*.

This early cycle of forest contraction may have been so extensive that Amazon forests were limited to the northern and upper parts of the basin. This would explain the distribution of the *Saguinus* radiation today, with the ancestral *Callithrix* later expanding, with the forests, throughout the southern Amazon, reaching its current distribution south of the Rio Amazonas-Solimões and on the right bank of the Rio Madeira (see Ferrari, 1993a, 1993b). *Cebuella* in this case achieved its extreme tree-gouging and gum-feeding adaptations independently of the *Callithrix* specialists, *C. jacchus* and *C. penicillata*, and probably in competition with syntopic *Saguinus* (Ferrari, 1993b). Interestingly, the phylogenetic trees arising from cytogenetic studies indicate *Callithrix emiliae* from Rondônia, geographically the nearest marmoset to *Cebuella*, as its closest relative (Nagamachi, 1995).

In summary, therefore, we propose three scenarios: 1) First half of the Tertiary with the Amazonian and Atlantic forests coalesced and populated by ancestors of *Alouatta, Callicebus*, a callitrichine, an ateline, and perhaps *Cebus apella*; 2) Mid-Tertiary with the first set of repeated climatic cycles giving rise to a separation of the Amazon and Atlantic forests, and resulting in evolutionary lineages leading to *Brachyteles* in the Atlantic forest, and *Ateles* in the Amazon (*Lagothrix* had already differentiated in the upper Amazon), *Saguinus* in the northern Amazon, and *Callithrix* and *Leontopithecus* in the southern and eastern portions of Brazil; 3) Late Tertiary - early Pleistocene with further cycles (the Atlantic forest already mostly separated from the Amazon) giving rise to species and subspecies in all of the Atlantic forest primates. This would argue for a progressively limited stretch of forest which has resulted in species extinctions as well as allopatric speciation/subspeciation.

Extinctions of large platyrrhines in areas which today lack forest, include that of *Protopithecus brasiliensis* Lund, 1938 (referred to as *Brachyteles brasiliensis* by Hill, 1962) from Lagoa Santa in the *cerrado* of Minas Gerais. Hill (1962) argued that *Protopi-*

thecus bonariensis Gervais and Ameghino, 1880, from the environs of Buenos Aires is synonymous. Cartelle (1993) has also since found two large late Pleistocene platyrrhines (probably not *Brachyteles*, as was originally reported; Cartelle, pers.comm.) to the north, near to the Rio São Francisco, in the state of Bahia (predicted by Hill, 1962). The analysis of this recent finding of two large primate species will be important in clarifying ateline origins and phylogeny.

2) Atlantic Forest Fragmentation

Largely due to the mountainous topography and resulting orographic precipitation, there is little doubt that at least the southeastern Atlantic forest remained as a forest refuge through climatic fluctuations during the Quaternary (Simpson and Haffer, 1978; Brown, 1987). Kinzey (1982) discussed the evidence for Pleistocene forest refuges in the Atlantic forest, and applied it to the current distributions and speciation patterns of the primates. Based on his analysis of the current distribution and taxonomy, he argued for the importance of three centers of endemism as reflecting the isolation of primate communities and their subsequent differentiation in Pleistocene forest refuges: the Bahia Center, Rio Doce Center, and the Paulista Center. He argued that the various forms differentiated in these centers, and subsequently, and to differing degrees, dispersed throughout the geographic distributions recognized today. The primates which characterize each of these centers are shown in Table 2. Taxonomic changes, and some new data on the distribution of the Atlantic forest primates, have modified somewhat the species' composition of these centers: the recognition of *Callithrix kuhli* as distinct from *C. penicillata* (v. Rylands *et al.*, 1993), and the possibility that *Alouatta fusca fusca* is restricted to southern Bahia (Rylands *et al.*, 1988) add another species and subspecies to the Bahia Center; the existence of two subspecies (or even species) of *Brachyteles* (v. Lemos de Sá and Glander, 1993; Lemos de Sá *et al.*, 1993; Coimbra-Filho *et al.*, 1993b) adds further distinct forms to the Paulista and Bahia Centers (Table 2).

While agreeing with Kinzey (1982) regarding the importance of Pleistocene fragmentation of the Atlantic forest, it is evident that the relatively small centers which he outlines provide an oversimplified picture. As argued by Haffer (1993) and Bush (1994) the distributions observed today result from the interaction of numerous processes of differing scale and importance in time and space. They include fragmentation and complex mosaics of vegetation depending on altitude, soils, climate on yearly and longer-term time scales, and the effect of rivers and other natural boundaries. Kinzey (1982) is correct in pointing out that there are basically four sets of primate communities in the Atlantic forest (Table 2), but most especially in the more mountainous south-east, the scenario of Pleistocene forest refuges was probably complex, with effects on any one species being dependent on the type and extent of their habitat specialization (or non-specialization), and the refuge(s) being perhaps highly dendritic and complex, depending on subtle altitude and local climate differences.

The existence of *Leontopithecus rosalia* in the state of Espírito Santo, a component of Kinzey's Rio Doce Center in the northern part of the state and north of the Rio Doce, has never been substantiated (Kierulff, 1993), and the historical and current distribution of this species is midway between his Rio Doce and Paulista Centers. *C. flaviceps*, likewise, lacks a center of endemism, occurring as it does in southern Espírito Santo and eastern Minas Gerais. Due to its great similarity to *C. aurita* (see Hershkovitz, 1977, p.492; Coimbra-Filho, 1986a, 1986b), Kinzey (1982) argued that *C. flaviceps* arose from a subsequent, recent parapatric (sub)speciation event following the dispersal of *C. aurita* north from the

Table 2. Atlantic forest primates communities and the primate species components of the centers of endemism identified by Kinzey (1982)

Species Groupings in the Atlantic Forest	Kinzey's (1982) Centers of Endemism	Observations
Northeast Atlantic forest, north of the Rio Paraguaçu	**Pernambuco Center**	
Callithrix jacchus	*Callithrix jacchus*	Originally restricted to north of the Rio São Francisco. Introduced and hybridizing in numerous regions to the south.
Cebus apella libidinosus		See Hill (1960)
Callicebus personatus barbarabrownae		See Hershkovitz (1990)
Callicebus ssp.		See Kobayashi and Langguth (1994)
Alouatta belzebul belzebul		See Bonvicino *et al.* (1989)
Southern Bahia	**Bahia Center**	
Callithrix kuhli	*Callithrix j. penicillata*	Considered to be a species: *Callithrix kuhli v.* Rylands *et al.* (1993)
Leontopithecus chrysomelas	*Leontopithecus r. chrysomelas*	Considered a species, *Leontopithecus chyrsomelas*
Callicebus personatus melanochir	*Callicebus p. melanochir*	See Hershkovitz (1990)
Cebus apella xanthosternos	*Cebus a. xanthosternos*	See Santos *et al.* (1987), Oliver and Santos (1991)
Brachyteles arachnoides hypoxanthus	*Brachyteles arachnoides*	See Lemos de Sá *et al.* (1993), Coimbra Filho *et al.* (1993b)
Alouatta fusca fusca		Kinzey (1982) did not recognize that *A. fusca* occurred in southern Bahia.
Rio Doce - Espírito Santo, Minas Gerais	**Rio Doce Center**	
Callithrix geoffroyi	*Callithrix j. geoffroyi*	Considered a species: *C. geoffroyi*
Callithrix flaviceps		*C. j. flaviceps* not considered by Kinzey (1982) as an element of the Rio Doce Center.
	Leontopithecus r. rosalia	Considered a species: *L. rosalia*
Callicebus personatus personatus	*Callicebus p. personatus*	See Hershkovitz (1990)
Cebus apella robustus	*Cebus a. robustus*	See Hill (1960), Kinzey (1982)
Alouatta fusca clamitans	*Alouatta f. fusca*	See Rylands *et al.* (1988)
South-east and southern Atlantic forest	**Paulista Center**	
Callithrix aurita	*Callithrix j. aurita*	See Rylands *et al.* (1993)
Leontopithecus rosalia		Attributed by Kinzey (1982) to the Rio Doce Center
Leontopithecus chrysopygus	*Leontopithecus r. chrysopygus*	Considered a species: *L.chrysopygus*
Leontopithecus caissara		Described by Lorini and Persson (1990)
Callicebus personatus nigrifrons	*Callicebus p. nigrifrons*	See Hershkovitz (1990)
Cebus apella nigritus	*Cebus a. nigritus*	See Hill (1960)
Alouatta fusca clamitans	*Alouatta f. clamitans*	
Brachyteles arachnoides arachnoides	*Brachyteles arachnoides*	See Lemos de Sá *et al.*(1993), Coimbra-Filho *et al.* (1993b)

Paulista Center. In the case of *Brachyteles*, Kinzey (1982) himself supposed that the refuges were of insufficient duration to permit specific or subspecific differentiation. He argued that the isolation of *Brachyteles* in the Bahia Center, and its lack in the Rio Doce Center (northern Minas Gerais) explains the lacuna in its range on the north bank of the Rio Doce (Fig. 6). Recent morphological and genetic studies, however, are demonstrating differences between the Paulista *Brachyteles* (*B. a. arachnoides*) and those north of the Serra da Mantiqueira (*B. a. hypoxanthus*) which extend into southern Bahia, and probably in the past as far north as the Rio Paraguaçu. The lacuna in its range north of the Rio Doce is a mystery but could be explained more parsimoniously by the fact that *Brachyteles* is typically a montane, upland forest species (600 to 1,800 m altitude; see Aguirre, 1971), and the predominance of lowland coastal forests in the region north of the Rio Doce to southern Bahia. Differentiation in this genus, therefore, can be explained in terms of isolation by forest refuges either side of a mountain range and habitat preference.

What is perhaps more significant is the lack *Leontopithecus* in Espírito Santo. The distribution of this genus is the most highly fragmented of any of the Atlantic forest primates (Fig. 2) and yet adequate habitat exists today, for example, along the lowland coastal region extending from southern Bahia (*L. chrysomelas*) to Rio de Janeiro (*L. rosalia*). *Leontopithecus* is restricted to lowland forests, and the possible explanation for the lack of lion tamarins in Espírito Santo would be the opposite of the supposition of Kinzey (1982) of the existence of forest refuge north of the Rio Doce: that is, a hiatus was caused by the loss of forests in the lowland regions, then restricted to the mountainous interior, supposedly during a Pleistocene dry period. Populations of *L. chrysopygus* and *L. caissara* were evidently isolated in the inland plateau of São Paulo, and the small area of coastal lowland forest in southern São Paulo and northern Paraná, respectively. Natori (1989), studying dental and cranial characters, found *L. rosalia* and *L. chrysopygus* to be more closely related to each other than either is to *L. chrysomelas*. This would argue that the Bahia refuge was separated first and that refuges in the western part of the state of São Paulo and in Rio de Janeiro were more recent (Natori, 1989).

The existence of two subspecies of *Alouatta fusca* is still doubted, but recent evidence concerning its distribution (Rylands *et al.*, 1988) would suggest that *A. f. fusca*, if valid, is a Bahian form, with *A. f. clamitans* not restricted to the south of the Rio Doce as suggested by Kinzey (1982), but extending north as far as the Rio Jequitinhonha. Kinzey (1982) did not recognize that *A. fusca* also occurs in southern Bahia, and still does in some remoter regions.

In summary, Kinzey's refuges or centers of endemism, would account for the differentiation of all the Atlantic forest primates, except for *Leontopithecus, Callithrix aurita/flaviceps, Alouatta* and *Brachyteles*. There is indisputable evidence for centers of endemism in southern Bahia (Bahia Center) and São Paulo (Paulista), but that for the Rio Doce Center, north of the Rio Doce, is weak. Whereas southern Bahia probably formed a fairly uniform and coherent center, that in São Paulo was probably larger and more complex. A reasonable assumption would be that the Rio Doce center would have extended along the mountains of the interior of the state of Espírito Santo and eastern Minas Gerais, and having as their principal components which differentiated, *Callithrix flaviceps* (south of the Rio Doce, *Cebus apella robustus* (isolated north of the Rio Doce), and *Callicebus personatus personatus* (both sides of the Rio Doce). Isolated secondary refuges of lowland forest enclaves in certain parts are necessary to account for the existence of *Callithrix geoffroyi* and *Leontopithecus rosalia*, and likewise in the case of the Paulista Center, *L. chrysopygus* and *L. caissara*. Whether the Pleistocene refuges, such as those postulated by Kinzey (1982) have resulted in speciation and subspeciation or merely the distributions observed today is still, however, uncertain. A number of leading specialists argue that pre-Quaternary refuges rather than Pleistocene refuges are the cause of speciation patterns observed today (see, for example, Hershkovitz, 1969; Heyer and Maxson, 1982; Cracraft and Prum, 1988; Silva and Patton, 1993).

3) Rivers, Topography, and Habitat Specialization

Haffer (1993 and earlier publications; see also Simpson and Haffer, 1978) discussed the role of rivers as barriers to dispersion and consequently in speciation. Hershkovitz (1977, 1983) has emphasized the importance of rivers in delimiting the ranges of especially the Amazonian primates, and consequently as an important driving force in speciation. River dynamics has been suggested as a major force in allopatric speciation and, through forest disturbance and riparian succession, partially responsible for high biological diversity, particularly in the upper Amazon (Salo *et al.* 1986; Salo, 1990). In the Atlantic forest such large rivers as the Rio São Francisco, Rio Paraguaçu, Rio Jequitinhonha, the Rio Doce, Rio Paraná, Rio Grande, Rio Paraíba, and Rio Paranaiba have undoubtedly played important roles as barriers in the past. The river barrier effect is most strong along the lower reaches where they are wider. However, forest destruction and the silting-up of the majority of the Atlantic forest rivers means that it is very difficult to identify their past or even present role as barriers, and in some cases the demise of once broad and fast-flowing rivers has resulted in range expansion (A.F.Coimbra-Filho, pers. comm.). The effects of lateral erosion and changing river courses in creating mosaics of variously-aged forests, an important feature in the Amazon (Salo *et al.*, 1986), is not evident in the Atlantic forest, where vegetational heterogeneity is maintained more by altitudinal differences and orographic effects.

In the Atlantic forest, therefore, topography rather than river dynamics is undoubtedly the most important in determining the physiognomy and floristic composition of the habitats available to primates. Vegetation types are determined by climate (seasonality), topography and soils. Too little is known concerning Pleistocene and Holocene changes in the distribution and floristic composition of the vegetation types available, but, as discussed in the previous section, they undoubtedly had a fundamental role in speciation and subspeciation. At the simplest level, mosaics of fragments of lowland and upland forest will have affected the distributions and speciation of the primates adapted to each in different ways.

The Atlantic forest marmosets demonstrate three main ecological and taxonomic groupings: 1) *C. aurita* and *C. flaviceps*, occurring in montane forest in the south-east; 2) *C. geoffroyi* and *C. kuhli*, occurring in lower montane, coastal lowland, semideciduous and gallery forests in the eastern Atlantic forest in the states of Minas Gerais, Espírito Santo and Bahia; and 3) *C. jacchus* and *C. penicillata* which have extended their ranges into gallery forest and forest patches in the, today, dry north-east (caatinga), and the bush savanna (cerrado) of central Brazil, respectively (Rylands and Faria, 1993). *C. jacchus* and *C. penicillata* are able to occupy relatively hostile habitats, in terms of severe seasonal or even annual lack of fruits, through their extreme specialization for tree-gouging to obtain gums (Rylands, 1984, 1993, 1996; Natori and Shigehara, 1992). The widespread distribution of *Callithrix* throughout the Atlantic forest is related to their ability to occupy secondary and successional forests, whereas the restricted distributions of the *Leontopithecus* species results from their specialization on mature lowland forest. *Brachyteles* is an upland, montane forest inhabitant, whereas the remaining genera, *Callicebus*, *Cebus*, and *Alouatta*, occur in all forest types.

PHYLOGENY OF THE ATLANTIC FOREST PRIMATES

The high degree of endemism found in the Atlantic forest primates indicates that speciation processes have been active following the colonization of the region by the following forms: a callitrichine ancestral to both *Callithrix* and *Leontopithecus*, *Callicebus*, *Cebus*, *Alouatta*, and a form ancestral to *Brachyteles*. There are only two species which occur in both the Atlantic and Amazonian forests; *Cebus apella* which has undergone subspeciation, and *Alouatta belzebul*, which, as described above, probably entered the Amazon during a post-glacial period *c.* 5,000–6,000 B.P., along the north-east of Brazil. To date it has been described as the subspecies *A. b. belzebul*, indicating its relatively recent colonization, but the possibility remains that further morphological and genetic studies may indicate that it is subspecifically different from the Amazon forms. Both endemic genera, and the most diverse communities occur in the south-east of the region, and it is presumed that the primate fauna originated from a broad connection with the Amazonian forests extending from the region of São Paulo-Paraná probably at least as far north as southern Bahia.

Atelidae

Following the growing consensus, reviewed by Rosenberger and Strier (1989) and Strier (1992b), that the atelines (*Alouatta*, *Brachyteles*, *Lagothrix* and *Ateles*) are the monophyletic descendants of a single common ancestor, the occurrence of the endemic *Brachyteles* alongside *Alouatta* in the Atlantic forest has interesting implications in terms of their phylogeny. *Alouatta* is generally accepted as the first extant offshoot of the group, but there is some dispute as to the relations between *Brachyteles*, *Lagothrix* and *Ateles*. In terms only of their distributions, however, the best fit would follow the phylogeny suggested by Zingeser (1973), in which an ancestral form, common to both the Amazonian and Atlantic forests, gave rise to *Brachyteles* on the one hand, and *Lagothrix* and *Ateles* on the other, following the isolation of the two forests. This is contrary to Rosenberger and Strier's (1989; Strier, 1992b) evaluation in terms of morphology, behavior and ecology, where they place *Brachyteles* as a sister-taxon to *Ateles*, with *Lagothrix* having differentiated earlier. Kay *et al.* (1987), studying dental morphology, and Ford (1986), who studied shoulder, elbow, knee and lower ankle joints, also found that *Alouatta* and *Brachyteles* shared derived traits, although they were indecisive and proposed alternative phylogenies. The phylogeny proposed by Rosenberger and Strier (1989; see also Rosenberger 1992a; Strier, 1992b) implies that the high degree of folivory shown by *Brachyteles* is secondarily evolved and the assumption that either: 1) *Lagothrix*, or the ancestral form to it became extinct in the Atlantic forest; or 2) *Lagothrix* arose in the upper Amazon independent of a separation from the Atlantic forest, and *Ateles* and *Brachyteles* subsequently differentiated following the separation. Considering the second alternative, the closest relative to *Brachyteles* should be *Ateles belzebuth chamek*. *Alouatta fusca* and *Alouatta caraya* of central Brazil, may have differentiated from other members of the genus after the separation of *Lagothrix*. The possibility that an ancestral form to the highly frugivorous *Ateles*, or even *Brachyteles*, became extinct in areas outside of the Amazon during the Pleistocene is possible considering the arguments of Brown (1987; see also Fonseca *et al.* in press) that forest contraction may have might have resulted in more extinction than speciation. In geographic terms, Zingeser's (1973) phylogeny is the best (simplest) fit, but that of Rosenberger and Strier (1989) is still plausible.

The platyrrhine lineage leading to *Callicebus*, grouped with the subfamily Pitheciinae by Rosenberger (1981; see also Schneider *et al.*, 1993), is considered to have diverged during the Oligocene, some 25 million years ago according to Rosenberger (1984) or about 15 million years according to Schneider *et al.* (1993). The diversity of *Callicebus* in the Atlantic forest, five subspecies of *C. personatus*, can be explained in terms of differentiation in isolated forest refuges: *C. p. nigrifrons* in the Kinzey's Paulista refuge, *C. p. personatus* in the region of the Rio Doce, *C. p. melanochir* in southern Bahia, and *C.p.barbarabrownae* and the new *C. personatus* referred to by Kobayashi and Langguth (1994) are relict, scattered, and scarce populations surviving in today's remnants of the Atlantic forest in the interior and northern coastal regions of Bahia and Sergipe.

Cebidae

Like *Callicebus*, the lineage of the capuchin monkeys *Cebus*, along with the squirrel monkeys, *Saimiri*, is believed to have begun in the Oligocene, about 27 million years ago, although the divergence times estimated through the molecular genetics of the genera indicate it would have occurred in the early Miocene, about 17 million years ago (Schneider *et al.*, 1993). *Cebus apella* evidently colonized the Atlantic forest as a full-blown species, and shows a degree of subspecific differentiation not present in the Amazon, where only one subspecies is currently recognized, although Torres (1988) indicated the likelihood of two. The reasons for the greater diversity of Atlantic forest *Cebus apella* may lie in the greater degree of altitudinal variation (variation in forest types as well as mountain ranges acting as geographic barriers), along with fragmentation through Pleistocene forest refuges as proposed by Kinzey (1982).

The phylogeny and radiation of the Callitrichinae (*Callimico, Cebuella, Callithrix, Saguinus,* and *Leontopithecus*) is the subject of considerable controversy (for reviews see Hershkovitz, 1977; Ford, 1980, 1986; Martin, 1992; Ferrari, 1993a, 1993b; Snowdon, 1993; Garber, 1994). The presence of *Leontopithecus* in the south-east Atlantic forest, with *Saguinus* restricted to the northern Amazon, argues for the latter being an early offshoot of the lineage which gave rise to *Callithrix*, with the Amazonian *Callithrix* radiation having its origins in the south of the basin (see Ferrari, 1993a, 1993b) and the Atlantic forest *Callithrix* having their origins in south-east Brazil. It is necessary to imagine that the callitrichine ancestor, once widespread throughout a broad belt connecting the Amazon to the Atlantic forest (at least 30 million years ago in the Oligocene, see Rosenberger, 1984; Fanning *et al.*, 1989; Martin, 1992), was separated in the north, giving rise to *Saguinus* (see Ferrari, 1993a, 1993b), with ancestral populations in the south, possibly in subtropical conditions, giving rise to *Callithrix*. *Callithrix* later spread to the north in the Atlantic forest, and north in the Amazon reaching the Rio Madeira and Rio Amazonas. This would argue for the most primitive forms of *Callithrix* being the most southerly, certainly true for the Atlantic forest (see Natori, 1986, 1990, 1994). The most primitive Atlantic forest species are *C. aurita* and *C. flaviceps* (considered to be only subspecifically distinct by Coimbra-Filho, 1986a, 1986b; see also Coimbra-Filho *et al.*, 1993a), and these may be considered a separate grouping from the remaining species which branched off the *Callithrix* lineage in the following order, *C. geoffroyi, C. kuhli,* and the most specialized sister-taxa *C. penicillata* and *C. jacchus* (see Natori, 1986; Natori and Shigehara, 1992). A south to north dispersal would also seem to be the case for the Amazonian *Callithrix*, where differentiation increases in a northerly direction (Natori, 1986; Ferrari, 1993a; Rylands *et al.*, 1993; Nagamachi, 1995). Ferrari (1993b) argued that ancestral *Saguinus* must have been excluded from the present-day range of Amazonian *Callithrix* through competition, but it is equally possible that *Callithrix* simply colonized an area which had previously been uninhabitable, and the cause for the separation of the ancestor which gave rise to the two genera.

In summary, there are two disputed scenarios for the origin of the Atlantic forest cal-litrichinae: 1) *Saguinus* branched away first, and *Leontopithecus* and *Callithrix* are sister-taxa (Ford, 1986; Rosenberger, 1984; Rosenberger and Coimbra-Filho, 1984; Natori, 1989; Snowdon, 1993); and 2) *Leontopithecus* branched away first and *Saguinus* and *Callithrix* are sister-taxa (Garber, 1994). The present-day distributions of the Atlantic forest genera fit with either of these phylogenies, and both would indicate a dispersal (going from most primitive to more advanced) in a south-north direction rather than the reverse (see Hershkovitz, 1977). The strongest arguments for the second scenario (*Saguinus* and *Callithrix* as sister taxa), include the fact that lion tamarins lack the phenomenon of physiological reproductive inhibition common to all other genera (Abbott *et al.*, 1993; Martin, 1992; Garber, 1994). If lion tamarins are the sister-taxon to *Callithrix*, and arose after the separation of *Saguinus*, the implication would be that *Leontopithecus* has lost this trait (evolving a number of other traits associated with behavioral reproductive inhibition, see Garber, 1994), or, less likely, that it evolved twice (Martin, 1992; Garber, 1994). A supposed loss of physiological reproductive inhibition in lion tamarins could be related to an ecological niche quite different from the other callitrichinae: the occupation of mature, lowland forests, whereas both *Saguinus* and *Callithrix* are associated with successional forest (Rylands, 1993; Rylands, 1996).

The Isolation of the Atlantic Forest

The separation of the Atlantic forest from the Amazonian forests in the southeastern part was evidently pre-Quaternary (Cracraft and Prum, 1988; Bush, 1994). Following the time-scale for the differentiation of New World primates as suggested by Rosenberger (1984, p.738), and presuming that *Brachyteles* arose following the Atlantic forest's isolation, the lack of *Aotus*, *Saimiri*, and least one form of the lineage which gave rise to the seed-predators, *Pithecia*, *Chiropotes* and *Cacajao* requires explanation. *Saimiri* and *Cacajao* are specialists occupying principally floodplains and inundated forests not characteristic of the Atlantic forest. Likewise, it would seem by the present-day distribution of the pitheciin genera *Cacajao*, *Chiropotes* and *Pithecia*, that their origin and radiation was essentially in the northern part of the Amazon. Why *Chiropotes* of southern Pará and western Maranhão failed to follow *Alouatta belzebul* into the northern Atlantic forest is not known, but one can surmise that it is related to its more specialized feeding habitats. The lack of *Aotus* in the Atlantic forest is surprising considering its occurrence south of the Amazon as far as northern Paraguay.

A recent review of the molecular phylogeny of the Platyrrhini by Schneider *et al.* (1993) provided an estimate of the time of the divergence of *Alouatta* from the remaining Atelinae at about 13 million years ago. Rosenberger (1984), however, compared paleontological evidence of the platyrrhines with that of molecular clocks and demonstrated that the latter consistently underestimate the divergence times. Paleontological evidence based on the existence of *Stirtonia tatacoensis*, a form closely related to *Alouatta* and from the Colombian Miocene dated to about 15 million years ago, places the *Alouatta* lineage as about five million years older than the molecular clock would indicate. Rosenberger (1984) argued, therefore, that the separation of *Alouatta* from the Atelines occurred around 20 million years ago. Whereas Rosenberger (1984) places *Stirtonia* firmly in the *Alouatta* lineage, Schneider *et al.* (1993) suggest it might be more correctly a representative of the ateline lineage prior to the divergence of the extant forms. The molecular phylogeny provided by Schneider *et al.* (1993) indicates that *Ateles*, *Lagothrix* and *Brachyteles* diverged a little over 10 million years ago, but according to Rosenberger

(1984) this would have occurred in the mid-Miocene, some 14–15 million years ago. If *Brachyteles* arose as a result of the isolation of an ancestral form in the Atlantic forest, this would argue for the break occurring at one of these two times.

The separation of the atelines pre-dates the radiation of the extant callitrichins which is believed to have occurred in the late Miocene, early Pliocene, some five million years ago according to Rosenberger (1984), although later Rosenberger (1992a) indicates that the radiation began much earlier, 16 million years ago, with *Saguinus* branching off about 10 million years ago. The molecular clock provided by Schneider *et al.* (1993), gives a similar date, of nearly 10 million years ago. This would reinforce the argument that the Atlantic forest became largely isolated at this time, leaving necessarily the *Leontopithecus* lineage and *Callithrix* in the Atlantic forest and *Saguinus* and *Callithrix/Cebuella* in the Amazon basin.

CONSERVATION STATUS OF THE ATLANTIC FOREST PRIMATES

Due to the widespread destruction of the Atlantic forest, estimates today indicate that it has been reduced to less than 10% of its original extent (Fonseca, 1985; Câmara, 1991). Forest destruction has been almost total in the northeastern states (north from Bahia) where less than 0.3% of the forest still exists. In the south-east less than 6% remains, and in the south approximately 10% (Câmara, 1991). Deforestation is still continuing despite recent legislation for its protection (Anon. 1993). Between 1985 and 1990 an estimated 144,240 ha of forests were cut in the state of Paraná. Atlantic forest formations originally covered 85% of the state, reduced by 1990 to 7.7% (Brazil, SOS Mata Atlântica-INPE, 1992). Likewise 87% of the state of Espírito Santo was originally covered by forest, which in 1990 was reduced to 8.3% (Brazil, SOS Mata Atlântica-INPA, undated).

Deforestation and hunting, and to some extent capture for commerce, have resulted in the decline of numerous species, and the threatened status of most of the Atlantic forest primates (Mittermeier *et al.*, 1982). A recent evaluation using the Mace-Lande categorization for threatened species, adopted by the World Conservation Union (a slightly modified version (2.3) of Mace and Stuart, 1994) of the all the Neotropical Primates, carried out by the IUCN Species Survival Commission Primate Specialist Group (Mittermeier, Rylands and Rodríguez-Luna, in prep.), lists seven species and subspecies as critically endangered, five as endangered, six as vulnerable, and six in the low risk category (Table 3). One of the "low risk" species is *Alouatta belzebul belzebul*, which, however, although not immediately threatened in the Amazon, is undoubtedly critically endangered in the Atlantic forest (Bonvicino *et al.*, 1989).

Conservation measures for the majority of species and subspecies are limited to their survival in protected areas. However, there are some significant conservation programs which should be mentioned. The golden lion tamarin, *L.rosalia*, has been the subject of a major research, reintroduction, and environmental education program since 1983 (Kleiman *et al.*, 1986). Captive and field conservation efforts, and the status and threats to both wild and captive populations of lion tamarins, were reviewed and evaluated during a population viability analysis workshop in 1990 (Seal *et al.*, 1990) and resulted in the formalization and official recognition by the Brazilian Institute for the Environment (Ibama) of international management committees for each of the four species (Mallinson, 1989, 1994). *L. rosalia* and *L. chrysomelas* are supported by successful captive breeding programs, with the managed populations numbering around 500 individuals (Ballou, 1993;

Table 3. Conservation status of the Atlantic forest primates according to the
Official Brazilian List of Threatened Fauna (Bernardes *et al.*, 1990),
the *1994 IUCN List of Threatened Animals* (Groombridge, 1993),
and the Mace-Lande Categories adopted by IUCN in 1994 (Mace and
Stuart, 1994; Mittermeier *et al.*, in prep.). V = vulnerable, E = endangered

Species/subspecies	Brazil	IUCN 1994	Mace-Lande
Callithrix aurita	X	E	Endangered
Callithrix flaviceps	X	E	Endangered
Callithrix geoffroyi	—	V	Vulnerable
Callithrix kuhli	—	V	Low Risk
Callithrix penicillata	—	—	Low risk
Callithrix jacchus	—	—	Low risk
Leontopithecus rosalia	X	E	Critical
Leontopithecus chrysomelas	X	E	Endangered
Leontopithecus chrysopygus	X	E	Critical
Leontopithecus caissara	X	E	Critical
Cebus apella			
C. a. nigritus	—	—	Low risk
C.a.robustus	—	V	Vulnerable
C.a.xanthosternos	X	E	Critical
C.a.libidinosus	—	—	Low risk
Callicebus personatus	X	V	—
C.p.personatus	X	E	Vulnerable
C.p.melanochir	X	E	Vulnerable
C.p.nigrifrons	X	E	Vulnerable
C.p.barbarabrownae	X	E	Critical
C.personatus ssp.[1]	X	E	Critical
Alouatta fusca		V	
A.f.fusca	X	E	Critical
A.f.clamitans	X	V	Vulnerable
Alouatta belzebul belzebul	X	—	Low risk
Brachyteles arachnoides			
B.a.arachnoides	X	E	Endangered
B.a.hypoxanthus	X	E	Endangered

[1] A subspecies described but not named by Kobayashi and Langguth (1994).

De Bois, 1994). Wild populations of *L.rosalia* are now severely reduced, numbering approximately 500, the majority in the Poço das Antas Biological Reserve (Kierulff, 1993), and a translocation program is underway for single isolated groups located during a major survey of its entire geographic range (Kierulff and Oliveira, 1994). *L. chrysomelas* in southern Bahia is spread over a wider area, and a recent survey estimated between 4,000 and 6,000 individuals (Pinto and Tavares, 1994). The captive breeding of *L. chrysopygus* is still limited, largely due to a very small founder population (Valladares-Padua and Simon, 1992; Mansour and Ballou, 1994) but more success is expected in the near future, and considerable investment has been made in promoting the conservation of populations in the wild, which are estimated to total about 1,000 individuals in five areas (principally the Morro do Diabo State Park, São Paulo) where they are known to survive (Valladares-Padua *et al.*, 1994; Padua, 1994). The recently discovered *L. caissara* is the most endangered of the lion tamarins. There are none in captivity, and the population estimate is of about 260 animals (Lorini and Persson, 1994). The integrity of the Superagui National

Park covering part of their range is itself seriously threatened (Câmara, 1994; Vivekananda, 1994).

Conservation of the muriqui, *Brachyteles arachnoides*, has to date concentrated on surveys, and ecological and behavioral research (Mittermeier *et al.*, 1987; Milton, 1984; Oliver and Santos, 1991; Lemos de Sá and Glander, 1993; Lemos de Sá *et al.*, 1990, 1993; Martuscelli *et al.*, 1994). The survey by Lemos de Sá *et al.* (1990, 1993; Lemos de Sá and Glander, 1993) involved morphometric and genetic analyses of individuals captured in localities in Minas Gerais and São Paulo, and provided evidence indicating the existence of two forms, which, pending further study, may even be specifically distinct (see also Coimbra-Filho *et al.*, 1993b). The surveys of Martuscelli *et al.* (1994) provided 14 new localities, although the majority of these have minimal or even recently extinct populations. Most important has been the preservation of a small area of Atlantic forest in the Fazenda Montes Claros where research and demographic monitoring have been carried out since 1982 (Nishimura *et al.*, 1988; Strier, 1991, 1992a, 1992b). Due to this study, *Brachyteles* is now, along with the lion tamarins, one of the best known of the Atlantic forest primates. Captive breeding has been successfully initiated at the Rio de Janeiro Primate Center (Coimbra-Filho *et al.*, 1993b), but active conservation measures are still pending (Mendes, 1994).

Following the field surveys of south-east Brazilian Atlantic forests by Santos *et al.* (1987) and Oliver and Santos (1991), the wild populations of *Cebus apella xanthosternos* and *C. a. robustus* were recognized as critically endangered and vulnerable, respectively. This resulted in the formation of an international management committee in 1992, similar to those for the lion tamarins, and efforts are underway to organize a captive breeding program for each (Santos and Lernould, 1993).

Some important initiatives concerning the protection of the Atlantic forest remnants should also be mentioned. A large part of the Atlantic forest is now recognized as a Biosphere Reserve, part of the network organized by the Man and the Biosphere Program (MAB) of the United Nation's (UNESCO). This has resulted in important collaboration between the governments of the Atlantic forest states and the development of action plans for the region (see Brazil, Consórcio Mata Atlântica - UNICAMP, 1992; Guatara *et al.*, 1994). The non governmental organization SOS Mata Atlântica is also collaborating with the National Institute for Space Research (INPE) in a major project involving the mapping and monitoring of the Atlantic forest fragments using satellite imagery (Brazil, SOS Mata Atlântica - INPE, 1992, 1993, undated). Finally, the Brazil Program of the Washington D. C. based Conservation International is organizing a series of workshops on priority areas for biodiversity conservation in the principal Brazilian biomes. The northern Atlantic forest was the focus of such a workshop in December 1993, and involved the identification and mapping of key areas in terms of biological importance throughout the Atlantic forest north of the Rio Doce in the state of Espírito Santo (Conservation International, 1995). A further workshop for the southern Atlantic forest is planned for the near future, and will provide a data base on current knowledge of the areas and orientation for research and conservation efforts.

REFERENCES

Abbott, D.H., Barrett, J. and George, L.M. 1993. Comparative aspects of the social suppression of reproduction in female marmosets and tamarins. In: *Marmosets and Tamarins: Systematics, Ecology, and Behaviour*, A.B. Rylands, (ed.), pp.152–163. Oxford University Press, Oxford.

Aguirre, A.C. 1971. *O mono* Brachyteles arachnoides *(E.Geoffroy): situação da espécie no Brasil*. Academia Brasileira de Ciências, Rio de Janeiro.

Andrade, G.O. de, and Lins, R.C. 1964. Introdução ao estudo dos "brejos" pernambucanos. *Arquivos*, Universidade do Recife, Instituto de Ciências da Terra, Recife, (2)21–34.

Anonymous. 1993. Legal protection for Brazil's Atlantic coastal forest. *Neotropical Primates*, 1(2): 7–9.

Ballou, J.D. 1993. *1992 International Studbook Golden Lion Tamarin* Leontopithecus rosalia. Smithsonian Institution, Washington, D.C.

Bernardes, A.T., Machado, A.B.M. and Rylands, A.B. 1990. *Fauna Brasileira Ameaçada de Extinção*. Fundação Biodiversitas, Belo Horizonte.

Bigarella, J.J. 1991. Aspectos físicos da paisagem / Physical landscape features. In: *Mata Atlântica / Atlantic Rain Forest*, I. de G. Câmara (ed.), pp.63–93. Editora Index Ltd. and Fundação SOS Mata Atlântica, São Paulo.

Bigarella, J.J. and Andrade-Lima, D. de. 1982. Paleoenvironmental changes in Brazil. In: *Biological Diversification in the Tropics*, G.T.Prance (ed.), pp.27–40. Columbia University Press, New York.

Bigarella, J.J., Andrade-Lima, D.de and Riehs, P.J. 1975. Considerações a respeito das mudanças paleoambientais na distribuição de algumas espécies vegetais e animais no Brasil. *An. Acad. Brasil. Ciênc.*, 47(supl.): 411–463.

Bitetti, M.S.di, Placci, G., Brown, A.D. and Rode, D.I. 1994. Conservation and population status of the brown howling monkey (*Alouatta fusca clamitans*) in Argentina. *Neotropical Primates*, 2(4): 1–4.

Bonvicino, C.R., Langguth, A., and Mittermeier, R.A. 1989. A study of the pelage color and geographic distribution in *Alouatta belzebul* (Primates: Cebidae). *Rev. Nordestina Biol.*, 6(2): 139–148.

Brazil, Consórcio Mata Atlântica - UNICAMP. 1992. *Reserve da Biosfera da Mata Atlântica. Plano de Ação. Vol. 1*. Consórcio Mata Atlântica, São Paulo, and Universidade Estadual de Campinas (UNICAMP), Campinas.

Brazil, SOS Mata Atlântica - INPE. 1992. *Mata Atlântica do Estado do Paraná. Evolução dos Remanescentes Florestais e Ecossistemas Associados do Domínio da Mata Atlântica. Período: 1985–1990*. Map. Scale 1:1,700,000. 1st Edition. SOS Mata Atlântica, São Paulo, and Instituto Nacional de Pesquisas Espaciais (INPA), São José dos Campos.

Brazil, SOS Mata Atlântica - INPE. 1993. *Evolução dos Remanescentes Florestais e Ecossistemas Associados do Domínio da Mata Atlântica no Período 1985–1990 - Relatório*. SOS Mata Atlântica, São Paulo, and Instituto Nacional de Pesquisas Espaciais (INPA), São José dos Campos. 46pp.

Brazil, SOS Mata Atlântica - INPE. Undated. *Mata Atlântica, Estado do Espirito Santo. Evolução dos Remanescentes Florestais da Mata Atlântica. Período: 1985/1990*. Map. Scale 1:1,000,000. 1st Edition. SOS Mata Atlântica, São Paulo, and Instituto Nacional de Pesquisas Espaciais (INPA), São José dos Campos.

Brown K.S., Jr. 1987. Conclusions, synthesis, and alternative hypotheses. In: *Biogeography and Quaternary History in Tropical America*, T.C.Whitmore and G.T.Prance, pp.175–191. Oxford University Press, Oxford.

Bush, M.B. 1994. Amazonian speciation,: a necessarily complex model. *J.Biogeog.*, 21: 5–17.

Câmara, I. de G. 1991. *Plano de Ação para a Mata Atlântica*. Fundação SOS Mata Atlântica, São Paulo.

Câmara, I. de G. 1994. Conservation status of the black-faced lion tamarin, *Leontopithecus caissara*. *Neotropical Primates*, 2(suppl.): 50–51.

Cartelle, C. 1993. Achado de *Brachyteles* do Pleistoceno Final. *Neotropical Primates*, 1(1): 8.

Coimbra-Filho, A.F. 1986a. Sagui-da-serra *Callithrix flaviceps* (Thomas, 1903). *FBCN/Inf.*, Rio de Janeiro, 10(1): 3.

Coimbra-Filho, A.F. 1986b. Sagui-da-serra-escuro *Callithrix aurita* (E.Geoffroy, 1812). *FBCN/Inf.*, Rio de Janeiro, 10(2):3.

Coimbra-Filho, A.F. and Mittermeier, R.A. 1973. New data on the taxonomy of the Brazilian marmosets of the genus *Callithrix* Erxleben 1777. *Folia Primatol.* 20:241–264.

Coimbra-Filho, A.F., Rocha e Silva, R da and Pissinatti, A. 1991. Acerca da distribuição geográfica original de *Cebus apella xanthosternos* Wied, 1820 (Cebidae, Primates). In: *A Primatologia no Brasil - 3*, A.B.Rylands and A.T.Bernardes (eds.), pp.215–224. Sociedade Brasileira de Primatologia and Fundação Biodiversitas,. Belo Horizonte.

Coimbra-Filho, A.F., Pissinatti, A. and Rylands, A.B. 1993a. Experimental multiple hybridism among *Callithrix* species from eastern Brazil. In: *Marmosets and Tamarins: Systematics, Ecology, and Behaviour*, A.B.Rylands, (ed.), pp.95–120. Oxford University Press, Oxford.

Coimbra-Filho, A.F., Pissinatti, A. and Rylands, A.B. 1993b. Breeding muriquis *Brachyteles arachnoides* in captivity: the experience of the Rio de Janeiro Primate Center (CPRJ-FEEMA). *Dodo, J. Wildl. Preserv. Trusts*, 29: 66–77.

Conservation International. 1995. *Prioridades para Conservação da Biodiversidade da Mata Atlântica do Nordeste*. Map. Scale 1:2,500,000. Conservation International, Washington, D.C., Fundação Biodiversitas, Belo Horizonte, and Sociedade Nordestina de Ecologia, Recife.

Cracraft, J. and Prum, O. 1988. Patterns and processes of diversification: speciation and historical congruence in some Neotropical birds. *Evolution* 42:603–620.

Daegling, D.J. 1992. Mandibular morphology and diet in the genus *Cebus*. *Int. J. Primatol.*, 13(5): 545–570.

De Bois, H. 1994. Progress report on the captive population of golden-headed lion tamarins, *Leontopithecus chrysomelas* - May 1994. *Neotropical Primates*, 2(suppl.): 28–29.

Fanning, T.G., Seuánez, H.N. and Forman, L. 1989. Satellite DNA sequences in the neotropical marmoset *Callimico goeldii* (Primates, Platyrrhini). *Chromosoma (Berl.)*, 98: 396–401.

Ferrari, S.F. 1993a. Ecological differentiation in the Callitrichidae. In: *Marmosets and Tamarins: Systematics, Ecology, and Behaviour*, A.B.Rylands, (ed.), pp.314–328. Oxford University Press, Oxford.

Ferrari, S.F. 1993b. The adaptive radiation of Amazonian callitrichids (Primates, Platyrrhini). *Evolución Biológica*, 7: 81–103.

Ferrari, S.F. and Mendes, S.L. 1991. Buffy-headed marmosets 10 years on. *Oryx*, 25(2): 105–109.

Fonseca, G.A.B. da. 1985. The vanishing Brazilian Atlantic forest. *Biol. Conserv.*, 34: 17–34.

Fonseca, G.A.B. da. 1994. Mono-carvoeiro, muriqui, *Brachyteles arachnoides* (E.Geoffroy, 1806). In: *Livro Vermelho dos Mamíferos Brasileiros Ameaçados de Extinção*. G.A.B. da Fonseca, A.B.Rylands, C.M.R. Costa, R.B.Machado and Y.L.R.Leite (eds.), pp.191–199. Fundação Biodiversitas, Belo Horizonte.

Fonseca, G.A.B.da, Herrmann, G. and Leite, Y.L.R. In press. Macrogeography of Brazilian mammals. In: *Mammals of the Neotropics, Vol. 3*, J.F.Eisenberg (ed.), University of Chicago Press, Chicago.

Ford, S.M. 1980. Callitrichids as phyletic dwarfs, and the place of the Callitrichidae in Platyrrhini. *Primates*, 21: 31–43.

Ford, S.M. 1986. Systematics of the New World Monkeys. In: *Comparative Primate Biology, Vol.1, Systematics, Evolution and Anatomy*, D.R.Swindler and J. Erwin (eds.), pp.73–135. Alan R. Liss, New York.

Froehlich, J.W., Supriatna, J. and Froehlich, P.H. 1991. Morphometric analyses of *Ateles*: systematic and biogeographic implications. *Am. J. Primatol.*, 25:1–22.

Garber, P.A, 1994. Phylogenetic approach to the study of tamarin and marmoset social systems. *Amer. J. Primatol.*, 34: 199–219.

Groombridge, B. 1993. *1994 IUCN Red List of Threatened Animals*. World Conservation Monitoring Center, Cambridge, UK.

Guatara, I. S., Costa, J.P. de O., Corrêa, F. and Azevedo, P.U.E. de. 1994. *Reserva da Biosfera da Mata Atlântica MAB-UNESCO. A Questão Fundiária: Roteiro para Solução dos Problemas das Áreas Protegidas*. Consórcio Mata Atlântica, São Paulo. 31pp.

Haffer, J. 1969. Speciation in Amazon forest birds. *Science*, 165: 131–137.

Haffer, J. 1993. Time's cycle and time's arrow in the history of Amazonia. *Biogeographica*, 69(1): 15–45.

Hershkovitz, P. 1969. The evolution of mammals on southern continents. IV. The recent mammals of the Neotropical region: a zoogeographical and ecological review. *Quart. Rev. Biol.*, 44: 1–70.

Hershkovitz, P. 1977. *Living New World Monkeys (Platyrrhini) With an Introduction to Primates. Vol. 1*. University of Chicago Press, Chicago.

Hershkovitz, P. 1983. Two new species of night monkeys, genus *Aotus* (Cebidae, Platyrrhini): a preliminary report on *Aotus* taxonomy. *Am. J. Primatol.*, 4: 209–243.

Hershkovitz, P. 1984. Taxonomy of squirrel monkeys genus *Saimiri* (Cebidae, Platyrrhini): a preliminary report with description of a hitherto unnamed form. *Am. J. Primatol.*, 7: 155–210.

Hershkovitz, P. 1990. Titis, New World monkeys of the genus *Callicebus* (Cebidae, Platyrrhini): a preliminary taxonomic review. *Fieldiana Zoology, New Series*, (55): v + 109pp.

Heyer, W. R. and Maxon, L. R. 1982. Distributions, relationships, and zoogeography of lowland frogs: the *Leptodactylus* complex in South America, with special reference to Amazonia. In: *Biological Diversification in the Tropics*, G.T.Prance (ed.), pp.375–388. Columbia University Press, New York.

Hill, W.C.O. 1960. *Primates. Comparative Anatomy and Taxonomy. IV. Cebidae, Part A*. Edinburgh University Press, Edinburgh.

Hill, W.C.O. 1962. *Primates. Comparative Anatomy and Taxonomy. V. Cebidae, Part B*. Edinburgh University Press, Edinburgh.

Joly, C.A., Leitão-Filho, H.F. and Silva, S.M. 1991. O patrimônio florístico/ The floristic heritage. In: *Mata Atlântica / Atlantic Rain Forest*, I. de G. Câmara (ed.), pp.95–125. Editora Index Ltd. and Fundação SOS Mata Atlântica, São Paulo.

Kay, R.F., Madden, R.H., Plavcan, J.M., Cifelli, R.L. and Diaz, J.G. 1987. *Stirtonia victoriae*, a new species of Miocene Colombian primate. *J. Hum. Evol.*, 16: 173–196.

Kierulff, M.C.M. 1993. Status and distribution of the golden lion tamarin in Rio de Janeiro. *Neotropical Primates*, 1(4): 23–24.

Kierulff, M.C.M. and Oliveira, P.P.de. 1994. Habitat preservation and the translocation of threatened groups of golden lion tamarins, *Leontopithecus rosalia*. *Neotropical Primates*, 2(suppl.): 15–18.

Kinzey, W. G. 1982. Distribution of primates and forest refuges. In: *Biological Diversification in the Tropics*, G.T.Prance (ed.), pp.455–482. Columbia University Press, New York.

Kleiman, D.G., Beck, B.B., Dietz, J.M., Dietz, L.A., Ballou, J.D., and Coimbra-Filho, A.F. 1986. Conservation program for the golden lion tamarin: captive research and management, ecological studies, educational strategies, and reintroduction. In: *Primates: The Road to Self-Sustaining Populations*, K.Benirschke (ed.), pp.959–979. Springer Verlag, New York.

Kobayashi, S. and Langguth, A.L. 1994. New titi monkey from Brazil. In: *Handbook and Abstracts, XV Congress of the International Primatological Society*, p.166. Kuta, Bali, Indonesia, 3–8 August, 1994. (Abstract).

Langguth, A., Teixeira, R.A. and Mittermeier, R.A. 1987. The red-handed howler monkey on northeastern Brazil. *Primate Conservation*, 8: 36–39.

Lemos de Sá, R.M. and Glander K.E. 1993. Capture techniques and morphometrics for the woolly spider monkey, or muriqui (*Brachyteles arachnoides*, E.Geoffroy 1806). *Am. J. Primatol.*, 29: 145–153.

Lemos de Sá, R.M., Pope, T.R., Glander, K.E., Struhsaker, T.T. and Fonseca, G. A. da. 1990. A pilot study of genetic and morphological variation in the muriqui (Brachyteles arachnoides). Primate Conservation (11): 26–30.

Lemos de Sá, R.M., Pope, T.R., Struhsaker, T.T. and Glander, K.E. 1993. Sexual dimorphism in canine length of woolly spider monkeys (*Brachyteles arachnoides*, E. Geoffroy 1806). *Int. J. Primatol.*, 14(5): 755–763.

Lorini, M.L. and Persson, V.G. 1990. Nova espécie de *Leontopithecus* Lesson, 1840, do sul do Brasil (Primates, Callitrichidae). *Bol. Mus. Nac., Nova Série*, Rio de Janeiro, (338): 1–14.

Lorini, M.L. and Persson, V.G. 1994. Status of field research on *Leontopithecus caissara*: the Black-Faced Lion Tamarin Project. *Neotropical Primates*, 2(suppl.): 52–55.

Mace, G. and Stuart, S. 1994. Draft IUCN Red List categories, Version 2.2. *Species*, (21–22), 13–24

Mallinson, J.J.C. 1989. A summary of the work of the International Recovery and Management Committee for Golden-headed Lion Tamarin *Leontopithecus chrysomelas*, 1985–1990. *Dodo, J. Jersey Wildl. Preserv. Trust*, 26: 77–86.

Mallinson, J.J.C. 1994. The Lion Tamarins of Brazil Fund: with reference to the International Management Committees for *Leontopithecus*. *Neotropical Primates*, 2(suppl.): 4–7.

Mansour, J. A. and Ballou, J. D. 1994. Capitalizing the ark: the economic benefit of adding founders to captive populations. *Neotropical Primates*, 2(suppl.): 8–11.

Martin, R.D. 1992. Goeldi and the dwarfs: the evolutionary biology of the small New World monkeys. *J. Hum. Evol.*, 22: 367–393.

Martuscelli, P. and Rodrigues, M.G. 1992. Novas populações do mico-leãocaissara, Leontopithecus caissara (Lorini and Persson, 1990) no sudeste do Brasil (Primates-Callitrichidae). Rev. Inst. Flor., São Paulo, 4:920–924.

Martuscelli, P., Petroni, L.M. and Olmos, F. 1994. Fourteen new localities for the muriqui *Brachyteles arachnoides*. *Neotropical Primates*, 2(2): 12–15.

Mendes, F.D.C. 1994. Muriqui conservation: the urgent need of an integrated plan. *Neotropical Primates*, 2(2): 16–19.

Milton, K. 1984. Habitat, diet, and activity patterns of free-ranging woolly spider monkeys (*Brachyteles arachnoides* E.Geoffroy 1806). *Int. J. Primatol.*, 5(5): 491–514.

Mittermeier, R.A., Coimbra-Filho, A.F., Constable, I.D., Rylands, A.B. and Valle, C. 1982. Conservation of primates in the Atlantic forest region of eastern Brazil. *Int. Zoo. Ybk.*, 22: 2–17.

Mittermeier, R.A., Valle, C.M.C., Alves, M.C., Santos, I.B., Pinto, C.A.M., Strier, K.B., Young, A.L., Veado, E.M., Constable, I.D., Paccagnella, S.G. and Lemos de Sá, R.M. 1987. Current distribution of the muriqui in the Atlantic forest region of eastern Brazil. *Primate Conservation*, 8: 143–149.

Mittermeier, R.A., Rylands, A.B. and Coimbra-Filho, A.F. 1988. Systematics: species and subspecies - an update. In: *The Ecology and Behavior of Neotropical Primates, Vol. 2*, R.A.Mittermeier, A.B.Rylands, A.F.Coimbra-Filho and G.A.B.da Fonseca (eds.), pp.13–75. World Wildlife Fund, Washington, D.C.

Mori, S.A., Boom, B.M. and Prance, G.T. 1981. Distribution patterns and conservation of eastern Brazilian coastal forest tree species. *Brittonia*, 33(2): 233–245.

Nagamachi, C.Y. 1995. Relações cromossômicas e análises filogenética e de agrupamento na família Callitrichidae (Platyrrhini, Primates). Unpublished Doctoral thesis, Universidade Federal do Rio Grande do Sul, Porto Alegre.

Natori, M. 1986. Interspecific relationships of *Callithrix* based on dental characters. *Primates*, 27(3): 321–336.

Natori, M. 1989. An analysis of cladistic relationships of *Leontopithecus* based on dental and cranial characters. *J. Anthropol. Soc. Nippon*, 97(2): 157–167.

Natori, M. 1990. Numerical analysis of the taxonomical status of *Callithrix kuhli* based on the measurements of the postcanine dentition. *Primates*, 31(4): 555–562.

Natori, M. 1994. Craniometrical variations among eastern Brazilian marmosets and their systematic relationships. *Primates*, 35(2): 167–176.

Natori, M. and Shigehara, N. 1992. Interspecific differences in lower dentition among eastern Brazilian marmosets. *J. Mammal.*, 73(3): 668–671.

Nishimura, A., Fonseca, G.A.B. da, Mittermeier, R.A., Young, A.L., Strier, K.B. and Valle, C.M.C. 1988. The muriqui, genus *Brachyteles*. In: *Ecology and Behavior of Neotropical Primates, Vol. 2*, R.A.Mittermeier, A.B.Rylands, A.F.Coimbra-Filho and G.A.B. da Fonseca (eds.), pp.577–610.

Oliver, W.L.R. and Santos, I.B. 1991. Threatened endemic mammals of the Atlantic forest region of south-east Brazil. *Wildlife Preservation Trust, Special Scientific Report 4*, 126pp.

Padua, S.M. 1994. Environmental education and the black lion tamarin, *Leontopithecus chrysopygus*. *Neotropical Primates*, 2(suppl.): 45–49.

Pinto, L.P.S. and Tavares, L.I. 1994. Inventory and conservation status of wild populations of golden-headed lion tamarins, *Leontopithecus chrysomelas*. *Neotropical Primates*, 2(suppl.): 24–27.

Pinto, L.P.S., Costa, C.M.R., Strier, K.B. and Fonseca, G.A.B.da 1993. Habitat, density and group size of primates in a Brazilian tropical forest. *Folia. Primatol.*, 61: 135–143.

Pinto, O.M.O. 1941. Da validez de *Cebus robustus* Kuhl e de suas relações com as formas mais afins. *Papéis Avulsos, Departamento de Zoologia, Secretaria da Agricultura, São Paulo*, 1:111–120.

Rizzini, C.T. 1979. *Tratado de Fitogeografia do Brasil: Aspectos Sociológicos e Florísticos. Vol.2*. Editora da Universidade de São Paulo, São Paulo.

Rizzini, C.T., Coimbra-Filho, A.F. and Houaiss, A. 1988. *Ecossistemas Brasileiros / Brazilian Ecosystems*. Editora Index, Rio de Janeiro.

Rosenberger, A.L. 1981. Systematics: the higher taxa. In: *Ecology and Behavior of Neotropical Primates, Vol.2*, A.F.Coimbra-Filho and R.A.Mittermeier (eds.), pp.9–27. Academia Brasileira de Ciências, Rio de Janeiro.

Rosenberger, A.L. 1984. Fossil New World monkeys dispute the molecular clock. *J. Hum. Evol.*, 13: 737–742.

Rosenberger, A.L. 1992a. Evolution of feeding niches in New World monkeys. *Amer.J.Phys.Anthropol.*, 88: 525–562.

Rosenberger, A.L. 1992b. Evolution of the Mata Atlantica primates. In: *Abstracts. XIVth Congress of the International Primatological Society*, August 16–21, 1992, Strasbourg. p.93. (Abstract).

Rosenberger, A.L. and Coimbra-Filho, A.F. 1984. Morphology, taxonomic status and affinities of the lion tamarins, *Leontopithecus* (Callitrichinae, Cebidae). *Folia Primatol.*, 42: 149–179.

Rosenberger, A.L. and Strier, K.B. 1989. Adaptive radiation of the ateline primates. *J. Hum. Evol.*, 18: 717–750.

Rosenberger, A.L., Setoguchi, T. and Shigehara, N. 1990. The fossil record of callitrichine primates. *J. Hum. Evol.*, 19: 209–236.

Rylands, A.B. 1984. Exudate-eating and tree-gouging by marmosets (Callitrichidae, Primates). In: *Tropical Rain Forest: The Leeds Symposium*. A.C. Chadwick and S.L. Sutton (eds.), pp.155–168. Leeds Philosophical and Literary Society, Leeds.

Rylands, A.B. 1987. Primate communities in Amazonian forests: their habitats and food resources. *Experientia*, 43: 265–279.

Rylands, A.B. 1989. Sympatric Brazilian callitrichids: the black tufted-ear marmoset, *Callithrix kuhli*, and the golden-headed lion tamarin, *Leontopithecus chrysomelas*. *J. Hum. Evol.*, 18: 679–695.

Rylands, A.B. 1993. The ecology of lion tamarins, *Leontopithecus*: some intrageneric differences and comparisons with other callitrichids. In: *Marmosets and Tamarins: Systematics, Ecology, and Behaviour*, A.B.Rylands, (ed.), pp.296–313. Oxford University Press, Oxford.

Rylands, A.B. 1994a. Guariba-preto, guariba-de-mãos-ruivas, *Alouatta belzebul* (Linnaeus, 1766).In: *Livro Vermelho dos Mamíferos Brasileiros Ameaçados de Extinção*. G.A.B. da Fonseca, A.B.Rylands, C.M.R. Costa, R.B.Machado and Y.L.R.Leite (eds.), pp.153–159. Fundação Biodiversitas, Belo Horizonte.

Rylands, A.B. 1994b. Bugio, guariba, barbado, *Alouatta fusca* (E.Geoffroy, 1812). In: *Livro Vermelho dos Mamíferos Brasileiros Ameaçados de Extinção*. G.A.B. da Fonseca, A.B.Rylands, C.M.R. Costa, R.B.Machado and Y.L.R.Leite (eds.), pp.161–170. Fundação Biodiversitas, Belo Horizonte.

Rylands, A.B. 1994c. Guigó, sauá, *Callicebus personatus* (E. Geoffroy, 1812). In: *Livro Vermelho dos Mamíferos Brasileiros Ameaçados de Extinção*. G.A.B. da Fonseca, A.B.Rylands, C.M.R. Costa, R.B.Machado and Y.L.R.Leite (eds.), pp.211–218. Fundação Biodiversitas, Belo Horizonte.

Rylands, A.B. 1996 Habitat and the evolution of social and reproductive behavior in Callitrichidae. *Am. J. Primatol.*38: 5–18.

Rylands, A.B. and Faria, D.S.de. 1993. Habitats, feeding ecology, and home range size in the genus *Callithrix*. In: *Marmosets and Tamarins: Systematics, Ecology, and Behaviour*, A.B.Rylands, (ed.), pp.262–272. Oxford University Press, Oxford.

Rylands, A.B. and Pinto, L.P.S. 1994.. Macaco-prego-de-peito-amarelo, Cebus apella xanthosternos (Wied, 1820). In: Livro Vermelho dos Mamíferos Brasileiros Ameaçados de Extinção. G.A.B. da Fonseca, A.B.Rylands, C.M.R. Costa, R.B.Machado and Y.L.R.Leite (eds.), pp.219–226. Fundação Biodiversitas, Belo Horizonte.

Rylands, A.B., Spironelo, W.R., Tornisielo, V.L., Sá, R.L. de, Kierulff, M.C.M. and Santos, I.B. 1988. Primates of the Rio Jequitinhonha valley, Minas Gerais, Brazil. *Primate Conservation*, 9: 100–109.

Rylands, A.B., Coimbra-Filho, A.F. and Mittermeier, R.A. 1993. Systematics, geographic distribution, and some notes on the conservation status of the Callitrichidae. In: *Marmosets and Tamarins: Systematics, Ecology, and Behaviour*, A.B.Rylands, (ed.), pp.262–272. Oxford University Press, Oxford.

Salo, J. 1990. External processes influencing origin and maintenance of inland water-land ecotones. In: *The Ecology and Management of Aquatic-Terrestrial Ecotones*, R.J.Naiman and H. Décamps (eds.), pp.37–64. UNESCO, Paris.

Salo, J., Kalliola, R., Häkkinen, I., Mäkinen, Y., Niemelä, P., Puhakka, M. and Coley, P.D. 1986. River dynamics and the diversity of Amazon lowland forest. *Nature, Lond.*, 322: 254–258.

Santos, I.B. and Lernould, J.-M. 1993. A conservation program for the yellow-breasted capuchin, *Cebus apella xanthosternos*. *Neotropical Primates*, 1(1): 4–5.

Santos, I.B., Mittermeier, R.A., Rylands, A.B. and Valle, C. 1987. The distribution and conservation status of primates in southern Bahia. *Primate Conservation*, 8: 126–142.

Schneider, H., Schneider, M.P.C., Sampaio, I., Harada, M.L., Stanhopes, M., Czelusniak, J. and Goodman, M. 1993. Molecular phylogeny of the New World monkeys (Platyrrhini, Primates). *Molecular Phylogenetics and Evolution*, 2(3): 225–242.

Seal, U.S., Ballou, J.D. and Valladares-Padua, C. (eds.).1990. Leontopithecus - *Population Viability Analysis Workshop Report*. IUCN/SSC Captive Breeding Specialist Group (CBSG), Apple Valley, Minnesota.

Seuánez, H.N., Armada, J.L., Freitas, L., Rocha e Silva, R., Pissinatti, A. and Coimbra-Filho, A.F. 1986. Intraspecific chromosome variation in *Cebus apella* (Cebidae, Platyrrhini): the chromosomes of the yellow breasted capuchin *Cebus apella xanthosternos* Wied, 1820. *Am. J. Primatol.*, 10: 237–247.

Silva, A.F. da and Shepherd, G.S. 1986. Comparações florísticas entre algumas mata brasileiras utilizando análise de agrupamento. *Rev. Brasil. Bot.*, 9: 81–86.

Silva, M.N.F. da and Patton, J.L. 1993. Amazonian phylogeography: mtDNA sequence variation in arboreal echimyid rodents (Caviomorpha). *Molecular Phylogenetics and Evolution*, 2: 243–255.

Simpson, B.B. and Haffer, J. 1978. Speciation in the Amazonian forest biota. *Ann. Rev. Ecol. Syst.*, 9: 497–518.

Snowdon, C.T. 1993. A vocal taxonomy of the callitrichids. In: *Marmosets and Tamarins: Systematics, Ecology, and Behaviour*, A.B.Rylands, (ed.), pp.78–94. Oxford University Press, Oxford.

Strier, K.B. 1991. Demography and conservation of an endangered primate, *Brachyteles arachnoides*. *Conserv. Biol.*, 5: 214–218.

Strier, K.B. 1992a. *Faces in the Forest: The Endangered Muriqui Monkeys of Brazil*. Oxford University Press, New York.

Strier, K.B. 1992b. Atelinae adaptations: behavioral strategies and ecological constraints. *Am. J. Phys. Anthropol.*, 88: 515–524.

Terborgh, J. 1983. *Five New World Primates: A Study in Comparative Ecology*. Princeton University Press, Princeton, NJ.

Torres, C. 1988. Resultados preliminares de reavaliação das raças do macaco-prego *Cebus apella* (Primates: Cebidae). *Rev. Nordest. Biol.*, 6(1): 15–28.

Valladares-Padua, C. and Simon, F. 1992. *International Studbook Black Lion Tamarin* Leontopithecus chrysopygus. Fundação Parque Zoológico de São Paulo, São Paulo.

Valladares-Padua, C., Padua, S.M. and Cullen Jr., L. 1994. The conservation biology of the black lion tamarin, *Leontopithecus chrysopygus*: first ten year's report. *Neotropical Primates*, 2(suppl.): 36–39.

Vivekananda, G. 1994. The Superagüi National Park: problems concerning the protection of the black-faced lion tamarin, *Leontopithecus caissara*. *Neotropical Primates*, 2(suppl.): 56–57.

Vivo, M. de 1991. *Taxonomia de Callithrix Erxleben, 1777 (Callitrichidae, Primates)*. Fundação Biodiversitas, Belo Horizonte.

Zingeser, M.R. 1973. Dentition of *Brachyteles arachnoides* with reference to alouattine and atelinine affinities. *Folia Primatol.*, 20: 351–390.

3

PRIMATE POPULATIONS IN EASTERN AMAZONIA

Stephen F. Ferrari[1] and M. Aparecida Lopes[2]

[1]Departamento de Genética and Departamento de Psicologia Experimental
Universidade Federal do Pará
Caixa Postal 8607, 66.075–150 Belém - PA, Brazil
[2]Departamento de Biologia
Universidade Federal do Pará
Belém do Pará, Brazil

ZOOGEOGRAPHY OF EASTERN AMAZONIAN PRIMATES

Eastern Amazonia is defined here as the region to the east of the Rio Xingu, including the Marajó archipelago in the Amazon delta (Figure 1). This region's habitats can be divided into three main types: lowland and submontane tropical rainforest (*terra firme*), flooded forests (including mangrove, *várzea* and *igapó*) and *cerrado* (savanna). A fourth major category that is becoming increasingly dominant encompasses disturbed and secondary forest, and land cleared for agriculture (SUDAM, 1988). In some areas, such as the 30,000 km^2 Zona Bragantina to the east of Belém, the original cover is now virtually nonexistent (Johns and Ayres, 1987; Salomão, 1994), while a large part of the lowland forest between the Xingu and the Tocantins remains intact.

Despite being the most densely-populated region of the Amazon basin and home to the Goeldi Museum, whose zoological collection was established more than a century ago, surprisingly few studies of the fauna of eastern Amazonia were available until very recently, and our knowledge of the zoogeography of this region's primates is still far from complete. This is emphasized most clearly by the recent discovery of the Ka'apor capuchin (*Cebus kaapori* Queiroz, 1992) in western Maranhão. Subsequent fieldwork has revealed that this monkey, which belongs to the "untufted" capuchin group, occurs at least as far west as the Rio Tocantins (Lopes and Ferrari, 1993, in press), implying that it originally occurred throughout much of eastern Pará (see also Goeldi and Hagmann, 1906).

Cebus nigrivittatus has recently been recorded (Fernandes et al., submitted) from Caviana island in the Marajó archipelago (Figure 1), and there is also unconfirmed evidence that untufted capuchins occur west of the Tocantins (Ferrari and de Souza Jr., 1994). As Queiroz's original classification (1992) of *Cebus kaapori* as a true species was based primarily on the disjunct nature of its known distribution in relation to that of *Cebus*

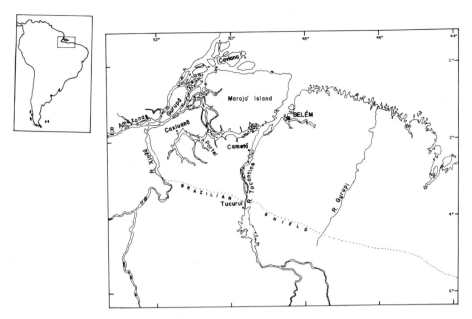

Figure 1. Eastern Amazonia.

nigrivittatus rather than distinct morphological characteristics, these data support the need for a re-appraisal of the former's status (Anon., 1993). This is reinforced by a recent study of molecular genetics (Harada, 1994), which has shown that *kaapori* may in fact be differentiated from *Cebus nigrivittatus* at no more than the subspecific level. Definition of the taxonomic status of the *kaapori* form is far more than a merely academic question, given that it is clearly one of, if not the most endangered of Amazonian primates (Lopes and Ferrari, 1993, in press; Ferrari and Queiroz, 1994).

The occurrence of the white-fronted spider monkey, *Ateles belzebuth marginatus*, in the region also remains unclear. The type locality of this taxon is Cametá (Kellogg and Goldman, 1944), on the west bank of the lower Tocantins (Figure 1), but no further specimens have been observed or collected in the Xingu-Tocantins interfluvium, and recent fieldwork (e.g. Egler, 1985; Mascarenhas and Puorto, 1989; Ferrari and Lopes, 1990; present study) indicates that *Ateles* is absent from this area. Of course, confirmation of a species's absence from an area is far more difficult than that of its occurrence, so this possibility cannot yet be discounted altogether.

The zoogeography of the region's other primates is more straightforward (Table 1, Figure 2). Tufted capuchins (*Cebus apella apella*), owl monkeys (*Aotus infulatus*), squirrel monkeys (*Saimiri sciureus sciureus*) and black-handed tamarins (*Saguinus midas niger*) are ubiquitous, at least where appropriate habitat occurs (e.g. Peres, 1989; Silva Jr. et al., 1993) and/or is still available.

Howler monkeys (*Alouatta*) are equally widespread, although Fernandes (1994) recently found *Alouatta seniculus*, and not *Alouatta belzebul*, on Gurupá island in the Amazon delta (Figure 2). Analyzing the highly variable pelage coloration in *Alouatta belzebul*, Bonvicino et al. (1989) recognized two subspecies in eastern Amazonia, *Alouatta belzebul belzebul* on the mainland, and *Alouatta belzebul discolor* in the Marajó archipelago. A single specimen from Portel on the mainland east of the Rio Xingu (Figure 1) is classified by

Table 1. Primate species and subspecies known to occur in eastern Amazonia

Sub/species	Known distribution
Callithrix argentata	Lowland floodplain (± sea level) in the Xingu-Tocantins interfluvium.
Saguinus midas niger	Widespread.
Cebus apella apella	Widespread.
Cebus kaapori	Mainland east (possibly also west) of the Rio Tocantins, and north of the Brazilian Shield.
Cebus nigrivittatus	Caviana and possibly other islands in the Amazon delta.
Saimiri sciureus sciureus	Widespread.
Aotus infulatus	Widespread.
Callicebus moloch	Mainland west of the Rio Tocantins.
Chiropotes satanas satanas	Mainland east of the Rio Tocantins.
Chiropotes satanas utahicki	Mainland west of the Rio Tocantins.
Alouatta belzebul belzebul	Mainland.
Alouatta belzebul discolor	Marajó archipelago, except Gurupá Island.
Alouatta seniculus	Gurupá Island in the Amazon delta.
?Ateles belzebuth marginatus	If present, mainland west of the Rio Tocantins, and probably north of the Brazilian Shield.

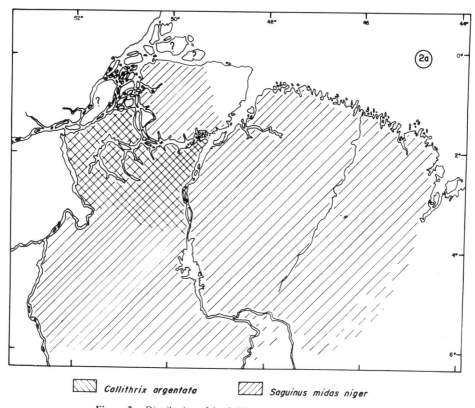

Figure 2a. Distribution of the Callitrichinae in eastern Amazonia.

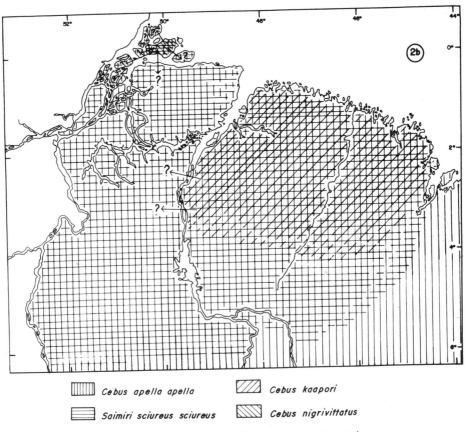

Figure 2b. Distribution of the Cebinae in eastern Amazonia.

these authors as *Alouatta b. discolor*, but its color pattern is within the range of that of *Alouatta b. belzebul*, and in fact, Fernandes (1994) has questioned the validity of pelage coloration for the classification of subspecies. As *Alouatta b. belzebul* has been recorded from all other localities in the Xingu-Tocantins interfluvium (Schneider et al., 1991), it seems reasonable to identify the Portel specimen as *Alouatta b. belzebul* until further data are available.

The three remaining genera are less widely distributed (Figure 2). Bearded sakis, *Chiropotes satanas*, are found everywhere on the mainland (i.e. not in the Marajó archipelago), with *Chiropotes satanas satanas* restricted to forested habitats east of the Tocantins (Lopes, 1993) and *Chiropotes satanas utahicki* to the west (Hershkovitz, 1985). Titi monkeys, *Callicebus moloch*, are found on the mainland west of the Tocantins (Hershkovitz, 1990).

In eastern Amazonia as defined here, the silvery marmoset (*Callithrix argentata*) is not only restricted to the west of the Tocantins, but to the lowland floodplain of the Xingu-Tocantins interfluvium (Ferrari and Lopes, 1990). Given the absence of geographic barriers, Ferrari and Lopes (1990) have argued that the restricted distribution of *Callithrix argentata* east of the Xingu (it is more widespread west of this river) is related to two principal factors: habitat differences and sympatry with a second callitrichine, *Saguinus m. niger* (absent west of the Xingu).

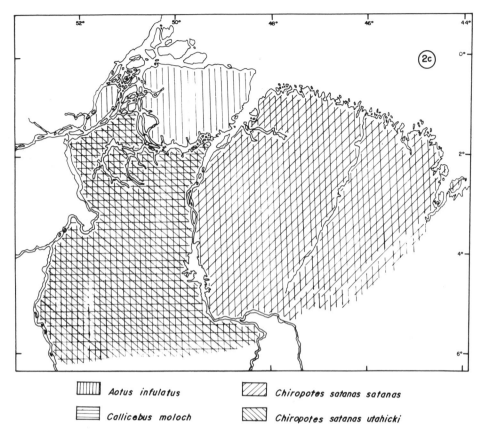

Figure 2c. Distribution of the Pitheciinae in eastern Amazonia.

Habitat differences would be underpinned by contrasts in fertility between the sediment-rich floodplain and the nutrient-poor soils associated with the geologically ancient uplands of the Brazilian Shield further south (Figure 1). Given the distribution of callitrichines in southern Amazonia, Ferrari (1993) has interpreted the present-day situation in the Xingu-Tocantins interfluvium as the result of a relatively recent (possibly Holocenic) expansion of the range of *Callithrix argentata* into that of *Saguinus m. niger*. Under present conditions, the resident tamarin population would appear to have the upper hand competitively in the habitats associated with the poorer soils further south.

Similar factors may determine the restricted distribution of *Cebus kaapori*, which is sympatric with *Cebus apella* throughout its known range (Lopes, 1993; Lopes and Ferrari, in press). While the Ka'apor capuchin has now been at least reported from a number of localities in eastern Pará and western Maranhão (Queiroz, 1992; Lopes, 1993), it is almost certainly absent from the region of Tucuruí (Mascarenhas and Puorto, 1989), which lies towards the northern extreme of the Brazilian Shield (Figure 1).

DIVERSITY GRADIENTS

Two basic gradients of primate diversity thus exist within the region's rainforest habitats: increasing diversity from east to west, and declining diversity moving upriver along the

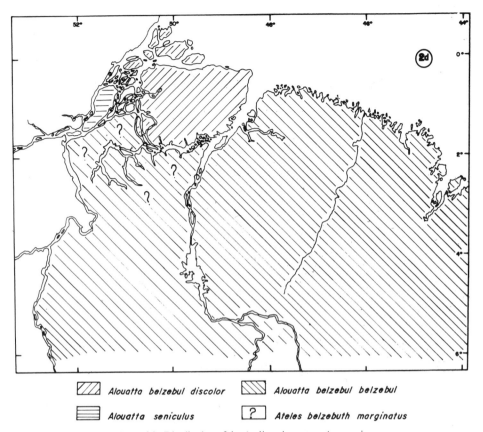

Figure 2d. Distribution of the Atelinae in eastern Amazonia.

main Amazon tributaries (Xingu, Tocantins). The east-west gradient is typical not only of Amazonian primates as a whole (see chapters in this volume), but also of other faunal and floral groups, and is probably related to a combination of factors including climate and geomorphology (Emmons, 1984; Gentry, 1988; Prance, 1992; Ayres, 1993). The second gradient is associated with similar factors, and in fact an abrupt decline in primate diversity is encountered where the Amazon Forest intergrades with central Brazilian *cerrado* habitats.

A third gradient is also habitat-related, that is, reduced diversity in savanna, mangrove and disturbed/secondary habitats in comparison with undisturbed rainforest (e.g. Johns and Ayres, 1987; Ayres et al., 1989; Peres, 1989; Fernandes and Aguiar, 1993; Lopes, 1993). In general, the primates most seriously affected by habitat disturbance are the medium-sized sakis and capuchins. The much larger *Alouatta belzebul* frequently persists at sites from which these monkeys have disappeared, while populations of the small-bodied *Saguinus m. niger*, in particular, may actually be benefitted by habitat disturbance.

An interesting aspect of the second diversity gradient is that of the transition between Amazonian and "Atlantic Forest" communities, in particular the callitrichines. Despite their ability to thrive not only in the Atlantic Forest, but also in the *cerrado* and *caatinga* habitats of central and northeastern Brazil, the distribution of both common (*Callithrix jacchus*) and pencil-tufted (*Callithrix penicillata*) marmosets appears to stop short of neighboring areas of Amazon rainforest, inhabited by *Saguinus m. niger*.

Table 2. Inventories of Amazonian tree communities (DBH ≥10 cm)

Site	Habitat type[1]	Sample size (ha)	Inds/ha	Families	Genera	Species	Source[2]
East of Rio Gurupi							
Açailândia	VA	1.0	456[3]	31	79	100	1
Alto Turiaçu	TF	1.0	498	35	78	116	2
Buriticupú	TF	1.0	533[3]	38	95	133	1
Buriticupú	TF	1.0	425[3]	23	57	66	1
Rio Tocantins/Rio Gurupi							
Alto Guamá	TF	1.0	456	35	85	138	3
Belém	TF	1.0	423	31	65	87	4
Belém	TF	2.0	449	—	—	—	5
Belém	IG	1.0	564	28	51	60	4
Castanhal	TF	3.5	423	—	—	—	6
Guamá	VA	3.8	483	—	—	—	7
Irituia	TF	0.5	462	—	—	—	8
Quiandeua	TF	1.0	484	39	—	151	9
Tailândia	TF	1.0	483	38	88	147	8
Rio Tocantins/Rio Xingu							
Caxiuanã	TF	1.0	649	43	—	196	10
	TF	1.0	527	40	—	191	10
	TF	1.0	727	37	—	147	10
	TF	1.0	538	38	—	179	10
O Deserto	TF	1.0	393	33	76	133	11
	TF	1.0	460	33	72	118	11
	TF	1.0	567	33	83	162	11
	VA	0.5	440	—	—	—	11
Marajó Island							
Breves	TF	1.0	516	36	—	157	12
Western Amazonia							
Ji-Paraná	TF	1.0[3]	564	41	125	164	1
Mamirauá	VA	1.0	416	36	84	109	13
	VA	1.0	580	36	95	135	13
	IG	1.0	546	36	91	118	13
Manu	TF	1.0	584	—	—	153	14
	TF	1.0	673	43	—	210	15
Mishana	TF	1.0	859	—	—	295	16
Cabeza de Mono	TF	1.0	544	—	—	185	16

[1]TF = *terra firme*; VA = *várzea*; IG = *igapó*.
[2]1=Salomão et al. (1988); 2=Balée (1986); 3=Balée (1987); 4=Black et al. (1950); 5=Cain et al. (1956); 6=Pires et al. (1953); 7=Pires and Koury (1958); 8=Lopes (1993); 9=Shanley (in prep.); 10=Almeida et al. (1993); 11=Campbell et al. (1986); 12=Pires (1966); 13=Ayres (1993); 14=Hartshorn (1980); 15=Gentry (1985); 16=Gentry (1986).
[3]DBH ≥9.55 cm.

Current knowledge of the extremes of the distribution of these three species in Maranhão and southern Pará/northern Mato Grosso is scant (Hershkovitz, 1977; de Vivo, 1991; Silva Jr. et al., 1992), but there appears to be no obvious geographic barrier between them. Factors determining the absence of these marmosets from Amazonian rainforest habitats may thus be similar to those that restrict the sympatry of *Callithrix argentata* and *Saguinus m. niger* further north. Of the other eastern Amazonian primates, only two (*Alouatta belzebul* and *Cebus apella*) make the transition to central Brazilian and Atlantic Forest habitats, although there is some evidence that *Aotus* may also occur in the former (A. Silva, personal communication).

HABITATS

Within Amazonia, plant communities are also characterized by an east-west gradient in tree (Gentry, 1982) and species density (Ducke and Black, 1954; Gentry, 1986, 1988; Ayres, 1993), correlating with those in soil fertility and precipitation. Species density is also lower in flooded habitats - *várzeas* and *igapós* - in comparison with *terra firme* forest in the same area, although western Amazonian *várzea* may be as species-rich as eastern Amazonian *terra firme* (Ayres, 1993). There are local exceptions, however, such as Caxiuanã, in the northwestern extreme of eastern Amazonia, whose *terra firme* forest is among Amazonia's richest (Almeida et al., 1993).

Once again, relatively few botanical studies are available for eastern Amazonia, and methodological differences tend to limit systematic comparisons between data sets. Nevertheless, an east-west gradient in both species and tree density is again apparent within eastern Amazonia, and there are clear contrasts with western Amazonian communities (Table 2, Figure 3). According to these data, an average hectare of *terra firme* forest in western Amazonia contains over forty percent more species than that of eastern Amazonia, and almost thirty percent more trees.

The families Leguminosae and Sapotaceae account for approximately a third of tree species at five sites in eastern Amazonia (Table 3), with Burseraceae, Chrysobalanaceae, Moraceae and Lecythidaceae accounting for another fifteen to twenty percent. In western Amazonia, by contrast, whereas the Leguminosae is still the major family, others, such as the Lauraceae, Annonaceae, Rubiaceae, Myristicaceae and Meliaceae are also relatively well represented (Gentry, 1986).

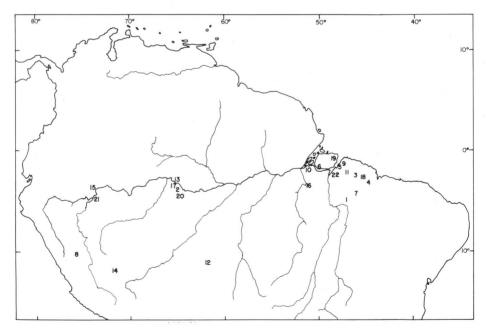

Figure 3. Study sites mentioned in the text: 1=Açailândia, 2=Açaituba, 3= Alto Guamá, 4=Alto Turiaçu, 5=Belém/Guamá, 6=Breves, 7=Buriticupú, 8=Cabeza de Mono, 9=Castanhal, 10=Caxiuanã, 11=Irituia, 12=Ji-Paraná, 13=Mamirauá, 14=Manu, 15=Mishana, 16=O Deserto, 17=Ponta da Castanha, 18=Quiandeua, 9=Rio Jutuba, 20=Rio Urucu, 21= Tahuayo, 22=Tailândia.

Table 3. Plant families represented by five or more tree species (DBH ≥10 cm) in 1 ha plots at five sites in eastern Amazonia

Family	Number of tree species (% of total) in 1 ha plot at				
	Tailândia	Irituia[1]	Alto Guamá	Alto Turiaçu	Quiandeua
Leguminosae	30 (20.4)	10 (12.5)	25 (18.1)	16 (13.8)	34 (22.5)
Sapotaceae	28 (19.1)	14 (17.5)	18 (13.0)	17 (14.7)	25 (16.6)
Burseraceae	13 (8.8)	6 (7.5)	13 (9.4)	10 (8.6)	6 (4.0)
Chrysobalanaceae	10 (6.8)	5 (6.3)	8 (5.8)	7 (6.0)	13 (8.6)
Lecythidaceae	8 (5.4)	5 (6.3)		7 (6.0)	9 (6.0)
Moraceae	8 (5.4)	8 (10.0)		5 (4.3)	
Euphorbiaceae			8 (5.8)	5 (4.3)	
Annonaceae			5 (4.3)		5 (3.3)
Lauraceae			6 (4.4)		
Anacardiaceae			5 (3.6)		
Meliaceae			5 (3.6)		

[1] 0.5 ha plot.

With a fifth to a third of individual trees at any one site (Table 4) the Lecythidaceae is by far the most abundant family in eastern Amazonia *terra firme* forests, although the Leguminosae and Sapotaceae are relatively well represented at most sites. Together, the eight most frequently encountered families account for at least 80% of the trees at each of the study sites, re-emphasizing the relatively homogeneous nature of this region's forests (Ducke and Black, 1954).

Perhaps unsurprisingly, these families, in particular Leguminosae, Sapotaceae, Moraceae and Lecythidaceae, are also among the most prominent in the diets of Amazonian primates (e.g. Terborgh, 1983; Ayres, 1989; Ferrari and Lopes, 1990). In eastern Amazonia, important primate food sources, such as *Eperua bijuga* and *Hymenaea* (Leguminosae), *Lecythis pisonis* (Lecythidaceae) and *Manilkara* spp. (Sapotaceae) are valued for their timber. In addition to habitat disturbance, then, selective logging may have seri-

Table 4. Relative frequency of individuals (DBH ≥10 cm) of the eight most commonly encountered plant families in 1 ha plots at five sites in eastern Amazonia

Family	Number of trees (% of total) in 1 ha plot at:				
	Tailândia	Irituia[1]	Alto Guamá	Alto Turiaçu	Quiandeua
Lecythidaceae	105 (21.8)	77 (33.3)	83 (18.2)	162 (32.5)	88 (18.2)
Leguminosae	73 (15.1)	29 (12.6)	62 (13.6)	28 (5.6)	86 (17.8)
Sapotaceae	70 (14.5)	23 (10.0)	54 (11.8)	41 (8.2)	101 (20.9)
Burseraceae	34 (7.0)	17 (7.4)	63 (13.8)	72 (14.5)	20 (4.1)
Chrysobalanaceae	25 (5.2)	10 (4.3)	30 (6.6)	21 (4.2)	42 (8.7)
Euphorbiaceae	21 (4.4)		52 (11.4)	72 (14.5)	13 (2.7)
Violaceae	56 (11.6)	12 (5.2)			44 (9.1)
Moraceae	15 (3.1)	12 (5.2)			
Meliaceae			15 (3.3)		13 (2.7)
Annonaceae			17 (3.4)		
Anacardiaceae			12 (2.6)		
Palmae				12 (2.4)	
Lauraceae		5 (2.2)			

[1] 0.5 ha plot.

Table 5. Population densities of diurnal primates in eastern Amazonia

	Tailândia	Rio Capim	Irituia	Gurupi Biological Reserve	Rio Jutuba, Marajó Island
Habitat[1]	TF	TF	TF	TF	GF
Hunting pressure	intense	high	moderate	mod./low	low
Km surveyed	216	205	408	480	36.8
Population density (inds/km^2)[2] of					
Alouatta belzebul	0.14	0.80	1.86	5.09	16.80
Cebus apella	1.40	4.03	8.16	8.76	a
Cebus kaapori	p[3]	p	p	0.98	a
Chiropotes satanas	10.08[4]	1.80	7.21	7.15	a
Saguinus midas	10.40	23.33	19.02	12.71	a
Saimiri sciureus	p	a	p	0.15	54.20
All species	22.02	29.96	36.24	34.84	72.00
Total primate biomass (kg/km^2)	38.31	32.76	63.05	83.41	122.60
Source[5]	1	1	1	1	2

[1] TF = *terra firme* forest; GF = gallery forest within *cerrado*.
[2] Estimated by Fourier series expansion (see Brockelman and Ali, 1987).
[3] p = present, but not observed in censuses; a = reported absent.
[4] Probable overestimate due to sampling problems (see Lopes, 1993), density estimate using the Kelker method was 3.24 inds/km^2.
[5] 1 = Lopes (1993); 2 = Peres (1989).

ous consequences for local primate populations, especially for species such as *Chiropotes satanas* (Johns and Ayres, 1987) and *Cebus kaapori* (Lopes, 1993).

Unfortunately, the available data do not yet permit the identification of keystone taxa or evaluation of the habitat variables that may contribute to the restricted distribution and limited population density of certain eastern Amazonian primates.

PRIMATE POPULATIONS IN EASTERN AMAZONIA

Reflecting these differences in habitat structure and diversity, population densities and biomass of primates in eastern Amazonia (Table 5) are also relatively low in comparison with the more diverse western Amazonian communities (Table 6). Mainland *terra firme* communities are dominated by four diurnal species, *Alouatta belzebul*, *Cebus apella*, *Chiropotes satanas* and *Saguinus midas*, whose relative densities vary in accordance with hunting pressure and habitat disturbance (Lopes, 1993). While rarely, if ever hunted, the population density of *Saguinus midas*, for example, was lowest at Tailândia (Table 5), where habitat remains intact, and highest in the logged forest at Rio Capim. The abundance of *Alouatta belzebul*, by contrast, is inversely related to hunting pressure, irrespective of habitat disturbance.

The low densities or absence of the remaining species at most sites are due to a variety of factors. The predominantly nocturnal habits of *Aotus* make sightings in daytime censuses unlikely (but not impossible: Peres, 1989; Lopes, 1993), but if records from other regions are typical (e.g. Terborgh, 1983; Aquino and Encarnación, 1988), *Aotus infulatus* may be one of, if not the region's most abundant primate. In the case of *Saimiri sciureus*, the principal factor would appear to be habitat specialization (e.g. Terborgh, 1983; Peres, 1989; Lopes, 1993). Thus, while the species apparently reaches relatively high ecological densities in riparian habitats, absolute densities are low at most sites.

Table 6. Primate densities and biomass in western Amazonia

Site	Number of diurnal primate species	Total population density (inds/km^2)	Total primate biomass (kg/km^2)	Source
Açaituba	11	67	283	Johns (1986)
Manu	11	249	625	Terborgh (1983)
Rio Urucu	12	137	372	Peres (1993)
Ponta da Castanha	11	156	223	Johns (1986)[1]
Tahuayo	12	82	112	Bodmer et al. (1988)[1]

[1]Hunted site.

By contrast, interspecific competition with closely-related taxa may be the principal factor determining the restricted geographical distribution and low population densities of both *Callithrix argentata* and *Cebus kaapori* (Lopes, 1993; Lopes and Ferrari, in press). The density of untufted capuchin populations also appears to relatively low in comparison with sympatric *Cebus apella* in other regions (Baal et al., 1988; Peres, 1993), and sympatry with *Saguinus fuscicollis* in southwestern Amazonia appears to have a similar influence on the distribution and abundance of *Callithrix emiliae* (Lopes and Ferrari, 1994; Ferrari et al., 1995).

Preliminary data from Caxiuanã National Forest (Table 7) indicate that *Callithrix argentata* is rare or absent from undisturbed *terra firme* forest occupied by *Saguinus m. niger*, but may be relatively abundant in neighboring areas of disturbed and/or secondary forest. As at other sites, *Saguinus m. niger* was also more abundant in disturbed habitat at Caxiuanã (Table 7). Ferrari and Lopes (1990) encountered *Callithrix argentata* and *Saguinus m. niger* groups with equal frequency during informal surveys in highly disturbed forest habitat close to the southern limit of the distribution of the former species. Even so, groups of *Callithrix argentata* appeared to be smaller than those of *Saguinus m. niger*, suggesting a lower population density. At Caxiuanã, in addition, gum-producing trees (e.g. *Parkia* sp.) with typical marmoset gouge holes were encountered in disturbed forest, but not in primary habitat.

It is usually possible to confirm the presence of *Callicebus* in a given area on the basis of its characteristic early-morning duetting (e.g. Ferrari and Lopes, 1990). At Caxiuanã, however, *Callicebus moloch* was neither seen (Table 7) nor heard during more than two weeks at each of the two study sites, although some local residents reported its occurrence within the area of the National Forest. *Callicebus moloch* is relatively abundant at sites further south (Mascarenhas and Puorto, 1988; Ferrari and Lopes, 1990), so why it

Table 7. Preliminary records of primate densities at the Caxiuanã National Forest, Pará. Straight-line transects were walked at 1–2 km/h at 07:00–13:00 h

Species	Groups sighted per 10 km	
	IBAMA - Sede[1]	Ferreira Penna Research Station[2]
Alouatta belzebul	p[3]	2.0 (1-6)[4]
Callithrix argentata	1.2 (3-5)	—
Cebus apella	0.4 (≥6)	0.7 (5-6+)
Chiropotes satanas	—	0.3 (≥7)
Saguinus midas	2.4 (3-7+)	0.7 (4-5)

[1]25 km of disturbed and edge habitat walked in 1990.
[2]30 km of undisturbed *terra firme* forest walked in 1994.
[3]Feces observed.
[4]Number of animals observed.

should be apparently so rare at Caxiuanã, especially given the habitat characteristics at this site (Almeida et al., 1993), is something of an enigma.

DIVERSITY AND CONSERVATION

With no more than seven species in a given area, the *terra firme* primate community in the region east of the Rio Tocantins is Amazonia's least diverse. It is thus probably little surprise that both research and conservation efforts in this region have been relatively limited in comparison with those further west (e.g. Terborgh, 1983; Ayres, 1989, 1993; Peres, 1993). In fact, beyond the surveys and anecdotal records cited here, virtually nothing is known of the ecology or behavior of any of the primates of eastern Amazonia.

While understandable, this relative neglect not just of the region's primates but of many of its other faunal and floral groups, is more than a little problematic for the understanding - and conservation - of Amazonian biodiversity as a whole. Of the fourteen taxa mentioned here (Table 1), for example, only one (*Alouatta seniculus*) occurs in western Amazonia, i.e. between the Madeira and Amazonas/Japurá rivers, where primate diversity is greatest (Peres, 1993), and two genera (*Callithrix* and *Chiropotes*) are absent altogether. Significant contrasts in the primate community are apparent even in the neighboring Xingu-Tapajós interfluvium, where *Saguinus* and untufted capuchins are absent, *Ateles b. marginatus* is present, *Chiropotes albinasus* replaces *Chiropotes satanas*, *Callithrix* is more widespread, and so on (Martins et al., 1988).

A second, related question is the relatively low density of most of the region's primate populations, which further emphasizes its marginal position within the ecosystem. This characteristic is equally relevant to the understanding of Amazonian biodiversity as a whole, while further increasing the vulnerability of most species to the effects of human colonization and habitat degradation (Terborgh and Winter, 1978). Even where habitat disturbance is minimal (e.g. Tailândia, Table 5), hunting pressure alone can be highly deleterious (Baal et al., 1988; Lopes, 1993).

The primate community of eastern Amazonia is not only unique, but is probably also its most vulnerable. The situation to the east of the Rio Tocantins is the most critical, given that less than half of the original forest cover now remains and colonization continues unchecked. While a conservation unit, the Gurupi Biological Reserve, has recently been established in the region (Oren, 1988), it receives no practical protection, and encroachment by ranchers, loggers and prospectors is frequent (Ferrari & Queiroz, 1994).

However serious, the current situation in this region is not yet anywhere near as critical as that of the Atlantic Forest (Mittermeier et al., 1982). Lopes (1993) found that even the rarest primates, such as *Chiropotes*, are still far more abundant in the region than was previously thought (e.g. Johns and Ayres, 1987), although the management of populations isolated in forest fragments will almost certainly be required over the long term. A second ray of hope is the growing interest of local *latifundiários* in conservation issues (Lopes, 1993; Lopes and Ferrari, 1993). While not necessarily ideal, protection of native habitat by these landowners may be a far more practical solution for the region's problems than the establishment of official reserves that remain totally unprotected in practice.

SUMMARY

While encompassing a total of at least thirteen subspecific taxa, *terra firme* primate communities to the east of the Rio Xingu are among Amazonia's least diverse, reflecting the east-west diversity gradient common to many other faunal and floral groups. Population densities and biomass are also relatively low in comparison with western Amazonian communities, which may in part be due to differences in habitat quality. Diversity also declines in marginal habitats such as savanna formations, mangrove and, in particular, secondary or disturbed forest, even though the population density of some species may increase. Four diurnal species, *Alouatta belzebul*, *Cebus apella*, *Chiropotes satanas* and *Saguinus midas*, dominate in most areas. Habitat quality and interspecific competition appear to limit the geographic distribution and population density of other species, especially *Callithrix argentata* and *Cebus kaapori*. Throughout the region, but to the east of the Rio Tocantins in particular, widespread and largely uncontrolled deforestation, selective logging and hunting together constitute a major conservation problem, exacerbated by the current lack of data on the ecology of the region's primate species.

ACKNOWLEDGMENTS

Original fieldwork presented here was supported by the National Environment Fund of the Brazilian government, Conservation International, the A.H. Schultz-Stiftung, Wildlife Conservation International, the Federal University of Pará (UFPa), Conglomerado Real, the Goeldi Museum and IBAMA-PA. We would like to thank Marilyn Norconk, Claudio Emidio, Luís Alves, Andréa Nunes, Cazuza Júnior, Arlindo Júnior, Pedro Lisboa, Aline de Azevedo, Carlos Leôncio, Evandro Moreira, Jaldecy Pancieri, Josaphá Azevedo, Livia Gasbarra, Olga de Oliveira, Tarcísio Magalhães Sobrinho, Ana Cristina Oliveira, Ana Lúcia Pina, Cristina Fontella, Denis Sana, Luciane Souza, Márcia Jardim, Moira Adams, Osvaldo de Carvalho Jr., Paulo Coutinho, Urbano Bobadilla, and José Maria Cardoso.

REFERENCES

Anon., 1993, A new species of untufted capuchin from the Brazilian Amazon, *Neotrop. Primates* 1(2):5–7.

Aquino, R., and Encarnación, F., 1988, Population densities and geographic distribution of night monkeys (*Aotus nancymae* and *Aotus vociferans*) (Cebidae: Primates) in northeastern Peru, *Am. J. Primatol.* 14:375–381.

Ayres, J.M., 1989, Comparative feeding ecology of the uakari and bearded saki, *Cacajao* and *Chiropotes*, *J. Hum. Evol.* 18:697–716.

Ayres, J.M., 1993, *As Matas de Várzea do Mamirauá*, MCT-CNPq, Rio de Janeiro.

Ayres, J.M., Bonsiepe, J.I., and Clare, T.T., 1989, A preliminary survey of monkeys and habitats in northeastern Marajó island, *Primate Conserv.* 10:21–22.

Baal, F.L.J., Mittermeier, R.A., and van Roosmalen, M.G.M., 1988, Primates and protected areas in Suriname, *Oryx* 22:7–14.

Balée, W., 1986, Informe preliminar sobre inventário florestal e a etnobotânica Kaapor (MA), *Bol. Mus. Para. E. Goeldi, Bot.* 2:141–167.

Balée, W., 1987, A etnobotânica quantitativa dos índios Tembé (Rio Gurupi, Pará), *Bol. Mus. Para. E. Goeldi, Bot.* 3:29–50.

Black, G.A., Dobzhansky, T., and Pavan, C., 1950, Some attempts to estimate species diversity and population density of trees in Amazonian forests, *Bot. Gaz.* 111:413–425.

Bodmer, R.E., Fang, T.G., and Ibañez, L.M., 1988, Primates and ungulates: a comparison of susceptibility to hunting, *Primate Conserv.* 9: 79–83.

Bonvicino, C.R., Langguth, A., and Mittermeier, R.A., 1989, A study of pelage color and geographical distribution in *Alouatta belzebul* (Primates: Cebidae), *Rev. Nordest. Biol.* 6:139–148.

Brockelman, W.Y., and Ali, R., 1987, Methods of surveying and sampling forest primate populations, in C.W. Marsh and R.A. Mittermeier (eds.) *Primate Conservation in the Tropical Rainforest*, Alan R. Liss, New York, pp. 21–62.

Cain, S.A., Castro, G.M.O., Pires, J.M., and Silva, N.T., 1956, Application of some phytosociological techniques to Brazilian rain forest, *Am. J. Bot.* 43:911–941.

Ducke, A., and Black, G.A., 1954, Phytogeographical notes on the Brazilian Amazon, *An. Acad. Bras. Cien.* 25:1–46.

Egler, S.G., 1985, *Levantamentos da fauna de vertebrados terrestres do Projeto Carajás*, Unpublished report to the Companhia do Vale do Rio Doce.

Emmons, L.H., 1984, Geographic variation in densities and diversities of non-flying mammals in Amazonia, *Biotropica* 16:210–222.

Fernandes, M.E.B., 1994, Notes on the geographic distribution of howling monkeys in the Marajó archipelago, Pará, Brazil, *Int. J. Primatol.* 15:919–926.

Fernandes, M.E.B., and Aguiar, N.O., 1993, Evidências sobre a adaptação de primatas neotropicais às áreas de mangue com ênfase no macaco-prego *Cebus apella apella*, in M.E. Yamamoto and M.B.C. Souza (eds.) *A Primatologia no Brasil - 4*, Sociedade Brasileira de Primatologia, Natal, pp. 67–80.

Fernandes, M.E.B., Silva, J.M.C., and Silva Junior, J.S., Submitted, The monkeys of the Amazon estuary, Brazil: a biogeographic analysis, *Mammalia*.

Ferrari, S.F., 1993, Ecological differentiation in the Callitrichidae, in A.B. Rylands (ed.) *Marmosets and Tamarins: Systematics, Ecology and Behaviour*, Oxford University Press, pp. 314–328.

Ferrari, S.F., and Lopes, M.A., 1990, A survey of primates in central Pará, *Bol. Mus. Para. E. Goeldi, Zool.* 6:169–179.

Ferrari, S.F., and Queiroz, H.L., 1994, Two new Brazilian primates discovered, endangered, *Oryx* 28:31–36.

Ferrari, S.F., and de Souza Junior, A.P., 1994, More untufted capuchins in southeastern Amazonia? *Neotrop. Primates* 2:9–10.

Ferrari, S.F., Lopes, M.A., Cruz Neto, E.H., Silveira, M.A.E.S., Ramos, E.M., Ramos, P.C.M., Tourinho, D.M., and Magalhães, N.F.A., 1995, Primates and Conservation in the Guajará-Mirim State Park, Rondônia, Brazil, *Neotrop. Primates*, 3:in press.

Gentry, A.H., 1982, Patterns of Neotropical plant species diversity, *Evol. Biol.* 15:1–84.

Gentry, A.H., 1985, Some preliminary results of botanical studies in Manu Park, in A. Tovar and M. Ríos (eds.) *Estudios Biológicos en el Parque de Manu*, Ministerio de Agricultura, Lima, pp.

Gentry, A.H., 1986, An overview of neotropical phytogeographic patterns with an emphasis on Amazonia, in *Anais do 1°. Simpósio do Trópico Úmido*, Vol. II, EMBRAPA/CPATU, Belém, pp.

Gentry, A.H., 1988, Changes in plant community diversity and floristic composition on environmental and geographical gradients, *Ann. Miss. Bot. Gard.*, 75:1–34.

Goeldi, E.A., and Hagmann, G., 1906, Prodromo de um catálogo crítico, comentado da colecção de mammíferos do Museu do Pará (1894–1903), *Bol. Mus. Goeldi (Mus. Para.) Hist. Nat. Enthog.* IV:38–122.

Harada, M.L., 1994, *Abordagem molecular para o esclarecimento da filogenia dos gêneros Aotus, Callicebus, Cebus e Saimiri (Platyrrhini, Primates)*, Unpublished Ph.D thesis, Universidade Federal do Pará, Belém.

Hartshorn, G.S., 1980, *Forest vegetation (of Manu National Park, Peru)*. Tropical Science Center, San José, Costa Rica.

Hershkovitz, P., 1977, *Living New World monkeys (Platyrrhini) with an Introduction to Primates. Vol. 1.*, University of Chicago Press, Chicago.

Hershkovitz, P., 1985, A preliminary taxonomic review of the south American bearded saki monkeys genus *Chiropotes* (Cebidae, Platyrrhini), with the description of a new subspecies, *Fieldiana, Zool.* 27.

Hershkovitz, P., 1990, Titis, New World monkeys of the genus *Callicebus* (Cebidae, Platyrrhini): a preliminary taxonomic review, *Fieldiana, Zool.* 55.

Johns, A.D., 1986, Effects of habitat disturbance on rain forest wildlife in Brazilian Amazonia. Unpublished report to the World Wildlife Fund-US, Washington, D.C.

Johns, A.D., and Ayres, J.M., 1987, Southern bearded sakis beyond the brink, *Oryx* 21:164–167.

Kellogg, R., and Goldman, E.A., 1944, Review of the spider monkeys, *Proc. Nat. Mus.* 96:1–45.

Lopes, M.A., 1993, Conservação do cuxiú-preto, Chiropotes satanas satanas (Platyrrhini, Primates), e de outros mamíferos na Amazônia oriental, Unpublished M.Sc dissertation, Universidade Federal do Pará, Belém.

Lopes, M.A., and Ferrari, S.F., 1993, Primate conservation in eastern Brazilian Amazonia, *Neotrop. Primates* 1(4):8–9.

Lopes, M.A., and Ferrari, S.F., 1994, Foraging behaviour of a tamarin group (*Saguinus fuscicollis weddelli*), and interactions with marmosets (*Callithrix emiliae*), *Int. J. Primatol.* 15:373–387.

Lopes, M.A., and Ferrari, S.F., In press, Preliminary observations on the Ka'apor capuchin, *Cebus kaapori*, from eastern Brazilian Amazonia, *Biol. Conserv.*

Martins, E.S., Ayres, J.M., and Valle, M.B.R., 1988, On the status of *Ateles belzebuth marginatus* with notes on other primates of the Iriri river basin, *Primate Cons.* 9:87–93.

Mascarenhas, B.M., and Puorto, G., 1988, Nonvolant mammals rescued at the Tucuruí dam in the Brazilian Amazon, *Primate Conserv.* 9:91–93.

Mittermeier, R.A., Coimbra-Filho, A.F., Constable, I.D., Rylands, A.B., and Valle, C.M.C., 1982, Conservation of primates in the Atlantic Forest of eastern Brazil, *Int. Zoo Ybk.* 22:2–17.

Oren, D.C., 1988, Uma reserva para o Maranhão, *Ciência Hoje* 44:36–45.

Peres, C.A., 1989, A survey of a gallery forest primate community, Marajó island, Pará, Brazil, *Vida Sylv. Neotrop.* 2:32–37.

Peres, C.A., 1993, Structure and spatial organization of an Amazonian terra firme forest primate community, *J. Trop. Ecol.* 9:259–279.

Pires, J.M., 1966, The estuaries of the Amazon and Oyapoque rivers, in *Proceedings of the Decca Symposium*, UNESCO, pp. 211–218.

Pires, J.M., Dobzhansky, T., and Black, G.A., 1953, An estimate of the number of species of trees in an Amazonian forest community, *Bot. Gaz.* 114:467–477.

Pires, J.M., and Koury, H.M., 1958, Estudo de um trecho de mata de várzea próximo de Belém, *Bol. Téc. I.A.N.* 36:3–44.

Prance, G.T., 1992, The diversity of the Amazon flora, *Royal Inst. Proc.* 64:169–195.

Queiroz, H.L., 1992, A new species of capuchin monkey, genus *Cebus* Erxleben 1777 (Cebidae: Primates), from eastern Brazilian Amazonia, *Goeldiana Zool.* 15:1–13.

Salomão, R.P., 1994, Estimativas da biomassa e avaliação do estoque de carbono da vegetação de florestas primárias e secundárias ("capoeiras") de diversas idades na Amazônia oriental, município de Peixe-Boi, Pará, Unpublished M.Sc dissertation, Universidade Federal do Pará, Belém.

Salomão, R.P., and Lisboa, P.L.B., 1988, Análise ecológica da vegetação de uma floresta pluvial tropical de terra firme, Rondônia, *Bol. Mus. Para. E. Goeldi, Bot.* 4:195–234.

Salomão, R.P., Silva, M.F.F., and Rosa, N.A., 1988, Inventário ecológico em floresta pluvial tropical de terra firme, Serra Norte, Carajás, Pará, *Bol. Mus. Para. E. Goeldi, Bot.* 4:1–46.

Schneider, H., Sampaio, M.I.C., Schneider, M.P.C., Ayres, J.M., Barroso, C.M.L., Hamel, A.R., Silva, B.T.F., and Salzano, F.M., 1991, Coat color and biochemical variation in Amazonian wild populations of *Alouatta belzebul*, *Am. J. Phys. Anthropol.* 85:85–93.

Silva Jr., J.S., Queiroz, H.L., and Fernandes, M.E.B., 1992, Primatas no Maranhão: dados preliminares (Primates: Platyrrhini), *Resumos do XIXº. Congresso da Sociedade Brasileira de Zoologia*, p. 19.

SUDAM, 1988, *Mapa de Alteração da Cobertura Florestal*, SUDAM, Superintendência de Desenvolvimento da Amazônia, Belém.

Terborgh, J., 1983, *Five New World Primates: A Study in Comparative Ecology*, Princeton University Press, Princeton, USA.

Terborgh, J., and Winter, B., 1978, Some causes of extinction. In M.E. Soulé and B.A. Wilcox (eds.), *Conservation Biology*, Sunderland, Mass., pp. 119–133.

de Vivo, M., 1991, *A taxonomia de Callithrix Erxleben, 1777 (Callitrichidae, Primates)*, Fundação Biodiversitas, Belo Horizonte.

4

PRIMATES OF GUAYANA SHIELD FORESTS

Venezuela and the Guianas

Marilyn A. Norconk,[1] Robert W. Sussman,[2] and Jane Phillips-Conroy[2]

[1]Department of Anthropology
Kent State University
Kent, Ohio 44242–0001
[2]Department of Anthropology
Washington University
St. Louis, Missouri 63130

INTRODUCTION

The Guayana Shield represents a land mass of 1,800,000 km^2 in northern South America (Kelloff & Funk 1995). Dating from the Precambrian, it has dominated the interior regions of the Guianas (Guyana, Suriname, and French Guiana), Venezuela, northern Brazil and southeastern Colombia (American Geographical Society 1978). Ten of the 16 platyrrhine genera (following the taxonomic arrangement of Schneider and Rosenberger, this volume) are known from forests either on or bordering the Shield. The most poorly represented subfamily is the Callitrichinae with only one of the five callitrichine genera present from this region. Only one species of tamarin, *Saguinus midas,* is found in the entire area. Both cebine genera (*Cebus* and *Saimiri*) are well-represented with interrupted distributions only in Venezuela. Of the three pitheciin genera (*Pithecia, Cacajao,*and *Chiropotes*), *Pithecia* has the broadest distribution and *Cacajao* has a very limited distribution. *Aotus* and *Callicebus* are found only in Bolívar and Amazonas states of southern Venezuela. Both species appear to be rare, but may also be under represented in surveys. *Brachyteles* and *Lagothrix* are absent from Guayana Shield forests, but both *Ateles* and *Alouatta* are broadly distributed. There is also a notable absence of endemic primate genera and species from these habitats suggesting that migration corridors into Shield forests from central or eastern Brazil have been open for thousands of years.

Spellings of political entities in the northern Neotropics abound due to regional occupations by the Spanish (Guayana), British (Guiana), French (Guyane), and Dutch (Suriname). We follow Lindeman & Mori (1989) and Berry et al (1995) in the following designations, Guayana: the physiographic region and geological formation referred to as "the Guayana Shield"; Guyana: the country (previously British Guyana); Guiana: the region including the three countries of Guyana, Suriname, and French Guiana.

Adaptive Radiations of Neotropical Primates
edited by Norconk *et al.* Plenum Press, New York, 1996

Relatively few long-term studies and extensive primate censuses have been conducted in Guayana Shield regions. Extensive surveys were conducted by Mittermeier (1977) and long-term field sites do exist in the extensive national park system of Suriname (Raleighvallen-Voltzberg: Kinzey & Norconk 1990, Mittermeier & Fleagle 1980, van Roosmalen 1985, van Roosmalen et al 1988). To date, no long term studies have been completed in Guyana, although two extensive surveys have been conducted (Muckenhirn et al 1975, Sussman & Phillips-Conroy 1995). A long-term study site "Les Nouragues" has been established in central French Guiana (Julliot & Sabatier 1993, Riera et al 1995). The best known long-term primate studies in Venezuela have not taken place in Guayana Shield forests south of the Orinoco River, rather they have been conducted in the "llanos" north of the Orinoco (Fig 1) (e.g. *Alouatta seniculus* - Crockett 1984 & this volume, Pope 1990, Rudran 1979 and *Cebus olivaceus* - Robinson 1986, Miller 1991, this volume). In addition, two long-term study sites have been established in Bolívar State, Venezuela (Guri Lake: Kinzey & Norconk 1993, Norconk this volume, Walker 1994 & this volume and Dedemay: Castellanos 1993 & this volume) (Table 1). Exact primate distributions are still being compiled for virtually all of the countries that border the Shield. Accessibility to interior localities is still quite limited where human populations tend to be low in density. For example, human population density in the Venezuelan state of Amazonas is only

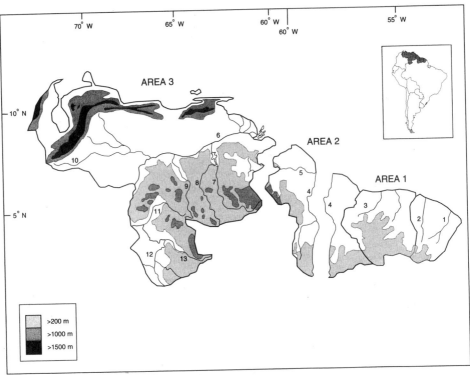

Figure 1. Countries of Venezuela, Guyana, Suriname and French Guiana with major geological formations, the Andes mountains and Guyana Shield represented. The region is divided into three separate areas for the purpose of discussing primate distributions. Area 1: eastern Guyana separated at the Essequibo River, Suriname, French Guiana; Area 2: western Guyana; and Area 3: Venezuela. The rivers indicated on the map are 1 - Approaugua, 2 - Mana, 3 - Coppename, 4 - Essequibo, 5 - Cuyuní, 6 - lower Orinoco, 7 - Caroni, 8 - Paragua, 9 - Caura, 10 - Apure, 11 - Ventuari, 12 - Casiquiare, 13 - upper Orinoco.

Table 1. Primates of Venezuela and the Guianas: characteristics of distribution; biogeographic regions of distribution after Eisenberg (1989); location of field studies. Geographical regions of distribution are 1: Guianas, east of the Essequibo river; 2: Guyana, west of the Essequibo River, 3: Venezuela

Primate species	Geographical region of distribution	Biogeographic areas	Localities of field studies[a]
Callitrichinae			
Saguinus midas	1,2	eastern Guianas	Raleighvallen-Voltzberg, Suriname[1]; Nourague Station, French Guiana
Cebinae			
Saimiri sciureus	1,2,3	southern Venezuela; eastern Guianas	Raleighvallen-Voltzberg, Suriname[2]
Cebus olivaceus	1,2,3	southern Venezuela; central Guyana highlands; eastern Guyanas; Venezuelan llanos; Maracaibo basin	Hato Masaguaral, Venezuela; Hato Piñero, Venezuela[3]; Raleighvallen-Voltzberg, Suriname[1]
Cebus apella	1,3	north coast range; southern Venezuela; eastern Guyanas	Raleighvallen-Voltzberg, Suriname[2]
Pitheciinae			
Aotus trivirgatus	3	southern Venezuela	
Aotus lemurinus	3	Maracaibo basin, Venezuela	
Callicebus torquatus	3	southern Venezuela	
Pithecia pithecia	1,2,3	southern Venezuela; central Guyana highlands; eastern Guyanas	Raleighvallen-Voltzberg, Suriname;[1] Guri, Venezuela[4]
Chiropotes satanas	1,3	southern Venezuela; central Guyana highlands; eastern Guyanas	Raleighvallen-Voltzberg, Suriname; Guri, Venezuela[5]
Cacajao melanocephalus	3	southern Venezuela	Río Baria, Venezuela[6]
Atelinae			
Alouatta seniculus	1,2,3	southern Venezuela; north coast Venezuela; Venezuelan llanos; central Guyana highlands; eastern Guyanas; Maracaibo basin	Hato Masaguaral, Venezuela; Hato Piñero, Venezuela; Nourague Station, French Guiana[7]; Raleighvallen-Voltzberg, Suriname[1]
Ateles paniscus paniscus	1	eastern Guyanas	Raleighvallen-Voltzberg, Suriname[1,8]
Ateles belzebuth hybridus	3	Maracaibo basin, Venezuela	
Ateles belzebuth belzebuth	3	southern Venezuela; central Guyana highlands	Estación Biológica Dedemay, Venezuela[9]

[a]References to Localities of Field Sites
[1]Fleagle & Mittermeier (1980); Mittermeier & van Roosmalen (1981)
[2]Fleagle et al. (1981)
[3]Miller (1991 + this volume); Robinson (1981, 1986)
[4]Kinzey and Norconk (1993)
[5]Kinzey & Norconk (1990); Mittermeier & van Roosmalen (1981); Norconk & Kinzey (1994); Kinzey & Norconk (1993); van Roosmalen et al. (1988)
[6]Lehman and Robertson (1994)
[7]Braza et al. (1983); Crockett (1984); Crockett & Rudran (1987a & b); Julliot & Sabatier (1993); Peetz et al. (1992); Pope (1990); Sekulic (1982);
[8]Kinzey & Norconk (1990); Norconk & Kinzey (1994); van Roosmalen (1985)
[9]Castellanos (1994)

0.057 individuals/km^2 (Huber 1995). Perhaps for this reason, the forests of the Guayana Shield provide some of the most pristine habitats remaining in Middle and South America.

GEOLOGY OF THE SHIELD AND ANTIQUITY OF ITS FORESTS

The Guayana Shield extends from the equator in Brazil north to include all but littoral habitats bordering the Caribbean Sea of French Guiana, Suriname, and Guyana. In Venezuela, the northern border of the Shield is delimited by the Orinoco River (Harrington 1956) (Fig 1). Exposures of the Guayana Shield in the northern Neotropics are part of the remnant geological core of the continent of South America (Clapperton 1993, Berry et al, 1995). Together with the Brazilian Shield, these regions represent the remaining exposed area of the western section of Gondwanaland (Huber 1995).

Geological processes (uplift and erosion) that began in the Precambrian have altered the substrate to the point that the earliest substrates no longer exist (Briceño and Schubert 1990; Clapperton 1993). The Venezuelan tepuis are the oldest surfaces on the Shield ranging from 900 m to 2,900 m in elevation (Briceño & Schubert 1990) (Fig 1). Current exposures date from the Mesozoic to very recent Holocene deposits and remodeling of the Shield and surrounding area is an on-going process. For example, slopes of the tepuis (table top mountains in Venezuela) were denuded during pluvial periods in the late Pleistocene, but now support forests (Clapperton 1993) and primate populations. Likewise, erosion of the Shield and subsequent deposition of intrusive rock has resulted in more recent surfaces in Guyana, French Guiana, Venezeula, and Brazil. In most areas of the Shield, intrusive rock has invaded the metamorphic basement resulting in soils that are more favorable for forest growth (lower in silica, higher in phosphate, calcium, and nitrogen than basement rock) (Huber 1995). The two surfaces supporting widespread tropical wet and dry forests in Venezuela are the Imataca surface (lower Tertiary - south of #6 on Fig 1) and Caroní-Ari surfaces (Oligocene-Miocene - Venezuelan states of Amazonas and Bolívar). The low-lying llanos of Venezuela, the Orinoco floodplain, and littoral regions of the Venezuelan Delta Amacuro and northern borders of the Guianas are Plio-Pleistocene or more recent (Briceño & Schubert 1990; Zonneveld 1993).

Today, forests cover much of the Shield in Brazil, French Guiana, Suriname and Guyana and there is little evidence that eastern Amazonia provided any barriers to primate dispersal. The disjunct distribution of primates in Venezuela and western Guyana however, suggests that a combination of geological processes and strong seasonal drying in the western Shield may have served as dispersal barriers. Quaternary deposits in the Gran Sabana and Caroní basin of Venezuela (region east of #7, Fig 1) suggest that climatic conditions were dryer than today, but climatic evidence prior to the Pleistocene is poor.

BIOGEOGRAPHY OF THE GUIANAS AND VENEZUELA

Of the nine biogeographic areas proposed by Eisenberg (1989) to characterize the northern Neotropics, six are applicable here (Table 1) although three are not specifically derived from the Guayana Shield. These three non-Shield areas are found in Venezuela: the "north coast range" of Venezuela, an arid area that includes Margarita Island; the strongly seasonal "llanos" of central Venezuela, a low, flat (< 100 m elevation) expanse of mixed gallery forest and savanna north of the Orinoco River; and the Maracaibo basin, sedimentary lowlands between the two northern extensions of the Andes in the far northwestern Venezuela and Lake Maricaibo.

The other biogeographical regions are more widespread and cross political boundaries: the wet, evergreen forests of "southern Colombia and adjacent Venezuela" of which only the southern tip of Amazonas State in Venezuela will be considered here; "central Guyana highlands" a complex set of habitats south of the Orinoco in Venezuela consisting of dry to transitional wet tropical forests interrupted by areas of tropical savanna near the Orinoco and high mountain ranges in the west and south; and "eastern Guyana" from the right bank of the Essequibo River eastward to French Guiana, a band of nearly continuous forest dotted with occasional elevated granite outcrops of the Guayana Shield.

PRIMATE DISTRIBUTIONS

Distributions of Guayana Shield primates in Venezuela and the Guianas are best understood by dividing the region into three parts (Fig 1): 1) Guianas east of the Essequibo River (eastern Guyana, Suriname, and French Guiana; 2) Guyana west of the Essequibo River; and 3) Venezuela. Much of area 2 and the Gran Sabana (area 3: far east and southeastern Venezuela) contain regions that are largely unexplored and in which primate distributions are still uncertain. A survey of the patchwork of highlands and savanna in the Brazilian territory of Roraima bordering southeastern Venezuela and extending northward into Venezuela and southwestern Guyana are critical to a complete record of primate distributions in the northern neotropics.

Area 1: Guianas, East of the Essequibo River

Eight primate species (*Ateles paniscus, Alouatta seniculus macconnelli, Cebus apella apella, Cebus olivaceus castaneus* (and *C.o. olivaceus* in Guyana), *Saimiri sciureus sciureus, Pithecia pithecia pithecia, Chiropotes satanas chiropotes,* and *Saguinus midas midas*) have widespread distributions in Suriname, French Guiana, and eastern Guyana (Rylands et al 1995) (Figs 2–5). We use Mittermeier's (1977) work in Suriname as an example of the kind of vegetation zones that might effectively provide habitat barriers and delimit primate populations in Area 1.

Mittermeier divided Suriname into four vegetation zones from north to south. The coast is geologically very recent (8,000 to 6,000 years BP, Clapperton 1993) consisting of mudflats and mangrove swamps of coastal Suriname and French Guiana. Abutting the young coastal plain is an older coastal plain (rain forest), a narrow belt of white sand savanna, and finally the interior (evergreen rain forests) (Mittermeier, 1977). Approximately 75% of Suriname is in the "interior", is well-forested with low human population density. Remnants of the Shield occur as granite outcroppings ("inselbergs") rising 200 to 350 m above the surrounding forest. Savannas are found in the interior, but are much smaller than the Gran Sabana of Venezuela and probably did not provide a deterrence to the spread of primate populations north from Amazonian Brazil. Mittermeier conducted surveys in all four vegetational zones in Suriname and although all eight species of primates can be found in the interior, they vary in the extent that they occupy the coastal and savanna belts. *Alouatta seniculus* and *Cebus apella* were abundant throughout the country, including the young coastal plain; *Pithecia pithecia* are found throughout, but were rare everywhere; *Saimiri sciureus* were found throughout and were very abundant in the old coastal plain; *Ateles paniscus, Cebus nigrivitattus,* and *Chiropotes satanas* were limited primarily to interior forests.

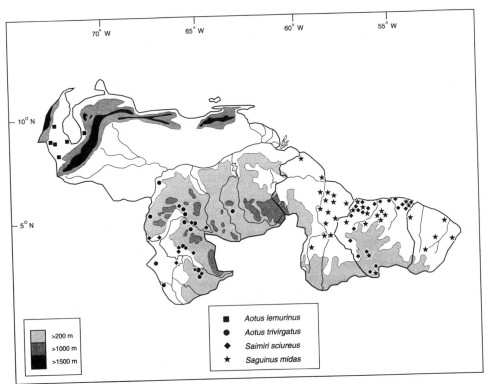

Figure 2. Distribution of the small-bodied platyrrhines in Venezuela and the Guianas: *Aotus* spp., *Saimiri sciureus*, and *Saguinus midas*. Data compiled from Bodini & Pérez-Hernández 1987, Eisenberg 1989, Hershkovitz 1977, Huson 1978, Julliot & Sabatier 1993, Mittermeier 1977, Muckenhirn et al 1975, Sussman & Phillips-Conroy 1995.

Primate distributions are not as well known from French Guiana, but all eight species found in Suriname have also been documented from French Guiana (Eisenberg 1989, Emmons 1990). Mittermeier (1977) provided some additional notes to his Suriname data from the village of Saül, French Guiana (185 km southwest of Cayenne). He confirmed the presence of *Saguinus midas* and *Alouatta seniculus* only and failed to find evidence of the other six species. *Chiropotes satanas* and *Saimiri sciureus* were not included in the primate species inventory at Norangue (Julliot & Sabatier 1993) which is midway between Cayenne and Saül, although the authors do indicate that these species have been documented from other parts of the country. Distribution of the eight species is evidently spotty in French Guiana as it is in Suriname, but there are no apparent barriers to dispersal and in general, all 8 species appear to occur in diverse habitats throughout these two countries.

The habitats of eastern Guyana (right bank of the Essequibo River) are similar to Suriname. The northern third of the region consists of a wide band of coastal plain followed by a wide band of savanna. Tropical forest replaces savanna in the interior (Huber et al, 1995). Of the sites surveyed by Muckenhirn et al (1975) and Sussman & Phillips-Conroy (1995), the following survey localities are found on the Essequibo or in the region east of the Essequibo (#4, Fig 1): east Berbice District on the Berbice River, Essequibo River, 24 Mile reserve is on the west bank of the Essequibo (non-flooded lowland forest on white sand) and Moraballi Re-

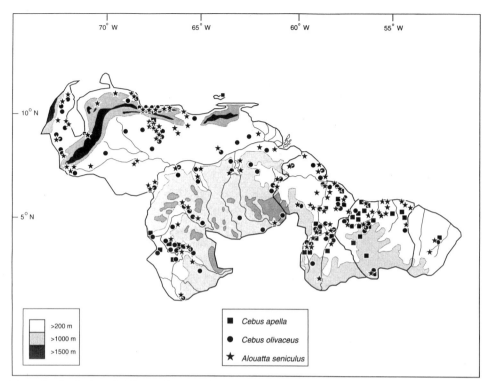

Figure 3. Distribution of *Cebus apella, C. olivaceus,* and *Alouatta seniculus* in Venezuela and the Guianas. Data compiled from Bodini & Pérez-Hernández 1987, Eisenberg 1989, Huson 1978, Julliot & Sabatier 1993, Kinzey et al 1988, Mittermeier 1977, Muckenhirn et al 1975, Sussman & Phillips-Conroy 1995.

east bank of the Essequibo (non-flooded lowland forest), Pakani and Apoteri settlements on the upper Essequibo River. All eight species were observed only at the Apoteri settlement at the junction of the Essequibo and Rupununi Rivers in forest habitat (at the bifurcation of the Essequibo River south of #4). Only *Alouatta seniculus* and *Cebus nigrivitattus(= olivaceus)* were found in common at the twin sites on either side of the Essequibo: Moraballi (non-flooded lowland forest on the east bank of the Essequibo) and 24 mile Reserve (nonflooded lowland forest on the west bank of the Essequibo) (see Sussman and Phillips-Conroy 1995). The Berbice River survey was conducted in tall evergreen flooded riparian forest and only *Cebus nigrivitattus(= olivaceus)* and *Pithecia pithecia* were not observed, although they were reported to have been seen by villagers.

Area 2: Guyana, West of the Essequibo River

There is no clear disruption in the distribution of the eastern 8 Guianan primate species until the Essequibo River is crossed to the west. Primate distribution is very sketchy in western Guyana. Terrain becomes more complex with the Koraima Mountain range on the border of Guyana and Venezuela, and the Pakaraima Mountains and widespread savanna of the Kanuku mountains in the southwestern part of the country (Huber et al 1995). Nevertheless, it is difficult to explain the apparent decline in primate species number and

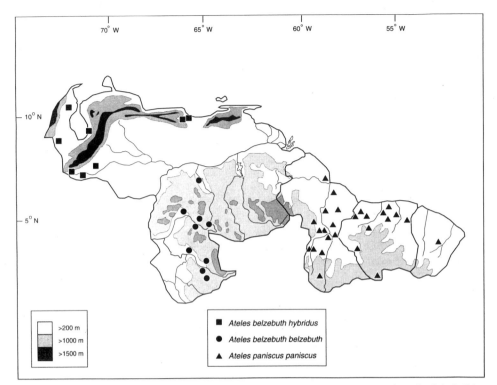

Figure 4. Distribution of *Ateles* spp. in Venezuela and the Guianas. Data compiled from Bodini & Pérez-Hernández 1987, Eisenberg 1989, Hershkovitz 1977, Huson 1978, Julliot & Sabatier 1993, Konstant et al 1985, Mittermeier 1977, Mondolfi & Eisenberg 1978, Muckenhirn et al 1975, Sussman & Phillips-Conroy 1995.

diversity in this region except that the southwest region is made up largely of savanna habitats to 400 m elevation.

Muckinhirn et al (1975) reported that four primate species (*Alouatta seniculus, Pithecia pithecia, Saimiri sciureus,* and *Cebus olivaceus*) were found throughout Guyana (Figs 2, 3, 5), but that *Ateles paniscus, Cebus apella, Chiropotes satanas,* and *Saguinus midas* (Figs 2, 3, 4, 5) had more restricted distributions. While *Cebus apella* and *Chiropotes satanas* also occur in Venezuela, their current distribution suggests that they entered southwestern Venezuela from Brazil or Colombia rather than moving directly westward from Guyana. Both *Chiropotes* and *Saguinus* appear to be largely geographically limited by the Essequibo River found only on the east bank, although we have reliable reports of *Saguinus* virtually on the Venezuelan border (Bourne, pers. comm.) and Napier (1976) reported on museum specimens from western Guyana. In a recent survey Sussman & Phillips-Conroy (1995) found neither species to be present west of the Essequibo and north of the Rupununi.

In the 1975 report of Muckenhirn et al., *Ateles paniscus* and *Cebus apella* were said to be absent from the northwest region, particularly north of the Potaro River (located at approximately # 4 on Fig 1). Although our findings generally concur with this, there is a report of *Ateles* in northwest Guyana along the mouth of the Supenaam River, just west of the Essequibo River (Fig 4). Thus, the western limits of the geographical ranges of *Ateles paniscus, Saguinus midas, Chiropotes satanas,* and *Cebus apella* are at present unknown, as are the factors limiting their distribution in the west. Furthermore, habitat preferences

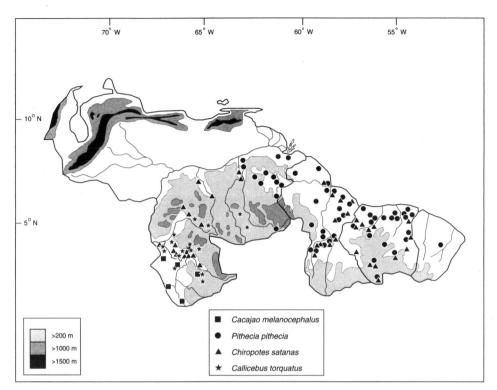

Figure 5. Distribution of the pitheciines Venezuela and the Guianas. Data compiled from Balbas personal communication, Bodini & Pérez-Hernández 1987, Eisenberg 1989, Huson 1978, Julliot & Sabatier 1993, Kinzey et al 1988, Lehman & Robertson 1994, Mittermeier 1977, Muckenhirn et al 1975, Rodrigues personal communication, Sussman & Phillips-Conroy 1995.

and the extent of sympatry of the *Cebus* spp. is as yet unknown. Consequently, further study of the biogeography of these species is sorely needed.

Tate (1939) had reported *Aotus* in Guyana, but Muckenhirn et al (1975) believed *Aotus* did not occur there. Two reliable observers in northern Guyana accurately described *Aotus* and recognized this species from color photographs. Thus, we feel the question of *Aotus* in Guyana remains open.

Area 3: Venezuela

Venezuela with its very diverse habitats supports the largest number of species (13) of the four Guayana Shield countries discussed here. There are only two endemic taxa found in Venezuela, *Ateles belzebuth hybridus* and *Aotus lemurinus*. Neither are found in Shield forests, but are limited to the intermontane region of Lake Maracaibo of northwestern Venezuela bordering Colombia.

Alouatta seniculus and *Cebus olivaceus* are the most broadly distributed primates in Venezuela occupying seasonally dry forests to wet forests from the llanos south to Amazonas state bordering Brazil and Colombia (Table 1; Fig. 2). Bodini & Péres Hernàndez (1987) identified 8 populations/subspecies of non - apella *Cebus* ssp. in Venezuela of which we have simplified to a single taxon, *C. olivaceus*. Both *Alouatta* and *C. olivaceus* inhabit a wide variety of habitats, although this perhaps characterizes *Alouatta* better than *C. olivaceus. Alouatta* inhabit diverse and relatively poor or disturbed habitats; *C. olivaceus* appears to be more limited to forest habitats that can provide relatively broad dietary diversity including reliable sources of ripe fruit. For example, small islands of tropical dry forest that range from 10 to 20 ha in Guri Lake, Venezuela, supported social groups of both *Pithecia pithecia* and *Alouatta seniculus*, but not *C. olivaceus*. Islands less than 10 ha supported only *Alouatta* ten years after flooding and creation of the reservoir (Kinzey et al 1988). Bodini & Péres-Hernàndez (1987) recognized four geographically distinct subspecies of *Alouatta seniculus* in Venezuela: south of the Orinoco in Amazonas state and in the state of Bolívar into Delta Amacuro (the Orinoco delta); north of the Orinoco in the Venezuelan llanos; in the basin of Lake Maricaibo with an intermontane extension into the western llanos; and a northern coastal population.

Aotus trivirgatus and *Callicebus torquatus lugens* have similar distributions in Venezuelan Amazonas and have been found as far east as the basin between the Paragua and Caroní rivers in southeastern Venezuela (Bodini & Péres-Hernàndes, 1987; Table 1, Figs 2 & 5). Habitats range from wet evergreen forest in the south and west of Amazonas and western Bolívar states to seasonal dry forests in eastern Bolívar state.

Two pitheciins, *Cacajao melanocephalus melanocephalus* and *Chiropotes satanas chiropotes*, have adjacent, but apparently non-overlapping distributions in southern Amazonas of Venezuela. The distribution of *Chiropotes* is much more extensive than that of *Cacajao* extending eastward to the Caroní River in Bolívar State (Fig 5). The problematic area with regard to distribution of these two genera continues to be in the state of Amazonas. Eisenberg (1989) and Bodini & Péres-Hernàndez (1987) cited only two collection localities for *Cacajao melanocephalus*, both south of the Orinoco in central and southern Amazonas. A recent survey by Lehman and Robertson (1994) provides an update to the distribution of *Cacajao*. They suggested that *Cacajao* are now limited to the southernmost tip of their previous distribution. There are more collection localities for *Chiropotes satanas* than for *Cacajao* in Amazonas. Although the upper Orinoco River (#13, Fig 1) appears to provide a geographic boundary, the right bank of the Orinoco supports terra firme and montane evergreen forests compared with the floodplains and seasonally flooded basins (igapó) of the Casequiare River and Río Negro. Thus the "barriers" to dispersal for these two species may be in the form of habitat preference or specialization.

The third pitheciin found in Venezuela, *Pithecia pithecia pithecia*, has a much broader distribution and is more tolerant of dry habitats than either *Cacajao melanocephaus* or *Chiropotes satanas*. They are known to occur in both evergreen forests and relatively open, savanna-like habitats of eastern Venezuela (Bodini and Péres-Hernàndez 1987, Kinzey et al 1988, Balbás, pers. comm.; Rodriguez, pers. comm.). The distribution of *Pithecia* is predominantly in the state of Bolívar, a huge area south of the lower Orinoco River (#6, Fig 1) in southeastern Venezuela. In addition, there are collection localities on the border of Brazil (Rodriguez, pers. comm.) as well as other reliable sightings very near the Delta in the north (Alvarez, pers. comm.), a few localities in the delta and one, outlying locality in Amazonas reported by Bodini & Péres-Hernàndez (1987) (Fig 5). Walker (1993) suggested that the distribution of *Pithecia* might be continuous from east to west in Venezuela south of the Orinoco and into the territory of Amazonas, but they were absent from the lower Caura River (#9, Fig 1) (Norconk, personal observation) and have not been seen during extensive surveys in the upper Caura (Balbás, personal communication). *Pithecia* distribution does appear to be very extensive throughout eastern Venezuela in habitats considered to be in the western portion of the Essequibo River basin (Huber 1995). We concur with Mittermeier however, that *Pithecia* appear to be widely distributed but relatively rare everywhere.

There are two subspecies of *Ateles belzebuth* in Venezuela, *Ateles belzebuth belzebuth* found south of the Orinoco River in Amazonas and across the central Guyana highlands to the Caura River (Bodini and Perés-Hernàndez, 1987) and *Ateles belzebuth hybridus* from the Maracaibo Basin, western llanos and coastal mountains in the northern state of Miranda (Mondolfi and Eisenberg, 1979) (Fig 4). The north coast distribution of *Ateles b. hybridus* has recently been confirmed in the mountainous terrain of the Guatopo National Park, state of Miranda (Ochoa et al, 1995).

The only endemic primate species in Venezuela (*Aotus lemurinus*) is reported to have a similar, but more limited distribution than *Ateles b. hybridus* in the narrow intermontane forests of western Venezuelan states of Zulia and Trujillo and in northeastern Columbia (Bodini & Peres-Hernandez, 1987; Konstant et al, 1985). Populations (or subspecies: Bodini & Perés-Hernàndez, 1987) of the more widespread species, *Alouatta seniculus* and *Cebus olivaceus*, are also found in the intermontane region and floodplain of Lake Maricaibo in northwestern Venezeula (north of #10, Fig 1).

Saimiri sciureus and *Cebus apella* have limited ranges in southern Venezuela (Figs 2 & 3) suggesting that a southern entrance into Venezuela was possible along the west edge of the Gran Sabana with dispersal northward as far as either the Orinoco river (#6, Fig 1) or the Sierra de Maigualida mountain range in the central Amazonas of Venezuela (Eisenberg, 1989). Only *Chiropotes satanas* has migrated further east into Bolívar State to the apparent eastern limit of their distribution at the west bank of the Caroní River (#7, Fig 1). *Cebus apella* has a disjunct distribution within Venezuela with a second population or subspecies (Bodini & Péres-Hernàndez, 1987) endemic to Margarita Island off the northeast coast (Fig 1).

Saguinus midas has a continuous distribution from French Guiana west across Suriname to the Essequibo River in Guyana and it is puzzling why *Saguinus* spp. have never been documented from Venezuela. Little is known specifically about either *Saguinus midas* or *Saguinus inustus* bordering Venezuela, but *Saguinus* spp. in general are not habitat specialists. Tamarins are reported to exploit secondary and disturbed habitats (Sussman & Kinzey 1984) and apparently do well in colonizing such habitats (Hershkovitz 1977). Sussman & Phillips-Conroy (1995) reported sightings of *Saguinus midas* west of the Essequibo and close to Delta Amacuro of Venezuela (Fig 2). Likewise, Hershkovitz (1977) reported a collection locality for *S. inustus* on the right bank of Río Negro in Brazil and apparently near the border of Venezuelan Amazonas. Since *Pithecia* may have dispersed into Venezuela via a northeastern route and several cebids moved into Venezuela from the southwestern Amazonia route, it is unclear why *Saguinus* did not do likewise.

POSSIBLE DISPERSAL ROUTES BETWEEN THE SHIELD AND LOWLAND FORESTS SOUTH OF THE SHIELD

Geomorphology, climatic and floristic sources provide evidence for the absence of dispersal barriers in the Guianas, east of the Essequibo River (Clapperton, 1993; Pires & Prance, 1978, Prance 1990) despite intermittent changes in elevation to 500 m and interruptions of continuous forest. In contrast, discontinuity of distribution characterizes many of the primates in Venezuela and in western Guiana. Venezuela offers a wide diversity of habitats (ranging from the foothills of the northern Andes and coastal ranges, the llanos and Orinoco delta, and the upland regions of the Guayana Shield), some of which may act as barriers to primate dispersal. For example, mapping the dispersal of primates into Venezuela involves by-passing the widespread savanna and tepuis of the Gran Sabana in southern Venezuela. Similarly, mountain and/or savanna barriers may have reduced dispersal capabilities for primates in western Guyana, but these regions of Guyana and Venezuela are virtually unknown from censuses.

Eisenberg (1989) proposed three dispersal routes into Venezuela and the Guianas, all of which appear to be barrier free today and it is doubtful if they would have been closed at any point in the past several thousand years: a) from the southwest (via western Amazonia of Brazil and Colombia), b) from the northeast (via Guyana), and c) a route across the northern Andes bordering Venezuela and Colombia. Contemporary barriers to primate dispersal in the northern Neotropics thus appear to reflect the distribution of widespread tropical savannas, particularly in the Gran Sabana area of southeastern Venezuela and adjacent regions of western Guyana. Distribution of some primates in Guyana may also have been limited by the Essequibo River as they apparently were also delimited by the Orinoco River in Venezuela.

CURRENT ISSUES IN CONSERVATION OF PRIMATES OF THE GUAYANA SHIELD FORESTS OF THE GUIANAS AND VENEZUELA

With the exception of Venezuela, little primate research has been conducted in the small countries the make up the Guianas and much of the research in Venezuela has been conducted on private ranches in the llanos (Table 1). Coastal areas of the Guianas are the most densely populated regions of human habitation in Suriname, French Guiana, and Guyana. Central and southern regions of the countries have not been developed and travel in the interior is often limited to access by river or by small plane. The interiors of these countries may support much of the remaining pristine tropical forests in South America, but conservation is more passive than active in these small countries. Protection is possible only as long as limited human access can be maintained. Suriname established a number of well-protected parks in the 1950's that were, perhaps, the first ecotourism camps in South America. Monetary support for the parks was reduced to a trickle when Suriname became independent of Holland in 1975, tourism declined, and support ground to a halt with the onset of a civil war in 1986.

A recent survey by Sussman & Phillips-Conroy (1995) in Guyana provided convincing evidence that primate populations are being subjected to increased threats from human encroachment (habitat destruction and hunting) regardless of whether they were near towns or in the interior. They reported a decline in the frequency of sighting primate groups since the previous survey by Muckenhirn et al. (1975), 20 years earlier. Furthermore, there is a threat of major concessions being granted to mining and timber companies. Recently Asian logging companies, having exhausted the forests of South East Asia, are threatening to move into the relatively untouched forests of western Guyana and Suriname (Colchester 1994).

Venezuela has a different history of both conservation efforts and habitat exploitation, but until very recently, remote regions remained relatively undisturbed. There is no shortage of interest and personal investment by Venezuelans in conservation and many individuals have literally devoted their lives to resource protection. Nor is there a shortage of national "parks". But active conservation is hindered by lack of support from government and private industry, poor coordination among the many conservation groups, and ironically by the very richness of the mineral and organic wealth of the country. Unlike the primate fauna that have migrated into Venezuela from Brazil, Colombia and Guyana, new endemic species of plants are still being discovered in the many diverse habitats of

Venezuela. Steyermark (1977) lamented the loss of the one of the largest rainforests north of the Orinoco River. In this instance, no attempt was made to collect even a vegetation inventory before the forest was reduced to a remnant. This forest was also the habitat of *Ateles belzebuth hybridus*, the only primate population in the countries of the northern Guayana Shield recognized to be in danger of extinction by Mittermeier (1986). Construction of hydroelectric plants, designation of huge areas of forest for logging concessions, and mining and hunting (both legal and illicit) are the major threats to successful habitat conservation in Venezuela.

SUMMARY

The distribution of extant primates and rare primate endemism in the northern Neotropics suggest that widespread exchange of populations has been possible from Amazonian Brazil despite the expanse of the Guayana Shield. Two corridors have apparently allowed primates to colonize the region: northward from eastern Amazonia into the Guianas and northward from western Amazonia (Brazil and Columbia) into southwestern Venezuela (Eisenberg, 1989). The continuous distribution of eight species of primates across the Guianas is interrupted in western Guyana by several possible barriers: the Essequibo River, savanna habitats, mountain ranges in the Guyana-Brazil border, and the Orinoco delta in the northern Guyana-Venezuela border. The most obvious barrier to primate dispersal in eastern Venezuela is the Gran Sabana, a vast savanna dotted with ancient tabletop mountains. Forests are discontinuous and are limited to the slopes of the tepuis and gallery forests along perennial streams.

The only endemic primates in the region, *Aotus lemurinus* and *Ateles belzebuth hybridus*, entered northwestern Venezuela through an Andean corridor and are very limited in distribution and very vulnerable as habitats continue to be threatened. Primate populations have probably been protected with low human population densities, but hunting, mining, and logging in Venezuela, as in Guyana and Suriname, and expansion of human populations into the interior of the Guianas appear to now cause a serious threat, particularly to atelines and pitheciins.

ACKNOWLEDGMENTS

The research of RWS and JP-C was supported in part by the National Science Foundation (BNS 9213532) and the National Geographic Society. The surveys in Venezuela were conducted with Warren G. Kinzey and the support of National Science Foundation (BNS 8719800). MN thanks Luis Balbás S., Eduardo Alvarez and Manuel Felipe Rodriguez for providing unpublished census data in Venezuela. We are grateful to Barbara Hammer, Dr. Ute Dymon, and Rick Zach of the Cartography Laboratory, Department of Geography at Kent State University for preparation of the maps.

REFERENCES

American Geographical Society. 1978. Guyana (map). New York: New York Botanical Garden.
Berry, P.E., Holst, B.K., Yatskievych, K. 1995. Introduction. In Flora of the Venezuelan Guayana. Volume 1. St. Louis: Missouri Botanical Garden.

Bodini, R., and Pérez-Hernández R. 1987. Distribution of the species and subspecies of cebids in Venezuela. *Fieldiana: Zoology* n.s. 39:231–244.

Braza, F. Alvarez, F., and Azcarate T. 1981. Behaviour of the red howler monkey (*Alouatta seniculus*) in the llanos of Venezuela. *Primates* 22:459–473.

Briceño, H.O., and Schubert, H. 1990. Geomorphology of the Gran Sabana, Guayana Shield, southeastern Venezuela. *Geomorphology* 3:125–141.

Castellanos, H. 1993. Feeding behaviour of *Ateles belzebuth* E. Geoffroy 1806 (Cebidae:Atelinae) in Tawadu Forest southern Venezuela. PhD dissertation, The University of Exeter.

Clapperton, C. 1993. Quaternary Geology and Geomorphology of South America. Amsterdam: Elsevier.

Colchester, M. 1994. The new sultans of the west: asian loggers move in on Guyana's forests. (unpub. ms.)

Crockett, C. 1984. Emigration by female red howler monkeys and the case for female competition, in: M.F. Small (ed). *Female Primates: Studies by Women Primatologists*. Alan Liss, New York.

Crockett, C., and Rudran, R. 1987a. Red howler monkey birth data I: Seasonal variation. *Amer. J. Primatol.* 13:347–368.

Crockett, C., and Rudran, R. 1987b. Red howler monkey birth data II: Interannual, habitat, and sex comparisons. *Amer. J. Primatol.* 13:369–384.

Eisenberg, J.F. 1989. Mammals of the Neotropics: The Northern Neotropics. Chicago University Press, Chicago.

Emmons, L. 1990. Neotropical Rainforest Mammals: A Field Guide. Chicago: University Press.

Fleagle, J.G., and Mittermeier R.A. 1980. Locomotor behavior, body size and comparative ecology of seven Suriname monkeys. *Amer. J. Phys. Anthropol.* 52:301–314.

Fleagle, J.G., Mittermeier, R.A., and Skopec, A. 1981 Differential habitat use by *Cebus apella* and *Saimiri sciureus* in central Suriname. *Primates* 22:361–367.

Harrington, H.J. 1956. Morphostructural regions of South America, in: W.F. Jenks (ed). Handbook of South American Geology, pp. xiii-xviii. Geological Society of America, New York.

Hershkovitz, P. 1977. Living New World Platyrrhines. Volume 1. Chicago University Press, Chicago.

Huber, Otto 1995. Geography and Physical Features. In Introduction In Flora of the Venezuelan Guayana. Volume 1. St. Louis: Missouri Botanical Garden. pp. 1–61. Timber Press.

Huber, O, Ghorbarran, G., Funk, V. 1995. Vegetation map of Guyana. Center for the Study of Biological Diversity, University of Georgetown, Georgetown Guyana.

Husson, A.M. 1978. The Mammals of Suriname. Leiden: E.J. Brill.

Julliot, C., and Sabatier, D. 1993. Diet of the red howler monkey (*Alouatta seniculus*) in French Guiana. *Int. J. Primatol.* 14:527–550.

Kelloff, C.L., and Funk, V.A. 1995. A preliminary study of the vegetation of Kaieteur National Park, Guyana. (abstract) *Measuring and monitoring forest biologicaldiversity: The international network of biodiversity plots*. Smithsonian / Man and the Biosphere Biodiversity Program International Symposium. Washington D.C., p. 95.

Kinzey, W.G., and Norconk, M.A. 1990. Hardness as a basis for food choice in two sympatric primates. *Am. J. Phys. Anthropol.* 81:5–15.

Kinzey, W.G., and Norconk, M.A. 1993. Physical and chemical properties of fruit and seeds eaten by *Pithecia* and *Chiropotes* in Suriname and Venezuela. *Int. J. Primatol.* 14:207–227.

Kinzey, W.G., Norconk, M.A., and Alvarez-Cordero, E. 1988. Primate survey of eastern Bolívar, Venezuela. *Primate Conservation* 9:66–70.

Konstant, W., Mittermeier, R.A., and Nash, S.D. 1985. Spider monkeys in captivity and in the wild. *Primate Conservation* 5:82–108.

Lehman, S.M., and Robertson, K.L. 1994. Survey of Humboldt's black head uakari (*Cacajao meianocephalus melanocephalus*) in southern Amazonas, Venezuela. *Int. J. Primatol.* 15:927–934.

Lindeman J.C., and Mori, S.A. 1989. The Guianas in D.G. Campbell and H.D. Hammond, eds. *Foristic Inventory of Tropical Countries: The status of plant systemaitcs, collections and vegetation, plus recommendations for the future* pp. 375–390. New York Botanical Garden.

Miller, L. 1991. The influence of resource dispersion on group size among wedge-capped capuchins (*Cebus olivaceus*). *Am. J. Primatol.* 24:123.

Mittermeier, R.A. 1977. Distribution, Synecology and Conservation of Surinam monkeys. PhD dissertation, Harvard University.

Mittermeier, R.A. 1986. Primate conservation priorities in the neotropical region in K.Benirschke, ed. *Primates: The road to self-sustaining populations*, pp. 221–240. New York: Springer-Verlag.

Mittermeier, R.A. and Fleagle, J.G. 1980. Locomotor behavior, body size and comparative ecology of seven Suriname monkeys. *Amer. J. Phys. Anthropol.* 52:301–314.

Mittermeier, R.A., and van Roosmalen M.G.M. 1981. Preliminary observations on habitat utilization and diet in eight Suriname monkeys. *Folia Primatol.* 36:1–39.

Mondolfi, E., and Eisenberg, J.F. 1978. New records for *Ateles belzebuth hybridus* in northern Venezuela, in J.F. Eisenberg (ed). *Vertebrate Ecology in the Northern Neotropics*, pp. 93–96. Smithsonian, Washington D.C.

Muckenhirn, N.A., Mortensen, B.K., Vessey, S., Fraser, C.E.O., Singh, B. 1975. Report on a primate survey in Guyana. Washington D.C.: Pan American Health Organization.

Norconk, M.A., and Kinzey, W.G. 1994. Challenge of neotropical frugivory: travel patterns of spider monkeys and bearded sakis. *Am. J. Primatol.* 34:171–183.

Ochoa, J.G., Aguilera, M., Soriano, P. 1995. The mammals from Guatopo National Park (Venezuela): Checklist and Community Study. (abstract) *Measuring and monitoring forest biological diversity: The international network of biodiversity plots*. Smithsonian / Man and the Biosphere Biodiversity Program International Symposium. Washington D.C., p. 160.

Peetz, A., Norconk, M.A., Kinzey, W.G. 1992. Predation by jaguar on howler monkeys (*Alouatta seniculus*) in Venezuela. *Am. J. Primatol.* 28:223–228.

Pires, J. Murça and Prance, G.T. 1977. The Amazon forest: A natural heritage to be preserved in G.T. Prance and T.S. Elias, eds. Extinction is Forever, pp. 158–194. New York: New York Botanical Garden.

Pope, T. 1990. The reproductive consequences of male cooperation in the red howler monkey: paternity exclusion in multi-male and single-male troops using genetic markers. *Behav. Ecol. Sociobiol.* 27:439–446.

Prance, G.T. 1990. Floristic similarities and differences between southern central America and upper and central Amazonia, in: A.H. Gentry (ed). *Four Neotropical Rainforests*, pp. 141–157. Yale, New Haven.

Riera, B., Poncy, O., Larpin, D., Joly, A., Belbenoit, P., Charles-Dominique, P., Hoff, M. 1995. Tree diversity, spatial distribution, structure, dynamics of the rainforest at the Nouragues permanent field research station, French Guiana. (abstract) *Measuring and monitoring forest biological diversity: The international network of biodiversity plots*. Smithsonian/ Man and the Biosphere Biodiversity Program International Symposium. Washington D.C., p. 94.

Robinson, J. 1986. Seasonal variation in the use of time and space by the wedge-capped capuchin monkey, *Cebus olivaceus*: Implications for foraging theory. *Smithsonian Contributions to Zoology*. Washington DC: Smithsonian Institution Press.

Rudran, R. 1979. The demography and social mobility of a red howler (*Alouatta seniculus*) population in Venezuela. in J.F. Eisenberg (ed.) *Vertebrate Ecology of the northern Neotropics*). Washington DC: Smithsonian Institution Press.

Rylands, A.B., Mittermeier, R.A., Luna, Ernesto Rodriguez 1995. A species list for the new world primates (Platyrrhini): Distribution by country, endemism, and conservation status according to the Mace-Land system. *Neotropical Primates* 3 (suppl): 113–164.

Sekulic, R. 1982. Daily and seasonal patterns of roaring and spacing in four red howler (*Alouatta seniculus*) troops. *Folia Primatol.* 39:22–48.

Steyermark, J. A. 1977. Future outlook for threatened and endangered species in Venezuela. In G.T. Prance and T.S. Elias, eds. *Extinction is Forever*, pp. 128–135. New York: New York Botanical Garden.

Sussman, R.W., Kinzey, W.G. 1984. The ecological role of the Callitrichidae: a review. *Am. J. Phys. Anthropol.* 64:419–449.

Sussman, R.W, Phillips-Conroy J. 1995. A survey on the distribution and density of the primates in Guyana. *International J. Primatol.* 16:761–792.

Tate, G.H.H. 1939. The mammals of the Guiana region. *Bull. Am. Mus. Nat Hist.* 76:151–229.

van Roosmalen 1985. Habitat preferences, diet, feeding strategy and social organization of the black spider monkey (*Ateles paniscus* Linnaeus 1758) in Suriname. *Acta Amazonica* 19:1–238.

van Roosmalen, M.G.M., Mittermeier, R.A., and Fleagle, J.G. 1988. Diet of the northern bearded saki (*Chiropotes satanas chiropotes*): a neotropical seed predator. *Am. J. Phys. Anthrop.* 14:11–35.

Walker, S. 1993. Positional adaptations and ecology of the Pitheciini. Unpublished PhD dissertation, City University of New York.

Walker, S. 1994. Habitat use by *Pithecia pithecia* and *Chiropotes satanas*. *Am. J. Phys. Anthropol.*, Suppl. 17:203.

Zonneveld, J.I.S. 1993. Planation and summit levels in Suriname (S. America). *Zeitschrift für Geomorphologie, N.F., Supplementband*, 93:29–46.

SECTION II

On Collitrichines

MARMOSET MISCONCEPTIONS

Paul A. Garber,[1] Alfred L. Rosenberger,[2] and Marilyn A. Norconk[3]

[1]Department of Anthropology
University of Illinois
Urbana, Illinois 61801
[2]National Zoological Park, Department of Zoological Research
Smithsonian Institution
Washington, DC 20008
and Department of Anthropology
University of Illinois at Chicago
Chicago, Illinois 60680
[3]Department of Anthropology
Kent State Univerity
Kent, Ohio 44242

INTRODUCTION

Beginning with the early 16th century reports by Western scientists of tiny, primitive, clawed, squirrel-like monkeys inhabiting the forests of South America, misconceptions and bias regarding tamarin and marmoset phylogeny, classification, ecology, behavior, and anatomy have continued. For example, as recently as 1992, Martin tenaciously guarded the notion that callitrichines should be specially treated and proposed a scheme of classification that by his own admission was unlikely to represent the evolutionary history of this group. He advocated dividing New World monkeys into two major clades, the 'true' New World monkeys and the 'clawed' New World monkeys for systematic purposes. Tamarins, marmosets, and Goeldi's monkeys were assigned to the latter group, although Martin (1990) followed the tradition of Simpson (1945), Simons (1972) and others in aligning *Callimico* with noncallitrichine ceboids ("...because it lacks some of the defining features of marmosets and tamarins, such as reduction in the number of molar teeth and twinning; "pg. 714).

As we discuss below - and as Martin apparently agrees despite his systematic arrangement - not only do the genera *Callimico*, *Saguinus*, *Leontopithecus*, *Callithrix*, and *Cebuella* represent a monophyletic group (subfamily Callitrichinae) (Rosenberger, 1981, 1992), but recent immunological (Sarich & Cronin, 1980), biochemical (Seuanez et al., 1989), and molecular data (Schneider et al, 1993) tentatively place Goeldi's monkey as a sister group to the *Callithrix/Cebuella* clade. We doubt this later linkage will stand against the scrutiny of further research, but nevertheless it represents strong evidence against a classification scheme that would place *Callimico* anywhere but within the Callitrichinae.

Given the recent data available on callitrichine behavior, ecology, and anatomy, we use this paper to highlight and dispel several commonly held misconceptions about tamarins and marmosets. We occasionally exercise all too much license in stating points of view that are perhaps more prevalent as intellectual currents than published ideas. Our apologies. We assume there is a large, multidisciplinary audience interested in callitrichines and our intention is to move the field forward by crossing off overly simplistic ideas and ill-founded notions, even if this means stating the obvious or the unlikely as a way of making a point.

Misconception #1. Tamarins, Marmosets, and Goeldi's Monkeys Are a Systematic Enigma, Difficult to Classify

Despite unambiguous morphological evidence of the dentition, cranium, and post-cranial skeleton supporting monophyly for tamarins, marmosets and Goeldi's monkeys (Rosenberger, 1981; Ford 1986; Kay, 1990), several researchers still embrace the archaic taxonomic placement of *Callimico* outside the tamarin and marmoset clade. The dual effect of this is to endorse a platyrrhine classification scheme that overrepresents biodiversity by setting up a single-species family, *Callimico*nidae, and to openly accept taxonomic groups of mixed ancestry, as in Martin's decision to include *Callimico* in the Cebidae. We strongly advocate accepting a classification scheme that places the genera *Saguinus*, *Leontopithecus*, *Callimico*, *Cebuella*, and *Callithrix* in the Subfamily Callitrichinae.

Misconception #2. *Callimico* Is an Intermediate between Marmosets and Tamarins and Other Platyrrhines

The molecular and morphological evidence linking *Callimico* with tamarins and marmosets dispels the notion of 'intermediacy' an idea that must be applied cautiously. In a cladistic model, phylogenetic relationships are linked through ancestral- descendent affinities with one taxon or a collection of taxa. There is no intermediacy, no measure of shared/equal affinities with more than one group. Thus an intermediate set of characters has no bearing on classification. The concept of 'intermediacy' does have value in realizing a continuity of form or behavior between taxa, and this recognition of continuity is key to understanding or rationalizing how apparently disparate taxa may in fact be closely related via ancestry. *Callimico* is a case in point. Few researchers currently doubt its close taxonomic affinity with other callitrichines. Yet the 'extra' molar and the 'absence' of a twin offspring are reminders that we should be able to reconcile the root or origin of the callitrichine stock elsewhere among the platyrrhines, where three molars and singleton births are the norm. By the same token, we might expect to find a comparable 'intermediate' among non-callitrichine platyrrhines as a conceptual and phylogenetic bridge toward callitrichines. *Saimiri* may be this link.

Misconception #3. Callitrichines Are an Isolated Stock of New World Monkeys

Irrespective of *Callimico*'s place in the tamarin and marmoset clade, there is another entrenched view which holds that callitrichines are a lone radiation without ties to other platyrrhines. This view is wrong. The idea has been furthered for decades by reading too much into classification, and by a philosophy that emphasizes static gaps as opposed to phylogenetic and adaptive continuities. This is one reason why we prefer to move from a family-level allocation of marmosets, tamarins and *Callimico*s, to a subfamily rank. The subfamily distinction offers a framework in which each of the 4 or 5 major platyrrhine radiations can be defined (i.e. Atelinae, Pitheciinae, Callitrichinae, Cebinae, and possibly Aotinae), and then re-aligned with related subfamilies into the same family (i.e. include Cebinae and Callitrhichinae in the Cebidae).

It appears that virtually all systematists now recognize that as a group, callitrichines are closely related to another known lineage of living platyrrhines. There is a healthy debate about which non-callitrichines are actually their nearest relatives (Schneider & Rosenberger, this volume). Evidence is mounting from morphological and molecular studies that *Saimiri* and *Cebus* are callitrichine sister-taxa. Although the histories of each of these genera are not well known and are likely complex, this linkage should offer a phylogenetic perspective on the most important adaptive features of the callitrichine radiation. For example, we expect researchers may begin to recognize more continuity in form, function and behavior. How large is the gap in foraging adaptations between squirrel monkeys and the typically insectivorous-frugivorous callitrichines? Why should we assume the high-pitched vocalizations of cebus monkeys, squirrels and callitrichines are parallelisms rather than shared-derived traits? Another view is that callitrichine and cebines are monophyletically related, part of a broader adaptive sub-radiation of platyrrhines (Rosenberger, 1980, 1992). Knowing that callitrichines are part of a larger group also justifies classifying them at a level below the family.

Misconception #4. Callitrichines Are either Primitive or Derived

Here we overstate the case in our effort to make a point. While this debate has basically polarized views on callitrichines for a century, in modern terms such expressions only serve as shorthand caricatures. Characters are primitive or derived, not lineages or taxa. Taxa are always a mixture of ancestral and derived traits, and some lineages may be relatively more conservative than others. Therefore, whereas we are convinced that many well known callitrichine features are not primitive primate or platyrrhine features (e.g., claws, twining, tricuspid teeth; see Hershkovitz, 1977; Rosenberger, 1977; Ford, 1980; Garber, 1980), we must continue to reevaluate our interpretations of the derived or primitive nature of traits and trait complexes as new fossil and comparative data become available. Overall, we maintain that callitrichines have not retained the ancestral platyrrhine morphology and behavior, and in this respect, the radiation is best considered as derived. The *Callimico* lineage, bearing single infants instead of twins and having three molars rather than two, is the least derived branch (in terms of these characters). *Callithrix* and *Cebuella*, using the yardsticks of skulls, teeth, postcrania, and genetic evidence, are the most derived forms. By the same token, *Leontopithecus* and *Saguinus* each present their own unique features and evolutionary trajectories.

We include another example where caution must be used in assessing the primitive or derived nature of callitrichine biology, namely adult body size. Although we believe that many tamarins and marmosets are secondarily reduced in body size, this does not conflict with the possibility that early platyrrhines were small. Those early forms would have been part of an initial radiation, one that may not be directly ancestral to all living platyrrhines. That is, in general terms, we would not expect them to be monophyletically related to cebids. Takai & Anaya (1996) have recently described extremely small platyrrhine teeth from the oldest primate site in South America. Early Old World anthropoids were also small. As discussed below, platyrrhine groups have experienced increases and decreases in body size several times in parallel. The challenge to paleoanthropologists is to identify which size-shifts (and features among the taxa) are homologous.

Misconception #5. Callitrichines Are a Recent, Derived Group

There is no direct linkage between time of origin and preponderance of derived traits. Based on cladistic evidence and related fossils 18–20 million years old, Rosenberger (1979) inferred that callitrichines were an ancient group in spite of their derived morphology. He also argued that *Mohanamico* was definitively callitrichine, possibly part of the *Callimico* lineage (Rosenberger, 1992). Interesting fossils recovered from La Salla, Bolivia, about 25 million years ago, also are very callitrichine-like (Takai & Anaya, 1996). Thus, although there it may have been a suggestion some years ago to link the derived aspect of callitrichine anatomy with a recent origin, perhaps in connection with Pleistocene refugia, this now is an unlikely scenario.

Misconception #6. Marmosets and Tamarins Represent Two Natural and Ecologically Distinct Adaptive Radiations

Critical to this idea are two assumptions: one, that these are natural, phylogenetic groups, and two, that based on their dentitions there is a clear ecological division between marmoset gum-eaters and tamarin fruit- insect eaters. These dichotomies are not supported by the evidence (Garber, 1992; Ferrari, 1993). There is universal agreement that marmosets (*Callithrix* and *Cebuella*) are a monophyletic group, but there is no evidence that *Saguinus* and *Leontopithecus* similarly represent a monophyletic group. The relatively large canines and small incisors that these two 'tamarin' genera share are ancestral callitrichine features and do not prove they are closely related. Although there is a continuing debate as to which one of these two genera is closer to *Callithrix/Cebuella* (Schneider & Rosenberger, this volume), there are no acceptable arguments supporting a close cladistic linkage between *Saguinus* and *Leontopithecus*.

Initial studies of callitrichine diet and dental morphology presented a simple ecological dichotomy with marmosets as gum- eaters and tamarins as fruit-eaters (Coimbra-Filho & Mittermeier, 1977). *Callithrix/Cebuella* have tall lower incisors combined with a set of incisor-like canines which form a dental scraper. The other callitrichines all have the primitive condition of low-crowned incisors and tall canine tusks. Clearly, *Saguinus* and *Leontopithecus* lack the scraping specialization and are more prone to eat prey, fruits and gums that do not require extensive chiseling with their front teeth. However, we are not at all certain that gum-eating alone, and not extractive foraging of insects under bark, or the two combined, has shaped the anatomy of ancestral marmosets via natural selection. Marmoset species show a range of dental and digestive morphologies, occupy a diversity of habitats, and most are larger in body size than *Callithrix jacchus* and certainly larger than *Cebuella*, which are reported to be the most dependent on gums as a dietary staple. It is possible that intense specialization on plant gums evolved locally in some forms of *Callithrix* and *Cebuella*.

Misconception #7. Callitrichines Are Dwarfs

We hope to redirect the discourse on this highly interesting issue. New World primates are unique in their extreme variation in body size. Given the constraints that smaller and larger body size place on positional behavior, feeding ecology, reproductive output, and susceptibility to predators, evolutionary changes in rates and patterns of growth and development likely represent fundamental changes in the manner in which a species exploits it environment. Among free-ranging living platyrrhines, adult body weight ranges from 120 grams in the pygmy marmoset (*Cebuella pygmaea*) to over 12,000 grams in the woolly spider monkey (*Brachyteles arachnoides*) (Ford & Davis, 1992; Rosenberger, 1992). This represents over two orders of magnitude and far exceeds the range found in extant cercopithecoids and pongids, groups that are distributed across much more landmass and many more ecozones.

Based on our current understanding of platyrrhine phylogeny, major increases and decreases in body size have occurred independently in several different lineages. These include *Cebus*, *Saimiri*, callitrichines, and atelines (Kay 1990, 1994; Rosenberger & Strier, 1990; Ford & Davis, 1992; Rosenberger, 1992; Cartelle & Hartwig, 1996). For example, Ford & Davis (1992:438–439) surmise that over the course of platyrrhine evolution capuchin monkeys have "nearly tripled in body size...independent of all other New World monkey lineages." *Saimiri*, in contrast, has probably undergone a significant body size reduction in comparison to its nearest early Miocene fossil relative *Dolichocebus* (Rosenberger, 1990). Relatives of spider and howler monkeys were, in the recent past, twice as large as any remaining alive today (Cartelle & Hartwig, 1996; Hartwig & Cartelle, 1996). Although it is unclear how frequently increases and especially decreases in body size have occurred in platyrrhine evolution, these events appear to have played a major role in shaping the reproductive, mating, and social systems of New World primates (Ford & Davis, 1992; Martin, 1992; Garber, 1994).

The picture of body size evolution among callitrichines is more complex and interesting than one might gather from the historical focus on the relatively narrow paradigm of the dwarfism hypothesis. For example, in addition to selection for size reduction connected with the origins of the group, among callitrichines there is evidence of several independent size-reduced lineages, perhaps going from a *Saguinus*-sized creature (400–600 gms) to a *Callithrix*-sized creature (400–250 gm) and, from a *Callithrix*-sized animal to *Cebuella* (125 gms). Related to the latter case, consider that Rosenberger & Coimbra-Filho (1984) and Rosenberger (1992) have also argued that *Leontopithecus*, possibly the sister-group of *Callithrix/Cebuella*, has undergone a body size increase since splitting from this clade. Thus, whether or not callitrichines arose as miniatures relative to the last common ancestor, they shared with other platyrrhines the body sizes of subsequent independent lineages continues to be an object of selection.

The notion of dwarfing also remains ill-defined and thus easily abused. Adult female squirrel monkeys (680 gm) are barely distinguishable in body weight from adult female golden lion tamarins (575–622 gm; Dietz et al, 1994) and adult female moustached tamarins (550–620; Garber et al., 1993). Thus there is nothing remarkable about callitrichine body size *per se*. Historically, the semantic implications of the dwarfism hypothesis became accepted despite limited biometric data of any kind, the strong influence of typology and orthogenesis in systematic thinking, and the lack of a sound cladistic framework for interpreting platyrrhine evolution. The impetus for a dwarfing theory as an evolutionary explanation was promoted by W.K. Gregory and R.I. Pocock in the 1920s, who thought callitrichine morphology was generally not primitive. One can imagine Gregory, a paleontologist, being enthusiastic about the idea as a counter example to Cope's Law of evolutionary size increase.

Given the present data, it cannot be stated with certainty that *Leontopithecus*, *Callimico*, and most species of *Saguinus* are smaller in body size than ancestral callitrichines; or that the extant forms as a group represent radically small, phylogenetic dwarfs. What we can do profitably is clarify the boundaries of the discussion. We suggest the term dwarf is best restricted to a special case of hypomorphosis (evolutionary size reduction) that results in maintenance of the same shape and form as a lineage evolves from a larger-bodied ancestor to a smaller descendant (proportioned dwarfism). Size reduction in an evolving lineage that produces shape changes relative to the ancestral condition results in hypomorphs, not dwarfs. Data presented by Garber & Leigh (in press) indicate that differences in adult body weight among extant callitrichines can be explained by ontogenetic changes in growth rates rather than by any significant decrease in the age at maturation. Differences in growth rates during particular developmental periods may result in significant size and shape differences among taxa. In addition, there is evidence of significant differences in limb proportions (Jungers, 1985; Garber, 1991), and hand size and shape among callitrichine species (Bicca-Marques, in prep), as well as reports of overscaling in the cheekteeth (Plavcan & Gomez, 1990) and the size of the eye (Martin, 1992). Together, these data do not support the contention that callitrichines are proportioned dwarfs.

Misconception #8. Claw-Like Nails in Callitrichines Are an Adaptation to Gum Feeding

All species of callitrichines have laterally compressed and elongated claw-like nails on all digits except the big toe, which bears a flattened nail. These claw-like nails are termed tegulae to distinguish them from the true claws of many nonprimate mammals (faculae) and nails shaped like ours and other catarrhines (ungulae). Histologically, the claw-like tegulae of tamarins, marmosets, and Goeldi's monkey are thought to be the same as the nails of other New World monkeys, many of which (e.g., *Saimiri, Aotus, Pithecia*) exhibit the compressed and pointed shape, but are not hooked like claws. Since many species of tamarins and marmosets are known to cling to large vertical trunks while feeding on plant gums (Garber, 1992), it has often been assumed that the evolution of claw-like nails is directly related to the evolution of a gum feeding habit.

There are several problems with this inference. One is theoretical: during gum feeding, all callitrichines embed their claw-like nails into the tree trunk to maintain support. Although gum feeding—vertical clinging—clawed digits are associated as a trait complex, this by itself does not establish causality. The other main problem is that there is considerable variability in the degree to which plant gums are exploited by different callitrichine taxa. Plant gums are critically important in the diet of most marmoset species (although as mentioned differences in anterior dental morphology and digestive physiology exist among and between marmosets of the *Callithrix*-jacchus group and marmosets of the *Callithrix*-argentata group suggesting differing degrees of gum feeding specializations). In contrast, Goeldi's monkey has not been observed to feed on plant gums, and for many *Saguinus* and *Leontopithecus* species, gum feeding may comprise only 1–8% of the annual diet. Overall, the feeding ecology of callitrichines is distinguished from other platyrrhines by the ability of these primates to exploit a range of resources that are associated with tree trunks in the forest understory. This includes plant gums, bark refuging insects, small vertebrates concealed in knotholes, prey hidden in bromeliads that grow along the main axis of the tree, as well as use of vertical trunks to scan for insects and small vertebrates located on the ground. Given the highly faunivorous diet of all callitrichines, the evolution of claw-like nails is best understood as a foraging adaptation enabling these small primates to exploit high protein and carbohydrate resources restricted to particular micro-habitats in the forest understory. In the absence of claw-like nails, access to large vertical trunks would be highly limited. Gum feeding and tree gouging in extant marmosets (*Ce-*

buella and *Callithrix*) represent a derived behavioral pattern related to an expansion of the original trunk foraging adaptation.

Misconception #9. Callitrichines Have Simple Social and Mating Systems

Initial reports of callitrichine social and mating systems were based first on lab studies and later on short-term field research on a few species. This work continued to build on the premise that tamarins and marmosets lived in small, monogamous social groups characterized by a pair bond between a single adult male and a single adult female. This characterization drew largely from captive studies in which (1) large groups were often found to be unstable, (2) only a single female in each group gave birth, (3) adult males and other group members helped care for the young, as well as the assumptions that (4) tamarins and marmosets were primitive platyrrhines and (5) the ancestral mating system for New World monkeys is monogamy. It is now apparent that among platyrrhines a monogamous pair bonded social system is found only in *Aotus* and *Callicebus*, and that monogamy in night and titi monkeys is best understood as a derived behavioral/social pattern (Garber,1994; Garber & Leigh, in press). In contrast, tamarins and marmosets live in multimale, multifemale groups of 5–15 animals. In all species for which data are available there is evidence of a extremely broad range of mating and grouping patterns including polyandry, polygyny, and less often monogyny. There is no evidence from the wild that a single male and a single female maintain an exclusive mating relationship over an extended period of time. Although only a single female in each group typically gives birth, groups with two breeding females have been reported in a few species (*Saguinus fuscicollis*, *Callithrix jacchus*, *Leontopithecus rosalia*).

Less is known regarding the social and mating system of *Callimico*. These callitrichines are reported to live in multimale, multifemale groups of at least 5–8 individuals. Observations in the wild indicate that more than a single female in a group may breed. Although *Callimico* is the only callitrichine that gives birth to single infants rather than twins, all species in the subfamily are characterized by an extremely high reproductive rate. Early age at maturation (approximately 2 years) coupled with the potential of female tamarins and marmosets to produce 2 litters of two offspring each year, and the potential of female Goeldi's monkeys to produce 2 litters per year each containing one offspring, results in intrinsic rates of population increase that are greater than those found in any other group of higher primates (Martin, 1992; Garber, 1994). The ability of individual callitrichines to achieve their reproductive potential is directly related to a range of proximate environmental, social and demographic factors. These include group size and composition, availability of helpers to care for young, age, kinship, opportunity to migrate with one or more peers, the presence of breeding vacancies in nearby groups, and the availability of suitable habitats within their range.

Callitrichine social groups are based on high levels of both competition and cooperation. Within each group, males and females compete for extremely limited reproductive opportunities. Intragroup aggression is rare, and competition appears to be mediated through olfactory cues, age-related dominance, and cooperation. Individuals may need to cooperate to insure infant survivorship, maintain range integrity and access to productive feeding sites, detect predators, and form social bonds that aid in paired migration or group fissioning. Behavioral options and behavioral tactics in callitrichine social groups appear to be extremely complex and dynamic, and any notion that tamarins, marmosets, or Goeldi's monkeys live in simple social or mating systems is completely in error.

In closing, we wish to highlight the social and ecological complexity of the callitrichine radiation, and underscore the importance of dispelling tamarin and marmoset misconceptions as a necessary step in understanding platyrrhine systematics and evolution. We hope this Introduction has indicated new directions of inquiry and debate.

REFERENCES

Cartelle , C. and Hartwig, W.C., 1996, A new extinct primate among the Pleistocene megafauna of Bahia, Brazil. *Proc. Nat. Acad Sciences* 93: 6405–6409.

Coimbra-Filho, A.F. and Mittermeier, R.A., 1977, Tree-gouging, exudate-eating, and the "short-tusked" condition in *Callithrix* and *Cebuella*. In: *The Biology and Conservation of the Callitrichidae*, D. Kleiman (ed.), Washington, DC: Smithsonian Institution Press, pp. 105–115.

Dietz, J.M, Baker, A.J., and Miglioretti, D., 1994, Seasonal variation in reproduction, juvenile growth, and adult body mass in golden lion tamarins (*Leontopithecus rosalia*). *Amer. J. Primatol.* 34: 115–132.

Ferrari, S.F., 1993, Ecological differentiation in the Callitrichidae. In: *Marmosets and Tamarins: Systematics, Behaviour, and Ecology*, A.B. Rylands (ed.) Oxford: Oxford University Press, pp. 314–328.

Ford, S.M. 1980, Callitrichids as phyletic dwarfs, and the place of the Callitrichidae in Platyrrhini. *Primates* 21:31–43.

Ford, S.M. 1986, Systematics of the New World monkeys. In: *Comparative Primate Biology Vol. 1.: Systematics, Evolution, and Anatomy*, D. Swindler and J. Erwin (eds.). New York: Alan R. Liss, Inc., pp. 73–135.

Ford, S.M. and Davis, L.C., 1992, Systematics and body size: implications for feeding adaptations in New World monkeys. *Am. J. Phys. Anthropol.* 88:415–468.

Garber, P.A., 1980, Locomotor adaptations and feeding ecology of the Panamanian tamarin (*Saguinus oedipus geoffroyi*, Callitrichidae, Primates). *Int. J. Primatol.* 1: 185–201.

Garber, P.A., 1991, A comparative study of positional behavior in three species of tamarin monkeys. *Primates* 32: 219–230.

Garber, P.A., 1992, Vertical clinging, small body size, and the evolution of feeding adaptations in the callitrichinae. *Am. J. Phys. Anthrop.* 88: 469–482.

Garber, P.A., 1994, Phylogenetic approach to the study of tamarin and marmoset social systems. *Amer. J. Primatol.* 34: 199–220.

Garber, P.A., Encarnación, F., Moya, L. and Pruetz, J.D., 1993, Demographic and reproductive patterns in moustached tamarin monkeys (*Saguinus mystax*): implications for reconstructing platyrrhine mating systems. *Amer. J. Primatol.* 29: 235–254.

Garber, P.A. and Leigh, S.R., in press, Ontogenetic variation in small-bodied New World primates: implications for patterns of reproduction and infant care. *Folia Primatol.*

Hartwig, W.C. and Cartelle, C. 1996, A complete skeleton of the giant South American primate *Protopithecus. Nature* 381:307–311.

Hershkovitz, P., 1977, *Living New World Primates (Platyrrhini), with an Introduction to Primates, Vol 1*. Chicago: University of Chicago Press.

Jungars, W.L., 1985, Body size and scaling of limb proportions in Primates. In: *Size and Scaling in Primate Biology*. W.L. Jungars (ed), New York: Academic Press, pp. 345–381

Kay, R.F. 1990, The phyletic relationships of extant and fossil Pitheciinae (Platyrrhini, Anthropoidea). *J. Hum. Evol.*. 19: 175–208.

Kay, R.F., 1994, "Giant" tamarin from the Miocene of Colombia. *Am. J. Phys. Anthropol.* 95: 333–353.

Martin, R.D., 1990, *Primate Origins and Evolution: A Phylogenetic Reconstruction*. New Jersey: Princeton University Press.

Martin, R.D. 1992, Goeldi and the dwarfs: the evolutionary biology of the small New World monkeys. *J. Hum. Evol.* 22: 367–393.

Plavacan, J.M. and Gomez, A.M., 1990, Phyletic dwarfing and dental scaling in callitrichines. *Am. J. Phys. Anthrop.* 81: 282.

Rosenberger, A.L., 1977, *Xenothrix* and ceboid phylogeny. *J. Hum. Evol.* 6: 461–481.

Rosenberger, A.L., 1979, *Phylogeny, Evolution, and Classification of New World Monkeys*. Ph.D. thesis, City University of New York.

Rosenberger , A.L.,1980, Gradistic views and adaptive radiation of platyrrhine primates. *Z. Morphol. Anthropol.* 71:157–163.

Rosenberger, A.L., 1981, Systematics: The higher taxa. In: Ecology and Behavior of Neotropical Primates, Vol. 1., A.F. Coimbra-Filho and R.A. Mittermeier (eds.). Rio de Janeiro: Academia Brasiliera de Ciencias, pp. 9–27.

Rosenberger, A.L., Setoguchi, T., and Shigerhara, N., 1990, The fossil record of callitrichine primates. *J. Hum. Evol.* 19: 209–236.

Rosenberger, A.L., 1992, Evolution of feeding niches in New World monkeys. *Am. J. Phys. Anthrop.* 88: 525–562.

Rosenberger, A.L. and Coimbra-Filho, A.F., 1984, Morphology, taxonomic status, and affinities of the lion tamarins, *Leontopithecus* (Callitrichinae, Cebidae). *Folia Primatol.* 42: 149–179.

Rosenberger, A.L., and Strier, K.B., 1989, Adaptive radiation of the ateline primates. *J. Hum. Evol.* 18: 717–750.

Sarich, V.M. and Cronin, J.E., 1980, South American mammal molecular systems, evolutionary clocks, and continental drift. In: *Evolutionary Biology of the New World Monkeys and Continental Drift*, R.L. Ciochon and A.B. Chiarelli (eds.), New York: Plenum Press, pp. 399–421.

Seuanez, H., Forman, L., Matayoshi, T., and Fanning T., 1989, The *Callimico goeldii* (Primates, Platyrrhini) genome: karyology and middle repetitive (LINE-1) DNA sequences. *Chromosoma* 98: 389–395.

Schneider, H., Schneider, M.P.C., Sampaio, I., Harada, M.L., Stanhope, M., Czelusniak, J., and Goodman, M., 1993, Molecular phylogeny of the New World monkeys (Platyrrhine, Primates). *Mol. Phylogen. Evol.* 2: 225–242.

Schneider, H., and Rosenberger, A.L., this volume, Molecules, morphology and platyrrhine systematics.

Simons, E.L., 1972, *Primate Evolution: An Introduction to Man's Place in Nature*. New York: Macmillan.

Simpson, G.G., 1945, The principles of classification and classification of mammals. *Bull. Amer. Mus. Nat. Hist.* 85: 1–350.

Takai, M. and Anaya, F., 1996, New specimen of the oldest fossil platyrrhine, *Branisella boliviana* from Salla,Bolivia. *Am. J. Phys. Anthrop.* 99:301–317.

5

THE OTHER SIDE OF CALLITRICHINE GUMMIVORY

Digestibility and Nutritional Value

Michael L. Power

Department of Zoological Research
National Zoological Park
Smithsonian Institution
Washington, D.C. 20008

GUMMIVORY AND CALLITRICHINES

Gummivory is rare among mammals. Within primates, however, there are species from many different taxa that feed on gums and other plant exudates (Nash, 1986). Why primates appear to be predisposed to gummivory is an intriguing question. Gum would not be considered an intrinsically high quality food. It is difficult to obtain and presumed to be difficult to digest (Van Soest, 1982). It generally contains little protein, no fat and no vitamins. Gums can contain tannins, phenolic compounds, and other chemicals that have potential adverse effects for animals that ingest them (Wrangham and Waterman, 1981; Nash and Whitten, 1989). Gums do provide complex carbohydrates, and often contain significant quantities of nutritionally important minerals. But on balance, gums would not appear to be particularly good food.

How important has gummivory been in the evolution of the Callitrichinae? It is well accepted that marmosets and tamarins feed on gums and other plant exudates. There are documented instances of gummivory for every genus within the Callitrichinae (Kinzey et al., 1975; Coimbra-Filho and Mittermeier, 1977; Garber, 1980; Pook and Pook, 1981; Soini, 1982, 1987; Rylands, 1984; Peres, 1989). Many features of callitrichine biology have been proposed to be adaptations for exudate feeding, including their small body size, their claw-like nails, and aspects of their postural and locomotor morphology that appear related to clinging to large-diameter vertical substrates (Kinzey et al., 1975; Garber, 1980, 1992). Indeed, exudate feeding has been suggested to have been a prime adaptive feature in the radiation of this lineage (Sussman and Kinzey, 1984). However, the explosion of field studies beginning in the late 1970s and continuing to this day has revealed wide variation in the extent of gum-feeding among callitrichine species. Only a few marmoset species can truly be regarded as gummivore-faunivores. Most callitrichines are frugivore-faunivores, with gum serving at most as a seasonal food, often during times of fruit scarcity (Garber, 1984, 1988; Rylands ,1984; Soini, 1987).

Was gummivory an important aspect of the entire callitrichine radiation, or an adaptation peculiar to the marmoset lineage? In this chapter I will explore this issue primarily by examining the morphological and physiological differences among callitrichines that could affect the digestion of gum. I will first examine the potential value of gum as a food. Then I will present data on digestive parameters and the digestive response to dietary gum from five species of callitrichines. The variation in these parameters among the species will be examined to determine how it might impact the actual nutritional value of gum as a food.

GUM AS FOOD

Foods are often termed either high or low quality. What is meant by these terms, however, is not always explicitly stated. Not surprisingly then, different authors come to different conclusions regarding the "quality" of a species diet. For example, one might consider an animal that feeds predominantly on ripe fruits to have a "high" quality diet. This assessment would be based primarily on the supposed ease of digestion of ripe fruit, implying a high nutrient "reward." McNab (1986), however, considers frugivores to have "low" quality diets based on the patchiness of ripe fruits in space and time. In both cases, however, the assessment of food quality is made by estimating the rate of nutrient gain the food affords an ingesting animal.

Animals must ingest foods whose net rate of nutrient gain is capable of supporting their nutrient requirements. Net rates of assimilation of nutrients depend on the availability of the food (i.e., the rate the food can be harvested), the nutritional costs associated with obtaining and ingesting the food (Note: this ignores costs such as predation risk.), and the nutritional value of the food. The nutritional value of a food is determined both by its nutrient content and by the bioavailability of those nutrients. The bioavailability of the nutrients depends not only upon the food's chemical structure, but upon the digestive physiology of the ingesting animal. This latter factor has generally been ignored in discussions of "high" versus "low" quality foods, but it is crucial to understand that the nutritional value of a food does not solely depend on its own intrinsic properties. The physiology of the animal ingesting the food is a key component in determining what nutrients are available, and whether the rate of nutrient gain from that food is capable of meeting the animal's nutritional needs.

In this section I consider the potential value of gums as food for callitrichines. My assessment is based on available information concerning the chemical composition and structure of gums, the potential abundance of gum, and the ecology and physiology of the callitrichines. Variation in the extent of knowledge among these issues is huge, therefore my presentation will strive for the development of reasonable hypotheses rather than pretending to firm conclusions.

Chemical Composition

This section must start with an important qualification. Few of the gums that primates (let alone callitrichines) naturally feed on have been chemically analyzed. The best studied gums are from trees in the genus *Acacia,* with most data coming from African and Australian species. There are data for a few gums eaten by callitrichines, most notably for

Anacardium excelsum. But the extent of the chemical variation of gums, either among gum-producing species, within the same species, or even seasonally within the same tree, is largely unknown.

Gums are predominately composed of water and carbohydrate. They appear to contain little or no fat or vitamins. Gums do contain a small protein fraction, ranging from 1%-9% of the dry matter content, and they usually contain nutritionally significant quantities of minerals, especially calcium. Often the mineral content of a gum will exceed its protein content. Thus, gums, as food, largely have been considered to provide energy and minerals (Bearder and Martin, 1980; Nash, 1986).

Gums have been suggested to provide needed calcium to highly insectivorous primates (Bearder and Martin, 1980; Garber, 1984). Insects appear to be poor sources of calcium, and to have extremely low calcium:phosphorus ratios (Allen 1989). Gums that have been analyzed are the reverse (Anderson et al., 1983; Garber, 1984). Thus, the value of gum to a primate may depend, in part, on what proportion of nutrient needs are met by ingesting insects.

Many gums have been shown to contain tannins, phenolic compounds, and other potentially noxious substances. Wrangham and Waterman (1981) demonstrated that *Acacia* gums fed on by vervet monkeys were significantly lower in total tannins than those that were avoided. Nash and Whitten (1989) proposed that flavonoids in gums might deter ingestion by primates. Gums are chemically complex substances, and are certain to contain both nutritive and non-nutritive components.

Animals consuming gum might also ingest insects that become trapped in the gum. Whether this does indeed happen, and whether these incidently ingested insects have nutritional significance is unclear. Gum-feeding animals probably also ingest dirt and tree bark, which most likely provide no nutrition.

Chemical Structure

Gums are multi-branched, β-linked polysaccharides. As such, they should be resistant to endogenous mammalian digestive enzymes (Van Soest, 1982) and thus provide something of a digestive challenge to mammals. Fermentation by microorganisms within the gut appears to be required if animals are to utilize the energy in gums. This appears to be true of their mineral component as well (Power, unpublished data).

Gums vary greatly in water solubility. The more water soluble a gum, the higher the expected rate of fermentation. Thus, water insoluble gums would provide a much lower rate of nutrient gain than would highly water soluble gums. In all cases, some of the energy and other nutrients utilized by the fermenting microorganisms are lost to the ingesting animal.

Abundance

The abundance of gums in callitrichine natural environments is difficult to ascertain. Callitrichines certainly feed from a large number of gum-producing plant species. From data published through 1991, Power (1991) produced a list of over 120 plant species that were fed on by at least one of the 16 callitrichine species. These plant species include some of the more common trees found in callitrichine field sites. Thus, there is potentially a large amount of gum in the callitrichine environment. How much of that gum is available to be eaten, however, is a major unanswered question.

Among callitrichines, the marmosets (the genera *Callithrix* and *Cebuella*) have at least partially solved the problem of gum availability via morphological adaptations to their lower anterior dentition that allow them to gouge holes in trees (Coimbra-Filho and Mittermeier, 1977; Rosenberger, 1978). The tree-gouging abilities of marmosets allow them to rely on gum as a food resource. In contrast, tamarins and lion tamarins (the genera *Saguinus* and *Leontopithecus*) can only be opportunistic gum-feeders. Marmosets have traded the costs of gouging for an effective increase in the abundance of gum, a likely explanation for why marmosets typically live in smaller home ranges and have smaller daily path lengths than other callitrichines (Rylands, 1984; Ferrari and Lopes, 1989; Power, 1991). Thus marmosets probably have lower energetic costs and predation risks associated with travel compared with tamarins and lion tamarins. Although the extent of either of these reductions is currently unknown, they are important considerations in evaluating the value of gum as a food.

DIGESTIVE PHYSIOLOGY OF CALLITRICHINES

To fully evaluate the value of gum as food for different callitrichines requires an understanding of their digestive physiology. How does digestion vary among callitrichines, and what implications does that variation have for the value of gum as a food? To investigate this issue, animals from five species of callitrichines were fed a controlled, homogeneous diet, both with and without added gum. Transit time (time elapsed from the ingestion of a marker to its first appearance in the feces, Warner, 1981) and the completeness of digestion of the diet (as measured by the coefficients of apparent digestibility of dry matter and energy) were measured for each animal under both dietary regimens.

Table 1 summarizes the available field data on exudate feeding for the five species in this study. These five species span the range of body sizes within the Callitrichinae. The two marmoset species are consistent gum-feeders; indeed the pygmy marmoset (*Cebuella pygmaea*) would appear to be a true gum specialist. The common marmoset (*Callithrix jacchus*) appears to be able to survive on either a fruit- or gum-dominated diet (Stevenson and Rylands, 1988; Alonso and Langguth, 1989). The golden lion tamarin (*Leontopithecus rosalia*) and the cotton-top tamarin (*Saguinus oedipus*) are primarily frugivore-faunivores, for which plant exudates probably are minor dietary constituents. The saddle-back tamarin (*S. fuscicollis*) would appear to be intermediate in the potential dietary importance of gums.

Digestion trials were conducted at three different institutions (National Zoological Park, Washington D.C., National Institutes of Health, Bethesda MD, and the Marmoset

Table 1. Percentage of monthly feeding time devoted to eating plant exudates based on field studies

Species	Average (%)	Range (%)	Sources
Cebuella pygmaea	67	50 - 75	1,2
Callithrix jacchus	30	20 - 70	3,4
Saguinus fuscicollis	12	5 - 58	5,6,7
Saguinus oedipus	<5	0 - 5	8
Leontopithecus rosalia	<5	0 - 5	9

1: Ramirez et al. 1977, 2: Soini 1982, 3: Alonso and Langguth 1989, 4: Stevenson and Rylands 1988, 5: Terborgh 1983, 6: Soini 1987, 7: Garber 1988, 8: A. Savage pers. comm., 9: Peres 1989

Research Center, Oak Ridge TN). All animals were singly-housed adults. Two diets were used: a basal diet consisting of a standardized, homogeneous mixture of canned marmoset diet (Hills Pet Foods, Topeka, KS), banana, and gelatin, and a modification of the basal diet adding gum arabic powder at a level of 9% of the dry matter of the diet. The diet ingredients were blended together, and the resulting mixture poured into a container and refrigerated overnight to gel. The purpose of making homogeneous diets was to ensure that animals could not self-select particular ingredients.

Indigestible markers were used to measure passage rates and to mark the feces from the beginning and end of a digestion trial. Markers used in this study were polystyrene and cellulose acetate beads (1 mm in diameter) and chromic oxide. Within species there were no differences in transit times measured by the different markers (Power, 1991).

Total collection of feces and refused food enabled the calculation of two measures of digestive efficiency, the coefficients of apparent digestibility of dry matter (ADDM) and energy (ADE). These are defined as:

$$ADDM = (1 - F_{DM}/I_{DM}) * 100, \tag{1}$$

$$ADE = (1 - FE/GE) * 100, \tag{2}$$

where F_{DM} is total fecal dry matter, I_{DM} is total food dry matter ingested, FE is total energy of feces, and GE is gross energy intake (total energy of food ingested) (Van Soest, 1982). The energy of food and feces was determined by adiabatic bomb calorimetry.

The standard procedure was to conduct two consecutive digestion trials on each diet, starting with the non-gum diet. For a small number of animals this was not possible. These include a golden lion tamarin that refused the gum diet after one trial. Data were available from 37 animals that completed at least one trial on each diet (four pygmy marmosets, eight common marmosets, ten cotton-top tamarins, seven saddle-back tamarins, and eight golden lion tamarins). The digestive parameters on the gum-free diet was examined across species using correlation analysis and analysis of covariance. Within each species, the differences in digestive parameters between the diets were tested by paired sample t-tests.

Details of these methods can be found elsewhere (Power, 1991; Power and Oftedal, 1996).

The Use of Gum Arabic in Callitrichine Research

Gum arabic comes from Old World *Acacia* species (*Acacia senegal* or related species). Thus there is the danger that the digestive responses of callitrichine species to gum arabic is irrelevant to interpretations of their evolutionary history. However, biochemically, gum arabic should represent a digestive challenge similar to that of gums eaten in the wild by callitrichines. Gum arabic is a heterogeneous, complex polysaccharide containing galactose, arabinose, rhamnose, and glucuronic acid as the most abundant sugar constituents (Meer, 1980; Adriani and Assoumani, 1983). Gum arabic is highly water soluble, which should facilitate fermentation, making it a conservative test of the abilities of these species to digest gum. Although gum arabic is resistant to mammalian digestive enzymes, it is readily fermentable by bacterial action (Salyers et al., 1978; McLean Ross et al., 1984; Eastwood et al., 1986; Wyatt et al., 1986, 1988; Walter et al., 1986, 1988). Gum arabic is useful as a test substance as it has no known toxic levels in humans or other animals (Adrian and Assoumani, 1983; Anderson et al., 1983), and it is readily available.

Table 2. Species averages for digestive parameters on the gum-free diet

Species	n	BWt (gm)	TFA (min)	ADDM (%)	ADE (%)	DM Intake (% BWt)
Cebuella pygmaea	4	127	388	84.5	83.8	5.4
Callithrix jacchus	8	355	198	77.2	75.1	6.3
Saguinus fuscicollis	7	310	159	74.3	71.0	9.7
Saguinus oedipus	10	472	233	83.0	81.5	6.1
Leontopithecus rosalia	8	673	266	85.4	85.6	4.6

BWt = body weight; TFA = transit time; ADDM = apparent dry matter digestibility; ADE = apparent energy digestibility; DM Intake = dry matter intake.

Digestive Function on the Gum-Free Diet

Four of the five species in this study showed a pattern of decreasing transit time and digestive efficiency with a decline in body weight when fed the non-gum diet. Only the pygmy marmosets did not fit this pattern (Table 2). Although they were the smallest animals, pygmy marmosets had the absolutely longest transit times (Fig. 1), and had coefficients of apparent digestibility equivalent to those of tamarins four times their body

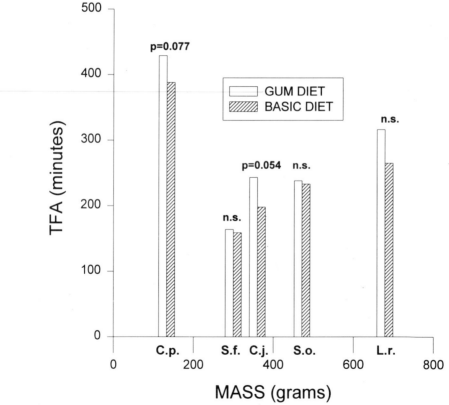

Figure 1. Transit time = time to first appearance of marker (TFA) for the basic (gum-free) and gum diets. C.p. = *Cebuella pygmaea*, C.j. = *Callithrix jacchus*, S.f. = *Saguinus fuscicollis*, S.o. = *Saguinus oedipus*, L.r. = *Leontopithecus rosalia*.

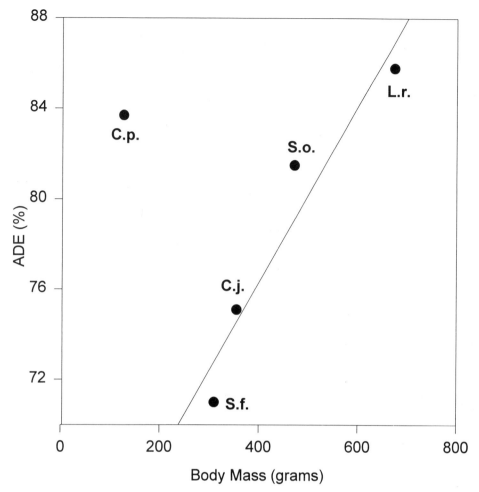

Figure 2. Plot of the coefficient of apparent energy digestibility (ADE) against body weight for the basic (gum-free) diet. C.p. = *Cebuella pygmaea*, C.j. = *Callithrix jacchus*, S.f. = *Saguinus fuscicollis*, S.o. = *Saguinus oedipus*, L.r. = *Leontopithecus rosalia*.

weight (Fig. 2). In contrast, common marmosets did not differ from the tamarins or lion tamarins in the allometry of any of these digestive parameters (p>.1, ANCOVA).

Excluding the pygmy marmoset, transit time and both coefficients of apparent digestibility were significantly correlated with body weight (p<0.01 in all cases). Body weight appeared to have a substantial effect on digestion in these four species. The average transit time of saddle-back tamarins was 42% less than that of golden lion tamarins when both were fed the gum-free diet, and the apparent digestibility of energy was lower by 14 percentage points.

Digestive Responses to Gum Arabic

The presence of gum arabic in the diet did not significantly affect transit time for any species (Fig. 1), although both marmoset species tended to have longer transit times

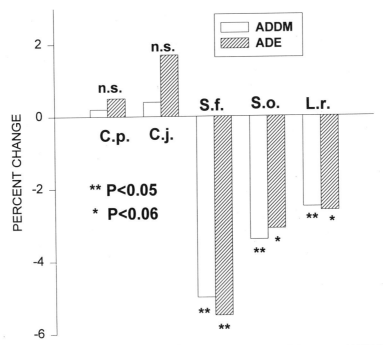

Figure 3. The percent change in the coefficients of apparent digestibility of dry matter (ADDM) and energy (ADE) between the two diets, calculated as 100 * (gum diet - basic diet)/basic diet. C.p. = *Cebuella pygmaea*, C.j. = *Callithrix jacchus*, S.f. = *Saguinus fuscicollis*, S.o. = *Saguinus oedipus*, L.r. = *Leontopithecus rosalia*.

on the gum diet (0.05<p<0.1). Gum did not affect digestion in pygmy or common marmosets. Neither marmoset species differed in ADDM or ADE between the diets. In contrast, both tamarin species and the golden lion tamarins had significantly lower ADDM when fed the gum arabic diet compared to the basal diet, although ADE were significantly different only for the saddle back tamarins (Fig. 3). The average percent reduction in digestive efficiency was greatest in *Saguinus fuscicollis*.

In both *Saguinus* species, the reductions in ADDM and ADE from the gum-free diet were significantly less in the second trial on the gum diet (Power 1991). The results reported above are for the second trial only, and thus are a conservative estimate of the reduction in digestion due to dietary gum in these species. The second trial on the gum diet took place 13 to 16 days after the animals had been fed exclusively on the gum diet, which would seem to be a sufficiently long period for animals with such rapid passage rates of digesta to adapt physiologically. Marmosets appeared not to need an adaptation period (Power, 1991). Nevertheless, further trials on cotton-top and saddle-back tamarins might have resulted in further reductions in the digestive deficit due to gum.

CONCLUSIONS

The dietary strategy of a species cannot be accurately assessed by taxonomic lists of ingested foods, nor by lumping those foods into broad categories such as leaves, fruit, gum, and animal matter. Those are not the parameters that evolution has operated on to produce the observed behaviors. To understand the value of a food to an animal requires an understanding of the nutritional needs of the animal, the relative bioavailabilities of nutrients in the different foods that the animal eats (and doesn't eat), and the physiological ecology of the species (Belovsky and Schmitz, 1994). There are many constraints, both external and internal, that act to modify a species dietary strategy from the theoretical "optimum" based on the nutritional qualities of the available foods.

Although much information pertinent to assessing the dietary strategy of a species can be collected by observation in the field, the opportunities to assess the effects of different nutritional regimens are severely restricted, and are at best opportunistic. The ability of the captive animal researcher to control diet brings the power of the experimental method to the research. However, although the results of captive research will always have physiological validity, the evolutionary implications can be problematic. Captivity can impose conditions that are outside the historical parameters affecting the species. Thus the observed physiological responses could be evolutionarily irrelevant. Collaboration between field ecologists and animal nutritionists, combining studies of wild animals with captive animal research, is needed. The input of the field ecologist would enable the captive animal researcher to design more realistic physiological and nutritional experiments. In turn, the results of the captive research will enable the field ecologist to better comprehend the physiological consequences of the behavior of wild animals.

What Is the Value of Gum as a Food?

Gum is a problematic food. Although there is potentially an enormous quantity of gum within tropical forests, the rate at which gum can actually be harvested may be quite limited. Gums are, on average, likely to be much more readily fermentable than leaves. Yet they probably still do require fermentation to yield nutrition. Gums are certainly not a balanced food. Animals that feed on gum must have other dietary sources of protein, fat and vitamins. When fruit is plentiful, marmosets reduce their gum intake in favor of fruit (Rylands, 1984; Alonzo and Langguth 1989). Even pygmy marmosets are reported to feed preferentially on fruit (Moynihan, 1976; Soini, 1982). The scarcity of non-primate gum-feeding mammals is a further indication that gum is a difficult food resource on which to survive.

The available information indicates that marmosets likely gain proportionally more nutrition from gums than do other callitrichines. Their ability to stimulate the flow of gum by gouging increases the relative availability and dependability of gum as a food. In addition to their dental adaptations, marmosets differ from other callitrichines in intestinal morphology. Common marmosets have larger and more complex caeca than do golden lion tamarins (Coimbra-Filho et al., 1980). Marmoset species appear to have a relatively greater proportion of both intestinal length and surface area accounted for by the caecum and colon than do tamarins (Power, 1991; Ferrari and Martins, 1992; Ferrari et al., 1993; Power and Oftedal, submitted b). The results from Power (1991) further suggest that lion tamarins also differ from marmosets in intestinal proportions, and resemble *Saguinus*. Thus the caecum and colon, the regions of the intestine where fermentation of gum would most likely occur, account for a larger proportion of the intestinal tract in marmosets than they do in other callitrichines. The results of these digestion trials are consistent with these morphological differences, and support the hypothesis that the marmoset lineage has derived adaptations to gummivory that the rest of the callitrichines lack.

For callitrichines outside of the marmoset lineage, gum likely is less available, less dependable a food resource, and when ingested probably provides a lower nutritional return. For these species, gum has all the characteristics of a low quality food: limited in quantity, patchy in space and time, and difficult to digest.

The above discussion refered to marmosets in a rather monolithic fashion. However, the extent of gummivory among *Callithrix* species appears quite variable (Rylands, 1984). It is plausible that the above conclusions only apply to the more gummivorous marmosets. More research, on both captive and wild animals, is needed to explore this issue.

Saddle-Back Tamarins: What Are They Doing?

In apparent contradiction to the above, saddle-back tamarins (*Saguinus fuscicollis* spp.) feed on gums at levels significantly higher than those for larger, sympatric *Saguinus* species, and comparable with Amazonian *Callithrix* species (Soini 1987). My research certainly does not indicate that they have the ability to efficiently digest gum. What aspects of saddle-back tamarin biology differ from the larger, sympatric *Saguinus* species that might increase the value of gum? One possibility is their habit of consistently ingesting gum throughout the year (Terborgh 1983, Soini 1987, Garber 1988) enables them to maintain a greater ability to digest gum than the captive animals in my study. My research suggests that tamarins improve their apparent ability to digest gum over the short term. In addition, there is evidence to suggest that wild callitrichines have larger caeca than captive conspecifics (Power, 1991). It is plausible that wild saddle-back tamarins have greater gum-digesting capabilities than captive tamarins maintained on gum-free diets. Wild saddle-back tamarins still resemble other tamarins and lion tamarins in gut proportions, however, and thus differ from wild or captive *Callithrix* species (Ferrari and Martins, 1992; Ferrari et al., 1993; Power, 1991). Therefore, the expectation is that *Callithrix* species would still be more efficient at digesting gum.

Another possibility is that energy is not the primary nutritional "reason" why saddle-back tamarins ingest gum. Saddle-back tamarins are highly insectivorous. It is quite likely that saddle-back tamarins catch and eat as many or more insects in a day than can sympatric, larger *Saguinus* species. It also is reasonable to hypothesize that their absolute energy requirements are lower than those for the larger species. Therefore, insects most likely account for a significantly greater proportion of the diet of saddle-back tamarins, while fruit is more prevalent in the diets of the larger *Saguinus*. Thus, saddle-back tamarins likely ingest a diet that has an imbalance in the calcium-phosphorus ratio. For saddle-back tamarins, gums might serve primarily as a mineral source, especially for calcium, as opposed to a primary energy source, as is suspected for marmosets.

Gummivory in the Callitrichine Radiation

All available information is compatible with the hypothesis that gum-feeding is an important adaptive aspect of marmoset biology. The ability of marmosets to rely on exudates as food potentially allows them to lead a less energetically stressful life than other callitrichines. They can live in smaller home ranges, and travel shorter daily path lengths than most other callitrichines (Rylands, 1984; Ferrari and Lopes, 1989; Power, 1991). They presumably suffer less from seasonal declines in food availability, although the seasonal variations in gum abundance and nutrient content are not known. They appear to be able to survive in drier habitats where fruit is relatively scarce (Fonseca and Lacher, 1984). These differences, all related to gummivory, have been suggested to underlie the differences between them and the other callitrichine species in social structure (Ferrari and Lopes, 1991) and possibly in energy metabolism (Power, 1991).

There is little support for the hypothesis that gummivory played a substantial role in the evolution of the other callitrichines, although it is likely that the ancestral callitrichine opportunistically fed on gums. The common callitrichine features such as small body size and digital claws could have evolved in response to insect foraging strategies that are still fundamental to the lifestyles of tamarins and lion tamarins (Garber, 1992). Small body size and claws could have served as preadaptations that enabled the proto-marmoset lineage to make a dietary shift to gummivory, but is not evidence for gummivory *per se*.

The fact that common marmosets are similar to tamarins and lion tamarins in the allometry of digestive parameters suggests that a rapid passage rate of digesta is the likely ancestral condition. A rapid passage rate is a digestive strategy that maximizes food intake by quickly eliminating indigestible bulk. For most callitrichines, the indigestible bulk in the diet largely consists of seeds that are swallowed, but are passed through the digestive tract largely unchanged. Many callitrichine defecations consist almost entirely of seeds (Garber, 1986, 1995). Seeds might represent a significant cost for the frugivorous callitrichines, both as ballast that must be carried around, and by potentially restricting food intake by filling the gut.

I propose that the physiological evidence supports the interpretation that the callitrichine common ancestor was a frugivore-faunivore. Although plant exudates could have been nutritionally important, digestive function was probably most influenced by bulk-fill limitations imposed by ingested seeds. Only in the lineage leading to *Cebuella* has this constraint been relaxed, allowing a slowing of the passage rate of digesta, and an apparent increase in the efficiency of digestion relative to body size.

Changes in the lower anterior dentition and the shift in gut proportions in the marmoset lineage were likely related to an increased importance of gum in the diet. Unlike some aspects of the tooth morphology (Rosenberger and Coimbra-Filho, 1984), the shift in gut proportions appears to have occured after the marmosets diverged from the lineage leading to the lion tamarins. The more recent divergence of the pygmy marmoset from the other marmoset species is marked by a slowing of the passage rate of digesta and a further reduction in body size, both most probably linked to that species becoming a gum-feeding specialist. Pygmy marmosets thus would appear to have the most derived digestive physiology of the callitrichines.

SUMMARY

Gum-feeding is a rare dietary adaptation in mammals. Within primates, however, it is a relatively common behavior. Among New World primates, members of the Callitrichinae are the most gummivorous. Marmosets (the genera *Callithrix* and *Cebuella*) appear to be more gummivorous than other members of the subfamily, although within the genus *Callithrix* there is wide variation in the extent of gum-feeding. Marmosets have a suite of morphological adaptations that aid in gum-feeding. These features include modifications of the lower anterior dentition (Coimbra-Filho and Mittermeier, 1977; Rosenberger, 1978) and an enlargement of the caecum and colon relative to the small intestine (Coimbra-Filho et al., 1980; Power, 1991, Ferrari and Martins, 1992; Ferrari et al., 1993; Power and Oftedal, submitted b).

Despite the lack of these, or any other adaptations clearly related to gum-feeding within the genera *Saguinus* and *Leontopithecus*, gummivory has been suggested to have played a major role in the callitrichine radiation (Garber, 1980; Sussman and Kinzey,

1984). To explore this issue, I conducted research on digestion and the digestive response to dietary gum in five species (*Cebuella pygmaea*, *Callithrix jacchus*, *Saguinus oedipus*, *S. fuscicollis*, and *Leontopithecus rosalia*). The animals were fed identical, homogenous single-item diets, both with and without added gum arabic powder. Transit time and the coefficients of apparent digestibility of dry matter and energy were measured for each animal on each diet.

Among all species except the pygmy marmoset, these digestive parameters followed a positive allometry under a gum-free dietary regimen. The pygmy marmosets alone differed, having the absolutely longest transit times, and coefficients of apparent digestibility equal to those of tamarins four times their size. In contrast, common marmosets were not different from the tamarins and lion tamarins. This implies that a rapid passage rate of digesta is the likely ancestral condition, and that pygmy marmosets are the most derived in terms of digestive parameters among these species.

The addition of gum to the diet had no effect on digestion in either marmoset species, but significantly decreased the digestion of the diet in the other three species. The results of these digestion trials offer further evidence that marmosets have derived adaptations to gum-feeding that other callitrichines lack.

I propose that the ancestral callitrichine was most likely a frugivore-faunivore. Although the small size of callitrichines possibly predisposes them to gummivory, only in the marmoset lineage has there been any significant adaptations to a gummivorous dietary strategy.

ACKNOWLEDGMENTS

This research would have been impossible without the support and cooperation of many people from the National Zoological Park, Washington D.C., the University of California at Berkeley, and the Marmoset Research Center, Oak Ridge Associated Universities, Oak Ridge, TN, and the National Institute of Mental Health, Bethesda, MD. The criticisms and encouragement of Dr. Katharine Milton are gratefully acknowledged. My many conversations with Dr. Suzette Tardif helped me to develop and sharpen my ideas. Dr. Elizabeth McClure, Neuroscience Section, National Institute for Mental Health, kindly allowed me to conduct digestion trials on common marmosets in her colony. Dr. Olav Oftedal provided invaluable technical, logistical, intellectual and moral support. Financial support during this research was provided by the Office of Fellowships and Grants, Smithsonian Institution, the Friends of the National Zoo, and Elaine Power. The author was supported during the preparation of this manuscript by NIH grant RR02022.

REFERENCES

Adriani, J.; Assoumani, M. 1983. Gums and hydrocolloids in nutrition. In CRC Handbook of Nutritional Supplements: vol II Agricultural Use. M. Reicheigl (ed.), pp. 301–333, CRC Press, Boca Rotan.

Allen, M.E. 1989. Nutritional aspects of insectivory in geckos and other small animals. Unpublished doctoral dissertation, Michigan State University.

Alonso, C.; Langguth, A. 1989. Ecologia e comportamento de *Callithrix jacchus* (Primates: Callitrichidae) numa ilha de floresta Atlantica. *Revista Nordestina de Biologia* 6:105–137.

Anderson, D.M.W.; Bridgeman, M.M.E.; Farquhar, J.G.K.; McNab, C.G.A. 1983. The chemical characterization of the test article used in toxicological studies of gum arabic (*Acacia senegal* (L.) Willd). *International Tree Crops Journal* 2:245–254.

Bearder, S.K.; Martin, R.D. 1980. Acacia gum and its use by bushbabies, *Galago senegalensis* (Primates: Lorisidae). *International Journal of Primatology* 1:103–128.

Belovsky, G.E.; Schmitz, O.J. 1994. Plant defenses and optimal foraging by mammalian herbivores. *Journal of Mammalogy* 75(4):816–832.

Coimbra-Filho, A.F.; Mittermeier, R.A. 1977. Tree-gouging, exudate-eating, and the short-tusked condition in *Callithrix* and *Cebuella*. In The Biology and Conservation of the Callitrichidae. D.G. Kleiman (ed.), pp. 105–115, Smithsonian Institution Press, Washington, D.C.

Coimbra-Filho, A.F.; Rocha, N.D.C.; Pissinatti, A. 1980. Morfofisiologia do ceco e sua correlacao com o tipo odontologico em Callitrichidae (Platyrrhini, Primates). *Rev. Brasil. Biol.* 40:177. .

Eastwood, M.A.; Brydon, W.G.;Anderson, D.M.W. 1986. The effect of the polysaccharide composition and structure of dietary fibers on cecal fermentation and fecal excretion. *American Journal of Clinical Nutrition* 44:51–55.

Ferrari, S.J.; Lopes, M.A. 1989. A re-evaluation of the social organization of the Callitrichidae, with reference to the ecological differences between genera. *Folia Primatologica* 52:132–147.

Ferrari, S.J.; Martins E.S. 1992. Gummivory and gut morphology in two sympatric callitrichids (*Callithrix emiliae* and *Saguinus fuscicollis weddelli*) from Western Brazilian Amazonia. *American Journal of Physical Anthropology* 88:97–103.

Ferrari, S.J.; Lopes, M.A.; Krause, E.A.K. 1994. Gut morphology of *Callithrix nigriceps* and *Saguinus labiatus* from Western Brazilian Amazonia. *American Journal of Physical Anthropology* 90:487–493. 1994.

Fonseca, G.A.B.; Lacher, T.E. 1984. Exudate feeding by *Callithrix jacchus penicillata* in semideciduous woodland (cerradão) in central Brazil. *Primates*, 25:441–450.

Garber, P.A. 1980. Locomotor behavior and feeding ecology of the Panamanian tamarin (*Saguinus oedipus geoffroyi*, Callitrichidae, Primates). *International Journal of Primatology* 1:185–201.

Garber, P.A. 1984. Proposed nutritional importance of plant exudates in the diet of the Panamanian tamarin, *Saguinus oedipus geoffroyi*. *International Journal of Primatology* 5:1–5.

Garber, P.A. 1986. The ecology of seed dispersal in two species of callitrichid primates (*Saguinus mystax* and *Saguinus fuscicollis*). *American Journal of Primatology* 10:155–170.

Garber, P.A. 1988. Diet, foraging patterns, and resource defense in a mixed species troop of *Saguinus mystax* and *Saguinus fuscicollis* in Amazonian Peru. *Behaviour* 105:18–34.

Garber, P.A. 1992. Vertical clinging, small body size, and the evolution of feeding adaptations in the Callitrichinae. *American Journal of Physical Anthropology* 88:469–482.

Garber, P.A. 1995. Fruit feeding and seed dispersal in two species of tamarin monkeys (*Saguinus geoffroyi* and *Saguinus mystax*). *American Journal of Primatology* Supplement 20:95–96.

Kinzey, W.G.; Rosenberger, A.L.;Ramirez, M. 1975. Vertical clinging and leaping in a neotropical anthropoid. *Nature* 255:327–328.

McClean Ross, A.H.; Eastwood, M.A.; Brydon, W.G.; Busuttil, A.; McKay, L.F. 1984. A study of the effects of dietary gum arabic in the rat. *British Journal of Nutrition* 51:47–56. 1984.

McNab, B. 1986. The influence of food habits on the energetics of eutherian mammals. *Ecological Monographs*, 56:1–19.

Moynihan, M. 1976. Notes on the ecology and behavior of the pygmy marmoset, *Cebuella pygmaea*, in Amazonian Columbia. In *Neotropical Primates - Field Studies and Conservation*. R.W. Thorington and P.G. Heltne (eds.), pp. 79–84, National Academy Press, Washington D.C.

Nash, L.T. 1986. Dietary, behavioral, and morphological aspects of gummivory in primates. *Yearbook of Physical Anthropology* 29:113–137.

Nash, L.T.; Whitten, P.L. 1989. Preliminary observations of the role of *Acacia* gum chemistry in *Acacia* utilization by *Galago senegalensis* in Kenya. *American Journal of Primatology*, 17:27–39.

Peres, C. 1989. Exudate-eating by wild golden lion tamarins, *Leontopithecus rosalia*. *Biotropica* 21:287–288.

Pook, A.G ; Pook, G. 1981. A field study of the socio-ecology of the Goeldi's monkey (*Callimico goeldii*) in northern Bolivia. *Folia Primatologica* 35:288–312.

Power, M.L. 1991. Digestive Function, Energy Intake and the Digestive Response to Dietary Gum in Captive Callitrichids. Unpublished doctoral dissertation, University of California at Berkeley.

Power, M.L.; Oftedal, O.T. 1996. Differences among captive callitrichids in the digestive responses to dietary gum. American Journal of Primatology 40:131–144.

Ramirez, M.F.; Freese, C.H.; Revilla, C.J. 1977. Feeding ecology of the pygmy marmoset, *Cebuella pygmaea*, in northeastern Peru. In The Biology and Conservation of the Callitrichidae. D.G. Kleiman (ed.), pp. 91–104, Smithsonian Institution Press, Washington, D.C.

Rosenberger, A.L. 1978. Loss of incisor enamel in marmosets. *Journal of Mammalogy* 59:207–208.

Rosenberger, A.F.; Coimbra-Filho, A.F. 1984. Morphology, taxonomic status and affinities of the lion tamarin, *Leontopithecus* (Callitrichinae, Cebidae). *Folia Primatologica* 42:149–179.

Rylands, A.B. 1984. Exudate-eating and tree-gouging by marmosets (Callitrichidae, Primates). In Tropical Rain-forest: The Leeds Symposium. A.C. Chadwick and S.L. Sutton (eds.), pp. 155–168, Leeds Philosophical Society, Leeds.

Salyers, A.A.; Palmer, J.K; Wilkins, T.D. 1978. Degradation of polysaccharides by intestinal bacterial enzymes. *American Journal of Clinical Nutrition* 31:S128-S130.

Soini, P. 1982. Ecology and population dynamics of the pygmy marmoset, *Cebuella pygmaea. Folia Prima-tologica* 39:1–21.

Soini, P. 1987. Ecology of the saddle-back tamarin *Saguinus fuscicollis illigeri* on the Rio Pacaya, Northeastern Peru. *Folia Primatologica* 49:11–32.

Stevenson, M.F.; Rylands, A.B. 1988. The marmosets, genus *Callithrix*. In Ecology and Behavior of Neotropical Primates. Vol. 2. R.A. Mittermeier, A.B. Rylands, A. Coimbra-Filho, and G.A.B. Fonseca (eds.), pp. 131–222, World Wildlife Fund, Washington D.C.

Sussman, R.W.; Kinzey, W.G. 1984. The ecological role of the Callitrichidae: a review. *American Journal of Physical Anthropology* 64:419–449.

Terborgh, J. 1983. Five New World Primates. Princeton University Press, Princeton, N.J.

Van Soest, P.J. 1982. Nutritional Ecology of the Ruminant. O and B books Inc. Corvalis, Oregon.

Walter, D.J.; Eastwood, M.A.; Brydon, W.G.; Elton, R.A. 1986. An experimental design to study colonic fibre fer-mentation in the rat: duration of feeding. *British Journal of Nutrition* 55:465–479.

Walter, D.J.; Eastwood, M.A.; Brydon, W.G.; Elton, R.A. 1988. Fermentation of wheat bran and gum arabic in rats fed on an elemental diet. *British Journal of Nutrition* 60:225–232.

Warner, A.C.I. 1981. Rate of passage of digesta through the gut of mammals and birds. *Nutrition Abstracts and Reviews Series 'B'* 51:699–820.

Wrangham, R.W.; Waterman, P.G. 1981. Feeding behavior of vervet monkeys on *Acacia tortilis* and *Acacia xan-thophlea* with special reference to reproductive strategies and tannin production. *Journal of Animal Ecol-ogy* 50:715–731.

Wyatt, G.M; Bayliss, C.E.;Holcroft, J.D. 1986. A change in human faecal flora in response to inclusion of gum arabic in the diet. *British Journal of Nutrition* 55:261–266.

Wyatt, G.M.; Horn, N.; Gee, J.M.; Johnson, I.T. 1988. Intestinal microflora and gastrointestinal adaptation in the rat in response to non-digestible dietary polysaccharide. *British Journal of Nutrition* 60:197–207.

6

LOCOMOTION OF GOLDEN LION TAMARINS
(*Leontopithecus rosalia*)

The Effects of Foraging Adaptations and Substrate Characteristics on Locomotor Behavior

Brian J. Stafford,[1][*] Alfred L. Rosenberger,[2] Andrew J. Baker,[3]
Benjamin B. Beck,[2] James M. Dietz,[4] and Devra G. Kleiman[2]

[1]Department of Anthropology
City University of New York
33 West 42nd Street
New York, New York 10036
and New York Consortium in Evolutionary Primatology
[2]Department of Zoological Research
National Zoological Park
Smithsonian Institution
Washington, DC 20008
[3]Curator of Small Mammals and Primates
Philadelphia Zoological Gardens
Philadelphia, Pennsylvania 19104
[4]Department of Zoology
University of Maryland at College Park
College Park, Maryland 20742

INTRODUCTION

Our study of the locomotor behavior of golden lion tamarins (*Leontopithecus rosalia*) was initiated because these unique, highly endangered primates, were perceived to possess locomotor deficiencies upon reintroduction to the wild. The critical status of the wild population (Coimbra-Filho and Mittermeier, 1978, Kleiman *et al.*, 1986) led to the establishment of the Poço das Antas Biological Reserve 70 km outside of Rio de Janiero in 1974. The reserve consists of approximately 5000 ha of disturbed lowland rainforest

---

* Current address to which correspondence should be addressed: Department of Zoological Research, National Zoological Park, Smithsonian Institution, Washington, DC 20008.

Adaptive Radiations of Neotropical Primates
edited by Norconk *et al.* Plenum Press, New York, 1996

(Kleiman *et al.* 1986, 1991; and Rylands, 1993 for details on reserve condition and environment). A program of reintroductions designed to resupply the declining wild population (Beck *et al.*, 1991; Kleiman, 1989; Kleiman *et al.*, 1986, 1991) by culling social groups from the world's captive stock was initiated in 1984. The first reintroductions, although successful, raised concerns that captive animals released into the forest may exhibit locomotor, and other behavioral deficiencies resulting from their lack of experience in such a complex environment (Kleiman *et al.*, 1986). Thus, a program of prerelease and postrelease training was designed to aid in the transition of captive-born animals into the wild. The research program reported in this paper was conceived to describe and quantify locomotion in *L. rosalia* with these issues in mind.

This report summarizes the first phase of this project, consisting of three separate but interrelated studies. First, a description and quantification of the locomotor behavior of captive animals housed in conventional enclosures; second, a companion study of captive animals newly released into a free-ranging setting at the National Zoological Park; and third, an initial field study of locomotion in wild *L. rosalia*. These studies were designed to allow a comparison of locomotion across these groups to determine the degree to which positional behavior of captive individuals differs from that of wild animals. During the first study (Rosenberger and Stafford, 1994), comparisons were made with captive Goeldi's monkeys, *Callimico goeldii*, housed in the same enclosures with the *L. rosalia* in order to gain taxonomic perspective on locomotion in callitrichines (*sensu* Rosenberger, 1979). Only data on *L. rosalia* are included here.

One of our main goals has been to separate behaviors related to the adaptations of the wild population from those resulting from the effects of captivity. Another was to evaluate the effects of differences in substrate structure on locomotor behavior in order to determine how these variables affect the locomotor profile. Throughout this study we consider morphology to be constant across our groups of *L. rosalia* since our investigations (Stafford and Rosenberger, in prep) do not indicate morphological differences between captive and wild animals for the characters considered here.

We were able to identify two factors that affect locomotion in *L. rosalia*. The first of these was related to substrate structure. These effects are difficult to evaluate between wild and captive groups because of the different ontogenetic experiences of captive and wild animals, as well as the differences in substrate availability between captive and wild settings. The second set of effects relates to the expression of locomotor patterns which we hypothesize are circumscribed by morphology, and therefore related to the phylogenetic experiences of the species.

METHODOLOGY

We studied four social groups of *L. rosalia* (Table 1), two captive-born and two wild-born. Details of group composition, housing and substrate setting for the two captive groups were discussed in detail elsewhere (Rosenberger and Stafford, 1994; Stafford, Rosenberger, and Beck 1994) and will only be summarized here. The CRC group lived in conventional cinder block enclosures, in mixed housing with several groups of *Callimico*, at the Smithsonian Institution's Conservation and Research Center (CRC) in Front Royal, Virginia. They occupied indoor and outdoor cages, furnished by a substrate network of mostly horizontal branches, roofed and fronted with standard cyclone fencing. Supports were arranged in a grid pattern four feet off the floor with one or two vertical or diagonal supports providing access to the cage floor. Animals were only observed in the outdoor

Table 1. Study groups

	Number of Adults	Number of Bouts
Conservation and Research Center (CRC)		
Leontopithecus rosalia	4	825
Callimico goeldii	14	1197
Beaver Valley, National Zoological Park (NZP)	5	3338
Poço das Antas Biological Reserve (PDA)	8	3795

enclosures for consistency across study groups. The Beaver Valley group consisted of five *L. rosalia* newly released into a forested area in the National Zoological Park (NZP) for the first time. This area consists of about 0.2 hectares of mature beech and oak forest forming a continuous canopy, and an understory below three meters of shrubs and bushes. A network of hemp ropes was strung up in the subcanopy, connecting the centrally located nest box to the perimeter of the site. Wild *L. rosalia* were studied at the Poço das Antas Biological Reserve (PDA) in Brazil and were observed over 19 days for a total of 76 hours. This resulted in the collection of 3795 locomotor bouts.

Data were recorded using a modified focal animal sampling method (Altmann, 1974; see Rosenberger and Stafford, 1994; or Stafford *et al.*,1994 for details of our particular method) for all groups. Visual observations were supplemented by videotapes recorded under the same protocols as visual sampling. Our unit of observation was a locomotor sequence, which we defined as a string of locomotor bouts proceeding without a postural interruption of more than 3–5 seconds. The locomotor bout, in turn, was defined by the maintenance of a single locomotor pattern across a single class of supports. This convention is required because locomotor behaviors (e.g., walking) may be performed differently on supports of different size or orientation. Locomotor categories were based on detailed observations of how the animals moved, and considered within the framework of discrete behaviors as described by Hildebrand (1967, 1977, 1980). This methodology allowed us to distinguish gait patterns between species, and to discern differences within gait categories.

We find this approach most instructive due to the transient nature of the taxonomy of primate locomotion (see Prost, 1965; Martin, 1990; Napier and Walker, 1967; Rose, 1973; Fleagle, 1988 for examples of how the terminology associated with primate locomotor studies has changed over the years). A functional and kinematic approach to defining locomotor behaviors (as advocated by Prost, 1965; and Hildebrand, 1967) should ensure relative constancy in the delineation of discrete behaviors over time. Also, since we are interested in discovering anatomical correlates of locomotion that will be useful in interpreting the fossil record, we believe that a kinematically based definitional system (as advocated by Prost, 1965) provides the most powerful methodology for linking behavior and morphology.

We have argued (Rosenberger and Stafford, 1994) that some features of the skeleton in *L. rosalia* are linked to gross interspecific differences in locomotor behavior while other characters affect the system more subtly. For this reason it is important to understand the information content of one's observations and throughout this report we consider our locomotor variables at two levels. The first level is that of the generalized locomotor profile in which locomotor behavior is grouped into larger conceptual categories (Table 2). The second is a finer grained analysis, in which the components of the generalized profile are broken down into more specifically defined subcategories for consideration. A

Table 2. Locomotor categories and substrate classes

Locomotor Categories

Specific Locomotor Categories

Quadrupedal Walking: Pronograde quadrupedal progression using a diagonal sequence gait, including "running".

Quadrupedal "Transaxial" Bounding: A transverse gallop with extended suspension characterized in *L. rosalia* by unique hand and foot placements. See Rosenberger and Stafford (1994) for a detailed discussion of this behavior.

Quadrumanous Climbing: Quadrupedal progression among small terminal branches where an animal's weight is spread across more than one support.

Saltatory Leaping: Saltation from a stationary posture.

Bounding-leap: Saltational extensions of quadrupedal walking or bounding, as when crossing between supports, or passing bends or obstructions.

Vertical Climbing: Ascent of a steeply inclined ($>60^0$) support.

Suspension: Walking suspended below a support, or hindlimb suspension when it is used to cross between supports.

Gap Bridging: Crossing between two discontinuous supports by placing some combination of limbs in contact with the target support before transfering the body accross the gap.

General Locomotor Categories[1]:

Quadrupedalism-g: Walking + Bounding + Climbing.

Leaping-g: Leaping + Bounding Leaping.

Vertical Climbing-g: Vertical Climbing.

Suspensory-g: Suspension + Gap Bridging.

Substrate Classes

No. 1: Vertical trunks below the canopy, too wide for the animals to reach halfway around with their forelimbs (>30 cm in diameter).

No. 1a: Vertical trunks below the canopy which the animals can reach halfway around ($_12.5 - 30$ cm in diameter).

No. 2: Boughs within the canopy of any angular orientation ($_12.5 - 30$ cm in diameter).

No. 3: Canopy or subcanopy branches approximately the same diameter as the animals shoulder width ($_5 - _12.5$ cm in diameter).

No. 4: Canopy or subcanopy branches that a tamarin can encircle with the hand ($_1.5 - _5$ cm in diameter).

Rope: The 2.5 cm diameter manila rope.

No. 5: Canopy or subcanopy supports about which the animals could curl their fingers, generally a terminal branch ($<_1.5$ cm in diameter).

Terrestrial: Cage or forest floor.

Substrate Orientation

Horizontal: Substrates between 0^0 and 30^0 inclination.

Diagonal: Substrates between 30^0 and 60^0 inclination.

Vertical: Substrates between 60^0 and 90^0 inclination.

Forest Level

Canopy: Locomotion above the level at which branches begin to spread from the trunks of the trees forming an interlocking layer.

Subcanopy: Locomotion between the canopy and above two meters from the ground.

Below 2 Meters: Locomotion within two meters of the ground.

Terrestrial: Locomotion on the forest floor, or floor of the cage.

---

[1]General locomotor categories are distinguished from specific categories of the same name by adding the suffix -g.

comparison of these two levels of resolution offers valuable insight into the application of functional analyses based on extant taxa to the fossil record.

Table 2 provides definitions of our locomotor and substrate categories. Some familiar categories were lumped together here when we could not distinguish between them consistently, or because of their usage in earlier phases of this project. For example, "vertical climbing" as presented here is actually composed of two distinct behaviors, vertical climbing and vertical bounding. In vertical climbing the animal is ascending or descending a vertical support with diagonal hindlimbs and forelimbs moving in synchrony. In vertical bounding, however, the forelimbs move in synchrony with each other, as do the hindlimbs. Because these behaviors were not distinguished during our initial study at CRC due to substrate availability at this site (i.e., only a few short supports were available for this behavior), we lump them into one category for comparison. A second instance of combining categories involves our walking category, which includes both walking and running. Both of these gaits utilize a diagonal couplets, diagonal sequence footfall formula. The difference between walking and running relates to the amount of time each limb retains contact with the substrate (see Hildebrand, 1967 for more complete descriptions). Operationally, this translates into how fast the animal is moving, i.e., walking is slower than running. Because we could not distinguish the transition between these two gaits during observations, we classified them together as walking. General locomotor categories are composites of more rigidly defined specific locomotor categories, and are identified by the suffix -g.

Substrate diameters were determined in relation to the size of the animal and the manner in which the animal used the support, especially how the animal grasped a support. We decided on this approach because of theoretical expectations that an animal walking on a 10 cm diameter support that cannot be grasped with the hand will move differently than it would when walking on a 2 cm diameter support that can be grasped. To what degree this is true has yet to be determined through kinematic analysis, but analysis of videotapes indicates that hand and foot placements and general body orientation are different on substrates of different sizes. This method of estimating substrate size also offers the observer a built-in scale when collecting data, namely the animal itself. As a result, we are very confident in our assignment of substrate sizes between sites presenting very different viewing conditions.

Our forest level categories deserve special mention. We divided the habitats of the animals into four levels based, in part, on geometry and continuity of supports (see Table 2). The canopy and subcanopy can be distinguished by the presence of interconnected branches in the canopy. The subcanopy, however, is dominated by the vertical trunks of the trees and presents a less continuous environment. The terrestrial level is self explanatory, and our "below two meters" category reflects the fact that wild *L. rosalia* appear to spend a good deal of time close to the forest floor, scanning the leaf litter for invertebrate prey items. The specific height of two meters was chosen because it could be reliably and repeatedly identified by observers. In fact, throughout our entire study, *L. rosalia* were seldom seen scanning for terrestrial prey from a height above two meters. Therefore, only this category and terrestriality legitimately represent the height of the animal. The habitat structure at the Poço das Antas reserve is extremely variable, ranging from areas similar in structure to old growth forest to areas of open grasslands. As a result, in some parts of the reserve the structural subcanopy may extend above the height of the canopy found in other parts of the forest. Tall stands of bamboo are one such example which we would consider not to have a canopy level. Therefore, our concept of forest level is one of structure, geometry, and substrate continuity, and has little or nothing to do with how high above the ground the animals actually were.

Table 3. Results of analysis of variance

| | *General Locomotor Categories* | | | |
	Quadrupedalism	Leaping	Vertical Climbing	Suspension
CRC vs NZP	NS	**	*	NS
CRC vs PDA	**	NS	**	**
NZP vs PDA	**	**	NS	*

| | *Specific Locomotor Categories* | | | | | | | |
	Walking	Bounding	Climbing	Leaping	Bounding Leaping	Vertical Climbing	Suspension	Gap Bridging
CRC vs NZP	**	**	**	NS	**	*	**	NS
CRC vs PDA	NS	**	**	*	**	**	**	NS
NZP vs PDA	**	NS	NS	**	NS	NS	NS	NS
NZP vs PDA-rope	**	**	NS	**	NS	NS	NS	NS

| | *Substrate Classes* | | | | | | | |
	Trunks >30 cm	Trunks 12.5-30 cm	Branch 12.5-30 cm	Branch 5-12.5 cm	Branch 1.5-5 cm	Rope	Branch <1.5 cm	Terrestrial
CRC vs NZP	*	**	**	**	**	**	**	NS
CRC vs NZP-rope	*	**	**	**	**	NA	**	NS
CRC vs PDA	*	**	**	**	**	NA	**	*
NZP vs PDA	**	NS	NS	**	**	**	NS	NS
NZP vs PDA-rope	**	NS	NS	**	NS	NA	NS	NS

| | *Substrate Orientation* | | | |
	Horizontal	Diagonal	Vertical	Terminals
CRC vs NZP	NS	*	**	*
CRC vs PDA	**	NS	**	**
NZP vs DA	**	**	**	**

| | *Forest Level* | | | |
	Canopy	Subcanopy	Below 2 meters	Terrestrial
CRC vs NZP	NS	**	**	*
CRC vs PDA	**	**	**	**
NZP vs PDA	*	**	NS	NS

*p<0.01; **p<0.001; NS = not significant; NA = not applicable.

Spearman's rank correlation (r) was used to test for significant correlation between locomotor, substrate size, substrate orientation, and forest level profiles between groups. Correlations were considered to be significant when p<0.01. To test for difference between individual components of these profiles we used pairwise single classification analysis of variance. Here, categories were considered to be significantly different when p<0.01.

THE MAJOR FEATURES OF LOCOMOTION IN *Leontopithecus rosalia*

Figure 1 shows the general locomotor profiles of *L. rosalia*. At CRC we find that *L. rosalia* is basically quadrupedal with leaping of secondary importance. Suspension and

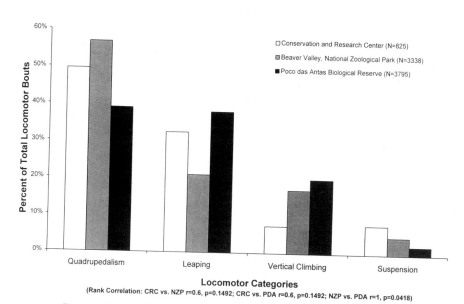

Figure 1. Locomotion of *Leontopithecus rosalia* (general locomotor categories).

vertical climbing represented less significant components of the locomotor profile. These generalizations are consistent with differences in limb indices between *L. rosalia* and *Callimico* (Rosenberger and Stafford, 1994) (Table 4), with forelimb and hindlimb lengths more nearly equal in *Leontopithecus* while *Callimico* has appreciably longer hindlimbs. Given this, one would expect more quadrupedalism in *Leontopithecus*. Data collected from the videotapes (Figure 2) showed that *L. rosalia* typically utilize the transverse gallop, a gait which involves a marked overstriding of the forelimbs by the hindlimbs. Such a condition is not unique for primates (see Hildebrand, 1967; Tuttle, 1969; and Vilensky, 1989; Vilensky and Larson, 1989) or even for mammals that gallop, but it does appear to be unique to *Leontopithecus* among the callitrichines. This may be the result of the incorporation of elongate forelimbs into the locomotor system. As shown in Figs. 2 and 3, *L. rosalia* use a unique pattern of hand and foot placement when galloping. Rather than positioning the hands and feet on either side of a support, both forelimbs and hindlimbs are set to one side. We have proposed the term "transaxial bounding" (Rosenberger and Stafford, 1994) to describe this pattern because although the gait is technically a gallop it fulfills the role of rapid quadrupedal locomotion in *L. rosalia* where other callitrichines utilize a half-bound.

We first identified transaxial bounding at CRC, and initially thought it represented an artifact of captivity but since then we have confirmed the occurrence of this pattern in wild-born *L. rosalia* at Poço das Antas and also in captive *L. chrysomelas*. To date, we have found no differences in gait patterns between the captive-born zoo animals and the wild animals, although we continue to investigate this possibility. Therefore, transaxial bounding appears to be a normal locomotor pattern in *Leontopithecus*.

Kinematically this gait is quite distinctive. The hindquarters are displaced lateral to the midline of the support before the hind feet contact it (Figure 2b&c). During the next phase of the stride, when the animal is extending the spine (Figure 2a), the shoulders

Table 4. Selected limb indices in callitrichines

Taxon	Mass[2]	STL	IMI[3]	BI[4]	CI[5]	Forelimb[6]	Hindlimb[7]	McI[8]	MtI[9]	TotFore[10]	TotHind[11]
Callimico goeldii	482 g	163 mm	71	92	101	63	89	8.52	14.14	72	103
Saguinus sp.	472 g	167 mm	76	90	101	60	79	8.94	14.39	69	94
Leontopithecus rosalia[12]	495 g	173 mm	88	96	100	74	83	13.25	16.73	87	100
Callithrix jacchus	294 g	139 mm	75	90	102	62	84	8.90	15.78	71	99
Cebuella pygmaea	116 g	93 mm	82	90	102	69	84	9.14	15.24	78	98
t-test[13]		$p<.01$	$p<.01$	$p<.05$	$p=.13$	$p<.01$	$p=.47$	$p<.01$	$p<.01$	$p<.01$	$p=.05$

[1] Data based on selected records from Dykyj (1982) unless otherwise noted. N = 4 in all cases.

[2] Data taken from Rosenberger (1992). More recent data on body weights in wild *L. rosalia* are available in Dietz *et al.* (1994). Ford and Davis (1992) and Ford (1994) provide data for a wider range of platyrrhines.

[3] Intermembral Index = ((Humerus length + Radius length)/(Femur length + Tibia length))*100.

[4] Brachial Index = (Radius length / Humerus Length)*100.

[5] Crural Index = (Tibia length / Femur length)*100.

[6] Forelimb Index = ((Humerus length + Radius length) / STL)*100.

[7] Hindlimb Index = ((Femur length + Tibia length) / STL)*100.

[8] Metacarpal Index = (Metacarpal III length / STL)*100.

[9] Metatarsal Index = (Metatarsal III length / STL)*100.

[10] Total Forelimb Index = ((Humerus length + Radius length + Metacarpal III length) / STL)*100.

[11] Total Hindlimb Index = ((Femur length + Tibia length + Metatarsal III length) / STL)*100.

[12] Data collected from USNM#s 546317, 546320, 546321, and 546322.

[13] t-test is 1 tailed for samples with unequal variance testing *L. rosalia* against all other taxa. Level of significance is $p<.05$.

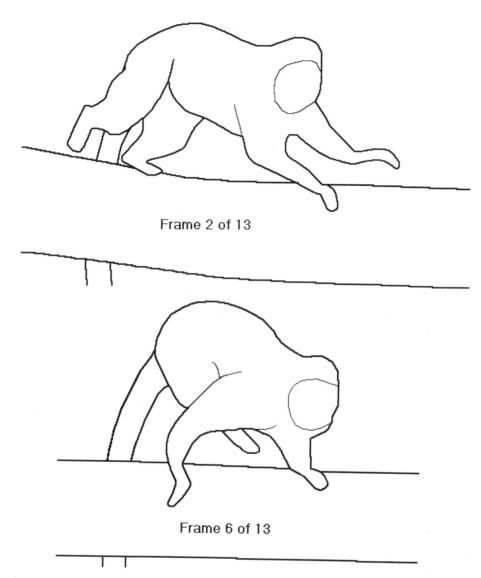

Frame 2 of 13

Frame 6 of 13

Figure 2a and b. Galloping in *L. rosalia* at CRC taken from a sequence of video frames.

straighten out in the direction of travel (Figure 2c) but turn laterally again as the forelimbs contact the substrate (Figure 2a, see Figure 5 Rosenberger and Stafford, 1994). The resulting pattern is visually very distinctive compared to the half-bounding of *Callimico* and produces a situation where the animal's body oscillates back and forth over the midline of the support.

We have suggested (Rosenberger and Stafford, 1994) that the specific pattern of forelimb elongation in *Leontopithecus* contributes to this gait pattern. The proportions of the antebrachium and manus in *Leontopitecus* are unique among callitrichines (Table 4), and indices clearly show that the longer forelimbs in *Leontopithecus* result from the elongation of distal limb segments. Such a condition, with long forelimbs leading to more

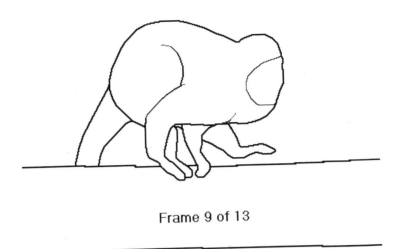

Frame 9 of 13

Figure 2c. Galloping in *L. rosalia* at CRC taken from a sequence of video frames.

equal relative limb indices, may impart an advantage to galloping over bounding in *Leontopithecus*. Certainly, this condition would lead to increased arcs of excursion of the body over the forelimb, producing an overstriding gait and increased stride length. These advantages probably explain the common occurrence of overstriding among mammals (i.e., equids, canids, felids, bovids) and primates (*Gorilla*, *Pan*, *Pongo*, or *Cercopithecus*; see Hildebrand, 1967; or Vilensky and Larson, 1989 for examples) that gallop. However, these examples involve either terrestrial cursorial mammals, or primates moving terrestrially. Our observations indicate that other callitrichines (*Callimico goeldii*, *Saguinus oedipus*, *Callithrix argentata*, *Cebuella pygmaea*) never enlist overstriding during bounding (*sensu stricto*) on arboreal supports and, to our knowledge, the only other arboreal mammal documented to utilize hindlimb overstriding on arboreal supports is the northern flying squirrel (*Glaucomys sabrinus*), which overstrides while half-bounding (Hampson, 1965). These animals are also notable for having relatively longer forelimbs than their arboreal nongliding relatives (Thorington and Heaney, 1981), although this elongation does not extend to the manus (Stafford, unpublished data).

We propose that the elongate hands of *Leontopithecus* necessitate transaxial placement as a means of reducing shearing stresses on the elongate manus. This placement may also allow overstriding to occur on arboreal supports because it reduces the possibility of interference between the forelimbs and hindlimbs by displacing the hindlimbs lateral to the midline of the support. Additionally, the oblique placement of the manus may provide enhanced stability through frictional forces and compensate for the loss of pollical grasping in *Leontopithecus*, a consequence of the highly specialized nature of the hand (i.e., manual elongation and interdigital webbing; see Garber, 1992; Hershkovitz, 1977; Rosenberger and Stafford, 1994).

Our observations on other captive callitrichines (*Callimico goeldii*, *Saguinus oedipus*, *Callithrix argentata*, *Cebuella pygmaea*) suggest that half-bounding is the ancestral pattern for callitrichines, and that forelimb elongation is linked with a change from bounding to galloping in *Leontopithecus*. Such an evolutionary transformation incorporates transaxial bounding as an integral aspect of the locomotor system in *Leontopithecus*. We do not deny the possibility that transaxial bounding confers some selective benefit in locomo-

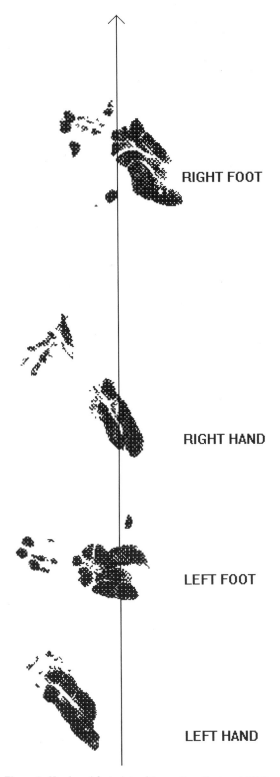

Figure 3. Hand- and footprints of *L. rosalia* galloping at CRC.

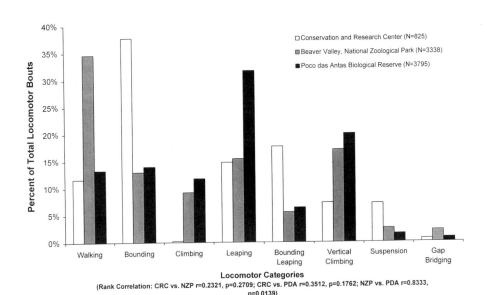

Figure 4. Locomotion of *L. rosalia* (specific locomotor categories).

tor or positional behaviors, but would propose that transaxial bounding is a byproduct of foraging adaptations that are incorporated as integral parts of the skeletal system, not a product of selection for this specific gait pattern.

We prefer the hypothesis that selection for extractive foraging is a better causal explanation of forelimb and manual elongation in *Leontopithecus* (e.g., Hershkovitz, 1977; Rosenberger, 1992) because there is ample biological evidence to support this hypothesis (Peres, 1986; Rylands, 1989). *L. rosalia* rely heavily on concealed, embedded prey which are removed from crevices, holes and the boles of large epiphytes by forceful extraction. Biomechanically, longer forelimbs increase reach and improve leverage. There is also comparative evidence suggesting that the elongation of anatomical components associated with extractive foraging behavior (i.e., phalangeal elongation in *Daubentonia madagascarensis* and *Dactylopsila* sp.; and lingual elongation in *Orycteropus afer*, *Manis* sp., *Myrmecophaga tridactyla*, and *Tamandua* sp., for example) is widespread.

Within the Callitrichinae, forelimb elongation is associated with extractive foraging, galloping, and transaxial hand and foot placement. Extractive foraging provides the most strongly supported hypothesis for the origin of forelimb elongation in *Leontopithecus*, until it can be shown that galloping or transaxial hand and foot placement confer some selective advantage. However, the influence of an elongate forelimb on other aspects of an animal's behavior (i.e., vertical clinging *sensu* Jungers, 1977; Cartmill, 1985; Thorington and Thorington, 1989: leaping Garber, 1991: or quadrupedal locomotion Rosenberger and Stafford, 1994) can have equally important secondary consequences in the daily life of the individual and this may explain the origins of galloping and transaxial hand and foot placement in *Leontopithecus*. It is within this context that we consider the locomotion of *L. rosalia* constrained by morphology.

It should be noted that we may never have identified transaxial bounding had we not studied the animals at CRC. Given the size of these animals, and observation conditions in the wild, it simply is not possible to determine what the animals are doing with their feet. In fact, it required the transplanting of our captive experimental setup into the wild for us to identify transaxial bounding conclusively in wild animals. Furthermore, our ability to recognize transaxial bounding when collecting behavioral observations in the wild relied upon our comparisons of the kinematics of this gait with half-bounding callitrichines. This allowed us to identify characteristics of transaxial bounding that do not rely upon being able to see the animal's feet, and therefore allowed us to identify this behavior reliably in the wild.

CAPTIVE *L. rosalia* IN CONVENTIONAL AND FREE-RANGING ENVIRONMENTS

Locomotor, substrate size, substrate orientation, and forest level usage profiles were not significantly correlated between CRC and Beaver Valley (Figures 1, 4–8), and the differences in locomotion between the two sits were not in accord with our predictions based on our initial work at CRC. We expected that the more discontinuous and flexible substrate conditions at Beaver Valley would elicit more leaping-g and less quadrupedalism-g than we observed at CRC. However, in examining our gross locomotor categories (Figure 1, Table 3) we found exactly the opposite pattern. There was proportionately more leaping-g at CRC, while the frequencies of quadrupedalism-g remained unchanged between the sites.

A more detailed look at locomotion at the two sites (Figure 4, Table 3) also seemed discordant with our predictions. Although quadrupedalism-g was the most frequent category of locomotor behavior at each site, different components of quadrupedalism-g predominated. Animals at CRC showed more ttransaxial bounding, whereas walking dominated at Beaver Valley. Also, bounding leaps occurred with a higher frequency at CRC. This may be explained by the fact that bounding leaping is, by definition, an extension of rapid quadrupedal locomotion (i.e. rapid walking or running, and bounding). We would, therefore, expect more bounding leaping to occur at a site where the animals are doing more bounding. Unexpectedly, saltatory leaping did not differ between the two sites, nor did gap bridging. We would have expected these behaviors to be more frequent in the more varied, unstable, and discontinuous environment at Beaver Valley.

A consideration of substrate character, however, leads to the interpretation (Stafford *et al.*, 1994), that each of the groups was, in fact, behaving in accordance with available substrate options. The animals at CRC were presented with a uniformly continuous, stable, and barrier-free network of supports and as a result they employed more transaxial bounding. In Beaver Valley the animals walked more because they had to negotiate more complex and unstable supports. Figure 5 shows the substrate usage profiles for both groups and illustrates the fact that the Beaver Valley group was using smaller, more flexible supports more than was the CRC group. Note specifically that the most commonly used support at Beaver Valley is the rope, which is small in diameter and highly flexible. Consequently, these animals may have been less able to employ transaxial bounding and bounding leaping and opted for walking, a gait which maintains the limbs in contact with the support for a greater percentage of the stride cycle, and is therefore more secure for the animal.

Figure 5 and Table 5 examined together further clarify the differences between the two captive sites. The absence of large vertical trunks, small branches or terminal sup-

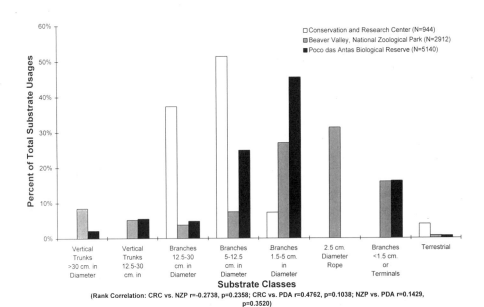

Figure 5. Substrate use by *Leontopithecus rosalia.*

ports, and ropes at CRC can immediately account for some differences in behavior. The animals at CRC used larger supports for a majority of their locomotor behaviors while the Beaver Valley animals used the smaller supports in most cases. At CRC there were no close associations between specific locomotor behaviors and specific supports, while in Beaver Valley animals did appear to use specific locomotor behaviors on certain classes of supports. For example, quadrumanous climbing did not occur at CRC but it accounted for 9% of the locomotor profile at Beaver Valley. This can be attributed to the fact that the proper substrates (terminal supports of less than 1.5 cm in diameter) were not present at CRC, while these kinds of substrates were used for 75% of all climbing activity at Beaver Valley (Table 3). Similarly, the greater frequencies of vertical climbing seen at Beaver Valley can be attributed to the presence of large vertical supports in the understory (tree trunks), a class of substrate not available to the CRC animals. The differing character of quadrupedalism between the two sites, however, seems to have less to do with substrate size than with substrate stability (see above). The Beaver Valley animals conducted most of their quadrupedal behaviors on the ropes that were strung between trees in the sub-canopy layer (47% for walking, and 77% for bounding).

LOCOMOTION IN WILD AND CAPTIVE *L. rosalia*

The most salient difference between the captive groups and the two wild groups studied (Figure 1, Table 3) is that the captive animals are more quadrupedal than the wild

Table 5. Associations between locomotion and substrate for *Leontopithecus rosalia*

	Walking	Bounding	Climbing	Leaping	Bnd. Leap	VCb	Su&GB
Conservation and Research Center (N=825)							
Vert. trunk >30 cm dia.	0%	0%	0%	0%	0%	0%	0%
Vert. Trunk 12.5-30 cm dia.	0%	0%	0%	0%	0%	0%	0%
Branch 12.5-30 cm dia.	39%	39%	0%	38%	44%	7%	0%
Branch 5-12.5 cm dia.	49%	57%	0%	54%	47%	59%	0%
2.5 cm Rope	0%	0%	0%	0%	0%	0%	0%
Branch 1.5-5 cm dia.	12%	5%	0%	7%	9%	33%	0%
Branch <1.5 cm dia	0%	0%	0%	0%	0%	0%	0%
Beaver Valley, National Zoological Park (N=3338)							
Vert. trunk >30 cm dia.	0%	0%	0%	7%	12%	35%	0%
Vert. Trunk 12.5-30 cm dia.	0%	0%	0%	7%	3%	32%	0%
Branch 12.5-30 cm dia.	1%	0%	1%	4%	2%	8%	0%
Branch 5-12.5 cm dia.	8%	7%	3%	12%	13%	7%	5%
2.5 cm Rope	47%	77%	3%	29%	28%	12%	35%
Branch 1.5-5 cm dia.	27%	13%	19%	17%	34%	0%	19%
Branch <1.5 cm dia	18%	3%	75%	23%	7%	6%	41%
Beaver Valley Minus Ropes (N=2912)							
Vert. trunk >30 cm dia.	0%	0%	0%	10%	17%	40%	0%
Vert. Trunk 12.5-30 cm dia.	0%	0%	0%	10%	4%	36%	0%
Branch 12.5-30 cm dia.	2%	0%	1%	6%	3%	9%	0%
Branch 5-12.5 cm dia.	15%	30%	3%	17%	18%	8%	8%
Branch 1.5-5 cm dia.	50%	57%	20%	24%	47%	0%	29%
Branch <1.5 cm dia	33%	13%	77%	32%	10%	7%	63%
Poço das Antas, Combined Groups (N=3795)							
Vert. trunk >30 cm dia.	0%	0%	0%	2%	1%	8%	0%
Vert. Trunk 12.5-30 cm dia.	0%	0%	0%	6%	1%	16%	0%
Branch 12.5-30 cm dia.	5%	9%	0%	4%	5%	7%	7%
Branch 5-12.5 cm dia.	22%	37%	3%	23%	26%	34%	21%
Branch 1.5-5 cm dia.	59%	48%	31%	46%	54%	33%	41%
Branch <1.5 cm dia	14%	5%	66%	18%	13%	2%	31%
Poço das Antas, Dois Femmes (N=2535)							
Vert. trunk >30 cm dia.	0%	0%	0%	2%	1%	7%	0%
Vert. Trunk 12.5-30 cm dia.	0%	0%	0%	6%	1%	15%	0%
Branch 12.5-30 cm dia.	4%	10%	0%	6%	6%	11%	8%
Branch 5-12.5 cm dia.	20%	35%	4%	24%	26%	33%	20%
Branch 1.5-5 cm dia.	58%	46%	31%	42%	50%	31%	40%
Branch <1.5 cm dia	16%	8%	64%	20%	16%	2%	32%
Poço das Antas, Cacador (N=1260)							
Vert. trunk >30 cm dia.	0%	0%	0%	2%	1%	9%	0%
Vert. Trunk 12.5-30 cm dia.	0%	0%	0%	7%	0%	18%	0%
Branch 12.5-30 cm dia.	6%	9%	0%	1%	5%	0%	0%
Branch 5-12.5 cm dia.	26%	38%	0%	22%	26%	35%	25%
Branch 1.5-5 cm dia.	61%	49%	29%	52%	61%	36%	50%
Branch <1.5 cm dia	7%	3%	71%	14%	7%	2%	25%

animals, while the wild animals use quadrupedalism-g and leaping-g with equal frequency. Although leaping-g appears to occur with comparable frequencies at CRC and PDA, we will see that the nature of this leaping-g is quite different between these two sites. However, vertical climbing-g occurs with comparable frequencies in the wild and at Beaver Valley, but at much lower frequencies at CRC. Surprisingly, suspensory-g behaviors were highest at CRC, lowest in the wild, and intermediate at Beaver Valley. These differences may be the result of the lack of diverse, flexible, and complex substrates at CRC, and reliance of the Beaver Valley animals on the rope bridges. The ropes provide highways through the subcanopy, strung between large vertical supports, a situation that may artificially enhance the frequencies of quadrupedal-g and vertical climbing-g behaviors at the expense of leaping-g behaviors.

A more detailed look (Figure 4) at the locomotor profiles of the captive and wild groups indicates that certain differences between these groups are quite marked (Table 3). Each group of *L. rosalia* exhibited a unique, predominant locomotor preference; transaxial bounding at CRC, walking in Beaver Valley, and leaping at Poço das Antas. The fact that both captive groups predominantly utilize some form of quadrupedalism-g largely explains the differences between the generalized locomotor profiles of the captive sample as a whole and the wild groups. In spite of such large scale differences, the specific locomotor profile of the free-ranging animals in Beaver Valley most closely resembles that of the wild animals. Only two of the eight specific locomotor categories (walking and leaping) differ by more than 3% between the Beaver Valley animals and the wild sample (Table 3). However, between the CRC group and the wild animals only two locomotor categories (walking and gap bridging) are within 3% of each other (Table 3). Furthermore, we now see that the apparent similarity of leaping-g between CRC and PDA is the result of inflated levels of bounding leaping at CRC. In fact, the frequencies of both saltatory leaping and bounding leaping are different between these two sites (Figure 4, Table 3). To this extent, we can confirm that locomotion in the free-ranging environment better approximates the wild than the locomotor behavior exhibited in the cages at CRC.

It is interesting to note that when the effects of rope use are removed from the analysis of Beaver Valley locomotion (Table 3), this group's locomotor profile diverges from that of the wild groups even more. We would have expected the Beaver Valley animals to have more closely approximated the wild groups in such a comparison. However, the locomotor profile of the Beaver Valley animals remains essentially unchanged (i.e. within a few percent of the values for the wild groups in all categories) except that bounding, which now drops to only 3%, is now also significantly different between these groups. Obviously, the fact that 77% of all bounding at Beaver Valley occured on the ropes has a lot to do with this. We interpret these data as evidence that the captive animals are utilizing more conservative gaits on unstable natural supports because even with the exclusion of rope use walking remains disproportionately high at Beaver Valley, while leaping remains low. In effect, the inclusion of rope use in the substrate profile of the Beaver Valley group "artificially" raises the frequency of bounding for this group but does not dramatically affect other behaviors.

Substrate usage profiles (Figure 5, Table 3) were very different between all groups, with the CRC animals showing highly restricted substrate selection and Beaver Valley animals strongly preferring the ropes. The CRC situation is explained by the fact that the animals did not have a diversity of substrates available to them. Because of this, the Beaver Valley substrate profile resembles that at Poço das Antas in indicating a wider selection of substrate types. When the overall profiles are recomputed with the rope category eliminated (Figure 6), the Beaver Valley and the wild samples appear even more similar,

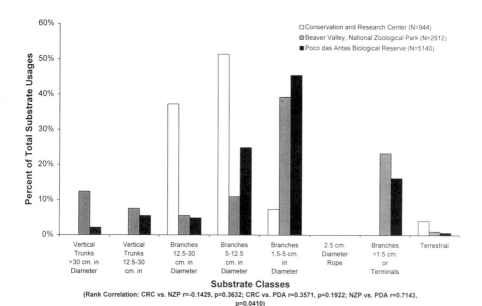

Figure 6. Substrate use by *Leontopithecus rosalia* (calculated without ropes).

with branches between 1.5 and 5 cm now no longer statistically different as compared to the wild sample. Interestingly, the exclusion of the ropes from the analysis does not change the relationship between the CRC sample and the Beaver Valley sample, providing further evidence that the Beaver Valley environment elicits behaviors that more closely approximate behaviors seen in the wild.

In both the Beaver Valley and PDA groups, branches that can be grasped in the hand (i.e. 1.5–5 cm in diameter) were the most often used substrates. The Beaver Valley animals used large vertical (>30 cm) trunks much more than did the wild animals and this may be an artifact of the positioning of the ropes, which were strung between large vertical trunks. It is interesting that the Beaver Valley animals appear to use small and terminal branches (<1.5 cm) relatively more than the wild animals once rope use is removed from the analysis, although this change is not statistically significant (Table 3). It is also interesting that the wild animals used larger supports (i.e., ones that are roughly the same diameter as the animal's trunk) more often.

There are two likely explanations for this pattern. The first is that substrate availability is different between the two sites and that the animals are randomly selecting supports based on availability. Testing this hypothesis would require data on relative abundance of substrate types between sites, and such data are not available for these groups. The second hypothesis is somewhat more complex, and proposes that the differences exhibited by our study groups reveal some "deficit" in the locomotor skills of captive *L. rosalia*. A greater frequency of small and terminal branch use by the Beaver Valley animals coupled with their higher frequencies of walking and lower frequencies of saltatory leaping may reflect their preferred means of crossing gaps in the canopy that the wild animals would cross by leaping. A greater ability, or willingness, of wild animals to cross gaps in the forest by leaping while captive animals find alternate routes is indicated by the greater frequency of leaping in wild animals and more walking in the Beaver Valley Group. Routes used by the Beaver Valley animals typically include more small and terminal supports. The data presented in Table 5 support this hypothesis. The Beaver Valley animals used branches smaller than 1.5 cm in diameter for 33% of all their walking bouts, and for 77% of all their climbing bouts. By contrast, the wild groups only used branches smaller than 1.5 cm in diameter for 14% of all walking bouts, although 66% of all climbing bouts used these supports. This may indicate that the Beaver Valley animals are walking further out into the periphery of the tree crowns than are the wild groups. In contrast, the greater reliance on larger supports in the wild groups may reflect the need for stable launching and landing platforms during leaping. Thirty-two percent of all leaping at Beaver Valley involved the smallest class of supports, as opposed to 18% at PDA. Conversely, leaping from larger supports (i.e. 1.5–12.5 cm in diameter) is more common in the wild (41% vs. 69%). We propose that the Beaver Valley animals are walking and climbing further out into the canopy before leaping to cross gaps, while the wild animals are leaping greater distances from more stable supports.

When substrates are broken down according to inclination (Figure 7), it is apparent that captive animals strongly prefer horizontal supports (Table 3). Although wild animals appear to show a slight preference for vertical supports, they also use horizontal and diagonals with similar frequency. It is interesting that under the protocols for which these data were collected, the wild animals show a greater preference for terminal supports than the Beaver Valley animals, somewhat contrary to the discussion above. This is because

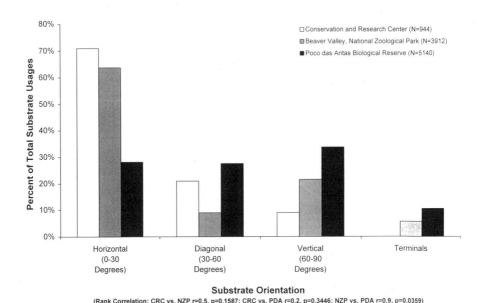

Figure 7. Substrate orientation use by *Leontopithecus rosalia.*

the category "terminals" is a structural, and not a size, category. Terminal branches are, by definition, smaller than 1.5 cm in diameter, but not all branches less than 1.5 cm in diameter are terminal branches. In order to qualify as "terminal branches" such small supports must have a geometrically complex structure. If the support that the animal is traveling on can be identified as a single support with a given orientation, then it is not classified as a terminal. Another point is that the concentrated use of large (>30 cm) tree trunks, which only occurs in Beaver Valley, contributes disproportionately to the vertical category in this group, while the wild animals spend relatively more time on smaller diameter vertical supports. Table 5 indicates that while 85% of all vertical climbing in Beaver Valley occurs on supports greater than 12.5 cm in diameter, in the wild 69% of all vertical climbing occurs on supports smaller than 1.5 cm in diameter.

Forest level usage also differs between our study groups (Figure 8). The animals at CRC were constrained by the fact that their cages were only just a little over two meters high and they will not be further discussed. The more evident similarities between the Beaver Valley and PDA groups are somewhat difficult to evaluate because the relatively greater use of the subcanopy at Beaver Valley relates to the arrangement of the ropes which were attached below the tree crowns. To what degree the Beaver Valley animals would have used the canopy in the absence of the ropes is difficult to say. We have noted (Stafford *et al.*, 1994) that the removal of two of the peripheral ropes in Beaver Valley did seem to elicit locomotion in the canopy along paths previously crossed by the ropes, but to what degree this was a result of rope removal, or increased locomotor competency in the animals is unknown (see Price, 1994 for an excellent study of *Saguinus oedipus* under similar but better controlled circumstances). In general, Beaver Valley offers a closer approximation of wild conditions in terms of the levels of the forest used.

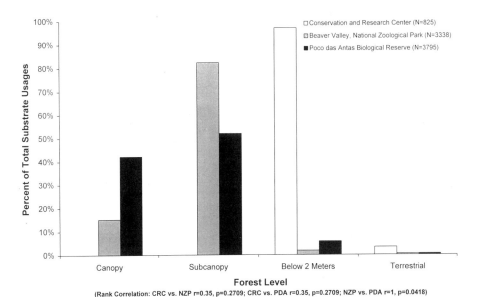

Figure 8. Use of forest level by *Leontopithecus rosalia*.

CONCLUSIONS

The results of this study lead us to several conclusions.

1. The locomotor profiles of *L. rosalia* in the wild and captivity are different. This difference appears to be manifested in a greater reliance on horizontal supports, a reduced degree of leaping-g behavior, and an increased reliance on a single form of quadrupedal locomotion in captive animals. In comparison, wild groups used leaping and vertical climbing as their primary locomotor choices, followed by variations of quadrupedalism-g.
2. The Beaver Valley environment seems to approximate the wild conditions better than the "conventional" enclosures at CRC. Even so, we would urge caution when positioning substrates into exhibits because of the apparent associations between specific locomotor categories and certain substrates.
3. The unique nature of quadrupedalism in *L. rosalia* is best explained as the result of incorporating foraging adaptations into the locomotor system, and not of locomotor adaptations *per se*. This unique pattern of cheiridial placement, now confirmed as the common pattern of *L. rosalia* in the wild, is constrained by forelimb morphology and is never seen to vary between settings.
4. The use of generalized locomotor categories to characterize the locomotor profile of any species may underrepresent the diversity of behaviors being used. Narrowly defined locomotor categories based on footfall patterns and kinematics may offer more acute insight into the specific correlations between locomotor behavior and morphology.
5. Captive studies are important because they offer opportunities to study behaviors without the obstructions present in the field. They allow experimentation, allowing the researcher to eliminate substrate as a variable between taxa, and to evaluate the effects of substrate size and inclination on locomotion.

Finally, as a general comment regarding the characterization of locomotion in *L. rosalia*, we point out that the locomotor profile of the wild animals suggests caution concerning our earlier hypothesis (Rosenberger and Stafford, 1994) which cast *L. rosalia* as a basically quadrupedal species based on our comparison with *Callimico*. In the wild, *L. rosalia* apparently use equal amounts of quadrupedalism and leaping. This makes their classification as "basically quadrupedal" somewhat problematic. A critical piece of missing information is comparative locomotor data on other callitrichines living in the same environment. Such studies could easily answer the question of whether or not *L. rosalia* are relatively more quadrupedal than other callitrichines as their morphology suggests.

SUMMARY

We report here on the locomotor behavior of three groups of golden lion tamarins (*Leontopithecus rosalia*); one in conventional captive enclosures, another in a captive free-ranging setting, and a third in the wild in Brazil. We find that *L. rosalia* appear to be more quadrupedal than other callitrichines in captivity, and that they use a unique mode of quadrupedal progression probably related to the elongate manus of this genus. We propose that the unique character of locomotion in *L. rosalia* is the result of incorporating foraging specializations into the locomotor system, rather than consider it a locomotor adaptation in the strictest sense. We found that locomotion did differ significantly between our groups, and that there were close associations between certain substrate types and certain locomotor behaviors for all groups. We propose that the different environments between the study sites drives this difference in locomotor behavior. While the locomotor behavior of the wild groups studied was different than that of the captive groups studied, there were few differences in the locomotion between wild groups in spite of the fact that there appear to be substantial differences in utilized substrates. We propose that morphological constraints explain this relative constancy in locomotor behavior between the wild sites.

ACKNOWLEDGMENTS

For access to the animals housed at CRC we thank Chris Wemmer and Larry Collins. Kim Pojeta was always helpful, and tolerant of our presence, during this phase of the project; especially when it came time to put ink on the hands and feet of the monkeys. The hand- and footprints presented here were collected by Doug C. Broadfield, who also assisted in the early development of the CRC study and data collection. Similarly, the staff at the National Zoological Park responsible for the daily management of the animals at Beaver Valley, under the direction of Lisa Stevens, were always helpful. At Poço das Antas, special thanks go to Andrea Martins and Denise Rambaldi for solving numerous logistical problems for us over the years. We would also like to thank the members of the ecology and reintroduction teams working in Brazil. They have proved a valuable and reliable source of information on golden lion tamarins.

Financial support has been provided by the Smithsonian Institution International Environmental Sciences Program, the Smithsonian Institution Scholarly Studies Program, the National Zoological Park (Research Development Award), Friends of the National Zoo, Brookfield Zoo (The Chicago Zoological Society), World Wildlife Fund for Nature - WWF, National Geographic Society, National Science Foundation (grant # DBS9008186), TransBrasil Airlines, the Brazilian Institute of the Environment and Natural resources (IBAMA), and the Brazilian National Council for Scientific and Technological Development (CNPq).

REFERENCES

Altmann, J (1974) Observational study of behavior: sampling methods. Behaviour 49:227–265.

Beck BB, Kleiman DG, Dietz JM, Castro I, Carvalho C, Martins A, and Rettberg-Beck B (1991) Losses and reproduction in reintroduced golden lion tamarins *Leontopithecus rosalia*. Dodo 27:50–61

Cartmill, M (1985) Climbing. Pp. 73–389 in M Hildebrand, DM Bramble, HF Liem, and DB Wake (eds.) Functional Vertebrate Morphology. The Belknap Press of Harvard University Press, Cambridge, massachusetts.

Coimbra-Filho AF, and Mittermeier RA (1978) Reintroduction and translocation of lion tamarins: a realistic appraisal. Pp. 41–46 in H Rothe, H Woters, and JP Hearn (eds.) Biology and Behavior of Marmosets. Eigenverlag Hartmut Rothe, Gottingen.

Dietz JM, Baker AJ, and Miglioretti D (1994) Seasonal variation in reproduction, juvenile growth, and adult body mass in golden lion tamarins (*Leontopithecus rosalia*). American Journal of Primatology 34:115–132.

Dykj D (1982) Allometry of the trunk and limbs in New World Monkeys. Doctoral Dissertation, City University of New York.

Fleagle JG (1988) Primate Adaptation and Evolution. Academic Press, Inc., New York.

Ford SM (1994) Evolution of sexual dimorphism in body weight in platyrrhines. American Journal of Primatology 34:221–244.

Ford SM and Davis LC (1992) Systematics and body size: implications for feeding adaptations in New World monkeys. American Journal of Physical Anthropology 88:415–468.

Garber PA (1980) Locomotor behavior and feeding ecology of the Panamanian tamarin *Saguinus oedipus geoffroyi* (Callitrichidae, Primates). International Journal of Primatology 1:185–201.

Garber PA (1991) A comparative study of positional behavior in three species of tamarin monkeys. Primates 32:219–230

Garber PA (1992) Vertical clinging, small body size, and the evolution of feeding adaptations in the Callitrichinae. American Journal of Physical Anthropology 88:469–482.

GeboDL, and Chapman CA (1995) Habitat, annual, and seasonal effects on positional behavior in red colobus monkeys. American Journal of Physical Anthropology 96: 73–82.

Grand, TI (1967) A mechanical interpretationof terminal branch feeding. Journal of Mammalogy 53:198–201.

Hampson, CG (1965) Locomotion and some associated morphology in the Northern flying squirrel. PhD Dissertation, University of Alberta.

Hildebrand M (1967) Symmetrical gaits of primates. American Journal of Physical Anthropology 26:119–130.

Hildebrand M (1977) Analysis of asymmetrical gaits. Journal of Mammalogy 58:131–156.

Hildebrand M (1980) The adaptive significance of tetrapod gait selection. American Zoologist 20:255–267.

Jungers WL (1977) Hindlimbs and pelvic adaptations to vertical climbing and clinging in *Megaladapis*, a giant subfossil prosimian from Madagascar. Yearbook of Physical Anthropology 20:508–524.

Kleiman DG (1989) Reintroduction of captive mammals for conservation. BioScience 39:152–161.

Kleiman DG, Beck BB, Dietz JM, Dietz LA, Ballou JD, and Coimbra-Filho AF (1986) Conservation program for the golden lion tamarin: captive research and management, ecological studies, educational strategies, and reintroduction. Pp. 959–979 in K Benirschke (ed.) Primates the Road to Self-Sustaining Populations. Springer-Verlag, New York.

Kleiman DG, Beck BB, Dietz JM, and Dietz LA (1991) Costs of a re-introduction and criteria for success: accounting and accountability in the golden lion tamarin conservation program. Symposium of the Zoological Society of London 62:125–142.

Martin RD (1990) Primate Origins and Evolution. Princeton University Press, Princeton.

Napier JR, and Walker AC (1967) vertical clinging and leaping - a newly recognized category of locomotor behavior of primates. Folia Primatologica 6:204–219.

Peres CA (1986) Costs and benefits of terretorial defense in golden lion tamarins, *Leontopithecus rosalia*. MS Thesis, University of Florida.

Price EC (1994) Adaptation of captive-bred cotton-top tamarins (*Saguinus oedipus*) to a natural environment. Zoo Biology 11:107–120.

Prost JH (1965) A definitional system for the classification of primate locomotion. American Anthropologist 67:1198–1214.

Ripley S (1967) The leaping of langurs: a problem in the study of locomotor adaptation. American Journal of Physical Anthropology 26:149–170.

Rollinson J, and Martin RD (1981) Comparative aspects of primate locomotion, with special reference to arboreal cercopithecines. Symposium of the Zoological Society of London 48:377–427.

Rose MD (1973) Quadrupedalism in Primates. Primates 14:337–357.

Rosenberger AL (1992) Evolution of feeding niches in New World monkeys. American Journal of Physical Anthropology 88:525–562.

Rosenberger AL, and Stafford BJ (1994) Locomotion in captive *Leontopithecus* and *Callimico*: a multimedia study. American Journal of Physical Anthropology 94:379–394.

Rylands AB (1989) Sympatric callitrichids: the black tufted-ear marmoset, *Callithrix khuli*, and the golden-headed lion tamarin, *Leontopithecus chrysomelas*. Journal of Human Evolution 18:679–695.

Rylands AB (1993) The ecology of the lion tamarins, *Leontopithecus*: some intrageneric differences and comparisons with other callitrichids. Pp. 296–313 in AB Rylands (ed.) Marmosets and Tamarins, Systematics, Behavior, and Ecology. Oxford University Press, Oxford.

Stafford BJ, Rosenberger AL, and Beck BB (1994) Locomotion of free-ranging golden lion tamarins (*Leontopithecus rosalia*) at the National Zoological Park. Zoo Biology 13:333–344.

Thorington RW, and Heaney LR (1981) Body proportions and gliding adaptations of flying squirrels (Petauristinea). Journal of Mammalogy 62:101–114.

Thorington RW, and Thorington EM (1989) Postcranial Proportions of *Microsciurus* and *Sciurillus*, the american pygmy tree squirrels. Advances in Neotropical Mammalogy 1989:125–136.

Vilensky JA (1989) Primate Quadrupedalism: how and why does it differ from that of typical quadrupeds? Brain behavior and Evolution 34:357–364.

Vilensky JA, and Larson SL (1989) Primate locomotion: utilization and control of symmetrical gaits. Annual review of Anthropology 18:17–35.

7

FUNCTIONAL AND PHYLOGENETIC IMPLICATIONS OF ANKLE MORPHOLOGY IN GOELDI'S MONKEY (*Callimico goeldii*)

Lesa C. Davis

Department of Anthropology
Southern Illinois University
Carbondale, Illinois 62901-4502

INTRODUCTION

Callimico goeldii maintains a unique position among platyrrhine primates. Due to its intriguing assortment of anatomical and behavioral traits, *Callimico* has figured prominently in the controversial issues of platyrrhine phylogeny and systematics (for modern discussions see Ford, 1980a, b, 1986a, b; Ford and Davis, 1992; Hershkovitz, 1977; Martin, 1990, 1992; Rosenberger, 1979, 1981, 1984; Rylands *et al.*, 1993). Field studies of *Callimico* (Buchanan-Smith, 1991; Cameron *et al.*, 1989; Christen and Geissmann, 1994; Izawa, 1979; Izawa and Yoneda, 1981; Masataka, 1981a, b; Pook and Pook, 1979, 1981, 1982) have documented a number of specific ecological and behavioral patterns that further distinguish this monkey. Much less is known of its postcranial anatomy, although some data are available (Davis, 1994; Ford, 1980a, 1986a, b; Hershkovitz, 1977; Hill, 1957, 1959). The present study examines the functional significance and phylogenetic affinities of morphological features in the ankle of *Callimico*.

It has become increasingly apparent (c.f. Davis *et al.*, 1993; Davis *et al.*, in prep; Fleagle, 1977; Ford and Hobbs, 1994; Garber, 1991; Rodman, 1979; Ward and Sussman, 1979) that significant morphological and behavioral diversity exists between closely related species. Primate form / function analyses conducted at the genus level can and do hide these significant species-level morphological and behavioral distinctions. In the present study, analysis is based at the level of the species to gain a more comprehensive, and accurate, picture of *Callimico* ankle adaptations by comparison to 14 other platyrrhine species.

Callimico's Place within Platyrrhini

Upon its discovery 90 years ago, *Callimico goeldii* (Thomas, 1904) was immediately recognized as an phylogenetic link between the small, clawed, two-molared cal-

Adaptive Radiations of Neotropical Primates
edited by Norconk *et al.* Plenum Press, New York, 1996

litrichines and the larger platyrrhines characterized by three molars, a hypocone, nails, and single births. Anatomically, *Callimico* shares a relatively small body size, clawed digits, and numerous postcranial features (Ford, 1980a, 1986a; Hill, 1957, 1959) with the callitrichines, and retains the presence of a hypocone, a third molar (although greatly reduced in size), and single births in common with the larger taxa. Behaviorally, *Callimico* is aligned with the callitrichines in its preferred mode of locomotion and postures (Pook and Pook, 1981).

Taxonomically, *Callimico* has been variously placed in Cebidae with the larger platyrrhine taxa (Martin, 1990; Simons, 1972; Simpson, 1945); in its own family, Callimiconidae (Hershkovitz, 1977; Hill, 1957); and most commonly with the marmosets and tamarins in the family Callitrichidae (Fleagle, 1988; Ford, 1980a, 1986a; Szalay and Delson, 1979; c.f. Rosenberger, 1979, 1981, 1984), with various subfamily and tribe designations (for review, see Rylands *et al.*, 1993). The latter designation is followed here. Specifically, *Callimico goeldii* is considered to be the sole representative of the subfamily Callimiconinae, and the subfamily Callitrichinae is reserved for *Cebuella*, *Callithrix*, *Saguinus*, and *Leontopithecus*. These two subfamilies make up the family Callitrichidae.

MATERIALS AND METHODS

Materials

A total of 53 quantitative and 10 qualitative traits of the ankle and foot regions were measured or scored for *Callimico* and a comparative outgroup. Only those found to be significantly distinct in *Callimico* are reported here. These traits were concentrated in the distal tibia and astragalus and the quantitative features are shown in Figure 1. All measurements were taken to the nearest tenth of a millimeter using Helios 160mm dial calipers.

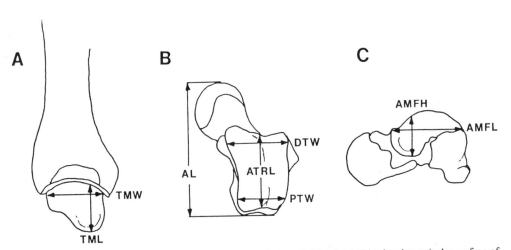

Figure 1. Measurements discussed in this study. A, lateral view of right distal tibia showing articular surface of malleolus. TML, tibial malleolar length; TMW, tibial malleolar width. B, Superior view right astragalus. AL, total astragalar length; ATRL, astragalar length; DTW, distal trochlear width; PTW, proximal trochlear width. C, medial view right astragalus showing medial astragalar facet for tibial malleolus. AMFH, medial facet height; AMFL, medial facet length.

Table 1. Study sample

Taxon	N	Sex			Origin		Body Weight[1] (g)
		Male	Female	Unknown	Wild	Captive	
Callitrichids							
Cebuella pygmaea	15	6	7	2	10	5	123
Callithrix penicillata	6	2	4	0	5	1	249
Callithrix jacchus	12	6	6	0	3	9	257
Saguinus fuscicollis	13	4	9	0	9	4	373
Saguinus nigricollis	15	7	7	1	11	4	435
Saguinus labiatus	10	4	6	0	0	10	497
Saguinus midas	12	3	5	4	10	2	523
Saguinus mystax	12	5	6	1	8	4	542
Saguinus geoffroyi	13	6	3	4	9	4	545
Callimico goeldii	13	6	5	2	3	10	492
Leontopithecus rosalia	12	5	7	0	4	8	628
Non-Callitrichids							
Saimiri sciureus	9	4	4	1	6	3	796
Callicebus torquatus	8	2	6	0	7	1	1303
Pithecia pithecia	9	5	3	1	7	2	1682
Ateles geoffroyi	10	1	6	3	8	2	7704

[1]Body weights taken from Ford and Davis (1992).

The comparative outgroup consists of ten callitrichine species and four non-callitrichid platyrrhines: *Leontopithecus rosalia, Saguinus geoffroyi, S. mystax, S. midas, S. fuscicollis, S. labiatus, S. nigricollis, Callithrix jacchus, C. penicillata, Cebuella pygmaea, Ateles geoffroyi, Pithecia pithecia, Callicebus torquatus,* and *Saimiri sciureus.* Table 1 provides a summary of the sample. Only non-pathological adults were used in this study. While an attempt was made to include primarily wild-caught individuals where possible, it is noted that the *Callimico* sample is drawn largely from captive specimens and the *Saguinus labiatus* sample is drawn entirely from captive individuals. Since wild-caught *Callimico* skeletons are poorly represented in museum collections, only three of the thirteen specimens examined here were wild-caught. However, Student's t-test was performed to test for significant differences between the captive and wild-caught specimens and no significant differences (at p = .05) were found. Approximately 58% of the comparative sample was wild-caught. Locomotor and postural data for *Callimico* and the comparative outgroup are drawn from the literature.

Methods

Both multivariate and bivariate analyses were performed. First, principal components analysis on the original 53 non-transformed skeletal variables was performed to grossly explore the placement of *Callimico* within Platyrrhini. Bivariate analyses were then conducted to identify differences between species and to identify which species were distinct from *Callimico.*

In order to compare morphological data across species which differ in body size, a standardizing variable was identified. Several recent works favor various body size surrogates (see references in Damuth and MacFadden, 1990; and in Jungers, 1985), and body weight was chosen as the target body size variable in the present study. Since individual body weight is often missing from museum specimen tags, an indirect measure of body weight that was available for each specimen was chosen. Two techniques were used to

HHH

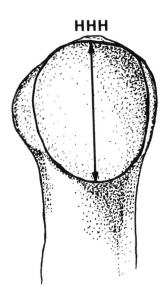

Figure 2. Body weight variable. Medial view of humeral head. HHH, humeral head superoinferior height.

identify this standardizing variable. First, measurements of specimens with associated body weights were regressed directly against their reported body weight. Second, species mean skeletal measurements were regressed against average species body weights taken from the literature (Ford and Davis, 1992, their Table 3) (all values were logrithmically transformed). Both techniques revealed that superoinferior height of the humeral head (Fig. 2) had a very high correlation (e.g. Technique 2: $r = .986$, $r^2 = .985$, P< .0001) with body weight and a slope which closely approached isometry (e.g. Technique 2: least squares estimate = .34). Thus, humeral head height (HHH) was chosen as the body weight variable. Analysis of residuals was not used in this study due to the substantial size extremes (ie. *Ateles geoffroyi* and *Cebuella pygmaea*) in the comparative sample; instead, indices were used (see Corruccini, 1987, 1995).

The standardizing variable HHH was incorporated as the denominator in ratios comparing specific quantitative ankle traits between species. Analysis of variance (ANOVA) tests were performed to determine if there were significant differences between species. If ANOVA was significant at the .05 level, the conservative Tukey-Kramer test, which controls for the experiment-wise error rate in large samples to be compared (see discussion under PROC GLM, SAS Institute Inc., 1989; also Dunnett, 1980; Hayter, 1984; Kramer, 1956; Miller, 1981), was used to identify significantly different taxa at the .05 level. All Tukey-Kramer analyses and the principal components analysis were performed using JMP Version 3.0.1 on a Macintosh IIsi (SAS Institute Inc., 1994).

Callimico POSITIONAL BEHAVIOR

Field studies documenting the habits of *Callimico* in the wild have been relatively short-term and have primarily focused on distribution, ecology, and social behavior (Buchanan-Smith, 1991b; Cameron *et al.*, 1989; Christen and Geissmann, 1994; Izawa, 1979; Izawa and Bejarano, 1981; Izawa and Yoneda, 1981; Masataka, 1981a,b; Moynihan, 1976; Pook and Pook, 1979, 1981, 1982). Our current understanding of the positional repertoire of *Callimico* comes from qualitative and sometimes anecdotal descriptions found within these sources. Although data on *Callimico* is incomplete, the positional activities of this species is more completely documented than most other callitrichid species, including 11 *Callithrix* species, all species of *Leontopithecus*, and almost half of the commonly recognized *Saguinus* species.

Callimico appears to prefer the dense lower forest levels (Buchanan-Smith, 1991; Cameron *et al.*, 1989; Christen and Geissmann, 1994; Izawa, 1979; Izawa and Yoneda, 1981; Moynihan, 1976; Pook and Pook, 1981). Most field reports note *Callimico* 's preference for thick patches of bamboo (Cameron *et al.*, 1989; Buchanan-Smith, 1991; Izawa, 1979; Izawa and Yoneda, 1981; Pook and Pook, 1981), although it was absent from this habitat in the brief survey of Christen and Geissmann (1994). Its diet consists primarily of fruit and insects, but unlike many other callitrichids, *Callimico* has not been observed feeding on gum or sap (Pook and Pook, 1981).

Quadrupedal walking, running, and bounding, vertical climbing with use of the claws, and leaping have been documented for *Callimico* (Christen and Geissmann, 1994; Izawa, 1979; Moynihan, 1976; Pook and Pook, 1981, 1982). Hanging by the hindlimbs is employed when feeding on fruit (Pook and Pook, 1981).

A unifying theme of most all reports of *Callimico* positional behavior is its common use of vertical trunks (Buchanan-Smith, 1991b; Izawa, 1979; Moynihan, 1976; Pook and Pook, 1981, 1982). These trunks are frequently used for vertical clinging as a resting posture (Pook and Pook, 1981), as well as the launch and landing posture between leaps (Buchanan-Smith, 1991b; Izawa, 1979; Moynihan, 1976; Pook and Pook, 1981, 1982). Izawa (1979) and Pook and Pook (1981) observed that leaping between horizontal branches and between terminal branches is used less frequently than leaping between vertical trunks. Conversely, in their five sightings of *Callimico*, Christen and Geissmann (1994) reported seeing no vertical trunk leaping. Instead, horizontal branch leaping and terminal branch leaping were the exclusive leaping behaviors noted. The intriguing results of the Christen and Geissmann study may either indicate the versatility of this species in habitat choice and positional behavior, or may be a function of the brevity of their observations of non-habituated animals. Further elucidation of the ecology and positional behavior of *Callimico* must await more intensive study.

Callimico ANKLE MORPHOLOGY

Results from the principal components analysis computed for the 53 quantitative ankle and foot traits indicate that *Callimico* clusters with the majority of the callitrichines (Fig. 3). *Callithrix penicillata*, *Cebuella pygmaea*, and *Saguinus nigricollis* are absent from this analysis due to missing data. Tukey-Kramer analyses on bivariate ratios of the 53 traits also indicate that *Callimico* is not significantly different from most callitrichines for most of these traits. These results agree with previous work documenting a distinctive suite of postcranial features shared by the callitrichines and *Callimico* (Ford, 1980a, 1986a, 1994). The discussion below focuses on morphological traits in *Callimico* that were found to be distinctive of the callitrichines as a group. These features are concentrated in the distal tibia and astragalus.

In the ankle, the distal tibial trochlear surface and its projecting malleolus articulate with the astragalus. The tibial trochlea and the corresponding astragalar trochlea form the superior tibioastragalar joint (Fig. 4a) The inner or medial surface of the tibial malleolus

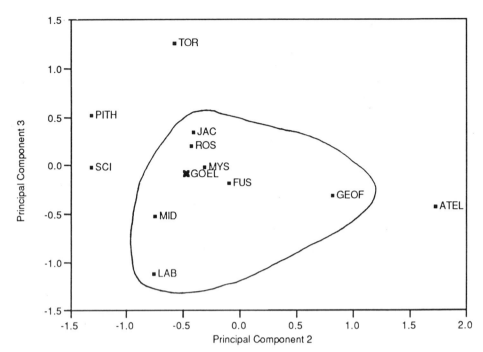

Figure 3. Plot of principal components analysis on the 53 ankle variables (principal component 3 by principal component 2). *Callithrix penicillata*, *Cebuella pygmaea*, and *Saguinus nigricollis* are absent from the plot due to missing data. Species abbreviations are as follows: JAC, *Callithrix jacchus*; FUS, *Saguinus fuscicollis*; LAB, *Saguinus labiatus*; MID, *Saguinus midas*; MYS, *Saguinus mystax*; GEOF, *Saguinus geoffroyi*; GOEL, *Callimico goeldii*; ROS, *Leontopithecus rosalia*; SCI, *Saimiri sciureus*, TOR, *Callicebus torquatus*, PITH, *Pithecia pithecia*; ATEL, *Ateles geoffroyi*. Circle encloses all callitrichids.

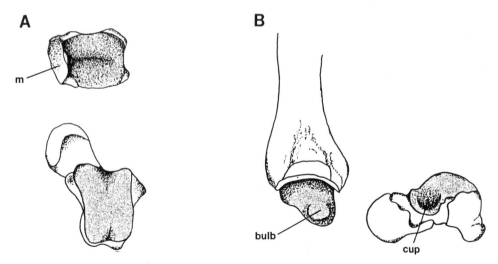

Figure 4. Joint surfaces examined in this study. A, Superior tibioastragalar joint. Top, inferior view of right distal tibia. Shaded area indicates tibial trochlea; m, malleolus. Bottom, superior view right astragalus. Shaded area indicates astragalar trochlea. B, Medial tibioastragalar joint. Left, lateral view of right distal tibia. Shaded area indicates medial or inner facetal surface of malleolus; bulb, rounded prominence on malleolar facetal surface. Right, medial view of right astragalus. Shaded area indicates medial astragalar facet; cup, depressed area of articular surface that receives malleolar bulb of tibia.

Table 2. Mean values for tibial malleolar area and medial astragalar facet area

	Tibial Malleolar Area $\sqrt{(\text{TML} \times \text{TMW})}/\text{HHH}$			Medial Astragalar Facet Area $\sqrt{(\text{AMFL} \times \text{AMFH})}/\text{HHH}$		
	n	Mean	SD	n	Mean	SD
Cebuella pygmaea	12	.360*	.038	6	.380*	.042
Callithrix penicillata	4	.380	.011	3	.395*	.070
Callithrix jacchus	12	.387*	.023	8	.422*	.031
Saguinus fuscicollis	10	.402	.027	5	.408*	.023
Saguinus nigricollis	13	.389*	.024	6	.417*	.056
Saguinus labiatus	9	.377*	.019	5	.405*	.031
Saguinus midas	10	.373*	.044	11	.406*	.051
Saguinus mystax	11	.387*	.019	10	.436*	.030
Saguinus geoffroyi	11	.375*	.038	7	.417*	.018
Callimico goeldii	10	.442	.059	8	.514	.025
Leontopithecus rosalia	11	.405	.025	10	.403*	.051
Saimiri sciureus	7	.477	.021	6	.484	.047
Callicebus torquatus	5	.422	.017	2	.479*	.017
Pithecia pithecia	8	.503*	.030	8	.480	.036
Ateles geoffroyi	4	.436	.020	4	.418*	.024

*Significantly different from *Callimico* at the .05 level.

articulates with the facet or cup on the medial side of the astragalus (Fig. 4b). This articulation forms the medial tibioastragalar joint. Laterally, the distal fibula articulates with the lateral facet of the astragalus.

Medial Tibioastragalar Joint

In the medial tibioastragalar joint, the size and morphology of the tibial malleolus and medial astragalar facet are distinct in *Callimico*. Examination of the relative area of the malleolus, computed as $\sqrt{(\text{TML} \times \text{TMW})}/\text{HHH}$, reveals that *Callimico* has a large malleolus (Table 2, Fig. 5). A large malleolus is shared by all the non-callitrichid platyrrhines, as well as *Callithrix penicillata*, *Leontopithecus rosalia*, and *Saguinus fuscicollis*. While *Callimico* is not statistically significantly larger than these three callitrichines, its mean value for this trait is higher than the callitrichines and is embedded within the range for non-callitrichids. Separate analysis of mean malleolar length and mean malleolar width indicate that the large malleolar area in *Callimico* and the non-callitrichid platyrrhines is driven by both a relatively long malleolus and a relatively wide malleolus. Interestingly, a relatively wide malleolus alone is primarily responsible for the large malleolar area in *C. penicillata*, *S. fuscicollis*, and *L. rosalia*. *Callimico* was not found to be significantly different from the comparative sample in the malleolar shape ratio (malleolar length / malleolar width) used by others (Dagosto, 1986; Ford, 1980a, 1986a, 1994).

The morphology of the medial malleolar articular surface was scored qualitatively as to the degree of rounding or convexity (slightly convex, moderately convex, or distinctly bulbous) of the articular surface (Fig. 4b). All callitrichines are polymorphic in this feature, although the majority of individuals in each species display a moderately convex tibial malleolar surface. Conversely, *Callimico* specimens consistently have a distinctly bulbous malleolar articular surface. Among the non-callitrichid platyrrhines, *Saimiri sciureus*, *Callicebus torquatus*, and *Ateles geoffroyi* were each polymorphically moderately

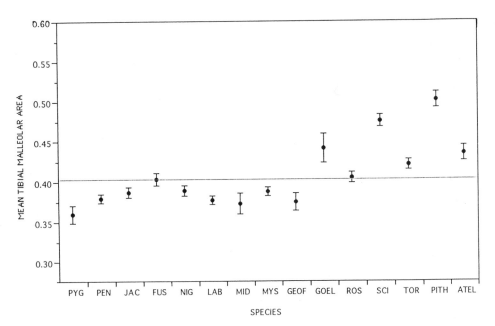

Figure 5. Plot of mean tibial malleolar area [expression 07/4] by species. Measurement abbreviations as in Figure 1, species abbreviations as in Figure 3, with the additions of: PYG, *Cebuella pygmaea*; PEN, *Callithrix penicillata*; NIG, *Saguinus nigricollis*. Dots represent species means, black horizontal bars represent standard error of the mean, grey horizontal line across plot represents sample mean.

convex and distinctly bulbous, with no clear pattern. *Pithecia pithecia*, like *Callimico*, consistently exhibits a distinctly bulbous tibial malleolar articular surface.

To fully assess the cohesion of the medial tibioastragalar joint, the area of the medial astragalar facet was also computed as $\sqrt{(\text{AMFL} \times \text{AMFH})/\text{HHH}}$ (Table 2, Fig. 6). *Callimico* emerges as having the largest medial astragalar facet area of the entire sample. It is significantly larger than all callitrichids, *Callicebus torquatus*, and *Ateles geoffroyi*. While *Callimico* 's mean value for this trait is larger than that for *Saimiri sciureus* and *Pithecia pithecia*, it is not statistically significantly larger. Analysis of the ratio of tibial malleolar area to medial astragalar facet area revealed no significant differences in the sample. In short, *Callimico* has among the largest medial tibioastragalar joint surfaces but the size relationship between the tibial portion and astragalar portion of this joint is statistically consistent for the entire sample.

Analyzed separately, the features of malleolar area, malleolar morphology, and medial astragalar facet area reveal little to clarify the relationship between anatomy and behavior in platyrrhines. For example, a large malleolar area is found among the smallest (*Callithrix penicillata*) and largest (*Ateles geoffroyi*) taxa in the sample, as well as several taxa intermediate in body size (*Saguinus fuscicollis*, *Leontopithecus rosalia*, *Callimico*, *Saimiri sciureus*, *Callicebus torquatus*, and *Pithecia pithecia*). Available positional behavior data for these species indicate a full range of locomotor and postural behaviors, including suspensory locomotion, vertical clinging, leaping, and quadrupedalism (e.g. Coimbra-Filho and Magnanini, 1972; Easley, 1982; Garber, 1991; Kinzey, 1976; Mittermeier, 1977; Norconk, 1986; Oliveira *et al.*, 1985; Pook and Pook, 1981; Walker, 1993, this volume), none of which characterize this sample to the exclusion of the entire comparative sample. Similarly, while *Callimico* shares an enlarged malleolar area with the entire non-callitrichid platyrrhine sample, it is not to the exclusion of all callitrichines.

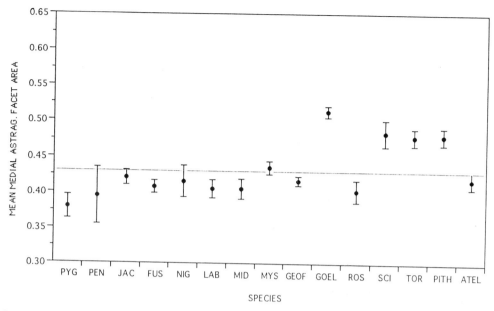

Figure 6. Plot of mean medial astragalar facet area [expression 07/6] by species. Measurement abbreviations as in Figure 1, species abbreviations as in Figure 5, key to features in the plot as in Figure 5.

Three taxa share both an enlarged malleolus and enlarged medial astragalar joint: *Callimico*, *Pithecia pithecia*, and *Saimiri sciureus*. Of these, *Callimico* and *P. pithecia* also have a distinctly bulbous medial tibial malleolar surface. An enlarged tibioastragalar articular surface area would seem to provide a stable displacement-resistant brace on the medial astragalus during plantar- and dorsiflexion excursions. Coupled with a protruding, bulbous malleolar surface, these features may form a morphological complex that serves to further stabilize the ankle during the closed-packed, fully dorsiflexed position, such as that which occurs during vertical clinging and / or leaping from a vertical platform. In addition, an enlarged tibial malleolus may also indicate a larger attachment area for the deltoid or medial ligament. Although this ligament has not been described for platyrrhines, in humans, it passes from the malleolus to the medial surface of the astragalus and calcaneal sustentaculum and stabilizes the ankle by resisting twisting forces (Basmajian, 1982). Available field studies of *Callimico* and *P. pithecia* report that vertical clinging and vertical leaping are common positional behaviors in both these taxa (Buchanan-Smith, 1991b; Heltne *et al.*, 1981; Izawa, 1979; Moynihan, 1976; Oliveira *et al.*, 1985; Pook and Pook, 1981; Walker, 1993, this volume). Interestingly, this complex is absent in the smaller-bodied callitrichines that frequently engage in vertical clinging and / or vertical leaping, such as *Cebuella pygmaea* (Kinzey *et al.*, 1975; Terborgh, 1983, 1985), *Callithrix jacchus* (Maier *et al.*, 1982), and *Saguinus fuscicollis* (Garber, 1992; Norconk, 1986; Terborgh, 1983, 1985). It is possible that *Callimico* and *P. pithecia* retained this complex and the associated vertical clinging and vertical leaping behaviors from the ancestral platyrrhine. However, reconstructions of the ancestral platyrrhine morphotype suggest that quadrupedalism is the predominant ancestral platyrrhine locomotor mode (Ford, 1988, 1990; cf. Gebo, 1989), and that *Aotus*, *Callicebus*, or possibly *Saimiri* closely approximate the primitive platyrrhine condition in body size and certain morphological aspects (Dunlap *et al.*, 1985; Ford, 1980b, 1986a ; Kay, 1980, 1990; Rose and Fleagle, 1981). Although *Aotus* was not included in the present study, the morphological complex is completely absent in *Callice-*

bus torquatus and only partially present in *Saimiri sciureus*. Furthermore, if vertical clinging and vertical leaping were a significant part of the positional repertoire of the platyrrhine ancestor, it is unclear why these behaviors and the associated morphological complex would be retained in one dwarfed callitrichid (*Callimico*), while others (e.g. *C. pygmaea*, *C. jacchus*, and *S. fuscicollis*) lose the morphological complex but retain the behaviors. A more plausible explanation would be that vertical clinging, vertical leaping, and the associated morphological complex were convergently acquired in *Callimico* and *P. pithecia*, derived from a more quadrupedal ancestor. It is possible that a form / function body size threshold exists in which relatively large species require a more stable medial tibioastragalar joint for habitual ankle dorsiflexion, while smaller species may maintain this activity without the supportive features in the ankle. In this scenario, *Callimico* at 492g (Ford and Davis, 1992) falls above the threshold while the smaller callitrichines are below the threshold. The evolutionary development of vertical clinging in the small callitrichines is clearly related to the gummivorous component of their diet and their use of vertical platforms to locate prey (Coimbra-Filho and Mittermeier, 1977; Crandlemire-Sacco, 1986; Garber, 1988, 1992; Maier *et al.*, 1982; Ramirez *et al.*, 1977; Soini, 1987). Gummivory is absent in both *Callimico* (Pook and Pook, 1981) and in *P. pithecia* (Mittermeier, 1977; Mittermeier and van Roosmalen, 1981). *Leontopithecus rosalia* poses an interesting problem in this issue. As the largest callitrichid, *L. rosalia* (628g, Ford and Davis, 1992), could conceivably provide a test for the size threshold hypothesis. However, positional behavior data in the wild for this species are almost entirely absent, and what little anecdotal data are available indicate that the movements of *L. rosalia* tend to be oriented horizontally as opposed to vertically (Coimbra-Filho and Magnanini, 1972). Stafford *et al.* (this volume) indicate that wild-born *L. rosalia* at the Poco das Antas Biological Reserve in Brazil use vertical trunks only 8% of the time. Similarly, vertical supports accounted for less than 15% of the substrates used by a group of cage-reared but free-ranging *L. rosalia* released on the grounds of the National Zoological Park (Stafford *et al.*, 1994, their Table 1). Even if future positional studies of *L. rosalia* indicate a common use of vertical clinging and or leaping from vertical platforms in the wild, there is evidence to suggest that this species has secondarily increased in size from a previously dwarfed condition (Ford and Corruccini, 1985; Ford and Davis, 1992; Garber, 1992, 1994). The effect of a possible secondary size increase from a phyletic dwarfing event would have to be addressed for this morphological complex.

Saimiri sciureus exhibits part of this complex (large tibial malleolus and large medial astragalar facet), but has a smaller malleolar bulb. While leaping constitutes a significant portion of *S. sciureus'* positional repertoire (Fleagle and Mittermeier, 1980; Mittermeier, 1977), leaping between vertical supports and vertical clinging are relatively rare (Mittermeier, 1977; Yoneda, 1988). Evolutionarily, *S. sciureus* is derived from the ancestral platyrrhine in its slightly smaller body size and the increased frequency of branch leaping. It is suggested that enlarged medial tibioastragalar joint surfaces were also convergently acquired in *S. sciureus*, as compared to *Callimico* and *P. pithecia*.

Superior Tibioastragalar Joint

Several features were examined in the superior tibioastragalar joint, including total astragalar length, astragalar trochlear length, degree of astragalar trochlear wedging, and the presence of a tibial stop at the distal end of the astragalar trochlea. Total astragalar length is greatest in *Callimico* and the non-callitrichid platyrrhines (Table 3, Fig. 7). However, within the callitrichids, *Callimico*'s longer astragalus is only statistically significantly different from that of *Cebuella pygmaea*, *Callithrix jacchus*, *Saguinus fuscicollis*, and *Leontopithecus rosalia*.

Table 3. Mean values for astragalar length, astragalar trochlear length, and ration of trochlear length to astragalar length

	Astragalar Length (AML/HHH)			Astragalar Trochlear Length (ATRL/HHH)			Trochlear Length/Total Length (ATRL/AL)		
	n	Mean	SD	n	Mean	SD	n	Mean	SD
Cebuella pygmaea	5	1.12*	.036	6	.56*	.051	6	.51	.029
Callithrix penicillata	3	1.25	.026	3	.59*	.017	3	.47	.024
Callithrix jacchus	8	1.22*	.046	8	.60*	.047	8	.49	.046
Saguinus fuscicollis	6	1.22*	.061	6	.61*	.027	7	.50	.040
Saguinus nigricollis	6	1.26	.111	6	.63*	.062	6	.50	.032
Saguinus labiatus	5	1.27	.061	5	.62*	.047	5	.49	.022
Saguinus midas	11	1.26	.098	11	.62*	.051	11	.49*	.023
Saguinus mystax	9	1.27	.075	11	.64*	.038	9	.50	.018
Saguinus geoffroyi	7	1.28	.103	9	.64*	.050	7	.51	.037
Callimico goeldii	8	1.37	.082	8	.74	.052	9	.54	.024
Leontopithecus rosalia	10	1.25*	.036	10	.62*	.029	11	.50	.025
Saimiri sciureus	6	1.40	.068	7	.74	.030	7	.53	.021
Callicebus torquatus	2	1.36	.089	2	.77	.056	3	.57	.011
Pithecia pithecia	7	1.36	.064	8	.77	.052	7	.58	.030
Ateles geoffroyi	4	1.34	.079	4	.77	.044	6	.59*	.030

*Significantly different from *Callimico* at the .05 level.

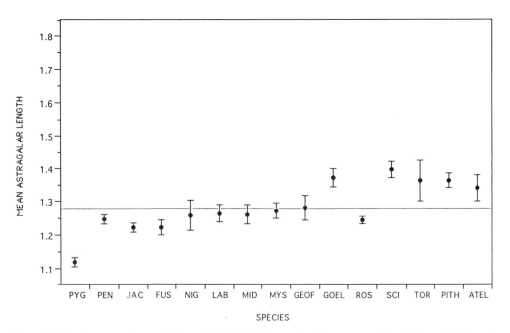

Figure 7. Plot of mean total astragalar length (AL / HHH) by species. Measurement abbreviations as in Figure 1, species abbreviations as in Figure 5, key to features in the plot as in Figure 5.

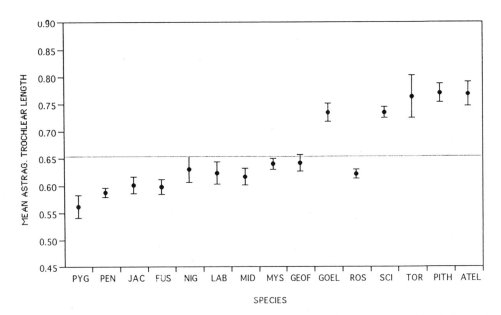

Figure 8. Plot of mean astragalar trochlear length (ATRL / HHH) by species. Measurement abbreviations as in Figure 1, species abbreviations as in Figure 5, key to features in the plot as in Figure 5.

Callimico also has among the longest astragalar trochleae (Table 3, Figs. 8 and 11). It is significantly longer than all callitrichines and is within the range for the non-callitrichid platyrrhines. The astragalar trochlea articulates with the tibial trochlea. The antero-posterior length of the *Callimico* tibial trochlea is not significantly different from the comparative sample. In order to gage the relative length of the astragalar trochlea to total astragalar length, a ratio of trochlear length / total astragalar length was computed (Table 3, Fig 9). Unlike the callitrichines, *Callimico* and the non-callitrichid platyrrhines have a trochlea that is more than half the length of the astragalus, suggesting that their elongated astragalus is not simply a function of an elongated trochlea.

The functional significance of an elongated astragalus is difficult to determine, as its length incorporates several distinct joint surfaces. Secondary to being associated with a relatively long astragalus, an elongated astragalar trochlea would facilitate an increased plantar- and dorsiflexion excursion at the ankle joint. Several positional behaviors could benefit from such an arrangement, including vertical clinging, leaping, climbing, and hindlimb suspension, etc. However, these behaviors are frequently used by many of the callitrichines which lack a relatively elongated trochlea. Conversely, the marked diversity of predominant positional behaviors employed by the non-callitrichids in the sample preclude the identification of any strict behavioral correlate to an elongated astragalar trochlea, or an elongated astragalus. This feature may better represent a phylogenetic character rather than a strong functional trait. Given the distribution of a relatively long astragalus and astragalar trochlea in *Callimico* and the non-callitrichid platyrrhines, it is suggested here that a relatively long astragalus (and trochlea) characterized the last common ancestor of callitrichines and non-callitrichid platyrrhines. This feature is retained in *Callimico* and the non-callitrichid platyrrhines, and secondarily reduced among callitrichines.

The degree of astragalar trochlear wedging, a posterior narrowing of the trochlea, has been correlated with various locomotor behaviors in primates (Gebo, 1986, 1988, 1989;

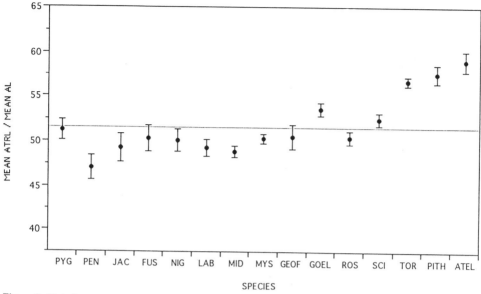

Figure 9. Plot of mean astragalar trochlear length to total astragalar length (ATRL / AL) by species. Measurement abbreviations as in Figure 1, species abbreviations as in Figure 5, key to features in the plot as in Figure 5.

Meldrum, 1990, 1993). A lower index reflects a more parallel-sided astragalar trochlea, reported by Gebo and by Meldrum as consistent with increased leaping behaviors, while a wedged trochlea is said to correlate with climbing. Second only to *Ateles geoffroyi*, *Callimico* has the highest mean wedging index (Table 4, Figs. 10 and 11). It is statistically significantly greater than all callitrichines, *Saimiri sciureus* and *Pithecia pithecia*.

Based on previous studies (see above), platyrrhines with the highest index should exhibit the greatest frequency of climbing. Using data collected by Paul A. Garber, Davis *et al.* (in prep) found that climbing increases among *Saguinus geoffroyi*, *S. fuscicollis*, and *S. mystax*, respectively, and this matches their increasing values for the trochlear wedging index reported here. Similarly, *Ateles geoffroyi* uses climbing very frequently (Cant, 1986; Mittermeier, 1978) and has the least parallel-sided trochlea of the entire sample. While these individual examples support previously held views of a strong correlation between trochlear wedging and locomotor mode, just as many examples, if not more, confound these associations. For example, *Pithecia pithecia* is a predominant leaper (75% of the time [Fleagle and Mittermeier, 1980]; and approximately 40% of all positional behaviors [Walker, 1993]), yet it has significantly greater trochlear wedging compared to that of *Saimiri sciureus* and *Saguinus geoffroyi*, both of which leap approximately 42% of the time (Fleagle and Mittermeier, 1980; Garber, 1991). In addition, *Saguinus midas* is intermediate between *S. mystax* and *S. fuscicollis* in trochlear wedging but leaps less (24%, [Fleagle and Mittermeier, 1980]) than either of the other species (31% and 33%, respectively, [Garber, 1991]). More central to the topic of this study, *Callimico* has one of the most wedged astragalar trochleae and is not significantly different in degree of wedging from a frequent climber (*Ateles geoffroyi*), nor a quadruped / leaper, *Callicebus torquatus* (Easley, 1982; Kinzey, 1976; Kinzey *et al*, 1977). It is possible that a future intensive field study of *Callimico* will reveal climbing as the predominant locomotor behavior, however inconsistent with all currently available reports on its behavior, but the wedging index still fails to predict positional behavior in many other platyrrhines.

Table 4. Mean values for astragalar trochlear wedging, distal trochlea width, and proximal trochlear width

	Wedging Index (DTW/PTW) × 100			Distal trochlear width/HHH (DTW/HHH)			Proximal trochlear width/HHH (PTW/HHH)		
	n	Mean	SD	n	Mean	SD	n	Mean	SD
Cebuella pygmaea	7	117*	9.721	6	.49*	.025	6	.42	.030
Callithrix penicillata	3	121*	2.301	3	.62	.043	3	.52*	.038
Callithrix jacchus	8	121*	3.325	8	.58	.033	8	.48	.027
Saguinus fuscicollis	7	118*	6.625	6	.57	.036	6	.48	.021
Saguinus nigricollis	6	123*	5.487	6	.57	.042	6	.46	.031
Saguinus labiatus	5	115*	6.751	5	.60	.018	5	.52*	.033
Saguinus midas	11	121*	6.924	12	.59	.044	11	.49*	.032
Saguinus mystax	11	124*	9.266	11	.58	.035	11	.47	.031
Saguinus geoffroyi	10	113*	4.410	10	.56*	.040	10	.49*	.029
Callimico goeldii	10	145	12.685	10	.63	.034	10	.44	.035
Leontopithecus rosalia	10	125*	8.627	10	.59	.031	9	.48	.034
Saimiri sciureus	8	123*	9.311	7	.58	.030	7	.47	.023
Callicebus torquatus	3	133	8.181	2	.60	.003	2	.46	.037
Pithecia pithecia	8	129*	10.929	8	.63	.058	8	.49	.053
Ateles geoffroyi	7	147	19.455	4	.74*	.070	4	.48	.042

*Significantly different from *Callimico* at the .05 level.

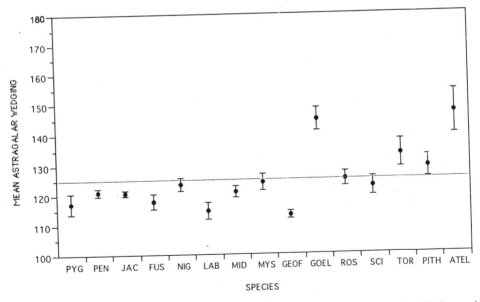

Figure 10. Plot of mean distal trochlear width to proximal trochlear width ([DTW / PTW] x 100) by species. Measurement abbreviations as in Figure 1, species abbreviations as in Figure 5, key to features in the plot as in Figure 5.

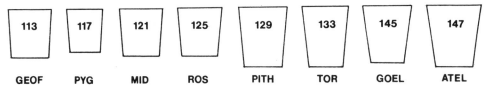

Figure 11. Schematic drawing of astragalar trochlear wedging and trochlear length of selected species. Each wedge shape is based on the following computations: DTW / HHH; PTW / HHH; ATRL / HHH.

In an attempt to clarify the significance of trochlear wedging in platyrrhines, the two measures that make up the index, proximal and distal width of the astragalar trochlea, were examined separately (Table 4). While *Callimico*, together with *P. pithecia*, and *Ateles geoffroyi*, has the widest distal trochlea, it is not significantly wider than most of the sample. The differences in the mean values for distal trochlear width, despite their lack of statistical significance, are interesting. It is possible that a widened distal astragalar trochlea is advantageous for a relatively large-sized callitrichid which frequently engages in vertical clinging with the ankle highly dorsiflexed, such as in *Callimico*. A widened distal trochlea would serve to increase the weight-transmitting surface of the ankle joint during extreme dorsiflexion. This may also explain the identically wide distal trochlea in *P. pithecia*, probably the largest platyrrhine to habitually engage in vertical clinging (Oliveira *et al.*, 1985; Walker, 1993). A widened distal trochlea may also be advantageous for stability in *A. geoffroyi*, which holds its foot in a dorsiflexed position during climbing. However, this does not consistently explain the range of values for distal trochlea width found in the sample. For example, *Callicebus torquatus* also has a relatively wide distal astragalar trochlea but climbing constitutes only 9% of its locomotor behaviors, and only 4% of its postural behaviors are in a vertical clinging position (Easley, 1982). Conversely, *Cebuella pygmaea*, a virtual expert at vertical clinging (Kinzey *et al*, 1975; Terborgh, 1985), has the narrowest distal astragalar trochlea of the entire sample.

The relatively marked wedging index in *Callimico* appears to be driven primarily by a greatly narrowed proximal trochlea. This proximal narrowing would seem to reduce the superior tibioastragalar joint surface contact area during plantarflexion, resulting in decreased stability and / or increased mobility. Analysis of anterior and posterior width of the tibial trochlea and tibial trochlear wedging revealed no significant differences between the species. Increased mobility at the proximal end of the superior tibioastragalar joint may be adaptive for the hindlimb hanging posture in which the inverted body is supported by the pedal digits alone grasping a branch. This feeding posture has been described for *Callimico* (Pook and Pook, 1981), *Saguinus fuscicollis* (Norconk, 1986), *S. geoffroyi* (Garber, 1984), *S. labiatus* (Pook and Pook, 1981; Yoneda, 1984), *S. mystax* (Castro and Soini, 1977; Norconk, 1986), *Ateles geoffroyi* (Mittermeier and Fleagle, 1976), *Pithecia pithecia* (Mittermeier, 1977), and *Saimiri sciureus* (Mittermeier, 1977). However, these species together represent almost the full range of values for proximal astragalar trochlear width.

Among platyrrhines, the proximal and distal astragalar trochlear widths and their incorporation into the wedging index do not consistently separate species with regard to positional behavior, body size, nor phylogenetic relationship. Previous platyrrhine studies similarly failed to identify a locomotor correlation with degree of trochlear wedging (Davis, 1987; Ford *et al.*, in prep). The significance of astragalar trochlear wedging in platyrrhines remains unclear at this time.

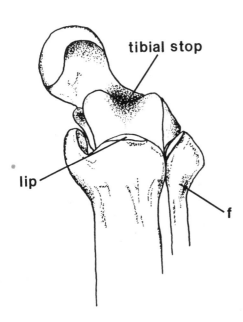

Figure 12. Superior view of right plantarflexed superior tibioastragalar joint, showing location of tibial stop on distal astragalus and location of anterior extension or '"lip" of the anterior tibial trochlear margin. "f" indicates fibula.

The tibial stop, a cup-like indentation or concavity on the distal edge of the astragalar trochlea, receives the anterior tibial trochlear margin during extreme dorsiflexion and acts to halt the distal progression of the tibia on the astragalus (Conroy, 1976; Dagosto, 1986; Ford, 1980a) (Fig. 12). This feature has previously been interpreted as an indicator of vertical clinging in primates (Fleagle and Meldrum, 1988; Meldrum, 1990). Although no species consistently lacked a tibial stop, considerable polymorphism in both presence / absence and depth of the tibial stop was found for this feature in platyrrhines (Table 5). Both the presence of a tibial stop and its depth (slight, moderate, deep) when present are most variable in *Callimico* and *Ateles geoffroyi*. In five species, the stop is invariably present and consistently moderate to deep: *Callithrix penicillata*, *C. jacchus*,

Table 5. Presence and depth of astragalar tibial stop

	N	Tibial Stop		Depth if Present		
		Absent	Present	Slight	Moderate	Deep
Cebuella pygmaea	7	3	4	4	0	0
Callithrix penicillata	2	0	2	0	0	2
Callithrix jacchus	9	0	9	0	4	5
Saguinus fuscicollis	7	0	7	2	1	4
Saguinus nigricollis	6	0	6	0	3	3
Saguinus labiatus	5	0	5	0	2	3
Saguinus midas	12	3	9	2	7	0
Saguinus mystax	12	2	10	5	5	0
Saguinus geoffroyi	8	1	7	1	6	0
Callimico goeldii	11	3	8	3	3	2
Leontopithecus rosalia	11	1	10	3	6	1
Saimiri sciureus	7	0	7	0	0	7
Callicebus torquatus	3	0	3	3	0	0
Pithecia pithecia	8	0	8	2	4	2
Ateles geoffroyi	7	3	4	2	1	1

Saguinus nigricollis, S. labiatus, and *Saimiri sciureus*. The tibial stop is also consistently present in *Pithecia pithecia* and *S. fuscicollis*, but exhibits the full range of depth in both taxa.

In some primates, an occluding facet is present on the anterior face of the anterior tibial trochlear margin (Fig 12). This facet contacts the astragalar tibial stop in extreme dorsiflexion (Conroy, 1976; Dagosto, 1986; Ford, 1980a). The apparent logic of a consistent co-distribution of these two features among platyrrhines fails to materialize morphologically. Previous platyrrhine studies have shown that the tibial facet is variable within a given genus (Davis, 1987; Ford, 1980a, 1994). Results from the present study further indicate that this feature is variably present at the species level as well. While it is not disputed that the anterior tibial trochlear margin contacts the tibial stop in extreme dorsiflexion, the presence or absence of a tibial stop and anterior extension of the tibial trochlea in the present sample are deemed too variable to be reliable indicators of a distinct posture or movement.

CONCLUSIONS

Given the strong relationship between body size and positional behavior (cf. Fleagle and Mittermeier, 1980), one would expect a relatively small-bodied platyrrhine to exhibit postures and movements characteristic of the callitrichines, as opposed to the larger atelines and pitheciines. Available field data for *Callimico* suggest that this is in fact the case. In addition, the discipline of functional anatomy predicts a reasonable correlation between form and function. In other words, a relatively small bodied platyrrhine should not only act like a callitrichine, but look like one too. However, relatively recent morphofunctional studies of primates in general, and platyrrhines specifically, have continued to serve as a reminder that morphology and behavior do not consistently form an exclusive one-to-one relationship (Davis, 1987, 1994; Davis *et al.*, in prep; Ford, 1980a, 1988, 1990; Rose, 1993a). A third critical factor in morphological analyses is that of phylogenetic heritage. The compelling relationship between size, function, and evolutionary history defies a "cook book" approach to functional anatomy. The form / function correlations that fit for one taxonomic group cannot be assumed to hold for another taxonomic group (c.f. Rose, 1993b). Accordingly, species-level analyses provide a more accurate picture of the complex morphology / behavior relationship.

Despite its callitrichid affinities and callitrichine-like size and positional behavior, *Callimico* shared several distinct features of the ankle with the larger-bodied non-callitrichid platyrrhines. In the medial tibioastragalar joint, the features of an enlarged malleolus, a distinctly bulbous malleolar surface, and an enlarged medial astragalar facet together form a morphological complex shared by *Callimico* and *Pithecia pithecia*. This complex enhances ankle stability during full dorsiflexion and is consistent with these species' habits of vertical clinging and of leaping from a vertical platform. The complex is absent from the smaller vertical clinging callitrichines, suggesting that a body size threshold of approximately 500g governs the presence of this morphological complex. As the only platyrrhines at or above this threshold who habitually engage in vertical clinging and vertical leaping, *Callimico* and *P. pithecia* have convergently acquired this complex. Given the apparent locomotor and postural similarities of *P. pithecia* and *Callimico*, future comparative analyses of other anatomical regions may identify additional convergent form / function correlations. If so, *Callimico* may prove to be key in understanding the distinctive locomotor anatomy of the fossil pitheciine, *Cebupithecia sarmientoi*. Slightly larger than *P. pithecia* in its reconstructed body size (1800 - 2100g, Gingerich *et al.*, 1982), the *Cebupithecia* skeleton exhibits a puzzling mix of pitheciine and callitrichid traits, including several that are consistent with vertical clinging and / or vertical leaping (Ciochon and Corruccini, 1975; Davis, 1987, 1988; Fleagle and Meldrum, 1988; Ford, 1980a, 1986a, 1988, 1990; Kay, 1990; Meldrum, 1993; Meldrum and Fleagle, 1988; Meldrum *et al.*, 1990; Meldrum and Kay, 1990, Meldrum and Lemelin, 1991). Indeed, with regard to the features discussed in the present study, measurements taken from casts of the tibia and astragalus indicate that *Cebupithecia* shares an enlarged tibial malleolar area ((Ã[TML x TMW]) / HHH = .446) and an enlarged medial astragalar facet area ((Ã[AMFL x AMFH]) / HHH = .451), neither of which are statistically significantly different from either *Callimico* or *P. pithecia*. Due to minor erosion of the medial articular surface of the malleolus, it is difficult to ascertain the convexity of the malleolar bulb. However, Davis (1987) noted that the corresponding cup on the medial astragalus is distinctly deep. It would appear that the fossil also exhibits the morphological complex for ankle stability during full dorsiflexion shared by *Callimico* and *P. pithecia*. An in depth comparative analysis of these three species is clearly warranted.

Callimico shares an elongated astragalar trochlea with the non-callitrichid platyrrhines. *Callimico* and the larger platyrrhines also have relatively long astragali, as do *Leontopithecus rosalia*, the *Callithrix* species and *Cebuella*. Given the lack of positional behaviors exclusively shared by these taxa, an elongated astragalus and elongated trochlea may represent traits retained from the last callitrichid / non-callitrichid platyrrhine ancestor.

This analysis of the *Callimico* ankle also examined features previously shown to correlate well with specific locomotor behaviors. However, the degree of astragalar trochlear wedging in this platyrrhine sample failed to produce a reliable correlation to any specific positional behavior with the expected consistency. While *Callimico*'s high wedging index may indicate a yet to be described propensity for climbing, the range of wedging values in the comparative sample did not consistently match the known frequency of leaping and climbing behaviors for many species. In addition, the presence of a tibial stop on the distal astragalus and an opposing facet on the anterior tibial trochlear margin were found to be highly variable in *Callimico* and the comparative sample. This polymorphism may indicate the lack of strong selection for either trait, and therefore they are poor indicators of function.

Finally, it is an important reminder that in the majority of morphological traits examined, remarkable consistency was noted between the callitrichines and *Callimico*. These results agree with previous studies documenting a distinctive suite of features in the postcranium shared by callitrichines and *Callimico* (Ford, 1980a, 1986a, 1994), and add further support for platyrrhine classification schemes which place *Callimico* within the family Callitrichidae.

SUMMARY

Due to its intriguing assortment of callitrichine and non-callitrichid platyrrhine traits, *Callimico goeldii* has figured prominently in the controversial issues of platyrrhine phylogeny and systematics. Field studies of the ecology and behavior of *Callimico* have documented a number of specific ecological and behavioral patterns that further distinguish this monkey. While our understanding of this species' ecology and behavior is incomplete, several field studies have noted its propensity for using vertical supports as platforms for leaping and vertical clinging, despite its lack of gummivory. Fewer studies of its skeletal anatomy are available. The present study examines the functional significance and phylogenetic affinities of skeletal features in the ankle region of *Callimico,* as compared to 14 other platyrrhine species.

For a more accurate appreciation of the complex relationship between morphology and behavior, analysis was conducted at the species level. The comparative sample is comprised of *Cebuella pygmaea, Callithrix jacchus, C. penicillata, Saguinus fuscicollis, S. geoffroyi,S. labiatus, S. midas, S. mystax, S. nigricollis, Leontopithecus rosalia, Saimiri sciureus, Callicebus torquatus, Pithecia pithecia,* and *Ateles geoffroyi.* A total of 53 quantitative and 10 qualitative traits of the distal tibia, astragalus, and pedal elements were examined. Both multivariate and bivariate analyses were conducted. First, principal components analysis on the 53 quantitative features was performed to grossly explore the placement of *Callimico* within Platyrrhini. To facilitate comparisons between species of different body sizes, all quantitative measures were then size-corrected. ANOVA and Tukey-Kramer tests were used to identify significant morphological differences between species and to identify which species were distinct from *Callimico* . Discussion focuses on those features in *Callimico* found to be distinct from the callitrichines. Information on the locomotor and postural behaviors of *Callimico* and of individual species in the comparative sample were taken from the literature.

Analysis of the majority of traits indicate that *Callimico* is closely aligned with the callitrichines. These results agree with previous studies documenting a distinctive suite of postcranial features shared by the callitrichines and *Callimico.* However, despite its callitrichid affinities and callitrichine-like body size and positional behavior, *Callimico* shared several features with the non-callitrichid platyrrhines. *Callimico* has a distinctly large tibial malleolar area and corresponding medial astragalar facet area. These two features, coupled with a distinctly bulbous tibial malleolar articular facet, form a morphological complex that is exclusively shared with *Pithecia pithecia.* In relatively large species, this anatomical complex facilitates increased ankle stability in a fully dorsiflexed, close-packed ankle position, and is consistent with these species' shared propensity for vertical clinging. These shared features may help in elucidating the positional adaptations of a fossil platyrrhine, *Cebupithecia sarmientoi. Callimico* is also found to have the greatest mean astragalar trochlear length of all callitrichines, and it falls within the range for the non-callitrichids in the sample. Preliminary analysis suggests that the elongated trochlea in *Callimico* represents a primitive retention. Analysis of two astragalar features, index of trochlear wedging and presence of a tibial stop distal to the trochlea, fails to confirm previous suggestions of an exclusive form / function relationship. The wedging index, predicted in previous studies to be low in leapers, was significantly high in *Callimico* and *P. pithecia*, both frequent leapers. In addition, the wedging index in *Callimico* and *P. pithecia* is not significantly different from that of *Ateles geoffroyi,* a frequent climber. The significance, if any, of trochlear wedging in platyrrhines remains unclear at this time. The presence of a tibial stop distal to the astragalar trochlea has previously been suggested to correlate with the vertical clinging posture. However, in the present study, this feature is found to be far too variable within most platyrrhine species to represent a consistent anatomical correlate to vertical clinging.

Results from this study emphasize that the relative contributions and correlations of body size, positional behavior and phylogenetic heritage must be taken into account in the functional analysis of skeletal elements. The form / function correlations that fit for one taxonomic group cannot be assumed to hold for another taxonomic group. Accordingly, species-level analyses provide a more accurate picture the complex relationship between morphology and behavior. While *Callimico* shares several morphological features with the non-callitrichid platyrrhines, overall, a remarkable morphological consistency was found between *Callimico* and the callitrichines. These results add further support for platyrrhine classification schemes which place *Callimico* within the family Callitrichidae.

ACKNOWLEDGMENTS

I wish to thank Marilyn Norconk, Paul Garber, and Alfred Rosenberger for inviting me to participate in this volume. This work has greatly benefited from the comments of Susan Ford, Marian Dagosto, Paul Garber, David Hobbs, Alfred Rosenberger, and one anonymous reviewer. I would especially like to thank Susan Ford for numerous discussions of this work. I would also like to express my appreciation to the following people who provided access to primate collections in their care: Bruce Patterson and Melissa Morales, Field Museum of Natural History; Richard Thorington, Linda Gordon, and David Schmidt, U.S. National Museum of Natural History, Smithsonian Institution; Neil Tappen, University of Wisconsin, Milwaukee; Wolfgang Fuchs, American Museum of Natural History; and Maria Rutzmoser, Museum of Comparative Zoology, Harvard University. This research was supported by NSF Dissertation Improvement Award (DBS 9203884), and grants from the Smithsonian Institution, Sigma XI, and the Graduate School, Southern Illinois University-Carbondale.

REFERENCES

Basmajian, J.V., 1982, *Primary Anatomy*, 8th Edition, Baltimore: Williams and Wilkins.
Buchanan-Smith, H., 1991, Field observations of Goeldi's monkey *Callimico goeldii*, in Northern Bolivia. Folia Primatol., 57: 102–105.
Cameron, R., Wiltshire, C., Foley, C., Dougherty, N., Aramayo, X., and Rea, L., 1989, Goeldi's monkey and other primates in Northern Bolivia, *Primate Conservation* (10): 62–70.
Cant, J.G.H., 1986, Locomotion and feeding postures of spider and howling monkeys: Field study and evolutionary interpretation, *Folia Primatol*. 46: 1–14.
Castro, R. and Soini, P., 1977, Field studies on *Saguinus mystax* and other callitrichids in Amazonian Peru. In D.G. Kleiman (ed.): *Biology and Conservation of the Callitrichidae*. Washington D.C.: Smithsonian Institution Press, pp. 102–105.
Christen, A. and Geissmann, T., 1994, A primate survey in Northern Bolivia, with special reference to Goeldi's monkey, *Callimico goeldii*, *Int. J. Primatol*. 15: 239–274.
Ciochon, R.L. and Corruccini, R.S., 1975, Morphometric analysis of platyrrhine femora with taxonomic implications and notes on two fossil forms, *J. Hum. Evol*. 4: 193–217.
Coimbra-Filho, A.F. and Magnani, A., 1972, On the present status of *Leontopithecus*, and some data about new behavioral aspects and management of *L. rosalia rosalia*. In D.D. Bridgewater (ed.), *Saving the Lion Marmoset*, Wheeling, WV: Wild Animal Propagation Trust, pp. 59–69.
Coimbra-Filho, A.F. and Mittermeier, R.A., 1977, Tree-gouging, exudate-eating, and the "short-tusked" condition in *Callithrix* and *Cebuella*. In D.G. Kleiman (ed.): *Biology and Conservation of the Callitrichidae*. Washington D.C.: Smithsonian Institution Press, pp. 105–115.
Conroy, G.C., 1976, Primate postcranial remains from the Oligocene of Egypt, *Contrib. Primatol*. 8: 1–134.
Corruccini, R.S., 1987, Shape in morphometrics: comparative analyses, *Am. J. Phys. Anthropol*. 73: 289–303.
Corruccini, R.S., 1995, Of ratios and rationality, *Am. J. Phys. Anthropol*. 96: 189–191.
Crandlemire-Sacco, J., 1986, The ecology of the saddle-backed tamarin, *Saguinus fuscicollis*, of southeastern Peru, Ph.D. dissertation, University of Pittsburgh.
Dagosto, M., 1986, The joints of the tarsus in the strepsirhine primates: Functional, adaptive, and evolutionary implications, Ph.D. dissertation, City University of New York.
Damuth, J. and MacFadden, B.J., 1990, *Body Size in Mammalian Paleobiology: Estimation and Biological Implications*, Cambridge: Cambridge University Press.

Davis, L.C., 1987, Morphological evidence of positional behavior in the hindlimb of *Cebupithecia sarmientoi* (Primates: Platyrrhini), M.A. thesis, Arizona State University.

Davis, L.C., 1988, Morphological evidence of locomotor behavior in a fossil platyrrhine, *Am. J. Phys. Anthropol.* 75: 202.

Davis, L.C., 1994, Locomotor and postural adaptations in an unusual platyrrhine, *Callimico goeldii*, *Am. J. Phys. Anthropol.* (Suppl. 18): 76–77.

Davis, L.C., Ford, S.M., and Garber, P.A., 1993, Functional anatomy and positional behavior in three *Saguinus* species, *Am. J. Phys. Anthropol.* (Suppl. 16): 78–79.

Davis, L.C., Ford, S.M., and Garber, P.A., in prep, Postcranial anatomy and positional behavior in three *Saguinus* species.

Dunlap, S.S., Thorington, R.W. Jr., and Aziz, M.A., 1985, Forelimb anatomy of New World monkeys: Myology and the interpretation of primitive anthropoid models, *Am. J. Phys. Anthropol.* 68: 499–517.

Dunnett, C.W., 1980, Pairwise multiple comparisons in the homogeneous variance, unequal sample size case, *J. Am. Statis. Assoc.* 75: 301–314.

Easley, S.P., 1982, Ecology and behavior of *Callicebus torquatus*, Cebidae, Primates, Ph.D. dissertation, Washington University, St. Louis.

Fleagle, J.G., 1977, Locomotor behavior and skeletal anatomy of sympatric leaf-monkeys (*Presbytis obscura* and *Presbytis melalophos*), *Yrbk. Phys. Anthropol.* 20: 440–453.

Fleagle, J.G., 1988, *Primate Adaptation and Evolution*, New York: Academic Press.

Fleagle, J.G. and Meldrum, D.J., 1988, Locomotor behavior and skeletal morphology of two sympatric pitheciine monkeys, *Pithecia pithecia* and *Chiropotes satanas*, *Am. J. Primatol.* 16: 227–249.

Fleagle, J.G. and Mittermeier, R.A., 1980, Locomotor behavior, body size, and comparative ecology of seven Surinam monkeys, *Am. J. Phys. Anthropol.* 52: 301–314.

Ford, S.M., 1980a, A systematic revision of the Platyrrhini based on features of the postcranium, Ph.D. dissertation, University of Pittsburgh.

Ford, S.M., 1980b, Callitrichids as phyletic dwarfs, and the place of the Callitrichidae in Platyrrhini, *Primates*, 21: 31–43.

Ford, S.M., 1986a, Systematics of New World monkeys, In D.R. Swindler and J. Erwin (eds.), *Comparative Primate Biology, Volume 1. Systematics, Evolution, and Anatomy*, New York: Alan R. Liss, pp. 73–135.

Ford, S.M., 1986b, Comment on the evolution of claw-like nails in callitrichids (marmosets/tamarins), *Am. J. Phys. Anthropol.* 70: 25–26.

Ford, S.M., 1988, Postcranial adaptations of the earliest platyrrhine, *J. Hum. Evol.* 17: 155–192.

Ford, S.M., 1990, Locomotor adaptations of fossil platyrrhines, *J. Hum. Evol.* 19: 141–173.

Ford, S.M., 1994, Primitive platyrrhines? Perspectives on anthropoid origins from platyrrhine, parapithecid, and preanthropoid postcrania, In J.G. Fleagle and R.F. Kay (eds.), *Anthropoid Origins*, New York: Plenum Press, pp. 595–673.

Ford, S.M. and Corruccini, R.S., 1985, Intraspecific, interspecific, metabolic, and phylogenetic scaling in platyrrhine primates. In W.L. Jungers (ed.), Size and Scaling in Primate Biology, New York: Plenum Press, pp. 401–435.

Ford, S.M. and Davis, L.C., 1992, Systematics and body size: implications for feeding adaptations in New World monkeys, *Am. J. Phys. Anthropol.* 88: 415–468.

Ford, S.M., Davis, L.C., and Kay, R.F., in prep, New platyrrhine astragalus from the Miocene of Colombia.

Ford, S.M. and Hobbs, D.G., 1994, Species differentiation in the postcranial skeleton of *Cebus*, *Am. J. Phys. Anthropol.* (Suppl. 18): 88.

Garber, P.A., 1984, Use of habitat and positional behavior in a Neotropical primate *Saguinus oedipus*. In P.S. Rodman and J.G.H. Cant (eds.): *Adaptations for Foraging in Nonhuman Primates*. New York: Columbia University Press, pp. 112–133.

Garber, P.A., 1988, Diet, foraging patterns, and resource defense in a mixed species troop of *Saguinus mystax* and *Saguinus fuscicollis* in Amazonian Peru, *Behaviour*, 105: 18–33.

Garber, P.A., 1991, A comparative study of positional behavior in three species of tamarin monkeys, *Primates* 32: 219–230.

Garber, P.A., 1992, Vertical clinging, small body size, and the evolution of feeding adaptations in the Callitrichinae, *Am. J. Phys. Anthropol.* 88: 469–482.

Garber, P.A., 1994, Phylogenetic approach to the study of tamarin and marmoset social systems, *Am. J. Primatol.* 34: 199–219.

Gebo, D.L., 1986, The anatomy of the prosimian foot and its application to the primate fossil record, Ph.D. dissertation, Duke University.

Gebo, D.L., 1988, Foot morphology and locomotor adaptation in Eocene primates, *Folia Primatol.* 50: 3–41.

Gebo, D.L., 1989, Locomotor and phylogenetic considerations in anthropoid evolution, *J. Hum. Evol.* 18: 201–233.

Gingerich, P.D., Smith, B.H., and Rosenberg, K., 1982, Allometric scaling in the dentition of primates and prediction of body weight from tooth size in fossils, *Am. J. Phys. Anthropol.* 58: 81–100.

Hayter, A.J., 1984, A proof of the conjecture that the Tukey-Kramer Method is conservative, Ann. Statis. 12: 61–75.

Heltne, P.G., Wojcik, J.F., and Pook, A.G., 1981, Goeldi's monkey, genus *Callimico*, In A.F. Coimbra-Filho and R.A. Mittermeier (eds.), *Ecology and Behavior of Neotropical Primates, Vol. 1*, Rio de Janeiro: Academia Brasileira de Ciencias, pp. 169–209.

Hershkovitz, P., 1977, *Living New World Monkeys (Platyrrhini), with an Introduction to Primates, Vol. 1*, Chicago: Univ. of Chicago Press.

Hill, W.C.O., 1957, *Primates: Comparative Anatomy and Taxonomy, Vol. 3: Pithecoidea, Platyrrhini (Families Hapalidae and Callimiconidae)*, Edinburgh: Edinburgh University Press.

Hill, W.C.O., 1959, The anatomy of *Callimico goeldii* (Thomas), a primitive American Primate, *Trans. Am. Phil. Soc.* 49: 1–116.

Izawa, K., 1979, Studies on the peculiar distribution pattern of *Callimico*, *Kyoto Univ. Overseas Res. Rep. of New World Monkeys* I: 1–9.

Izawa, K. and Bejarano, G., 1981, Distribution ranges and patterns of nonhuman primates in Western Pando, Bolivia, *Kyoto Univ. Overseas Res. Rep. of New World Monkeys* II: 1–11.

Izawa, K. and Yoneda, M., 1981, Habitat utilization of nonhuman primates in a forest of the Western Pando, Bolivia, *Kyoto Univ. Overseas Res. Rep. of New World Monkeys* II: 13–22.

Jungers, W.L., 1985, *Size and Scaling in Primate Biology,* New York: Plenum Press.

Kay, R.F., 1980, Platyrrhine origins: a reappraisal of the dental evidence, In R.L. Ciochon and A.B. Chiarelli (eds.), *Evolutionary Biology of the New World Monkeys and Continental Drift*, New York: Plenum Press, pp. 159–188.

Kay, R.F., 1990, The phyletic relationships of extant and fossil Pitheciinae (Platyrrhini, Anthropoidea), *J. Hum. Evol.* 19: 175–208.

Kinzey, W.G., 1976, Positional behavior and ecology in *Callicebus torquatus*, *Yrbk. Phys. Anthropol.* 20: 468–480.

Kinzey, W.G., Rosenberger, A.L., and Ramirez, M., 1975, Vertical clinging and leaping in a Neotropical anthropoid, *Nature* 255: 327–328.

Kinzey, W.G., Rosenberger, A.L., Heisler, P.S., Prowse, D.L., and Trilling, J.S., 1977, A preliminary field investigation of the yellow-handed titi monkey, *Callicebus torquatus torquatus*, in northern Peru, *Primates* 18: 159–181.

Kramer, C.Y., 1956, Extension of multiple range tests to group means with unequal numbers of replications, *Biometrics* 12: 307–310.

Maier, W. Alonso, C., and Langguth, A., 1982, Field observations on *Callithrix jacchus jacchus*, L.Z. *Saugetierkunde* 47: 334–346.

Martin, R.D., 1990, *Primate Origins and Evolution: A Phylogenetic Reconstruction*, Princeton: Princeton University Press.

Martin, R.D., 1992, Goeldi and the dwarfs: the evolutionary biology of the small New World monkeys, *J. Hum. Evol.* 22: 367–393.

Masataka, N., 1981a, A field study of the social behavior of Goeldi's monkey (*Callimico goeldii*) in Northern Bolivia. I. Group composition, breeding cycle, and infant development, *Kyoto Univ. Overseas Res. Rep. of New World Monkeys* II: 23–32.

Masataka, N., 1981b, A field study of the social behavior of Goeldi's monkey (*Callimico goeldii*) in Northern Bolivia. II. Grouping patterns and intergroup relationship, *Kyoto Univ. Overseas Res. Rep. of New World Monkeys* II: 33–41.

Meldrum, D.J., 1990, New fossil platyrrhine tali from the Early Miocene of Argentina, *Am. J. Phys. Anthropol.* 83: 403–418.

Meldrum, D.J., 1993, Postcranial adaptations and positional behavior in fossil platyrrhines, In: D.L. Gebo (ed.), *Postcranial Adaptation in Nonhuman Primates*, DeKalb: Northern Illinois Univ. Press, pp. 235–251.

Meldrum, D.J. and Fleagle, J.G., 1988, Morphological affinities of the postcranial skeleton of *Cebupithecia sarmientoi*, *Am. J. Phys. Anthropol.* 75: 249–250.

Meldrum, D.J., Fleagle, J.G., and Kay, R.F., 1990, Partial humeri of two Miocene Colombian primates, *Am. J. Phys. Anthropol.* 81: 413–422.

Meldrum, D.J. and Kay, R.F., 1990, A new partial skeleton of *Cebupithecia sarmientoi* from the Miocene of Colombia, *Am. J. Phys. Anthropol.* 81: 267.

Meldrum, D.J. and Lemelin, P., 1991, Axial skeleton of *Cebupithecia sarmientoi* (Pitheciinae, Platyrrhini) from the Middle Miocene of La Venta, Colombia, *Am. J. Primatol.* 25: 69–89.

Miller, R.G. Jr., 1981, *Simultaneous Statistical Inference*, New York: Springer-Verlag.

Mittermeier, R.A., 1977, Distribution, synecology, and conservation of Surinam monkeys, Ph.D. dissertation, Harvard University.

Mittermeier, R.A ., 1978, Locomotion and posture in *Ateles geoffroyi* and *Ateles paniscus, Folia Primatol.* 30: 161–193.

Mittermeier, R.A. and Fleagle, J.G., 1976, The locomotor and postural repertoires of *Ateles geoffroyi* and *Colobus guereza*, and a re-evaluation of the locomotor category semibrachiation, *Am. J. Phys. Anthropol.* 45:235–255.

Mittermeier, R.A. and van Roosmalen, M.G.M., 1981, Preliminary observations on habitat utilization and diet in eight Surinam monkeys, *Folia primatol.* 36: 1–39.

Moynihan, M., 1976, *The New World Primates: Adaptive Radiation and the Evolution of Social Behavior, Languages, and Intelligence*, Princeton: Princeton Univ. Press.

Norconk, M.A., 1986, Interactions between primate species in a neotropical forest: Mixed-species troops of Saguinus mystax and S. fuscicollis (Callitrichidae), Ph.D. dissertation, University of California, Los Angeles.

Oliveira, J.M.S., Lima, M.G., Bonvincino, C., Ayres, J.M., and Fleagle, J.G., 1985, Preliminary notes on the ecology and behavior of the Guianan saki (*Pithecia pithecia*, Linnaeus 1766; Cebidae, Primate), *Acta Amazonica* 15: 249–263.

Pook, A.G. and Pook, G., 1979, The conservation status of the Goeldi's monkey *Callimico goeldii* in Bolivia, *Dodo* (16): 40–45.

Pook, A.G. and Pook, G., 1981, A field study of the socioecology of the Goeldi's monkey (*Callimico goeldii*) in Northern Bolivia, *Folia Primatol.* 35: 288–312.

Pook, A.G. and Pook, G., 1982, Polyspecific associations between *Saguinus fuscicollis, Saguinus labiatus, Callimico goeldii*, and other primates in northwestern Bolivia, *Folia Primatol.* 38: 196–216.

Ramirez, M.R., Freese, C.H., and Revilla, C.J., 1977, Feeding ecology of the pygmy marmoset, *Cebuella pygmaea*, in northeast Peru. In D.G. Kleiman (ed.): *Biology and Conservation of the Callitrichidae*, Washington D.C.: Smithsonian Institution Press, pp. 91–104.

Rodman, P.S., 1979, Skeletal differentiation of *Macaca fascicularis* and *Macaca nemestrina* in relation to arboreal and terrestrial quadrupedalism, *Am. J. Phys. Anthropol.* 51: 51–62.

Rose, M.D., 1993a, Locomotor anatomy of Miocene hominoids, In D.L. Gebo (ed.), *Postcranial Adaptation in Nonhuman Primates*, DeKalb: Northern Illinois University Press, pp. 252–272.

Rose, M.D., 1993b, Functional anatomy of the elbow and forearm in primates, In D.L. Gebo (ed.), *Postcranial Adaptation in Nonhuman Primates*, DeKalb: Northern Illinois University Press, pp. 70–95.

Rose, K.D. and Fleagle, J.G., 1981, The fossil history of nonhuman primates in the Americas, In A.F. Coimbra-Filho and R.A. Mittermeier (eds.), *Ecology and Behavior of Neotropical Primates, Vol. 1*, Rio de Janeiro: Academia Brasileira de Ciencias, pp. 111–168.

Rosenberger, A.L., 1979, Phylogeny, evolution, and classification of New World monkeys (Platyrrhini, Primates), Ph.D. dissertation, C.U.N.Y.

Rosenberger, A.L., 1981, Systematics: the higher taxa, In A.F. Coimbra-Filho and R.A. Mittermeier (eds.), *Ecology and Behavior of Neotropical Primates, Vol. 1*, Rio de Janeiro: Academia Brasileira de Ciencias, pp. 9–27.

Rosenberger, A.L., 1984, Aspects of the systematics and evolution of the marmosets, In M. Thiago de Mello (ed.), *A Primatologia no Brasil*, Brasilia: Sociedade Brasileira de Primatologia, pp. 159–180.

Rylands, A.B., Coimbra-Filho, A.F., and Mittermeier, R.A., 1993, Systematics, geographic distribution, and some notes on the conservation status of the Callitrichidae, In A.B. Rylands (ed.), *Marmosets and Tamarins: Systematics, Behaviour, and Ecology*, Oxford: Oxford University Press, pp. 11–77.

SAS Institute Inc., 1989, *SAS/STAT User's Guide, Version 6, Fourth Edition, Vol. 2*, Cary, N.C.: SAS Institute Inc.

SAS Institute Inc., 1994, *JMP, Version 3.0.1: Statistics Made Visual*, Cary, N.C.: SAS Institute Inc.

Simons, E.L., 1972, *Primate Evolution: An Introduction to Man's Place in Nature*, New York: Macmillan Press.

Simpson, G.G., 1945, The principles of classification and a classification of mammals, *Bull. Am. Mus. Nat. Hist.* 85: 1–350.

Soini, P., 1987, Ecology of *Cebuella. Int. J. Primatol.* 8: 437.

Stafford, B.J., Rosenberger, A.L., and Beck, B.B., 1994, Locomotion of free-ranging golden lion tamarins (*Leontopithecus rosalia*) at the National Zoological Park, *Zoo Biol.* 13: 333–344.

Stafford *et al.*, this volume.

Szalay, F.S. and Delson, E., 1979, *Evolutionary History of the Primates*, New York: Academic Press.

Terborgh, J., 1983, *Five New World Primates: A Study in Comparative Ecology*. Princeton: Princeton University Press.

Terborgh, J., 1985, The ecology of Amazonian primates, In G.T. Prance and T.E. Lovejoy (eds.), *Amazonia*, New York: Pergamon Press, pp. 284–304.

Thomas, O., 1904, New *Callithrix, Midas, Felis, Rhipidomys*, and *Proechimys* from Brazil and Ecuador, *Ann. Mag. Nat. Hist.* 14: 188–196.

Walker, S.E., 1993, Positional adaptations and ecology of the Pitheciini, Ph.D. dissertation, City University of
    New York.
Ward, S.C. and Sussman, R.W., 1979, Correlates between locomotor anatomy and behavior in two sympatric spe-
    cies of *Lemur*, *Am. J. Phys. Anthropol.* 50: 575–590.
Yoneda, M., 1984, Comparative studies on vertical separation, foraging behavior, and travel mode of saddle-
    backed tamarins (*Saguinus fuscicollis*) and red-chested moustached tamarins (*Saguinus labiatus*) in North-
    ern Bolivia, *Primates* 25: 414–422.
Yoneda, M., 1988, Habitat utilization of six species of monkeys in Rio Duda, Colombia, *Field Studies of New
    World Monkeys, Colombia* 1: 39–45.

8

ECOLOGY OF THE "SOUTHERN" MARMOSETS
(*Callithrix aurita* AND *Callithrix flaviceps*)

How Different, How Similar?

Stephen F. Ferrari,[1] H. Kátia M. Corrêa,[2] and Paulo E. G. Coutinho[3]

[1]Departamento de Genética & Departamento de Psicologia Experimental
Universidade Federal do Pará
Caixa Postal 8607, 66.075-150 Belém - PA, Brazil
[2]Departamento de Zoologia
Universidade Federal de Minas Gerais
Belo Horizonte, Brazil
[3]Centro de Ciências Biológicas
Universidade Federal do Pará, Belém, Brazil

Endemic to the Atlantic Forest of southeastern Brazil, buffy-headed (*Callithrix flaviceps* Thomas, 1903) and buffy-tufted-ear (*Callithrix aurita* E. Geoffroy, 1812) marmosets face harsh seasonal extremes of temperature and rainfall, together possibly the most severe conditions regularly experienced by any callitrichine species in the wild.

The recently-discovered golden-handed lion tamarin, *Leontopithecus caissara*, does occur slightly further south than either of these marmosets (> 25°S: Persson and Lorini, 1993), but in areas at or near sea level with a relatively amenable climate. By contrast, *C. aurita* and *C. flaviceps* occupy altitudes of up to at least 1200 m, where winter (dry season) temperatures often fall close to, or below freezing. The winter is also characterized by a significant decline in precipitation, especially at sites located in the rainshadow of the main coastal range, where mean annual rainfall may be little more than 1000 mm.

Virtually nothing was known of the ecology of either species prior to Mittermeier et al.'s (1980) unexpected confirmation of the occurrence of *C. flaviceps* in the Brazilian state of Minas Gerais, an important expansion of its known range. Our knowledge has grown considerably since then, through both short-term studies and surveys (Torres de Assumpção, 1983; Muskin, .1984a, 1984b; Alves, 1986; Milton and Lucca, 1986; Stallings and Robinson, 1991; Diego et al., 1993; Mendes, 1993) and detailed long-term studies of both species (Ferrari, 1988; Ferrari and Diego, 1992; Corrêa, 1995; Coutinho, in prep.). Equivalent methods were followed in the latter studies, creating highly comparable data sets for the analysis of behavioral and ecological parameters.

Adaptive Radiations of Neotropical Primates
edited by Norconk *et al.* Plenum Press, New York, 1996

These data highlight a number of interesting differences and similarities, not just between these "southern" marmosets, but also in comparison with other Atlantic Forest forms. Overall, while one or two patterns have begun to appear in the growing body of data on marmoset ecology (e.g. Ferrari and Digby, 1996), even more questions have been raised, showing that our understanding of these fascinating monkeys is still far from complete.

SPECIES OR SUBSPECIES?

The status of *C. aurita* and *C. flaviceps* remained unchanged until Hershkovitz's classic revision of callitrichine taxonomy (1977), in which he recognized only a single species of Atlantic Forest marmoset, *Callithrix jacchus*, with five intergrading subspecies, including *aurita* and *flaviceps*. Their original classification as true species was subsequently upheld by de Vivo's detailed review (1988, 1991) of the genus. More recently, Coimbra-Filho (1990, 1991) has proposed that both forms be considered subspecies of *C. aurita*.

All three viewpoints are supported by equally convincing evidence (primarily morphological) and arguments, and in fact a definitive taxonomy of the Atlantic Forest marmosets has yet to be reached (Rylands et al., 1993). If and when such a classification may be available is unclear, especially as adequate data on the genetic and zoogeographic relationships between the two forms are currently lacking. Worse still, satisfactory reconstruction of the latter may now be impossible, given the present-day reduction of the Atlantic Forest of southeastern Brazil to a series of relatively small, isolated fragments (Ferrari and Mendes, 1991; Diego et al., 1993; Mendes, 1993).

The present paper follows the more traditional, and possibly more widely-accepted (Rylands et al., 1993) classification of the *aurita* and *flaviceps* forms as true species. Hopefully, continued collection of behavioral and ecological data, such as those presented here, will eventually contribute to a more definitive classification, essential for the effective conservation of these endangered primates.

STUDY SITES AND METHODS

Buffy-headed marmosets, *C. flaviceps*, were studied at the 880 ha Caratinga Biological Station in eastern Minas Gerais (19°50'S 41°50'W, altitude 318–682 m: Fig. 1) between 1985 and 1991 (Ferrari, 1988; Ferrari and Diego, 1992). In the rainshadow of the Mantiqueira mountains, whose peaks include the highest in South America south of the Amazon and west of the Andes, the climate at Caratinga is characterized by a marked dry season between April and September when monthly precipitation is frequently near zero, and minimum temperatures reach 5–10°C. Mean annual rainfall at the site is 1150 mm.

Observations of buffy-tufted-ear marmosets, *C. aurita*, have been carried out since 1991 (Corrêa, 1995; Coutinho, in prep.) in the 2584 ha Cunha Nucleus of the Serra do Mar State Park in southeastern São Paulo (23°14'-23°18'S, 45°03'-45° 05'E; altitude 1075–1200 m: Fig. 1). While wet and dry seasons coincide with those at Caratinga, the climate at Cunha, located in the coastal Serra do Mar range, is more humid (mean annual rainfall approximately 2000 mm), but also cooler, with minimum temperatures falling below 0°C.

Equivalent study methods were used throughout, beginning with the full habituation of study groups to the presence of human observers, and the individual identification of all

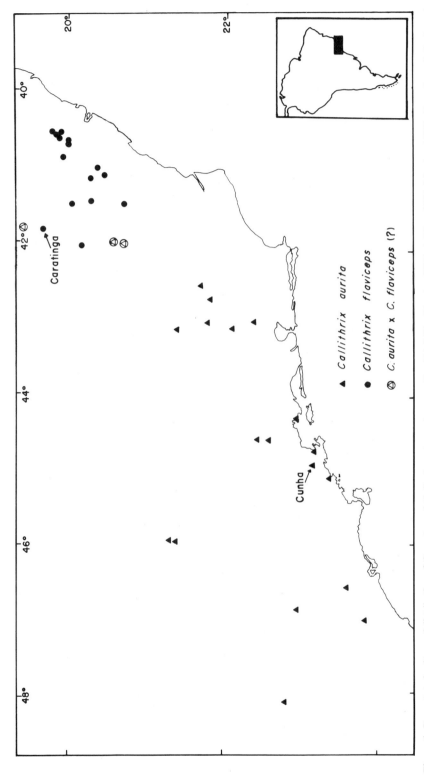

Figure 1. Present study sites and distribution of *C. aurita* and *C. flaviceps* in southeastern Brazil (cf. Hershkovitz, 1977; de Vivo, 1991; Mendes, 1993). *C. aurita* x *C. flaviceps* hybrids were identified by Mendes (1993).

group members through natural variations in pelage markings. The same scan sampling method (Altmann, 1974) and schedule was used for the collection of quantitative behavioral data in the main study periods analyzed here. One-minute scans of group members were carried out at five-minute intervals throughout the day for thirteen consecutive months (8–10 days per month) at Caratinga in 1985/86 and for a continuous seventeen-month period (5–12 days per month) at Cunha in 1992/94. The quantitative data included in the present study are from the period December 1992 to August 1993 (Corrêa, 1995).

The first activity of each group member observed during each scan was recorded, together with the animal's identity, details of support use, feeding behavior, interactions with other group members and so on. Scan sample data were complemented by the collection of both *ad libitum* observations and all-events records of categories such as scent marking, social interactions and prey feeding. At both Cunha and Caratinga, the main study groups occupied marginal areas of disturbed and/or secondary forest habitat located in the lowest part of the reserve. The use of space was recorded with the help of 50 m x 50 m trail grids. The location of the group (estimated central point) was recorded in each scan and additional records of group movements allowed detailed mapping of day ranges.

For the analysis of seasonal patterns, observation periods are divided into wet season (October to March) and dry season months (April to September), in accordance with precipitation patterns at both study sites.

GROUP COMPOSITION AND DYNAMICS

The *C. aurita* study group was smaller than that of *C. flaviceps*, but both were highly typical of *Callithrix* marmosets in terms of size, composition and stability (Table 1). As in other marmosets, two or more adults of either sex were present during at least part of the study period, and all group members participated in the rearing of offspring, irrespective of sex, age or reproductive status. Only one immigration was observed, that of an adult male *C. flaviceps*, which became the breeding male in the study group following a change in breeding female.

In both groups, recruitment through births was balanced overall by emigrations and other losses, although patterns differed. Assuming that disappearances of immature marmosets are normally due to death rather than emigration (which seems likely, given group

Table 1. Observed changes in the composition of the *C. aurita* and *C. flaviceps* study groups

	C. aurita	C. flaviceps
Total members	6-11	11-15
Adult males	1-2	3-6
Adult females	2-3	1-6
Litters/infants born	5/7	≥10/20
Immigrations	0	1
Known emigrations	0	9
Disappearances		
Adults	4	≥7
Immatures	1	2
Known deaths	1	0
Months of monitoring	24	±72

stability: Ferrari and Digby, in press), infant/juvenile mortality was three times higher in *C. aurita* (two of seven infants) than in *C. flaviceps* (two of twenty infants). This may be related, at least partly, to differences in mating systems (see below).

Similarly, assuming that most, if not all disappearances of adult *C. aurita* are attributable to emigrations, the data also suggest a contrast between the two groups in dispersal patterns. In *C. flaviceps*, the four adults emigrating during the main study period did so virtually simultaneously, whereas in *C. aurita*, the four adults dispersed singly at intervals of three to nine months, coinciding, to a certain extent, with the timing of births.

BREEDING PATTERNS

One of the most interesting contrasts between the two groups is in the number of breeding females. Never more than one of the adult females in the *C. flaviceps* study group bred at any one time between 1985 and 1991, in "typical" - up to now - callitrichine fashion. By contrast, the two adult female members of the *C. aurita* group reproduced simultaneously, a pattern also observed in *C. jacchus* (Roda, 1989; Digby and Ferrari, 1994). Polygyny has thus now been reported for half of the marmoset species for which detailed long-term field data are available. Polyandry has been reported for *Callithrix intermedia* (Rylands, 1986a) and, at least potentially, for *C. flaviceps*, while monogamous breeding has been observed in *C. jacchus* (Alonso and Langguth, 1989) and *C. flaviceps* (Ferrari and Diego, 1992). Polygyny is actually the most commonly recorded pattern in terms of the number of groups observed.

The small number of social groups and/or species studied to date may obscure possible patterns, e.g. a lack of polygyny in Amazonian marmosets (represented by a single group of *C. intermedia*), but as with other aspects of marmoset ecology, the observed variability would seem to be intrinsic to the genus rather than specific taxa. The stable polygynous associations observed in both *C. aurita* and *C. jacchus* nevertheless indicate that both the factors and the mechanisms determining reproduction in free-ranging females are more complex than was previously thought (e.g. Abbott, 1984).

Digby and Ferrari (1994) concluded that the only major ecological difference between polygynous *C. jacchus* and other marmosets was the former's high population density. Dietz and Baker (1993) concluded that similar factors may contribute to polygyny in *Leontopithecus rosalia*. At Cunha, however, the population density of *C. aurita* is even lower than that of *C. flaviceps* at Caratinga. Moreover, other aspects of the ecology of these two species are more similar to each other than either is to *C. jacchus*. Given this, an even more pertinent question might be why the *C. flaviceps* study group should have only a single breeding female?

In both studies, the minimum inter-birth interval was a little over five months, indicating postpartum ovulations. The dominant females in both groups reproduced almost continuously, the only difference being that in *C. flaviceps*, breeding was occasionally delayed for up to three months, resulting in a maximum inter-birth interval of eight months. The two births recorded for the second *C. aurita* breeding female were separated by just over twelve months, however, contributing to its lower reproductive output, as observed in subordinate *C. jacchus* females (Digby & Ferrari, 1994).

Breeding seasonality is also apparent in both species. Four of the five births (involving six of the seven infants) recorded at Cunha between May 1992 and April 1994 occurred at the beginning of the wet season (September-November), when resources - in particular insects - are most abundant. While there was also a tendency for wet season

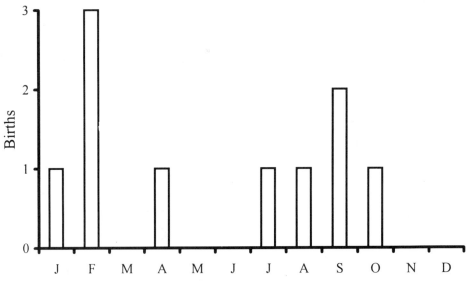

Figure 2. Timing of recorded births in the *C. flaviceps* study group (1985–1991).

births in *C. flaviceps*, their distribution was bimodal (Fig. 2). The more marked seasonality in *C. aurita* is at least partly due to both the smaller number of births recorded and the influence of the second breeding female. On the basis of the three births observed for the dominant *C. aurita* female, a bimodal distribution might also be expected for this species.

FEEDING ECOLOGY

Like other marmosets (Coimbra-Filho and Mittermeier, 1978; Ferrari et al., 1993), *C. aurita* and *C. flaviceps* are specialized morphologically for gummivory, and overall, plant exudates provided the majority of feeding records for both study groups (Table 2). The proportion of animal material in the diets of the two species was highly similar. However, *C. aurita* was more far more frugivorous, and less gummivorous, than *C. flaviceps*, and reproductive plant parts accounted for a majority of feeding records during the wet season at Cunha (Table 2).

Although *C. flaviceps* spent less time overall than *C. aurita* feeding on fruit, the study group spent 53.5% of feeding time in January consuming *Allophyllus* drupes when these were abundant at Caratinga. *C. flaviceps* also exploited the fruit of more plant species than *C. aurita* (Appendix). This suggests that overall differences in fruit feeding are related to contrasts in the abundance of this resource at the respective study sites rather than to any significant interspecific difference in feeding behavior.

Seasonal changes in the diets of both species were similar (Table 2), with the consumption of animal material and fruit declining during the dry season, reflecting the decreasing abundance of these resources (Fig. 3a&b). *C. aurita* exploited the fruit of eighteen different plant species during the four wet season months analyzed here. In contrast, only eleven species were exploited during the five dry season months. Similarly, *C. flaviceps* used twenty-two species in the wet season at Caratinga as against nine during the dry season.

Table 2. Diets of the *C. aurita* and *C. flaviceps* study groups

Species/period	Percentage of feeding records attributed to			
	Animal material	Exudates	Reproductive plant parts	N
C. aurita				
Wet season	26.83	13.80	59.37[1]	2102
Dry season	14.07	60.15	25.78[1]	3887
All records	18.55	43.88	37.57[1]	5989
C. flaviceps				
Wet season	22.96	53.80	23.24	5193
Dry season	16.14	80.30	3.56	4334
All records	19.89	65.72	14.39	9527

[1]Includes fungi.

Both species also demonstrated dietary peculiarities. *C. aurita* fed regularly on at least three different forms (as yet unidentified) of fungi throughout the study period, while *C. flaviceps* consumed the mature seeds of *Siparuna arianea* when these were available (February to April). Neither active seed predation nor such systematic fungivory has previously been recorded for callitrichines. In addition, the main source of exudates for the *C. aurita* group, *Prunus sellowii*, is a member of the Roseaceae, a family not previously recorded as a marmoset gum source. Nectar feeding was recorded for *C. flaviceps*, but not *C. aurita*.

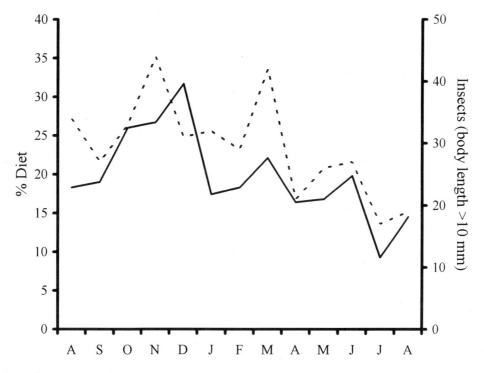

Figure 3a. Percentage of animal material in the diet of *C. flaviceps* (solid line), and number of insects (body length >10 mm) collected at Caratinga, August 1985 to August 1986 (for methods, see Ferrari, 1988).

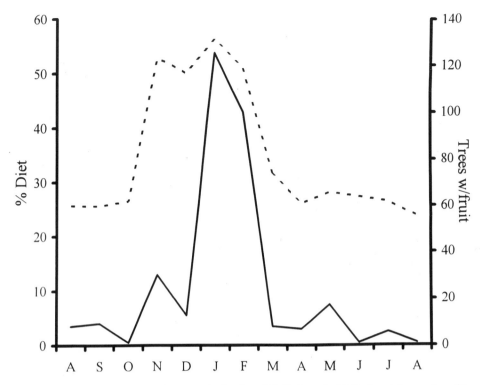

Figure 3b. Percentage reproductive plant parts in the diet of *C. flaviceps* (solid line), and the number of trees bearing fruit in phenology quadrats, August 1985 to August 1986.

Other, broader aspects of the diets of the two species were both highly similar to each other and typical of callitrichines in general. Grasshoppers (Orthoptera: Acrididae, Tetigoniidae) represented by far the major type of insect prey for both marmoset species, and a majority of gum sources were legumes (Leguminosae: Mimosoidea). The fruits exploited were mostly small, succulent drupes and berries.

Detailed identification of insect prey was limited, but most plant sources are known to genus, if not species. Here, similarities between the two marmosets break down (Appendix). Of the thirty plant families identified as food sources, only six were recorded for both marmoset species, and only two genera (*Croton* and *Inga*) were shared by *C. aurita* and *C. flaviceps*. As for the dietary "peculiarities" mentioned above, it seems likely that these contrasts are related to differences in habitat composition at the two study sites rather than to systematic interspecific differences in feeding behavior.

FORAGING BEHAVIOR

As might be expected from the similarities in their diets, the foraging behavior of the two species was also broadly similar. Basic patterns, such as the stealthy "scan-and-pounce" mode of insect predation in canopy foliage appear to be common to these and all other marmoset species (see Stevenson and Rylands, 1988), as is, of course, the systematic gouging and use of plant exudate sources.

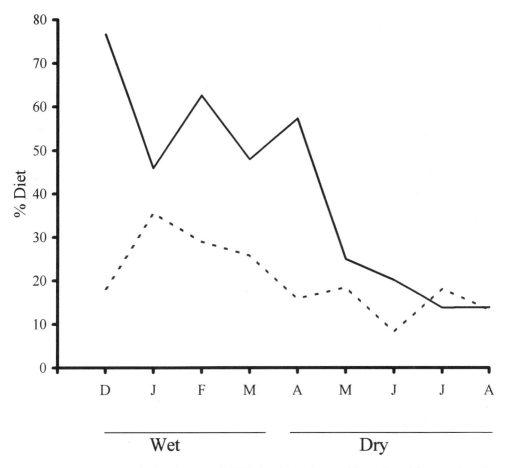

Figure 3c. Percentage reproductive plant parts (solid line) and animal material in the diet of *C. aurita*, December 1992 to August 1993.

Both *C. aurita* and *C. flaviceps* also frequently utilized gum sources created by damage other than gouging, in particular insect bore holes. These sources actually constituted a majority of the species used, but not of total feeding records. Use of these "opportunistic" sources by *C. flaviceps* nevertheless increased systematically as their abundance at the Caratinga study site increased, a seasonal pattern related to the reproductive cycle of wood-boring insects (Ferrari, 1988). These sources contributed a majority of gum feeding records in some months.

While gum was the major component of the diet overall, a second pattern common to both species was decreasing exudate feeding with increasing fruit availability. When fruit was abundant, during wet season months at both sites, it substituted gum as the principal component of the diet (Table 2). As gouging ensures an adequate supply of exudates year-round, fruit is clearly the preferred source of plant material, principally carbohydrates. The predominance of gum-feeding in both studies would thus appear to be related to the paucity of alternatives at the two sites, i.e. habitat composition and seasonality.

Seasonal patterns also marked foraging for animal material. As at most tropical sites, insect abundance reaches its peak at the beginning of the wet season in southeastern

Table 3. Foraging success of the *C. aurita* and *C. flaviceps* study groups

Species/period	Index of Foraging Success[1]
C. aurita	
Wet season	12.3 (10.3-15.2)[2]
Dry season	9.0 (8.7-10.3)[2]
All records	10.4
C. flaviceps	
Wet season	16.5 (11.8-17.2)[2]
Dry season	7.0 (3.7-9.4)[2]
All records	11.0

[1]Index of Foraging Success (IFS): IFS = (records of feeding animal material/records of foraging for animal material) × 100 (cf. Ferrari, 1991).
[2]Range of monthly values.

Brazil (Fig. 3c), and falls off significantly during the dry season. This is reflected in a decline in the efficiency of the marmosets' foraging behavior (Table 3). While the overall IFS values for the two marmoset species are remarkably similar, seasonal variation, in particular the dry season minimum, is more extreme in *C. flaviceps*, as might be expected from the much drier climate at Caratinga.

C. aurita was also observed foraging in bromeliads - a behavior more characteristic of *Leontopithecus* (Rylands, 1989) - more frequently during the dry season, accounting for 8% of foraging records in June, from a minimum of 0.4% in December. Bromeliads were scarce at Caratinga, reflecting once again differences in humidity at the two sites, and *C. flaviceps* was never observed foraging in this fashion.

Correlating with these changes was a significant decrease in the mean daily activity period, from almost eleven hours in November to less than nine in June-August. As the daily activity period was always much shorter than the period of daylight, both species appeared to compensate for decreasing prey abundance by reducing overall activity time rather than increasing foraging efforts, in typical "time-minimizing" fashion (Schoener, 1971). This strategy would be reinforced by the typical callitrichine night-time torpor (Morrison and Middleton, 1967), through which metabolic expenditure may be reduced significantly.

RANGING BEHAVIOR

The home ranges of the *C. aurita* and *C. flaviceps* study groups were virtually equal in size (35.3 ha and 35.5 ha, respectively), but while *C. aurita* actively defended a territory almost equivalent to its home range (15% overlap), that of *C. flaviceps* overlapped with its neighbors by more than 80%. Given differences in group size, this implies that mean range per individual in *C. aurita* was approximately double that for *C. flaviceps*, and, consequently, population density much lower. While population density at Caratinga appears to be of the order of 40 individuals per square kilometer, it is less than 30 at Cunha.

Mean daily paths were almost a third longer at Caratinga than at Cunha (Table 4). Given the similarity of home range size, one possible reason for this difference is that the much smaller *C. aurita* group has comparatively reduced daily foraging requirements. On the other hand, both species exhibited as much variation in daily path length within as between months, and there was no clear seasonal pattern in either case.

Table 4. Day ranging of the *C. aurita* and *C. flaviceps* study groups

Species/period	Mean daily path (m)	N
C. aurita		
Wet season	927.1 (580 - 1350)[1]	32[2]
Dry season	978.7 (595 - 1400)	54
All records	958.8	86
C. flaviceps		
Wet season	1250.6 (650 - 2200)	58
Dry season	1195.1 (755 - 2670)	67
All records	1222.5	125

[1]Range.
[2]Excludes December 1992.

Given the abundance of exudate sources at Caratinga, Ferrari (1988) concluded that the relatively large size of the home range of the *C. flaviceps* study group was related primarily to the distribution of insect prey. As potential exudate sources are also relatively abundant at Cunha, it seems reasonable to conclude that the size of the *C. aurita* home range at this site is also linked to insect availability. The extent to which local differences in the availability of insect prey may also contribute to the much higher population density and smaller home ranges of other Atlantic Forest marmosets (Lacher et al., 1984; Alonso and Langguth, 1989; Faria, 1989; Rylands, 1989; Scanlon et al., 1989; Digby, this volume) remains unclear.

Despite habitat and other ecological differences (Rylands and Faria, 1993), home range size in *C. aurita* and *C. flaviceps* is actually more similar to values recorded for the Amazonian marmosets *C. intermedia* (Rylands, 1986b) and *Callithrix argentata* (Albernaz, 1993). Albernaz's study nevertheless revealed as much variation in home range size within a single population as that found in the genus as a whole, further reinforcing the importance of local environmental factors rather than taxon-specific characteristics.

HOW DIFFERENT, HOW SIMILAR?

The present comparison of behavior patterns in free-ranging groups of *C. aurita* and *C. flaviceps* leaves little doubt that the two forms are as similar to each other behaviorally and ecologically as they are morphologically. However, this comparison appears to have done more to re-emphasize the enormous flexibility underlying the behavioral ecology of the marmosets as a whole than clarify the relationships between these two taxa in particular.

Overall, the evidence suggests that observed differences in feeding, foraging and ranging behavior are related to contrasts in habitat composition, climatic conditions, and the temporal/spatial distribution of resources at the respective study sites, although far more data will be required if the exact effects of each of these factors are to be defined. Given the stability of the marmoset resource base (Ferrari and Digby, 1996), contrasts with other Atlantic Forest marmosets in characteristics such as population density may be similarly related to local conditions rather than any significant contrasts in foraging adaptations. In addition to the collection of basic information on behavior patterns, then, future ecological studies of Atlantic Forest marmosets should also emphasize the quantification of factors such as habitat composition and resource abundance, in order to permit a more systematic evaluation of their influence on behavioral and demographic variables.

Despite the conditions encountered in the highlands of southeastern Brazil, both study groups exhibited what now seems to be a universal (or almost universal - see Scanlon et al., 1989) characteristic of marmoset social organization, that is, the marked stability of the "extended family" group (Ferrari and Digby, 1996). The occurrence of polygyny in *C. aurita*, by contrast, re-emphasizes the flexibility of the marmoset mating system, but appears to contradict population density as a possible determining factor (Dietz and Baker, 1993; Digby and Ferrari, 1994). Once again, far more data will be required before callitrichine breeding patterns can be related systematically to environmental factors.

As altitude (and concomitant conditions) appears to play a role in the distribution of both *C. aurita* and *C. flaviceps* in relation to that of neighboring forms such as *Callithrix geoffroyi* (Coimbra-Filho and Mittermeier, 1973; Ferrari and Mendes, 1991), it is possible that these marmosets have certain morphological and/or physiological specializations not found in other Atlantic Forest forms. If they do exist, however, such specializations have yet to be identified. On the basis of the behavioral data alone, the ability of the "southern" marmosets to survive under relatively harsh seasonal conditions may be based on little more than the ecological flexibility common to all *Callithrix* species.

SUMMARY

Endemic to the mountain ranges of southeastern Brazil, the "southern" marmosets, *Callithrix aurita* and *Callithrix flaviceps*, face what are probably the most extreme climatic conditions encountered by any callitrichine species in the wild. Results from two long-term field studies, presented here, show that the general characteristics of the ecology of the two species are very similar, and that most contrasts are apparently due to differences in habitat composition, conditions and group size at the two study sites rather than to interspecific differences in foraging adaptations. Despite this, a major contrast was found in breeding patterns in the two study groups. Whereas only a single female was reproductively active at any one time in the *C. flaviceps* group, two *C. aurita* females bred throughout the study period. As the *C. flaviceps* group contained as many as six adult females, reasons for this difference are unclear. Overall, this and other patterns observed in the two studies appear to re-emphasize the ecological flexibility of the marmosets as a whole rather than clarify species-specific characteristics or adaptations.

ACKNOWLEDGMENTS

Fieldwork at Caratinga and Cunha was supported by the Medical Research Council, London University Central Research Fund, the A.H. Schultz-Stiftung, the Leakey Trust and the Boise Fund, and the São Paulo State Forestry Institute and the National Research Council (CNPq) of the Brazilian government, respectively. We would like to thank Marilyn Norconk for inviting this contribution and Paul Garber and two anonymous reviewers for providing helpful comments on the text, and extend our deepest gratitude to the many people who made these studies possible, in particular Cida Lopes, Anthony Rylands, Vânia Diego and José Luíz de Carvalho.

REFERENCES

Abbott, D.H., 1984, Behavioral and physiological suppression of fertility in subordinate marmoset monkeys, *Am. J. Primatol.* 6:169–186.

Albernaz, A.L.K.M., 1993, *Área de uso de* Callithrix argentata: *habitats, disponibilidade de alimento e variação entre grupos*, M.Sc dissertation, Fundação Universidade do Amazonas/INPA, Manaus.

Alonso, C., and Langguth, A., 1989, Ecologia e comportamento de *Callithrix jacchus* (Primates: Callitrichidae) numa ilha de floresta atlântica, *Rev. Nordest. Biol.* 6:107–137.

Altmann, J., 1974, Observational study of behavior: sampling methods, *Behaviour* 49:227–267.

Alves, M.C., 1986, Observações sobre o *Callithrix flaviceps* (Thomas, 1903) na Estação Ecológica de Caratinga - EBC/FBCN, Minas Gerais (Callitrichidae, Primates), in M.T. de Mello (ed.) *A Primatologia no Brasil - 2*, Sociedade Brasileira de Primatologia, Brasília, pp. 205–206.

Coimbra-Filho, A.F., 1990, Sistemática, distribuição geogrica e situação atual dos símios brasileiros (Platyrrhini - Primates). *Rev. Brasil. Biol.* 50:1063–1079.

Coimbra-Filho, A.F., 1991, Apontamentos sobre *Callithrix aurita* (E. Geoffroy, 1812), um sagui pouco conhecido, in A.B. Rylands and A.T. Bernardes (eds.), *A Primatologia no Brasil - 3*, Sociedade Brasileira de Primatologia, Belo Horizonte, pp. 145–158.

Coimbra-Filho, A.F., and Mittermeier, R.A., 1973, New data on the taxonomy of the Brazilian marmosets of the genus *Callithrix* Erxleben, 1777, *Folia Primatol.* 20:241–264.

Coimbra-Filho, A.F., and Mittermeier, R.A., 1978, Tree-gouging, exudate-eating and the "short-tusked" condition in *Callithrix* and *Cebuella*, in D.G. Kleiman (ed.) *Biology and Conservation of the Callitrichidae*, Smithsonian Institution Press, Washington D.C., pp. 105–115.

Corrêa, H.K.M., 1995. *Observações sobre a dieta e padrão de atividade de um grupo silvestre de sagui-da-serra-escura (*Callithrix aurita *E. Geoffroy, 1812) no Parque Estadual Serra do Mar, Núcleo Cunha, São Paulo*, MSc dissertation, Universidade Federal de Minas Gerais, Belo Horizonte.

Coutinho, P.E.G., In Preparation, Comportamento reprodutivo de Callithrix aurita (Platyrrhini, Primates) no Parque Estadual Serra do Mar, Ndcleo Cunha, Sno Paulo, MSc dissertation, Universidade Federal do Par<, BelJm.

Diego, V.H., Ferrari, S.F., and Mendes, F.D.C., 1993, Conservação do sagüi-da-serra (*Callithrix flaviceps*) o papel de matas particulares, in M.E. Yamamoto and M.B.C. de Sousa (eds.), *A Primatologia no Brasil - 4*, Sociedade Brasileira de Primatologia, Natal, pp. 129–137.

Dietz, J.M., and Baker, A.J., 1993, Polygyny and female reproductive success in golden lion tamarins, *Leontopithecus rosalia, Anim. Behav.* 46:1067–1078.

Digby, L.J., and Ferrari, S.F., 1994, Multiple breeding females in free-ranging groups of *Callithrix jacchus, Int. J. Primatol.* 15:389–397.

Faria, D.S., 1989, O estudo de campo com o mico-estrela no planalto central brasileiro, in C. Ades (ed.), *Etologia de Animais e de Homens*, EDICON/EDUSP, São Paulo, pp. 109–121.

Ferrari, S.F., 1988, *The behaviour and ecology of the buffy-headed marmoset,* Callithrix flaviceps *(O. Thomas, 1903)*, Ph.D thesis, University College London.

Ferrari, S.F., and Diego, V.H., 1992, Long-term changes in a wild marmoset group, *Folia Primatol.* 58:215–218.

Ferrari, S.F., and Digby, L.J., 1996, Wild *Callithrix* groups: stable extended families? *Am. J. Primatol.* 38:19–28

Ferrari, S.F., Lopes, M.A., and Krause, E.A.K., 1993, Gut morphology of *Callithrix nigriceps* and *Saguinus labiatus* from western Brazilian Amazonia, *Am. J. Phys. Anthropol.* 90:487–493.

Ferrari, S.F., and Mendes, S.L., 1991, Buffy-headed marmosets 10 years on, *Oryx* 25:105–109.

Hershkovitz, P., 1977, *Living New World Monkeys (Platyrrhini) with an Introduction to Primates. Vol. 1.*, University of Chicago Press, Chicago.

Lacher, T. Jr., Fonseca, G.A.B., Alves, C. Jr., and Magalhães-Castro, B., 1984, Parasitism of trees by marmosets in a central Brazilian gallery forest. *Biotropica* 16: 202–209.

Mendes, S.L., 1993, Distribuição geogrica e estado de conservação de *Callithrix flaviceps* (Primates: Callitrichidae), in M.E. Yamamoto and M.B.C. de Sousa (eds.), *A Primatologia no Brasil - 4*, Sociedade Brasileira de Primatologia, Natal, pp. 139–154.

Milton, K., and Lucca, C., 1986, Population estimate for *Brachyteles* at Fazenda Barreiro Rico. *IUCN/SSC Primate Spec. Group Newsl.* 4:27–28.

Mittermeier, R.A., Coimbra-Filho, A.F., and Constable, I.D., 1980, Range extension for an endangered marmoset, *Oryx* 15:380–383.

Morrison, P., and Middleton, E.H., 1967, Body temperaure and metabolism in the pygmy marmoset, *Folia Primatol.* 6:70–82.

Muskin, A., 1984a, Preliminary field observations of *Callithrix aurita* (Callitrichinae, Cebidae), in M.T. de Mello (ed.), *A Primatologia no Brasil*, Sociedade Brasileira de Primatologia, Brasília, pp. 79–82.

Muskin, A., 1984b, Field notes and geographic distribution of *Callithrix aurita* in eastern Brazil, *Am. J. Primatol.* 7:377–380.

Persson, V.G., and Lorini, M.L. 1993, Notas sbre o mico-leão-de-cara-preta, *Leontopithecus caissara* Lorini and Persson, 1990, no sul do Brasil (Primates, Callitrichidae), in M.E. Yamamoto and M.B.C. de Sousa (eds.), *A Primatologia no Brasil - 4*, Sociedade Brasileira de Primatologia, Natal, pp. 169–181.

Roda, S.A., 1989, Ocorrência de duas fêmeas reprodutivas em grupos selvagens de *Callithrix jacchus* (Primates, Callitrichidae), in M.L. Cristofferson and D.S. Amorim (eds.), *Resumos do XVI° Congresso Brasileiro de Zoologia*, Universidade Federal da Paraíba, João Pessoa, p. 122.

Rylands, A.B., 1986a, Infant carrying in a wild marmoset group *Callithrix humeralifer*: evidence for a polyandrous mating system, in M.T. de Mello (ed.), *A Primatologia no Brasil - 2*, Sociedade Brasileira de Primatologia, Brasília, pp. 131–144.

Rylands, A.B., 1986b, Ranging behaviour and habitat preference in a wild marmoset group, *Callithrix humeralifer* (Callitrichidae, Primates), *J. Zool., Lond.* 210: 489–514.

Rylands, A.B., 1989, Sympatric Brazilian callitrichids: the black tufted-ear marmoset, *Callithrix kuhli*, and the golden-headed lion tamarin, *Leontopithecus chrysomelas*, *J. Hum. Evol.* 18:679–695.

Rylands, A.B., Coimbra-Filho, A.F., and Mittermeier, R.A., 1993, Systematics, geographic distribution, and some notes on the conservation status of the Callitrichidae, in A.B. Rylands (ed.), *Marmosets and Tamarins: Systematics, Behaviour, and Ecology*, Oxford University Press, Oxford, pp. 11–77.

Rylands, A.B., and Faria, D.S., 1993, Habitats, feeding ecology, and home range size in the genus *Callithrix*, in A.B. Rylands (ed.), *Marmosets and Tamarins: Systematics, Behaviour, and Ecology*, Oxford University Press, Oxford, pp. 262–272.

Scanlon, C.E., Chalmers, N.R., and da Cruz, M.A.O.M., 1989, Home range use and the exploitation of gum in the marmoset *Callithrix jacchus jacchus*. *Int. J. Primatol.* 19:123–136.

Schoener, T.W., 1971, Theory of feeding strategies. *Ann. Rev. Ecol. Syst.* 2:369–404.

Stallings, J.R., and Robinson, J.G., 1991, Disturbance, forest heterogeneity and primate communities in a Brazilian Atlantic Forest state park, A.B. Rylands and A.T. Bernardes (eds.), *A Primatologia no Brasil - 3*, Sociedade Brasileira de Primatologia, Belo Horizonte, pp. 357–368.

Stevenson, M.F., and Rylands, A.B., 1988, The marmosets, genus *Callithrix*, in R.A. Mittermeier, A.B. Rylands, A.F. Coimbra-Filho and G.A.B. da Fonseca (eds.), *Ecology and Behavior of Neotropical Primates, vol. 2.*, World Wildlife Fund, Washington D.C., pp. 131–222.

Torres de Assumpção, C., 1983, *An ecological study of primates in southern Brazil, with a reappraisal of* Cebus apella *races*, Ph.D thesis, University of Edinburgh.

de Vivo, M., 1988, *Sistemática de* Callithrix *Erxleben, 1777 (Callitrichidae, Primates)*, Ph.D thesis, University of São Paulo.

de Vivo, M., 1991, *A Taxonomia de* Callithrix *Erxleben, 1777 (Callitrichidae, Primates)*, Fundação Biodiversitas, Belo Horizonte.

APPENDIX

A plant species exploited by *C. aurita* and *C. flaviceps*

Family	Species	Habitus	Part used	Marmoset
Annonaceae	*Rollinia sylvatica*	Tree	Fruit	*C. aurita*
Araceae	*Anthurium* sp.	Epiphyte	Fruit	*C. aurita*
Boraginaceae	*Tournefortia bicolor*	Vine	Fruit	*C. flaviceps*
Bromeliaceae	*Aechmea* sp.	Epiphyte	Fruit	*C. aurita*
Cactaceae	*Hatiora* sp.	Epiphyte	Fruit	*C. aurita*
	Rhipsalis spp. (4)[1]	Epiphyte	Fruit	*C. aurita*
Chrysobalanaceae	*Hirtella sprucei*	Tree	Fruit	*C. flaviceps*
Combretaceae	*Terminalia* sp.	Tree	Gum	*C. aurita*
Eleocarpaceae	*Sloanea stipitata*	Tree	Gum	*C. flaviceps*
Erythroxylaceae	*Erythroxylon subracemosum*	Shrub	Fruit	*C. flaviceps*
Euphorbiaceae	*Croton* sp.	Tree	Gum	*C. aurita*
	Croton sp.	Tree	Gum	*C. flaviceps*
	Mabea fistulifera	Tree	Nectar	*C. flaviceps*
Flacourtiaceae	*Casearia decandra*	Tree	Fruit	*C. aurita*
Gesneriaceae	Undet.	Tree	Fruit	*C. aurita*
Guttiferae	Undet.	Tree	Fruit	*C. flaviceps*

Family	Species	Habitus	Part used	Marmoset
Leguminosae				
(Faboidea)	*Dalbergia nigra*	Tree	Gum	*C. flaviceps*
	Dalbergia sp.	Tree	Gum	*C. flaviceps*
(Mimosoidea)	*Acacia paniculata*	Vine	Gum	*C. flaviceps*
	Anadenanthera peregrina	Tree	Gum	*C. flaviceps*
	Inga barbata	Tree	Gum/Fruit	*C. aurita*
	Inga marginata	Tree	Gum/Fruit	*C. aurita*
	Inga sellowiana	Tree	Fruit	*C. aurita*
	Inga sessilis	Tree	Gum	*C. aurita*
	Inga sp.	Tree	Gum	*C. flaviceps*
	Inga sp.	Tree	Fruit	*C. flaviceps*
	Piptadenia gonocanthus	Tree	Gum	*C. flaviceps*
	Undet. (3)[1]		Gum	*C. flaviceps*
Loranthaceae	*Phoradendron* sp.	Epiphyte	Fruit	*C. aurita*
Melastomataceae	*Miconia fasciculata*	Tree	Fruit	*C. aurita*
	Undet. (2)[1]	Tree	Fruit	*C. flaviceps*
Monimiaceae	*Siparuna arianea*	Tree	Seeds	*C. flaviceps*
Moraceae	*Acantinophyllum ilicifolia*	Tree	Fruit	*C. flaviceps*
	Ficus enormis	Tree	Fruit	*C. aurita*
	Sorocea guilleminiana	Tree	Fruit	*C. flaviceps*
Myrtaceae	*Eugenia* sp.	Tree	Fruit	*C. aurita*
	Gomidesia sp.	Tree	Fruit	*C. aurita*
	Myrcia cf. *fallax*	Tree	Fruit	*C. aurita*
	Undet. (3)[1]	Tree	Fruit	*C. flaviceps*
Nyctiginaceae	*Bougainvillea spectabilis*	Vine	Gum	*C. flaviceps*
Palmae	Undet.	Tree	Fruit	*C. flaviceps*
Rosaceae	*Prunus sellowii*	Vine	Gum	*C. aurita*
Rubiaceae	*Alseis* sp.	Tree	Gum	*C. flaviceps*
	Coffea arabica	Tree	Fruit	*C. flaviceps*
	Coussarea sp.	Tree	Fruit	*C. flaviceps*
Rutaceae	*Zanthoxylum* sp.	Tree	Gum	*C. flaviceps*
Sapindaceae	*Paullinia carpopodia*	Tree	Gum	*C. aurita*
	Allophyllus sp.	Tree	Fruit	*C. flaviceps*
Sapotaceae	*Pouteria* sp.	Tree	Fruit	*C. flaviceps*
Simaroubaceae	*Picramnia* cf *glaziviana*	Tree	Fruit	*C. aurita*
Solanaceae	*Solanum* cf *excelsum*	Tree	Fruit	*C. aurita*
Theophrastaceae	*Clavija spinosa*	Shrub	Fruit	*C. flaviceps*
Vochysiaceae	*Qualea* sp.	Tree	Gum	*C. aurita*
	Vochysia magnifica	Tree	Gum	*C. aurita*
Undet.	Undet. (2)[1]	Tree	Fruit	*C. flaviceps*
Undet.	Undet.	Vine	Fruit	*C. flaviceps*
Undet.	Undet.	Shrub	Fruit	*C. flaviceps*
Undet.	Undet. (3)[1]		Fruit	*C. flaviceps*
Undet.	Undet.	Tree	Gum	*C. aurita*
Undet.	Undet.		Gum	*C. flaviceps*

[1]Number of species.

9

ACTIVITY AND RANGING PATTERNS IN COMMON MARMOSETS (*Callithrix jacchus*)

Implications for Reproductive Strategies

Leslie J. Digby[1] and Claudio E. Barreto[2]

[1]Department of Biological Anthropology and Anatomy
Duke University
Durham, North Carolina 27708
[2]Setor de Psicobiologia
Universidade Federal do Rio Grande do Norte
Natal, RN 59072-970, Brazil

INTRODUCTION

The geographic range of the marmosets and tamarins (tribe Callitrichini; *sensu* Rosenberger et al. 1990) spans from Panama to northern Bolivia and throughout much of Colombia, Peru, Ecuador, and Brazil (Rylands et al., 1993; Hershkovitz, 1977). In these areas they exist in a variety of habitats including the primary rain forests of the Amazon basin, the Atlantic forests of southeastern Brazil, and the xeric thorn scrub (*caatinga*) and savannah forests (*cerrado*) of northeastern Brazil (Rylands, 1993; Rylands & de Faria, 1993). In these and other habitats, field studies have documented considerable variability in the social organization and ecology of callitrichin species. It has been hypothesized that some aspects of callitrichin social organization, in particular mating and infant care patterns, are related to the high cost of rearing infants in these species (see below). Few studies have attempted to quantify these costs under natural conditions (Goldizen, 1987; see also Tardif, 1997), and it remains unclear whether variation in ecological factors may result in species differences in the costs of infant care.

Reproduction in callitrichin species is believed to require a heavy energy expenditure by breeding females. For instance, females typically give birth to twin infants that, together, are heavy relative to maternal weight (Leutenegger, 1973, 1979; Kleiman, 1977). In addition, female marmosets and tamarins lack a post-partum anovulatory period and can become pregnant as soon as ten days after giving birth (Hearn, 1983; note that under natural conditions, biannual births are more common in *Callithrix* species), thus breeding females may be simultaneously lactating and pregnant. Goldizen (1987) suggested that the resulting energetic costs make it unlikely that a female could successfully raise twin infants alone (see also Garber et al., 1984; Terborgh & Goldizen, 1985; Sussman & Garber, 1987; Ferrari, 1992; Price, 1992a). Instead, a system of cooperative infant care distributes rearing responsibilities among group members (reviewed in Tardif, 1997). Participation in the care of young infants (primarily carrying) is expected to be costly for helpers, but we also need to take into account the possibility that individuals may adjust their behaviors in ways that minimize the energetic costs of helping behaviors (Goldizen, 1987; Price, 1992a; Tardif, 1997).

The home range size and daily path length of a group may put constraints on the ability of a helper to reduce the energetic costs of infant transport. In order to try and quantify the interactions between ecological variables and the costs of infant care we present new data on the ranging and activity patterns of three groups of common marmosets. Ideally, we would want to compare data taken from carriers and non-carriers recorded during the same time period. Unfortunately these data are not yet available. Instead we present comparisons of activity and ranging patterns from periods when there were and were not infants less than 2 months old in a group. We then compare our data on activity and ranging patterns with similar data from other marmoset and tamarin populations. Our aim is to provide a preliminary examination of the interactions between ecological variables and the costs of infant care, and to explore the ways in which these differences may influence the reproductive strategies of individuals in various callitrichin populations.

METHODS

Study Site and Animals

We conducted this study at EFLEX-IBAMA, an experimental forestry station run by the Brazilian Institute for the Environment (IBAMA) in the municipality of Nísia Floresta in the northeastern Brazilian state of Rio Grande do Norte (06° 05'S, 35° 12'W). The station encompasses 154 ha, including an approximately 70 ha reserve of semi-deciduous Atlantic forest. Surrounding the reserve are plantations consisting primarily of coconut (*Cocos nucifera*) and eucalyptus (*Eucalyptus citriodora*). Common marmosets, *Callithrix jacchus*, inhabit both forests and plantations and are the sole primate species found in the reserve and surrounding areas.

We focused on three marmoset groups. Group A inhabited a plantation area which included blocks of mahogany (*Swietenia mahogoni*), eucalyptus, coconut, "azeitona" (*Syzygium jambolanum*), and assorted fruit and nut trees including mango (*Mangifera indica*) and cashew (*Anacardium occidentale*). Groups B and C resided in adjacent ranges in the forest reserve. Group B was known to cross occasionally into a neighboring coconut plantation that also contained several mango and cashew trees. Group C was never observed outside the forest reserve.

Following an initial period of habituation and general monitoring, we captured individuals from all three groups using manually operated live traps baited with fruit. Animals were lightly sedated with ketamine hydrochloride (< 5 mg/kg intramuscularly), and those above 200 grams were fitted with ball-chain collars strung with color-coded plastic beads. To identify smaller animals, we clipped the tail fur in unique patterns.

During the course of the study, group compositions changed due to births, emigrations, or disappearances. No immigrations were observed. Group A (monitored for 18 months) was the smallest of the three groups (5 to 9 individuals) and contained a single adult male and two adult females for the majority of the study period (Table 1). Groups B

Table 1. Group compositions

	Group A	Group B	Group C
Study Period (mo)	18	12	12
Group Size	5-9	11-15	10-14
Adult Females	2-3	4-6	3-4
Adult Males	1-2	2-5	3-4

and C (monitored for 12 months) were equivalent in size (10 to 15 individuals), and contained multiple adult males and females throughout the study (Digby & Barreto, 1993). All three groups contained two reproductively active females (Digby & Ferrari, 1994).

Thirteen births occurred between January 1991 and June 1992. Systematic data on infant care were recorded for 9 of these litters (6 twins, 3 singletons), with 3 litters in each of the three study groups. Births occurred throughout the year (details in Digby, 1995a). In order to control for possible seasonal effects, the core study period (September 1991 to June 1992) was broken down into 5 "dry" months (September through January; range 9.7 to 87.3 mm rain/month) and 5 "wet" months (February through June; range 107.2 to 441 mm rain/month). Four litters were born during the dry months and another 5 (including one infant born in late January) during the wet months.

Activity Patterns

Systematic behavioral observations were collected during the core study period (10 months). We used two different data collection protocols depending on the presence or absence of young infants in the group. During periods when there were no infants less than 2 months old, we conducted focal animal sampling of adult group members on 3 to 5 complete days (sleep-tree to sleep-tree) each month. Focal animal samples follow those used by Goldizen (1987). Using only animals that were adults at the beginning of the study, we followed focal animals for 30 minute time blocks during which we recorded their activity and location (see below) at 2 minute intervals. Records of activity were broken down into four general categories: rest (inactivity), travel (locomotion), forage/feed (including scanning for insects, manipulating food, and opening gouge holes in addition to the consumption of foods), and social behavior (primarily grooming but also including play, chase, scent marking, etc.; behaviors are defined in Digby, 1995b). The sampling order of focal animals was changed each day, and care was taken to insure that focal samples for each individual were evenly distributed throughout the day and the month. Intervals of at least 90 minutes separated repeated observations of a focal individual. We completed a total of 452 focal samples for Group A, 364 samples for Group B, and 425 samples for Group C.

During periods when there were infants less than 2 months old, we focused behavioral observations on infant care, and replaced focal animal samples with group scan samples [note: infant carrying declines rapidly after the first 8 weeks, and infants are rarely carried following week 12]. Following a birth, we conducted all day group follows (sleep-tree to sleep-tree) for 2 consecutive days at 10 day intervals until the infants were approximately 2 months old (a total of 7 two-day periods). At 5 minute intervals throughout the observation day, we scanned the group (for a maximum of 1 minute) until the infants had been located. The activity of the infant (on or off a carrier) was noted, and, if carried, the identity and activity of the carrier were also recorded.

We calculated activity budgets (the proportion of records in each of the four activity categories) for each adult in the three groups. Statistical comparisons used the Wilcoxon

Matched Pairs test. For the core study period and the "dry" months comparisons, there were a total of 19 adults for which data for both the "no infants" and "infant carrying" conditions were available. There were 13 adults for which data were available for the "wet" months comparison due to emigrations, disappearances, and a few individuals (from Group C) that did not participate in the care of infants born during these months (in part due to early infant deaths in this group; see Digby, 1995a). Possible biases in the results due to the two data collection protocols will be discussed below.

We calculated the daily activity periods for all three groups by noting the time between when the first animal left the night nest each morning until the last animal had entered the night nest in the evening. Only those days for which night nests were confirmed the next morning are included.

Ranging Patterns

Home Range. Areas used by the three marmoset groups were broken down into 25 m x 25 m quadrants. We recorded all quadrants entered by the animals during both focal and scan sampling. We estimated home ranges for each group by calculating the number of quadrants entered by at least one group member per day and over the core study period. Note that this procedure is likely to result in slight overestimates in home ranges as the proportions of edge quadrants used were not determined. We estimated overlap between groups by calculating the proportion of quadrants entered by members of more than one group over the core study period.

Path Length. Exact measurements of daily path length are not available for these groups. We have estimated path length by summing the number of times the group crossed quadrant boundaries and multiplying by the length of the quadrant (25 m). We took care not to include crossings where animals entered a quadrant and immediately crossed back into the previous quadrant. Overestimates of path length (animals crossing between quadrants in such a way that less than 25 m was covered) should be roughly balanced by underestimates (animals circling within a quadrant or crossing diagonally). Again, path lengths should be considered rough estimates, and are only intended to provide data for general comparisons between overall group and infant daily path lengths. Statistical comparisons for both day range and daily path length used the Mann-Whitney U Test.

RESULTS

Activity Patterns

Activity budgets were generally similar between the three groups and between individuals despite differences in habitat types and group size. One exception to this was a smaller proportion of records indicating foraging and feeding in the plantation group (Group A; 35%) compared to groups in the forest reserve (Groups B and C; 41% and 48%)(Mann Whitney; $p = .03$). There were no significant differences in the proportion of records indicating rest, travel, or social behaviors between the plantation and forest groups.

During periods when there were no young infants in the group, the majority of an adult's time was spent foraging and feeding (an average of 43% of records/individual; Groups A, B and C combined). Gum feeding accounted for 75% of plant feeding records

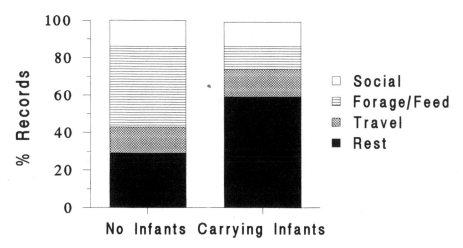

Figure 1. Activity budgets of adult animals during periods when their were no young infants in the group and while carrying infants (Groups A, B & C combined). Data cover the core study period (dry and wet months combined).

(10% of the overall activity budget). Resting accounted for an average of 30% of activity records, and both travel and social behaviors each accounted for an average of 14% of records (Figure 1). In contrast, adults who were carrying infants had significantly more resting records (59%) and less foraging and feeding records (12%)(Wilcoxon Matched Pairs; p < .01). There was no significant change in the proportion of records during the two periods for travel (15%) or social behaviors (13%)(Figure 1).

Comparisons between wet and dry months showed a significant decrease in the proportion of resting records (Wilcoxan Matched Pairs p < .10), but no significant differences for the proportion of travel, forage and feeding, or social behavior records. Within both dry and wet periods, results were the same as those above. The proportion of resting records was still significantly higher for adult infant carriers than for adults during periods with no young infants in the group (dry: .31 vs. .63; wet: .23 vs. .54; both p < .01). Foraging and feeding records again showed a significant decrease between the two conditions (dry: .42 vs. .11; wet: .45 vs. .12; both p < .01). As above, there were no significant differences in travel or social behaviors.

Differences in data collection protocol between the two periods may have resulted in some biases in the proportions of behaviors recorded, but scan sampling is typically associated with biases toward more active behaviors (e.g. travel) which tend to catch the observers eye while scanning (Fragaszy et al., 1992). The change in the proportion of resting records was opposite that expected from a procedural bias, making it unlikely that the difference was due to the change in data collection technique. In addition, the proportion of resting records also increased along with infant age (and thus weight) during the two month period of the scan samples, independent from any change in sampling protocol (Figure 2). As with the above comparisons, the increase in resting is compensated primarily with a decrease in foraging and feeding with little change to travel and social behaviors.

The average daily period of activity was similar for all three groups (11 hours 40 minutes to 11 hours 48 minutes) (Table 2). Variation in monthly means (min = 11 hours 15 minutes, max = 12 hours 02 minutes) was associated with changes in day length, rather than the presence or absence of young infants in the group.

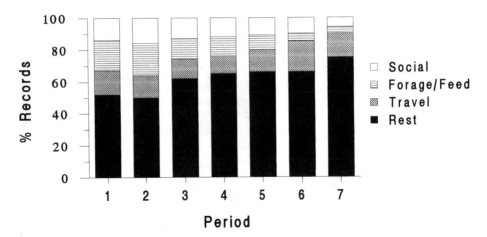

Figure 2. Activity budgets for all infant carriers (Groups A, B & C combined) over time. Periods represent two-day samples taken at 10-day intervals beginning the first few days following a birth (Period 1) and ended when the infant is approximately 2 months old (Period 7).

Ranging Patterns

Home Range. Home range over the core study period (10 months) for Group A was 5.2 ha, and day ranges varied from 0.6 to 2.8 ha (avg = 1.2 ha) (Table 2). Groups B and C from the forest reserve had somewhat smaller overall home ranges (B = 4.6 ha; C = 3.9 ha), but day ranges were comparable to those of Group A (Table 2).

Day ranges recorded for infants and their carriers tended to be smaller than the ranges recorded for the group as a whole when there were no young infants present (Figure 3). In Group A, day ranges averaged 1.4 ha/day during periods when there were no young infants present compared 1.1 ha for infants and their carriers. Group B's day ranges averaged 1.8 ha compared to 1.5 ha for infants and carriers, and Group C's day ranges averaged 1.6 ha compared to 1.3 ha for infants and carriers. Only Group B showed a significant difference between these two conditions (Mann-Whitney; $p < .01$).

Table 2. Ecological characteristics of the study groups

	Group A	Group B	Group C
Core Study Period			
Home Range (ha)	5.2	4.6	3.9
Range Overlap (%)	46	72	86
Daily Means			
Day Range (ha)	1.2	1.6	1.4
[min-max]	[0.6-2.8]	[0.9-2.4]	[1.0-2.4]
(n)	(50)	(41)	(37)
Path Length (m)	912	1181	1243
[min-max]	[525-1875]	[550-1975]	[525-1950]
(n)	(48)	(41)	(36)
Activity Period (hours)	11:45	11:48	11:45
[min-max]	[11:08-12:18]	[11:01-12:35]	[10:58-12:13]
(n)	(62)	(50)	(50)

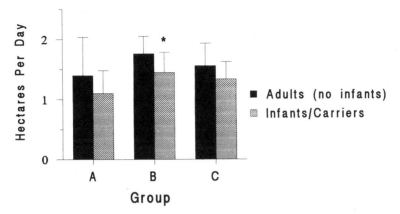

Figure 3. Day ranges (hectares used per day) for (1) the group as a whole during periods when there were no young infants in a group and for (2) infants and their carriers for two months following a birth (* = p < .01). Data cover the core study period (dry and wet months combined).

To control for possible seasonal differences, data were also compared between and within dry and wet months. Group A showed no significant difference in day range between wet and dry periods. Both Groups B and C did show significant differences in area use, but Group B had a decrease (dry = 1.8 ha, wet = 1.5 ha; Mann Whitney p < .02) and Group C an increase (dry = 1.4 ha, wet = 1.7 ha; p < .01) in day range. Within both dry and wet periods, again only Group B showed a significant decrease between the two conditions (dry months: p = .10; wet months: p < .01). All groups showed a trend toward a smaller area of use by infants and their carriers compared to the group as a whole during periods with no young infants present.

Path Length. The average path lengths varied between groups, but the range of variation within groups was comparable (range: 912–1243 m, avg = 1100 m) (Table 2). In all three groups estimated path lengths were significantly shorter for infants and their carriers compared to those for adults when there were no young infants in the group (Figure 4). In Group A, daily path lengths estimated for the group as a whole averaged 1192 m compared to the infant/carrier daily path length of 765 m (Mann-Whitney; p < .01). Similarly, estimated path lengths for Group B averaged 1322 m compared to 1018 m for infants/carries (p < .01), and Group C averaged 1356 m compared to 1046 m (p < .01).

Comparisons between dry and wet months produced a significant difference in daily path length in Group B only (dry = 1267 m, wet = 1091 m; Mann Whitney p < .01). Within both dry and wet periods all 3 groups again showed a significant decrease in daily path length (p < .06) for carriers and infants compared to the group as a whole, with the exception of Group B during the dry period.

DISCUSSION

Costs of Infant Care

Infant care has inherent costs. For mothers this may include the cost of both nourishing and carrying an infant either during pregnancy or following birth. The increased en-

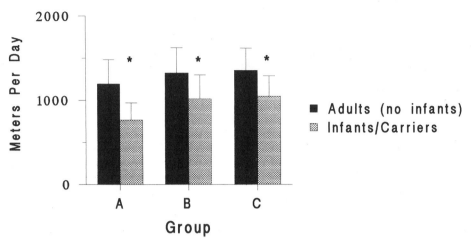

Figure 4. Daily path lengths (meters traveled per day) for (1) the group as a whole during periods when there were no young infants in a group and for (2) infants and their carriers for two months following a birth (* = p < .01). Data cover the core study period (dry and wet months combined).

ergy expenditure may necessitate greater energy input and thus an increase in the proportion of time spent feeding. Such is the case with female yellow baboons (*Papio cynocephalus*) and gelada baboons (*Theropithecus gelada*), both of which increase time spent feeding during pregnancy and while carrying their infants (Altmann, 1980; Dunbar, 1984, 1988). But these mothers are the primary caretakers of their young and are in almost constant contact with their infants for the first several weeks of life.

In species with cooperative rearing of young, the costs of infant care (other than that due to lactation) are distributed across several group members (Tardif, 1994, 1997). Also, because each caretaker spends only a portion of its time caring for infants, it can adjust its behavior such that costs are minimized. For example, in both wild dogs (*Lycaon pictus*) and dwarf mongooses (*Helogale parvula*) helpers contribute to infant care by "babysitting" (Rood, 1986; Malcolm & Marten, 1982). This behavior requires the helper to stay near the nest and be vigilant for predators. While babysitters often eat less than others in the group (Malcolm & Marten, 1982), they reduce energetic needs by engaging in relatively little activity.

In marmosets and tamarins, helping in the early weeks following a birth consists primarily of carrying infants. Transporting infants is estimated to increase the caloric expenditure of a helper by as much as 21% compared to the costs of traveling without infants (based on travel speed in natural populations and relative infant and adult weights; Tardif, in press). The results of this study suggest that marmoset helpers behave in ways that minimize this energy expenditure. Carriers spent significantly more time resting and less time foraging and feeding. Similar patterns have been reported for free-ranging saddle back tamarins (*Saguinus fuscicollis*; Goldizen, 1987) and captive cotton-top tamarins (*Saguinus oedipus*; Price, 1992b). Though the marmosets in this study showed no significant decrease in the proportion of time spent traveling, the reduction in the day ranges and path lengths of infants and their carriers do suggest that they traveled over shorter distances and stayed in more central areas as other group members foraged around them.

In addition to minimizing energetic costs, helpers may remain relatively immobile for other reasons, such as predator avoidance (Price, 1992b; Harrison & Tardif, 1994). By remaining still, the carrier maintains a greater crypticity, a practical strategy if the extra weight of the infant reduces the speed and manuverabilty of the carrier (Terborgh & Goldizen, 1985; Caine, 1987; Price, 1992b). But, an increase in time spent resting may also decrease opportunities to forage. To test if this is the case, we need to determine whether or not helpers are able to compensate for lost foraging time when they are not carrying. Ideally we would want to examine an individual's activity budget while carrying an infant compared to not carrying on a given day. Unfortunately such data are not yet available in detail (but see Baker, 1991; p. 145).

In summary, carriers in both marmosets and tamarins appear to adjust their behavior in ways that minimize energetic demands and possibly predation risk. These behaviors suggest that helping may be energetically costly, but that compensatory behaviors may result in minimal overall costs for shared infant care. The degree to which a helper can minimize costs is likely to vary between species depending on ecological variables such as range size and daily path length. Such constraints may play a role in the variability of callitrichin social organization.

Species Differences in the Cost of Infant Care

Callithrix species tend to have smaller home ranges and daily path lengths than those described for *Saguinus* species, despite the fact that they also tend to have larger group sizes (reviewed in Sussman & Kinzey, 1984; Ferrari & Lopes Ferrari, 1989; Rylands & de Faria, 1993). These differences are likely to be related to the marmoset's morphological specializations that enable them to exploit plant exudates as a major food source (Coimbra-Filho & Mittermeier, 1978; Ferrari & Lopes Ferrari, 1989; Ferrari, 1993; Harrison & Tardif, 1994). Within the *Callithrix* genus, common marmosets are an extreme with gum making up as much as 70 to 80% of the plant diet, home ranges as small as half a hectare, and groups containing up to 15 individuals (reviewed in Rylands & de Faria, 1993).

The small home ranges of the groups studied at EFLEX-IBAMA (3.9 to 5.2 ha) are consistent with those described for other common marmoset sites (.72 to 6.5 ha)(Hubrecht, 1985; Alonso & Langguth, 1989; Scanlon et al., 1989). Our estimated daily path lengths (912 to 1243 m) also fall within the range of other populations (528 to 1300 m; Hubrecht, 1985; Alonso & Langguth, 1989). Interestingly, the path length of common marmosets are not as small as might be expected given their small home range sizes (Table 3).

If the cost of carrying infants is proportional to the group's path length for a given day, then we might predict only a small reduction in these costs for common marmosets when compared to other callitrichin species. But, because common marmosets travel within a small area, an infant carrier can greatly reduce its own travel (e.g., resting in a central area) while still maintaining contact with the group. It also allows for infant "parking", where infants are left in a central area (generally with an adult "babysitter" nearby) while other group members forage around them (also seen in *Cebuella*; Soini, 1988 and some prosimians; Richard, 1987). These tactics would not be possible in other species where groups travel over a larger area, forcing infants and their carriers to move along with the group in order to avoid losing contact.

The reduction in energy expenditure associated with infant carrying is only one of several factors that may make helping less costly in common marmosets. This species also

Table 3. Ranging patterns for selected callitrichin species

Species	Home Range (ha)	Daily Path length (m)	Reference
Cebuella pygmaea	.1-.5	280-300	Soini 1988
Callithrix jacchus	.72-6.5	528-1300	Alonso & Langguth 1989, Hubrecht 1985
Callithrix kuhli	10	830-1120	Rylands 1989
Callithrix humeralifer	28.25	772-2115	Rylands 1986
Callithrix flaviceps	35.5	1222	Ferrari 1988
Saguinus fuscicollis (various subspecies)	16-40	1220-1849	Garber 1993
Saguinus mystax	40	1946	Garber 1989
Saguinus imperator	30+	1420	Terborgh 1983
Saguinus oedipus	7.8-12.4	1500-1900	Neymann 1978, Savage 1990
Leontopithecus chrysomelas	36	1410-2175	Rylands 1989
Leontopithecus rosalia	24.6-40	1339-1533	Peres 1986, Rylands 1993

has a shorter dependency period than the tamarin species to which has been compared. For example, under captive conditions, common marmoset infants are carried only 11% of the time by 8 weeks of age, compared to 60% of the time in cotton-top tamarins (Tardif et al., 1986, 1993; Harrison & Tardif, 1994). In addition, common marmosets, like other *Callithrix* species, tend have large groups, resulting in a greater number of potential helpers (Ferrari & Lopes Ferrari, 1989). Under these circumstances helping costs (including not only carrying but also the provisioning of older infants) may be distributed over a greater number of individuals, and energetic costs per helper may be reduced.

The small home range size (which may allow for infant parking and increased resting by carriers), shorter infant dependency period, and greater availability of helpers may reduce the costs of infant care in common marmosets. These factors may, in turn, play a role in the ability of breeding females in this species to give birth to twins twice a year and could increase the chances that some subordinate females can successfully raise young (see also Ferrari et al. this volume). Differences in the costs of infant care due to ecological, demographic, and morphological variables are likely to play an important role in the reproductive strategies of marmosets and tamarins. We suggest that these costs, as well as an individual's ability to reduce or compensate for them, may be a key to deciphering the causes of variability in the social organizations of callitrichin species. The comparisons of activity budgets and range use presented here should be interpreted as preliminary. Nevertheless, it is our hope that the patterns presented here will encourage future research projects to test these hypotheses with additional data from various callitrichin populations.

SUMMARY

We present data on the activity and ranging patterns of three groups of free-ranging common marmosets (*Callithrix jacchus*). During periods when there were no young infants in a group, adults spent approximately 43% of their time foraging or feeding, 30% resting, 14% traveling, and 14% engaged in social behaviors. In contrast, during the first 2 months following a birth, individuals who were carrying infants significantly increased their proportion time of resting (59%) and decreased time spent foraging and feeding (12%). The ranging patterns of these groups were comparable to those reported for other common marmosets (home range: 3.9 to 5.2 ha; path length: 912 to 1243 ha). Both day ranges and estimates of daily path length were reduced for infants and their carriers compared to those recorded for the group as a whole (during periods when there were no young infants). We suggest that changes in the activity and ranging patterns are indicative of individuals minimizing energetic expenditures while carrying infants. We also suggest that the ability to compensate for and minimize the costs of infant care is likely to be related to such factors as range size, group size and infant development. Each of these factors needs to be considered in our attempt to decipher the causes of both intra and interspecific variability in marmoset and tamarin social organization.

ACKNOWLEDGMENTS

This study was authorized by the Science Research Council (CNPq) of the Brazilian government and was supported by National Science Foundation Grant No. BNS-89–20448, the Wenner-Gren Foundation for Anthropological Research, the University of California at Davis and IBAMA. We thank Steve Ferrari, Peter Rodman, Fátima Arruda, Emilia Yamamoto, José Juarez da Costa, João Dantas, and the Universidade Federal do Rio Grande do Norte for their help in various aspects of this project. We are grateful to Dwain Santee for botanical names and to Paul Garber, Suzette Tardif, Steve Ferrari, and an anonymous reviewer for their helpful comments on earlier drafts of this paper. We would also like to thank Marilyn Norconk for inviting us to contribute to this volume.

REFERENCES

Alonso, C. & Langguth, A., 1989, Ecologia e comportamento de *Callithrix jacchus* (Callitrichidae, Primates) numa ilha de floresta Atlântica, *Rev. Nordestina Biol.* 6:105–137.
Altmann, J., 1980, *Baboon Mothers and Infants.* Cambridge: Harvard University Press.
Baker, A.J., 1991, *Evolution of the social system of the golden lion tamarin (Leontopithecus rosalia): mating system, group dynamics and cooperative breeding.* PhD Thesis, University of Maryland, College Park.
Caine, N., 1987, Vigilance, vocalizations, and cryptic behavior at retirement in captive groups of red-bellied tamarins (*Saguinus labiatus*), *Am. J. Primatol.* 12:241–250.
Coimbra-Filho, A.F. & Mittermeier, R.A., 1978, Tree gouging, exudate eating, and the "short tusked" condition in *Callithrix* and *Cebuella*. In D.G. Kleiman (ed.) *The Biology and Conservation of the Callitrichidae*. Washington D.C.: Smithsonian Institution Press, pp. 105–115.
Digby, L.J., 1995a, Infant care, infanticide, and female reproductive strategies in polygynous groups of common marmosets (*Callithrix jacchus*). *Behav. Ecol. Sociobiol.* 36:000–000.
Digby, L.J., 1995b, Social Organization in a Wild Population of *Callithrix jacchus*: II. Intragroup Social Behavior, *Primates.* 36(3):361–375.
Digby, L.J. & Barreto, C.E., 1993, Social Organization in a wild population of common marmosets (*Callithrix jacchus*). Part I: group composition and dynamics, *Folia Primatol.* 61:123–134.
Digby, L.J. & Ferrari, S.F., 1994, Multiple breeding females in free-ranging groups of *Callithrix jacchus, Int. J. Primatol.* 15(3):389–397.
Dunbar, R.I.M., 1984, *Reproductive Decisions: An Economic Analysis of Gelada Baboon Social Strategies.* Princeton: Princeton University Press.
Dunbar R.I.M., 1988, *Primate Social Systems.* New York: Comstock Publishing Associates.
Ferrari, S.F., 1988, *The Behaviour and Ecology of the Buffy-Headed Marmoset, Callithrix flaviceps (O. Thomas, 1903).* Unpublished PhD Thesis, University College London.
Ferrari, S.F., 1992, The care of infants in a wild marmoset (*Callithrix flaviceps*) group, *Amer. J. Primatol.* 26:109–118.
Ferrari, S.F., 1993, Ecological differentiation in the Callitrichidae. In: A.B. Rylands (ed.) *The Marmosets and Tamarins: Systematics, Behaviour, and Ecology.* London, Oxford University Press, pp. 314–328.

Ferrari, S.F. & Lopes Ferrari, M.A., 1989, A re-evaluation of the social organisation of the Callitrichidae, with reference to the ecological differences between genera, *Folia Primatol.* 52:132–147.

Fragaszy, D.M., Boinsky, S. & Whipple, J., 1992, Behavioral sampling in the field: comparison of individual and group sampling methods. *Amer. J. Primatol.* 26:259–275.

Garber, P.A., 1989, Role of spatial memory in primate foraging patterns: *Saguinus mystax* and *Saguinus fuscicollis*, *Am. J. Primatol.* 19:203–216.

Garber, P.A., 1993, Feeding ecology and behavior in the genus *Saguinus*. In: A.B. Rylands (ed.) *The Marmosets and Tamarins: Systematics, Behaviour, and Ecology.* London, Oxford University Press, pp. 273–295.

Garber, P.A., Moya, L. & Málaga, C., 1984, A preliminary field study of the moustached tamarin monkey (*Saguinus mystax*) in northeastern Peru: questions concerned with the evolution of a communal breeding system, *Folia Primatol.*, 42:17–32.

Goldizen, A.W., 1987, Facultative polyandry and the role of infant-carrying in wild saddle-backed tamarins (*Saguinus fuscicollis*), *Behav. Ecol. Sociobiol.* 20:99–109.

Harrison, M.L. & Tardif, S.D., 1994, Social implications of gummivory in marmosets, *Amer. J. of Phys. Anthro.* 95(4):399–408.

Hearn, J., 1983, The common marmoset (*Callithrix jacchus*). In J. Hearn (ed.) *Reproduction in New World Primates.* Lancaster, MTP Press Ltd, pp. 181–215.

Hershkovitz, P., 1977, *Living New World Monkeys (Platyrrhini).* Chicago, University of Chicago Press.

Hubrecht, R.C., 1985, Home-range size and use and territorial behavior in the common marmoset, *Callithrix jacchus jacchus*, at the Tapacura field station, Recife, Brazil, *Int. J. Primatol.* 6:533–550.

Kleiman, D.G., 1977, Monogamy in mammals, *Q. Rev. Biol.* 52:39–69.

Leutenegger, W., 1979, Evolution of litter size in primates, *Am. Nat.* 114:525–531.

Leutenegger, W., 1973, Maternal fetal weight relationships in primates, *Folia Primatol.* 20:280–293.

Malcolm, J.R. & Martin, K., 1982, Natural selection and the communal rearing of pups in African wild dogs (*Lycaon pictus*), *Behav. Ecol. Sociobiol.* 10:1–13.

Neyman, P.F., 1978, Aspects of the ecology and social organization of free-ranging cotton-top tamarins (Saguinus oedipus) and the conservation status of the species. In D.G. Kleiman (ed.) The Biology and Conservation of the Callitrichidae. Washington, D.C., Smithsonian Institution Press, pp. 39–71.

Peres, C.A., 1986, Golden lion tamarin project II. Ranging patterns and habitat selection in golden lion tamarins *Leontopithecus rosalia* (Linnaeus, 1766)(Callitrichidae, Primates). In M. Thiago de Mello (ed.) *A Primatologia no Brasil - 2*, Brasilia, Sociedade Brasileira de Primatologia, pp. 223–241.

Price, E.C., 1992a, The benefits of helpers: effects of group and litter size on infant care in tamarins (*Saguinus oedipus*), *Am. J. Primatol.* 26:179–190.

Price, E.C., 1992b, The costs of infant carrying in captive cotton-top tamarins, *Am. J. Primatol.* 26:23–33.

Richard, A., 1987, Malagasy prosimians: female dominance. In B. Smuts, D. Cheney, R. Seyfarth, R. Wrangham & T. Struhsaker (eds.) *Primate Societies.* Chicago, University of Chicago Press, pp. 24–33.

Rood, J.P., 1986, Ecology and social evolution in the mongooses. In D. Rubenstein & R. Wrangham (eds.) *Ecological Aspects of Social Evolution.* Princeton, Princeton University Press, pp. 131–152.

Rosenberger, A.L., Setoguchi, T, & Shigehara, N, 1990, The fossil record of callitrichine primates, *J. Human Evol.* 19:209–236.

Rylands, A.B., 1986, Ranging behaviour and habitat preference of a wild marmoset group, *Callithrix humeralifer* (Callitrichidae, Primates), *J. Zool, Lond, (A),* 210:489–514.

Rylands, A.B., 1989, Sympatric Brazilian callitrichids: the black tufted-ear marmoset, *Callithrix kuhli*, and the golden-headed lion tamarin, *Leontopithecus chrysomelas, J. Human Evol.* 18:676–695.

Rylands, A.B., 1993, The ecology of the lion tamarins, *Leontopithecus*: some intrageneric differences and comparisons with other callitrichids. In: A.B. Rylands (ed.) *The Marmosets and Tamarins: Systematics, Behaviour, and Ecology.* London, Oxford University Press, pp. 296–313.

Rylands, A.B., Coimbra-Filho, A.F. & Mittermeier, R.A., 1993, Systematics, geographic distribution, and some notes on the conservation status of the Callitrichidae. In: A.B. Rylands (ed.) *The Marmosets and Tamarins: Systematics, Behaviour, and Ecology.* London, Oxford University Press, pp. 11–77.

Rylands, A.B. & de Faria, D.S., 1993, Habitats, feeding ecology, and home range size in the genus *Callithrix*. In: A.B. Rylands (ed.) *The Marmosets and Tamarins: Systematics, Behaviour, and Ecology.* London, Oxford University Press, pp. 262–272.

Savage, A., 1990, *The Reproductive Biology of the Cotton-Top Tamarin (Saguinus oedipus oedipus) in Colombia.* PhD Thesis, University of Wisconsin, Madison.

Scanlon, C.E., Chalmers, N.R. & Monteiro da Cruz, M.A.O., 1989, Home range use and the exploitation of gum in the marmoset *Callithrix jacchus jacchus, Int. J. Primatol.* 10:123–136.

Soini, P., 1988, The pygmy marmoset, genus *Cebuella*. In R.A Mittermeier, A.B. Rylands, A.F. Coimbra-Filho, & G.A.B. Fonseca (eds.) *Ecology and Behavior of Neotropical Primates, Volume 2.* Washington, D.C., World Wildlife Fund, pp. 79–129.

Sussman, R.W. & Kinzey, W.G., 1984, The ecological role of the Callitrichidae: a review, *Am. J. Phys. Anthrop.* 64:419–449.

Sussman, R.W. & Garber, P.A., 1987, A new interpretation off the social organization and mating system of the Callitrichidae, *Int. J. Primatol.* 8:73–92.

Tardif, S.D., 1994, Relative energetic costs of infant care in small-bodied neotropical primates and its relation to infant-care patterns, *Am. J. Primatol.* 34:133–143.

Tardif, S.D., 1997, The bioenergetics of parental behavior and the evolution of alloparental care in marmosets and tamarins. In N.G. Solomon & J. French (eds.) *Cooperative Breeding in Mammals.* Cambridge, Cambridge University Press.

Tardif, S.D., Carson, R. L. & Gangaware, B.L., 1986, Comparison of infant care in family groups of the common marmoset (*Callithrix jacchus*) and the cotton-top tamarin (*Saguinus oedipus*), *Am. J. Primatol.* 11:103–110.

Tardif, S.D., Harrison, M.L. & Simek, M.A., 1993, Communal infant care in marmosets and tamarins: relation to energetics, ecology, and social organization. In: A.B. Rylands (ed.) *The Marmosets and Tamarins: Systematics, Behaviour, and Ecology.* London, Oxford University Press, pp. 220–234.

Terborgh, J., 1983, *Five New World Primates: a study in comparative ecology.* Princeton, Princeton University Press.

Terborgh, J. and Goldizen, A.W., 1985, On the mating system of the cooperatively breeding saddle-backed tamarin (*Saguinus fuscicollis*), *Behav. Ecol. Sociobiol.* 16:93–99.

10

PARENTAL CARE PATTERNS AND VIGILANCE IN WILD COTTON-TOP TAMARINS (*Saguinus oedipus*)

Anne Savage,[1] Charles T. Snowdon,[2] L. Humberto Giraldo,[3] and Luis H. Soto[3]

[1]Roger Williams Park Zoo
1000 Elmwood Avenue
Providence, Rhode Island 02907
[2]Department of Psychology
1202 W. Johnson Street
University of Wisconsin
Madison, Wisconsin 53706
[3]INDERENA
Proyecto Primates
Coloso, Colombia

INTRODUCTION

The relationship between callitrichid infants and their caregivers has been a topic of extensive study in both the field and the laboratory for the past decade. Although there is extensive information on infant care in captive callitrichids, in only a few cases is direct comparison of the behavior of caregivers and infants in natural habitats and captivity possible for the same species.

In captive callitrichids, reproduction is generally limited to one adult female in each social group and infants are cared for cooperatively by group members of both sexes (Abbott, 1984; Epple & Katz, 1984; Price, 1990a;b;c, 1992; Snowdon et al., 1985; Tardif et al., 1990). Adult callitrichids are small in size (100–700g) and generally produce twins that weigh between 15–20% of the female's body weight at birth (Leutenegger, 1973; 1979). Most captive females experience a post-partum ovulation that results in a new pregnancy shortly after parturition (Ziegler et al., 1987). Thus, it is common for breeding females to continue nursing their young offspring during their subsequent pregnancy. These females are rarely the sole caretakers of their offspring, given the extraordinary energetic demands placed on them. Females are rarely capable of rearing their offspring without assistance from other group members (Epple, 1978; Hoage, 1977; Tardif et al., 1990).

Secondly, parental care in callitrichids is not instinctual, it is learned. Several studies have shown that if a callitrichid male or female has not had previous experience carry-

Adaptive Radiations of Neotropical Primates
edited by Norconk *et al*. Plenum Press, New York, 1996

ing someone else's infants prior to the birth of its own infants, its own infants will be rejected and often abused. Animals that have carried infants prior to caring for their own infants are successful in rearing their offspring (Epple, 1978; Hoage, 1977; Snowdon et al., 1985; Tardif et al., 1986, 1992) and infant care is not necessarily confined to genetically related infants (Box, 1977; Dronzek et al., 1986; Ferrari 1992; Rylands, 1986; Savage, 1990).

Finally, additional helpers may assist in predator detection and increasing foraging efficiency. Goldizen (1987) found that free-ranging saddle-back tamarins (*Saguinus fuscicollis*) were less likely to engage in foraging activities when they were carrying an infant. Price (1991) found similar results for captive cotton-top tamarins (*Saguinus oedipus*) and also noted that cotton-top tamarins carrying infants were more likely to be still and hidden than were non-carriers. Caine (1993) has argued that the need for multiple infant carriers may be directly related to predation pressures. As group size increased, there would be a greater level of predator detection as well as additional individuals to assist in infant care. Caine (1993) suggested that the high level of vigilance required for infant care and the need to develop efficient foraging strategies has formed the basis for the relatively tolerant and flexible social system observed in tamarins. Thus, the evolution of cooperative care in callitrichids appears to have evolved based on: 1) the high energetic demands facing reproductive females, 2) the need for both males and females to learn parental care skills prior to reproduction to ensure the survival of their own offspring and 3) the increase in group foraging efficiency and predator detection.

Several captive studies have shown an age and sex bias in caring for offspring. Arruda et al., (1986) found that older common marmosets (*Callithrix jacchus*) were more involved in carrying infants and that males carried infants more than females. Studies on captive saddle-back tamarins have found a similar pattern in infant care. The adult male tends to carry the infants longer and more frequently during the first month of life than other group members (Epple, 1975; Vogt, et al., 1978). Vogt et al., (1978) found that males accounted for 70% of infant carrying. Age and sex differences in infant care has also been documented for cotton-top tamarins (Cleveland & Snowdon, 1984; Price, 1990b;c, 1992) with older siblings and fathers carrying infants more than mothers and other adult females. In contrast, golden lion tamarin (*Leontopithecus rosalia*) mothers were the principal infant carriers through the first three weeks of an infant's life but became secondary carriers in weeks 4–12. Fathers were the secondary carriers for the first three weeks but became the primary infant carriers during weeks 4–12. Older siblings carried infants primarily during weeks 2–8, with female siblings carrying earlier than males (Hoage, 1977).

Few studies have examined infant care patterns in wild callitrichids yet, the results appear similar to what has been found in captive studies. Adult male saddle-back tamarins carried infants (40%) about twice as often as did the infant's mother (20%) (Goldizen, 1987). The amount of infant-carrying by non-reproductive members of the group varied, depending on the age of the individuals. The eldest animals in the group (3 years) carried more than the youngest (1 year). Juveniles were rarely observed to carry infants in the study groups. Ferrari (1992) found that adult male buffy-headed marmosets (*Callithrix flaviceps*) carried infants more relative to females, and that younger animals carried less than adults. There was however, considerable variation in carrying behavior within each age-sex class and through time. Potentially reproductive males exhibited a relatively high degree of interest and carrying of the infants during the first week following birth. Based on inter-birth intervals (5–8 months) and a gestation of approximately 150 days, Ferrari (1992) suggested that reproductively active females may be ovulating as early as one week post-partum. Thus, infant care may be a component and/or a consequence of the competition between males for access to the breeding female during periods of highest fertility (see also Price, 1990b;c).

There has been considerable controversy regarding callitrichid group size and infant survival. Garber et al., (1984) suggested a positive correlation between infant survival and group size in wild moustached tamarins (*Saguinus mystax*). Moreover, Garber et al., (1984) and Sussman and Garber (1987) have suggested that since males are more involved in infant carrying the number of adult males in a group, rather than overall group size, may be the most important factor influencing infant survival in tamarins. Goldizen and Terborgh (1989) found that pairs of wild saddle-back tamarins were unable to raise twins successfully without helpers. In contrast, captive pairs of callitrichids given proper early infant caretaking experience, are capable of rearing their offspring without additional helpers (Rothe et al., 1993; Snowdon et al., 1985, Tardif et al., 1984). Interestingly, Snowdon et al., (1985) found a high infant mortality in pairs of captive cotton-top tamarins which was attributed to primiparous females regardless of early rearing history. In contrast Tardif et al., (1986) found that primiparous females had a higher probability of rearing their infants when they were paired with an experienced male.

Cotton-top tamarins have been studied extensively in captivity. Cleveland & Snowdon (1984) and Price (1990a;b; 1992) present a comprehensive overview of development and infant care patterns of captive-born cotton-top tamarins. Such information provides a basis from which to compare the development and social interactions of wild-born cotton-top tamarins. This study presents information on infant care patterns, survival, and vigilance behavior of wild-born cotton-top tamarins in Colombia. This information will be compared to studies in captivity as well as with the available information on wild callitrichids.

METHODS

Study Site

Our studies of wild cotton-top tamarins have been conducted in Colombia at IN-DERENA's (Instituto Nacional de los Recursos Renovables de Ambiente) research station (Estación Experimental de Fauna Silvestre de Colosó). The research station is adjacent to a protected reserve (la Reserva Forestal Protectora Serranía de Coraza-Montes de María). The reserve covers approximately 4,000 ha and is one of the principal refuges for the cotton-top tamarins in the northern Atlantic coast of Colombia. Observations were conducted during March - July, 1988; January - April, 1989 and January, 1991- December 1993.

Subjects

Twelve litters of cotton-top tamarins were studied between February 1988 and December 1993. Each litter is treated as independent since group composition varied with each new birth.

Using compartmentalized traps, the animals were captured twice each year. For a complete review of the trapping, individual identification, radio tracking techniques and anesthesia procedures see Savage et al., (1993). Each individual animal was tattooed on the abdomen or thigh and in 1992 a transponder chip (Trovan, Infopet Identification Systems, Burnsville, MN) was implanted in each individual for permanent identification. Animals less than 7 months of age were not permanently identified until the next trapping session. Individuals were classified as adult (>14 months of age), juvenile (7–14 months), or infant (< 7 months). Sections of the head, chest, arms and legs were dyed in various color combinations (Redken DecoColors hair dye, Redken Laboratories, Inc., Canoga Park, CA) to permit observers to identify individual tamarins from a distance.

To locate groups accurately in the forest, an adult male in each group was fitted with a radio transmitter package (MOD-070) (Telonics, Inc., Mesa, AZ). Using a Telonics TR-2 Receiver, RA-2A 2-element directional antenna, and headphones, tamarins were located by the observer moving in the direction of the signal frequency, which emanated from the target male's transmitter, until the group was sighted. Battery life was approximately 6 months for each transmitter. Each group was recaptured every 5–7 months to replace the transmitter on the male prior to battery expiration.

Observations

The paths through the study site were marked every 100 m. Once a habituated group was located using the radiotracking equipment, its location in the study site was noted and the group was followed either until they became nervous (defined as running away from the observer or type E chirp chatter and mobbing (Cleveland & Snowdon, 1982) for more than 5 min) or until 4.0 hr of observations on that group had been completed for the day. Two teams of observers collected data from 0700–1930 hr 4–5 days/week. Attempts to contact each study group was made each day of observation and two-minute scan samples were used to note the activity of each group member. Although extensive behavioral data were collected on this species, we report here only the information relative to infant care and development (see Savage et al., 1996, for additional information).

Infant Care and Development

Scan samples of the group were used only to locate the infant and/or caregiver. Once an infant(s) was located, focal sampling was used to collect data on infant care and behavior. Only observations where there was direct visual contact with the infant(s) and/or caregiver were used in this analysis. Using this strict criteria, we have over 391 hrs of direct observation (minimum of 32 hours/group) on infant care and development. Observations were distributed relatively equally by week during the first 8 weeks of an infant's life. Detailed infant observations vary after 8 weeks due to the end of a field season, infant death or disappearance, or difficulty in observing groups due to disturbances at the field site. Sample sizes for each analysis performed will be presented. To examine male contribution to infant care, we analyzed the data using the combined average of all males in the group as well as data from the "best male" (ie. carried the most).

Vigilance as a Mechanism for Group Defense

Two minute scan samples were conducted when the group was located and the individual adopting a vigilant role was noted. All vigilance data are reported as a percent of scan samples. An individual was classified as vigilant if he/she was separated from the group, but was positioned between his/her group and the observers and was observed scanning. Our observations of vigilance are surely underestimated using this method, since it is quite likely that there may have been other animals out of our sight exhibiting similar behavior. We have chosen a conservative, yet reliable, method for determining vigilance so that we may compare the relative proportion of scans contributed by animals at different age/sex classes.

Table 1. Composition of the study groups at parturition

	Adult		Juvenile		Litter Size	Survival	
	Males	Females	Males	Females	(M.F.?)	to 1 year	Parity
1988							
Group 1	2	2	1	1	(1.1)	100%	Multiparous
Group 2	2	2		1	(2.0)	?*	Multiparous
1989							
Group 1	2	2	2	1	(1.1)	100%	Primiparous/Polygnous
Group 2	2	2	1	2	(1.1)	?*	Multiparous
1991							
Group 1	2	2	1	1	(2.0)	100%	Multiparous
Group 2	2	2		2	(0.2)	100%	Multiparous
Group 4	2	1			(1.1)	50%	Multiparous
Group 5	2	1			(0.0.2)	0	Multiparous
1992							
Group 3	1	2	1		?	?	Multiparous/Polygynous
Group 4	2	1		1	(0.2)	100%	Multiparous
Group 5	2	1			(0.1)	100%	Multiparous
1993							
Group 2	3	1			(0.0.1)	0	Primiparous
Group 4	2	2		2	(1.0.1)	50%	Multiparous

*Observations terminated after 6 months

RESULTS

Demography

All study groups contained at least one reproductively active male and female (Table 1). We have not been able to determine the paternity of the infants born in this study, therefore, all adult males present in the group 6 months prior to parturition should be considered as potential sires. Several groups (N=9) had at least one adult male and female, as well as juveniles (7–14 mo of age) of either sex. Most groups experienced modest growth due to the immigration of animals into existing groups and births from previous years (Savage et al., 1996). Immigrants were defined in this study as having entered the group 6 months or less prior to the birth of the infant(s). Number of animals in each group at birth of the infants varied from 3–6 individuals. Of the 58 animals in our study groups, no sex ratio differences were found (28:30).

Females generally gave birth between March-June each year (see Savage et al., 1996, for a complete review). In 1992, northern Colombia experienced a severe drought and none of the five females that were palpated and presumed to be pregnant during the January trapping of the study groups gave birth in March-June. However, two of the five females conceived a second time during the year and gave birth during the months of October and November when the rains began. We have also observed two pregnant females of similar gestational length in the same group. In one case, only one female delivered live offspring and in the second case, no live births were observed in the group. Of the 12 births, we observed ten twins and two single infants. Infant sex ratio at one year of age was equal (9:9:4). An 82% survival rate to one year of age based on successful births was observed in this study.

Figure 1 illustrates the development of independence of wild cotton-top tamarin infants. Infants were carried by caregivers exclusively for the first four weeks of life. During

Carry

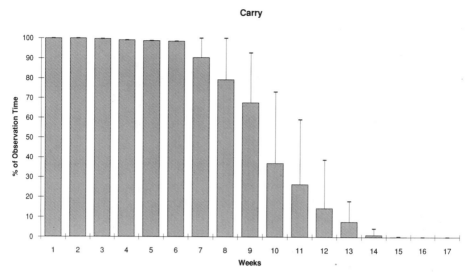

Figure 1. Development of independence of cotton-top tamarin infants.

weeks 5–9 there was a gradually decrease in the amount of time infants were carried, so that by Week 10 the infant was solo for nearly 50% of the observation time. Infants were rarely observed carried by Week 14.

Infant Care Patterns

All individuals in a group were observed to carry infants. However, adults were more likely to carry infants (88.3%) than juveniles (11.7%). In groups containing adults and juveniles (N=8), mothers (23.2 ± 12%) carried infants more than juveniles (11.7 +15.3%) and at equal rates to other adult females (25.7 ± 17.5%) in the group. Mothers were observed to carry less in groups when compared with total adult male carrying (50.3 ± 21.4%) (t(12) = -3.38, p≤.005), and less than the male that carried most in the group (ie. "best male") (34 ± 15.6%) (t(15) =-1.73, p≤ .05), but no significant differences were found between rates of carrying by the mother and the rate that the average adult male (27.7 ± 12.5%) carried.

Figure 2 illustrates that mothers carried their infants the most during the first week of life and gradually decreased the amount of time they carried during the first 8 weeks. As males increased the amount of time they carried infants, mothers significantly reduced the amount of time they carried their infants during the first 8 weeks (r=0.71, p≤.05, df=6). Other individuals in the group did not significantly change their contribution to infant care during the first 8 weeks.

Group Size Effects

Our data suggests that infant survival to one year of age was higher in larger groups than in smaller groups (r=0.62, p≤.06, df =8) (Table 2). There was a significant correlation between group size and male carrying of infants. As the size of the group increased, the overall contribution of the adult male in carrying the infant decreased. There was a significant negative correlation between group size and the percentage of time all adult males

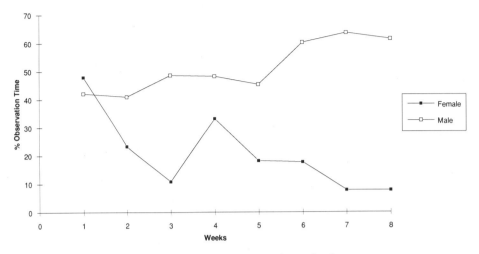

Figure 2. Infant carrying by mothers and males.

carried infants (r=-0.88, p≤ .01, df=8), the average amount of time infants were carried by males (r=-0.76, p≤ .05, df=8) and the "best male" contribution to carrying of infants (r=-0.87, p≤.01, df=8). However, the contribution of the mother in carrying the infants remained constant and independent of group size.

Primiparous vs. Multiparous Females

The previous experience of caring for infants prior to first reproduction could not be assessed in our study animals. In the two cases where females assumed a breeding position in a group, both females had early infant caretaking experience. Both of these females' infants survived to at least one year of age. We have yet to witness a female that we could positively identify as one lacking early infant caretaking experience assume a breeding position and rear offspring. Although we could document that several potentially breeding males had early infant caretaking experience, without the availability of paternity tests for this species, we cannot interpret this information further.

There was however an interesting difference between primiparous and multiparous females. Primiparous females were observed to carry their infants much more during the first two weeks than multiparous females (Figure 3). However, given this small data set, we cannot yet determine whether parity of female is positively correlated with infant survival.

Table 2. Infant survival as a function of group size

Group Size (N)	Infant Survival Rate
3	40%
4	66.7%
5	100%
6	87.5%

r=0.62, p≤.06, df=8

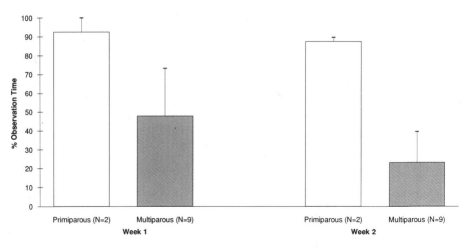

Figure 3. Carrying of infants by primiparous and multiparous females.

Immigrants

There was a significant difference in the percentage of carrying infants in groups that had long-term resident and recent immigrant males. Males residing in a group at the time of conception were observed to carry infants more than males immigrating into the group after possible conception ($F(15,5)=29.3\%$, $p \leq .05$).

Vigilance

Adult animals were involved in vigilance more than juveniles ($t(5)=9.29$, $p \leq .001$) (Figure 4). To examine the effects of infant independence and sex of vigilant individual, the data were divided into two categories for analysis. Phase I encompassed the first 9 weeks of an infants life. During this period the infant was carried for at least 50% of the observation period. Phase II included Weeks 10–15, a time in which the infant was spending more than 50% of its time locomoting independently. Vigilant behavior by adult males was significantly greater than that of adult females during Phase I ($t(16)=3.60$, $p \leq .002$) and even greater in Phase II ($t(16)=6.03$, $p \leq .00002$). There were no significant differences between residents and recent immigrants in vigilant behavior during Phase I (resident X = 35.97% ± 19.91, immigrant X = 30.67% ± 21.98, N=3) or Phase II (resident X = 36.67% ± 27.54, immigrant X = 41.2% ± 13.24, N=3).

It was rare that an animal was vigilant while carrying an infant (16.6%), yet males were more likely to be vigilant while carrying an infant (12.1%) than females (4.5%). There was a positive correlation between group size and carrying an infant while vigilant. Individuals living in smaller groups were more likely to be vigilant while carrying an infant than those individuals living in larger groups ($r=0.69$, $p \leq .05$, df=7).

DISCUSSION

The patterns of reproduction in wild cotton-top tamarins differ from their captive counterparts. Wild tamarins generally give birth to twins once a year, in contrast captive

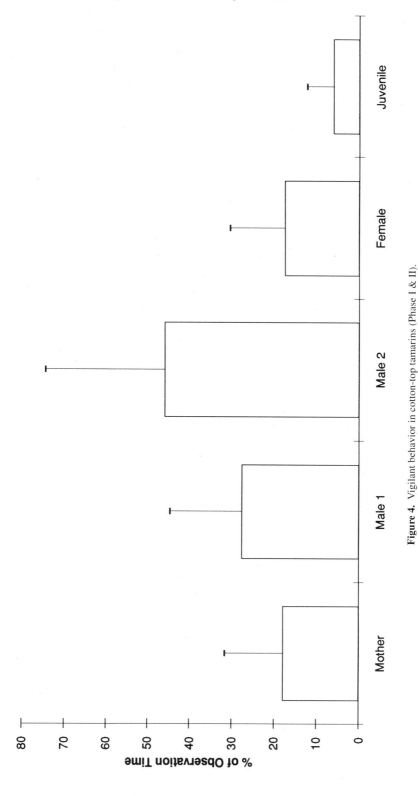

Figure 4. Vigilant behavior in cotton-top tamarins (Phase I & II).

females can either give birth every 28 weeks (Snowdon et al., 1985) or exhibit a bimodal distribution of interbirth intervals (Kirkwood et al., 1983;1985). Fertility in the wild can be affected by environmental changes as evidenced by a high fetal loss during the drought of 1992. Yet, when environmental conditions are appropriate, females will produce live offspring that have a high probability of surviving (82%). Fertility was also affected by the presence of more than one reproductively active female in a group. Although the incidences of two pregnant females in one group was low in our study, fetal loss was high (75%) in polygynous groups.

Cotton-top tamarin infants born in the forests of Colombia exhibited similar patterns of development as those infants born in captive colonies. Infants born in a natural environment took their first steps away from their caregiver during Weeks 4–5 similar to those infants born in captive U.S. colonies (Cleveland & Snowdon, 1984; Tardif et al., 1990). By Week 9 wild and captive infants spend more than 50% of their time locomoting independently. In wild groups, infants were no longer carried by Week 15, yet captive infants were still carried for very brief periods (less than 5% of the observation period) until Week 20 (Cleveland & Snowdon, 1984).

Infant survival in wild groups of tamarins was considerably greater than has been observed in captive colonies (see Snowdon et al., 1985). This may be due to several factors. In captivity, there is a greater incidence of triplet births in callitrichids (Tardif & Jaquish, 1994). Since few females are capable of rearing all three infants, survival values are greatly deflated. No triplet births have been observed in wild cotton-top tamarins to date. In captivity, the highest infant mortality usually occurred within the first week of life. In wild tamarins, infant mortality was observed between 5–10 weeks of age. This developmental stage coincided with the period of increased independence of the infants. Infants were first observed locomoting independently from their caregiver during Week 5. Captive and wild tamarin infants frequently fall during this period. Although these falls are rarely fatal in captivity, a fall of 15–20 m in the forest may in fact cause death in the wild. Moreover, infants that are learning to locomote and function independently of their caregivers are likely to be more vulnerable to predation.

Several factors influence the survival of infants in captivity. Previous early infant caretaking experience for both male and females as well as parity of the female have been demonstrated to be critical for the successful rearing of cotton-top tamarin infants (Snowdon et al., 1985). In the wild, the opportunity for individuals to gain early infant caretaking experience appears readily available, as immigrants (non-kin) and kin were frequently observed to carrying infants. Although adults were more likely to carry infants than juveniles, all individuals in the group participated in infant carrying. Even recent immigrant males were observed to carry infants. Thus, the lack of parental care skills typically found in captive-born tamarins, appears to be a function of poor captive management practices rather than what is likely the norm in wild groups of tamarins.

Regardless of early infant caretaking experience, primiparous captive females have a much lower infant survival rate than multiparous females (Snowdon et al., 1985). In contrast, the infants born to wild primiparous cotton-top tamarins in this study all survived. Interpreting this data is somewhat difficult, since parity may in fact be confounded by other variables that influence infant survival. In captivity, most reports of primiparous births occur in groups with only 2 animals (Snowdon et al., 1985). There are no published reports of a primiparous female giving birth in a large captive group. Since Price (1992) found that infants born to large families were carried more than infants born to small families and also suggested that group size may influence infant survival, it appears that the captive data needs to be re-evaluated. Group size, in relation to parity and previous infant caretaking experience should be examined to determine the factors involved in infant survival in captivity.

Similar to observations of captive golden-lion tamarins (Hoage, 1977), wild, cotton-top tamarin mothers carried their infants more during the first week of life and gradually decreased the amount of time they carried during the first 8 weeks. As the adult male(s) contribution increased the amount of time mothers carried their infants decreased. Primiparous cotton-top tamarin females were observed to carry their infants more frequently than multiparous females during the first two weeks of life. Although it has been suggested that first time mothers tend to be more cautious, additional observations of the behavior of primiparous females are needed to evaluate this claim.

The factor that appears to be most important in influencing infant survivorship in wild cotton-top tamarin groups is group size. Large groups of tamarins were more successful at rearing offspring than smaller groups. Although all individuals in a group participated in infant carrying, in large groups the overall adult male contribution was less than in smaller groups yet the adult female contribution in carrying infants remained constant regardless of group size. Not only do large groups have more individuals to assist in carrying infants, thereby reducing the energetic costs of carrying, but they are more likely to increase their foraging efficiency and predator detection. Similar results have been found in captive cotton-top tamarins (McGrew, 1988; Tardif et al., 1990).

Individuals that adopted a vigilant role, tended to separate themselves from the group and were generally in fairly open areas making them potentially more vulnerable to predation. In general, adult males were more likely to be vigilant than other individuals in the group. It was rare that adult males in large groups were vigilant while carrying an infant. However, males in small groups were more likely to be vigilant while carrying an infant. Thus, males in small groups may be placing their infants at a greater risk, by adopting a vigilant role while carrying infants since they may be exposed to predation. Although the exact cause of death could not be determined, it is likely that a predator killed an adult male and his infants in one of our small study groups. No such observations were made in our larger groups.

Moreover, males were more involved in vigilance during the time of highest infant vulnerability. Males were more likely to be vigilant during the first 9 weeks of an infants life when an infant was carried for at least 50% of the observation period. Male vigilance increased during Weeks 10–15, when infants were spending more than 50% of their time locomoting independently. Since infant mortality in the wild was highest during this period, males may be adopting a vigilant role to assist in protecting the infants.

Thus, infant survivorship in wild cotton-top tamarins appears to be directly related to the size of the group. The larger the group, the more individuals there are to assist in carrying and protecting the infant, thereby distributing the energetic load of infant care, group foraging efficiency and defense among several individuals (Caine, 1993; Tardif, 1994). It may be advantageous for small groups to accept immigrants to assist in carrying infants and group defense thereby increasing the probability of infant survival.

SUMMARY

Studies of wild tamarins were conducted at Estación Experimental de Fauna Silvestre de Colosó on the northern Atlantic coast of Colombia during 1988, 1989, and 1991–1993. Individuals of five groups have been permanently marked for identification (both tattoo and beginning in 1992 using transponder chips - Trovan, Infopet Identification System, Burnsville, MN) and one male in each group was fitted with a radio transmitter package (MOD-070, Telonics, Inc,. Mesa, AZ). Individuals were retrapped every 5 to 7 months to replace batteries in the transmitters.

The reproductive patterns of wild and captive cotton-top tamarins differed in both reproductive rates and infant care. Wild tamarins gave birth once a year, while captive tamains generally give birth every 28 weeks (Snowdon et al., 1985). More than one mature female in some of the wild groups also had an adverse effect on fetal survival. Wild cotton-top tamarin infants exhibited similar patterns of development as those infants born in captive colonies and had a high probability of surviving to one year of age (82%). An important factor that appeared to influence infant survivorship in wild cotton-top tamarin groups was group size. Large groups were more successful at rearing offspring than smaller groups. Large groups not only had more individuals to assist in carrying and protecting the infants, but were also more likely to increase their foraging efficiency and predator detection thus distributing the energetic load of infant care, group foraging efficiency, and defense among several individuals in the group.

ACKNOWLEDGMENTS

We greatly appreciate the assistance of Felix Medina, Pacho Ochoa, the staff of Proyecto Primates, and Dr. Evan Blumer for their continued support and assistance in this program. Drs. Bernardo Ortiz and Jose Antonio Lopera have provided us with the necessary logistical support for our studies in Colombia. Redken Laboratories, Inc. kindly supplied us with hair dye for the tamarins. This study was supported in part by National Science Foundation grants (IBN-9222313 & IBN-8922741), National Geographic Research and Exploration, Rhode Foundation, an Institute for Museum Services General Operating Support to the Roger Williams Park Zoo and Fossil Rim Wildlife Center.

REFERENCES

Abbott, D. H., 1984, Behavioral and physiological suppression of fertility in subordinate marmoset monkeys. *Am. J. Primatol.* 6:169–186.

Arruda, M.F., Yamamoto, M. E., and Bueno, O. F. A., 1986, Interactions between parents and infants, and infants-father separation in the common marmoset (*Callithrix jacchus*), *Primates* 27:215–228.

Box, H. O., 1977, Quantitative data on the carrying of young captive monkeys (*Callithrix jacchus*), *Primates* 18:475–484.

Caine, N. G., 1993, Flexibility and co-operation as unifying themes in *Saguinus* social organization and behavior: the role of predation pressures., In: *Marmosets and tamarins: systematics, behaviour and ecology.*, Ed. A. B. Rylands, Oxford:Oxford University Press.

Cleveland, J., and Snowdon, C. T., 1982, The complex vocal repertoire of the adult cotton-top tamarin (*Saguinus oedipus*). *Ziet. fur Tier.* 58:213–270.

Cleveland, J., and Snowdon, C. T., 1984, Social development during the first twenty weeks in the cotton-top tamarin (*Saguinus o. oedipus*), *Anim. Behav.* 32:432–444.

Dronzek, L. A., Savage, A., Snowdon, C. T., Whaling, C. S., and Ziegler, T. E., 1986, Techniques for handrearing and reintroducing rejected cotton-top tamarin infants, *Lab. Anim. Sci,* 36:243–247.

Epple, G. and Katz, Y., 1984, Social influences on estrogen excretion and ovarian cyclicity in saddle-back tamarins (*Saguinus fuscicollis*), *Am. J. Primatol.* 6:215–227.

Epple, G., 1975, Parental behavior in *Saguinus fuscicollis* ssp. (Callitrichidae). *Folia Primatol.* 24:221–238.

Epple, G., 1978, Reproductive and social behavior of marmosets with special reference to captive breeding. *Primate Medicine* 10:50–62.

Ferrari, 1992, The care of infants in a wild marmoset (*Callithrix flaviceps*) group. *Am. J. Primatol.* 26:109–118.

Garber, P. A., 1984, Proposed nutritional importance of plant exudates in the diet of the Panamanian tamarins, *Saguinus oedipus geoffroyi. Int. J. Primatol.* 5:1–15.

Goldizen, A. W., 1987, Facultative polyandry and the role of infant-carrying in wild saddle-back tamarins (*Saguinus fuscicollis*). *Behav. Ecol. and Sociobiol.* 20:99–109.

Hoage, R. J., 1977, Parental care in *Leontopithecus rosalia rosalia*: sex and age differences in carrying behavior and the role of prior experience, In: *The Biology and Conservation of the Callitrichidae* pp. 293–305., Ed. D. G. Kleiman, Washington, D. C., Smithsonian.

Kirkwood, J.K., Epstein, M.A. and Terlecki, A.J., 1983, Factors influencing the population growth of a colony of cotton-top tamarins. *Lab. Anims.* 17:35–41.

Kirkwood, J.K., Epstein, M.A., Terlecki, A.J. and Underwood, S.J. 1985, Rearing a second feneration of cotton-top tamarins (*Saguinus oedipus oedipus*) in captivity. *Lab. Anims.* 19:269–272.

Leutenegger, W., 1973, Maternal-fetal weight relationships in primates. *Folia primatol.* 20:280–293.

Leutenegger, W., 1979, Evolution of litter-size in primates. *Am. Nat.* 114:525–531.

McGrew, W.C. 1988, Parental division of infant caretaking varies with family composition in cotton-top tamarins. *Anim. Behav.* 36:285–286.

Price, E. C., 1990a, Parturition and perinatal behaviour in captive cotton-top tamarin (*Saguinus oedipus*). *Primates*, 31:523–535.

Price, E. C., 1990b, Infant-carrying as a courtship strategy of breeding male cotton-top tamarins. *Anim. Behav.* 40:784–786.

Price, E. C., 1990c, Reproductive Strategies of Cotton-top Tamarins. Ph.D. Dissertation, University of Stirling, Scotland.

Price, E. C., 1991, The costs of infant carrying in captive cotton-top tamarins. *Am. J. Primatol.* 26:23–33.

Price, E. C., 1992, Contributions to infant care in captive cotton-top tamarins (*Saguinus oedipus*): The influence of age, sex, and reproductive status. Int. J. Primatol. 13:125–142.

Rothe, H., Darms, K., Koenig, A., Radespiel, U., and Juenemann, B., 1993, Long-term study of infant-carrying behavior in captive common marmosets (*Callithrix jacchus*) - Effect of nonreproductive helpers on the parents carrying performance. *Int. J. Primatol.* 1:79–93.

Rylands, 1986, Ranging behaviour and habitat preference of a wild marmoset group, *Callithrix humeralifer* (Callitrichidae, Primates)., *J. Zool.* 210:489–514.

Savage, A., 1990, *The reproductive biology of the cotton-top tamarin (Saguinus oedipus oedipus) in Colombia.* Ph.D. Thesis, University of Wisconsin-Madison.

Savage, A., Giraldo, L.H., Blumer, E.S., Soto, L.H., Burger, W., Snowdon, C.T.1993, Field techniques for monitoring cotton-top tamarins (*Saguinus oedipus oedipus*) in Colombia. *Am. J. Primatol.* 31:189–196.

Savage, A., Giraldo, L.H., Soto, L.H., Snowdon, C.T. 1996, Demography, group composition, and dispersal in wild cotton-top tamarins. *Am. J. Primatol.* 38: 85–100.

Snowdon, C. T., Savage, A., McConnell, P. B., 1985, A breeding colony of cotton-top tamarins (*Saguinus oedipus oedipus*). *Lab. Anim. Sci.* 35:477–480.

Sussman, R. W. and Garber, P.A. 1987, A new interpretation of the social organization and mating system of the Callitrichidae. *Int. J. Primatol.* 8:73–92.

Tardif, S. D. 1994, Relative energetic cost of infant care in small-bodied neotropical primates and its relation to infant-care patterns. *Am. J. Primatol.* 34:133–143.

Tardif, S. D. and Jaquish, C.E. 1994, The common marmoset as a model for nutritional impacts upon reproduction. *Ann. N.Y. Acad. Sci.* 709:214–215.

Tardif, S. D., Carson, R. L., and Gangaware, B. L., 1992, Infant-care behavior of non-reproductive helpers in a communal-care primate, the cotton-top tamarin (*Saguinus oedipus*). *Ethology,* 92:155–167.

Tardif, S. D., Carson, R. L., and Gangaware, B. L., 1990, Infant-care behavior of mother and fathers in a communal-care primate, the cotton-top tamarin (*Saguinus oedipus*). *Am. J. Primatol.* 22:73–85.

Tardif, S. D., Carson, R. L., and Gangaware, B. L., 1986, Comparison of infant care in family groups of the common marmoset (*Callithrix jacchus*) and the cotton-top tamarin (*Saguinus oedipus*). *Am. J. Primatol.* 11, 103–110.

Tardif, S. D., Richter, C. B., and Carson, R. L., 1984, Effects of sibling-rearing experience on future reproductive success in two species of Callitrichidae. *Am. J. Primatol.* 6:377–380.

Vogt, J. L., Carlson, H., and Menzel, E., 1978, Social behavior of a marmoset (*Saguinus fuscicollis*) Group I: Parental care and infant development., *Primates*, 19:715–726.

Ziegler, T.E., Bridson, W.E., Snowdon, C.T., Eman, S. 1987, Urinary gonadotropin and estrogen excretion during the postpartum estrus, conception and pregnancy in the cotton-top tamarin (*Saguinus oedipus oedipus*). *Am. J. Primatol.* 12:127–140.

11

TESTING LEARNING PARADIGMS IN THE FIELD

Evidence for Use of Spatial and Perceptual Information and Rule-Based Foraging in Wild Moustached Tamarins

Paul A. Garber[1] and Francine L. Dolins[2]

[1]Department of Anthropology
University of Illinois
Urbana, Illinois 61801
[2]Department of Psychology
Centre College
600 N. Walnut Street
Danville, Kentucky 40422

INTRODUCTION

An animals' foraging efficiency may be enhanced by generating rules or strategies of behavior that increase the probability of encountering suitable prey or food patches (Kamil, 1984; Parker, 1986). The information upon which these strategies are based is acquired during a process of exploration and sampling, and is associated with both random encounters and expectations regarding the productivity and distribution of previously visited feeding sites. As resources become depleted and change in spatial and temporal availability, animals must incorporate new or updated information in order to generate more effective foraging patterns. The ability of a forager to exploit resources efficiently is dependent, therefore, on its ability to remember and integrate information concerning: (1) direction, position, and distance between multiple feeding sites; (2) food availability in a patch prior to and after a foraging bout (rates of renewal); (3) number of visits to a patch and time interval since last visit; and, (4) food type associated with a specific feeding site (Dolins 1993; see also Krebs et al, 1977, 1978; Krebs, 1981; Shettleworth & Krebs, 1982; Vander Wall, 1982, 1990; Balda & Turek, 1984; Sherry, 1984; Stephens & Krebs, 1986; Armstrong et al, 1987; Garber, 1988, 1989; Milton, 1988; Brown & Gass, 1993).

In deciding when and where to search for food, a forager is likely to use a number of behavioral 'rules' as a guide (Krebs, 1981). In the case of primates and other higher vertebrates, these rules appear to be based on the application of past experience to classes of problems that consistently reappear within the environment. The ability to apply past experience to novel problems enables animals to generate solutions without re-learning cause and effect relationships involved in every new situation.

Rule-Based Learning

In the experimental analysis of decision making or problem solving, Krechevsky (1932, 1938) suggested that animals develop 'rules' on which to base decisions during the trial-and-error learning phase prior to solution. In trial-and-error learning, it has been found that animals do not respond 'randomly' with chance responses in order to generate the correct solution. Instead, based on previous experience and learning capacity, an animal develops a number of different 'hypotheses' or pre-solutions that it can apply in attempting to solve a new task, or a similar task in a novel context. Krechevsky (1932, 1938) stipulated that an animal will apply a number of these hypotheses systematically until reaching the correct solution to a particular problem. When the correct solution has been achieved it is then reinforced through reward. Evidence for the existence of 'hypotheses' is "response biases and systematic error-producing strategies...[in which the] nonrandomness may reveal important aspects of the learning processes that could not be detected by analysis of only percentage correct" (Fobes & King, 1982, pg. 316). An example of rule-based behavior is when an animal consistently chooses fruits on a tree that are of a certain hue but not hardness. In this case the animal is making foraging decisions based on visual information associated with the degree of ripeness of the fruit. The assumption here is that the animal has learned that fruit color signals ripeness and readiness for eating, whereas hardness does not predict edibility of fruit. Thus, in attempting to solve present problems by testing and applying previously successful behavioral responses, an animal will build up a number of rules by which it will be able to respond effectively when confronted with similar problems in the future.

Despite the fact that monkeys and apes have been the focus of numerous field and laboratory investigations, little is presently known regarding how primates generate foraging rules, and encode and use spatial and temporal information during foraging (Boesch & Boesch, 1984, 1989; Garber, 1987; Meador et al, 1987; Dolins, 1993; Byrne, in press). This reflects limitations in the ways in which many captive and field projects are designed. In general, traditional primate field studies have not focused on testing hypotheses of learning and spatial memory. Moreover, captive experiments have often disregarded factors of ecological validity when presenting animals with problem solving tasks, and in doing so, have failed to stimulate the full range of complex behavioral responses observed in the wild (Dolins, 1993). An alternative approach to the study of primate learning and foraging behavior involves the use of naturalistic field experiments. Experimental field studies offer an opportunity to examine species differences in problem solving and perceptual skills, and to identify a set of adaptive challenges that have shaped the evolution of primate sensory systems.

Experimental Field Study

This paper presents an experimental field study of spatial and temporal rule-based foraging behavior in free-ranging groups of Peruvian moustached tamarins (Saguinus mystax). Moustached tamarins are small bodied New World monkeys (550–600 gm) that exploit a diet composed principally of insects, fleshy fruits, the arrilate seed coat of certain legume species, floral nectar, and plant exudates (Garber, 1989, 1993). These resources exhibit a scattered and patchy distribution within a groups' home range and are characterized by marked changes in seasonal availability. The challenges that tamarins face in exploiting these resources include *where* to forage (use of spatial information and expectations concerning food availability) and *when* to return to a previous feeding site (temporal information, rates of resource renewal, and memory of prior visits to each site).

Data collected in the field (Garber, 1989, 1993) indicate that moustached tamarins use a limited set of foraging rules to successfully exploit spatially scattered but temporally predictable feeding sites. Trees visited by the tamarins were characterized by intraspecific fruiting synchrony. During the course of each day, these primates concentrated their feeding efforts on a large number of trees from a small number of plant species. Trees of the same species were visited in succession (a behavior reminiscent of trap-lining) and particular trees may be revisited many times over a 1–2 week period. Many of these trees exhibited a piece-meal fruiting pattern, with each tree producing only a small amount of ripe fruit each day. The spatial location of fruiting trees relative to each other, and expectations regarding the amount of food available on a particular tree were primary factors in selecting feeding sites (Garber, 1988, 1993). Despite seasonal changes in the spatial distribution of food patches and the types of food items eaten, moustached tamarin foraging and ranging patterns were characterized by a high degree of behavioral consistency. Rather then continuously adopting new foraging patterns in response to local or temporal variation in food availability, these primates appeared to employ and re-employ a set of foraging rules throughout the year. The specific environmental information used in developing and applying these foraging rules has not been systematically investigated.

The aims of this study are to test directly the ability of wild moustached tamarins to use spatial, temporal, and perceptual information to predict the distribution and availability of food resources at multiple feeding sites. In a series of controlled and naturalistic field experiments, the tamarins were presented with select sets of experimentally manipulated information. The monkeys were tested on (1) their ability to localize the spatial positions of 16 novel baited feeding platforms; (2) attentiveness to local (nearby) spatial information; and, (3) differential use of perceptual cues (i.e., visual and/or olfactory) to discriminate between bait-types (banana [food reward] and plastic banana [sham - no reward]) at feeding sites.

METHODS

The Field Site

This research on moustached tamarins was conducted during an 8 week field study (July - August 1993) on Padre Isla, a protected biological reserve maintained by the Proyecto Peruano de Primatología in northeastern Peru (3° 44'S, 73° 14'W). The island is composed of a series of long narrow strips of gallery forest, secondary forest, and planted orchards separated by shallow lakes (Moya et al, 1980). A census conducted in 1990 (Garber et al, 1993a) indicated a total of 123 moustached tamarins living in 17 established groups on the island. *Saguinus mystax* is the only species of nonhuman primate on Padre Isla (For further information on the behavioral ecology of tamarins on Padre Isla see: Norconk 1986; Garber et al., 1993 a,b; Garber & Pruetz, 1995).

Research Design

The research design used in these field experiments involved the construction of four Feeding Stations located 50 meters apart, along a transect of the forest in which moustached tamarins had been encountered previously. A Feeding Station consisted of four visually identical platforms located 10 meters north, south, east, and west from a central point. On each day of the experiment, two of four platforms per Station contained 2 real bananas and two contained 2 plastic bananas.

In order to ensure that the presence of the observer did not effect tamarin feeding choices, a blind was constructed at the midpoint of each Feeding Station. Once in the blind, it was possible to observe undetected, the tamarins' movement patterns and foraging behavior at or near the platforms.

Platforms were placed approximately 1.5 meters above the ground and consisted of a fixed wooden board measuring 43 cm x 30 cm covered by a dark plastic sheet. Attached to each board were two plastic bags. During each test session carbon paper and white paper where placed in these plastic bags. Impressions of tamarin hand and foot prints on the white paper served as one method of verifying whether the tamarins had visited a feeding platform. Direct observations of tamarins' visits to platforms, however, served as the primary method of data collection. A visit was scored when at least one tamarin was observed to contact or orient to an experimental platform, or if evidence of tamarin foot prints were found on the paper. Orientation was defined as an individual climbing to within 1–2 feet of a platform and visually scanning for food. Over 400 person-hours were spent in the blind monitoring the Feeding Stations. During this period, no species other than moustached tamarins were observed to visit a feeding platform.

The research design included a 3 day pretest period in which all 16 feeding platforms were baited with real or sham bananas. This period was used to acquaint and habituate the animals to the test conditions. The tamarins visited several baited platforms on each of these 3 days.

Two experiments divided into six test conditions were presented to the tamarins. The order of presentation was as follows (summarized in Table 1): In the first condition, (1A) the positions of real and sham bananas were kept constant at each Feeding Station over a 6 day test period (i.e. spatial position was predictable); in (1B) locations of real and sham bananas remained constant and two local red cues (sticks covered with red flagging) were added to all platforms containing real bananas (4 days test period), and in (1C) the conditions of 1B were repeated except that the positions of real and sham bananas were rotated by 90° on the first day of the test session. For the remaining 5 days of condition 1C the positions of real and sham bananas remained constant. In the second set of experiments: the positions of real and sham bananas were changed randomly and visual cues were eliminated by covering the bait with leaves in condition 2A; the positions of real and sham bananas were changed randomly, bait covered, and olfactory cues were equivalent at all sites (banana skins were placed to plastic bananas) in condition 2B; and the positions of real and sham bananas were changed randomly, bait covered, olfactory cues equivalent, and two visual associative cues (sticks covered with red flagging that could be used as local landmarks) were placed at real banana sites in condition 2C. Conditions 2A, 2B, and 2C each lasted for 6 consecutive days.

During the course of our investigation tamarins from two groups were observed at the experimental feeding sites. The East group contained 10 animals. Members of this group visited platforms on all days of the experiment. The West group was composed of 11 animals. Based on information of the location of each groups' home range, the direc-

Table 1. Summary of field experiments

Experiment	Test	Conditions
1A	Associate spatial information with presence/absence of food rewards at multiple feeding sites	Place kept constant
		Visual cues and olfactory cues available
1B	Attend to local landmark cues to locate real banana sites	Place kept constant
		2 red cues placed only at platforms containing real bananas
		Other visual and olfactory cues available
1C	Dominance of spatial information over visual and olfactory cues in selecting feeding sites	Rotate positions of real and sham bananas 90°
		Visual and olfactory cues available
		Conflicting spatial vs. visual-olfacotory information
2A	Role of olfactory cues in locating feeding sites	Place random in AM and predictable in PM
		Bait covered, visual cues unavailable
		Olfactory cues available
2B	Control Condition	Place random in AM and predictable in PM
		Bait covered, visual cues unavailable
		Olfactory cues equivalent at real and sham sites
2C	Attend to local landmark cues to locate real banana sites	Place random in AM and predictable in PM
		Bait covered, visual cues unavailable
		Olfactory cues equivalent at real and sham sites
		2 red cues placed only at platforms containing real bananas

In Experiment 2, spatial-temporal sequence of baiting also permitted testing of rule-based foraging

tion of group travel and the behavior of individuals as they approached the study area, it appeared that the West group fed only at Feeding Station 1 and did so on less than 15% of experimental days.

None of the animals in our study groups were marked and therefore it was not possible to identify individual, age, or sex-specific patterns of learning, or to control for or quantify differences in individual experience. Moustached tamarins feed and forage as a cohesive social unit. In our experiments it was common for several individuals to feed simultaneously at the same platform, and therefore it is likely that many members of a group had similar experiences at each feeding site. In addition, based on idiosyncratic patterns of behavior and movement it was possible to determine that the same individual(s) consistently visited particular Feeding Stations.

It was assumed that once an animal had located a Feeding Station it had an equal probability of visiting any of the 4 feeding platforms. Visits to real banana sites were scored as correct choices. Visits to sham sites were scored as incorrect choices. Visits by more than one animal to a platform or repeated visits to the same platform were scored as a single visit. Given uncertainties (data from foot prints) in determining whether members of the East group or the West group had visited Station 1, data are pooled for analysis. Chi Square Tests were used to determine if the number of correct choices differed significantly from chance level.

RESULTS

General Results

Over the course of 34 days of data collection, the tamarins visited an average of 8.2 ± 3.4 feeding platforms daily (5.9 ±2.6 reward and 2.3 ±1.9 sham sites). The largest number of platforms visited in one day was 18 (during Experiment 2 platforms were baited in the morning and rebaited in the afternoon and therefore an animal could revisit a platform in the same day) and the fewest was 4. Platforms were visited during all days of the experiments. In total, 72% (N=201) of tamarin visits were to reward platforms and 28% (N=78) were to sham platforms.

Experiment 1: Differential Reliance on Distant and Local (Nearby) Spatial Information. In Experiment 1A, the position of the real and sham bananas remained constant at each station. The monkeys rapidly learned to distinguish between the locations of the eight real and eight sham banana platforms. During the 6 day test period, the tamarins visited the feeding platforms on 50 occasions, selecting the real banana sites 78% and the plastic banana sites only 22% of the time ($X^2 = 7.84$, df=1, p<.01). There was evidence of an increase in correct responses from Days 1 - 3 (69.2% ,18/26) to Days 4 - 6 (87.5%, 21/24; see Figure 1).

In Experiment 1B the monkeys were presented with a set of local visual association cues (red sticks) at platforms containing real bananas. This information could be used as a signal of food availability. The positions of real banana and local cues were the same as in 1A and remained constant over the 4 day test period. Beginning with the first day the

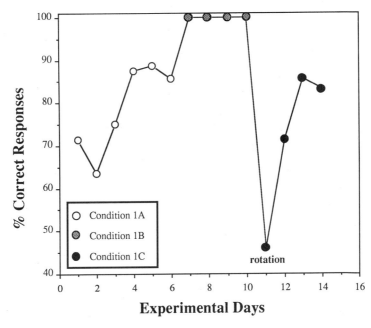

Figure 1. Tamarin visits to experimental platforms in Experiment 1. Note the increase in performance during Condition 1A, the high level of performance in Condition 1B (introduction of local landmark cues - red flagging), and the drop in performance in Condition 1C on the day of the rotation.

monkeys' search accuracy increased from 86% in the previous test condition to 100%. Search accuracy remained at 100% throughout the entire test period (Figure 1).

In Condition 1C, the real bananas (plus red local cues) and sham bait positions were altered from their previous positions by a 90° clockwise rotation. On the first day of the rotation the monkeys' accuracy in locating real banana platforms decreased to 46% (from 100% previously), a value indistinguishable from random performance (chance level was equivalent to 50% - see Figure 1). This occurred despite the presence of the red local cues on the platforms containing real bananas and the fact that olfactory and other visual cues (e.g. the bait type) were present. By Day 2 however, the monkeys returned to the real banana sites on 71% of their foraging choices suggesting that they had re-learned the new positions of the real bananas sites. On Day 3 the correct choices reached 86%, similar to that for the initial condition (1A). Figure 1 summarizes tamarin foraging choices in Experiment 1A, 1B and 1C, highlighting both the decrease in correct responses during the rotation condition and the gradual increase in correct responses over the next three days to a level approximately equivalent with their previous performance in Condition 1A.

The behavior of the monkeys in selecting feeding sites immediately after the rotation (in Condition 1C) appeared to be similar to that of random sampling or exploration. Although they were able to locate six reward sites, they also visited seven sham sites. The average number of sham sites visited on all other days of this test condition was only 1.3. These data suggest that the tamarins' were relying principally on the predictability of spatial position, possibly associated with a single or set of landmark cues, to locate the food rewards rather than olfactory cues or visual differences in the appearance of real and plastic bananas.

Experiment 2: Evidence for Rule Based Foraging. In Experiment 2 the tamarins were introduced to a new set of conditions in which the positions of real and sham bananas were less predictable and visual cues were eliminated by covering the bait with leaves. The goal was to identify what other types of sensory information the tamarins would use to locate feeding sites, and whether these primates would generate a set of behavioral rules relating to time and place that could be used to enhance foraging efficiency. This was accomplished by introducing a "resource renewal schedule" into the experimental design in which each platform was baited twice during the day.

The first baiting occurred at 6:00am, when the tamarins were still in their sleeping tree. As in previous experiments, 2 of the 4 platforms at each Feeding Station were baited with real bananas and 2 baited with plastic bananas. At the 6:00am baiting, the positions of real and sham bananas were assigned randomly; that is, the spatial positions of the real bananas were not predictable in the mornings. The second baiting occurred at approximately 10:00am. Extreme care was taken to ensure that no tamarins were in the vicinity of the Feeding Station when the sites were rebaited. At the 10:00am baiting, the positions of real and sham bananas were changed systematically such that: (a) if a platform had real bananas in the morning it would have plastic bananas in the afternoon; and, (b) if a platform had plastic bananas in the morning it would have real bananas in the afternoon.

Condition 2A (6 days) was a test of the monkeys' ability to discriminate between visually concealed real bananas and visually concealed sham banana feeding platforms on the basis of olfactory information. The results (Table 2) indicate that the tamarins did not rely on available olfactory cues in choosing feeding sites. Their performance in identifying platforms containing real bananas (23/37) versus plastic bananas (14/37) failed to deviate from chance levels ($X^2 = 1.09$, df=1, p>.10).

Table 2. Comparison of AM and PM foraging efficiency

	AM			PM			Total		
Condition	Reward	Sham	% Correct	Reward	Sham	% Correct	Reward	Sham	% Correct
2A	9	6	(60%)	14	8	(63%)	23	14	(62%)
2B	16	19	(46%)	20	4	(83%)	36	23	(61%)
2C	19	11	(63%)	38	5	(88%)	57	16	(78%)
Total	44	36	(55%)	72	17	(73%)	116	53	(69%)

For Condition 2B (6 days), the bait was again covered with leaves to eliminate visual cues and the positions of real and sham bananas were changed according to the morning/afternoon resource renewal schedule. In addition, banana skins were placed next to plastic bananas in order to equalize olfactory information between real and sham platforms. This represents a baseline condition. Given the absence of cues by which to discriminate between sites containing real and sham rewards, the tamarins were not expected to locate bananas at a frequency greater than random. This was indeed the case (Table 2). The number of visits to real banana sites (36/59) was statistically indistinguishable from the number of visits to plastic banana sites (23/59) ($X^2=1.43$, df=1, p>.10).

In the final experiment (2C) the baseline condition was repeated except that during both morning and afternoon trials, 2 red sticks serving as local landmark cues were placed on all platforms containing real bananas. During this 6 day test period the tamarins' accuracy in selecting real banana sites increased significantly ($X^2=11.51$, df=1, p<.001). As indicated in Table 2, 78% of visits were to platforms containing bananas + red local cues. In contrast, only 22% were to platforms containing sham rewards and no local cues.

Evidence for Spatial-Temporal Rule-Based Foraging

The resource renewal schedule employed in Experiment 2 introduced an element of predictability that could be used by the tamarins to increase their efficiency in locating feeding sites. Although the locations of real and sham sites were *random between days* (afternoon of Day 1 to the morning of Day 2), they were *predictable within a day* (morning on Day 1 to afternoon on Day 1). In order to take advantage of this element of predictability the tamarins would be required to:

1. differentiate between Feeding Stations and feeding platforms;
2. encode spatial and/or other sensory information to recall which of 4 Feeding Stations and which of 16 feeding platforms were visited in the morning; and,
3. use a combination of behavioral rules associated with a WIN-SHIFT (finding a real banana at a platform in the morning signifies not to return to that site in the afternoon), and LOSE-RETURN foraging strategy (finding plastic banana in the morning signifies to return in the afternoon to obtain a reward).

The activities of free-ranging tamarins at the feeding sites offered strong evidence for flexible use of rule-based foraging strategies. Although each of the three 6-day test conditions in Experiment 2 differed in the availability of olfactory and/or local spatial cues that could be used to discriminate between feeding platforms, the tamarins were consistently more accurate at locating reward sites in the afternoon test sessions than in the morning test sessions. As indicated in Table 2, during Experiment 2 the tamarins made 169 visits to the platforms, 69% to reward sites and 31% to sham sites. The frequency of correct choices (visits to reward sites) in the morning trials was 55% (44/80). The frequency of correct choices in the afternoon trials was 80% (72/89). Overall, differences in the ability of the tamarins to discriminate reward from sham sites in the morning and the afternoon trials were significant ($X^2=6.36$, df=1, p<.02).

An examination of the data by experimental condition offers additional insight into the ability of tamarins to solve problems associated with variation in the spatial/temporal availability of food resources. In the first condition (2A), morning and afternoon performances were approximately the same (60% in the morning and 63% in the afternoon; Table 2). A comparison of afternoon performances during Days 1–3 with afternoon performance during Days 4–6 however, shows a marked level of improvement from 54.5% correct to 72.7% correct. Although these differences are not statistically significant, they may indicate that the tamarins were beginning to test pre-solutions and apply a WIN-SHIFT/LOSE-RETURN foraging strategy in order to locate banana feeding platforms.

During the next two test conditions (Conditions 2B and 2C), the tamarins did significantly better in locating reward sites in the afternoon than in the morning (Table 2). It appears that by Day 7, the monkeys had successfully adopted a strategy of site selection that took advantage of temporal predictability in the location of food rewards. In Condition 2B, 83% of sites selected in the afternoon were reward sites. In Condition 2C, 88% of sites selected in the afternoon were reward sites.

Further evidence supporting a marked increase in afternoon foraging success is presented in Table 3. In this table those cases are examined in which the tamarins visited only two platforms at a Feeding Station and both contained a food reward. Given that two of four platforms per station were baited, the probability of encountering bananas on each of the first two platform visits to a given Feeding Station is 16.25% (.5 x .33). During Experiment 2, the tamarins visited only two platforms at a Feeding Station on 32 occasions. In 62.5% of these cases (N=20), two reward and no sham sites were visited (Table 3). Of the 20 cases discussed above, 18 (90%) occurred in the afternoon. A comparison of the morning and afternoon feeding behavior indicates two very different behavioral patterns. Of the 11 cases in the morning in which only two platforms at a station were visited, both were reward sites only 18.2% of the time (2/11). In contrast, of the 21 times only two platforms at a station were visited in the afternoon, 85.7% of the time (18/21) both sites selected were reward sites (see Table 3). Thus, under conditions in which the spatial position of food rewards was systematic but not constant, the tamarins used spatial and temporal information obtained in the morning to predict the location of food in the afternoon.

Is there evidence that the tamarins encode spatial/temporal information in a way that enables them to differentiate information associated with food availability within a day from food availability between days? This question was examined in Experiment 2 by analyzing tamarin foraging patterns on the afternoon of one day and the morning of the following day. If

Table 3. AM and PM Feeding Patterns: Cases in which only 2 platforms were visited at a station

Condition	AM			PM		
	No. Cases	2 reward	%	No. Cases	2 reward	%
2A	2	0	(0%)	4	2	(50%)
2B	5	1	(20%)	5	4	(80%)
2C	4	1	(25%)	12	12	(100%)
Total	11	2	(18.2%)	21	18	(85.7%)

Table 4. Between day foraging patterns in *Saguinus mystax* (comparison of feeding choices from PM of day 1 to AM of day 2)

	WIN/RETURN	WIN-SHIFT	LOSE-RETURN	LOSE/SHIFT
No. of Cases	12	15	4	3
	(44.4%)	(55.5%)	(57.1%)	(42.9%)

these monkeys continued to use a WIN-SHIFT and LOSE-RETURN foraging strategy between days, then they would be expected to (a) avoid visiting sites that offered a banana reward the previous afternoon and (b) return to sites that contained plastic bananas the previous afternoon. Retention of the same foraging pattern within and between days would not support a hypothesis of complex discrimination of spatial-temporal information.

An analysis of the data indicate that tamarins did not continue to adopt a WIN-SHIFT and LOSE-RETURN foraging pattern between days. Of the 27 platforms visited in the afternoon that offered a food reward, the monkeys exhibited a WIN-RETURN strategy 44.4% of the time (N=12) and a WIN-SHIFT strategy 55.5% of the time (N=15; see Table 4). Similarly, platforms containing plastic bananas the previous afternoon were no more likely to be visited the next morning (LOSE-RETURN; 57.1%, N=4) than not to be visited (LOSE-SHIFT; 42.9%, N=3; see Table 4). It appears that tamarin foraging patterns in the morning were considerably less predictable or more random than in the afternoon, and that 'rules' used for foraging efficiency in the afternoon were not applied by the monkeys the next morning. Given the random placement of real and sham feeding sites between days, there was in fact no consistent behavioral rule (although there were local landmark cues and olfactory cues) that the tamarins could have used to increase the morning foraging efficiency.

DISCUSSION

Under natural conditions, resources exploited by many species of rainforest primates exhibit a patchy distribution in time and space. For example, fruiting and flowering schedules of tropical trees may vary on a time scale measured in months (analogous to fruiting phenology), days (analogous to fruit ripening rates) or even with a few hours (analogous to rates of nectar renewal in flowers). Similarly, foliage disturbance associated with insect foraging may cause prey to temporarily flee areas of refuge, only to return a short time later. Thus, to enhance foraging success an animal must extract and remember information about the location of a food patch, the availability of food within a visited patch, and apply that knowledge comparatively across alternate food sites and food types in the environment. Decisions made by the forager may include: which patches to select; what prey items to consume in that patch (i.e., between different food items with different nutritional value); where to search within a patch (e.g. search strategy); when to leave one patch for another; when to return to a previously visited patch, and, what route(s) to follow to locate subsequent food patches (Dolins, 1993). The bases for these decisions are constrained by social and environment factors associated with within-group feeding competition, the energetic costs of acquiring and processing different foods types, susceptibility to predation, as well as the ability of the forager to encode and associate resource availability with temporal and spatial information (Vander Wall, 1982, 1990; Janson, 1985; Terborgh & Janson, 1986; Garber, 1987, 1993; van Schaik & van Noordwijk, 1988; Henzi et al, 1992; Brown & Gass, 1993; Dolins, 1993).

Decisions of when to visit and when to leave a feeding site depend on the acquisition of information about the quality of patches. This has been tested formally in the laboratory using concurrent variable ratio schedules in the 'two-armed bandit' reward paradigm (e.g. Krebs et al., 1978). In these studies the forager is presented simultaneously with a choice of food rewards that differ in reinforcement schedules by varying the rate or amount of food available per unit time (foraging effort). The model assumes that an optimal trade-off exists between the time an animal spends sampling and assessing food availability in a patch and the time an animal spends consuming food in that patch. The expectation is that the forager will learn a problem solving strategy (i.e. WIN-STAY or LOSE-SHIFT) that enables it to most efficiently predict and exploit the spatial/temporal availability of food rewards. It has been shown that birds respond to these experiments by developing 'hypotheses' to increase their rate of food intake over time (Krebs et al., 1978; Balda, 1980; Gass & Sunderland, 1985; Armstrong et al, 1987). These hypotheses include, returning to the most productive patch of the previous day, fully depleting one patch before moving to a second patch, and/or returning to cache sites that minimize time spent in travel.

Data presented in this experimental field study indicate that free-ranging Peruvian moustached tamarins (*Saguinus mystax*) learned several complex spatial and temporal foraging 'rules' that enabled them to predict the location of concealed food rewards. These rules represent behavioral responses to food acquisition problems that are frequently encountered in their environment, such as tracking spatial and temporal changes in the availability and location of feeding sites. Moustached tamarins were presented with four Feeding Stations and 16 feeding platforms that varied experimentally in the location of food rewards, presence/absence of olfactory cues and nearby visual cues (local landmarks), and temporal availability of food rewards. The use of Feeding Stations simulated a set of choices that tamarins face regarding which patch to exploit (patch choice). The use of feeding platforms simulated an ecological problem related to 'within patch' foraging or, when to leave one patch and move to another.

Each experimental condition required that the tamarins learned to distinguish between the location of platforms containing real bananas and platforms containing plastic bananas. Overall the results indicate that these primates were able to rapidly associate spatial information with resource information and adopt a Win-Return foraging pattern when the location of resources was constant (Experiment 1). Under conditions in which the location of reward sites changed systematically (Experiment 2), the tamarins learned to encode accurately spatial information, temporal information, and presence/absence of food rewards at multiple feeding sites. Our results are consistent with laboratory research conducted on cotton-top tamarins (*Saguinus oedipus*) by Dolins (1993). In a series of experiments in which visual cues (landmarks) were systematically rotated and translated on a board containing an 8 x 8 matrix of holes, it was found that the cotton-top tamarins learned to use the spatial relationship between two or more visual cues to locate hidden food items (Dolins, 1993). These results suggest that the cotton-tops developed complex spatial rules representative of cognitive and computational abilities in spatial problem solving. The computational skills required to maintain a mental representation of the distance, angle, and direction between salient objects or landmarks in the environment represent a form of complex cognitive learning that, although present in humans (Rovee-Collier, 1990), has been difficult to document experimentally in nonhuman primates.

In Experiment 1B, red local cues were added to each platform containing bananas. Within one day the monkeys' performance increased to 100%. These results can be interpreted in several ways. One possibility is that the tamarins readily paid attention to small changes in the environment immediately surrounding the feeding platforms, and exhibited considerable behavioral flexibility in learning the association of cues that predicted food availability. However, their response to the 900 clockwise rotation of the 'local cues plus food' condition (Experiment 1C), suggests that if local cues were used to assist accuracy in locating hidden food items, then they were not the primary basis used by the tamarins to locate 'place'. It is also possible that the tamarins did not attend to the red color cues, and that their enhanced ability to select real banana sites in Experiment 1B reflected an overall trend of increased accuracy over time.

Despite these uncertainties in interpretation, the tamarins' response during the rotation condition (Experiment 1C) offers strong support for the predominance of distant spatial information over that of local visual-olfactory cues in selecting feeding sites. Given contradictory information on the day of the rotation, the tamarins consistently based their decision making on spatial cues. Research conducted on gerbils (*Meriones unguiculatus*) indicates that when presented with conflicting local and distal spatial information, these rodents relied on the distal cues over those of the local cues to locate hidden food items (Collett et al., 1986). It is likely that distal spatial landmarks (e.g. permanent landmarks such as river boundaries, large trees, rocks, or forest margins) offer more salient and reliable information over time, whereas local spatial landmarks (e.g. a birds' nest, the presence of flowers or fruits nearby) offer more ephemeral information. Additional research examining the manner in which prosimian and simian species rely on distant versus local cues in locating feeding sites would provide important insight into relationships between navigation, learning and foraging in primates.

In both Experiment 1 and Experiment 2 there was no evidence that the monkeys used olfactory cues to navigate to or select banana-baited versus sham-baited feeding sites. These findings are consistent with those reported by Dolins (1993) on the use of perceptual cues during foraging in captive cotton-top tamarins. In this study, cotton-top tamarins were found to make use of the visual information over that of the olfactory information in locating feeding sites. Although the experimental design did not eliminate entirely the possibility that olfactory cues were being used, based on the results this seemed highly unlikely.

In Experiment 2, a pattern of systematic baiting was implemented that introduced a set of spatial and temporal rules that the tamarins could use to predict the location of resources in the afternoon trials. Predictability of food location was dependent on the tamarins' ability to remember and associate temporal information (over the course of several hours) with knowledge of the locations of earlier successful and unsuccessful foraging attempts. These associations would have to be updated daily and coupled with a second set of spatial behavioral rules, namely a WIN-SHIFT and LOSE-RETURN foraging strategy. Many animals exploit a WIN-SHIFT foraging strategy to avoid returning to resources that have been previously depleted or as a basis for determining optimal return times to renewable food patches (i.e. for foraging on floral nectar or refuging insects). A LOSE-RETURN foraging pattern is also likely to be a common element of a species' foraging pattern, especially in the context of resource sampling and updating information on the future availability or ripening schedule of food within a patch (Krebs, 1981; Garber and Hannon, 1993). Behavioral choices associated with a WIN-SHIFT and LOSE-RETURN pattern differ markedly from those associated with a WIN-RETURN and LOSE-SHIFT pattern.

The application of a learning rule associated with a WIN-SHIFT and LOSE-RETURN (cf. lose-stay) foraging strategy was used by the tamarins in Experiment 2 to increase their ability to accurately locate and differentiate food and nonfood items. In doing so, they exhibited flexible behavioral responses to the conditions specified by the experimental design, and foraged in a goal-directed manner in which time and energy associated with trial-and-error search processes were minimized. In order to accomplish this the tamarins learned information regarding not only where the food was located in the mornings, but also where nonfood (nonreward) items were located. This meant that they learned and applied rules in order to follow the temporal pattern of resource renewability/availability that occurred within a day. As food rewards were not predictable between days, the tamarins showed no evidence of attempting to locate food rewards in positions that had contained plastic bananas on the previous day. Our results compliment those of Menzel and Juno (1982, 1984, 1985) on one-trial learning of the learning set 'WIN-STAY and LOSE-SHIFT' foraging patterns in saddle-back tamarins (*Saguinus fuscicollis*). These authors have shown that tamarins residing in family groups learn to associate the location of objects with a food reward rapidly, and retain this information over a period of several months.

In conclusion, the results of our natural field experiments indicate that moustached tamarins exhibited a high level of learning flexibility and were able to switch from one set of foraging rules to another over the course of three to six days. This rapid change in behavior suggests that these monkeys are highly sensitive to small differences in foraging efficiency resulting from natural changes in the spatial and temporal availability of food resources. Improvement in the afternoon foraging success went from 63% in Experiment 2A to 83% (Day 7–12) in Experiment 2B to 88% (Day 13–18) in Experiment 2C. Although the process through which tamarins assess rates of foraging success remains unknown, it is clear that information sampled in the morning was used in the afternoon to predict the location of feeding sites. It is likely that the tamarins went through an initial process of trial-and-error (sampling behavior), applying behavioral solutions used to locate and acquire resources commonly found in their environment to solve the particular foraging problems posed by our field experiments. This could be tested in a set of expanded field experiments to determine if these or other primates consistently apply a similar sequence of patterned responses (through an examination of their error types and rates) to changes in rates of resource renewal, abundance, and distribution.

SUMMARY

Based on a series of natural field experiments, there is evidence that wild Peruvian moustached tamarins (*Saguinus mystax*) learned to distinguish between the spatial positions of 16 individually baited feeding platforms and use temporal information from previous visits to predict the present position of food rewards. These monkeys were sensitive to even small alterations in environmental information and rapidly learned behavioral foraging rules, applying these to solve novel foraging problems. The tamarins exhibited a high level of learning flexibility and adopted a set of foraging rules such as WIN-RETURN, LOSE-SHIFT, LOSE-RETURN, WIN-SHIFT in response to differences in the spatial and temporal availability of food resources.

In the first set of experiments the locations of eight platforms containing real bananas and eight platforms containing sham banana were held constant over time. The tamarins learned the spatial positions of reward and nonreward platforms rapidly, and by Day 4 over 80% of platform visits contained real bananas. The introduction of two local landmark cues (red flagging) on each platform that contained real bananas (1B) resulted in an increase in tamarin search accuracy to 100%. Given certain limitations in the design of this experimental condition, the degree to which the red cues alone were responsible for this increase in performance remains unclear. In condition 1C, the location of sham and real banana sites were rotated clockwise 90° indicated. Despite the availability of the red local landmark cues and available olfactory cues, on the first day of the rotation the monkeys returned to feeding platforms that had previously contained banana rewards. These results strongly suggest that the tamarins relied on spatial information over red local landmark and olfactory cues as a primary factor in locating feeding sites.

In the second series of experiments a within day resource renewal schedule was introduced into the research design such that (a) platforms containing real bananas in the morning contained plastic bananas in the afternoon, and (b) platforms containing plastic bananas in the morning contained real bananas in the afternoon. The moustached tamarins exhibited a high level of behavioral flexibility and rapidly learned to remember and associate temporal information with information regarding the locations of successful and unsuccessful foraging activities earlier in the day. These primates adopted a WIN-SHIFT/LOSE RETURN foraging strategy in order to predict the location of afternoon baited feeding sites. Between days, however, the resource renewal schedule was random and the tamarins applied a different set of foraging rules. The results from these experiments suggest that the tamarins did not use olfactory cues to navigate to feeding sites.

In conclusion, controlled studies in a field setting provide an excellent method to examine the ways in which nonhuman primates use environmental information when navigating to and selecting feeding sites. Field experiments offer an opportunity to explore hypotheses regarding species-specific differences in spatial learning, the development of foraging rules, and the hierarchy of perceptual cues used by free-ranging primates in making foraging decisions.

ACKNOWLEDGMENTS

This study was conducted with the permission and assistance of the Proyecto Peruano de Primatología "Manuel Moro Sommo" and the Instituto Veterinario de Investigaciones Tropicales y de Altura (IVITA). We thank Dr. Enrique Montoya Gonzales, Executive Director of the Proyecto Peruano de Primatología, Filomeno Encarnación, and Carlos Ique for their support in this project. Assistance in the field was provided by Walter Mermao and Eriberto Mermao. We are also indebted to the Jose Oscanoa Lagunas for his help and friendship. Comments on earlier drafts of this manuscript were provided by Marilyn Norconk, Leslie Digby, and anonymous reviewers. P.A. Garber wishes to thank Lynette, Sara, and Jenni for their encouragement and support. F.L. Dolins would like to acknowledge the support and encouragement of Merelyn Dolins in conducting fieldwork in Peru and in writing this paper. Financial assistant was provided by a William and Flora Hewlett Summer International Research Grant (P.A.G).

REFERENCES

Armstrong, D.P., Gass, C.L., and Sutherland, G.D., 1987, Should foragers remember where they've been? explorations of a simulation model based on the behavior and energetics of territorial humming birds. In: *Foraging Behavior*, A.C. Kamil, J.R. Krebs, and H.R. Pulliam (eds.). New York: Plenum Press, pp. 563–586.

Balda, R.P., 1980, Recovery of cached seeds by a captive *Nucifraga caryocatactes*. *Z. Tierpsychol.* 52: 331–346.

Balda, R.P., and Turek, R.J., 1984, The cache-recovery system as an example of memory capabilities in Clark's nutcracker. In: *Animal Cognition*, H.L Roitblat, T.G. Bever, and H.S. Terrace (eds.). Hillsdale, N.J.: Lawrence Erlbaum Associates, Publishers, pp. 513–532.

Boesch, C., and Boesch, H., 1984, Mental map in wild chimpanzees: an analysis of hammer transports for nut cracking. *Primates* 25: 160–170.

Boesch, C., and Boesch, H., 1989, Hunting behavior of wild chimpanzees in the Tai National Park. *Am. J. Phys. Anthropol.* 78: 547–573.

Brown, G.S., and Gass, C.L. 1993, Spatial association learning by hummingbirds. *Anim. Behav.* 46(3): 487–497.

Byrne, R.W., in press, Primate cognition: comparing problems and skills. *Am. J. Primatol.*

Collett, T.S., Cartwright, B.A., and Smith, B.A., 1986, Landmark learning and visuo-spatial memories in gerbils. *J. Comp. Physiol. A.* 158: 835–851.

Dolins, F.L., 1993, *Spatial Relational Learning and Foraging In Cotton-top Tamarins*, Unpublished Ph.D. Thesis, University of Stirling, Stirling, Scotland.

Fobes, J.L., and King, J.E., 1982, Measuring primate learning abilities. In: *Primate Behavior*, J.L. Fobes and J.E. King (eds.). London: Academic Press, pp. 289–326.

Garber, P.A., 1987, Foraging strategies among living primates. *Ann. Rev. Anthro.* 16: 339–364.

Garber, P.A., 1988, Foraging decisions during nectar feeding by tamarin monkeys (*Saguinus mystax* and *Saguinus fuscicollis*, Callitrichidae, Primates) in Amazonian Peru. *Biotropica* 20(2): 100–106.

Garber, P.A., 1989, Role of spatial memory in primate foraging patterns: *Saguinus mystax* and *Saguinus fuscicollis. Amer. J. Primat.* 19: 203–216.

Garber, P.A., 1993, Seasonal patterns of diet and ranging in two species of tamarin monkeys: Stability versus variability. *Int. J. Primat.* 14(1): 1–22.

Garber, P.A., Encarnación, F., Moya, L., and Pruetz, J.D., 1993a, Demographic and reproductive patterns in moustached tamarin monkeys (*Saguinus mystax*): implications for reconstructing platyrrhine mating systems. *Am. J. Primatol.* 29:235–254.

Garber, P.A., Pruetz, J.D., and Issacson, J., 1993b, Patterns of range use, range defense, and intergroup spacing in moustached tamarin monkeys (*Saguinus mystax*). *Primates* 34: 1–22.

Garber, P.A. and Hannon, B., 1993, Modeling monkeys: a comparison of computer generated and naturally occurring foraging patterns in 2 species of Neotropical primates. *Int. J. Primatol.* 14(6): 827–852.

Garber, P.A., and Pruetz, J.D., In Press, Positional behavior in moustached tamarin monkeys: effects of habitat on locomotor variability and locomotor stability. *J. Human Evol.*

Gass, C.L., and Sunderland, G.D., 1985, Specialization by territorial hummingbirds on experimentally enriched patches of flowers: energetic profitability and learning. *Canad. J. Zool.* 63: 2125–2133.

Henzi, S.P., Byrne, R.W., and Whiten, A., 1992, Patterns of movement by baboons in the Drakensberg mountains: primary responses to the environment. *Int. J. Primatol.* 13: 601–630.

Janson, C.H., 1985, Aggressive competition and individual food intake in wild brown capuchin monkeys. *Beh. Ecol. Sociobiol.* 18: 125–138.

Kamil, A.C., 1984, Adaptation and cognition: Knowing what comes naturally. In: Animal Cognition, H.L. Roitblat, T.G. Bever, and T.S. Terrace (eds.). Hillsdale, N.J.: Lawrence Erlbaum Associates, Publishers, pp. 533–544.

Krebs, J.R., 1981, Optimal foraging decision rules for predators. In: *Behavioural Ecology: An Evolutionary Approach*, J.R. Krebs and N.B. Davies (eds.). Sunderland, MA.: Sinauer Associates, Inc., pp. 23–63.

Krebs, J.R., Erichsen, J.T., Webber, M.I., and Charnov, E.L., 1977, Optimal prey selection in the great tit (*Parus major*). *Anim. Behav.* 25: 30–38.

Krebs, J.R., Kaclenik, A., and Taylor, P., 1978, Test of optimal sampling by foraging great tits. *Nature* 275:27–31.

Krechevsky, I., 1932, "Hypotheses" versus "chance" in the presolutional period in sensory discrimination-learning. *Univ. Calif. Publ. in Psych.* 6: 27–44.

Krechevsky, I., 1938, A study of the continuity of the problem-solving process. *Psych. Rev.* 45: 107–133.

Meador, D.M., Rumbaugh, D.M., Pate, J.L., and Bard, K.A., 1987, Learning, problem solving, cognition, and intelligence. In: *Comparative Primate Biology Vol. 2, Part B, Behavior, Cognition, and Motivation*, G. Mitchell and J. Erwin (eds.). New York: Alan R. Liss, pp. 17–84.

Menzel, E.W., Jr., and Juno, C., 1982, Marmosets (*Saguinus fuscicollis*): Are learning sets learned? *Science* 217: 750–752.

Menzel, E.W., Jr., and Juno, C., 1984, Are learning sets learned? Or: Perhaps no nature-nurture issue has any simple answer. *Anim. Learn. & Behav.* 12(1): 113–115.

Menzel, E.W., Jr., and Juno, C., 1985, Social foraging in marmoset monkeys and the question of intelligence. *Phil. Trans. Royal Soc. London* B308: 145–157.

Milton, K., 1988, Foraging behaviour and the evolution of primate intelligence. In: *Machiavellian Intelligence: Social Expertise and the Evolution of Intellect in Monkeys, Apes, and Humans*, R.W. Byrne and A. Whiten (eds.). Oxford: Oxford University Press, pp. 285–305.

Moya, L., Trigoso, M., and Heltne, P.G., 1980, Manejo de fauna silvestre en semicautiverio en la Isla de Iquitos y Padre Isla - Año 1980. Informe Ordeloreto Dirección Regional de Agricultura y Dirección Forestal y de Fauna, Iquitos, Perú.

Norconk, M.A., 1986, *Interactions between primate species in a Neotropical forest: mixed-species troops of Saguinus mystax and S. fuscicollis (Callitrichidae)*. Ph.D. Thesis, University of California at Los Angeles.

Parker, G., 1986, Evolutionary stable strategies. In: *Behavioural Ecology: An Evolutionary Approach*, J.R. Krebs and N.B. Davies (eds.). Oxford: Blackwell Scientific Publications, pp. 30–61.

Rovee-Collier, C., 1990, The "memory system" of prelinguistic infants. *Ann N.Y., Acad. Sci.* 608: 517–542.

Shettleworth, S.J., and Krebs, J., 1982, How marsh tits find their hoards: the roles of site preference and spatial memory. *J. Exp. Psych.: Anim. Behav. Proc.* 8: 354–375.

Sherry, D.F., 1984, What food-storing birds remember. *Canad. J. Psych.* 38(2): 304–321.

Stephens, D.W., and Krebs, J.R., 1986, *Foraging Theory*. Princeton: Princeton University Press.

Terborgh, J.W. and Janson, C.H., 1986, Socioecology of primates. *Ann Rev. Ecol. Syst.* 17: 111–135.

van Schaik, C.P., and van Noordwijk, M.A., 1988, Scramble and contest in feeding competition among female long-tailed macaques (*Macaca fascicularis*). *Behaviour* 105: 77–98.

Vander Wall, S.B., 1982, An experimental analysis of cache recovery in Clark's nutcracker. *Anim. Behav.* 30: 84–94.

Vander Wall, S.B., 1990, *Food Hoarding In Animals*. London: The University of Chicago Press.

SECTION III

Critical Issues in Cebine Evolution and Behavior

CRITICAL ISSUES IN CEBINE EVOLUTION AND BEHAVIOR

Linda Marie Fedigan,[1] Alfred L. Rosenberger,[2] Sue Boinski,[3]
Marilyn A. Norconk,[4] and Paul A. Garber[5]

[1]Department of Anthropology
University of Alberta
Edmonton, Alberta, T6G 2H4 Canada
[2]Department of Zoological Research
National Zoological Park
Washington, D.C. 20008
[3]Department of Anthropology and Division of Comparative Medicine
University of Florida
Gainesville, Florida 32611
[4]Department of Anthropology
Kent State University
Kent, Ohio 44242
[5]Department of Anthropology
University of Illinois
Urbana, Illinois 61801

Cebus and *Saimiri*, together with *Callicebus* and *Aotus*, represent four genera of New World monkeys whose unresolved taxonomic position has served to muddle platyrrhine cladistics. Their affinities remain somewhat problematic (Schneider and Rosenberger, this volume), although we argue that in the past 20 years, new molecular, genetic, morphological and behavioral analyses have narrowed the range of possible explanations regarding *Cebus* and *Saimiri*, in particular. Here, we attempt to clarify cebine evolutionary relationships and outline some interesting and relevant directions for future studies in behavior and ecology.

PART I: SYSTEMATICS AND MORPHOLOGICAL EVOLUTION
(A.L. Rosenberger, M.A. Norconk, and P.A. Garber)

As is usually the case in systematics, without the proper frame of reference it is often easier to understand what a taxon is not, phylogenetically, rather than what it is. For example, *Cebus* has a grasping tail, but it is not an ateline. *Cebus* is said to have a partially

Table 1. Genus level classification of cebines

Family Cebidae
  Subfamily Cebinae
    Tribe Cebini
      Cebus - Cebus monkey
    Tribe Saimiriini
      (*)*Saimiri* - Squirrel monkey; Middle Miocene, Colombia
      *Laventiana - Middle Miocene, Colombia
      *Dolichocebus - Early Miocene, Argentina
  Other cebines
      *Chilecebus - Early Miocene, Chile
      *Antillothrix - Pleistocene/Recent, Dominican Republic

*Extinct genus. (*) Living genus which includes *Neosimiri* as a subgenus. See Schneider and Rosenberger (this volume) and Rosenberger (1992) for references and discussion. "Other cebines" include fossils whose relationships *within* Cebinae are uncertain.

opposable thumb, but it is not a catarrhine. Nor is it likely that the relatively large capuchin brain will conjure up fantasies of a special evolutionary relationship with hominids.

The same would hold for *Saimiri*. With a round head, short face, agouti coloration, long tail and insectivorous diet, one might mistake it for a talapoin, which it is not. Like *Cebus*, *Saimiri* has a relatively large brain, highly sexually dimorphic canines, and shares a long, novel sequence of the IRBP and epsilon globin genes, with *Cebus,* to the exclusion of other platyrrhines. What are we to make of these similarities? The null hypothesis should be that *Cebus* and *Saimiri* are closely related.

As Schneider and Rosenberger (this volume) relate, there are dichotomous views on the relationships of *Cebus*. One view nests capuchins within a group that also includes callitrichines and *Saimiri* - all cebids (Table 1). The other view places *Cebus* quite outside the radiation of most modern forms, albeit with *Saimiri* again appearing as a potential sister-taxon. This latter view can be termed the "outlier" hypothesis, and argues that capuchins represent an ancient platyrrhine radiation isolated from all other genera for perhaps 20 million years. The fossil record is of little help here, and the crucial single step to resolving this conflict rests with understanding the linkage between *Cebus* and *Saimiri*. To us, capuchins and squirrel monkeys represent a pair of closely related genera and this makes the outlier hypothesis patently untenable. The hypothesis that *Cebus* and *Saimiri* are sister taxa has been tested often, at least implicitly, and it has been rarely if at all refuted.

Every point of similarity (either primitive or derived) found between *Cebus* and *Saimiri* is a corroboration of the null hypothesis. Every potential point of derived similarity found between either *Cebus* or *Saimiri* and a taxon outside this pair must be demonstrated to be homologous if it is to weaken the null hypothesis. In general, the modern radiation of extant capuchins is characterized by relatively large brain size, enhanced manual dexterity and tool use, elaborate visual system, semi-prehensile tail, complex system of social communication and group coordination, thickly enameled teeth, premolar dominance, hyper-short face, and narrow inter-orbital distance. Although it is possible that each of these traits could be interpreted as autapomorphic, unique add-ons that accumulated since the genus split from the stem of the platyrrhine radiation, we feel that this is highly unlikely. Moreover, if capuchins do represent an old, isolated lineage, then it is necessary to posit that *Cebus* and *Saimiri* have convergently evolved short faces, broad premolars, minuscule third molars, narrow nasal bones, rounded braincases containing relatively large brains, and highly dimorphic canines honing on a *Cebus*-like premolar an-

vil. Given that these traits are distributed across the face, cranium, and dentition supporting a prey-based foraging strategy, arguments for evolutionary convergences in each of
these traits are difficult to reconcile.

What is the genesis of the "outlier hypothesis" This result appeared in three numerical cladistic studies (Dunlap et al., 1985; Ford, 1986 et seq.; Kay, 1990), all of which
shared the same set of built-in constraints. In each of these studies, catarrhines were used
as the principle source of cladistic information (the out-group). No *a priori* study was undertaken to specify homologies shared by platyrrhines and catarrhines. Reconstructing the
ancestral platyrrhine pattern was left to the algorithms. We suggest their approach biased
the analyses to search for platyrrhines with the highest frequency of catarrhine-like features, based on the available sample. *Cebus* molars were anatomically likened to those of
Apidium; *Cebus* ankle joints were said to resemble early Fayum parapithecid anthropoids;
and *Cebus* forearm muscles were compared favorably with extant Old World monkeys.
Thus, the outlier hypothesis was driven by the limits of a methodology.

The null hypothesis regarding the systematic position of *Cebus*, as a member of a
lineage linked with *Saimiri*, remains the most compelling (Schneider and Rosenberger,
this volume). It will be strengthened as we continue to explore differences in cebine ontogeny (Hartwig, 1995 ; Armstrong and Shea, in press) and how that influences patterns of
behavior and ecology. We argue that *Cebus* and *Saimiri* are closely related genera, separated for millions of years but still bound to the pre-catching guild of cebids (Table 1). It
is not surprising that over time they would accumilate morphological differences that
might lead to questions of ancestry. However, it is a set of unique similarities that unite
these lineages phylogenetically.

PART II. BEHAVIOR AND ECOLOGY ISSUES IN *Cebus* AND *Saimiri* (L.M. Fedigan and S. Boinski)

To most casual observers *Saimiri* and *Cebus* are strikingly similar in their general
appearance and demeanor. Whether in cages or a neotropical forest, these beasts are usually recalled as busily moving about, poking, prying, peeling, and scraping substrates
looking for tasty bits, and bustling about fruit sources, bumping and jostling each other
like a litter of puppies at a food bowl. Although questions exist regarding their precise
taxonomic affinities (but see new molecular data in Schneider and Rosenberger, this volume), *Saimiri* and *Cebus* are readily pooled into an ecological 'clade' of insectivorous primates. Despite these similarities, many of issues relevant to the behavioral ecology of
Cebus and *Saimiri* are most obvious when their many differences are noted.

First, the systematics and genetic structure of these genera are dramatically different. *Cebus* has four well-defined species. *C. albifrons*, *C. olivaceous*, and *C. capucinus* replace each other geographically and together form a tidy clade which is clearly
morphologically and behaviorally different from *C. apella* (Mittermeier and Coimbra-
Filho,1981; Groves, 1987). In contrast, *Saimiri* presents systematists with a messy pattern
of parapatric and allopatric populations and with evidence of species distinctions. The me-
ticulous and herculean efforts of Hershkovitz (1984) and Thorington (1985) in sorting out
Saimiri taxonomy, based largely on pelage, osteological, and chromosomal characters, are
now in the process of being refined to include more recent behavioral and molecular data
(i.e., Costello et al., 1993; Silva et al., 1993; Garcia et al., 1995). Consensus on the spe-
cies- and subspecies-level taxonomy of *Saimiri*, however, is unlikely to be achieved even
within the next decade.

Second, adult capuchins can easily weigh four to six times more than an adult squirrel monkey. Although body proportions are not markedly dissimilar, the consequences from the size differences reverberate throughout the biology of both genera. First, all else being equal, *Saimiri* are much more vulnerable to predation than *Cebus*. A much broader range of potential predators can capture a *Saimiri* than a *Cebus*. For example, 50% of infant *S. oerstedi* are lost to confirmed or probable predation by avian predators by six months of age (Boinski, 1987). One probable consequence of their enhanced vulnerability to predation, is that *Saimiri* troop sizes can easily be three to six or seven times larger than a *Cebus* troop. Extremely large troop sizes appear to be an anti-predator adaptation in *Saimiri* (Boinski, 1988a), as do numerous peculiarities in *Saimiri* reproduction, including their remarkably synchronous seasonal birth peaks and the extended, unusually variable duration of gestation (Boinski, 1987; Hartwig, 1995).

Third, both genera are highly vocal in the wild, and individual troop members may produce more than a 1000 vocalizations each day in the course of normal activities (Boinski, 1991; 1993; Boinski and Mitchell 1992, 1995; Boinski and Campbell 1995, In press). In *Saimiri*, however, a much larger proportion of calls can be described as 'contact' calls with the function of exchanging positional information among visually isolated troop members. In *Saimiri*, the enhanced susceptibility to predation due to their smaller body size appears responsible for the greater emphasis of contact calls. The number of contact calls produced by a squirrel monkey is positively related to the extent of spatial separation between the squirrel monkey and its nearest neighbor. Only in infant capuchins are such 'security-blanket' vocalizations found.

Another repercussion of the body size difference is that *Cebus* have greater bite force and manual strength than do *Saimiri* (Janson and Boinski,1992). *Saimiri* are foliage gleaners, extracting arthropods and small vertebrates off leaf and bark surfaces or from within leaf curls. *Cebus* can twist, rip, bite, and crunch open hard substances to extract grubs, and other social insects unavailable to *Saimiri*. Even when foraging in mixed-species groups, the two genera overlap little in the sites in which they forage for arthropods. *Cebus* can also harvest high-quality fruit sources, such as dense clusters of hard-husked palm fruits, which are completely inaccessible to sympatric *Saimiri* because the latter cannot penetrate the husks.

Fifth, although both species have anomalously large brains relative to body size compared to other primates, the concomitant developmental trajectories that produce the large brains are markedly different and appear to reflect very different selective regimes (Hartwig, 1995, 1996). *Cebus* has more postnatal brain growth and slower motor skill development than other New World primates. Neonates are highly precocial in *Saimiri* in terms of both brain growth and motor skill development. Another developmental difference is that *Saimiri* evidences much more marked geographic variation in development than has yet been reported in *Cebus*. Infants are in great part weaned by 4.5 months in *S. oerstedi* in Costa Rica and are rarely in close vicinity to their mothers by 8 months of age (Boinski and Fragaszy, 1989). In contrast, *S. sciureus* in Peru are weaned by about 19 months of age (Mitchell, 1990; Boinski and Mitchell, 1995).

Finally, in regard to the extent and breadth of field studies the positions of the two genera are reversed. *Cebus* has been the subject of detailed long-term behavioral and ecological field studies since Oppenheimer's field work on Barro Colorado Island in the mid 1960's (see Freese and Oppenheimer (1981) for historical review). The number of field observational and experimental studies shows no signs of diminishing (see below). The long-term field legacy for *Saimiri* is far different. Squirrel monkeys were one of the five monkey species studied in Terborgh's (1983) and his associates year-long ecological study of the primate community at Manu, Peru. Fortuitous field conditions facilitated Boinski's (1986) studies of squirrel monkeys in Costa Rica, the first with detailed social observations of individually recognized troop members. Mitchell (1990) quickly followed with her superb investigation of the ecology and complex social behavior of *Saimiri* in Manu. The third, and only other population of squirrel monkeys studied for an extended period (although the results remain largely unpublished) is of an artificially stocked population on Isla de Santa Sofia, a 400-ha island in Amazonas, Colombia (Bailey et al., 1974; Sponsel et al., 1974). To our knowledge, no other researchers have undertaken long-term behavioral field studies of *Saimiri*. Quite a few workers, however, mention *Saimiri* in reports on other neotropical primates (e.g. Peres, 1994).

Clearly, the main challenge awaiting further insights into *Saimiri* are additional long-term field studies of behavior and ecology at new sites. Squirrel monkeys have the most geographically variable social organization of any group of closely related primate populations (Mitchell et al., 1991; Boinski, In press). *S. oerstedi* arguably exhibits the most egalitarian, least aggressive social organization of primates with large multi-female, multi-male social organizations (Boinski, 1988a, 1994; Boinski and Mitchell, 1994). Moreover, *S. oerstedi* is one of the minority of primate taxa in which female dispersal is the rule and negligible female-female bonds are evident. In contrast, Peruvian females are dominant to males and female-female social bonds are strong (Mitchell, 1990,1994). Colombian *Saimiri* are reminiscent of those in Peru, with male transfer and female-female bonds, but during the four month-long dry season, food becomes extremely scarce and troops fission into small subgroups (R. C. Bailey, pers. comm). In Suriname, a fourth type of social organization occurs. Males are fully integrated into the social group, most males are dominant to most females and much time is allocated to dyadic dominance displays in the wild (Boinski, unpublished data). The Peru-Costa Rica contrast has been explained by differing levels of within-group food competition. Studies begun by Boinski in Suriname aim to extend and test this model on a squirrel monkey population that exhibits a very different social organization.

Unlike *Saimiri* studies, recent studies of *Cebus* have been spurred by a surge of interest in primate cognition. The investigation of cognitive abilities is a hot topic throughout primatology, and fundamental to this issue is the study of social and ecological pressures that were likely to have selected for intelligence. Both social and foraging intelligence are now being investigated in captive and field studies. Among the topics under investigation are formation of coalitions (Perry, 1995a, 1996b), reconciliation (Perry, 1995b), cognitive capacities under captive, experimental conditions (Visalberghi, 1988, 1990; Anderson and Roeder, 1989; Fragaszy and Visalberghi, 1990; Visalberghi and Fragaszy, 1990; Fragaszy et al., 1994; Marchal and Anderson, 1993), spatial memory and rule-based foraging (Garber and Paciulli, 1996; Janson, this volume), social interactions and vocal behavior leading to troop travel decisions (Boinski 1993, this volume; Boinski and Campbell, 1995), alarm calls (Norris, 1990), choice of plants used for medicinal purposes (Baker, 1996), hunting (Fedigan, 1990, Rose, 1994a,1996; Perry and Rose, 1994), ontogeny of foraging skills (MacKinnon, 1995), food sharing (de Waal et al., 1993), tool making (in captivity: Anderson, 1990; Westergaard and Suomi, 1994a,b; in the wild: Boinski, 1988b, Chevalier-Skolnikoff, 1990, Fernandes, 1991).

We are also just starting to piece together the picture of social dynamics in capuchins. We know that males disperse and that females are usually philopatric. Does this mean that females form matrilines and that female relatedness underlies much of the affinitive interaction patterns? Capuchins readily form coalitions, they often allonurse and alloparent each others' young, and they engage in frequent triadic interactions (O'Brien

1988, 1991, 1993; O'Brien and Robinson, 1991; Perry, 1996b; Robinson, 1993). Are these patterns based on kinship, rank, friendship, or some other factor? Years ago, Bernstein (1966) conducted experiments in captivity to show that capuchins, unlike macaques, do not form linear dominance hierarchies, and capuchin field workers experience some difficulties in determining rank other than that of alpha individuals. How is dominance rank acquired and is it linear? All four capuchin species exhibit a pattern of prominent alpha males, but in *C. albifrons* and *C. capucinus*, adult males within the same group associate strongly with each other, they cooperate actively in group defense, they look for and retrieve lost males, and they sometimes transfer groups together (Fedigan, 1993, Fedigan et al., this volume; Perry, 1996a; Rose, 1994b). How far and how often males disperse is unknown, but in the white-faced capuchins that have been under observation for 13 years in Santa Rosa National Park, Costa Rica, small parties of adult males invade groups every few years, fighting with the resident males, and injuring females and infants in the process. At other times, single males join groups quietly and inconspicuously over an extended period of time. Resident males of a group sometimes exhibit extensive male care, protecting, carrying, and retrieving infants, and even allowing them to suckle. What factors underlay these highly variable patterns of male social behavior are as yet unknown, but field studies focused on males are underway. Field studies have also investigated female social behavior, particularly the feeding and traveling costs of female *Cebus olivaceus* in groups of different sizes (Miller, 1992, this volume).

As with *Saimiri*, we still have much to learn about *Cebus* mating systems. Some capuchin species mate cryptically (e.g., *C. capucinus*, see Parish et al., 1996), whereas in others the females clearly and overtly choose the alpha male for mating (e.g., *C. apella*, see Janson,1984; Phillips et al., 1994). Two of the four *Cebus* species (*C. albifrons* and *C. capucinus*) live in groups that are decidedly multi-male. The other two species (*C.olivaceus* and *C. apella*) live in what might best characterized as "age-graded male" (or functionally speaking, unimale) systems. In these cases, the top-ranking male is the only reproductively active male, and is highly conspicuous socially (Izawa, 1980; Janson, 1984; Robinson, 1988; O'Brien, 1991). Male capuchins in at least two of the *Cebus* species exhibit strong male-male bonds, and female kinship and dominance systems do not seem as clear cut as in cercopithecines.

Capuchins in captivity live very long lives, up to 47 years, which is much longer than expected for a primate of their body size, but less surprising in terms of their brain-to-body weight ratio. Are members of this genus similarly long-lived in the wild? The entire pace of life seems slower than expected in capuchins - weaning age, age at first birth, interbirth intervals, estrous cycle length (Fedigan and Rose, 1995). Is this "slow" life history pattern related only to their large brains, or are there other factors involved? We need more long-term life history and ecological data from field studies on both this genus and other primates exhibiting large brain to body ratios, such as squirrel monkeys. Birth rates are highly variable from year to year, however, we have yet to document the factors that might affect annual variation in reproduction. Capuchins have long been thought to be nonseasonal breeders, but at least one study found significantly more infants born in the dry than the wet season (Fedigan et al., this volume). Is it possible, as Susan Perry has suggested, that females of the same group exhibit some loose form of breeding synchrony, resulting in clusters of births within a troop over a several month-long period, but not strict seasonality? Capuchins appear to rely on pheromones and olfactory communication. This seems to be indicated by their neurophysiology and by their frequent use of behaviors with an olfactory component, such as urine-washing (Robinson, 1979), fur-rubbing with odoriferous substances (Ludes and Anderson, 1995; Baker, 1996), and mutual hand-sniffing (Perry, 1996b). Plausibly, olfactory communication might underlie the phenomenon of birth clusters within a troop.

A number of factors about the capuchins' relationship to their environment are also distinctive. As mentioned above, these monkeys are famous as extractive, manipulative foragers. Capuchins seem to specialize in food that "fights back", that is, flora and fauna with highly evolved defense mechanisms. What ecological factors might have selected for this pattern and what are the repercussions for the capuchins' ability to adapt and survive under rapidly changing ecological conditions? For example, how do these opportunistic, omnivorous feeders fare under the all too prevalent conditions in Central and South America of forest destruction and fragmentation? And how do they fare under the less common conditions of forest regeneration? In many parts of their range, capuchins are reported to prey on small vertebrates (e.g. birds, lizards, small mammals, see Newcomer and DeFarcy, 1985; Fedigan, 1990; Perry and Rose, 1994). Is this a recent adaptation or do they exhibit behaviors indicating that they have long been effective at vertebrate predation? They are also renowned for their ability to mount an impressive, effective, and cooperative anti-predator display by vocalizing, mobbing, and breaking branches on the source of the disturbance (e.g., Boinski, 1988b), and thus driving off creatures many times their size (cats, coyotes, boa constrictors, ecotourists). Finally, capuchins exhibit considerable variability from group to group and species to species in their diet and the way they manipulate and extract food — is this variation due to food availability or local traditions (Chapman and Fedigan, 1991)?

This brief review of issues in the behavioral ecology of *Cebus* and *Saimiri* make it clear that we need more long-term intensive studies of known individuals, groups and populations before we can fully understand the mechanisms that underpin many of the patterns discussed here. The last decade has seen a great increase in our knowledge of cebine behavioral ecology. With the increasing attention paid to the neotropical primates and the availability of new non-invasive techniques for biological sampling in the field, the next decade should prove to be an even greater leap forward in our understanding of these animals.

REFERENCES

Anderson, J.R. 1990. Use of objects as hammers to open nuts by capuchin monkeys (*Cebus apella*). *Folia primatol.* 54: 138–145.

Anderson, J.R., and Roeder, J. 1989. Responses of capuchin monkeys (*Cebus apella*) to different conditions of mirror-image stimulation. *Primates* 30: 581–587.

Armstrong, E., and Shea, M. A. In press. Brains of New World and Old World monkeys, in: W.G. Kinzey (ed.), *New World Primates*. Aldine de Gruyter, New York.

Bailey, R.C., Baker, R.S., Brown, D.S., von Hildebrand, P., Mittermeier, R.A., Sponsel, L.E., and Wolk, K.E. 1974. Progress of a breeding project for non-human primates in Colombia. *Nature* 248:453–455.

Baker, M. 1996. Fur-rubbing: use of medicinal plants by capuchin monkeys (*Cebus capucinus*). *Am J Primatol.* 38: 263–270.

Bernstein, I.S. 1966. Analysis of a key role in a capuchin (*Cebus albifrons*) group. *Tulane Studies in Zoology* 13: 49–54.

Boinski, S. 1986. The ecology of squirrel monkeys in Costa Rica. Unpub. Ph.D. Thesis. The University of Texas at Austin.

Boinski, S. 1987. Birth synchrony in squirrel monkeys (*Saimiri oerstedi*): a strategy to reduce neonatal predation. *Behav. Ecol. Sociobiol.* 21:393–400.

Boinski, S. 1988a. Sex differences in the foraging behavior of squirrel monkeys in a seasonal habitat. *Behav. Ecol. Sociobiol.* 23:177–186.

Boinski, S. 1988b. Use of a club by a wild white-faced capuchin (*Cebus capucinus*) to attack a venomous snake (*Bothrops asper*). *Am. J. Primatol.* 14:177–179.

Boinski, S. 1991. The coordination of spatial position: a field study of the vocal behaviour of adult female squirrel monkeys. *Animal Behaviour* 41:89–102.

Boinski, S. 1993. Vocal coordination of troop movement among white-faced capuchin monkeys, *Cebus capucinus*. *Am. J. Primatol.* 30: 85–100.

Boinski, S. 1994. Affiliation patterns among male Costa Rican squirrel monkeys. *Behaviour* 130:191–209.

Boinski, S. In press Stress Responses in Primates: Proximate Mechanisms in the Evolution of Social Organization, in: S. A. Foster and J.A. Endler (eds), *Geographic variation in behavior: An evolutionary perspective*. Oxford University Press, Oxford England.

Boinski, S., and Campbell, A. F. In press. The huh vocalization of white-faced capuchins: a spacing call disguised as a food call? *Ethology*

Boinski, S. and Campbell, A.F. 1995. Use of trill vocalizations to coordinate troop movement among white-faced capuchins: a second field test, *Behaviour* 132: 875–901.

Boinski, S., and Fragaszy, D.M. 1989. The ontogeny of foraging behavior in squirrel monkeys, *Saimiri oerstedi*. *Animal Behaviour* 47:415–428.

Boinski, S., and Mitchell, C.L. 1992. The ecological and social factors affecting adult female squirrel monkey vocal behavior. *Ethology* 92:316–330.

Boinski, S., and Mitchell, C.L. 1994. Male dispersal and association patterns in Costa Rican squirrel monkeys (*Saimiri oerstedi*). *Am. J. Primatol.* 34:157–170.

Boinski, S., and Mitchell, C. L. 1995. Wild squirrel monkey (*Saimiri sciureus*) "caregiver" calls: contexts and acoustic structure. *Am.J. Primatol.* 35:129–138.

Chevalier-Skolnikoff, S. 1990. Tool-use by wild *Cebus* monkeys at Santa Rosa National Park, Costa Rica. *Primates* 31: 375–383.

Chapman, C.A., and Fedigan, L.M. 1991. Dietary differences between neighboring *Cebus capucinus* groups: local traditions, food availability or responses to food profitability? *Folia primatol.* 54: 177–186.

Costello, R.K.; Dickinson, C.; Rosenberger, A.L.; Boinski, S.; Szalay, F.S. 1993. A multidisciplinary approach to squirrel monkey (genus *Saimiri*) species taxonomy, in: W.B. Kimbel and L. B. Martin (eds), *Species, Species concepts, and Primate evolution*, pp. 177–237. Plenum, New York.

Dunlap, S. S., Thorington, R. W. Jr., and Aziz, M. A. 1985. Forelimb anatomy of New World monkeys: myology, and the interpretation of primitive anthropoid models. *Am. J. Phys. Anthropol.* 68: 499–517.

Fedigan, L.M. 1990. Vertebrate predation in *Cebus capucinus*: meat-eating in a neotropical monkey. *Folia primatol.* 54: 196–205.

Fedigan, L.M. 1993. Sex differences and intersexual relations in adult white-faced capuchins (*Cebus capucinus*). *Int. J. Primatol.* 14: 1–25.

Fedigan, L. M., and Rose, L.M. 1995.Interbirth interval variation in three sympatric species of neotropical monkey. *Am. J. Primatol.* 37:9–24.

Fernandes, M.E.B. 1991. Tool use and predation of oysters (*Crassostrea rhizophorae*) by the tufted capuchins, *Cebus apella apella*, in brackish water mangrove swamp. *Primates* 32: 529–531.

Ford, S.M. 1986. Systematics of the New World monkeys, in: D.R. Swindler and J. Erwin (eds.), *Comparative Primate Biology, Volume 1: Systematics, Evolution and Anatomy*, pp. 73–135. Alan R. Liss, New York.

Fragaszy, D.M., and Visalberghi, E. 1990. Social processes affecting the appearance of innovative behaviors in capuchin monkeys. *Folia Primatol.* 54:155–165.

Fragaszy, D.M., Vitale, A.F., and Ritchie, B. 1994. Variation among juvenile capuchins in social influences on exploration. *Am. J. Primatol.* 32: 249–260.

Freese, C.H., and Oppenheimer, J.R. 1981. The capuchin monkey, genus Cebus, in: A.F. Coimbra-Filho and R.A. Mittermeier (eds.), *Ecology and Behavior of Neotropical Primates*, volume 1, pp. 331–390. Academia Brasiliera de Ciencias, Rio de Janeiro.

Garber, P.A., and Paciulli, L.M. In press. Experimental field study of spatial memory and learning in wild capuchin monkeys (*C. capucinus*). *Folia Primatol.*

Garcia, M., Borrell, A., Mudry, M., Egozcne, J., and Pansa, M. 1995. Prometaphase karyotype and restriction enzyme banding in squirrel monkeys, *Saimiri boliviensis boliviensis*. *J. Mammal.* 76:497–503.

Groves, C.P. 1987. *A theory of human and primate evolution*. Oxford University Press, Oxford UK.

Hartwig, W.C. 1995. Effect of life history on the squirrel monkey (Platyrrhini, *Saimiri*) cranium. *Am. J. Phys. Anthropol.* 97:435–449.

Hartwig, W.C. 1996. Perinatal life history traits in New World monkeys. *Am. J. Primatol.* 40:99–130.

Hershkovitz, P. 1984. Taxonomy of squirrel monkeys genus *Saimiri* (Cebidae: Platyrrhini): A preliminary report with description of a hitherto unnamed form. *Am. J. Primatol.* 6:257–312.

Izawa, K. 1980. Social behavior of the wild black-capped capuchin (*Cebus apella*). *Primates* 21: 443–467.

Janson, C.H. 1984. Female choice and mating system of the brown capuchin monkey *Cebus apella* (Primates: Cebidae). *Z. Tierpsychol.* 65: 177–200.

Janson, C.H., and Boinski, S. 1992. Morphological and behavioral adaptations for foraging in generalist primates: the case of the cebines. *Am. J. Phys. Anthropol.* 88:483–498.

Kay, R.F. 1990. The phyletic relationships of extant and fossil Pitheciinae (Platyrrhini, Anthropoidea). *J. Hum. Evol.* 19: 175–208.

Ludes, E., and Anderson, J.R. 1995. Peat-bathing by captive white-faced capuchin monkeys (*Cebus capucinus*). *Folia primatol.* 65: 38–42.

MacKinnon, K. 1995. Age differences in foraging patterns and spatial associations of the white-faced capuchin, *Cebus capucinus*, in Costa Rica. Unpubl. MA Thesis, Univ Alberta, Edmonton, Canada.

Marchal, P. and Anderson, J.R. 1993. Mirror-image responses in capuchin monkeys (*Cebus capucinus*): social responses and use of reflected environmental information. *Folia primatol.* 51:165–173.

Miller, L.E. 1991. Socioecology of the wedge-capped capuchin monkey (Cebus olivaceus). Unpub. PhD dissertation. University of California, Davis, CA.

Mitchell, C.L. 1990. The ecological basis for female social dominance: A behavioral study of the squirrel monkey (*Saimiri sciureus*) in the wild. Unpub. PhD dissertation. Princeton University, Princeton, NJ.

Mitchell, C.L. 1994. Migration alliances and coalition among adult male South American squirrel monkeys (*Saimiri sciureus*). *Behaviour* 130:169–190.

Mitchell, C., Boinski, S., and van Schaik, C.P. 1991. Competitive regimes and female bonding in two species of squirrel monkey (*Saimiri oerstedi* and *S. sciureus*). *Behav. Ecol. Sociobiol.* 28:55–60.

Mittermeier, R.A., and Coimbra-Filho, A.F. 1981. Systematics: species and subspecies, in: A.F. Coimbra-Filho and R.A. Mittermeier (eds.), *Ecology and Behavior of Neotropical Primates*. Volume 1, pp. 29–110. Academia Brasiliera de Ciencias, Rio de Janeiro.

Newcomer, M.W., and De Farcy, D.D. 1985. White-faced capuchin (*Cebus capucinus*) predation on a nestling coati (*Nasua narica*)" *J. Mammal.* 66: 185–186.

Norris, J.C. 1990. The semantics of *Cebus olivaceus* alarm calls: object designation and attribution. Unpubl. PhD dissertation University of Florida, Gainesville.

O'Brien, T.G. 1988. Parasitic nursing behavior in the wedge-capped capuchin monkey (*Cebus olivaceus*). *Am. J. Primatol.* 16: 341–344.

O'Brien, T.G. 1991. Female-male social interactions in wedge-capped capuchin monkeys. Benefits and costs of group living. *Anim. Behav.* 41:555–567.

O'Brien, T.G. 1993. Stability of social relationships in female wedge-capped capuchin monkeys, in: M. Pereira and L. Fairbanks (eds), *Juvenile Primates*, pp.197–210. Oxford University Press, Oxford, UK.

O'Brien, T.G., and Robinson, J.G. 1991. Allomaternal care by female wedge-capped capuchin monkeys: effects of age, rank and relatedness. *Behaviour* 119: 30–50.

Parish, A.R., Manson, J.H., and Perry, S.E. 1996. Nonconceptive sexual behavior in bonobos and capuchins. Abstract 001G of the XVIth Congress of the International Primatological Society and XIXth Conference of the American Society of Primatology, August 11–16, 1996, Madison, WI.

Peres, C.A. 1994. Primate response to phenological change in an Amazonian terre firme forest. *Biotropica* 26:98–112.

Perry, S.E. 1995a. Patterns of coalitionary aggression in wild white-faced capuchin monkeys, *Cebus capucinus*. *Am. J Primatol.* 36: 147.

Perry, S.E. 1995b. Social relationships in wild white-faced capuchin monkeys, *Cebus capucinus*. Unpubl. PhD dissertation, University of Michigan.

Perry, S.E. 1996a. Intergroup encounters in wild white-faced capuchins, *Cebus capucinus*. *Int. J. Primatol.* 17: 309–330.

Perry, S.E. 1996b. Female-female relationships in wild white-faced capuchin monkeys, *Cebus capucinus*. *Am. J. Primatol.* 40:167–182.

Perry, S.E., and Rose, L.M. 1994. Begging and transfer of coati meat by white-faced capuchin monkeys, *Cebus capucinus*. *Primates* 35: 409–415.

Phillips, K.A., Bernstein, I.S., Dettmer, E.L. Devermann, H., and Powers, M. 1994. Sexual behavior in brown capuchins (*Cebus apella*). *Int. J. Primatol.* 15: 907–917.

Robinson, J.G. 1979. Correlates of urine-washing in the wedge-capped capuchin, *Cebus nigrivittatus*," in J.F. Eisenberg (ed), *Vertebrate Ecology in the Northern Neotropics*, pp. 137–143. Smithsonian Institution Press, Washington D.C.

Robinson, J.G. 1988. Demography and group structure in wedge-capped capuchin monkeys, *Cebus olivaceus*. *Behaviour* 104: 202:232.

Robinson, J.G. 1993. Allogrooming behavior among female wedge-capped capuchin monkeys. *Anim. Behav.* 46: 499–510.

Rose, L.M. 1994a. Sex differences in diet and foraging in white-faced capuchins. *Int. J. Primatol.* 15: 95–114.

Rose, L.M. 1994b. Benefits and costs of resident males to females in white-faced capuchins. *Am. J. Primatol.* 32: 235–248.

Rose, L.M. 1996. Socio-ecology of meat eating and food sharing in *Pan* and *Cebus*. Abstract 001E of the XVIth Congress of the International Primatological Society and XIXth Conference of the American Society of Primatology, August 11–16, 1996, Madison, WI.

Rosenberger, A.L. 1992. Evolution of feeding niches in new world monkeys. *Am. J. Phys. Anthropol.* 88:525–562.

Silva, B.T.F., Sampaio, M.I.C., Schneider, H., Schneider, M.C., Montoya, E., Encarnacion, F., Callegiari-Jacques, S.M., and Salzano, F.M. 1993. Protein electrophoretic variability in *Saimiri* and the question of its species status. *Am. J. Primatol.* 29:183–193.

Sponsel, L.E., Brown, D.S., Bailey, R.C., and Mittermeier, R.A. 1974. Evaluation of squirrel monkey ranching in Santa Sophia Island, Amazonas, Colombia. *Int. Zoo Yrbk.* 14:233–240.

Terborgh, J. 1983. *Five New World Monkeys*. Princeton University Press, Princeton, N.J.

Thorington, R.W. 1985. The taxonomy and distribution of squirrel monkeys, in: L.A. Rosenblum and C. Coe (eds.), *Handbook of Squirrel Monkey Research*, pp. 1–33. Plenum Publishing, New York.

deWaal, F. Luttrell, L.M., and Canfield, M.E. 1993. Preliminary data on voluntary food sharing in brown capuchin monkey. *Am. J. Primatol.* 29: 73–78.

Visalberghi, E. 1988. Responsiveness to *objects in two social groups of tufted capuchin monkeys (Cebus apella).* *Am. J. Primatol.* 15: 347–360.

Visalberghi, E. 1990. Tool use in *Cebus*. *Folia Primatol.* 54: 146–154.

Visalberghi, E., and Fragaszy, D.M. 1990. Food-washing behavior in tufted capuchin monkeys, *Cebus apella*, and crab-eating macaques, *Macaca fascicularis*. *Anim Behav.* 40: 829–836.

Westergaard, G.C., and Suomi, S.J. 1994a. Hierarchical complexity of combinatorial manipulation in capuchin monkeys (*Cebus apella*). *Am. J. Primatol.* 32: 171–176.

Westergaard, G.C., and Suomi, S.J. 1994b. Asymmetrical manipulation in the use of tools by tufted capuchin monkeys (*Cebus apella*). *Folia Primatol.* 63: 96–98.

12

SPECIES DEFINITION AND DIFFERENTIATION AS SEEN IN THE POSTCRANIAL SKELETON OF *Cebus*

Susan M. Ford and David G. Hobbs

Department of Anthropology
Southern Illinois University
Carbondale, Illinois 62901

A critical aspect of any study of evolutionary process is an understanding of the nature and distinctiveness of species. While species are generally recognized as real entities, the identification of members of a species remains problematic, particularly for members of a paleospecies. Most researchers seek analogues in the defining differences among modern species, but little of this work has focused on the postcranial skeleton. Recent studies have shown that most species differences are manifested in the soft tissue, but several studies have documented some hard tissue differences between closely related species.

This study examines differences in numerous regions of the appendicular skeletons of species of the genus *Cebus*, which are generally characterized as cautious arboreal quadrupeds. Many aspects of the postcranial skeleton do not demonstrate significant differences across *Cebus* species. However, several significant differences were noted. *C. apella* has short forelimbs, short distal limb segments, and smaller joint surfaces. In addition, it differs markedly from the other species in diet and locomotion, utilizing more hard objects, including palm nuts, and being a more "deliberate" quadruped with more powerful foraging techniques. In contrast, *C. capucinus* has long limbs and large elbow and knee joint surfaces. *C. albifrons* is most like *C. capucinus* but smaller, while the untufted *C. olivaceus* shows some similarities to the tufted *C. apella*.

The ability of these differences to diagnose or support species identity differs under different species concepts, and these differing concepts lead to alternative interpretations of the implications of postcranial differences in speciation.

INTRODUCTION

Species have long been viewed as a basic unit in the study of evolutionary processes. While debates do exist about the nature of species, most researchers generally rec-

Table 1. Species concepts

Species concept	Species definition	How the concept is operationalized	Expectation of differentiation base on postcrania
Biological Species Concept (BSC)	Species are groups of actually or potentially interbreeding natural populations, which are reproductively isolated from other such groups	Determine reproductive isolation (in sympatry)	None necessary
Recognition Species Concept (RSC)	The most inclusive population of individual biparental organisms which share a common fertilization system	Identify distinct SMRS characters	Equivocal, Not necessary
Evolutionary Species Concept (ESC)	A lineage (ancestral-descendent sequence of populations) evolving separately from others and with its own evolutionary role and tendencies	Any morphological character	Yes
Phylogenetic Species Concept (PSC)	A diagnosable cluster of individuals within which there is a parental pattern of ancestry and descent, and which exhibits a pattern of phylogenetic ancestry and descent among units of like kind	Any apomorphic (derived) morphological characters	Yes, assuming species have apomorphic characters

ognize species as real entities. Thus envisioned, one should be able to define and recognize species within the biological world. Various concepts have been put forward in an attempt to define species and provide a framework in which species can be recognized (see Table 1), but none of them have proven themselves as a universal species concept applicable to all organisms living and fossil. Is this to imply that there are different types of species within the biological world, or is it merely the result either of our problems in identifying the evolutionary processes acting on the biological world or our difficulties in interpreting the data at hand? We believe the difficulties stem from the latter two, and that as we continue to gather more data on specific species radiations, we will improve our ability to identify species boundaries and understand the speciation process. Here, we examine whether data on a large number of features of the appendicular skeleton can distinguish between recognized species in an extant genus and, if so, if the distinctions are consistent with expectations of any of the current species concepts.

The genus Cebus contains five widely recognized species, *C. apella, C. albifrons, C. capucinus, C. olivaceus*, and the recently described *C. kaapori* (following Hershkovitz, 1949, and Queiroz, 1992; see also Kinzey, 1982; Mittermeier and Coimbra-Filho, 1981; and Mittermeier, et al., 1988; see Figure 1). Other valid species may exist which are currently designated to the subspecific level (Mittermeier, et al., 1988; Torres de Assumpçao, 1986, 1988). While field data are limited on capuchin species, ecological and behavioral differences have been documented between certain species (see Table 2). Dietary differ-

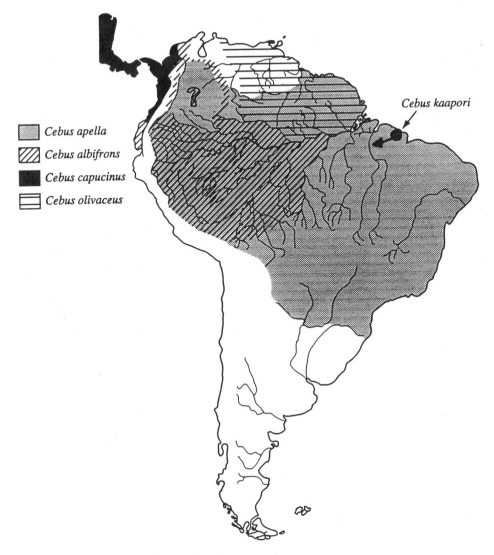

Figure 1. Map of distributions of species of *Cebus* (adapted from Emmons, 990 and Queiroz, 1992; contra Emmons' map #100, we follow her and others descriptions indicating no extension of the range of *Cebus albifrons* northeast of the Rio Negro).

ences have been noted, with *C. apella* eating harder seeds and nuts than any other species. In addition, while all species are considered quadrupeds, *C. apella* has been documented as being more quadrupedal and preferring larger supports during locomotion than *C. capucinus*. Based on the hypothesis that there are different adaptive strategies between the five currently recognized species of *Cebus*, we test the degree to which species differentiation in *Cebus* is reflected in the postcranium.

The central problem in all species concepts is found in the attempt to operationalize and apply them. A number of species concepts are current in the literature; these have been discussed in several recent volumes (e.g. Kimbel and Martin, 1993a; Lambert and Spencer, 1995; Otte and Endler, 1989; see also Godfrey and Marks, 1991; Froehlich,

Table 2. Body size, behavior, ecology by species

	C. apella	*C. albifrons*	*C. capucinus*	*C. olivaceus*	*C. kaapori*
Common name	tufted or brown capuchin	white-fronted capuchin	white-throated or black-capped	weeping or wedge-capped	ka'apor capuchin
Body weight	2645 g	2265 g	3212 g	2684 g	3050 g
Distribution	widespread	NW So. Amer. (W of Rio Negro)	Central Amer. to No. Colombia	NE So. Amer. (E of Rio Negro)	SE Brazil, so. of Amazon
Habitat	varied	mature & disturbed	varied	mature evergreen & rainforest, high forest	edge habitats & high forest
"tough" substrate foraging	44%	32%		21%	
palm foraging	40%	21%			
Diet	hard seeds palm nuts	less hard seeds big figs		lots of seeds	palm nuts
% fruit		80%	65%, 81%* 53-81%**	48%	
%leaves		0%	15%, 2%* 1-2%**	8%	
%insects		20%	20%, 17%* 18-45%**	36%	
Locomotion (during travel)	quadrupedal	quadrupedal	quadrupedal	quadrupedal	
% quad.	0.85		0.54		
% leap	0.1		0.15		
% climb	0.05		0.26		
Support Choice (during travel)	27.5% <2 cm 52.5% 2-10 cm 20% >10 cm		49% <5 cm 40% 6-20 cm 10% >25 cm		
Canopy level					
upper	least		1.0%		
middle	most		76.0%		
lower	frequent		21.0%		
under	little		2.0%		
ground	0.5%	10.0%	14.5%		

Sources: all data on C. kaapori: Queiroz, 1992; distribution and habitat: Emmons, 1990; habitat: Mittermeier & van Roosmalen, 1981; foraging substrate: Terborgh, 1983, Robinson, 1981; diet: seeds - Janson & Boinski, 1992; nuts/figs - van Schaik & van Noordwijk, 1989; frequencies -Rosenberger, 1992, except as noted: * Chapman, 1987, **Chapman & Fedigan, 1990; locomotion: C. apella - Fleagle & Mittermeier, 1980; C. capucinus - Gebo, 1992; support choice and canopy level: C. apella - Fleagle & Mittermeier, 1981, C. capucinus - Gebo, 1992; ground % - Janson & Boinski, 1992, Terborgh, 1983, Robinson, 1981.

1996). Many studies in these volumes (and elsewhere) attempt to identify features which diagnose extant species, often in the hope of refining lists of characters that may discriminate fossil species. However, selection of features and the interpretation of their possible significance in diagnosing species depends strongly on their likely importance under contrasting conceptualizations of what constitutes a species and how they differentiate. Most studies within the order Primates have tended to focus in descending frequency on craniodental morphology, pelage coloration, and vocalizations. The heavy reliance on craniodental morphology and pelage coloration is not historically due to any peculiar relationship of these features to concepts of species constitution but to the biases of both the fossil record, where teeth are the most commonly preserved element, and of museum collections, which are dominated by craniodental remains and skins. Few studies have addressed species discrimination in the postcrania of primates (notable exceptions are Teaford, et al. ,1993, and Turner & Chamberlain, 1989; see also Galdikas, et al.,1993).

Table 1 summarizes original formulations of the major competing species concepts (current workers continue to argue over the distinctiveness or overlap of these concepts; see, e.g., Eldredge, 1993; Kimbel and Martin, 1993b; Masters, 1993; Rose and Bown, 1993; Szalay, 1993). Within this context, predictions can be made about the likelihood of species differentiation to be evidenced in the postcranium under alternate species concepts. While certainly none of these precludes postcranial differentiation between species, the likelihood and potential significance of such differences varies.

The Biological Species Concept (BSC; Dobzhansky, 1937; Mayr, 1942) states that species are distinguished on the basis of reproductive isolation, requiring populations to be in sympatry for a test. (Only *C. apella* occurs sympatrically with other species of *Cebus*). The BSC makes no predictions about the nature and degree of divergence expected in any morphological features, including postcranial differences.

The Recognition Species Concept (RSC; Paterson, 1978) states that members of a species share a common fertilization system, predicting that differentiation will be evidenced through differences in characters of a Specific-Mate Recognition System (SMRS). Thus, the RSC predicts that the first and primary characters to differ will be those important for reproductive processes and behaviors (including pre- and post-mating; e.g. facial patterns or vocalizations might be very important). There does not have to be a correlation between non-SMRS characters and the speciation process. Although Galdikas, et al. (1993) have suggested that body form and associated postcrania may be a significant aspect of the human SMRS, they and others (e.g. Turner and Chamberlain, 1989) argue that this is likely not true for other primates. Thus, we would not expect to find any postcranial differences between *Cebus* species being maintained under tight stabilizing selection (as per Lambert and Paterson, 1982) under the RSC.

The Evolutionary Species Concept (ESC; Simpson, 1961; Wiley, 1978) states that species are recognizable because of membership in an evolutionary lineage; species differentiation may be reflected in any morphological characters related to genotype or phenotype. Under the ESC, postcranial differences between *Cebus* species could be predicted based on the fact that locomotor/foraging differences have been documented in the literature, thus implying distinct evolutionary lineages with different adaptive strategies.

The Phylogenetic Species Concept (PSC; Eldredge and Cracraft, 1980; Cracraft, 1983) is very similar to the ESC, with species members part of a diagnosable cluster. The PCS predicts that species can be identified through analysis of any morphological characters, phenotypic or genotypic, which are apomorphic. Based on current knowledge about the differences among *Cebus* species' behavior and morphology, we would predict that differences in the postcranium would exist.

Much has been written about the importance of genetic data and the relevance of facial patterns to species recognition. This paper explores the degree to which species differences are evidenced in the postcranium of related species and what relevance the expectations of different species concepts may have to the interpretation of these differences. Functional aspects are touched on but explored in more detail elsewhere (Ford and Hobbs, in prep.).

METHODS AND MATERIALS

Seventy-eight measurements, including eight angles, were measured on the humerus, femur, tibia, astragalus, and calcaneus. These measurements included bone lengths and diameters as well as joint surfaces and muscle attachment areas (see Ford, 1980, 1988), in an attempt to adequately characterize important taxonomic and functional aspects of a number of major joints in the appendicular skeleton. Eleven indices were computed from these data, including standard intermembral and intra-limb indices. All measurements were taken by SMF. Thirty-seven adult, healthy, wild-caught *Cebus* skeletons were measured, comprising 19 *Cebus albifrons*, 12 *Cebus apella*, 4 *Cebus capucinus*, and 2 *Cebus olivaceus*. No skeletons of *C. kaapori* were examined. Behavioral data, summarized in Table 2, were taken from the literature.

The data base is small, as wild-caught skeletons of many *Cebus* species are fairly rare. However, statistical techniques were chosen to identify only those features for which significant differences between species exist even with these small samples. All statistical analyses were done using SAS/STAT Version 6 (SAS Institute Inc., 1989) and JMP Version 3.0 (SAS Institute Inc., 1994).

Measurements were standardized for size in the following manner. First, an estimator of body size was chosen by regressing mean values of all 78 variables to average species body weights (taken from Ford and Davis, 1992) for a cross-platyrrhine sample, using a larger data set of 35 platyrrhine species. Body weights of *Cebus* species are similar enough that a within-*Cebus* regression was non-informative. By maximizing three parameters (Pearson's correlation coefficient squared [r^2] as close to 1 as possible, slope of the reduced major axis regression as close to isometry [0.33] as possible, and standard error of the slope as low as possible), a best predictor of body weight was chosen. Several features, all measures of joint surfaces through which large percentages of body weight

Figure 2. Illustrations of postcranial measurements exhibiting significant differences between species (see text): a) Right calcaneus, superior view. (b-d) Right astragalus: (b) distal, (c) lateral, (d) superior views. (e-f) Right tibia: (e) anterior, (f) lateral views. (g-h) Left femur: (g) posterior, (h) inferior views. (i-l) Left humerus: (i) ventral, (j) dorsal, (k) ventral, (l) inferior (distal) views. MAXL= maximum calcaneal length; W-CALC= maximum calcaneal width at sustentaculum; MASW= width of medial articular facet; PT-CAS= distance from the distal edge of the peroneal tubercle to the distal end of the calcaneus; CPT-CAS=distance from the center of the peroneal tubercle to the distal end of the calcaneus; LP= angle of the astragalar lateral malleolar facet (for the fibula) to the trochlear plane; FFL= antero-posterior length of the fibular facet on the astragalus; HW= width of the astragalar head; NHA= angle of the astragalar neck to the long axis of the trochlea; PMMA= angle of the posterior surface of the medial malleolus to the tibial shaft; ML= maximum medial femoral length; NSA= angle of the femoral neck to the shaft; LCH= height of the lateral condyle; LCW= width of the lateral condyle; MCH= height of the medial condyle; DED= antero-posterior depth of the femoral distal epiphysis; L-HUM= maximum humeral length; DELPECL= length of the deltopectoral crest; DLEW= dorsal width of the lateral epicondyle; ASW= ventral articular surface width of the humerus; MaxTD=maximum humeral trochlear diameter; MinTD= minimum humeral trochlear diameter. (Not illustrated: maximum radial length).

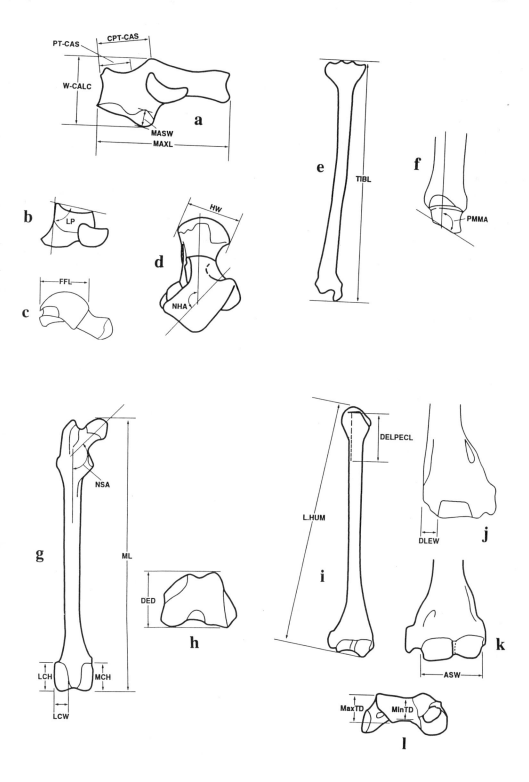

are transferred, ranked high. The best predictor was the width of the distal articular surface of the humerus measured dorsally (ASW, across trochlea and capitulum: $r^2=0.98$, slope=0.355, std. error=0.009, see Figure 2k). All raw data (linear measurements and angles) and indices were then log-transformed and regressed against log ASW; residuals to the reduced major axis best-fit line were then computed. Angles and indices were also analyzed separately without being size corrected.

Analysis of variance (ANOVA) tests were done on each variable to determine whether or not the means of the species are significantly different from one another. Only when the overall ANOVA F-test indicated that significant differences exist ($p < .05$) were post-hoc tests performed to determine which pairs of species differed significantly from one another. The more powerful Student's t-test was done, since the number of species (and thus the number of pair-wise comparisons) are fairly small and the likelihood of a false positive is not great, and because this test provides the strongest control for comparison-wise error. As a check, the more conservative post-hoc Tukey-Kramer test, or "honestly significant difference test," was also applied. This test controls for the experiment-wise error rate but not the comparison-wise error rate (see discussion under PROC GLM, SAS Institute Inc., 1989; also Dunnett, 1980; Hayter, 1984; Kramer, 1956; Miller, 1981). Pearson's correlation coefficients were computed to determine the degree of correlation between features which differ significantly between species.

RESULTS

Of the 78 features and 11 indices examined, 26 exhibited statistically significant differences between *Cebus* species as measured by ANOVA ($p < .05$), including four indices and three angles not size corrected, and four indices and three angles corrected for size (see Figure 2). Table 3 reports raw angles and indices, size-corrected ratio data (by division by ASW), and residuals to the reduced major axis, which were analyzed statistically, for those features which demonstrated significant differences between species. It also presents the F-value (from ANOVA) for differences between residuals and the species pairs differing at the .05 significance level in post-hoc tests. The size-corrected ratios are included for ease of comparison with other studies and additional material, and it is the original angles and size-corrected ratios which are plotted in Figures 3–5.

The features which differ significantly between *Cebus* species based on ANOVA include, on the calcaneus (see Figure 2a): width of the medial articular surface (MASW); distance from the distal edge of the peroneal tubercle to the distal end of the bone (PT_CAS); distance from the center of the peroneal tubercle to the distal end of the bone (CPT_CAS); relative position of the peroneal tubercle (PTPOSIT=CPT_CAS/MAXL [maximum length of the calcaneus]), both original index and corrected for body size; and shape of the calcaneus (CALSHP=W_CALC/MAXL), not size corrected. Significant differences in the astragalus (see Figure 2b-d) were found for: the width of the astragalar head (HW); length of the fibular facet (FFL); angle of the lateral projection (LP), both original and size corrected; and angle of the astragalar neck to the long axis of the trochlea (NHA), not size corrected. Significant differences in the tibia (see Figure 2f) were found only for angle of the posterior surface of the medial malleolus to the tibial shaft (PMMA), both original value and when corrected for body size. Significant differences in the femur (see Figure 2g,h) were found for: height of the medial condyle (MCH); antero-posterior depth of the distal epiphysis (DED); shape of the lateral condyle (LCSHAPE=LCW/LCH) when corrected for body size; and angle of the femoral neck to the shaft (NSA), corrected

Table 3. Features differing between *Cebus* species

Variable	Mean character value[1]				Mean residual value[2]				F[3]	Student's t[4]	Tukey-Kramer[4]
	apella	*albifrons*	*capucinus*	*olivaceus*	*apella*	*albifrons*	*capucinus*	*olivaceus*			
General											
Brachial Index (L_RAD/L_HUM)	93.8 [9]	97.3 [19]	97.7 [3]	— [0]	—	—	—	—	4.93	B-P C-P	B-P
Brachial Index (to body size) (L_RAD/L_HUM)	6.1 [9]	6.6 [19]	5.8 [3]	— [0]	-0.030	0.025	-0.025	—	7.24	B-C B-P	B-P
Crural Index (TIBL/ML)	92.9 [9]	95.8 [19]	95.3 [3]	— [0]	—	—	—	—	6.51	B-P	B-P
Crural Index (to body size) (TIBL/ML)	6.1 [8]	6.5 [19]	5.6 [3]	— [0]	-0.021	0.023	-0.024	—	9.10	B-C B-P	B-C B-P
Calcaneus											
Medial articular surface width (MASW)	22.4 [11]	24.1 [18]	25.6 [4]	21.5 [1]	-0.064	0.036	0.032	-0.125	3.00	B-P	
Peroneal tubercle to distal end (PT_CAS)	46.5 [11]	53.4 [18]	51 [4]	47.2 [2]	-0.129	0.132	-0.121	-0.070	3.13	B-P	B-P
Center of peroneal tubercle to the distal end (CPT_CAS)	79.2 [9]	82.8 [18]	65.3 [4]	— [0]	-0.002	0.094	-0.357	—	7.94	B-C P-C	B-C P-C
Peroneal tubercle relative position (PTPOSIT) PTPOSIT=CPT_CAS/MAXL	46.4 [10]	48 [18]	37.9 [3]	— [0]	—	—	—	—	4.39	B-C P-C	B-C
Peroneal tubercle relative position (PTPOSIT) to size	3 [9]	3.2 [18]	2.2 [3]	— [0]	-0.011	0.129	-0.488	—	9.77	B-C P-C	B-C
Calcaneal shape (CALSHP) CALSHP=W_CALC/MAXL	48.5 [12]	47.3 [18]	51.2 [4]	50.2 [2]	—	—	—	—	5.10	B-C B-P	B-C
Astragalus											
Head width (HW)	48.2 [11]	50 [17]	52 [4]	45.4 [2]	-0.030	0.012	0.046	-0.086	4.26	C-P C-O B-O	C-O
Fibular facet length (FFL)	62.4 [11]	65.6 [17]	68 [4]	64.4 [2]	-0.045	0.017	0.034	-0.011	4.92	C-P B-P	C-P B-P
Angle of lateral projection (LP)	86 [12]	93 [17]	93 [4]	83 [2]	—	—	—	—	6.38	B-O B-P C-O C-P	B-O B-P

Table 3. Features differing between *Cebus* species

Variable	Mean character value[1]				Mean residual value[2]				F[3]	Student's t[4]	Tukey-Kramer[4]
	apella	*albifrons*	*capucinus*	*olivaceus*	*apella*	*albifrons*	*capucinus*	*olivaceus*			
Angle of lateral projection (LP) to body size	549.9 [11]	622.9 [17]	553.6 [4]	535.7 [2]	-0.064	0.054	-0.035	-0.089	6.32	B-C, B-O, B-P	B-P
Angle of neck to trochlea (NHA)	32 [12]	29 [17]	29 [4]	39 [2]	—	—	—	—	6.91	B-O, B-P, C-O, P-O	B-O, C-O, P-O
Tibia											
Posterior medial malleolar angle (PMMA)	112 [12]	114 [19]	119 [4]	109 [2]	—	—	—	—	3.72	C-B, C-O, C-P	C-P
Posterior medial malleolar angle (PMMA) to body size	712.1 [11]	773.3 [19]	706.3 [4]	705.2 [2]	-0.032	0.022	0.002	-0.043	3.91	B-P	B-P
Femur											
Medial condyle height (MCH)	68.9 [11]	70 [19]	75.1 [4]	67.7 [2]	-0.018	0.009	0.067	-0.033	3.14	C-B, C-O, C-P	C-P
Distal epiphyseal depth anterior-posterior (DED)	88.7 [11]	94.1 [19]	99.6 [4]	88.5 [2]	-0.044	0.026	0.069	0.044	6.29	C-O, C-P, B-P	C-O, B-P
Lateral condyle shape (LCSHAPE=LCW/LCH) to body size	3.6 [11]	3.8 [19]	3.3 [4]	3.3 [2]	-0.014	0.053	-0.114	-0.101	2.97	B-C	
Neck/shaft angle (NSA) to body size	822.3 [11]	888.3 [19]	769.0 [4]	844.0 [2]	-0.035	0.045	-0.083	-0.004	3.26	B-C, B-P	
Humerus											
Humeral length (L_HUM)	677.7 [11]	715.8 [19]	703.2 [4]	662.1 [2]	-0.027	0.015	0.031	-0.050	3.29	B-P, C-P	
Length of deltopectoral crest (DELPECL)	285.1 [11]	314.9 [19]	298.9 [4]	242.9 [2]	-0.077	0.092	-0.074	-0.230	2.89	B-P, B-O	
Dorsal lateral epicondylar width (DLEW)	30.1 [11]	31.3 [19]	30.9 [4]	24.9 [2]	-0.029	0.059	-0.053	-0.212	4.03	B-O	B-O

Variable	Mean character value[1]				Mean residual value[2]				F[3]	Student's t[4]	Tukey-Kramer[4]
	apella	*albifrons*	*capucinus*	*olivaceus*	*apella*	*albifrons*	*capucinus*	*olivaceus*			
Maximum trochlear diameter (MaxTD)	44.6 [11]	47.4 [19]	46.1 [4]	43.3 [2]	-0.038	0.037	-0.013	-0.064	4.03	B-O B-P	B-P
Minimum trochlear diameter (MinTD)	38.4 [11]	37.6 [19]	41.4 [4]	34.7 [2]	0.001	-0.003	0.062	-0.095	3.70	B-C B-O C-O P-O	C-O

[1] Mean character values are given for all; indices and angles may be raw data or corrected for body size (by an index: [(measurement/ASW) X 100]) as noted. Mean values for all other variables are as an index: [(measurement/ASW) X 100], where ASW= body weight variable (see text). Sample size is given in [].

[2] Mean residuals are computed for raw variables regressed against ASW (as a measure of body size).

[3] ANOVAs were performed on residuals to the Reduced Major Axis where variables were regressed to the body size variable (ASW). ANOVAs were also performed directly on raw angles and indices (see text). All ANOVA F values have are significant at least at the .05 level. Those in bold are significant at the .01 level or lower.

[4] Significant pair-wise comparisons of species in post-hoc tests (p².05): B=C. *albifrons*, O=C. *olivaceus*, C=C. *capucinus*, P=C. *apella*

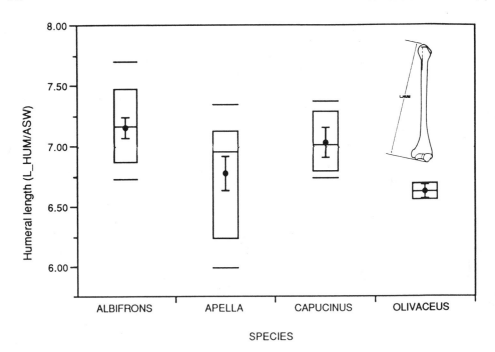

Figure 3. Box plots of the 10th, 25th, 50th, 75th, and 90th percentiles of maximum humeral length for four species of *Cebus*, with mean and standard error of mean (dot and whiskers); corrected for body size (L-HUM/ASW) (see text, Figure 2i).

for body size. Significant differences between species in the humerus (see Figure 2i-l) included: maximum humeral length (L_HUM); length of the deltopectoral crest (DELPECL); width of the dorsal epicondyle (DLEW); maximum trochlear diameter (MaxTD); and minimum trochlear diameter (MinTD). Finally, two additional indices were significantly different between species: the brachial index (maximum radial length [not illustrated]/ L_HUM), both original and size corrected; and the crural index (maximum tibial length [TIBL]/ maximum medial femoral length [ML]), both original and size corrected (see Figure 2e, g, i). Only these measurements and indices were further examined for specific inter-species comparisons using both Student's t-tests and Tukey-Kramer tests (see Table 3). Both are reported; given the small number of pairwise inter-species comparisons, we discuss the results of the less conservative t-tests. As a convention, for angles and indices examined in both size-corrected (regression to ASW) and non-corrected (raw angles or indices) versions, a suffix of "2" indicates the size-corrected analysis (e.g., "NSA" and "NSA2").

The 26 features differing significantly between species were examined for inter-correlations, using Pearson's correlation coefficient. Of 676 possible pairs of features, very few (36 pairs) were strongly correlated with r>.7, and even fewer (22 pairs) at r>.8. These included, not surprisingly, alternate ways of examining the position of the peroneal tubercle on the calcaneus (CPT_CAS2, PTPOSIT, and PTPOSIT2) or of femoral lateral condyle shape (LCSHAPE and LCSHAPE2). Additionally, while none of the standard limb indices correlated strongly with other features, all of them were strongly intercorrelated once corrected for body size (i.e., brachial index, crural index, and intermembral index each divided by ASW). More surprising were the moderate correlations (r=.7-.85) of these

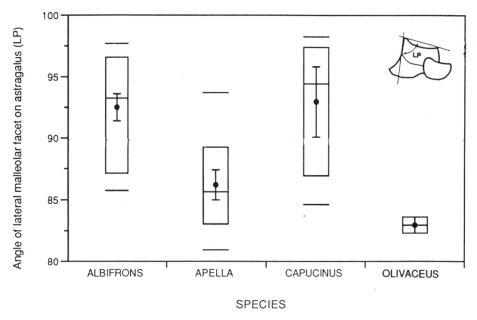

Figure 4. Box plots of the 10th, 25th, 50th, 75th, and 90th percentiles of astragalar lateral angle (LP) for four species of *Cebus*, with mean and standard error of mean (dot and whiskers) (see text, Figure 2b).

limb indices to calcaneal shape (CALSHP2) and several size-corrected angles throughout the hindlimb, including the angle of the femoral neck (NSA2), angle of the tibial posterior medial malleolus (PMMA2), and of the astragalar lateral malleolar facet for the fibula (LP2). These correlations may reflect the larger values for *C. albifrons* for all these characters (see below), although there is large overlap (see, e.g., Figure 4). Last, all of the individual long bone lengths (corrected for body size) are strongly correlated with one another and with the length of the deltopectoral crest (DELPECL), mirroring the distinctiveness of individual species in their limb dimensions.

ANOVA show that *C. apella* has short forelimbs, particularly the humerus (see Figure 3), and relatively shorter distal limbs (as evidenced in the brachial and crural indices, alone and controlled for body size; sample sizes for TIBL and L_RAD are smaller than for other measures and may obscure significant differences). While *C. capucinus* shares the relatively short distal (vs. proximal) hindlimb elements and approaches *C. apella* in brachial index (when controlled for body size), its humerus is quite long relative to body size and the distal forelimb is relatively elongated when the brachial index is not corrected for size (see Table 3). *C. albifrons* contrasts to both of these patterns in having relatively quite long humeri and long distal (relative to proximal) limb elements in all analyses. For most traits, the sample of *C. olivaceus* is too small to say anything with statistical confidence, but some interesting patterns are evident; e.g., it has the shortest humeri.

In addition, *C. apella* is distinctive in its shorter extension of the major forelimb "mover," the area of insertion of the deltoid and pectoral muscles distally on the shaft of the humerus, seen in the shorter deltopectoral crest relative to body size. *C. albifrons* in particular contrasts in the longer extension of these attachment areas. *C. olivaceus* falls close to the "robust" *C. apella* in this measure, having the shortest deltopectoral crest, while *C. capucinus* is intermediate but more like the extensive crest in *C. albifrons*. However, there is considerable overlap in all four samples.

In the ankle, several measures of peroneal tubercle position all point to a distinct contrast between *C. albifrons*, with a much more posterior placement of the tubercle, and *C. capucinus*, with a more anterior placement of the tubercle. *C. apella* is intermediate but closer to *C. albifrons*. Both *C. olivaceus* and *C. apella* have a more anterior distal endpoint for the peroneal tubercle. In addition, *C. capucinus* has a short, wide calcaneus (CALSHP) as does *C. olivaceus*, while *C. albifrons* is significantly different in its longer, narrower calcaneus. *C. apella* approaches the calcaneal shape seen in *C. albifrons*.

On the astragalus, *C. apella* and *C. olivaceus* share a relatively narrow astragalar head, while *C. albifrons* and especially *C. capucinus* contrast with wider heads. The lateral facet for the fibula is significantly shorter in *C. apella*. *C. apella* and *C. olivaceus* also share a far less projecting lateral shelf than that seen in *C. albifrons*, whose lateral shelf projection is extremely large both absolutely and relative to body size (see Figure 4). *C. capucinus* is similar. The angle of the astragalar head and neck (NHA) is significantly different in all species, being smallest in *C. albifrons* and *C. capucinus*, moderate in *C. apella*, and very large in *C. olivaceus*.

The knee joint also offers significant contrasts. *C. capucinus* has a tall, narrow lateral condyle, shared with *C. olivaceus*, in contrast to the other two species. In measures of the femoral medial condyle (height and width) and anterior-posterior depth of the distal end, *C. apella* and *C. olivaceus* are distinguished by being relatively small, and *C. capucinus* in particular by being quite large (see Figure 5a, b). *C. albifrons* is closest to *C. capucinus*. The pattern of a smaller knee joint in *C. apella* and *C. olivaceus* is found again in the elbow, where *C. apella* and *C. olivaceus* have a relatively narrow trochlea and *C. capucinus* in particular has a relatively wide trochlea, particularly for MinTD. *C. olivaceus* also has the smallest lateral epicondyle. And in the hip, *C. albifrons* and *C. olivaceus* share a large angle of the femoral neck (NSA), while *C. apella* and especially *C. capucinus* have a smaller angle, although there is considerable overlap.

Thus a consistent pattern of statistically significant differences appears, even in this small sample. *C. apella* has short forelimbs, short distal limb segments, a shorter extension of major forelimb muscle groups, and smaller joint surfaces. *C. capucinus* contrasts in having large joints and longer limbs. *C. albifrons* is in general more like *C. capucinus*, but with longer distal limb segments and a much more angled astragalar facet for the fibula. *C. olivaceus*, despite being an untufted capuchin, is more similar to the pattern seen in *C. apella*, although this varies.

C. apella is sympatric with all other species except *C. capucinus* (see Figure 1) and is consistently distinct in numerous features of the postcranium. Available data (see Table 2) suggest that it eats a tougher diet, including hard palm nuts (Janson and Boinski, 1992; van Schaik and van Noordwijk, 1989), utilizing "tougher" foraging substrates (Robinson, 1981; Terborgh, 1983), and it does less climbing, more "straight" quadrupedalism, perhaps using larger substrates (Fleagle and Mittermeier, 1980, 1981). In addition, it spends the least amount of time on the ground, almost never descending (Janson and Boinski, 1992; Robinson, 1981; Terborgh, 1983). In contrast, *C. capucinus* does more climbing, perhaps utilizing smaller substrates (Gebo, 1992); it is also the heaviest/largest capuchin (Ford and Davis, 1992). Dietary differences are less certain, although *C. capucinus* and *C. albifrons* do not appear to utilize hard palm nuts; they may differ from each other in *C. capucinus* utilizing less fruit and more leaves or insects than *C. albifrons* (reports vary on different groups - Chapman and Fedigan, 1990; Rosenberger, 1992); also, *C. albifrons* is the smallest of all species (Ford and Davis, 1992). Given the frequent similarities of the few *C. olivaceus* specimens examined here and *C. apella*, it is interesting to note that reports indicate much less ripe fruit foraging and more insect and seed foraging in *C. oli-*

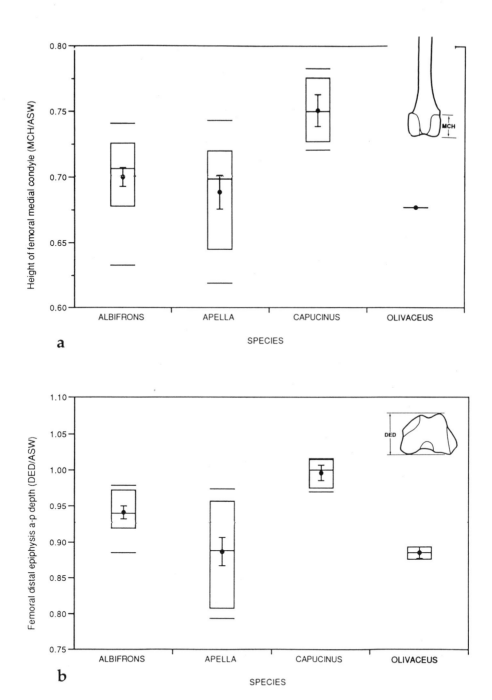

Figure 5. Box plots of the 10th, 25th, 50th, 75th, and 90th percentiles for four species of *Cebus* of (a) medial femoral condyle height, corrected for body size (MCH/ASW); (b) antero-posterior depth of the femoral distal epiphysis, corrected for body size (DED/ASW); with mean and standard error of mean (dot and whiskers) (see text, Figures 2g & h).

vaceus than the other untufted capuchins (Chapman and Fedigan, 1990; Janson and Boinski, 1992); these two are also fairly close in size, with *C. olivaceus* slightly larger. *C. kaapori* is especially intriguing here; preliminary reports suggest it is most like *C. olivaceus* (its neighbor to the north across the Rio Amazonas) including size, but that like *C. apella* it can crack and use hard palm nuts (Queiroz, 1992).

General reports all agree in calling the tufted capuchins, *C. apella*, the most "robust" (Jungers and Fleagle, 1980; Hershkovitz, 1949; Kinzey, 1982). In accord with this crude description, it does indeed have short limbs, but not significantly different from the two *C. olivaceus* specimens included here. The shorter length of muscle attachment in the hip and, especially, the shoulder (also shared with *C. olivaceus*) may provide greater strength at the expense of speed, although no differences in rapidity of travel have been documented. In addition, despite its apparent "robustness" and strength in utilizing tough feeding substrates (Janson and Boinski, 1992), which have clear repercussions in the masticatory apparatus (Daegling, 1992), the limb joint surfaces do not reflect stronger support or buttressing; just the opposite. It is the larger-bodied though apparently "more gracile" *C. capucinus*, with its long limbs, that has the relatively largest joint surfaces, coupled with more climbing activities and more descent to the ground. This cannot be a purely allometric effect, as the smallest capuchin, *C. albifrons*, shares many of the enlarged joint surfaces. No comparative data are available on locomotor frequencies and support choices for the *C. albifrons*, but one might predict more climbing and use of angled supports than is seen in *C. apella*. Likewise, one might predict more use of horizontal supports in *C. olivaceus*, although its comparatively light dependence on ripe fruits in the dry season is intriguing (Rosenberger, 1992). Clearly, and especially with the discovery of the new and similar *C. kaapori* with its ability to utilize hard palm nuts like *C. apella*, good field data on the diet and locomotion of *C. olivaceus* is called for, as it emerges in this study as an unexpectedly distinct untufted form. Of course, only two skeletons were examined; a larger sample must be examined to confirm its distinctive features, but these two animals definitely fall as separate from the other untufted capuchins.

DISCUSSION

This study has documented that clear differences exist in the postcrania of currently recognized species of *Cebus*. Preliminary study suggests these correlate to some degree to differences in body size or behavior, intimating distinctive adaptive niches (see also Ford and Hobbs, in prep.). However, in the absence of any good genetic data on the distinctiveness of these species, the nature and reality of each of the widely accepted species is open to question and interpretation, particularly given the conflicting criteria and expectations of competing conceptualizations of species. Our interest here is in documenting the degree and nature of differentiation in the postcranium and examining the implications of these differences for species diagnosis under competing views of species (see Figure 6).

Jungers and Fleagle (1980) examined growth allometry in two of these species. They found that *C. apella* and *C. albifrons* infants are roughly the same weight and limb lengths; the differences noted here emerge during growth, when *C. apella* increases in weight more rapidly than *C. albifrons*, attaining a heavier adult weight. Thus, while the limbs may remain the same length, those of *C. albifrons* are relatively longer in comparison to body weight at each growth stage; here, we found that to be true of adults as well. There is no comparable ontogenetic data set using body weight for the other species; we note, however, that *C. capucinus* not only has relatively longer limbs, like *C. albifrons*,

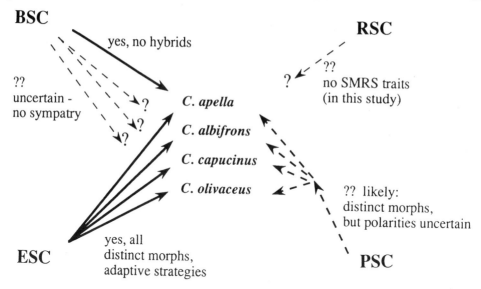

Figure 6. *Cebus* species diagnosis by postcranial features, under assumptions of differing species concepts (see text).

but also absolutely longer limbs and limb elements as well as a heavier body mass. Thus, its differences must begin prior to birth with a different body mass or through accelerated scaling of both mass and, extremely so, limb length. In either case, it is unlikely to be a precise parallel in ontogenetic growth to *C. albifrons* despite their adult similarities in relative dimensions.

The BSC does not make any predictions about species differentiation in particular morphological features; however, species in sympatry may diversify more due to competitive exclusion. In keeping with this, sympatric *C. apella* and *C. albifrons* are quite distinct from one another. However, the sympatric *C. apella* and *C. olivaceus* are much more similar in their postcrania. It may be that a larger sample of *C. olivaceus* would indicate more distinction, in keeping with the cranial and pelage differences noted by others (e.g., Hershkovitz, 1949; Masterson, 1995). On the other hand, the most isolated species, the Central American *C. capucinus*, is in many ways the most distinctive.

The RSC is unequivocal in its predictions about differentiating between species of *Cebus* solely on the basis of SMRS traits. If postcranial characters are not part of a species' SMRS, as seems likely in this case, they should have no special significance for defining or recognizing species. In addition, since there is no expected correlation between non-SMRS characters and the speciation process, these characters should show less intra-specific stability and more inter-specific overlap (Lambert and Paterson, 1982; Masters, 1993). Our data show that species of *Cebus* can be differentiated by postcranial (non-SMRS) characters, but overlap is often considerable except perhaps for *C. capucinus* (but our sample is small). Thus, if these are "good" species, the degree of differentiation in non-SMRS characters suggests that we might look further at characters related to reproduction which should differentiate first and most distinctly. This would seem like a good test scenario for the RSC - are SMRS traits present and more differentiated and stable than the non-SMRS traits in capuchin species?

The ESC and the PSC both are more easily applicable to any data set utilizing morphology; however, the interpretation and utility of these postcranial traits varies. Under the ESC, the postcranial differences suggest differing adaptive strategies which support the separate status of the four species of *Cebus* studied here as members of distinct evolutionary lineages. Further behavioral studies on the lesser known species should clarify their particular adaptive strategies.

Interpreting the potential significance of these traits under the PSC requires the identification of at least one apomorphic character for each group before we can state that it is in fact a valid species. There are currently insufficient data on *Cebus* species or on the phylogenetic affinities of *Cebus* (see review in Ford and Davis, 1992) to determine which of the morphs seen in the various species are plesiomorphic or apomorphic for the genus as a whole. For example, should the similarities between *C. apella* and *C. olivaceus* prove to be ancestral for the genus, they would tell us nothing about the potential distinctiveness of these species themselves. Thus, while it is theoretically possible to use postcranial traits in the study of speciation under the PSC, in reality its application in *Cebus* must await a clearer understanding of the polarity of traits within this genus.

Thus, we have shown that the four major recognized species of *Cebus* do differ significantly in their postcrania, and the little available behavioral data support the hypothesis that these differences are real and relate to basic differences in adaptive strategies, even with our small sample sizes. (The species with the smallest samples are often the most distinct and outside of the ranges of the other species). The ability to use these differences to diagnose and identify species varies depending on the species concept adopted; thus, it is important to clarify one's view on processes of species origination and differentiation before selecting and interpreting characters to diagnose individual species. This is true for characters outside the postcranium as well.

SUMMARY

The genus *Cebus* contains five generally recognized species, *C. apella, C. albifrons, C. capucinus, C. olivaceus*, and the recently described *C. kaapori*. Different behaviors have been documented between certain species, including diet and support choice. We test the degree to which species differentiation in *Cebus* is reflected in the postcranium.

The various species concepts current in the literature allow predictions to be made about the likelihood of species differentiation being evidenced in the postcranial skeleton. While none of these concepts precludes postcranial differentiation between species, the likelihood and potential significance of such differences varies.

Seventy-eight measurements, including 8 angles, and 11 indices were taken on 37 skeletons of *Cebus*, including all commonly recognized species except *C. kaapori*. Measurements were standardized for body size, and ANOVA, t-tests, and Tukey-Kramer tests were used to identify significant differences between species. These show that *C. apella* has short forelimbs, short distal limb segments, a shorter extension of major forelimb muscle groups, and smaller joint surfaces. In contrast, *C. capucinus* (even with the small sample here) has significantly larger joints and longer limbs; *C. albifrons* is similar, but with longer distal limb segments and a much more angled astragalar facet for the fibula. *C. olivaceus* shows a number of similarities to *C. apella* (the tufted capuchins), despite being in the "untufted" group. *C. apella* is also sympatric with all other species except *C. capucinus*; none of the others are known to overlap in their ranges.

Available behavioral data support the hypothesis of species differentiation in adaptations. *C. apella* forages for a tougher diet, including hard palm nuts, and does less climbing, uses larger substrates, and spends the least amount of time on the ground. In contrast, *C. capucinus* is much larger, does more climbing, and perhaps uses smaller substrates. *C.*

capucinus and *C. albifrons* do not eat hard palm nuts, and *C. capucinus* eats less fruit and more leaves or insects than *C. albifrons*. *C. olivaceus* differs behaviorally from the other untufted capuchins in eating much less ripe fruit. Preliminary reports of *C. kaapori* indicate that, unlike all other untufted capuchins, it cracks and eats hard palm nuts.

Despite the clear evidence of differences between the species, applying these data to issues of species distinctiveness and diagnosis is difficult under most current species concepts. The Biological Species Concept recognizes species solely on the basis of reproductive isolation and requires sympatry for evidence. The postcranial differences in *Cebus* are not precluded but also not particularly expected in association with speciation. The Recognition Species Concept (RSC) recognizes species by shared, common fertilization systems, and it argues that only characters associated with reproduction and mate recognition (SMRS) will necessarily change in direct association with speciation. Since limb features are likely not directly linked to mate recognition in non-human primates, species-specific, stable differences would not be expected in the postcranium of *Cebus*; thus, under the RSC, none are significant in diagnosing species. However, since non-SMRS differences have developed, the RSC does predict that SMRS traits not examined here should be already well differentiated. The Phylogenetic Species Concept necessitates the identification of apomorphies and thus a solid hypothesis on the polarity of traits under study. That is not yet established for this genus, but the potential is there in the future. The Evolutionary Species Concept recognizes species as members in an evolutionary lineage and requires only clear morphological / adaptive distinctions. Under this view, the data and analysis reported here would confirm that there are, at a minimum, four distinct species of *Cebus*.

ACKNOWLEDGMENTS

This paper is an outgrowth of a paper originally presented by SMF at the Conference of "Neotropical Primates: Setting the Future Research Agenda; Essays in Honor of Warren G. Kinzey," organized by A.L. Rosenberger and M.A. Norconk. SMF thanks them for this opportunity to exchange ideas, and in particular wishes to dedicate this work to Warren, who encouraged her interest in *Cebus* for many years. Helpful discussions or comments were provided by Drs. Jeffrey Froehlich, Thomas Masterson, Alfred Rosenberger, Robert Sussman, and 3 anonymous reviewers. We thank the many curators in the museums which provided access to specimens in their care, including those at the American Museum of Natural History, the Academy of Natural Sciences in Philadelphia, the Field Museum of Natural History, the United States National Museum, the University of Wisconsin-Milwaukee Neil Tappen Collection, and the University of Wisconsin-Madison Department of Zoology. Funds for data collection were provided at varying stages by a Smithsonian Predoctoral Fellowship, a Mellon Fellowships, and grants from Southern Illinois University to SMF. Mr. Philip Spielmacher greatly assisted in writing the SAS programs for reduced major axis computations. Ms. Karen Fiorino prepared Figures 1 and 2. We are grateful to all of these individuals and institutions for their support.

REFERENCES

Chapman, C.A., 1987, Flexibility in diets of three species of Costa Rican primates, *Folia Primatol.* 49:90–105.

Chapman, C.A., and Fedigan, L.M., 1990, Dietary differences between neighboring *Cebus capucinus* groups: Local traditions, food availability or responses to food profitability? *Folia Primatol.* 54: 177–186.

Cracraft, J., 1983, Species concepts and speciation analysis. *Current Ornithology* 1: 159–187.

Daegling, D.J., 1992, Mandibular morphology and diet in the genus *Cebus*. *Int. J. Primatol.* 13: 545–570.

Dobzhansky, T., 1937, *Genetics and the Origin of Species*, New York: Columbia University Press.

Dunnett, C.W., 1980, Pairwise multiple comparisons in the homogeneous variance, unequal sample size case. *Journal of the American Statistical Association* 75: 789–795.

Eldredge, N., 1993, What, if anything, is a species? In: W.H. Kimbel and L.B. Martin (eds.), *Species, Species Concepts, and Primate Evolution*, New York: Plenum Press, pp. 3–20.

Eldredge, N., and Cracraft, J., 1980, *Phylogenetic Patterns and the Evolutionary Process*, New York: Columbia University Press.

Emmons, L.H., 1990, *Neotropical Rainforest Mammals: A Field Guide*, Chicago: University of Chicago Press.

Fleagle, J.G., and Mittermeier, R.A., 1980, Locomotor behavior, body size, and comparative ecology of seven Surinam monkeys. *Amer. J. Phys. Anthropol.* 52: 301–314.

Fleagle, J.G., and Mittermeier, R.A., 1981, Differential habitat use by *Cebus apella* and *Saimiri sciureus* in Central Surinam. *Primates* 22: 361–367.

Ford, S.M., 1980, A systematic revision of the Platyrrhini based on selected features of the postcranium, Ph.D. dissertation, University of Pittsburgh.

Ford, S.M., 1988, Postcranial adaptations of the earliest platyrrhine. *J. Hum. Evol.* 17:155–192.

Ford, S.M., and Davis, L.C., 1992, Systematics and body size: Implications for feeding adaptations in New World monkeys. *Amer. J. Phys. Anthropol.* 88: 415–468.

Ford, S.M., and Hobbs, D.G., in prep, Intra-generic variability in form and function: the case of *Cebus*.

Froehlich, J.W., 1996, Primate species: the irreversible units in the evolution of our mammalian divergence. *Amer. J. Primatol.* 38:271–279.

Galdikas, B.M.F., Duffy, J.B., Odwak, H, Purff, R.C.M., Vasey, P., 1993, Postcrania and the specific mate recognition system. *Hum. Evol.* 8:281–289.

Gebo, D.L., 1992, Locomotor and postural behavior in *Alouatta palliata* and *Cebus capucinus*. *Am. J. Primatol.* 26: 277–290.

Godfrey, L. and Marks, J., 1991, The nature and origins of primate species. *Yearbook Phys. Anthropol.* 34:39–68.

Hayter, A.J., 1984, A proof of the conjecture that the Tukey-Kramer Method is conservative. *The Annals of Statistics* 12: 61–75.

Hershkovitz, P., 1949, Mammal of northern Colombia. Preliminary report no. 4: monkeys (Primates), with taxonomic revisions of some forms. *Proc. U.S. Nat. Mus.* 98: 323–427.

Janson, C.H., and Boinski, S. , 1992, Morphological and behavioral adaptations for foraging in generalist primates: The case of the cebines. *Amer. J. Phys. Anthropol.* 88: 483–498.

Jungers, W.L. and Fleagle, J.G., 1980, Postnatal growth allometry of the extremities in *Cebus albifrons* and *Cebus apella*: a longitudinal and comparative study. *Am. J. Phys. Anthropol.* 53:471–478.

Kimbel, W.H., and Martin, L.B. (eds.), 1993a, *Species, Species Concepts, and Primate Evolution.*, New York: Plenum Press.

Kimbel, W.H., and Martin, L.B. (1993b) Species and speciation: conceptual issues and their relevance for primate evolutionary biology. In: W.H. Kimbel and L.B. Martin (eds.), *Species, Species Concepts, and Primate Evolution.* New York: Plenum Press pp. 539–554.

Kinzey, W.G., 1982, Distribution of primates and forest refuges. In: G.J. Prance (ed.), *Biological Diversification in the Tropics*, New York: Columbia University Press, pp. 455–482.

Kramer, C.Y., 1956, Extension of multiple range tests to group means with unequal numbers of replications. *Biometrics* 12: 307–310.

Lambert, D.M., and Paterson,, H.E.H., 1982, Morphological resemblance and its relationships to genetic distance measures. *Evol. Theory* 5: 291–300.

Lambert, D.M., and Spencer, H.G., 1995, *Speciation and the Recognition Concept: Theory and Application.* Baltimore: The Johns Hopkins University Press.

Masters, J.C., 1993, Primates and paradigms: problems with the identification of genetic species. In: W.H. Kimbel and L.b. Martin (eds.), *Species, Species Concepts, and Primate Evolution*, New York: Plenum Press pp. 43–66.

Masterson, T.J., 1995, Morphological relationships between the Ka'apor capuchin (*Cebus kaapori* Queiroz, 1992) and other male *Cebus* crania: a preliminary report. *Neotropical Primates* 3(4):165–169.

Mayr, E., 1942, *Systematics and the Origin of Species.* New York: Columbia University Press.

Miller, R.G., Jr. , 1981, *Simultaneous Statistical Inference.* New York: Springer-Verlag.

Mittermeier, R.A. and Coimbra-Filho, A.F., 1981, Systematics: species and subspecies. In: A.F. Coimbra-Filho and R.A. Mittermeier (eds.), *Ecology and Behavior of Neotropical Primates, Vol. 1*, Rio de Janeiro: Academia Brasileira de Ciências, pp. 29–109.

Mittermeier, R.A. and van Roosmalen, M.G.M., 1981, Preliminary observations on habitat utilization and diet in eight Surinam monkeys. *Folia primatol.* 36: 1–39.

Mittermeier, R.A., Rylands, A.B., and Coimbra-Filho, A.F., 1988, Systematics: species and subspecies: an update. In: R.A. Mittermeier, A.B. Rylands, A.F. Coimbra-Filho, and G.A.B. da Fonseca (eds.), *Ecology and Behavior of Neotropical Primates, Vol. 2*, Washington D.C.: World Wildlife Fund, pp. 13–75.

Otte, D., and Endler, J.A. (eds.), 1989, *Speciation and its Consequences*, Sunderland: Sinauer Associates.

Paterson, H.E.H., 1978, More evidence against speciation by reinforcement. *South African Journal of Science* 74: 369–371.

Queiroz, H.L., 1992, A new species of capuchin monkey, genus *Cebus* Erxleben, 1777 (Cebidae: Primates) from Eastern Brazilian Amazonia. *Goeldiana Zoologia* 15: 1–13.

Robinson, J.G., 1981, Spatial structure of foraging groups of wedge-capped capuchin monkeys (*Cebus nigrivittatus*). *Anim. Behav.* 29: 1036–1056.

Rose, K.D. and Bown, T., 1993, Species concepts and species recognition in Eocene primates. In: W.H. Kimbel and L.b. Martin (eds.), *Species, Species Concepts, and Primate Evolution*, New York: Plenum Press, pp. 299–330.

Rosenberger, A.L., 1992, Evolution of feeding niches in New World monkeys. *Amer. J. Phys. Anthropol.* 88: 525–562.

SAS Institute Inc., 1989, *SAS/STAT User's Guide, Version 6, Fourth Edition, Volume 2*, Cary, N.C.: SAS Institute Inc.

SAS Institute Inc., 1994, *JMP 3.0: Statistics Made Visual*, Cary, N.C.: SAS Institute Inc.

Simpson, G.G., 1961, *Principles of Animal Taxonomy.*, New York: Columbia University Press.

Szalay, F.S., 1993, Species concepts: the tested, the untestable, and the redundant. In: W.H. Kimbel and L.B. Martin (eds.), *Species, Species Concepts, and Primate Evolution*, New York: Plenum Press, pp. 21–42.

Teaford, M.F., Walker, A., and Mugaisi, G.S., 1993, Species discrimination in *Proconsul* from Rusinga and Mfangano Islands, Kenya. In: W.H. Kimbel and L.B. Martin (eds.), *Species, Species Concepts, and Primate Evolution*, New York: Plenum Press, pp. 373–392.

Terborgh, J., 1983, *Five New World Primates. A study in comparative ecology*, Princeton: Princeton University Press.

Torres de Assumpçao, C., 1986, Resultados preliminares de reavaliação das raças do macaco-prego *Cebus apella* (Primates: Cebidae). In: M. T. de Mello (ed.), *A Primatologia no Brasil,Vol. 2*, Brasilia, DF: Sociedade Brasileira de Primatologia, pg. 369.

Torres, C., 1988, Resultados preliminares de reavaliação das raças do macaco-prego *Cebus apella* (Primates: Cebidae). *Revta. nordest. Biol.* 6(1): 15–28.

Turner, A. and Chamberlain, A., 1989, Speciation, morphological change and the status of African *Homo erectus*. *J. Hum. Evol.* 18: 115–130.

van Schaik, C.P., and van Noordwijk, M.A., 1989, The special role of male *Cebus* monkeys in predation avoidance and its effect on group composition. *Behav. Ecol. Sociobiol.* 24: 265–276.

Wiley, E.O., 1978, The evolutionary species concept reconsidered. *Systematic Zoology* 27: 17–26.

13

VOCAL COORDINATION OF TROOP MOVEMENT IN SQUIRREL MONKEYS (*Saimiri oerstedi* AND *S. sciureus*) AND WHITE-FACED CAPUCHINS (*Cebus capucinus*)

Sue Boinski

Department of Anthropology and Division of Comparative Medicine
University of Florida
Gainesville, Florida 32611

INTRODUCTION

Troops of New World monkeys in the wild are often first detected by field workers from the tumult of routine intra-group vocal communication. Prodigious numbers of calls can be produced by an individual, and especially a troop, within a brief time period. This vocal barrage usually represents a few types of vocalizations emitted at high rates (Smith et al.,1982; Boinski et al.,1994). Although a species' vocal repertoire usually comprises a limited set of distinct vocalizations, each type of vocalization may encompass much individual and population within-call variation in acoustic structure (Newman 1985; Snowdon 1982, 1989). Yet many vocalizations produced by wild New World monkeys are not associated with any overt social interaction, foraging activity, or predation threat. This prompts inquiry into why individual monkeys indulge in such varied, apparently non-essential vocal communication. The cumulative costs of these intra-group vocalizations are likely significant in terms of energy expenditure and enhanced exposure to predators, even if the cost of each individual vocalization is trivial (Krebs & Dawkins 1984; Jürgens & Schriever 1991).

Across many, if not most, primate species, those intra-group vocalizations with few obvious contextual associations are commonly described by the rubric "contact calls" and are conjectured to maintain cohesiveness among dispersed group members (Lindburg 1971; Gautier & Gautier-Hion 1977). Quantitative studies of the vocal behavior of primate troops in the wild have successfully identified more specific functions for these nominal contact calls in a few species (e.g. Robinson 1982; Harcourt et al., 1993). Further discoveries undoubtedly await close examination of intra-group vocal communication in other primates in the wild. For example, some of the common intra-group contact calls among several species of New World primates have only recently been shown to have an important role in the initiation and leading of troop travel (Boinski 1991; 1993; Boinski et al., 1994). Qualitative observations of other primate taxa in forested habitats also provide evidence that intra-group vocal signals are used to initiate travel in social groups of primates (Kudo 1987; Soini 1988).

In some respects vocalizations that coordinate troop movement can be considered contact calls in the purest sense as they act ultimately to maintain group cohesion. On the other hand, such vocalizations are distinguished from the general concept of contact call because the caller not only transmits and, perhaps, seeks information (Smith 1969), but also attempts to manipulate other group members to follow its movement decisions (Maynard Smith 1974; Krebs & Dawkins 1984). Travel coordination calls also differ from contact calls because the response to movement calls can be marked (i.e., extensive troop movement in a specific trajectory), and are almost certainly more notable than the subtle behavioral effects and associations suggested for many other intra-group vocalizations in primates (Green 1975; Smith et al.,1982; Dittus 1988). Moreover, usage of travel coordination calls is likely closely linked to individual knowledge of the spatial distribution of food sources and other critical environmental features (Garber 1988). These features of vocal travel coordination signals offer insights into the relative cognitive abilities of individuals, and have the potential to define how species communicate spatial, temporal, and social information in a manner similar to recent studies of predator alarm calls (i.e., Cheney & Seyfarth 1990; Hershek & Owings 1993).

In this chapter I consider the results and implications of three field studies addressing the vocal coordination of troop movement among species of two genera of Cebidae: two species of squirrel monkeys, *Saimiri oerstedi* in Costa Rica (Boinski 1991) and *S. sciureus* in Peru (Boinski & Mitchell 1992; see Costello et al.,1993 for discussion of *Saimiri* taxonomy); and the white-faced capuchin, *Cebus capucinus*, also in Costa Rica (Boinski 1993). The specific objectives of these studies were to use focal observations of individual behavior in association with troop movement patterns to identify (1) the vocal signal(s), if any, used to coordinate travel; (2) how such vocal signals are used; (3) which troop members were active in movement decisions; and (4) if within-group conflict over travel route is detected, how consensus for the selected route is achieved.

Long-term research on the foraging behavior and social organization of the study populations preceded each vocal study. These cebids are broadly similar in regard to diet (predominantly fruits, arthropods and the occasional small vertebrate), large to very large social groups (15 - 65 troop members) and ranging areas (1 - 5 km sq.), preference for arboreal, small branch, often heavily foliated, locomotor and foraging substrates, and lack of outstanding morphological adaptations for foraging that constrain behavioral flexibility (Janson & Boinski 1992; Fragaszy & Boinski, in press). In all three species, troops do not fission into subgroups and females allocate more time to foraging than do males (Boinski 1987, 1988a; Mitchell 1990; Rose 1994; Fedigan 1993).

For purposes of comparison, however, it is propitious that marked differences exist among the species in at least three behavioral features potentially affecting the expression of vocal coordination of troop movement: social organization, proclivity to form mixed-species troops, and intelligence. First, squirrel monkeys in Costa Rica, *S. oerstedi*, have an egalitarian social organization in which social aggression in any form is exceedingly rare, whereas Peruvian squirrel monkeys, *S. sciureus*, exhibit strong dominance hierarchies within each sex and females are dominant to males (Mitchell et al.,1991; Boinski & Mitchell 1994). Females are bonded within *C. capucinus* troops and a weak dominance hierarchy is evident (Fedigan 1993); troops have an unambiguous alpha male, and males are individually dominant over all but the alpha female. Second, South American squirrel monkey troops, including the *S. sciureus* in this study but not *S. oerstedi* in Costa Rica (Boinski 1989), usually follow capuchin monkey troops, especially *C. capella*, and are responsible for the formation of mixed-species groups (Terborgh 1983). Third, capuchins, including *C. capucinus*, display remarkably high levels of manipulative intelligence and problem solving abilities compared to squirrel monkeys, and nearly every other species of monkey (Parker & Gibson 1977; Westergaard & Fragaszy 1987; Boinski 1988b).

METHODS

Study Sites and Subjects

The study troops at each site were well habituated and focal subjects were individually identified. At Parque Nacional Corcovado, Costa Rica, 21 of the 24 adult females of the 65-member *S. oerstedi* troop were selected as study subjects (Boinski 1987; Boinski & Fowler 1989; Boinski 1991). Data were collected in January 1988. *Saimiri sciureus* were studied at Parque Nacional del Manu, Peru in March and April 1989 (Terborgh 1983, Mitchell 1990; Boinski & Mitchell 1992,1995). All 22 adult females in the 65-member study troop were study subjects. The study troop of capuchins at Parque Nacional Santa Rosa, Costa Rica, comprised 14 animals, 4 adult females and 3 adult males, 2 female and 2 male juveniles, and 3 infants, and all provided focal data (Chapman 1988, Boinski 1993; Rose 1994). This vocal study took place in January and February 1991.

Data Collection

To a great extent, the same protocol was used to collect, process, and analyze the vocal data in each field study.

Recording conditions were good at each of the three sites. All of the vocal recordings were taken within a 10 m radius of a focal animal, and most within a 5 m radius. A continuous sample of the vocalizations of the focal animal, termed a bout, was recorded onto one channel of a Marantz PMD 430 field recorder fitted with a Sennheiser ME88 directional microphone with a windscreen. Observations of all occurrences of individual and group behaviors were dictated as they occurred.

The non-vocal behavioral data obtained for each subject included descriptions of foraging and locomotor activity, and estimates of spatial position within stationary and travelling groups. Members of the *C. capucinus* troop when stationary had one of two positions: 1) *edge*, a 10 m wide peripheral zone with the outside edge defined by the perimeter of troop dispersion; or 2) *core*, the area interior to the edge. When squirrel monkey troops were stationary it was noted whether the focal subject was at or beyond the troop periphery. Two positions were defined for a travelling troop of all three species: 1) *vanguard*, a 10 m deep zone starting at the leading edge (5 m deep for *S. oerstedi*); and 2) *rearguard*, all positions within a travelling troop behind the vanguard. During preliminary observations of each species the depth of the vanguard was determined by my estimation of the positions within a travelling troop most critical in coordinating movement.

It was not feasible to locate individual group members on a random schedule. Thus a pseudo-random technique for selecting focal animals was employed (Fragaszy et al., 1992). From the animals observable at a given moment, the individual judged to be most 'undersampled' was selected. Preliminary observations indicated that female *C. capucinus* exhibited more variable vocal behavior than other age-sex classes. Therefore, to more efficiently achieve power in statistical analyses, focal samples in the *C. capucinus* study were biased toward adult females. Recording bouts were terminated when the focal animal moved out of recording range, and when callers could not be identified. Possible bias caused by these nonrandom protocols are reduced because many of the data are analyzed as vocal rates within contexts. Analyses do not depend on estimates of allocation of time across contexts by focal subject.

In each study, preliminary observations were used to identify the vocal signals, if any, that might function to coordinate (initiate and lead) troop movement. To quantify use of potential travel coordination calls within each study, a subject was described as making a 'start attempt' when it made at least one of the designated travel calls while the group was stationary. If the group began travelling within 10 min subsequent to the first travel call, the azimuth of the group movement relative to the center of the stationary group and the subject's position were determined by estimation and use of a Suunto compass. I selected a ten-min criterion based on my judgement that this interval represented a pragmatic balance between the minimum time necessary for group members to concur with or to reject the 'start attempt', but not so long that troop travel would occur independently. Travel calls emitted in the vanguard of a travelling group were interpreted as 'leading' the current troop travel. Similarly, a subject was identified as apparently attempting to 'change' troop direction when it made at least one travel call within the rearguard of a travelling group. The azimuth of the group movement relative to the azimuth predicted by the location of the monkey producing the travel call was determined, as was the azimuth of the troop's movement 10 min later.

Sample sizes were comparable across the studies. The data sets for *S. oerstedi, S. sciureus*, and *C. capucinus* consisted of, respectively, a cumulative 18.6 h, 17.1 h, and 33.7 h of continuous recordings. Each focal adult female squirrel monkey contributed an average (SE) of 53.0 (5.4) min and 146.7 (16.8) calls in the *S. oerstedi* study and 46.7 (4.8) min and 152.4 (19.0) calls in the *S. sciureus* study. Each adult female *C. capucinus* contributed an average (SE) of 288.8 min and 331.5 (24.3) calls, adult males 117.9 (31.6) min and 35.7 (14.7) calls, juveniles 75.8 (10.2) min and 40.5 (21.1) calls, and infants 70.6 (12.2) min and 99 (16.2) calls. The temporal sequence and duration of the individual and troop behavior and the time and context of each vocalization were coded from the tapes. Sonograms of the 3080 vocalizations recorded during the *S. oerstedi* study were made with a Kay Sonograph 700 (settings: narrow band, 0–16 kHz). Spectrograms of the vocalizations from the other two studies (*S. sciureus*, n = 3353; *C. capucinus*, n = 1892) were made with a Multigon Uniscan II (settings 5, 10, and 20 kHz; 256 FFT) and a Panasonic KY-P1180 printer.

Call Classification

Sonograms and spectrograms were assigned by visual inspection into call types. Context was not employed in assigning call types. For each species, the characteristic intra-group vocalizations, other than alarm calls, are illustrated (Fig. 1).

Although subtle variants and subcategories in the acoustic structure of the vocalizations of many primate species may convey dramatically different categories of information to conspecifics (Cheney & Seyfarth 1982; Boinski & Mitchell 1995), the analyses employed in these vocal field studies were insensitive to detection of this potential source of variation in the data. Robinson (1984), for example, subdivided the trill vocalizations produced by wedge-capped capuchins (*C. apella*) into four subcategories based on the extent and contour of the twitter's frequency modulation: descending trills, U trills, ascend-

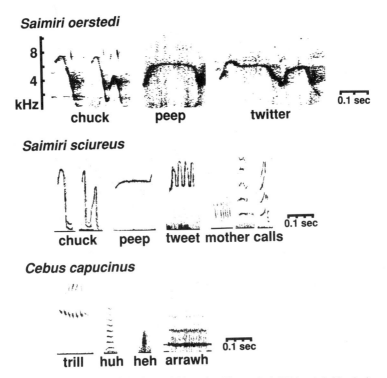

Figure 1. Examples of the major call types of the three cebid species. The vertical (kHz) axis is identical across all species, but the horizontal (time in sec) axis differs in scale between *Saimiri oerstedi* compared to *S. sciureus* and *Cebus capucinus*.

ing trills, and FM trills. He also determined that social communication functions, primarily indication of affiliation and relative status in potentially aggressive interactions, varied across these trill subcategories. The broad range of within-individual variation in acoustic structure typical of the trills produced by white-faced capuchins at Santa Rosa easily encompasses the trill subcategories defined by Robinson (1984) (See Figure 1 in Boinski, 1993). No association of specific contexts with variation in the acoustic structure of the trills has yet been found at Santa Rosa, despite many analyses that have sought unsuccessfully to identify such an association (Boinski 1993, unpublished data). Furthermore, no evidence suggests that the social communication functions of trills Robinson (1984) identified wedge-capped capuchins explained any of the variation in the usage among white-capped capuchins at Santa Rosa (Boinski 1993).

Analytical Techniques

The recording bouts for each subject were combined. Within-subject repeated-measure ANOVAs were used to examine the effects of position within stationary and travelling troops on the rate of travel coordination calls within each species. In the capuchin study, between-subject effects were also used to examine the effects of age-sex class. For each repeated-measure ANOVA, the raw data were first sorted by behavioral state (e.g., vanguard versus rearguard). The cumulative frequency of travel coordination calls and the duration of time a subject was recorded in each behavioral state were used to calculate the rate of calling per min for each behavioral state. This rate was then log 10 transformed to obtain normal distributions. Subadult capuchins were excluded from the analysis of trill rate within a travelling troop because of inadequate sample time within the vanguard position.

I used the V-test, a modified version of the Rayleigh test (Batschelet 1981) to determine whether the observed azimuths of troop movement were clustered about the position of the focal monkey that was apparently attempting to initiate or change the direction of movement. Instances in which the troop did not move subsequent to such an attempt were treated as if the troop had instead moved 180° from an angle predicted by an individual's position on the troop periphery. Similarly, if a monkey produced travel coordination calls from the core of the stationary troop, the troop was scored as having moved 180° from an angle predicted by a failed start attempt. The azimuths of troop movement 10 min after an attempt to change the trajectory were compared to the azimuths predicted by the subject's position on the side and back of a travelling troop. Instances in which a monkey emitted a travel coordination call in the center of a moving troop were treated as if the direction indicated was 180° from the current travel direction. The angular data were grouped into arcs of 90° prior to analysis for *C. capucinus*.

The data describing the average (SE) number of calls emitted in start attempts and the duration of start attempts are based on the individual means.

RESULTS

Calls Used to Coordinate Travel and Which Troop Members Produced Travel Calls

Each species relied upon a small set of call categories for nearly all intra-group vocal communication (Fig. 1, Table 1). Preliminary observations quickly suggested that the *twitter* call in *S. oerstedi* and the *trill* call in *C. capucinus* played important roles in the initiation and leading of troop movement. In both species intention movements (Kummer 1968; Rowell 1972) sometimes accompanied the production of the travel coordination calls. Yet the vocal signals alone probably contained sufficient information; in most situations for these forest primates, the majority of troop members were unlikely to be in visual contact with the signaller due to dense, intervening foliage (see Boinski 1987, 1993).

No vocalization or other non-vocal behavior, however, with comparable functions to coordinate troop movement was identified in the vocal repertoire of *S. sciureus*. In fact, the twitters of *S. oerstedi* have no apparent analogues in the repertoire of *S. sciureus*. All the data in the *S. sciureus* study were collected when the troop was in association with one of several troops of brown capuchin monkeys (*C. apella*). The latter species seems best described as 'leading' the mixed-species entourages; on an individual basis *S. sciureus* appeared to be loosely monitoring the location of capuchin monkeys both visually and by attending to their calls (SB and Carol Mitchell, pers. obs.; Terborgh 1983). Two types of intra-group calls important in maintaining spatial cohesion among subsets of *S. sciureus* troop members are chucks and peeps (Fig. 1; see also Boinski & Mitchell 1992), Observations during other study periods (Mitchell 1990) suggest that in those relatively rare instances when *S. sciureus* travel independently of capuchins, chucks and peeps are emitted at much higher rates, and that coordinated troop travel is accomplished awkwardly.

During extensive preliminary observations immediately prior to and within the *S. oerstedi* field study, no troop members other than adult females were observed to produce

Table 1. List of major call types, excluding alarm calls, within the vocal repertoire of *Saimiri oerstedi, S. sciureus,* and *Cebus capucinus,* and the absolute number and percentage of the total call sample of that species represented by each call type based on Boinski (1991, 1993) and Boinski and Mitchell (1992, 1995). Also included is a brief description of apparent function

	Absolute number	%	Apparent function
Saimiri oerstedi			
Chuck	2414	78.4	maintain spatial cohesion with neighbors; rate increases with increasing distance to nearest neighbor; chucks may have one, two, or rarely three major frequency modulations and subtle functional distinctions probably exist
Peep	173	5.6	separation call produced when caller is at risk of losing contact with troop
Twitter	355	11.5	travel coordination call
Saimiri sciureus			
Chuck	2287	68.2	maintain spatial cohesion with neighbors; rate increases with increasing distance to nearest neighbor; chucks may have one, two, or rarely three major frequency modulations and subtle functional distinctions probably exist
Peep	61	1.8	separation call produced when caller is at risk of losing contact with troop
Tweet	78	2.3	unclear; might function to notify other troop members of the location of a dense fruit patch to be shared
Mother calls	716	21.4	mother signals to infant her location, that she is willing to nurse, and when nursing bout will end
Cebus capucinus			
Trill	1258	66.4	travel coordination call in adults and contact call in subadults
Huh	428	22.6	most commonly produced within dense fruit patches; may also have role in negotiation of spacing between individual troop members
Heh	150	7.9	produced in mildly to moderately aggressive contexts
Arrawh	0	0	lost call emitted either when caller is completely separated from troop or by individuals within troop in response to an arrawh from a lost troop member (frequently documented in *ad lib* samples, but not focal samples)

twitters. Therefore, the focal subjects in this study were limited to adult females. In contrast, all age-sex classes of *C. capucinus* produced trills and were included in the study as focal subjects. The trills of infants and juveniles (subadult troop members less than 5 years old) were never associated with the coordination of troop travel, and over 90% of subadult trills were in vocal exchanges or social interactions with other subadults. Subadult capuchins were never observed in the vanguard of a travelling troop and only uncommonly in the edge of a stationary troop. Regardless of position, activity, or whether the troop was stationary or moving, trills were nearly exclusively produced by adults in contexts associated with the coordination of troop movement.

How Travel Coordination Calls Are Used

A) S. oerstedi. There were 23 instances in which a *S. oerstedi* female made an apparent attempt to initiate travel in a stationary troop. Twenty of these start attempts were termed 'successful' because within 10 min of the beginning of the start attempt the troop travelled in the direction predicted by the focal females' location relative to the core of the troop. The azimuths of troop movement subsequent to a start attempt were significantly clustered about the angle predicted by the focal female's position (u= 5.06, n=23, P < 0.0001). The troop travelled a minimum of 275 m in all 20 'successful' starts before becoming stationary again. On average (SE), 6.2 (3.2) twitters were emitted during a 3.6 min (1.3) period by the 12 females that successfully initiated travel at least once during focal samples.

During the 1988 field study, adult males never produced a twitter nor appeared to participate in any overt manner in the coordination of troop movement. However, during periods subsequent to predation attempts or when the troop passed through areas when susceptibility was high (i.e., little cover or in the vicinity of a raptor nest) the 3–4 adult males were usually clustered in the forefront of an extremely tightly coalesced troops during troop travel. In this situation, no troop members, including the adult males, emitted any vocalization, including twitters. Yet, in at least one instance in 3000+ hours of previous observation of this population (see Boinski 1987a, 1988, 1989) the four full adult males then resident in the study troop were extremely active in directing travel with twitters. The occasion was a late afternoon when the troop was in a seldom-frequented part of its range and far (> 1.5 km) from its habitual sleeping tree; normally the troop would be within 200 m of the sleeping tree at this time. The four males began twittering and leading the troop rapidly forward in the most direct trajectory to the sleeping tree. Shortly thereafter, two of the males dropped to the back of the troop. These latter two males continued to emit loud twitters while herding the laggards to join the main body of the troop in a manner reminiscent of a sheep dog rounding up stray sheep.

Sixteen of the 21 focal *S. oerstedi* females led a travelling troop at least once. In other words, these females produced twitters while in the vanguard position of a travelling troop. A total of 31 instances of a female leading a travelling troop were documented. Six of these were after the focal female had successfully initiated travel. Seven times a female moved forward from the rearguard of a travelling troop to a position within the vanguard where she then produced twitters. No significant difference was detected in the rate of twitters when a female was attempting to initiate travel versus leading a travelling troop (n=6, df =1,5, F = 2.55, P > 0.05, R^2 = 0.34; Fig. 2a). Twitters were produced at significantly higher rates, however, during the time females allocated to initiating and leading travel pooled compared to the time spent in the core of stationary troops or in the rearguard of travelling troops when the data were pooled (i.e. following the travel coordination efforts of other females) (n = 14, df = 1,13, F = 26.87, R^2 = 0.67; Fig. 2b).

No evidence suggested that the near-negligible rate (< 0.1 twitters / min) and absolute number of twitters (n= 32) emitted by *S. oerstedi* in the rearguard of a travelling troop were part of an attempt to alter the trajectory of troop travel. Instead, nearly all of these rearguard twitters were produced when a female was near or rapidly locomoting to the vanguard, or a female was at the forward edge of a second "wave" within a linearly dispersed troop. Similarly, there was little evidence that any female or set of females predominated in efforts to coordinate troop travel. All but three females led or initiated travel at least once, and no female was active in more than four instances of travel coordination in the focal samples. There might, however, have been a slight age effect. Females that appeared youngest (based on changes in pelage, skin pigmentation, and ease of movement) tended to be least active, and females that appeared older tended to be more active in the coordination of travel using vocalizations.

The outcome of these individual efforts to coordinate travel was that the daily movement patterns of the *S. oerstedi* troop consisted largely of straight travel segments punctu-

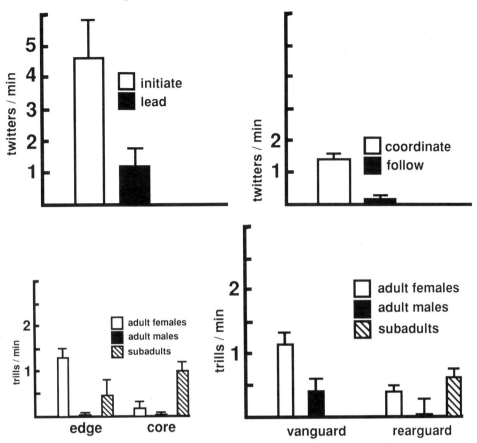

Figure 2. a) The mean rate (+SE) at which adult female *Saimiri oerstedi* produced twitters when initiating versus leading troop movement. b) The mean rate (+SE) at which adult female *Saimiri oerstedi* produced twitters when coordinating travel (time spent initiating and leading troop movement pooled) versus following (time spent in the core of a stationary troop and rearguard of a travelling troop pooled). c) The mean (+SE) rate at which *C. capucinus* age-sex classes produced trills within the edge and core positions of a stationary troop. d) The mean (+SE) rate at which *C. capucinus* age-sex classes produced trills within the vanguard and rearguard positions within a travelling troop. The data for subadults are given in Fig. 2d to document age-sex patterns more completely, but were not used in statistical analyses.

ated by intermissions where the troop was stationary and engaged in foraging activities or resting. Abrupt changes of trajectory when the troop was travelling were extremely rare. Instead, major alterations in trajectory were achieved after a troop stopped travelling and travel was re-initiated in a new direction. These sequences of starts and stops overlaid long-term (> 10 years), traditional travel routes.

B) C. capucinus. When a capuchin troop was stationary there was a significant age-sex class (between-subject) effect on the rate of trills (Table 2, Fig. 2c). No effect of position within the edge or core or position by age-sex interaction effects were detected. Adult females emitted trills 7 times more frequently on the edge than in the core. Males had very low rates of trills in both positions. Subadult capuchins (infants and juveniles pooled) trilled twice as often in the core as in the edge.

Table 2. Results of repeated measure ANOVAs on the effects of spatial position (within-subject effect), age-sex class (between-subject effect), and their interaction on the rates of trills when the *Cebus capucinus* troop was stationary and travelling

Troop movement	Effect	df	F	R^2
Stationary	Age-sex class	2,8	7.74 *	0.66
	Position	1,9	5.62 *	0.59
	Position by age-sex interaction	2,8	1.74	0.15
Travelling	Sex class	1,5	9.58 *	0.66
	Position	1,6	18.12 **	0.66
	Position by sex interaction	1,5	8.74 *	0.43

*P < 0.5; **P < 0.01

Significant sex (between-subject) effects and positional (within-subject) differences as well as an interaction effect were found on the rate of trills within a travelling troop (Table 2, Fig. 2d). Both males and females emitted trills at a significantly greater rate in the vanguard versus the rearguard positions, with females calling more than males in both positions.

The high rate of trills by adult females when in the edge of a stationary troop (1.3 trills / min) represents 36 instances in which an adult female made an apparent attempt to initiate movement in a stationary troop (Table 3). Adult males also made three apparent attempts to initiate troop movement. Thirty-six of the 39 total start attempts were termed 'successful' because within 10 min of the beginning of the start attempt the troop travelled in the direction predicted by the capuchin's location relative to the center of the troop. In at least 19 of the start attempts more than one adult produced trills in the same start attempt. The sex, but not always the individual identity, of others active in the start attempt were determined. No mixed-sex start attempts were observed. The azimuths of troop movement subsequent to a start attempt were significantly clustered about the angle predicted by the focal capuchin's position ($u = 7.67$, n=39, P < 0.0001).

Changes in direction during troop travel were also associated with trills produced by an adult at the sides or back of a travelling troop (Table 3). Ten 'successful' and 4 'unsuccessful' changes were documented. The azimuths of troop movement 10 min subsequent

Table 3. Tabulation of the number and type of instances in which adult females and males attempted to lead troop movement. Also given is the rate of successful attempts per hour of focal observation for each individual. Within each sex individuals are ranked by relative dominance as determined by Rose (1992)

Focal	Travel start in stationary troop		Change in travelling troop angle		Rate of successful attempts per hour of focal observation
	Successful	Failed	Successful	Failed	
Adult females					
Limp	10		4		3.41
Blanche	9		3		2.58
Chops	9	1		2	2.04
Patch	5	2	2	2	1.50
Adult males					
Junior	2		1		1.08
Winston	1				0.84
Blackie					0.00
Total	36	3	10	4	

to these attempts were significantly clustered about the position predicted by the calling animal's position ($u = 3.114$, n = 14, P < 0.001). For the 3 females and 1 male that successfully changed the azimuth of the troop trajectory, the change attempts represented an average (SE) of 11.6 (2.3) trills and 4.3 (1.5) min (n = 2). Two adult females each made two unsuccessful attempts with an average (SE) of 13.3 (2.9) trills and 3.2 (2.0) min per attempt (n = 2).

Within each sex there was a perfect concordance between an individual's dominance rank and the number of successful attempts to determine troop movement (Table 3); the greater the dominance rank, the greater the number of successful initiations and changes of troop movement. This relation is also evident when the number of successful attempts is converted to the rate of successful attempts per hour of focal observation.

Conflict and Consensus in Travel Route Selection

In two of the three 'failed' attempts to initiate travel that were documented for *S. oerstedi*, the troop did not start travelling within 10 min of the beginning of the start attempt. After 2 min at the periphery of a stationary troop in one instance, and three min in the other, with no overt movement response by the troop, the focal female ceased twittering and returned to the troop core. While I was recording from a focal female in the third 'failed' attempt, two other females were also engaged in a start attempt on the troop periphery 146° from the position of the focal female. Six minutes after the focal female began her start attempt, the troop travelled in the trajectory predicted by the position of the latter two females. No evidence suggested that any vocalization, including the twitter, of *S. oerstedi* (or *S. sciureus*) was ever employed in a deceptive manner.

The several instances of 'failed' start attempts observed in the *C. capucinus* study were reminiscent of those documented in *S. oerstedi*. In two of the three 'failed' *C. capucinus* start attempts, the troop did not start travelling within 10 minutes. One unsuccessful adult female produced 72 trills in an attempt lasting more than 9 minutes, and another emitted more than 93 trills in an attempt exceeding 10 minutes before apparently abandoning efforts to initiate travel. The third 'failed' attempt occurred when two adult females were simultaneously engaged in start attempts at positions 60° apart on the troop periphery. Nearly 9 min and 51 trills after the focal female initiated the start attempt, the troop travelled in the trajectory predicted by the other female's position.

Two instances of change in direction in the *C. capucinus* troop were noteworthy for their indication of strategic maneuvers. First, two females, A and B, simultaneously initiated start attempts at different positions on the periphery of a stationary troop; the troop soon began following the trajectory indicated by one of the trilling females. The 'unsuccessful' female, A, travelled in the rearguard for 2 min then ran in a purposeful manner to the vanguard while trilling very loudly and frequently, such that she 'overwhelmed' the 'successful' female's trills. Once A surpassed B in the vanguard, A continued the intense trilling and altered the trajectory to the route she had originally indicated. In another example, an adult female in the vanguard was producing trills and leading the capuchin troop on a trajectory that intersected with a favored fruit-foraging area. The alpha and beta males rapidly ran forward from the rearguard to a position facing the adult female such that her forward movement was blocked. Shortly after the vanguard female emitted another trill, the two males gave her highly aggressive threats. She ran back to the rearguard and produced no additional trills. The two adult males, now alone in the vanguard, immediately oriented 120° degrees away from the original trajectory, trilled, and led the troop directly to a water hole.

Based on the focal and ad libitum observations during the capuchin study, all but one instance in which a stationary troop initiated travel was subsequent to an adult emitting at least one trill from the troop periphery. The exception was when the alpha male appeared to initiate and lead troop travel solely with intention movements in an area with little foliage where he was likely visible to all troop members. Shortly after the completion of this field study, Lisa Rose (pers. comm.) twice observed a pair of adult females in an attempt to initiate troop movement direct their trills specifically to the alpha male, as if 'appealing' for his concurrence.

DISCUSSION

The array of mechanisms through which primates coordinate individual activities to travel as a cohesive unit are only beginning to be identified and the importance of this function appreciated (Norton 1986, Henzi et al.,1992). Somehow troops seldom fragment even when confronted with extensive and complex travel routes, broad dispersion of troop members, and for arboreal primates particularly, sporadic visual contact even among those troop members in close proximity. The challenges of coordinating travel are compounded further because troops typically comprise individuals with diverse foraging strategies, nutritional needs, susceptibilities to predation, and locomotor abilities. Potentially divisive factors are so numerous that it is surprising that fission-fusion social structures are not more prevalent among primates.

In at least two species of cebid, *S. oerstedi* and *C. capucinus*, the coordination of troop movement is mediated via specialized calls, the twitter and trill respectively. Use of these vocalizations by adults of both species are restricted largely to two major situations, the initiation of travel in a stationary troop and within the leading edge of a travelling troop. A troop typically initiates travel when a monkey (occasionally two, and rarely three) moves to a position on the periphery of the troop dispersion and produces the species-appropriate travel coordination call. Even one travel coordination call produced from the troop periphery significantly predicts that the troop will initiate travel within 10 min. Moreover, the trajectory of travel is also closely predicted by the calling monkey's position relative to the center of the troop. Within travelling troops the rates of travel coordination calls are significantly greater in the leading edge, the vanguard, compared to the rearguard positions. Perhaps trills and twitters might be best described as "start travel or continue travel in this specific direction" calls.

Another feature shared by the *C. capucinus* trill and the *S. oerstedi* twitter is that the acoustic structure of these calls accurately signals the location of the vocalizer. In my experience, these calls are reliably heard and the location of the caller easily discerned by human observers, and other troop members in nearly all instances in which they are produced. To be locatable would seem to be a requisite characteristic of a vocal signal employed to coordinate travel among animals living in forest habitats. Foliage and woody structures impede both visual contact and the reliable transmission of vocalizations due to reverberation and reflection of acoustic signals by these plant materials (Waser & Brown 1986).

The ease with which a recipient of a vocal signal can evaluate the angle and location of the vocalizer is related to the intensity and redundancy of time-of-arrival cues of the vocal signal (Thurlow, 1971). Frequency modulated calls and those with numerous time-of-arrival cues provide such information, as do signals with a broad band-width (Brown, 1982). Although the acoustic structures of the twitter and trill are not markedly similar at first glance, both appear adapted to be readily heard and located in forested habitats. Both of these calls are typically produced at very loud intensities compared to other intra-group vocalizations in these species' repertoires. The frequency contour of the acoustic signal of the *S. oerstedi* twitter is highly modulated. The *C. capucinus* trill, on the other hand, consists of a sequence of, on average, five pulses of sound during a 0.25 sec interval (Boinski 1993), a veritable firecracker of time-of-arrival cues.

Cognitive Implications

These data on the vocal coordination of troop movement from capuchins at Santa Rosa and squirrel monkeys at Corcovado provide strong evidence for the expression of individual preferences and decisions in this complex vocal behavior. Conflict and lack of consensus was frequently evident when simultaneous efforts to initiate travel in divergent trajectories occurred, and when the troop failed to follow efforts by group members to initiate travel. Qualitative observations during both studies suggest that individuals often attempted to lead their troop to preferred foraging patches using travel coordination vocalizations (Boinski, 1993, unpubl. obs.). In similar fashion, individuals foraging within preferred foraging patches often appear resistant to acquiesce to the travel initiatives of other troop members. Detailed knowledge of the distribution of resources and features within the home range appears to be a necessary precondition for this level of divergent individual preferences to be expressed.

The common observation of individuals at both sites actively supporting and reinforcing the efforts of other troop members to coordinate troop movement provide a second source of evidence for the importance of cognition, in this instance social cognition (Harcourt 1988). At Santa Rosa, pairs of adult female *C. capucinus* that formed alliances in aggressive food competition also reinforced each other's efforts to coordinate troop movement (Boinski 1993). Similarly, multiple adult female *S. oerstedi* at Corcovado often cooperated, or at least expressed concordant preferences, in joint efforts to direct troop travel. In many respects the process of initiation and leading of troop movement can be regarded as a social behavior on a very grand scale, particularly with a large social group. When an individual transmits a signal in an effort to determine troop movement a positive response may hinge on negotiation, concurrence, reciprocation, and acquiescence among a large proportion of group members.

Trills and twitters should probably be considered representational vocal signals. This classification is consistent with two criteria set forth by Macedonia and Evans (1992): (1) that the signal be closely associated with specific contexts, and (2) be sufficient in itself to provoke the appropriate response. Trills emitted by mature *C. capucinus* and twitters emitted by adult female *S. oerstedi* were nearly exclusively associated with coordination of travel. The second criterion was met because troop members were able to respond appropriately to these travel coordination calls within brief time periods, even though the typical broad dispersion of *C. capucinus* and *S. oerstedi* troops provided only limited opportunities for visual contact with the vocalizer. At both sites it was common to observe individuals that appeared to be totally isolated visually from other troop members orientating and moving in the appropriate direction after hearing travel coordination calls in the distance.

The use of these specialized calls in initiating and leading travel suggests a recognition of the effect on recipients, what Dennett (1983) and Cheney and Seyfarth (1990) describe as first-order intentionality. One way to assess whether full first-order intentionality can be invoked is to determine if the caller subsequently monitors the behavior of its audience. Monitoring would be expected if an individual gives a signal with the intent to inform and affect. Monitoring of the effect of trill or twitter calls on the other troop members was, in fact, ubiquitous, among *C. capucinus* at Santa Rosa and *S. oerstedi* at Corcovado (SB, pers. obs.). A monkey emitting calls to initiate travel in a stationary troop repeatedly looked towards the main body of the troop and scanned for evidence that the troop had begun travel. No capuchin or squirrel monkey gave travel coordination calls in the periphery of a stationary troop and then travelled onward more than a short distance by itself. In like fashion, and in both species, the individual vocally leading the troop in the vanguard commonly stopped and looked backward, especially if troop members travelling in the rearguard had slowed or even slightly deviated in trajectory.

To date, the sparse data available suggest that among primates the opportunity for individual trial and error learning is required for appropriate expression of most aspects of vocal behavior (Symmes & Biben 1992). Clearly, at least two types of learning will be involved in vocal coordination of troop movement (1) the association between travel coordination calls and subsequent movements (i.e., Thorpe 1963), and (2) the spatial learning necessary to orient and navigate over distances where the ultimate goal is not visible (Alyan & Jander 1994).

I suspect vocal coordination of travel may provide the first examples of social experience affecting the acquisition of a vocal behavior, including *social influence*, another individual stimulating or facilitating the expression of a vocal behavior, and the direct acquisition of knowledge from another via *social learning* (Whiten & Ham 1992). Future studies of capuchins, squirrel monkeys and other species of primate might profit from close attention to potential associations between early efforts by an individual to use vocalizations to coordinate movement and the active cooperation with and reinforcement of these efforts by older, and especially related, troop members. Perhaps a young individual might also 'practice' appropriate usage by reinforcing the travel coordination calls and leadership of experienced troop members before solo attempts.

Mixed-Species Troops

Despite ardent efforts by my colleague Carol Mitchell and myself, we were unable to identify any aspect of the vocal behavior of Peruvian *S. sciureus* in Manu that hinted at being specialized to coordinate travel in a manner similar to that evident in *S. oerstedi* and *C. capucinus*. The *S. sciureus* study troops did, however, swiftly and deftly range throughout large portions of its 6 km^2 home range with a facility that was truly staggering to the field workers trailing in their wake. This was accomplished by 'catching capuchin troops' (the sympatric *C. apella* and less often *C. albitrons*) much as surfers 'catch waves'.

'Parasitism' by *S. sciureus* of the capuchins' knowledge of the location of local fruit sources and their reliance upon capuchins to direct the movement of the mixed-species troop (Terborgh 1983; Boinski & Mitchell 1992) is apparently so complete (> 95% of diurnal activity) that no specialized call to coordinate travel independently of capuchins has evolved or been maintained in the repertoire of the study population. In the limited time *S. sciureus* travel apart from capuchins, they appear to determine travel paths and maintain a loose troop cohesion, albeit with apparent confusion and many false starts, by producing contact calls (more typically exchanged in response to increases in spatial separation with close neighbors) at exceptionally high rates and volume. In contrast, *S. oerstedi* avoids troops of the sympatric capuchin, *C. capucinus* (Boinski 1989), and always determines its travel route by a within-troop decision process.

Troop coordination calls are not the only category of vocalizations with a dichotomous presence in the vocal repertoires of the two squirrel monkey species. Mother to infant calls are common (22% of all adult female vocalizations) in *S. sciureus* vocal communication, but are absent among *S. oerstedi* (Table 1; Boinski & Newman 1988; Boinski & Mitchell 1995). That the vocal behavior of two closely related species, nearly indistinguishable in appearance and basic foraging behavior (Mitchell et al. 1991), may diverge so markedly on two axes, including use of specialized travel coordination calls, suggests that closely related species within other primate genera may easily have divergent means to coordinate travel. This is reinforced by the fact that *Cebus apella* differs from *C. capucinus* in lacking specialized travel coordination calls; instead troop members monitor closely the activity and location of the alpha male, the supremely dominant troop member (Boinski 1993; SB & C. Janson, unpubl. data).

We should expect, and already have evidence of, a diversity of adaptations to the challenges posed by the need to coordinate travel in primate troops. Phylogenetic constraints do not appear to be severe. Several factors seem predictive of the presence of vocal coordination of troop movement. The first is use of an arboreal habitat or other situation where visual contact among troop members is impeded. Second, within an egalitarian social organization or a relatively weakly ranked dominance hierarchy, as exhibited by *S. oerstedi* and *C. capucinus*, it would likely be advantageous for individuals to have a means to broadcast their preferences. Among social groups with one or a few clearly dominant individuals, as in *C. apella* and baboons (*Papio anubis*, Rowell 1972; *P. hamadryas*, Kummer 1968), these few individuals can control travel path decisions, and specialized vocalizations will be unlikely. Third, large group size might also tend to make calls more advantageous. Small family groups of titi monkeys (*Callicebus moloch*) do not employ travel coordination calls, and generally coordinate movement by closely following one another and restricting travel to customary routes (Menzel 1993). But golden lion tamarins (*Leontopithecus rosalis*) do rely upon a specialized call to coordinate travel, and their group size seldom exceeds eight (Boinski et al.,1994). Perhaps a better predictor than group size of the utility of travel coordination calls would be the extent of troop dispersion and nearest neighbor distances; the larger the typical separation and interference with visual contact among troop members, the more likely travel coordination calls will be in the vocal repertoire of a species.

Effects of Intelligence

Capuchins (*Cebus* spp.) display remarkable levels of manipulative intelligence and problem-solving abilities that appear to derive from their foraging specializations: extraction of food items encased within substrates, a precision grip, and the ability to integrate time and space in complex routes between foraging patches (Janson & Boinski 1992). The capuchins in this study also evinced unexpected ingenuity in two aspects of vocal behavior. First, capuchins were able to change the trajectory of an often rapidly travelling troop as much as 180° using trills. *S. oerstedi* did not display this flexibility, but travelled in a more or less straight line between points and changed direction only after the troop was stationary.

Second, the relation between a troop member trilling in the vanguard of a troop and the troop following is apparently exploited more fully by at least some individual *C. capucinus* than by *S. oerstedi*. The *C. capucinus* troop was successfully 'hijacked' by individual troop members usurping the vanguard position, once by aggression and in another instance by outdistancing the nominal leader and overwhelming her trills. These maneuvers suggest intentionality and the ability to anticipate behavioral effects and are reminiscent of instances of social manipulation documented in chimpanzees and other primates (de Waal, 1991).

Social Organization

Within those species that employed travel coordination calls, the extent to which troop members actively participated in troop movement decisions varied dramatically across age and sex classes. In *S. oerstedi,* adult females exclusively produced the twitters to initiate and lead travel during the study period, and among *C. capucinus* adult females vastly predominated in their efforts compared to adult males. The greater amount of time females allocate to foraging compared to other age and sex classes in both species probably underlies this adult female bias. Given that alternative travel routes to and through foraging areas are likely to differ in attainable foraging efficiency, the foraging costs of poor route selection would most immediately be borne by adult females.

In a similar fashion, individuals whose dominance accords more influence in the decision process should wield their dominance when advantageous. Therefore, within *C. capucinus* males and females separately, the relative success of initiating travel movements was exactly predicted by an individual's dominance rank. Among the egalitarian *S. oerstedi* adult females, no individual predominated, although the females that participated least also appeared to be the youngest individuals in this age-sex class. Some evidence suggests that the alpha male *C. capucinus* may on occasion decide the outcome of travel initiatives made by adult females. This possible role for alpha males agrees with the "initiative" and "decision" roles that Kummer (1968) and Rowell (1972) identified in the process of determination of travel routes within baboon troops. In *S. oerstedi*, there was no evidence that males actively participated in the decision process in other than extremely unusual situations. On the other hand, no data exclude the possibility that *S. oerstedi* males wield influence in a more subtle "decision" role.

Conclusion

The vocal behavior of New World primates should be considered much more explicitly within the context of the behavioral ecology of each species. In the future the phrase 'behavioral ecology of vocal communication' should become as commonplace as the phrase 'behavioral ecology of mating patterns'. This chapter emphasizes that social organization, cognitive abilities, and interspecific association patterns can be associated with specific features of intra-group vocal behavior. These conclusions are consistent with other recent findings that geographic variation in the usage of a vocalization in a social species can be identified with variation in specific ecological and social factors (East & Hofer 1991; Clark 1993). Moreover, these causal linkages between the nonvocal behavior and ecology of a species and its vocal behavior can and should be tested with future field studies with cebids and other species. More emphasis on quantitative studies of the patterns of vocal behavior among social groups whose social structure and ecology are well known will allow fresh insights into the functions and evolution of vocal communication.

SUMMARY

Although many social groups appear to remember where food is distributed because they move between foraging areas quite directly and efficiently, how coordinated movement is achieved has been obscure. Three recent field studies of the vocal behavior of Neotropical monkeys document that at least two species (white-faced capuchins and Costa Rican squirrel monkeys), but not all (Peruvian squirrel monkeys), rely on specialized calls to initiate and lead troop movement. In some respects, vocalizations that coordinate troop movement can be considered under the general rubric of contact calls as they act ultimately to maintain group cohesion. Yet such vocalizations are distinguished from the general concept of contact calls because the caller not only transmits information, but also manipulates others to follow its movement decisions. Use of travel coordination vocal signals in white-faced capuchins and Cost Rican squirrel monkeys is probably closely linked to the spatial knowledge of food patch distribution, foraging tactics, and social position of individual group members. As a result, this specialized vocal behavior provides useful vantage points from which to consider the relative cognitive abilities and social complexity of individuals and species. Peruvian squirrel monkey troops, although deftly traversing ranging areas much larger than the two other species, gave no evidence of any analogous travel coordination vocalization in its vocal repertoire. Instead, Peruvian squirrel monkeys form mixed-species troops with sympatric capuchin species to benefit from enhanced fruit foraging efficiency. The capuchin troops, in turn, appear to determine the travel route for the mixed species troop.

ACKNOWLEDGMENTS

I thank the Servicio de Parques Nacionales of Costa Rica and the Direccion General Forestal de Fauna of the Peruvian Ministry of Agriculture for the privilege of conducting field work with squirrel monkeys and capuchins. These vocal behavior field studies were supported by the Laboratory of Comparative Ethology, NICHD, NIH. Paul Garber, Linda Fedigan, and Gary Steck provided helpful comments on this manuscript.

REFERENCES

Alyan, S., and Jander, R., 1994, Short-range homing in the house mouse, *Mus musculus*: stages in the learning of directions. *Anim. Behav.* 48: 285–298.
Batschelet, E., 1981, *Circular Statistics in Biology*. London: Academic Press, 1981.
Boinski, S.,1987, Habitat use in squirrel monkeys (*Saimrii oerstedi*) in Costa Rica. *Folia primatol.*49: 151–167.
Boinski, S.,1988a, Sex differences in the foraging behavior of squirrel monkeys in a seasonal habitat. *Behav. Ecol. Sociobiol.* 23: 177–186.
Boinski, S., 1988, Use of a club by a wild white-faced capuchin (*Cebus capucinus*) to attack a venomous snake (*Bothrops asper*). *Amer. J. Primatol.*14: 177–179.
Boinski, S., 1989, Why don't *Saimiri oerstedii* and *Cebus capucinus* form mixed-species groups? *Int. J. Primatol.* 10: 103–114.
Boinski, S.,1991, The coordination of spatial position: a field study of the vocal behaviour of adult female squirrel monkeys. *Anim. Behav.* 41: 89–102.
Boinski, S., 1993, Vocal coordination of group movement among white-faced capuchin monkeys, *Cebus capucinus*. *Amer. J. Primatol.* 30:. 85–100.
Boinski, S., and Fowler, N. L.,1989, Seasonal patterns in a tropical lowland forest. *Biotropica* 21: 223–234.
Boinski, S., and Mitchell, C. L.,1992, Ecological and social factors affecting the vocal behavior of adult female squirrel monkeys. *Ethology* 92: 316–330.
Boinski, S., and Mitchell, C. L., 1994, Male residence and association patterns in Costa Rican squirrel monkeys (Saimiri oerstedi). *Amer. J. Primatol.* 34: 157–170.
Boinski, S., and Mitchell, C. L., 1995, Wild squirrel monkey (*Saimiri sciureus*) "caregiver" calls: contexts and acoustic structure. *Amer. J. Primatol.* 135: 129–137.
Boinski, S., and Newman, J. D.,1988, Preliminary observations on squirrel monkey (*Saimiri oerstedi*) vocalizations in Costa Rica. *Amer. J. Primatol.*14:. 329–343.
Brown, C. H., 1982, Auditory localization and primate vocal behavior. In: *Primate Communication*. C.T. Snowdon, C.H. Brown, and M. Peterson (eds.). Cambridge: Cambridge University Press, pp.144–164.

Chapman, C.A., 1988, Patterns of foraging and range use by three species of neotropical primates. *Primates* 29: 177–194.

Cheney, D. L., and Seyfarth, R. M.,1982., How vervet monkeys perceive their grunts: field playback experiments. *Anim. Behav.* 30: 739–751.

Cheney, D. L., and Seyfarth, R. M.,1990, *How Monkeys See the World*. Chicago: University of Chicago Press.

Clark, A. P.,1993, Rank differences in the production of vocalizations by wild chimpanzees as a function of social context. *Amer. J. Primatol.* 31: 159–179.

Costello, R. K., Dickinson, C., Rosenberger, A. L., Boinski, S., and Szalay, F. S., 1993, Squirrel monkey (Genus Saimiri) taxonomy: a multidisciplinary study of the biology of the species. In: Species, Species Concepts, and Primate Evolution, B. Kimbel and L. Martin (eds.). New York: Plenum Press, pp. 177–237.

Dennet, D. C.,1983, Intentional systems in cognitive ethology: the "Panglossian paradigm" defended. *Behav. Brain Sci.* 6: 343–355.

Dittus, W.,1988, An analysis of toque macaque cohesion calls from an ecological perspective. In: *Primate Vocal Communication,*. D. Todt, P. Goedeking, and D. Symmes (eds.). Berlin: Springer-Verlag, pp. 30–50.

East, M. L.,and Hofer, H.,1991, Loud calling in a female-dominated mammalian society: II. Behavioural contexts and functions of whooping of spotted hyaenas, *Crocuta crocuta. Anim. Behav.* 42: 651–670.

Fedigan, L.,1993, Sex differences and intersexual relations in adult white-faced capuchins, *Cebus capucinus. Int. J. Primatol.* 14: 853–877.

Fragaszy, D. M., and Boinski, S., In press, Patterns of individual choice and efficiency of foraging and diet in the wedge-capped capuchin, *Cebus olivaceus. J. Comp. Psych.*.

Fragaszy, D.M., Boinski, S., and Whipple, J., 1992, Behavioral sampling in the field: comparison of individual and group sampling methods. *Amer. J. Primatol.* 26: 259–275.

Garber, P.A., 1988, Foraging decisions during nectar feeding by tamarin monkeys (*Saguinus mystax* and *Saguinus fusciollis*, Callithrichidae, Primates) in Amazonian Peru. *Biotropica* 20: 100–106.

Gautier, J.-P., and Gautier-Hion, A.,1977, Communication in Old World monkeys. In: *How Animals Communicate*, T. Sebeok (ed.). Bloomington, Indiana University Press, pp. 890–964..

Green, S., 1975, Dialects in Japanese monkeys: vocal learning and cultural transmission of locale-specific behavior? *Z. Tierpsychol.* 38: 304–314.

Harcourt, A. H.,1988, Alliances in contests and social intelligence. In: *Machiavellian Intelligence: Social Expertise and the Evolution of Intellect in Monkeys, Apes, and Humans*, R. Byrne, and A. Whiten (eds.). Oxford: Clarendon Press, pp. 132–152.

Harcourt, A. H., Stewart, K. J., and Hauser, M., 1993, Functions of gorilla "close" calls. I. Repertoire, context, and interspecific comparison. *Behaviour* 124: 89–122.

Henzi, S. P., Byrne, R. W., and Whiten, A., 1992, Patterns of movement by baboons in the Drakensburg mountains: primary responses to the environment. *Int. J. Primatol.* 13:. 601–629.

Hershek, M. J., and Owings, D. H., 1993, Tail flagging by adult California ground squirrels: a tonic signal that serves different functions for males and females. *Anim. Behav.* 46: 129–138.

Janson, C.H., and Boinski, S., 1992, Morphological and behavioral adaptations for foraging in generalist primates: the case of the cebines. *Amer. J. Phys. Anthro.* 88: 483–498.

Jurgens, U., and Schriever, S., 1991, Respiratory activity during vocalization in the squirrel monkey. *Folia primatol.* 56: 121–132.

Krebs, J. R., and Dawkins, R.,1984, Animal signals: mind-reading and manipulation. In: *Behavioral Ecology: an Evolutionary Approach*, J.R. Krebs, and N. B. Davies (eds.). Sunderland, MA.: Sinauer Assoc, Inc., pp. 380–403.

Kudo, H., 1987, The study of vocal communication of wild mandrills in Cameroon in relation to their social structure. *Primates* 28: 289–308.

Kummer, H., 1968, *Social Organization of Hamadryas Baboons*. Chicago: University of Chicago Press, 1968.

Lindburg, D. G., 1971, The rhesus monkey in North India: an ecological and behavioral study. In: *Primate Behavior*, L. A. Rosenblum (ed.). New York, Academic Press, pp. 1–106..

Macedonia, J. M., and Evans, C. S.,1992, Variation among mammalian alarm call systems and the problem of meaning in animal signals. *Ethology* 93: 177–197.

Maynard Smith, J., 1974, The theory of games and the evolution of animal conflicts. *J. Theor. Biol.* 47: 209–221.

Menzel, C.,1993, Coordination and conflict in *Callicebus* social groups. In: *Primate Social Conflict*, W. A. Mason, and S. P. Mendoza (eds.). Albany: State University of New York Press, pp. 253–290.

Mitchell, C. L., 1990, *The Ecological Basis for Female Social Dominance: a Behavioral Study of the Squirrel Monkey (Saimiri sciureus)*. Unpublished Ph.D.Thesis, Princeton University, Princeton.

Mitchell, C. L., Boinski, S., and van Schaik, C. P., 1991, Competitive regimes and female bonding in two species of squirrel monkeys (*Saimiri oerstedi* and *S. sciureus*). *Behav. Ecol. Sociobiol.* 25: 55–60.

Newman, J. D., 1985, Squirrel monkey communication. In: *Handbook of squirrel Monkey Research*, L. A. Rosenblum, and C. L. Coe (eds.). New York, Plenum Press, pp. 99–126.

Norton, G. W., 1986, Leadership decision processes of group movement in yellow baboons. In: *Primate Ecology and Conservation*, J. G. Else, and P. C. Lee (eds.). Cambridge: Cambridge University Press, pp. 145–156.

Parker, S.T., and Gibson, K.R., 1977, Object manipulation, tool use and sensory motor intelligence as feeding adaptations in cebus monkeys and great apes. *J. Hum. Evol.* 6: 623–641.

Robinson, J.G., 1982, Vocal systems regulating within-group spacing. In: *Primate Communication*, C.T. Snowdon, C.H. Brown, and M. Peterson (eds.). Cambridge, Cambridge University Press, pp. 94–116 1982.

Robinson, J. G. 1984. Syntactic structures in the vocalizations of wedge-capped capuchin monkeys, *Cebus nigrivittatus*. *Behaviour* 90: 46–79.

Rose, L.M., 1992, *Sex Differences n Diet and Foraging Behavior and Benefits and Costs of Males to Females in White-faced Capuchins*. Unpublished M.A. thesis. Edmonton, University of Alberta.

Rose, L. M.,1994, Sex differences in diet and foraging behavior in white-faced capuchin monkeys, Cebus capucinus. *Int. J. Primatol.* 15: 63–82.

Rowell, T.E., 1972, *Social Behavior of Monkeys*. Baltimore: Penguin Books.

Smith, H. J., Newman, J. D., and Symmes, D.,1982, Vocal concomitants of affiliative behavior in squirrel monkeys (*Saimiri sciureus*). In: *Primate Communication*, C.T. Snowdon, C.H. Brown, and M. Peterson (eds.). Cambridge: Cambridge University Press, pp. 930–49.

Smith, W. J., 1969, Messages of vertebrate communication. *Science* 165: 145–150.

Snowdon, C. T., 1982, Linguistic and psycholinguistic approaches to primate communication. In: *Primate Communication,* C.T. Snowdon, C.H. Brown, and M. Peterson, eds. Cambridge: Cambridge University Press, pp. 212–238.

Snowdon, C. T., 1989, Vocal communication in New World monkeys. *J. of Human Evol.* 18: 611–633.

Soini, P., 1988, The pygmy marmoset, genus *Cebuella*. In: *Ecology and Behavior of New World Primates*, R.A. Mittermeier, A. B. Rylands, and A. F. Coimbra-Filho (eds.). Rio de Janeiro: Academia Brasileira de Ciencias, pp. 79–129

Symmes, D., and Biben, M., 1992, Vocal development of nonhuman primates. In: *Nonverbal Vocal Communication: Comparative and Developmental Approaches*, H. Papousek, U. Jurgens, and M. Papousek (eds.). Cambridge: Cambridge University Press, pp. 123–140.

Terborgh, J., 1983, *Five New World Primates: A Study in Comparative Ecology*. Princeton, Princeton University Press.

Thorpe, W. H., 1963, *Learning and Instinct in Animals*. London: Methuen.

Thurlow, W.R., 1971, Audition. In: *Experimental Psychology*, J.W. Kling and L.A. Riggs (eds.). New York: Holt Rinehart and Winston, pp. 223–259 .

de Waal, F.B.M., 1991, Complimentary methods and convergent evidence in the study of primate social cognition. *Behaviour* 18: 297–320.

Waser, P. M. and Brown, C. H., 1986, Habitat acoustics and primate communication. *Amer. J. Primatol.* 10: 135–154.

Westergaard, G.C., and Fragaszy, D.M., 1987, The manufacture and use of tools by capuchin monkeys (*Cebus apella*). *J. Comp. Psych.* 2: 159–168.

Whiten, A. and Ham, R., 1992, On the nature and evolution of imitation in the animal kingdom: reappraisal of a century of research. *Adv. Study Behav.* 21: 239–283.

14

THE BEHAVIORAL ECOLOGY OF WEDGE-CAPPED CAPUCHIN MONKEYS (*Cebus olivaceus*)

Lynne E. Miller

Department of Anthropology
Pitzer College
1050 N. Mills Avenue
Claremont, California 91711

The relationship between ecology and social structure has been an important area of primatological research since the 1960s (e.g., Crook and Gartlan 1966, Gartlan 1968, Kummer 1971, Eisenberg et al. 1972, Clutton-Brock 1974, Clutton-Brock and Harvey 1977a). Early work focused on group-level responses to the environment, employing the comparative approach to establish broad correlations between ecology and sociality (Crook and Gartlan 1966, Clutton-Brock and Harvey 1977b). Later, the sociobiological paradigm (Wilson 1975) stimulated investigation of how selection acts on individuals to maximize reproductive success and inclusive fitness within populations.

In light of the sociobiological model, socioecological research on primates has investigated, among other things, the costs and benefits to individuals of living in groups of different sizes (e.g., van Schaik 1983, Terborgh and Janson 1986). Additional research has examined the extent to which changes in the physical environment, such as resource dispersion, influence foraging and social behaviors of individuals within groups (e.g., Southwick 1967, Whitten 1983, Janson 1985).

This chapter explores behavioral responses to elements of both the social and physical environments. The analysis is structured around the hypothesis that (a) individuals experience costs associated with membership in either a large group or a small group (i.e., the social environment), (b) these costs are influenced by facets of the physical environment, in particular fluctuations in resource abundance, and (c) individuals are selected to employ behavioral strategies which mitigate these costs. Specific predictions are tested against data from a two-year study of wedge-capped capuchin monkeys (*Cebus olivaceus*) at Hato Piñero, Venezuela.

Life in a Small Group

One hypothetical cost of membership in a small group is reduced foraging success, relative to members of large groups (Krebs et al. 1972, Altmann 1974, Wrangham 1980).

Adaptive Radiations of Neotropical Primates
edited by Norconk *et al.* Plenum Press, New York, 1996

Possible mechanisms include the greater competitive ability of large groups over small groups during intergroup encounters (Wrangham 1980; for wedge-capped capuchins see Srikosamatara 1987, Miller 1992a), and the increased ability of large groups to locate food patches (Krebs et al. 1972, Pulliam and Caraco 1984; for wedge-capped capuchins see Miller 1992a). The first objective of this chapter is to test the prediction that individuals living in large groups ingest more food than do their small-group conspecifics.

The second objective of this investigation is to explore the influence of resource abundance upon the relative foraging success of individuals in large and small groups. When resources are scarce, members of small groups may experience disproportionately more difficulty locating and/or maintaining access to food patches. It is predicted that the disparity in food intake, between those in large and small groups, increases during times of food scarcity.

Third, this chapter examines strategies by which members of small groups mitigate (hypothetical) costs of reduced foraging success. For example, if individuals in small groups experience especially low levels of food intake during times of resource scarcity, as compared with those in large groups, they may respond by reducing time devoted to energetically costly behaviors, such as travel, and increasing time spent resting (Miller 1992b, Miller 1994). This study tests the prediction that members of small groups alter their daily activity budgets in response to seasonal changes in food availability, so as to maximize energy intake when food is abundant and accessible, and to minimize energy output when food is scarce.

It has been suggested that the reproductive success of females, more than of males, is influenced by net nutrient intake (Trivers 1972, Bradbury and Vehrencamp 1977, Emlen and Oring 1977) and so this section of the study emphasizes the activities of adult females.

Life in a Large Group

One of the hypothetical costs associated with membership in a large group is greater intragroup competition for limited resources, especially food (Janson and van Schaik 1988). Intragroup competition may result in a variety of behaviors which may, themselves, depress individual fitness. For example, among some primate species large groups travel longer day ranges, presumably to meet all members' nutritional requirements; this forces individuals in large groups to expend additional calories on travel (e.g., wedge-capped capuchins: de Ruiter 1986, Srikosamatara 1987, Miller 1992a). Intragroup feeding competition may also force individuals to spend more time foraging and less time resting (e.g., brown capuchins: Janson 1988). Competition may also result in frequent aggressive interactions as individual group members displace one another from important resources (e.g., wedge-capped capuchins: de Ruiter 1986, Srikosamatara 1987, Miller 1992a).

This analysis of the costs of large-group membership focuses upon intragroup aggressive interactions. The perspective is that such fights represent not just a measure of intragroup competition but are costly in and of themselves. Frequent fighting may increase an individual's likelihood of injury (Smuts 1987, Walters and Seyfarth 1987). Aggression increases levels of circulating cortisol (e.g., Sapolsky 1983), raises blood pressure (Smith et al. 1986) and reduces renal blood flow (Smith et al. 1986), all of which may have long-term, negative influence upon fitness. The fourth objective of this analysis is to test the prediction that individual members of large groups, on average, participate in fights more frequently than do those in small groups.

Feeding competition should increase with food scarcity for groups of all sizes, and hence the frequency of aggressive interactions should always be inversely correlated with resource abundance (e.g., Janson 1985). However, those in large groups may be especially hard hit by such competition. Therefore the fifth prediction is that the disparity in rates of aggression, between large and small groups, increases during times of resource scarcity.

The final goal of this chapter is to examine strategies by which individuals minimize intragroup aggression, thereby mitigating the (hypothetical) costs of large-group membership. One way that individuals might reduce the likelihood of being involved in a fight is to maintain a healthy distance from companions, for example by spreading out during foraging. This study tests the prediction that members of large groups maintain greater individual space than do those of small groups. Since rates of aggression may increase during periods of resource scarcity, individual space should also increase during these times.

Summary of Test Predictions

1. Adult females in large groups ingest more food than do their counterparts in small groups.
2. This disparity increases during times of resource scarcity.
3. Adult females in small groups adjust their daily activity budgets according to resource availability, maximizing energy intake when food is abundant and minimizing energy output when food is scarce.
4. Members of large groups engage more frequently in intragroup aggressive interactions than do those in small groups.
5. This disparity increases during times of resource scarcity.
6. Members of large groups maintain greater individual space (a) than do those of small groups, and (b) during times of resource scarcity.

METHODS

Subjects

Cebus olivaceus are small platyrrhines with moderate sexual dimorphism in body size and pelage (Robinson and Janson 1987). They are opportunistic foragers, exploiting a wide variety of resources (cf. Robinson 1986, Miller 1992a). Although the majority of the diet, by volume, comes from ripe fruit (Miller 1992a), a greater proportion of the daily activity budget is devoted to foraging for invertebrate material (Robinson 1986, Miller 1992a). *C. olivaceus* also capture and eat vertebrate prey such as small birds and baby squirrels (Robinson 1986, Miller 1992a). Mature leaves are rarely eaten, and the extent of seed use is currently under investigation.

Cebus olivaceus is an attractive species for this type of study because these monkeys live in groups of a wide range of sizes, from as small as eight to over 40 individuals (de Ruiter 1986, Srikosamatara 1987, Miller 1991). Furthermore, several groups may occupy overlapping home ranges, which controls the variable of habitat disparity. This investigation focuses on two groups, one large (n_{LG} = approximately 37) and one small (n_{SG} n = approximately 16) (see Table 1 for details of group composition). For the reasons cited above, collection of data on foraging and activity budgets focused upon the adult females in each group.

Study Site

Hato Piñero is a functioning cattle ranch in the *llanos* or plains of central Venezuela. It covers approximately 80,000 hectares lying between latitudes 8°40' and 9° 00' north

Table 1. Composition of study groups

Group	AM[2]	SAM	AF	SAF	J	I	Total
Large (min)[1]	5	0	12	1	17	0	35
Large (max)[1]	6	0	12	0	19	2	39
Small (min)	2	0	6	0	3	4	15
Small (max)	1	1	6	0	10	0	18

[1]Min/max=minimum/maximum number of individuals counted in the course of the two-year study.
[2]AM=adult male; SAM=subadult male; AF=adult female; SAF=subadult female; J=juvenile; I=infant.

and longitudes 69° 00' and 68° 18' west. Hato Piñero lies approximately 100 kilometers west of Hato Masaguaral, the site of most other published investigations of *Cebus olivaceus* (e.g., Robinson 1981, 1984, 1986, 1988; de Ruiter 1986; Srikosamatara 1987; Fragaszy 1990; O'Brien and Robinson 1993).

Like most of the *llanos*, Hato Piñero is a mosaic of open grassland and large stretches of semideciduous, dry tropical forest (Schuerholz and Demarchi, unpublished). The region is characterized by pronounced seasonality. In the wet season, from May through October, mean monthly rainfall is approximately 190 millimeters. During the dry season, from November through April, mean rainfall is approximately 20 millimeters per month (Miller 1992a).

The study site is a 270 hectare plot within a large stretch of contiguous forest. Approximately 45 kilometers of trails, running parallel and perpendicular at 125 meter intervals, form a grid over the site. Eleven groups of *Cebus olivaceus* were observed in this forest during the course of the study. Both of the focal study groups used all quadrats of the site at some point in the year, and thus — within this area — had completely overlapping annual home ranges.

Food Abundance

Food abundance is positively correlated with rainfall (cf. Robinson 1986 for nearby Hato Masaguaral). Seasonal indices of fruit abundance were developed from data on the spatial and temporal dispersion of ripe fruits provided by trees and bromeliads in sample plots throughout the study site. A fruit was considered "available" whenever it was exploited by the subjects. The index for a given season = the absolute number of plants of all "available" species X estimated fruit production per plant (on a geometric scale, cf. Leighton 1982) X estimated volume of food provided by each fruit (see Miller 1992a for detail). The wet season index of abundance is approximately twice that for the dry season (Miller 1992a). More species produce fruit during this time of year, and greater volumes of fruit are produced.

Invertebrate matter is also more abundant during the wet season than in the dry. Two month-long entomological surveys, one in August and one in March, revealed clear seasonal differences in both the density and diversity of arthropod species present (Dr. George McGavin, pers. comm.). During the wet season, flying arthropods and large caterpillars are abundant and readily accessible to the subjects. In the dry season, arthropods are available primarily as eggs or larvae developing under bark or inside twigs. They are thus smaller and less easily captured by the subjects (Miller 1992a).

Data Collection and Analysis

Field work took place between April, 1989, and June, 1991. Intensive data collection on the two focal study groups was conducted from June, 1990, through June, 1991.

This analysis is based upon 485 hours of observation, 265 with Large Group and 220 with Small Group, during which 2910 behavioral samples (of adult females) were recorded (LG wet season = 720, LG dry season = 1102; SG wet = 531, SG dry = 557).

Because of the density of vegetation and the extensive home ranges of the subjects, establishing and maintaining contact with the focal troops within the study site was unreliable (see Miller 1992a). Despite efforts to sample behaviors during all hours of the day, the subjects were more frequently sighted, and thus more data were collected, at midmorning and afternoon. Because the paucity of data for early morning and late evening hours represents a potential source of bias, data analyses were designed to minimize the influence of diurnal variation in behavioral patterns (cf., Robinson 1984; see also Harcourt and Stewart 1984).

Daily Activity Budgets. Data were collected via slow scan sampling. Subjects (primarily adult females) were observed as rapidly as they could be located, for 30 seconds; this was about the longest period of time during which a monkey could be consistently observed without interruption. Interrupted samples were discarded. The goal was to observe every troop member in each 30 minutes of contact time, throughout the 12 hours of daylight. Among the data collected with each sample, the following are pertinent to this analysis: (1) the time; (2) the subject's activity identified as *feeding* (gathering and ingesting plant matter), *foraging* (actively searching for and ingesting animal matter), *moving* (moving from one place to another without also foraging or engaging in some other activity), *moving and foraging* (moving along while also searching for prey), *resting* (sitting, lying down, sleeping) or *social behavior* (playing, fighting); (3) the type of any food item ingested (plant or animal, and species if known); and (4) the number of items ingested (e.g., of individual fruits, or bites of a very large fruit) (Miller 1992a).

To determine daily activity budgets, and to minimize the bias of irregular sampling, the data set was stratified by time. Samples (for all days and for all adult females) were grouped according to the half-hour time slot during which they were collected. The result was a 24-by-6 table — time slot versus activity — in which each cell contained the proportion of samples during that time slot in which subjects engaged in that activity. To determine the mean proportion of time devoted to each activity (hours per day for each group), the cell values for that activity were averaged across all 24 half-hour time slots (Miller 1992a). For this analysis, behaviors were grouped into three "larger activity categories": *food acquisition* (includes *feeding*, *foraging* and *moving and foraging*); *moving* (includes *moving* and *moving and foraging*); and *rest* (includes only *rest*). Grouping the activities in this manner may compromise the independence of samples, but provides a more appropriate picture of the subjects' behavior.

For statistical analysis of differences between groups, the data were converted to a 24-by-2 table — time slot versus study group (Large Group and Small Group) — with one such table for each of the "larger activity categories" (*food acquisition*, *moving* and *rest*). In each cell of a given table was the proportion of time devoted to that "larger activity category" during that time slot by that group. Large Group and Small Group were then compared with respect to the proportion of time spent in each "larger activity category" over all 24 half-hour time slots. Although it might seem appropriate to compare these data sets in a pair-wise fashion, there is no clear evidence that members of both troops followed the same diurnal pattern of activity (Miller 1992a). Both troops generally traveled less and rested more during the middle of the day, but there was a great deal of variability in activity during a given time slot between groups and from day to day. Therefore, a Mann-Whitney unpaired test of difference between two populations was used. Where appropriate, the data sets were also stratified by season and the above analyses were conducted separately for each data set (Large Group versus Small Group and wet season versus dry season) (Miller 1992a).

Food Intake. The collection of behavioral data described above provided an assessment of all food items ingested by the focal animal, including food type and number of items. These data were converted to volumetric estimates based upon the approximate dimensions of each food item, minus the volume of large seeds which were known to be discarded or passed undigested. For very large food items, the number of bites taken during the sample was recorded and the volume of each bite estimated. The result was a single value, a measure of volume, representing the food intake observed during each 30-second behavioral sample (Miller 1992a).

To estimate daily food intake, controlling for diurnal patterns in feeding (cf. Robinson 1984), the data set was stratified by time. Samples (for all days and for all adult females) were grouped according to the half-hour time slot in which they were collected, as above. Within each half-hour time slot, the volumetric estimates of food intake were averaged over all samples — including those samples in which the volume ingested was 0 — to determine the mean volume of food ingested per 30 seconds during that time slot. These means were then multiplied by 60 to determine the total volume of food ingested during each of the 24 half-hour time slots. Finally, these 24 values were totaled to determine the mean volume of food consumed per individual throughout the day. This procedure was followed for each study group for the year overall, for the wet season, and for the dry season. For statistical analyses, 24-by-2 tables (time slot versus group) were constructed as above, in which each cell contained the mean volume of food consumed during that half-hour by females in that group. The Mann-Whitney test was employed to compare the two groups' food intake (Miller 1992a).

Intragroup Aggressive Interactions. Intragroup aggressive interactions were rarely observable; instead, auditory cues were used to tally up the number of fights which occurred during contact time with each study troop (Miller 1992a). For statistical analysis, each day of observation served as a single sample, regardless of the number of hours spent with the troop in question. The frequency of aggressive interactions (number per hour) was calculated for each day, for each troop, for each season (Miller 1992a).

The absolute frequency of fights does not, however, accurately reflect the impact of intratroop aggression upon the individual group member. Of greater interest is the rate at which a single member participates in aggressive interactions. To determine the frequency of fights per individual per troop, the absolute frequencies were divided by the number of group members (LG=37, SG=16) and multiplied by two (because each fight must include at least two participants). The Mann-Whitney test was then used to assess the difference between the two "populations," stratified by troop and by season (Miller 1992a). This method of assessment assumes that all members of a troop are equally likely to engage in aggressive interactions, an assumption which is probably invalid (see Discussion). However, limited visibility of the subjects prevented finer analysis of fight participants.

Individual Space. When research assistance was available, the "spread" of the focal troop was measured by pacing off the distance from "front to back" and from "side to side" (Miller 1992a). Multiplying these two values gave an estimate of the area covered by each troop during different hours of the day and while engaged in different activities. These estimates were then divided by the number of group members (LG=37, SG=16) to

determine the mean space per individual in each troop. For statistical analysis, each estimate served as a single sample. Measurements were made at intervals of at least one hour
to enhance independence of samples. The Mann-Whitney test was employed to compare
the data sets, across troops and across seasons. This method provides only a two-dimen-
sional estimate of space, although it is clear that the monkeys exist in a three-dimensional
world. However, visibility often prohibited nearest-neighbor measures and thus a three-dimensional estimate of individual space is currently unavailable.

RESULTS

Life in a Small Group

The first objective of this research was to explore the prediction that adult females in
large groups ingest more food than do those in small groups. The results of this analysis
are shown graphically in the first pair of bars in Figure 1. For the year overall, an average
adult female in the Large Group ingested approximately 1800 cc of food per day. The average Small Group female ingested approximately 2136 cc of food per day. Although the
difference is greater than 16%, it is not statistically significant ($z = -1.25$, $p = 0.21$). Based
upon these results, Prediction 1 can be rejected: females in large groups apparently do not,
on average, eat significantly more than their small-group counterparts.

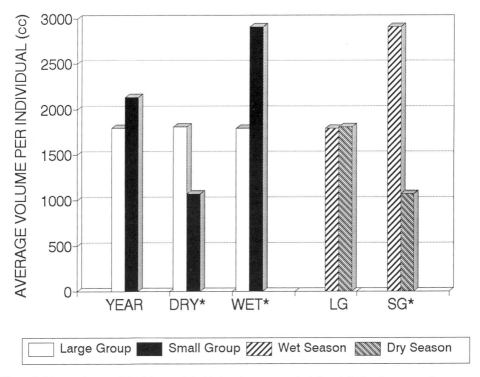

Figure 1. Mean volumes of food ingested (cubic centimeters per day) by adult females; comparisons across
groups and across seasons. Bar pairs marked with an asterisk (*) exhibit statistically significant differences; see
text for numerical values.

The second goal of this analysis was to explore the influence of resource abundance upon the foraging success of individuals in large and small groups. Specifically, it was predicted that any disparity in food intake, between females in large versus small groups, would increase during times of food scarcity, thus during the dry season.

The results of this analysis are presented graphically in the remainder of Figure 1. They reveal a pattern somewhat more complex than predicted. Indeed, Large Group females ingested significantly more than those in Small Group during the dry season (LG average = 1811 cc, SG average = 1070 cc; $z = 2.51$, $p = 0.0122$), thus supporting Prediction 2. However, the relationship was reversed during the wet season, when resources were abundant, with Small Group females ingesting significantly more (LG average = 1792 cc, SG average = 2911 cc; $z = -3.06$, $p = 0.0022$). What appears is a pattern in which females in the Large Group, despite great fluctuations in food availability, maintained consistent levels of food intake across seasons (wet season average = 1792 cc, dry season average = 1811 cc; $z = -0.26$, $z = 0.79$). In contrast, females in Small Group experienced extreme variance in food intake, with high levels when food was abundant (wet season average = 2911 cc) and low levels when food was scarce (dry season average = 1070 cc; $z = 4.60$, $p = 0.0000$). This pattern accords with the general hypothesis that the physical environment influences the costs of small-group membership, in particular that fluctuations in resource abundance correlate with foraging success among these adult females.

The third aim of this research was to explore ways in which members of small groups mitigate costs of reduced foraging success. The investigation tested the prediction that females in small groups alter their daily activity budgets in accordance with food availability, increasing energy intake when food is abundant, and reducing energy output when food is scarce. This prediction is especially pertinent in light of the pattern of food intake described above. If activity budget is correlated with food intake, then budgets for females in large groups should remain constant across seasons while budgets for those in small groups should fluctuate, and significant intergroup disparities should appear within each season.

The results of this analysis are presented graphically in Figures 2, 3 and 4. Three variables are of interest here. The first is the proportion of the daily activity budget devoted to food acquisition, that is, the sum of *feeding* time, *foraging* time, and time spent *moving and foraging*. Figure 2 shows that, when the data were averaged across the year overall, there was no significant difference between Large Group females and Small Group females in time devoted to food acquisition, thus matching the similarity in food intake. Those in the Large Group spent (on average over all days and all adult females) 58.1% of their time in food acquisition, compared with 60.8% for females in the Small Group ($z = -0.57$, $p = 0.57$). In the dry season, LG and SG females also spent similar amounts of time acquiring food (LG = 59.4%, SG = 53.5%; $z = 1.48$, $p = 0.14$), contrary to the contrast in food intake. However, during the wet season, there was a difference: Small Group females devoted significantly more time to food acquisition than their counterparts in the Large Group (LG = 56.0%, SG = 68.4%; $z = -2.48$, $p = 0.013$). Who, then, altered their activity budgets with the seasons? Figure 2 shows that food acquisition time among Large Group females remained essentially constant across seasons (wet season = 56.0%, dry season = 59.4%; $z = -0.47$, $p = 0.64$), while for Small Group females it increased significantly during the period of resource abundance (wet season = 68.4%, dry season = 53.5%; $z = 2.70$, $p = 0.007$), thus supporting the above prediction.

The next variable of interest is time spent moving. This is essentially a measure of energy output as directed movement is one of the more energetically costly activities in which adults participate. For this analysis, moving time was assessed via a combination of

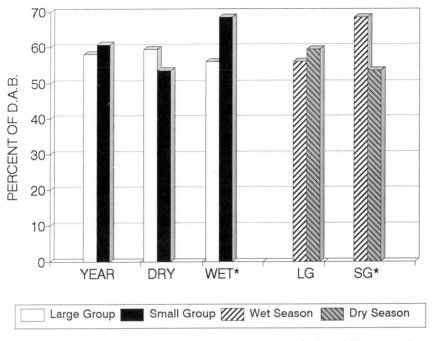

Figure 2. Percentages of daily activity budgets adult females devoted to food acquisition; comparisons across groups and across seasons. Bar pairs marked with an asterisk (*) exhibit statistically significant differences; see text for numerical values.

moving and *moving and foraging*. Figure 3 shows that, for the year overall, there was no significant difference between Large Group females and Small Group females in time spent moving (LG = 40.1%, SG = 36.8%; z = 1.15, p = 0.25). The same is true for the wet season (LG = 40.7%, SG = 43.8%; z = -0.21, p = 0.84). However, for the dry season, females in the Small Group spent considerably less time moving than did those in the Large Group (LG = 39.7%, SG = 30.0%; z = 1.97, p = 0.049). Who altered their behavioral patterns with the seasons? Figure 3 shows that, again, Large Group females were consistent across seasons (wet season = 40.7%, dry season = 39.7%; z = 0.54, p = 0.59), while Small Group females reduced moving time during the dry season (wet season = 43.8%, dry season = 30.0%; z = 2.22, p = 0.027). The results are in accordance with the prediction that females in small groups reduce energy output during times of resource scarcity.

The final variable of interest here is time spent resting, a measure of energy conservation. Figure 4 shows that, for the year overall, there was no difference between Large Group females and Small Group females in time spent resting (LG = 23.7%, SG = 23.7%; z = 0.30, p = 0.77). For the dry season, mean resting times were greatly different and in the direction predicted, with Small Group females resting much more than those in the Large Group, but the disparity is not statistically significant (LG = 22.7%; SG = 33.0%, z = -1.21, p = 0.23). In the wet season, however, Small Group resting was significantly less than Large Group resting (LG = 25.1%, SG = 13.9%; z = 2.52, p = 0.012). Who altered their activity budgets with the seasons? Figure 4 shows that, once again, Large Group females were consistent across seasons in time spent resting (wet season = 25.1%, dry season = 22.7%; z = -0.28, p = 0.78), while resting time decreased significantly in the wet season for females in the Small Group (wet season = 13.9%, dry season = 33.0%; z = -

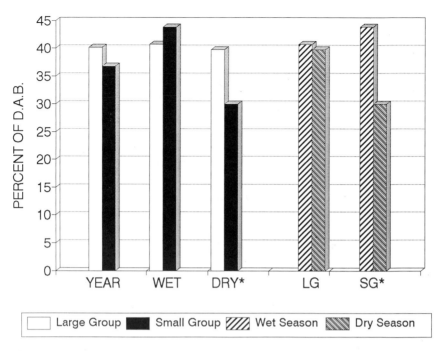

Figure 3. Percentages of daily activity budgets adult females devoted to moving; comparisons across groups and across seasons. Bar pairs marked with an asterisk (*) exhibit statistically significant differences; see text for numerical values.

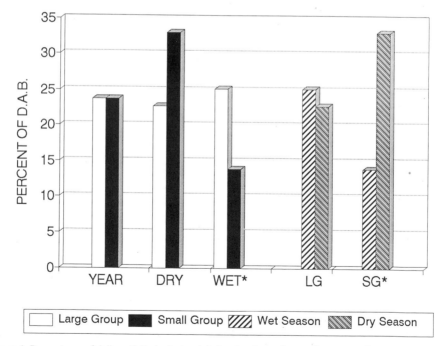

Figure 4. Percentages of daily activity budgets adult females devoted to resting; comparisons across groups and across seasons. Bar pairs marked with an asterisk (*) exhibit statistically significant differences; see text for numerical values.

2.52, p = 0.012). This third variable thus also behaves according to expectation. In general, therefore, the activity budgets of adult females correlate with their rates of ingestion. For those in small groups, activity budgets follow the predicted pattern of energy intake during the wet season and energy conservation during the dry season.

Life in a Large Group

One objective of this research was to highlight the costs of large-group membership by focusing on rates of aggression within groups. In addition, the investigation sought to explore the influence of resource abundance on such aggression. The study tested two related predictions: that members of large groups engage in fights more frequently than their counterparts in small groups (Prediction 4), and that this disparity increases during times of resource scarcity (Prediction 5).

The results of this analysis are summarized in Figure 5. The data support Prediction 4: individuals in Large Group participated in aggressive interactions significantly more frequently than did members of Small Group, for the year overall (LG median = 0.09 fights per individual per hour, n = 32, range = 0.02–0.39; SG median = 0.05, n = 22, range =0.00–0.31; z = 5.42, p = 0.0000), during the wet season (LG median = 0.08, n = 20, range = 0.02–0.21; SG median = 0.05, n = 12, range = 0.00–0.31; z = 3.64, p = 0.0003) and during the dry season (LG = 0.10, n = 12, range = 0.06–0.39; SG = 0.05, n = 10, range = 0.00–0.10; z = 3.96, p = 0.0001). However, the associations among group size, resource abundance and intragroup competition (Prediction 5) are more complex than expected. For Small Group, the food scarcity of the dry season did not result in more frequent fighting; instead, the rate of aggressive

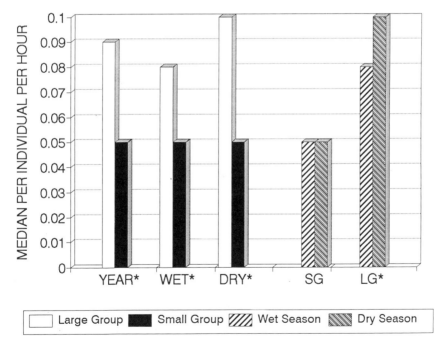

Figure 5. Median frequencies of aggressive interactions, per individual per hour; comparisons across groups and across seasons. Bar pairs marked with an asterisk (*) exhibit statistically significant differences; see text for numerical values.

interactions within this group remained virtually unchanged across seasons (wet season median = 0.05 fights per individual per hour, dry season = 0.05; z = 0.13, p = 0.90). For Large Group, though, as predicted, the frequency of fights rose significantly during the dry season (wet season = 0.08, dry season = 0.10; z = -1.99, p = 0.047). These data suggest that the costs of sociality, in this case rates of aggression, are greater within large groups than within small. Furthermore, the influence of resource scarcity on aggression is especially strong within large groups, presumably due to high levels of feeding competition.

The final goal of this research was to test the prediction that members of large groups might decrease the frequency of intragroup fights by increasing their individual space, (a) over those in small groups, and (b) during the dry season. The results of this analysis are summarized in Figure 6. Again, the results are more complex than originally expected. For the year overall, there was no significant difference between the two study groups in individual space (LG median = 391.7 m² per individual, n = 49, range = 69.4–2625.0; SG median = 293.8m², n = 28, range = 56.3–1618.8; z = 1.45, p = 0.15). Similarly, there was no significant difference for the wet season (LG = 213.9 m², n = 32, range = 72.2–2625.0; SG = 312.5 m², n = 21, range = 56.3–1618.8; z = -0.71, p = 0.48). However, in the dry season, Large Group members did spread out significantly more than their Small Group counterparts (LG = 597.2 m², n = 17, range = 69.4–1083.3; SG = 218.8m², n = 7, range = 75.0–350.0; z = 3.21, p = 0.0013). In this case, as for rates of aggression, it is Small Group which was consistent across seasons (wet season = 312.5 m², dry season = 218.8 m²; z = 1.35, p = 0.18). Members of Large Group, however, significantly increased their individual space during the dry season (wet season = 213.9 m², dry season = 597.2 m²; z = -3.10, p = 0.002), presumably as a response to food scarcity, and possibly as a strategy to minimize rates of aggression.

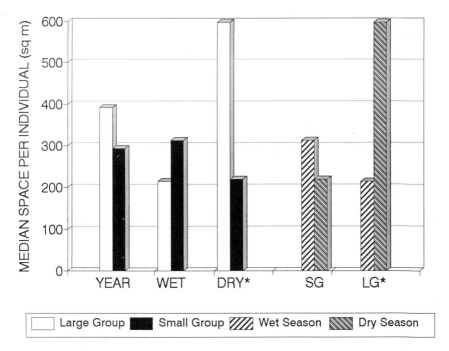

Figure 6. Median individual space; comparisons across groups and across seasons. Bar pairs marked with an asterisk (*) exhibit statistically significant differences; see text for numerical values.

DISCUSSION

As with most studies of nonhuman primates, this investigation is based upon an extremely limited sample, in this case, two years of observation, of approximately 53 individuals in two focal groups. Further research incorporating additional study groups would enhance the strength of the conclusions made here. Until such data become available, however, interpretations of the results should be made with care.

Life in a Small Group

Generalizing from observations of two study groups of *Cebus olivaceus* at Hato Piñero, these analyses reveal a complex pattern for the lives of adult females in small groups. Those in the small study group did not, as predicted, ingest less than females in large groups for the year overall. However, they did experience significant annual variance in rates of ingestion, correlating positively with resource abundance. The wet season, during which food is plentiful and intake is high, may normally provide small-group females with an opportunity to store fat and thus survive the months of food scarcity. However, an annual drop in food intake during the dry season may impose costs upon small-group females in terms of reproductive success. The dry season is the usual time of gestation for this species (Robinson and Janson 1987, and personal observation). Laboratory and field studies have demonstrated that, for various taxa, nutritional deprivation during gestation may result in low birth weight among neonates, a condition often associated with reduced viability (e.g., pigs: Pond et al. 1985; rats: Koski and Hill 1990; humans: Nutrition Review 1984). It is thus plausible that a restriction in food intake during this crucial time may adversely affect the annual reproductive success of small-group females. Although this investigation has not produced long-term data to test this hypothesis, a modicum of support comes from the fact that fecundity of females in the large study group was slighter higher than for those in the small group (average proportion of juveniles plus infants to adult females in LG = 1.59, in SG = 1.42, a difference of 12%). Data from over 15 years of research with another population of *Cebus olivaceus* further support this model by demonstrating that group size is, in fact, positively correlated with reproductive output (Robinson 1988). The results of the investigation presented here suggest a possible mechanism for this difference, namely, limited food intake during the months of gestation.

The results of this analysis are also significant from an evolutionary perspective. Under normal circumstances, resources are abundant for six months of the year (the wet season). Females in smaller groups may feed heavily at this time to compensate for the shortages they experience when resources are scarce (the dry season). However, environmental stochasticity can occasionally result in resource scarcity for extended periods of time. Under these circumstances, only individuals that are able to maintain adequate nutrient intake will survive and reproduce. If the dry season is taken as a model for periods of resource scarcity, when selective pressures are strongest, then the results of this analysis suggest that females in small groups would suffer more from food shortages. Those in large groups, however, would be better able to maintain adequate food intake for survival and reproduction. Therefore, both annually and on an evolutionary scale, there is a selective disadvantage to living in a small group.

For this study, activity budgets were generally correlated with patterns of food intake. Females in the large group achieved consistent levels of food intake across seasons, despite fluctuations in resource availability; similarly, they maintained consistent daily activity budgets. In contrast, those in the small group consumed a relatively large amount of food when resources were abundant (the wet season), but a small volume when resources were scarce (the dry season); their daily activity budgets followed suit. During the wet season, small-group females devoted considerable time to food acquisition and to travel (which may have enhanced their ability to locate food patches); time spent resting was very low. During the dry season, travel time dropped and resting time rose. This seasonal change in activity budget may have been a strategy to maximize energy intake during times of resource abundance and to minimize energy output when resources were scarce.

This correlation raises the issue of causality: Were variations in food intake the cause or consequence of the behavioral patterns of small-group females? That is, did low food intake encourage small-group females to conserve energy, or was low food intake a result of reduced foraging effort (cf. Janson 1988)? Intergroup comparisons may help to answer this question. During the wet season, females in the small group expended greater foraging effort and took in more food than did those in the large group. It therefore appears that increased foraging effort resulted in greater food intake. In the dry season, however, females in the large and small groups spent equal time gathering food, and yet food intake was significantly higher for those in the large group. Low food volume for small-group females was apparently not a result of reduced feeding time, but may instead reflect their inability to maintain access to food patches (cf. Wrangham 1980). This conclusion is supported by limited data on intertroop displacements. Of 23 encounters between groups of known size, 19 (83%) involved a larger troop displacing a small one. The exceptions involved two groups of similar size (35 versus 37) in which the smaller, with one or two more adult males, would displace the larger (Miller 1992a). If small groups are consistently displaced during intertroop encounters, and if this prevents females from gaining access to food patches, then low food intake would be a logical consequence. In turn, low energy intake may force these individuals to decrease energy expenditure by reducing their travel time and increasing their rest time, relative to wet season levels.

Life in a Large Group

Members of the large group participated in aggressive interactions more frequently than did those in the small group. This corroborates the results of previous studies on *Cebus olivaceus* (de Ruiter 1986, Srikosamatara 1987) and indicates that membership in a large group is relatively costly. Furthermore, rates of aggression within the small group were constant across seasons, while fighting increased significantly within the large group during the dry season. This suggests that food scarcity has greater impact upon members of large groups, presumably by raising disproportionately the level of intragroup feeding competition (Janson and van Schaik 1988).

The method used to determine the frequency of fights per group member assumes that all individuals are equally likely to engage in aggressive interactions. In fact, an individual's tendency to fight may be influenced by several variables, including age, sex, rank, reproductive status, and perhaps by temperament as well (cf. Janson 1985, Robinson 1981). These data only suggest that, on average, members of large groups are more likely to participate in fights, especially during the dry season. Given that fighting affects the cardiovascular system (Smith et al. 1986) and increases the likelihood of incurring serious injury (cf. Smuts 1987), this tendency may represent a significant cost to individuals in large groups.

Disparities in rates of aggression, across groups and across seasons, might conceivably be explained in terms of individual space. If members of the large group were foraging in closer proximity — for example, when a large number of individuals cluster to feed simultaneously in a fruit tree — then more frequent fights would be expected. However, the results of this investigation indicate that members of the large group were not in closer proximity than their small-group counterparts. In fact, during the dry season, when resources were scarce and rates of aggression were especially high within the large group, these members significantly increased their individual space, over that of small-group members and over their own wet season levels. If individual space had been maintained at constant levels throughout the year, it is plausible that dry season rates of aggression would have been still higher than observed. It is therefore conceivable that an increase in individual space during the dry season represents a behavioral strategy employed by those in the large group to minimize rates of aggression and thereby reduce the costs of large-group membership.

CONCLUSIONS

The results of this investigation suggest that, not only are there are costs to membership in both large and small groups, but also that the physical environment, especially fluctuations in resource abundance, may influence the magnitude of these costs. Members of small groups may suffer from a lack of competitive ability in intergroup encounters, particularly during periods of low food availability. Those in large groups may suffer from higher rates of aggression, especially when resources are scarce. However, selection should encourage individuals to develop strategies to reduce the costs of group membership. The data presented here support this hypothesis by elucidating changes in behavioral patterns correlated with rising costs. These results indicate that primates can and do develop adaptive strategies in response to the selective pressures of both their physical and social environments.

SUMMARY

Adult females in the small study group experienced a foraging disadvantage, relative to those in the large group, manifested in seasonal fluctuations in food intake. Rates of ingestion for small-group females were, in turn, positively correlated with resource availability. During the wet season, when food was abundant, daily food volume for females in the small group was high, significantly higher than for females in the large group. During the dry season, when food was scarce, intake was significantly lower, both lower than wet season levels and lower than the intake of large-group females. A restriction in food intake during the normal months of gestation may depress the reproductive output of small-group females, and thus membership in a small group may prove costly to individual fitness.

Annual variance in food intake for females in both groups was correlated with variance in activity budget. The budgets of large-group females remained virtually unchanged across seasons, thus matching their constant food intake. However, the budgets of small-group females shifted with the seasons. When food abundance and food intake were high, these individuals devoted considerable time to foraging and travel and very little time to rest, thus maximizing energy intake. When food was scarce and intake was low, these females significantly reduced travel time and increased rest time, thus conserving energy. Intertroop comparisons of activity budgets indicate that the low food intake of small-group females during the dry season was not a result of reduced foraging effort, but of an apparent inability to locate and/or maintain access to food resources, possibly because of periodic displacement by large groups.

The continuing existence of small groups, despite apparent foraging disadvantages faced by their members, may be facilitated by the behavioral strategies individuals employ to mitigate costs to fitness. By reducing their energy output during times of food scarcity, females in small groups may reduce the extent to which nutritional shortages suppress their reproductive success. Increasing their energy intake during times of food abundance may allow them to store fat and thus further improve their chances of coping with low food intake in drier months. Such a pattern represents a behavioral adaptation to aspects of the individual's physical environment, such as fluctuations in resource availability, and social environment, such as the inability to compete for food resources in intergroup encounters.

Individuals in the large study group experienced higher costs of intragroup competition as manifested in rates of aggressive interactions. The average member of the large group participated in fights significantly more frequently than did its small-group counterpart. Furthermore, the disparity between the two study groups increased during the dry season. While rates of aggression within the small group remained virtually unchanged throughout the year, fighting within the large group significantly increased when food became scarce. Since frequent fighting is likely to reduce individual fitness, membership in a large group is costly, especially during times of resource scarcity.

The relative rates of aggression for the study groups could, hypothetically, be explained by spacing; that is, if members of the large group were in closer proximity they would fight more frequently. However, the opposite was true. As food became scarce and rates of aggression rose within the large group, these members actually increased their individual space. They spread out signficantly more than they had in the wet season, and more than did those in the small group. For those in the small group, in contrast, frequency of fights and individual space were both constant across the seasons. Had members of the large group maintained close spacing into the dry season, rates of aggression might have been still higher than observed. The change in spacing within the large group may have been a means to control rising rates of aggression and thereby minimize the costs of large-group membership.

ACKNOWLEDGMENTS

First and foremost, I would like to thank Sr. Antonio Julio Branger for his hospitality and support during my two years at Hato Piñero. While in the field, I received valuable assistance from L. Aristiguieta, G.D. Cantrell, R. Dowhan, R.S.O. Harding, G.C. McGavin, and many members of the Hato Piñero staff. I am also grateful for comments on manuscripts made by L.M. Fedigan, A.H. Harcourt, R.S.O. Harding, C.H. Janson, S.A. Miller, S.F. Miller, M.A. Norconk, J.G. Robinson and P.S. Rodman. This research was supported in part by a U.C. Davis Humanities and Research Award, 1989, and by U.C. Regents' Fellowships, 1988 and 1989, which are gratefully acknowledged.

REFERENCES

Altmann, S.A. 1974. Baboons, space, time and energy. *American Zoologist* 14: 221–248.
Bradbury, J.W. and S.L. Vehrencamp. 1977. Social organisation and foraging in emballonurid bats. *Behavioral Ecology and Sociobiology* 2: 1–17.
Clutton-Brock, T.H. 1974. Primate social organisation and ecology. *Nature* 250: 539–542.
Clutton-Brock, T.H. and P.H. Harvey. 1977a. Primate ecology and social organisation. *Journal of Zoology, London* 183: 1–39.

Clutton-Brock, T.H. and P.H. Harvey. 1977b. Species differences in feeding and ranging behaviour in primates, pages 557–584 in, *Primate Ecology*, T.H. Clutton-Brock, ed. Academic Press. London.

Crook, J.H. and J.S. Gartlan. 1966. Evolution of primate societies. *Nature* 210: 1200–1203.

Eisenberg, J.F., N.A. Muckenhirn and R. Rudran. 1972. The relation between ecology and social structure in primates. *Science* 176: 863–874.

Emlen, S.T. and L.W. Oring. 1977. Ecology, sexual selection, and the evolution of mating systems. *Science* 197: 215–223.

Fragaszy, D. 1990. Sex and age differences in the organization of behavior in wedge-capped capuchins, *Cebus olivaceus*. *Behavioral Ecology* 1: 81–94.

Gartlan, J.S. 1968. Structure and function in primate society. *Folia Primatologica* 8: 89–120.

Harcourt, A.H. and K.J. Stewart. 1984. Gorillas' time feeding: aspects of methodology, body size, competition and diet. *African Journal of Ecology* 22: 207–215.

Janson, C.H. 1985. Aggressive competition and individual food consumption in wild brown capuchin monkeys (*Cebus apella*). *Behavioral Ecology and Sociobiology* 18: 125–138.

Janson, C.H. 1988. Food competition in brown capuchin monkeys (*Cebus apella*): Quantitative effects of group size and tree productivity. *Behaviour* 105: 53–76.

Janson, C.H. and C.P. van Schaik. 1988. Recognizing the many faces of primate food competition: Methods. *Behaviour* 105: 165–186.

Koski, K.G. and F.W. Hill. 1990. Evidence for a critical period during lab gestation when maternal dietary carbohydrate is essential for survival of newborn rats. *Journal of Nutrition* 120: 1016–1027.

Krebs, J., M. MacRoberts and J. Cullen. 1972. Flocking and feeding in the great tit *Parus major*: an experimental study. *Ibis* 114:507–530.

Kummer, H. 1971. *Primate Societies: Group Techniques of Ecological Adaptation*. AHM Publishing Corp. Arlington Heights, Ill.

Leighton, M. 1982. Fruit resources and patterns of feeding, spacing and grouping among sympatric Bornean hornbills (Bucerotidae). Doctoral dissertation. University of California, Davis, CA.

Miller, L.E. 1991. The influence of resource dispersion on group size among wedge-capped capuchins (*Cebus olivaceus*). *American Journal of Primatology* 17 (4): 123.

Miller, L.E. 1992a. Socioecology of the wedge-capped capuchin monkey (*Cebus olivaceus*). Doctoral dissertation. University of California, Davis.

Miller, L.E. 1992b. The association between group size and food intake in adult female wedge-capped capuchins (*Cebus olivaceus*). *Abstracts of the XIVth Congress of the International Primatological Society.*

Miller, L.E. 1994. Life's ups and downs: Activity budgets and feeding strategies of adult female wedge-capped capuchins (*Cebus olivaceus*). *Abstracts of the XVth Congress of the International Primatological Society.*

Nutrition Reviews. 1984. Nutrition intervention in pregnancy. *Nutrition Reviews* 42: 42–44.

O'Brien, T. and J.G. Robinson. 1993. Stability of social relationships in female wedge-capped capuchin monkeys, pages 197–210 in, *Juvenile Primates: Life History, Development and Behavior*, M. Pereira and L. Fairbanks, eds. Oxford University Press. New York.

Pond, W.G., H.J. Mersmann and J. Yen. 1985. Severe feed restriction of pregnant swine and rats: effects on postweaning growth and body composition of progeny. *Journal of Nutrition* 115: 179–189.

Pulliam, H.R. and T. Caraco. 1984. Living in groups: Is there an optimal group size?, pages 122–147 in, *Behavioural Ecology: An Evolutionary Approach*, second edition, J.R. Krebs and N.B. Davies, eds. Blackwell Scientific Publications. Oxford.

Robinson, J.G. 1981. Spatial structure in foraging groups of wedge-capped capuchin monkeys *Cebus nigrivittatus*. *Animal Behavior* 29: 1036–1056.

Robinson, J.G. 1984. Diurnal variation in foraging and diet in the wedge-capped capuchin *Cebus olivaceus*. *Folia Primatologica* 43: 216–228.

Robinson, J.G. 1986. Seasonal variation in use of time and space by the wedge-capped capuchin monkeys, *Cebus olivaceus*: Implication for foraging theory. *Smithsonian Contributions to Zoology* 431: 1–60.

Robinson, J.G. 1988. Group size in wedge-capped capuchin monkeys *Cebus olivaceus* and the reproductive success of males and females. *Behavioral Ecology and Sociobiology* 23: 187–197.

Robinson, J.G. and C.H. Janson. 1987. Capuchins, squirrel monkeys, and atelines: Socioecological convergence with Old World primates, pages 69–82 in, *Primate Societies*, B.B. Smuts, D.L. Cheney, R.M. Seyfarth, R.W. Wrangham and T.T. Struhsaker, eds. University of Chicago Press. Chicago, Ill.

Ruiter, J. de. 1986. The influence of group size on predator scanning and foraging behaviour of wedge-capped capuchin monkeys (*Cebus olivaceus*). *Behaviour* 98: 240–258.

Sapolsky, R.M. 1983. Endocrine aspects of social instability in the olive baboon (*Papio anubis*). *American Journal of Primatology* 5: 365–379.

Schaik, C.P. van. 1983. Why are diurnal primates living in groups? *Behaviour* 87: 120–144.

Schuerholz, G. and R. Demarchi. Unpublished. *Ecological Reconnaissance of Hato Piñero, Venezuela.*

O.A. Smith, C.A. Astley, M.A. Chesney, D.J. Taylor, and F.A. Spelman. 1986. Personality, stress and cardiovascular disease: human and nonhuman primates, pages 471–484 in, *Neural Mechanisms and Cardiovascular Disease,* B. Lown, A. Malliani and M. Prosdocimi, eds. Fidia Research Series, Volume 5. Liviana Press. Padova.

Smuts, B.B. 1987. Gender, aggression and influence, pages 400–412 in, *Primate Societies,* B.B. Smuts, D.L. Cheney, R.M. Seyfarth, R.W. Wrangham and T.T. Struhsaker, eds. University of Chicago Press. Chicago, Ill.

Southwick, C.H. 1967. An experimental study of intragroup agonistic behaviour in rhesus monkeys. *Behaviour* 28:182–209.

Srikosamatara, S. 1987. Group size in wedge-capped capuchin monkeys (*Cebus olivaceus*): Vulnerability to predators, intragroup and intergroup feeding competition. Doctoral dissertation. University of Florida, Gainesville, Fla.

Terborgh, J. and C.H. Janson. 1986. The socioecology of primate groups. *Annual Review of Ecology and Systematics* 17: 111–135.

Trivers, R.L. 1972. Parental investment and sexual selection, pages 136–179 in, *Sexual Selection and the Descent of Man,* B. Campbell, ed. Aldine. Chicago, Ill.

Walters, J.R. and R.M. Seyfarth. 1987. Conflict and cooperation, pages 306–317 in, *Primate Societies,* B.B. Smuts, D.L. Cheney, R.M. Seyfarth, R.W. Wrangham and T.T. Struhsaker, eds. University of Chicago Press. Chicago, Ill.

Whitten, P.L. 1983. Diet and dominance among female vervet monkeys (*Cercopithecus aethiops*). *American Journal of Primatology* 5: 139–159.

Wilson, E.O. 1975. *Sociobiology: The New Synthesis.* Harvard University Press. Cambridge, Mass.

Wrangham, R.W. 1980. An ecological model of female-bonded primate groups. *Behaviour* 75: 262–300.

15

SEE HOW THEY GROW

Tracking Capuchin Monkey (*Cebus capucinus*) Populations in a Regenerating Costa Rican Dry Forest

Linda M. Fedigan,[1]* Lisa M. Rose,[2] and Rodrigo Morera Avila[3]

[1]Department of Anthropology
University of Alberta
Edmonton, Alberta, T6G 2H4, Canada
[2]Department of Anthropology
Washington University
St. Louis, Missouri, 63130–4899
[3]Programa Regional en Manejo de Vida Silvestra
Universidad Nacional
Heredia, Costa Rica

INTRODUCTION

A major threat to world-wide conservation efforts is the loss and fragmentation of existing wildlife habitats (Boza 1993; Medley 1993; Mittermeier & Cheney 1987; Saunders *et al.* 1991; Shaffer 1981; Simberloff 1988). The expansion of habitat refuges through land restoration and subsequent forest regeneration is thus a critical tool in tropical conservation biology. The study of population dynamics of long-lived species in regenerating habitats can offer important insight into conservation research and behavioral ecology. Tropical forest primates are important indicator species because they have relatively slow life history patterns, are comparatively rare, and tend to attract public conservation interest. Very little is known about the long-term population dynamics of most primate species. Around the world, most non-human primate populations are in decline, and many are endangered or seriously threatened with extinction (Dobson & Lyles 1989; Dunbar 1987, 1988; Mittermeier & Cheney 1987). However, some populations, such as those of the relatively well-studied baboons, macaques and howling monkeys, show strong demographic fluctuations when studied over long periods of time. This study summarizes ten years of demographic data on the Central American white-faced capuchin monkey, *Cebus*

---

* Author to whom correspondence should be addressed.

Adaptive Radiations of Neotropical Primates
edited by Norconk *et al.* Plenum Press, New York, 1996

capucinus, with particular reference to ecological correlates of population growth, sex ratios, births rates, survivorship, and intergroup dispersal, in the early stages of forest regeneration. These data provide a context for our on-going studies of the behavioral ecology of this species (e.g. Boinski 1993; Chapman 1986, 1987, 1988*a;* Chapman & Fedigan 1990; Fedigan 1990, 1993; Rose 1994*a*, 1994*b*; Rose & Fedigan 1995), and also suggest how white-faced capuchins respond to changes in their environment.

White-faced capuchins are medium-sized, arboreal monkeys, recently ranked seventeenth in rarity among 100 Neotropical mammals (Dobson & Yu 1993). Both Arita *et al.* (1990) and Dobson & Yu (1993) classified this species as having "restricted distribution and low density". It is the only capuchin species present in Central America. The study area is Santa Rosa National Park in Costa Rica, the site of a regenerating tropical dry forest, and an experiment in creating a "megapark" from reclaimed ranch land (Janzen 1986*a*). On first consideration, one would expect the monkey populations in the park to have been on the increase over the past two decades, during which time cattle and hunters have been removed from their ranges, fires have been increasingly prevented or controlled, and trees have begun to grow again in abandoned pastures. On the other hand, the study area has been experiencing a continuing drying trend over the last 50 years (Fleming, 1986) and capuchins are highly dependent on waterholes during the dry season (Fedigan, unpubl. data, and see below). Furthermore, it remains to be determined whether newly regenerated forest provides sufficient food resources to attract and sustain monkey groups. By analyzing the demographic trends in the 28–29 capuchin monkey groups in the park during these crucial years of environmental change, we hope to shed some light on the potential for recovery of neotropical primate populations in reclaimed habitats (see also Chapman et al. 1989*a*). In addition, by presenting basic information on natality, mortality, and migration patterns in this population compared to other capuchins, we hope to augment our understanding of demographic processes in this little known neotropical monkey.

METHODS

Site Description

Santa Rosa National Park was established in 1971, and is located in northwestern Costa Rica in Guanacaste Province, near the Nicaraguan border. The original park covered 10,800 hectares of tropical dry forest in a series of stepped plateaus from the foothills of volcanic mountains down to the Pacific coastal plain. In the late 1980s, a project was begun to buy the ranchlands surrounding the park, using a "debt for nature" swap (Boza 1993; Liebow 1993; Wallace 1992), and the result is a greatly enlarged protected area (approximately 110, 000), which is now known as Area de Conservacion Guanacaste (ACG). The core of ACG remains Santa Rosa National Park, and the original park borders form the boundaries for our annual censuses. There are three monkey species in Santa Rosa: howlers *(Alouatta palliata)*, spiders *(Ateles geoffroyi)* and capuchins *(Cebus capucinus)*. Descriptions of Santa Rosa howler and spider monkey demography and behavioral ecology can be found in Chapman (1987, 1988*a*, 1988*b*, 1989), Chapman *et al.* (1989b), Chapman & Chapman (1991), Fedigan (1986), Fedigan *et al.* (1985, 1988), Fedigan & Rose (1995), and Freese (1976).

Originally, the Santa Rosa area was covered by a semi-deciduous, tropical dry forest, with patches of semi-evergreen oak forest *(Quercus oleoides)* on the upper plateaus (for detailed descriptions, see Janzen 1982, 1983*a*, 1983*b*, 1986*a*, 1986*b*). Over the past

300 years, 50% of the upper plateau was cleared for cattle pasture and planted with the African grass, *Hyparrhenia rufa*, and the forests were selectively logged, primarily for mahogany (*Swietenia macrophylla*). After the establishment of the park in 1971, cattle were removed, fires were gradually controlled, and the pastures have been slowly reverting to woody vegetation. As a result, the park is now a mosaic composed primarily of dry deciduous forest (dominated by *Spondias mombin*, *Luehea candida*, *Guazuma ulmifolia*, *Bursera simaruba* and *Ficus* species), along with fragments of semi-evergreen and riparian forest (dominated by *Hymenaea courbaril*, *Masticodendron capiri*, *Manilkara zapote*, *Sloanea terniflora*, *Brosimum alicastrum*) and of early secondary forest (succeeding in former pastures and usually dominated by wind-dispersed species, such as *Cochlospermum vitifolium* , *Tabebuia rosea* and *Luehea speciosa*) (Figure 1). Other forest types in the park which are sometimes used by the capuchins are fragments of oak, mangrove, and mixtures of deciduous and evergreen forests.

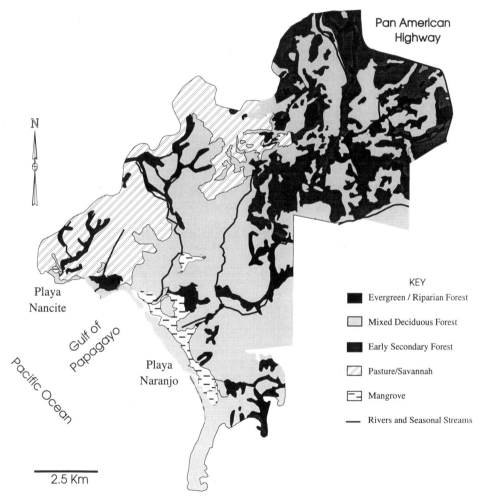

Figure 1. Vegetation map of Santa Rosa National Park, Area de Conservacion Guanacaste. Prepared by Cecilia Pacheco and Rodrigo Morera, based on field data (1993–1994) and aerial photographs (1987). Edited by Henry Chaves K and Jorge Fallas G., Laboratorio de Teledeteccion y Sistemas de Informacion Geografica (TELESIG), Universidad Nacional, Heredia, Costa Rica.

There are two distinct seasons at Santa Rosa. Almost all of the annual rainfall occurs in the wet season, from mid-May to mid-December (Janzen 1991). During the dry season the majority of non-riparian trees lose their leaves, and the streambeds and waterholes gradually dry up. Only a few spring-fed water sources last throughout the year.

The Study Sample: Study Groups and Census Population

The demographic research began with a brief survey of the monkeys in the park in 1982, followed by more complete censuses of the capuchin and howler groups throughout Santa Rosa National Park in 1983 and 1984 (Fedigan 1986; Fedigan *et al.* 1985). In 1984, we selected a few "study groups" of each species for intensive research and began to discriminate the individual monkeys in these groups, either through natural markings or by distinctive tags that we placed on the animals. In 1985, we began to observe individuals in two study groups of capuchins on a regular daily basis, and added a third group in 1989. We recorded births, deaths, disappearances and migrations, as well as various foraging and social behaviors, a practice that continues through the present (for life history and group composition data on our three intensively studied groups, see Fedigan & Rose, 1995).

Between 1983 and 1992, we conducted 8 annual censuses (see Table 1). The censuses were almost always carried out in the months of May and June, during which time as many as possible of the capuchin groups in the park were contacted and counted. In the early years, we chose one area of the park at a time and either walked transects, or walked all known trails and dry creek beds in the area to locate groups, which we then counted carefully and followed for the day. We used individual markings and distinctive age/sex compositions to help us recognize the same group on successive days for repeat counts. A targeted group was repeatedly counted by more than one observer until a stable count and

Table 1. Demographic trends in Santa Rosa Capuchins from 1983 to 1992

Year	No of monkeys counted	No. of groups counted	Average group size	Estimated population size	Mean proportion in each age/sex class[1]			
					Adult/sub Male	Adult/sub Female	Juvenile	Infant
1983	229	20	11.5	321	0.17	0.34	0.34	0.15
1984	393	28	13.6	393	0.17	0.37	0.36	0.10
1985	193	13	14.8	415	0.19	0.33	0.36	0.12
1986	311	19	16.4	458	0.20	0.29	0.41	0.10
1987	217	13	16.7	467	0.23	0.27	0.36	0.14
1988	164	10	16.4	459	0.23	0.27	0.37	0.12
1990	318	18	17.7	495	0.25	0.30	0.32	0.13
1992	526	29	18.1	526	0.22	0.31	0.29	0.18
Mean		19	15.7		0.21	0.31	0.35	0.13

Comparison of 1992 with 1983 (unpaired t-test, 2 tailed)

	Group size	Males	Females	Juveniles	Infants
t =	−4.209	−3.391	0.928	1.405	-1.329
p =	**<0.001***	**0.001***	0.358	0.167	0.190

[1]Group sizes and proportions are means across all groups counted in each year
*Statistically significant differences

composition was achieved. After establishing a stable count on one group, we located its closest neighboring group, and where possible, used simultaneous contact with both groups by different observers to establish their independence as two groups.

From our well-known study groups, we were able to determine average home range sizes (approximately 1–2 km^2), and we used this information to help estimate and map the ranges of the groups counted during the censuses. Over the years, it has become increasingly easy to relocate our census groups on successive years and to determine if new groups have appeared. Individual adult male capuchin monkeys are readily distinguishable from year to year because of scars, short tails, poorly-healed broken limbs, piebaldism, etc., and individual female capuchins are also identifiable with practice. Adults are easily sexed, and infants, immatures and adults are readily distinguished by size. The sexing of immatures and infants is more difficult, and we did so only for our well-known study groups. Clearly, the census data are not as detailed, complete or reliable as the daily information we collect on our study groups, but we believe that we have a reasonably accurate picture of the number of capuchin groups in the park, as well as their changing sizes and age/sex compositions over the past decade.

Data Analysis

Although the first census took place in 1983, there were a few areas of the park that we did not search adequately until 1984, and the latter is probably more representative of the population size at the beginning of our study. Thus, when considering growth in terms of overall numbers, we compared 1984 to 1992 , whereas for group composition analyses we compared 1983 and 1992. Individuals were assigned to one of four age/sex classes: infant (less than one year), juvenile (females 1–5 years; males 1–7 years), adult female, or adult/subadult male. We included subadult males (7–10 years) in the 'adult male' category because although slightly smaller than fully adult males, subadults are typically active in mating and group defense. (For further details on distinguishing individuals and age/sex classes, see Fedigan *et al.* 1985; Oppenheimer 1968). We examined infant survival and male tenure in our study groups with the use of SPSS-SURVIVAL. At the time we analyzed our data, some of the infants were still alive and some of the males were still resident in their groups. Analyses based only on completed (uncensored) intervals is known to create a bias toward shorter intervals. Thus, SPSS-SURVIVAL analysis requires the use of both completed (uncensored) intervals *and* the intervals that were still incomplete (censored) at the time the study terminated. SPSS-SURVIVAL generates median survivorship values for groups or subgroups, and performs a Wilcoxon-Gehan test for differences in interval distribution (Lee 1992). Significance was set at $p \leq 0.05$ for all tests, and two-tailed probabilities are reported.

RESULTS

Population Growth Rate

The total number of individuals counted increased from 393 in 1984 to 526 in 1992, an observed rate of increase (r) of 0.34 over 8 years (Table 1). Although we were not able to contact and count all the groups every year, it is clear that the population was growing steadily over this period, at a rate of approximately 4% per annum. Extrapolation from the number of groups counted in the partial censuses shows that growth across the years was fairly even, except for a small decline in 1988. Despite the increase in population size, the number of capuchin groups increased only marginally, from 28 in 1984 to 29 in 1992. However, average group size increased significantly, from 11.5 in 1983 to 18.1 in 1992 (t = 4.209, p < 0.001). Again the increase occurred steadily from year to year, except for a small decrease in 1988. As the increase in group size is independent of the number of groups counted, this result is unlikely to be an artifact of improved census taking.

Group Composition: Age/Sex Classes and Sex Ratio

Average group composition in each census year represents the proportion of individuals in each age/sex class within a group, averaged across all of the groups that were counted in a given year (Table 1). There was no significant difference between the overall distribution of proportions in 1983 and 1992 (X^2 = 1.499, p = 0.694). Although the population grew during this period, there was no significant increase in the proportion of juveniles (0.34 to 0.27) or infants (0.15 to 0.18; Table 1). The ratio of immature animals (juveniles plus infants) to adult females has fluctuated, but has not changed consistently over the course of the study in either the census population or in the study groups (Table 2). However, the proportion of males has increased significantly, from 0.17 in 1984 to 0.22 in 1992 (t = 3.391, p < 0.001; Table 1). The number of adult males in the population increased by 75%, from 65 in 1984 to 114 in 1992. By comparison, the number of adult females increased by only 9%, from 143 in 1984 to 156 in 1992, with no significant change in the proportion of adult females in each group (t = 0.928, p = 0.358).

The ratio of adult males to females in the census population also has increased significantly, from an average of 1.0M:1.9F per group in 1983, to 1.0M:1.3F in 1992 (t = 2.995, p = 0.004; Table 2). Averaged across years for which data are available, the sex ratio was higher in the study population (1.0M : 0.97F) than in the census population (1.0M :1.4 F), but the difference is not significant (t = 1.135, p = 0.191). The higher sex ratio in the study groups is mainly due to particularly high proportions of adult males in one group between 1985 and 1988.

Table 2. Comparison of capuchin study groups to Santa Rosa census population. (Average values across all groups counted in each year)

Year	Adult M:F (sex ratio)				Infants:adult fems (Birth rate)				Immatures[1]:adult fems			
	Census	Study	T	p	Census	Study	T	p	Census	Study	T	p
1983	0.53	—	—	—	0.53	—	—	—	1.30	—	—	—
1984	0.53	—	—	—	0.31	—	—	—	1.15	—	—	—
1985	0.61	2.25	3.03	**0.011***	0.38	0.31	−0.47	0.647	1.22	2.04	−0.93	0.373
1986	0.73	1.84	3.01	**0.008***	0.34	0.55	0.91	0.374	1.64	1.96	10.54	0.596
1987	0.81	1.42	3.11	**0.011***	0.41	0.29	−0.67	0.520	1.67	1.89	−0.33	0.747
1988	0.79	1.25	2.60	**0.036***	0.29	0.38	0.33	0.752	1.56	2.70	−1.55	0.165
1989	—	0.27	—	—	—	0.71	—	—	—	1.25	—	—
1990	0.96	0.66	−1.13	0.277	0.33	0.42	0.68	0.505	1.37	1.69	−0.79	0.441
1991	—	0.48	—	—	—	0.50	—	—	—	1.54	—	—
1992	0.76	0.63	−0.86	0.397	0.55	0.44	−0.94	0.354	1.30	0.93	1.00	0.327
1993	—	0.70	—	—	—	0.63	—	—	—	1.00	—	—
1994	—	0.80	—	—	—	—	—	—	—	2.94	—	—
Mean	0.72	1.03	1.356	0.191	0.39	0.47	1.266	0.225	1.40	1.79	1.607	0.128

*Statistically significant differences, p < 0.05
[1]"Immatures" includes juveniles and infants

We were able to determine sexes for 37 of the 44 infants born in the study groups between 1985 and 1994 (the remainder died before their sex could be unambiguously assigned). Of these 37 infants, 29 (78%) were male and 8 (22%) were female, indicating a mean neonatal sex ratio of 3.6M : 1.0F. The male bias in births is significant on a binomial test ($p = 0.001$). We considered the possibility that male births were over-estimated, as under field conditions, females are more difficult to sex than are males. However, juveniles are easier to sex, and a male bias also occurs in this age category (2.8M: 1.0F for small juveniles and 1.6 M: 1.0 F for large juveniles).

Natality

Almost all the census data were collected during the months of May and June. As births peak between January and April (see below), May-June censuses are the optimal time to include the majority of new infants. For each group in the census population, a birth rate was inferred from the number of infants counted at the time of census, divided by the number of adult females. The average number of infants per group is two (range 0 - 11), and the average proportion of infants per group is 0.13 (Table 1). In the three study groups, where known individuals are observed on a daily basis and all births throughout the year are recorded, the mean birth rate between 1985 and 1993 was 0.47 (Table 2). This was somewhat higher than the mean infant:adult female ratio (0.39) between 1983 and 1992 in the census population, but the difference is not significant (t = 1.266, $p = 0.225$).

Birth Rates and Group Size. We performed a regression analysis to examine the relationship between group size and birth rate over all census years. The size of a given group, and the birth rates of the group varied from year to year. For this analysis, we first calculated the average group size and the average birth rate for each groups across the years of the census, and then checked for a relationship between the two sets of average values. We found that average group size accounted for 39% of the variance in birth rates, and that the effect of group size was significant at 0.028. We therefore examined the relationship between group size and birth rates *within* each year, in both the census and study populations. We found a significant positive correlation in only one year, 1984, in the census population (r = 0.469, $p < 0.05$), This result must be interpreted with caution because multiple tests (n = 8) were performed, and only 25% of the variation was explained by the correlation. Birth rates were not significantly correlated with the number of adult females or males in a group in any year.

Birth Rates and Rainfall. Rainfall at Santa Rosa is recorded on a daily basis (Janzen 1991). The average annual rainfall between 1983 and 1991 was 1544 mm (range 915–2558 mm.) The median date of the first rain over 1 mm (excluding isolated showers in March) was May 14 (range April 26 to May 21), and the median date of the last rain over 1 mm was December 16 (range Dec. 10 to Feb. 27). We found no significant correlations between annual rainfall and births the following year in either the census population (rho = 0.167) or in the study groups (rho = 0.262). Birth rates were significantly correlated with the amount of 'early rain' (rain during May) in the census population (rho = 0.905, $p < 0.005$), but not in the study groups (rho = 0.276). In the study groups, where we know most birth dates to within a few days, there were significantly more births in the dry season than in the wet (Figure 2; binomial test, $p = 0.018$).

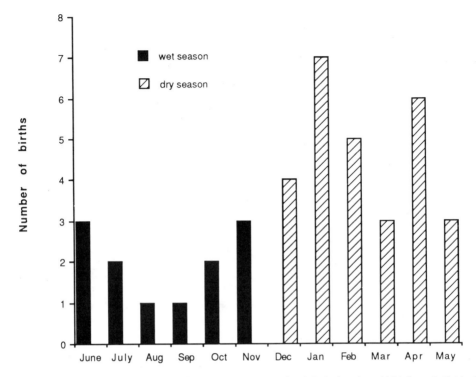

Figure 2. Birth seasonality in *Cebus capucinus* at Santa Rosa National Park, based on 44 births to individually known mothers on known dates.

Birthrates, Habitat Type, and Waterholes. Home range quality was categorized in terms of predominant forest type (evergreen/riparian or mixed deciduous) and the type of waterhole available (permanent or seasonal). We performed a multiple regression analysis using forest type, water type, and initial group size in 1984 as independent variables, with birth rate and growth rate (% increase between 1984 and 1992) as respective dependent variables. Groups in mixed deciduous forest *tended* to grow more than those in evergreen riparian forest (56% versus 19%) and those with permanent water *tended* to grow more than those with seasonal water (57% versus 21%). The three factors together (forest, water, and initial group size) explained 33% of the variation ($F = 3.948$, $p = 0.021$), but of the three independent variables in the equation, only "initial group size" had a significant effect on growth ($t = 2.522$, $p = 0.019$). Groups that were smaller in 1984 grew significantly more than did those that were larger in 1984. Thus, the effects of home range quality on growth rates may have been confounded by the effect of initial group size. When we ran the multiple regression, using only forest type and water type as independent variables, there were no significant effects on growth or birth rates.

Mortality and Survivorship

We performed survival analysis on the 44 capuchins that have been born in our study groups since 1985. Those that died or disappeared prior to age five (the approximate age of female sexual maturity) were treated as dead. Most disappearances were animals less than two years old that almost certainly would have died, but included are two 3–4

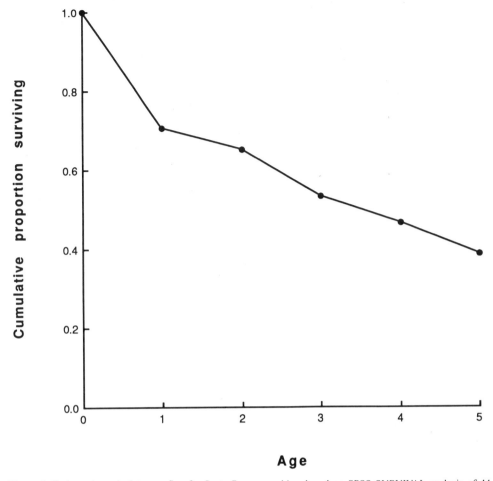

Figure 3. Estimated survival to age five for Santa Rosa capuchins, based on SPSS SURVIVAL analysis of 44 births in three intensively-studied groups, 1985–1994.

year old males that may have been unusually early emigrations. Those still alive in June 1994 were included in the analysis as incomplete ("censored ") intervals. We found survival to age five in the study groups to be 39%, indicating an infant/immature mortality rate of 61%. The highest mortality (29%) occurred in the first year of life (Figure 3). Seven out of twelve (58%) confirmed infant deaths and fourteen of twenty-three (61%) disappearances prior to age five occurred during months in which new males displaced previously resident males ("male takeovers"). We found no significant sex difference in survival to age five (Wilcoxon-Gehan = 0.577, p = 0.447). Infant males had higher mortality than infant females (22% versus 0%), but the difference was not significant (X^2 =2.08, $p > 0.10$).

Emigration and Immigration

Between 1985 and 1993, 31 males older than five years left our 3 study groups, and 14 males joined the study groups. (These figures exclude 9 cases in which males transferred between study groups.) If the difference between emigration and immigration can be assumed to reflect mortality, the adult male mortality rate is approximately 55%. Breaking this down by age classes; 23 adult males left and 9 joined (mortality 61%); while 8 subadult or large immature males left and 5 joined (mortality 37.5%). By comparison, 13 adult females disappeared and 3 appeared in our study groups, suggesting adult female mortality to be approximately 23%.

Based on the 19 males whose arrival and departure dates are known to within a few months, average male tenure in our study groups is 30 months (SD = 17.9 months). When we included all 32 individually identified males that have resided in our study groups since 1985 and assigned censored (incomplete) intervals to those whose arrival or departure dates were unknown, SURVIVAL analysis indicated the median length of male tenure to be 40.6 months. Multiple transfers in a male's life are suggested by these results, and are known to have occurred in our individually-recognized males.

Comparisons to Other Studies

Comparing our data with those from another long-term study site, Barro Colorado Island, (Mitchell 1989; Oppenheimer 1982), Santa Rosa capuchins have a higher ratio of males to females (average 0.76 compared to 0.57; t = 2.40, p = 0.031). However, the populations are otherwise very similar (Table 3). We found no significant differences in group size (t = 0.655, p = 0.523), birth rate (t = 0.261, p = 0.80), group composition (X^2 = 0.156, p = 0.984), ratio of immatures to females (t = 0.874, p= 0.397) or ratio of immatures to adults (t =0.694, p = 0.499). We also compared our data with those available for another capuchin species, *Cebus olivaceus* (Freese and Oppenheimer 1981; Miller 1992; Robinson 1988a, 1988b). Overall group composition was not significantly different (X^2 = 0.735, p = 0.865), but Santa Rosa capuchins have smaller groups (t = 4.15, p = 0.003), with a larger proportion of males (t = 2.259, p = 0.05) and a higher ratio of adult males to

Table 3. Comparison of Santa Rosa capuchins with other sites

Year	Average groups included	Average group size	Proportions in each age/sex class				Adult M:F sex ratio	Immature /adult female	Immature / adult	Annual Birth Rate
			Adult male	Adult female	Juvenile	Infant				
Cebus capucinus										
Santa Rosa National Park *										
1970's[1]	1?	17	0.18	0.35	0.41	0.06	0.50	1.33	0.88	
1983-1992[2]	18	26	0.21	0.32	0.33	0.12	0.76	1.80	1.01	0.41
Barro Colorado Island										
1966-1970[3]	2	15	0.17	0.31	0.36	0.16	0.54	1.70	1.41	
1986-1988[4]	4	20	0.19	0.32	0.37	0.12	0.60	1.59	1.00	0.42
Cebus olivaceus										
1973[1]	2	26	0.16	0.33	0.34	0.17	0.56	1.81	1.01	
1977-1985[5]	10	20	0.18	0.33	0.35	0.14	0.54	1.50	0.97	0.29 to 0.47
1989-1991[6]	8	22	0.16	0.33	0.47	0.04	0.51	1.61	1.07	

*Values for Santa Rosa include study groups and census data.
[1] Freese and Oppenheimer 1981 (Table 3).
[2] SR census data.
[3] Oppenheimer 1982 (Table 1).
[4] Mitchell 1989 (Table 3.1, 3.2).
[5] Robinson 1988a (Table 5); 1988b (Table 1).
[6] Miller 1992 (Table 1).

females (t = 2.41, p = 0.039). Birth rates are similar to those reported for young and middle-aged female *C. olivaceus* (Robinson 1988*b*), and we found no significant difference in the ratio of immatures to adult females (t = 0.883, p = 0.400) or adults (t = 0.137, p = 0.894).

DISCUSSION

From a conservation perspective, our most significant finding is that a primate population within a protected and regenerating habitat has increased in size over the past decade. The observed rate of increase (r) of 0.04 that we found in the Santa Rosa capuchins is relatively high for a non-provisioned population (e.g., see Richard's summary of demographic parameters in a variety of primate species, 1985: 244–7). This augurs well for the positive effects of habitat regeneration. However, much of the increase is due to increasing numbers of males. If we consider only adult females, the average annual growth rate is considerably lower at r = 0.015. It is also noteworthy that population growth has occurred mainly through increasing group size, and not through the proliferation of new groups. This contrasts with the howling monkey population in the park, which has grown faster over the same period (approximately 7.5% per annum), and in which the number of groups has increased more than has group size (Fedigan *et al.* in prep). What might be the ecological and demographic correlates of the capuchin pattern of population growth?

Ecological Factors Affecting Population Growth

Santa Rosa is a highly seasonal environment in which no appreciable rain falls for 5 months of the dry season. During this period, most streams stop running and the standing water remaining in tree holes and creek beds gradually dries up. Only a few spring-fed water sources remain available throughout the dry season, and even these have been drying up in recent years. Fleming (1986) has argued that annual rainfall in Guanacaste province has decreased over the last 50 years, leading to an earlier disappearance of standing water in the dry season. Capuchins prefer to drink water at least once and usually several times a day. During the dry season, groups with waterholes in their range forage over smaller areas than in the wet season, and center their movement patterns on any available water source. A study of foraging behavior in 1991 found that a group with a permanent spring-fed waterhole (since dried up) spent 48% of dry season observations in the 10% of its range containing the waterhole, compared with only 16% in the wet season (Rose, unpubl. data). During the dry season, capuchins with a single water source essentially become central place foragers (Orians & Pearson 1979). This is quite different from the behavior patterns found in howling and spider monkeys, which seldom drink water.

Some groups in the park are without water sources for the last few weeks of the dry season, but capuchins seem to work very hard to maintain access to water. Neighboring groups engage in territorial encounters over access to water, and those groups without access to water shift their ranges to encompass a water source. In the Santa Rosa study sample, larger groups, or groups with more males, are better able to displace smaller groups from water. One new group that appeared in 1990 gradually expanded its range to encompass a permanent waterhole, and within two years had displaced the smaller group that previously had maintained its range in that area.

Both Harding (1976) and Hamilton (1985) found that population growth and decline in baboons was largely determined by the availability of water sources, and this may also act as a constraint on white-faced capuchins. However, despite the obvious importance of water for capuchins, we were not able to document any significant effect of habitat type or water availability on growth of group sizes or birth rates. This may be because our measures were too crude to capture the variation between groups, because our sample sizes are too small, or because of the confounding effects of initial group size. Alternatively, we would argue that the lack of difference may be because capuchin groups are omnivorous and flexible, and manage to shift their ranging patterns and diet to compensate for constraints imposed by food and water availability in different seasons.

Food and drinking water likely represent different types of resources for capuchins. With respect to food, groups in evergreen and deciduous forests appear to use a similar resource base, composed primarily of fruit and small invertebrates. However, capuchins are able to exploit a wide variety of specific foods within these general categories in order to fulfill their dietary requirements, as evidenced by intergroup variation that appears to reflect food preferences as well as local resource availability (Chapman & Fedigan 1990). In contrast, water is not readily replaced by an alternate resource. Because water sources are discrete and limited, it may not be possible to distribute an increasing number of animals between water sources in any other pattern than that of increasing group size.

Large areas of Santa Rosa are newly regenerated secondary forest, growing up in abandoned pastures. One might predict that monkeys would do well in regenerating forest. Secondary forests often exhibit high plant diversity, and many colonizing species quickly produce foliage and fruit with few toxic defenses (Fleming *et al.* 1985; Hartshorn 1978; Janzen 1979; Webb *et al.* 1972). However, as Janzen (1988) and Gerhardt (1994) have documented, the new woody vegetation that grows in most regenerating pastures in Santa Rosa is made up of 90% wind-dispersed trees that do not offer fruits for vertebrates. Howling and spider monkeys have a diet primarily made up of fruits and flowers, and they seldom enter new secondary forest (although spiders will swing rapidly though a corridor of regenerating secondary forest to move from one forest fragment to another). Capuchins are able to make use of these new areas, but because of the lack of water sources and fruit trees, they do not do so as extensively as might have been anticipated.

Capuchins *do* regularly enter early secondary forest, largely in search of insects, but they do not reside in them exclusively. A newly-fissioned, small capuchin group residing in a patch of early secondary forest would have difficulty accessing large fruit trees or water sources; whereas a larger group could expand its range into regenerating forest and still maintain access to water and to fruit trees within its original range of older forest. Larger home ranges would, in turn, tend to support larger social groups that were more successful in intergroup competition. Thus there are social as well as ecological constraints on the distribution of capuchin groups, and the pattern of population growth. The major limiting resources (water sources and large fruit trees) are patchily distributed, and are subject to strong intergroup competition. By increasing group size rather than propagating new groups, Santa Rosa's capuchin population appears to be making the most effective use of the increasing resources offered by regenerating forest.

Demographic Variables Underlying Population Growth

Capuchin births peak in the dry season, and since gestation is approximately six months, this means that many females are becoming pregnant during the rainy season. The wet season typically begins with an intensive period of rainfall in late May/early June, followed by a small "dry period" in late July/early August, and heavier rains again in September and October. Soon after the onset of heavy rains in late May/early June, there is a intense upsurge in the insect population, marked especially by the appearance of caterpillars, on which the capuchins gorge themselves for a few weeks (Rose 1994*a*). The deciduous trees also leaf out rapidly. A similar peak in leaf flushing and arthropod abundance in the early wet season occurs in the Costa Rican lowland forest of Corcovada (Boinski & Fowler 1989). Although we did not find a correlation between the total amount of rainfall during the season and the birth rates in the subsequent year, we did find a significant relationship between the amount of "early rain" and birth rates in the following year in the census population. This suggests that conception and birth rates are affected by the timing, rather than the overall amount, of annual rainfall.

It is very common in primates to find high infant mortality in the first year of life (e.g., Dunbar 1988; Richard 1985). Twenty-nine percent of the infants in our study groups died in their first year, and 61% died before reaching reproductive maturity. Based on the average study group birthrate of 0.47 and assuming first parturition to occur at age 5, it would take a female 10.5 years to replace herself. If the sex ratio at birth is 3.6 males to 1 female, it would take a female more than 20 years to replace herself with another female. Female longevity is not known for free ranging *C. capucinus* populations, but it is likely to be considerably less than the 47 years reported in captivity (Ross 1988).

Although we have observed deaths to result from predation and disease, and inferred deaths to be caused by dry season stress and parasites, a large proportion of the deaths of immatures in our study groups have been associated with male take-overs. The typical social system for the genus *Cebus* is female-bonded with male dispersal (Fedigan 1993; Freese & Oppenheimer 1981; Robinson & Janson 1987). Males often emigrate singly. However, on three occasions in the history of our study, at approximately four year intervals, there has been a major movement of adult males between groups. Several new males entered a group together and over a course of days or weeks, fought with the resident males, wounding and very occasionally killing each other. During this time of social upheaval, females and immatures also died (Rose & Fedigan 1995). Furthermore, on one occasion a male was observed to kill an infant (Rose 1994*b*), and on two other occasions, strong inferential evidence suggested that adult males have wounded and killed infants that were conceived in the group prior to their arrival. Whether these events are interpreted as directed infanticide or not, it is clear that male take-overs are a source of mortality for immatures in our study groups.

The patterns of immigration and emigration in our study groups suggest that dispersal is also a considerable risk for males, and that adult mortality is higher for males than for females. Dunbar (1988) argues that monkeys are especially susceptible to *predation* when moving between groups, but we would add that capuchin males also run a higher risk of severe *agonism* during transfers between groups. Resident males may also be injured while attempting to exclude male immigrants, and the alpha male in one of our study groups recently died from wounds received during a male 'takeover' of his group.

The increasing proportion of adult males in the groups and in the population is one of our more puzzling findings. It is possible that additional males have been immigrating from outside the boundaries of Santa Rosa National Park, particularly as hunters and other forms of habitat disturbance were common outside the park borders until recent years when the adjacent ranches were purchased and converted to protected areas. If a significant component of the Santa Rosa capuchin population increase is due to male immigration, this carries the important implication that protected areas may attract the dispersing sex. In species where males rather than females disperse, this would result in a larger absolute population size, without a corresponding increase in effective population size.

It is also possible that the increasing proportion of males is due to a skewed birth sex ratio. Examination of the sex ratio at birth in our study groups shows it to be highly skewed toward male infants and small juveniles, and somewhat skewed toward males in large juveniles. This bias toward males at birth may have been introducing more males than females into the adult population, especially as we found no significant sex difference in mortality prior to age five. If we apply the average birth and mortality rates in our study sample to the larger population, a consistent 3:1 male bias in births since 1983 would account for the observed increase in adult male numbers. Until we have a larger sample size, and some other long term study of white-faced capuchins documents the sex ratio at birth, we cannot determine if this pattern is unique to our present study population. However, birth data from an ongoing study of one white-faced monkey group at Lomas Barbudol in Costa Rica suggests a similarly biased sex ratio, with 6 of 8 infants born between 1990 and 1994 being male (Susan Perry, pers. comm.)

Comparison to Other Capuchin Studies

One other site for which long-term demographic data are available for *Cebus capucinus* is the Smithsonian Tropical Research Institute station on Barro Colorado Island (BCI), Panama (Mitchell 1989, Oppenheimer 1968). The BCI *Cebus* population has expanded less rapidly than the Santa Rosa population, from an estimated 270 individuals in 1966 to 313 in 1988 (Mitchell 1989). This may be because BCI supports a fixed extent of mature forest, and the primate population is close to carrying capacity.

The most complete set of demographic data for *Cebus olivaceus* come from Hato Masaguaral in central Venezuela. Most published studies of wedge-capped capuchin demography and behavioral ecology come from this site (e.g., Fragaszy 1990; O'Brien 1991, 1993*a*, 1993*b*; O'Brien & Robinson 1993; Robinson 1981, 1984, 1986, 1988*a*, 1988*b*; Robinson & O'Brien 1991; de Ruiter 1986). Hato Masaguaral capuchins inhabit seasonally dry forest, but a permanent stream provides water throughout the year. The study population of 9–10 groups grew from 74 individuals in 1978 to 113 in 1985 (r = 0.075), and average group size increased from 14.8 to 23 with several group fissions (Robinson 1988a). The rate of population growth is thus somewhat faster than that found at Santa Rosa, but the pattern of increasing group size is similar.

Wedge-capped capuchins tend to live in larger groups with fewer adult males than do white-faced capuchins. Both the Santa Rosa and Hato Masaguaral capuchin populations are biased toward females, but the bias is less pronounced at Santa Rosa. The sex difference in group composition is heightened if we consider relative proportions of *breeding* males in groups, as there is typically only one breeding male per group in wedge-capped capuchins (Robinson 1986, 1988*a*). By contrast, our behavioral observations at Santa Rosa indicate that all adult and subadult males within a group may mate with estrous females. We do not have the data to address individual differences in male reproductive success, but it is clear that male white-faced capuchins have greater (and earlier) mating opportunities than do male wedge-capped capuchins. The shorter group tenure for Santa Rosa males (3.3 years compared with 6.6 years at Hato Masaguaral) also indicates greater intergroup mobility. Better mating opportunities (compared with *C. olivaceus*) may attract relatively more *C. capucinus* males into social groups, including immigrant males moving into Santa Rosa from outside the park boundaries.

Conclusions

Over the past decade, during an early stage of forest regeneration in Santa Rosa , the capuchin population has been steadily increasing. The relatively high rate of growth for a

non-provisioned primate population is likely due to the removal and control of disturbances, such as hunters, domestic animals, and anthropogenic fires from capuchin ranges
in the protected area of the park. The population increase has been expressed mainly
through increasing group size; almost no new groups are known to have formed. The most
likely explanation for the absence of new groups is the limited number of dry season water
sources. The scarcity of large fruit trees in newly-regenerating forest and patterns of intergroup competition, especially for dry season water, are also implicated. Newly regenerating patches of deciduous forest are visited and used by capuchins, particularly in search of
insects, but no group resides exclusively in such a habitat. Neotropical monkeys in general
rely on fruit trees and capuchins in particular rely on drinking water. These resources do
not reappear for many decades in reclaimed pastures. When a large ecological area has
been extensively deforested, probably to the extent of affecting the watershed and rainfall
pattern, it obviously cannot be expected to recover quickly. In addition, different resources may not increase proportionally in recently-reclaimed areas. When populations increase, competition for fixed or less-rapidly increasing resources may constrain patterns of
growth and distribution, particularly in group-living animals such as primates.

There are indications that in the Santa Rosa capuchin population, male numbers are
increasing faster than females, through skewed birth rates and/or immigration from outside the park's boundaries. One goal of our ongoing research at Santa Rosa will be to distinguish between these two factors. If migrating males are moving in to the protected area,
the implication that habitat refuges favor the dispersing sex is significant for other species
with sex biased dispersal patterns. Changing sex ratios can have important demographic
and behavioral consequences that warrant close attention. Finally, the increasing proportion of male capuchins at Santa Rosa illustrates the caveat that effective population size
does not necessarily increase at the same rate as overall population numbers. Habitat rec-
lamation and regeneration are invaluable measures for protecting vulnerable populations,
but the long process of recovery must take into account species-specific ecological and behavioral factors.

SUMMARY

The expansion of habitat refuges through land restoration and subsequent regeneration of tropical forest is a critical tool in conservation biology. Here we describe population growth in white-faced capuchins (*Cebus capucinus*) in a regenerating dry forest at
Santa Rosa National Park, Costa Rica. Data come from annual park-wide censuses, supplemented by ongoing observations of three intensively-studied social groups. Between
1984 and 1992, the population grew from 393 to 526 individuals. Growth was primarily
due to the increasing size of existing social groups, rather than the formation of new
groups. The proportion of males in the population and the ratio of adult males to females
increased significantly between 1983 and 1992. This may be due to immigration from outside the park, and/or to a male-biased birth ratio. A sample of 44 births showed a skew of
almost 3:1 toward males. Data from our study groups indicates an average birth rate of
0.47, with 39% survival to age five. Birth rates were not affected by group size, and neither birth rates, group size, nor increase in group size were affected by habitat type. However, birth rates were positively correlated with the amount of early rainfall in the
previous year, and births were significantly biased toward the dry season . We conclude
that the need for fruit and dry season drinking water prevents capuchins from living exclusively in newly regenerated forest, and inhibits the formation of new groups in response to

and neither birth rates, group size, nor increase in group size were affected by habitat type. However, birth rates were positively correlated with the amount of early rainfall in the previous year, and births were significantly biased toward the dry season . We conclude that the need for fruit and dry season drinking water prevents capuchins from living exclusively in newly regenerated forest, and inhibits the formation of new groups in response to habitat expansion. We suggest that, at least during the early stages of forest restoration, existing groups increase in size, expanding their ranges into newly-available habitat while maintaining access to essential fruit and water resources in areas of older forest. The growing proportion of males in the population suggests that the dispersing sex may benefit most from habitat protection and expansion.

ACKNOWLEDGMENTS

This paper is dedicated to the late Larry Fedigan, who both collected and oversaw the collection of the primate census data for ten years. We are grateful to the National Park Service of Costa Rica for allowing us to work in Santa Rosa National Park from 1983–1988, and to the administrators of the Area de Conservacion Guanacaste for permission to continue research in the park to the present day. We thank the many people who contributed to the collection of data for this project over the years, and the efforts of Colin Chapman are particularly acknowledged. Dan Janzen generously shared his expertise on the ecology and history of the park; John Robinson checked our values for *Cebus olivaceus* demographics; Terry Taerum and John Addicott advised on the analyses; and Cecilia Pacheco helped to prepare the vegetation map of Santa Rosa. We thank Sue Boinski, Colin Chapman, Tom Langen, Lynne Miller, Susan Perry and Sandra Zohar for many helpful suggestions that improved the manuscript. The research of LMF is funded by an ongoing grant (#A7723) from the Natural Sciences and Engineering Research Council of Canada (NSERCC).

REFERENCES

Arita, H.T., Robinson, J.G., and Redford, K.H., 1990, Rarity in Neotropical forest mammals and its ecological correlates, *Conserv. Biol.* 4:181–192.

Boinski, S., 1993, Vocal coordination of troop movement among white-faced capuchin monkeys, *Cebus capucinus, Am. J. Primatol.* 30:85–100.

Boinski, S., and Fowler, N.L., 1989, Seasonal patterns in a tropical lowland forest, *Biotropica* 21:223–233.

Boza, M.O., 1993, Conservation in action: past, present, and future of the national park system of Costa Rica, *Conserv. Biol.* 7:239–247.

Chapman, C.A., 1986, *Boa constrictor* predation and group response in white-faced cebus monkeys, *Biotropica* 18:171–177.

Chapman, C.A., 1987, Flexibility in diets of three species of Costa Rican primates, *Folia Primat.* 49:90–105.

Chapman, C.A., 1988a, Patterns of foraging and range use by three species of Neotropical primates, *Primates* 29:177–194.

Chapman, C.A., 1988b, Patch use and patch depletion by the spider and howling monkeys of Santa Rosa National Park, Costa Rica, *Behaviour* 105:99–116.

Chapman, C.A., 1989, Spider monkey sleeping sites: use and availability, *Am. J. Primatol.* 18:53–60.

Chapman, C.A. , and Chapman, L.J., 1991, The foraging itinerary of spider monkeys: when to eat leaves, *Folia Primat.* 56:162–166.

Chapman, C.A., and Fedigan, L.M., 1990, Dietary differences between neighboring *Cebus capucinus* groups: local traditions, food availability, or response to food profitability?, *Folia Primat.* 54:177–186.

Chapman, C.A., Chapman, L.J., and Glander, K.E., 1989a, Primate populations in Northwestern Costa Rica: potential for recovery, *Primate Conserv.* 10:37–44.

Chapman, C.A., Fedigan, L.M., Fedigan, L., and Chapman, L.J., 1989b, Post-weaning resource competition and sex ratios in spider monkeys, *Oikos* 54:315–319.

Dobson, A.P., and Lyles, A.M., 1989, The population dynamics and conservation of primate populations, *Conserv. Biol.* 3:362–380.

Dobson, F.S., and Yu, J., 1993, Rarity in Neotropical forest mammals revisited, *Conserv. Biol.* 7:586–591.

Dunbar, R.I.M, 1987, Demography and reproduction, pages 240–249 in Smuts, B.B., Cheney, D.L., Seyfarth, R.M., Wrangham, R.W., and Struhsaker, T.T., editors, *Primate Societies*, University of Chicago Press, Chicago.

Dunbar, R.I.M., 1988, *Primate Social Systems*, Cornell University Press, Ithaca, New York.

Fedigan, L.M., 1986, Demographic trends in the *Alouatta palliata* and *Cebus capucinus* populations of Santa Rosa National Park, Costa Rica, pages 287–293 in Else, J.G., and Lee, P.C., editors, *Primate Ecology and Conservation*, Volume 2, Cambridge University Press, Cambridge.

Fedigan, L.M., 1990, Vertebrate predation in *Cebus capucinus*: meat eating in a Neotropical monkey, *Folia Primat.* 54:196–205.

Fedigan, L.M., 1993, Sex differences and intersexual relations in adult white-faced capuchins, *Cebus capucinus*, *Int. J. Primatol.* 14:853–877.

Fedigan, L.M., and Rose, L.M., 1995, Interbirth intervals in three sympatric species of Neotropical monkey, *Am. J. Primatol.* (in press).

Fedigan, L.M., Fedigan, L., and Chapman, C.A., 1985, A census *of Alouatta palliata* and *Cebus capucinus* in Santa Rosa National Park, Costa Rica, *Brenesia* 23:309–322.

Fedigan, L.M., Fedigan, L., Glander, K.E., and Chapman, C.A., 1988, Spider monkey home ranges: a comparison of radio telemetry and direct observation, *Am. J. Primatol.* 16:19–29.

Fleming, T.H, 1986, Secular changes in Costa Rican rainfall: correlation with elevation, *J. Trop. Ecol.* 2:87–91.

Fleming, T.H., Williams, C.F., Bonaccorso, F.J., and Herbst, L.H., 1985, Phenology, seed dispersal, and colonization in *Muntingia calabura*, a Neotropical pioneer tree, *Am. J. Bot.* 72:383–391.

Fragaszy, D.M., 1990, Sex and age differences in the organization of behavior in wedge-capped capuchins, *Cebus olivaceus*, *Behav. Ecol.* 1:1–94.

Freese, C., 1976, Censusing *Alouatta palliata, Ateles geoffroyi*, and *Cebus capucinus* in the Costa Rican dry forest, pages 4–9 in Thorington, R.W., and Heltne, P.G., editors, *Neotropical Primates: Field Studies and Conservation*, National Academy of Sciences, Washington, DC.

Freese, C.H., and Oppenheimer, J.R., 1981, The capuchin monkey, genus *Cebus*, pages 331–390 in Coimbra-Filho, A.F., and Mittermeier, R.H., editors, *Ecology and Behavior of Neotropical Primates*, Volume 1, Academia Brasilia, Rio de Janeiro.

Gebhardt. K., 1994, *Seedling development of four tree species in secondary tropical forest in Guanacaste, Costa Rica*, Ph.D. Thesis, Uppsala University.

Hamilton, W.J., III, 1985, Demographic consequences of food and water shortages to desert chacma baboons, *Papio ursinus*, *Int. J. Primatol.* 6:451–462.

Harding, R.S.O., 1976, Ranging patterns in a troop of baboons (*Papio anubis*) in Kenya, *Folia Primat.* 24:143:185.

Hartshorn, G.S., 1978, Tree falls and tropical forest dynamics, pages 617–638 in Tomlinson, P.B., and Zimmerman, M.H., editors, *Tropical Trees as Living Systems*, Academic Press, New York.

Janzen, D.H., 1979, New horizons in the biology of plant defenses, pages 331–350 in Rosenthal, G.A., and Janzen, D. H., editors, *Herbivores. Their Interactions with Secondary Plant Metabolites*, New York: Academic Press.

Janzen, D.H., 1982, Natural history of guacimo fruits (Sterculiaceae: *Guazuma ulmifolia*) with respect to consumption by large mammals, *Am. J. Bot.* 69:1240–1250.

Janzen, D.H., 1983a, *Costa Rican Natural History*, University of Chicago Press, Chicago.

Janzen, D.H., 1983b, No park is an island: increase in interference from outside as park size decreases, *Oikos* 41:402–410.

Janzen, D.H., 1986a, *Guanacaste National Park: Tropical Ecological and Cultural Restoration*, Fundacion de Parques Nacionales, Editorial Universidad Estatal Distancia, San Jose, Costa Rica.

Janzen, D.H., 1986b, The external internal threat, pages 286–303 in Soule, M.E., editor, *Conservation Biology: the Science of Scarcity and Diversity*, Sinauer Associates, Sunderland, Massachusetts.

Janzen, D.H., 1988, Management of habitat fragments in a tropical dry forest: growth, *Ann. MO Bot. Gard.* 75:105–116.

Janzen, D.H., 1991, Daily temperature and rainfall records from the administration area of Santa Rosa National Park, Area de Conservacion Guanacaste, Northwestern Costa Rica, 1979–1991, unpublished manuscript.

Lee, E.T., 1992, *Statistical Methods for Survival Analysis*, Wiley and Sons, New York.

Liebow, E., 1993, The mobilization of resources for the conservation of Neotropical ecosystems: "Debt for nature" exchanges, pages 107–116 in Aramubulo III, P., editor, *Primates of the Americas: Strategies for Conservation and Sustained Use in Biomedical Research*, Battelle Press, Colombus, Ohio.

Medley, K.E., 1993, Primate conservation along the Tana River, Kenya: an examination of the forest habitat, *Conserv. Biol.* 7: 109–121.

Miller, L.E., 1992, *Socioecology of the Wedge-Capped Capuchin Monkey (Cebus olivaceus)*, Ph.D. dissertation, University of California, Davis.

Mitchell, B.J., 1989, *Resources, Group Behavior, and Infant Development in White-Faced Capuchin Monkeys, Cebus capucinus*, Ph.D. dissertation, University of California, Berkeley.

Mittermeier, R.A., and Cheney, D.L., 1987, Conservation of primates and their habitats, pages 477–490 in Smuts, B.B., Cheney, D.L., Seyfarth, R.M., Wrangham, R.W., and Struhsaker, T.T., editors, *Primate Societies*, University of Chicago Press, Chicago.

O'Brien, T., 1991, Female-male social interactions in wedge-capped capuchin monkeys: benefits and costs of group living, *Anim. Behav.* 41:555–567.

O'Brien, T., 1993*a*, Allogrooming behaviour among adult female wedge-capped capuchin monkeys, *Anim. Behav.* 46:499–510.

O'Brien, T., 1993*b*, Asymmetries in grooming interactions between juvenile and female wedge-capped capuchin monkeys, *Anim. Behav.* 46:929–938.

O'Brien, T., and Robinson, J.R., 1993, Stability of social relationships in female wedge-capped capuchin monkeys, pages 197–210 in Pereira, M.E. and Fairbanks, L.A., editors, *Juvenile Primates: Life History, Development, and Behaviour*, Oxford University Press, Oxford.

Oppenheimer, J.R., 1968, *Behavior and ecology of the white-faced monkey, Cebus capucinus, on Barro Colorado Island*, Ph.D. thesis, University of Illinois, Urbana.

Oppenheimer, J.R., 1982, *Cebus capucinus*: home range, population dynamics, and interspecific relationships, pages 253–272 in Leigh, A.S. R., and Windsor, D.M., editors, *The Ecology of a Tropical Forest: Seasonal Rhythms and Long-Term Changes*, Smithsonian Institute Press, Washington DC.

Orians, G.H., and Pearson, N.B., 1979, On the theory of central place foraging, pages 155–177 in Horn, D., Mitchell, R., and Stairs, G., editors, *Analysis of Ecological Systems*, Ohio State University Press, Columbus, Ohio.

Richard, A. F., 1985, *Primates in Nature*, W.H. Freeman, New York.

Robinson, J.G., 1981, Spatial structure in foraging groups of wedge-capped capuchin monkeys, *Cebus nigrivittatus*, *Anim. Behav.* 29:1036–1056.

Robinson, J.G., 1984, Diurnal variation in foraging and diet in the wedge-capped capuchin, *Cebus olivaceus*, *Folia Primat.* 43:216–228.

Robinson, J.G., 1986, Seasonal variation in use of time and space by the wedge-capped capuchin monkey, *Cebus olivaceus*: implications for foraging theory, *Smithson. Contrib. Zool.* 431, 60 pp.

Robinson, J.G., 1988*a*, Demography and group structure in wedge-capped capuchins, *Cebus olivaceus*, *Behaviour* 104:202–232.

Robinson, J.G., 1988*b*, Group size in wedge-capped capuchin monkeys *Cebus olivaceus* and the reproductive success of males and females, *Behav. Ecol. Sociobiol.* 23:187–189.

Robinson, J.G., and Janson, C.H., 1987, Capuchins, squirrel monkeys and atelines: socioecological convergence with Old World primates, pages 69–82 in Smuts, B.B., Cheney, D.L., Seyfarth, R.M., Wrangham, R.W., and Struhsaker, T.T., editors, University of Chicago Press, Chicago.

Robinson, J.G., and O'Brien, T.G., 1991, Adjustment in birth sex ratio in wedge-capped capuchin monkeys, *Am. Nat.* 138:1173–1186.

Rose, L.M., 1994*a*, Sex differences in diet and foraging behaviour in white-faced capuchins, *Cebus capucinus*, *Int. J. Primatol.* 15:63–82.

Rose, L.M. 1994*b*, Benefits and costs of resident males to females in white-faced capuchins, *Cebus capucinus*, *Am. J. Primatol.* 32:235–248.

Rose, L.M., and Fedigan, L.M., 1995, Vigilance in white-faced capuchins, *Cebus capucinus*, *Anim. Behav.* 49: 63–70.

Ross, C., 1988, The intrinsic rate of natural increase and reproductive effort in primates, *J. Zool.* 214:199–219.

de Ruiter, J.R., 1986, The influence of group size on predator scanning and foraging behaviour of wedge-capped capuchin monkeys (*Cebus olivaceus*), *Behaviour* 77:240–258.

Saunders, D.A., Hobbs, R.J., and Margules, C.R., 1991, Biological consequences of ecosystem fragmentation: a review, *Conserv. Biol.* 5:18–32.

Shaffer, M.L., 1981, Minimum population sizes for species conservation, *Bioscience* 31:131–134.

Simberloff, D., 1988, The contribution of population and community biology to conservation science, *Annu. Rev. Ecol. Syst.* 19:473–511.

Wallace, D.R., 1992, *The Quetzal and the Macaw: the Story of Costa Rica's National Parks*, Sierra Club Books, San Francisco.

Webb, L.J., Tracey, J.G., and Williams, W.T., 1972, Regeneration and pattern in the subtropical rainforest, *J. Ecol.* 60:675–696.

16

TOWARD AN EXPERIMENTAL SOCIOECOLOGY OF PRIMATES

Examples from Argentine Brown Capuchin Monkeys (*Cebus apella nigritus*)

Charles H. Janson

Department of Ecology and Evolution
State University of New York
Stony Brook, New York 11790

INTRODUCTION

Like many other fields, the study of primate social ecology has developed through at least three distinct phases (as reviewed in Terborgh and Janson, 1986). First and earliest, detailed *descriptive* studies added to our basic natural history knowledge of primates in general (e.g., Struhsaker, 1969; Kinzey, 1977). Explanatory models tended to be based on single species and many competing hypotheses were advanced based on distinct study systems. For instance, the ecology and social behavior of savannah-dwelling baboons were studied as analogies for the ecology and behavior of savannah-dwelling early hominids (e.g., DeVore and Washburn, 1963), or the thick molar enamel of capuchin monkeys was investigated as a model for the thick enamel of hominid dentitions (Kinzey, 1974). Second, as enough descriptive data were amassed, it became possible to begin *comparative* studies, at first either searching for broad patterns relating ecology and social organization (e.g., Crook and Gartlan, 1966) or testing very general hypotheses linking group size and ranging behavior to diet (Clutton-Brock and Harvey, 1977) . Third, across-species comprisons and detailed descriptive studies focused on *hypothetico-deductive* tests of specific theories, with the goal of distinguishing (and perhaps eliminating) some of the many competing hypotheses to explain social and ecological variation among primates, including the roles of predation and within- versus between-group competition (e.g., van Schaik, 1983; Terborgh, 1983; Janson, 1990).

There is still room for all three levels of analysis in primate socioecology, but we are now reaching the stage where the number of hypotheses used to explain species differences in social structure is rapidly becoming unwieldy. There are at least four distinct adaptive hypotheses regarding mating systems that could be applied to female mate choice in New World primates (Wrangham, 1979; Wrangham, 1980; Janson, 1984; van Schaik and van Noordwijk, 1989), and the differences between them are often subtle. Thus, even well-focused observational studies may prove incapable of distinguishing between alternate hypotheses. This problem is especially great when testing hypotheses about the ecological bases of social differences in primates, because (1) many ecological factors covary in nature and even sophisticated statistics may not be sufficient to disentangle the correlated effects of several variables on social structure, (2) it may be difficult to measure fitness differences among individuals under distinct social circumstances because animals may rarely voluntarily adopt suboptimal behaviors (such as living in excessively large groups: Janson, 1987), or because frequency-dependent selection may equalize the fitnesses of individuals adopting different social behaviors (Maynard Smith, 1982), and (3) many of the most easily-observed aspects of social structure (such as group size) are not simple traits, but emergent properties of many individual tendencies, the exact outcome of which may not respond in a linear way to changes in ecological pressures. The use of short-term field experiments can minimize several of these problems by (1) allowing independent control of a single ecological variable of interest, (2) changing ecological circumstances fast enough that the study animals are not likely to be well adapted to them, and (3) allowing the researcher to vary the magnitude of ecological change across broad ranges to detect non-linear social responses.

The advantages of controlled field experiments have previously been recognized by several primatologists. Starting in the mid-1970's, colleagues of Dr. Peter Marler began a series of vocalization playbacks to study diverse aspects of primate vocal and social behavior (e.g., intergroup spacing: Waser, 1977; semantic communication: Seyfarth et al., 1980; species recognition: Mitani, 1987). This methodology has been refined and adopted by many other researchers of primate communication (e.g., Kinzey and Robinson, 1983; Hauser and Wrangham, 1987). These refinements (many of which are described in Cheney and Seyfarth, 1990) include: systematic monitoring of pre-playback behavior to serve as a control, use of blank-tape playbacks to control for incidental response to machine noise, the use of habituation to investigate perceptual categorization or similarity (as routinely performed with human infants), matching test and control conditions within subjects, and randomization of the order of playback presentation.

The situation in primate ecology is less advanced, with only a small scattering of field experiments published over the past 3 decades, a couple using predator models (Kortland, 1967; van Schaik and van Noordwijk, 1989) and a number using single-site provisioning as a coarse manipulation of food availability (Fa, 1986; Altmann and Muruthi, 1988). While these experiments have yielded some important insights, these have been confounded, at least in the case of provisioning, by the inability to distinguish the effects of an increase in amount of resource from a change in its distribution in space and by a paucity of replicates.

In this chapter, I present evidence that: 1) large-scale manipulations of both food abundance and spatial pattern are feasible, at least in some primate populations; 2) at least some primates make excellent subjects for such experiments and can flexibly change behavior in response to short-term variations in ecological circumstances; 3) a broad variety of questions can be addressed with such experiments; and 4) the control available by such manipulations, even in the wild, allows far more confident conjectures about the ecological causality of the resulting changes in social behavior.

The results described below are derived from a single field experiment set up in 1992 in subtropical forest near Iguazu Falls, Argentina. The purpose of that experiment was to determine experimentally the important ecological factors responsible for high degrees of within-group aggressive (contest) competition and feeding success skewed toward dominant animals. In earlier non-experimental studies on brown capuchins in Peru, it was found that highly skewed feeding success was correlated with high levels of aggression in food patches (Janson, 1985). However, it could not be determined directly (1) if the high rates of aggression directly caused low feeding success of subordinates, or (2) what ecological factors were responsible for the high rates of aggression. Multiple regression analysis (Janson, 1988) suggested that high rates of aggression were linked to a lack of alternate food patches, but the many (10) independent variables and the sometimes strong correlations between them made it hard to draw strong conclusions. The 1992 experiments were set up to test the independent effects of resource 'size' (mean number of fruits per patch), distance to nearest alternate food patch, and spatial variance in resources (difference in resource size between adjacent food sources) on rates of aggression and amount of skew in feeding success among group members. Although not the focus of the experimental design, the platforms also provided a good opportunity to examine several facets of feeding decisions within and between platforms under relatively controlled conditions; in this paper, I report analyses of food dropping behaviors within platforms, and movement decisions between platforms. As the focus of this paper is the experimental method, I shall present preliminary analyses on these various topics rather than a complete analysis on any one of them.

MATERIALS AND METHODS

Study Species and Site

The study species is the brown capuchin monkey (*Cebus apella nigritus*). The study population lives in the Iguazu Falls National Park, a 60,000 ha reserve near the junction of Argentina, Brazil, and Paraguay (54° W, 26° S). Group sizes range from 7 to over 30 individuals, with a ratio of about 3 adult males to 4 adult females, and about as many adults as juveniles (Janson, in prep.). The study group used here was the Macuco group, which consisted of 3 adult males, 5 adult females, 5 juveniles, and 2 infants. They are omnivorous, with a diet showing strong seasonal variations from nearly exclusively frugivorous to a concentration on insects during the austral spring to almost totally folivorous during the austral winter when fruit is scarce (Brown and Zunino, 1990). Baseline data on time budgets and seasonal patterns of food availability were taken during a year-long study in 1991. The Macuco group maintained a typical group spread of 50m in the direction of travel by 30m wide, although on occasion the group would spread out up to 100m. Such group spreads often allowed individuals to feed in more than one food patch at a time (Janson and Brown, unpubl. data), but normally precluded feeding in more than one platform site (see below).

The observations for this study were made between May 15 and August 5, 1992, a time of fruit and insect scarcity in the study site (Janson, in prep.). The monkeys were already used to eating several commercially available exotic fruit species that had grown wild after the existing settlements in the study site were removed some 30 years ago. For this study, we used tangerines as the provisioning food. Densities of 'feral' tangerine trees in the group's home range were about 0.1 per hectare, and most such trees were shaded and not very productive; in any case, once the Macuco group learned to take tangerines from the platforms, they stopped visiting nearly all other available sources of tangerines. Tangerines were purchased from one commercial supplier with orchards outside the National Park; the fruits used were quite uniform in size and quality from the beginning of the study until July 20th, when we could only obtain a somewhat larger variety.

Platforms and Provisioning

Platforms were 1m-square structures of bamboo and wood with a raised wood railing around the edges to prevent fruits from falling off. Nylon cords were attached at the four corners and tied to a single heavier cord ca. 1.2m above the platform. The heavier cord led through a pulley (tied to a tree branch) and then to an attachment site on a nearby tree trunk at chest height. The tree branches used for platform pulleys varied in height from 4 to 15 m above the forest floor. Tangerines were placed on a platform after lowering it with the rope to ground level; the platform was then raised as close to the support branch as the rope and pulley would allow (typically 1.2–1.5m). At the start of the study period, tangerines were placed on a given platform 24 hrs. after the capuchin group last visited the platform (whether or not the capuchins were present), but by the beginning of data collection on June 16, coatimundis had learned to raid the platforms, so we changed to placing tangerines on the platform only when the capuchins had made a clear choice to visit that platform (i.e., they were less than 50 m from the platform and traveling rapidly toward it) *and* more than 24 hrs. had elapsed since their previous visit to that platform. Otherwise, we stayed near the center or back portion of the group spread to avoid providing any anticipatory cues to the group about the location or provisioning status of any platform. Visibility of animals on or near the platforms was very good to excellent at all platforms from at least one vantage point, although for several sites, a separate observer was needed to obtain data for each platform; with rare exceptions, observations at a site were always performed by pairs of observers.

Experimental Design

Platforms were arranged in 15 sites over an area of ca. 1 km^2 (A-O, Fig. 1), each at least 180 meters from the nearest other site, to prevent group members from using more than one site at a time. Platform sites were chosen to be within the area used by the study group in 1991, but no effort was made to place sites near familiar travel routes; indeed, two platform sites (not indicated in Fig. 1) were never discovered by the study group during the experimental period. The capuchin group took about 2 weeks to learn to enter the platforms and feed, and another 2 weeks to learn the locations of nearly all the sites. At each site were either one or two platforms; if two, these were spaced either 2, 10, or 40–50 meters apart (Table 1). The average numbers of fruits per platform was either 10 or 40, but when two platforms were present, the fruits could either be split evenly between the platforms or in a 1:4 ratio (e.g., for a mean of 10 fruits, one platform would have 4 tangerines, the other 16, see Table 1). After three weeks of observation (Period 1, June 15-July 5), the mean and/or variance of fruit numbers between platforms at a site were changed to allow detection of any differences between sites in capuchin behavior independent of treatment (Table 1); after five days, data collection continued under the new regime for an additional three weeks (Period 2, July 10–30). Every combination of distance between nearest platforms, mean number of fruits per platform and variance between adjacent paired platforms was present in at least one site for at least one period; a few treatments were repeated at different sites (Table 1). I could not control the prior feeding experience of the capuchins at each visit to a given site, but the ordering of visits to sites across the hours of the day was fairly uniform, so that temporal patterns (such as hunger or protein demands) should have little influence on the mean patterns of food intake at a given site.

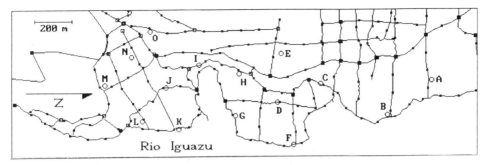

Figure 1. Spatial arrangement of platform sites at study site in 1992. The entire eastern edge of the trail system borders on 70-m tall cliffs bordering the Iguazu river.

Data Collection

We followed the group from first movement in the morning to final resting place at dusk, recording the spatial location of the group at least once every 15 minutes and the location and duration of all fruit-feeding activities, including those at natural fruit trees that partially coincided with feeding visits to platforms. Every day, the capuchins visited five to eight platform sites, for a total of nearly 250 site-visits over the course of the study. At each visit to a platform, we noted the time of first entry of each group member and how many tangerines it removed, ate, and/or dropped from the platform, until all tangerines were gone. Data on all agonistic interactions were noted, including the time of day, which animal initiated the aggressive behavior, which animals (if any) joined the initiator, the re-

Table 1. List of treatments at each platform site (see Fig. 1)

		Number of Fruits	
Site	Distance	Period 1	Period 2
A	2	16:4	64:16
B	10	10:10	16:4
C	0.5	10:0	40:0
D	0.5 (10)	40:0	64:16
E	40	40:40	10:10
F	10	40:40	64:16
G	0.5	40:0	10:0
H	40	16:4	10:10
I	2	16:4	64:16
J	0.5	10:0	40:0
K	40	64:!6	16:4
L	10	10:10	16:4
M	2	40:40	10:10
N	40	10:10	40:40
O	2	10:10	16:4

Distance is meters between the pair of platforms within a site, or 0.5m for solitary platforms (i.e., the only other feeding spot was within the 1-m confines of the lone platform); all sites were > 180 m apart. At site D, a second platform became usable between Period 1 (June 16-July 5) and Period 2 (July 10-July 30). The numbers of fruits are listed from richer to poorer platforms; for solitary platforms, the poorer platform = 0.

cipient(s) of the action, what behaviors were used (threat faces, lunges, chases, vocalizations), the duration of the interaction, and any subsequent behaviors of the recipient (leaving the area, returning to platform, etc.). Any rest, play, or grooming behaviors during platform visits were noted opportunistically.

Social dominance status was based on the outcomes of decided dyadic aggressive interactions and displacements at food sources; individuals were arranged into a matrix of agonistic interactions in such a way as to minimize the number of 'reversals', resulting in a linear hierarchy (cf. Janson, 1985). Behavioral data were combined from the baseline study of the Macuco group in 1991 and the experimental study of 1992. Dominants were the most aggressively successful male and female of the group. Non-tolerated adult males were mature, presumably non-natal, males of high dominance status that were routinely chased away from food sources by the dominant male and occasionally by the dominant female. Subordinate females were the two lowest-ranking adult females, and they were subordinate to all other adults and some juveniles. Juveniles are 1–5 years old; infants are less than one year old.

Statistical Analyses

For most analyses, feeding success of individuals was averaged across all visits to a given platform site to avoid problems of pseudo-replication. Exploratory data analysis was performed graphically and with conventional parametric regression; once likely trends were observed, more accurate analyses were performed to correct for multiple observed values for each level of the independent variable (Sokal and Rohlf, 1981). Log transformations of the dependent variable were used when needed to make the residuals from a regression homoscedastic; log transformations of the independent variable were used to make monotonic non-linear relationships more nearly linear. Analysis of covariance was used to test for equality of slopes and intercepts in regressions done on several social subsets within the study group.

To predict movement decisions between successive platform sites, logistic regression was performed using PROC CATMOD in SAS 6.0 with maximum-likelihood parameter estimation (see below for details of variable coding).

RESULTS

Aggressive Competition: Patterns of Feeding Success at Platforms

Several studies have hypothesized that dominants should feed better than subordinates only when resources are "monopolizable" (e.g., Harcourt, 1989; van Schaik, 1989), that is, when a dominant individual can defend access to the resource. Although intuitively reasonable, this conclusion leaves out other factors which theory asserts should be important in predicting the occurrence of aggressive defense of resources (such as benefit of access to the resource: Brown, 1964; Maynard Smith, 1982) and leaves open the question of what ecological parameters make a resource monopolizable. Numerous studies on primates suggest that defendability increases with resource patchiness (e.g., van Schaik and van Noordwijk, 1988) or spatial isolation (e.g., Symington, 1988), but only one field study (Whitten, 1983) has explicitly defined a measure of patchiness (a high variance/mean ratio of counts of food trees of a given species in point-quarter censuses) and found that it correlated positively with skewed feeding success. However, when I tried to use a similar use a similar measure to predict rates of aggression in Peruvian brown capuchin monkeys, I found that it correlated negatively (Janson, 1988). Because of the potential problem of disentangling the many intertwined ecological variables that might influence social food competition, I felt an experimental approach might produce clearer answers.

Two categories of individuals may find their food intake restricted in wild capuchin groups: non-tolerated animals (usually high-ranking adult males other than the dominant male) and adult or subadult animals of low dominance rank (Janson, 1985). Both in Argentina and Peru, juveniles and especially infants are generally tolerated by the dominant male and show little reduction in feeding success (Janson, 1985; unpubl. data). I calculated the mean food intake of two top dominants (the dominant male and the dominant female, hereafter referred to as DOM) and that of either: 1) the group's two (non-tolerated) subdominant adult males (NTOL), or 2) the group's two lowest-ranking (subordinate) adult females (SUBORD). As a measure of the ability of dominants to monopolize feeding access at a platform, I used the ratio of twice DOM's food intake, divided by the sum of the intakes of DOM plus NTOL or DOM plus SUBORD. This ratio varies from 2.0, if the subordinates eat nothing, through 1.0 if DOM's intake is equal to that of the subordinates, to 0.0, if the dominants eat nothing at a given platform site.

The variables manipulated in this experiment were mean amount of food per platform within a site, the difference in food amounts between platforms within a site, and the distance between platforms within a site. The effects of other feeding sites was not considered, as the individuals in the group rarely split up to use more than one site at a time. Within a site, food should be less defendable or monopolizable the farther apart the platforms are and the more evenly split the food is between platforms. Thus, I expected that the feeding skew toward dominants would increase the closer the platforms were to each other and the greater the relative difference in numbers of tangerines placed on adjacent platforms. From the perspective of game theory, aggression should be more prevalent (and thus feeding skew greater) as the benefit of exclusive access to a resource increases. The feeding benefit of exclusive access is likely to increase initially as food reward increases, but then should saturate when more food is present than an individual can ingest. In addition, a dominant's inclusive fitness may actually decrease by excluding subordinate females and juveniles when food is superabundant. Because a dominant rarely consumed more than 7 tangerines per visit to a platform, I expected that aggression (and feeding skew toward dominants) would increase with numbers of tangerines from 4 up to 16 per platform, but then saturate or even decrease at platforms with 40–64 tangerines.

The feeding success of DOMs relative to NTOLs depended primarily on the distance between the platform with the dominants on it and the nearest available alternative (Fig. 2). This result makes sense as the likelihood that DOMs would notice and/or chase away NTOLs was a function of distance between them (Janson, 1990; unpubl. data). Most adult group members used the platforms only to obtain food, but sat to feed on nearby tree branches; only the DOMs and the infants regularly fed while seated on the platform itself. The data (Fig. 2) suggest that DOMs have a depressing effect on food intake by NTOLs up to a distance of about 10 m. Neither the total amount of food at a site nor the difference in availability between platforms at a site had significant effects on the relative ingestion by DOMs versus NTOLs after controlling for distance between platforms. Relative ingestion by dominants was not affected by time of day, which should affect hunger levels, either by itself or in conjunction with the previous predictors. Not surprisingly, the absolute quantities ingested increased with total food availability for all individuals (Janson, unpubl. data).

In contrast to the above, the feeding success of DOMs relative to SUBORDs was not consistently affected by distance between platforms; instead, the relative ingestion of

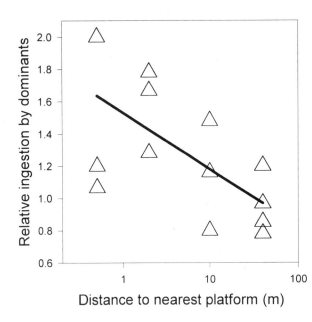

Figure 2. Ingestion of tangerines by top dominants relative to non-tolerated adult males as a function of distance between platforms within a site for Period 1; when only one platform was present, the distance was set at 0.5m to reflect the unavailability of alternate feeding platforms within the group spread. The slope of the regression is significantly different from zero (multiple Y per X value, $F[1,12] = 8.9$, $P = 0.011$, $r^2 = 0.43$).

DOMs was best predicted by the amount of food in the platform with the smaller quantity available (Fig. 3). This pattern appears to reflect the inability of SUBORDs to gain access to either platform, regardless of distance between them, when more dominant (even non-tolerated) animals are present. Because the dominant male tends to stay in the richer platform as long as juveniles and infants continue to feed there (cf. Janson, 1985), the SUBORDs generally feed at the poorer platform, along with or after the NTOLs have fed. Time of day had no significant effect on intake by DOMs relative to SUBORDs.

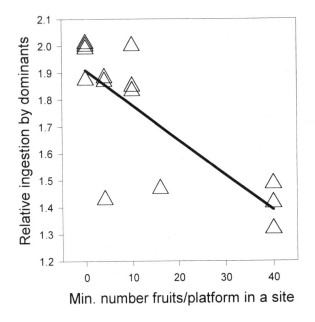

Figure 3. Ingestion of tangerines by top dominants relative to subordinate adult females in Period 1 as a function of the amount of food in the platform with the lesser amount of food at a given site; if only one platform was available at a site, the lesser amount was set to 0. The slope of the regression is significantly different from 0 (multiple Y per X value, unpooled mean squares, $F[1,3] = 8.99$, one-tailed $P = 0.028$, $r^2 = 0.61$). Deviations from the regression were not significant ($F[3,8] = 2.5$, $P = 0.13$).

Optimal Foraging: Fruit Rejection Frequency at Platforms

It is a common observation that primates drop considerable quantities of apparently edible fruit beneath their feeding trees, so much so that they are often considered 'wasteful' feeders (Howe, 1980). Such apparently non-adaptive behavior could have at least three explanations. First, fruit may be so abundant that they need not worry about being efficient foragers (Coelho et al., 1976). Second, frequent aggression between individuals within food patches (e.g., Janson, 1985) may incidentally knock down ripe fruits. Third, primates are choosy feeders that will reject even slightly imperfect (underripe, overripe, unusually small, or insect-damaged) fruits if alternatives are readily available; according to foraging theory, individuals should be more likely to reject fruits of a given quality as their future expectation of obtaining better fruits increases.

Each of these hypotheses predicts a distinct relationship between the frequency of dropping fruits and the amount of fruit on a platform in these experiments. If primates are simply clumsy or inefficient foragers, then the fraction of fruits dropped should not be related to the amount of fruit available in a given platform, and infants and juveniles should drop more than adults. If aggressive interactions are the cause of fruit loss from platforms, then rates of loss should be highest for individuals that received the highest rates of aggression (NTOLs and SUBORDs) and at platforms with the highest rates of aggressive interactions per unit time (because fruit intake is proportional to time as well). The platforms with the highest rates of aggressive interactions were those with intermediate amounts (10–16) of fruits (Janson, unpubl. data) — where only 4 fruits were available, feeding was so short that few interactions occurred, while at platforms with 40–64 fruits, dominants were able to satiate, thus allowing subordinates to feed without much interference (see Fig. 3). Finally, if primates are choosy feeders, then the fraction of fruits dropped should increase with the number of fruits on a platform; also, the fraction of fruits dropped should be highest in the most dominant or most tolerated individuals that have best access to food, and lowest in the most subordinate or least tolerated individuals.

The fraction of fruits dropped by an individual was calculated as the number of fruits it picked up and dropped uneaten to the ground, divided by the total number of fruits it dropped plus the number it ate (entirely or partially). Some fruits undoubtedly were dropped back into the platform, some may have been rejected without handling, and still others may have been dropped deliberately after some eating; however, I did not feel competent to much such judgements in the field. The pattern of dropping across platforms and social subsets (Fig. 4) best matches the predictions made by foraging theory for choosy feeders: overall, the fraction of fruits dropped increased more than twice between the platforms with the lowest and greatest amounts of fruits, and the social subsets with the greatest access to the platforms (dominants and infants) had the highest dropping rates, whereas the social subset with the least access to platforms (SUBORDs) had the lowest dropping rate. The increase of dropping rate with fruit amount clearly contradicts the hypothesis that dropping is a random act of clumsiness or inefficiency, as does the fact that dominant adults had dropping rates as high as did the least coordinated infants. The aggression hypothesis is not supported by the distribution of dropping among social subsets — NTOLs and SUBORDs receive the most aggression and should therefore have dropped fruits more often, but did not. However, there is a suggestion (Fig. 4) that the platforms with intermediate amounts of fruit (particularly those with 10 fruits) did have higher rates of dropping than expected if there were a monotonic increase of dropping rate with amount of fruit in the platform. Thus, a small fraction of the pattern may be attributable to incidental consequences of aggressive competition.

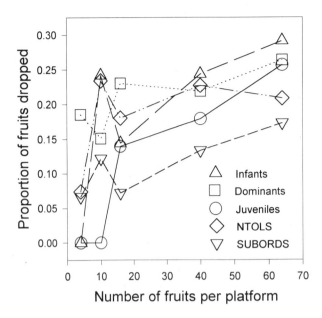

Figure 4. The proportion of fruits dropped by different social and age classes of capuchin monkeys while feeding on tangerines in provisioning platforms. The proportion dropped increases with the number of fruits per platform and at the same time varies significantly across social classes (Two-Way ANOVA, effect of fruit number: $F[4,16] = 9.38$, $P = 0.0004$; effect of social class: $F[4,16] = 3.19$, $P=0.042$).

Cognitive Maps: Patterns of Sequential Spatial Use of Platform Sites

Some excellent observational work strongly implies that movements of fruit-eating monkeys groups within their home ranges is determined by the distribution of fruit trees (e.g., Terborgh, 1983; Robinson, 1988) and that wild monkeys know the locations of their food trees and prefer to move from one tree to its nearest neighbor (e.g., Garber, 1989). However, these conclusions need to be tempered by the fact that the investigators did not know the locations and amounts of fruit in all resources available to the monkeys at a given time. Achieving this level of knowledge is almost certainly impossible in most cases, yet to understand what rules a group uses to choose food patches, it is important to know for every choice what resources the monkeys could have used but did not. In the Argentine winter, when this study was done, the capuchins had virtually no other fruit available, and the platforms became practically their only fruit trees. Thus, I could analyze the extent to which their ranging behavior depended on fruit at all, and on the specific spatial distribution of fruit sources in particular; some preliminary results are given here.

From a given starting site, the next site visited by the group was clearly unevenly distributed across the potential sites available (Table 2). Using only sites with at least 14 total visits (to yield an expected value of 1 per cell), the G-statistic suggests a strong rejection of the null hypothesis that all other sites are visited with equal frequency from a given site (G = 659.9, df = 154, P < 0.0001). Given the sparseness of observations across the matrix as a whole and the strong effect of distance on visitation probability (see below), it is not surprising that there is no detectable difference between the transition matrixes for Period 1 and Period 2.

However, the choices are hardly constant, with as many as seven or eight other sites visited from a given site (e.g., sites C,F,G,I,J). To investigate what factors predict the sequence of platform sites actually used, sequences of site visits were analyzed by logistic regression. Each move by the group from one platform to another was considered one choice among the all the platform sites. The chosen site was given an outcome of 1, the

Table 2. Transition matrix of movements between platform sites

		A	B	C	D	E	F	G	H	I	J	K	L	M	N	O
								TO								
	A		6													
	B	1		6												
	C	1	1		4	8	4	1								
	D		2			2		4								
	E	2	4	1				1	8			1	1			
	F	2	6	1	1			6	1			1				
	G		1	1	1	9			4	2		3				
FROM	H					3	2	3		13		1				1
	I					3		5	6		5	6		1	6	3
	J					1		2		4		4	2	6	3	
	K								1		5		17			1
	L							1			1	2		16	2	
	M									2		1	2		22	2
	N				1			1		15	8			9		1
	O							1			1	1	1		4	

There were no significant differences in transition probabilities between Periods 1 and 2, so data are combined here. The starting platform is listed in the 'FROM' rows, and the frequency of observed movements to each possible destination sites is listed in the 'TO' columns; for instance, from site B, the group moved once to site A and 6 times to site C.

'rejected' sites an outcome of 0. Logistic regression was used to predict the probability of 1 vs. 0 outcome for a potential choice site given three independent predictor variables: distance of each choice site to the starting site, the total number of tangerines available in the platforms at the potential choice site, and whether or not the monkeys should be rein-forced upon arriving at the next site (yes = 1 = at least a 24-hour delay since last visit; no = 0 = delay < 24 hours). Combining data across both experimental Periods (with appropri-ate changes in the independent variables associated with each choice platform), the logis-tic regression showed significant effects of all three variables (Table 3). The group visited closer platforms in preference to more distant ones, more rewarding sites over less well-provisioned ones, and rarely revisited platforms in less than 24 hours, the reinforcement delay established by us in the experiment. Figure 5 shows the effects of both total food re-ward at a choice site and distance to that site on the movement choices of the Macuco

Table 3. Results of logistic regression of probability of movement from a given starting site to any of 14 alternate sites

Effect	Slope	se(slope)	Probability
Intercept	-3.26	0.68	<0.0001
Distance	-0.0065	0.0014	<0.0001
Number of Fruits	0.014	0.0071	<0.05
Availability	3.76	0.73	<0.0001
Goodness of Fit (Deviance) = 75.99 on 81 df, P = 0.64			

Results are combined from two starting sites (G and K, chosen arbitrarily). Move-ment decisions to choice sites are predicted as a function of distance between start-ing and choice sites, the number of tangerines in the various choice sites, and expected reinforcement at the choice sites (as determined by the experimental pro-tocol: reinforcement only if delay since previous visit to that site was > 24 hrs). The dependent variable is the $\log_{10}$ of $p/(1-p)$, where p is the observed probability of visiting a choice platform.

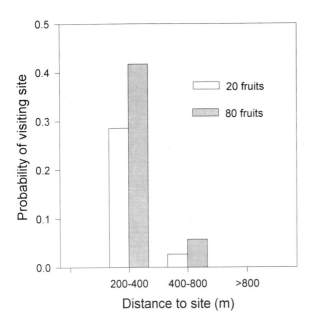

Figure 5. Probability of moving next to a potential feeding site from a given starting site (either site J or K) as a function of distance to the potential site and the number of tangerines provided there. Data only for potential sites last visited at least 24 hrs previously; more recently used sites were almost never visited (see Table 3).

group. The analysis suggests that the group should be equally likely to choose two sites, one of which is farther away by 150 m, but provides 70 tangerines more than the other site. Analysis of the nutritional reward of the fruits themselves suggests that the extra reward is far in excess (ca. 10–15-fold) of that needed to pay for the added travel effort (Janson, unpubl. data), thus suggesting that they prefer closer platforms even when they are less rewarding per unit distance travelled.

DISCUSSION

The Effect of Food Distribution and Abundance on Individual Feeding Success

The results of the experiments reported here are consistent with the results reported elsewhere for capuchin monkeys (Janson, 1985; 1988). However, in previous observational studies using correlational tests, I could not rule out the possibility that the observed correlations were due to confounding variables, such as different nutritional chemistry among food types or changing availability of alternate food patches across time and space. In the experimental design used here, the effects of confounding variables is probably small because: 1) the type of food reward (tangerines) was the same across sites and across time; 2) the possible confounding effects of uncontrolled variables at particular sites were reduced by replicating treatments across sites when possible and changing the treatment at a given site over time; 3) the seasonal near-absence of fruit meant that essentially no alternate food sources were used when the group visited a given platform.

The relative feeding success of non-dominant animals appears to depend on distinct ecological factors depending on their dominance rank and degree of tolerance by dominants (see also Janson, 1985). Subdominant high ranking, but non-tolerated, adult males (NTOLs) could feed as successfully as the top dominants (DOMs) as long as there existed a minimum separation of 10m between the feeding sites. The total amount of food, its spatial variance, and time of day had no significant effects on relative food intake by NTOLs after the effects of distance were removed. The lack of effect of difference in food availability between nearby platforms on the relative feeding success of NTOLs may be an artifact of the relatively small ratio of availabilities used in the experiment (4:1) — even the smallest number of fruits on a platform (4) could provide the two subdominant males with only slightly less than the average number of fruits ingested by DOMs at the nearby platform with 16 fruits (2 versus 2.45 fruits each). More extreme ratios of food availability would almost certainly have biased feeding success more in favor of DOMs.

Low-ranking adult females (SUBORDs) fed well only if there was sufficient food in the platforms so that some was left over after more dominant adults had left. Distance between platforms did not affect relative food intake for SUBORDs, probably because they were equally subordinate to both DOMs in the richer platform and to NTOLs in the poorer platform. The fact that SUBORDs never achieved average food intakes equal to DOMs is likely due to the restriction of the experiment that a maximum of two platforms were used per feeding site; if more platforms with less food each had been available per feeding site, it is likely that SUBORDs would have reached food intakes similar to that of DOMs.

The experimental results reported here help clarify some confusion about what food resources are likely to be monopolized by dominants in nature. First, monopolization is not a single concept, but rather may be determined by somewhat distinct ecological factors depending on the social relations between dominants and particular subordinates. Thus, distance between rich food patches may be the major variable when subordinates are relatively high-ranking but are chased preferentially by dominants, whereas for low-ranking subordinates, feeding success depends more on the evenness of food availability. Second, the natural spatial scale within which to measure food patchiness is that of the group spread. For instance, feeding success of NTOLs relative to DOMs increased with increasing distance between nearest platforms, but only up to 40–50m apart, the approximate spread of the group (Fig. 2). When the nearest platform was > 180 m away (coded as 0.5m in Fig. 2), NTOLs fed poorly because they had no alternate food sources within the group spread. Because the ability of DOMs to monopolize access to food will likely depend on the group's spread, the strength of within-group aggressive competition should depend on group cohesion, which may depend on predation risk, among other ecological variables. In any case, measuring patchiness of food sources at spatial scales much different from that of the group spread may produce results of little relevance to explaining social behavior.

Although the results of this analysis clarify which ecological variables are likely to be important in determining the relative feeding success of different social subsets within a brown capuchin group, the results presented so far do not resolve if aggression is the immediate causal factor affecting relative food intake or if aggression and biased feeding success are both linked to some other variable, such as food patch isolation. This question can be resolved by analysing the variation in rates of agonism and skew in feeding success among different visits to the same platform site (Janson, in prep.).

Tests of Foraging Theory

The evidence presented above (Fig. 4) suggests that at least one aspect of primate feeding behavior, frequency of dropping fruits, is best explained as deliberate fruit rejection behavior, rather than random or socially-induced accidents. If correct, this result has some interesting implications for seed dispersal by primates. In particular, primates should waste little in plants that produce few ripe fruits at a time, either because of small total crop size or slow ripening rate. Because large-seeded plants may have to rely on primates as dispersers (Foster and Janson, 1985), they might also benefit from having a protracted ripening phenology. However, a small effective crop size has its disadvantages as well — as seen in the movement patterns between platforms (Fig. 5), monkeys will skip over resources that offer too small a reward relative to the alternatives. One possible solution to this dilemma might be for plants to make relatively few large fruits with multiple seeds in them, so that it is worth at least one individual's time to visit the tree, yet the effective number of alternate fruits is small as well. This hypothesis might help explain why large-seeded monkey fruits typically contain multiple seeds whereas bird-dispersed fruits of comparable seed size usually have only one seed, at least in Peru (Janson, unpubl data).

Other topics in primate foraging that could be addressed with these platforms are: 1) when is food intake *rate* at a platform more important than total food *amount* in affecting visitation preference; 2) can group-level discovery of 'new' resources offset the costs of food competition within resources either by increasing the mean or reducing the variance of food intake; 3) how do capuchins trade off mean vs. variance in reward amount in making spatial movement choices; 4) do primates obey the marginal value theorem when resources are set up specifically to have declining intake rates over time within a platform (cf. Grether et al., 1992)?

Spatial Movement Rules

This experiment, although not designed specifically to test spatial movement patterns, provides suggestive evidence that capuchin monkeys know where their resources are, and when and how much food to expect when they get there. Although some of the actual foraging routes used by the monkeys may have been dictated by physical barriers (such as cliffs, see Fig. 1), the observed transitions among platforms (Table 2) are not likely to emerge simply from habitual long-term ranging patterns. First, observational study during 16 months prior to the experiments showed that ranging patterns were highly variable seasonally and could change markedly even within a season depending on the major food sources available (Janson and Brown, unpubl. data). Second, the choices of which site to visit from a given starting site were far from constant (Table 2) and apparently depended on the recent history of use of the alternate sites (Table 3). The fact that the transition matrix did not change significantly between the two experimental periods with different food reward (but not spatial) structures across sites might be viewed as evidence for long-term habitual foraging routes. However, the strong domination of distance over reward amount in predicting movement decisions (Table 3) means that the expected transition matrices for the two periods were very similar (Janson, unpubl. data).

Some unresolved questions remain and await further experimental manipulations specifically to test spatial movement rules and knowledge. For instance, the significant effect of distance would occur even if the group moved at random, stopping to feed at the first platform at which they were reinforced. To test whether non-random movement is occurring, the group's movement decisions must be compared to those predicted by random search (cf. Garber and Hannon, 1992). Even if the group's movements are not random, it is possible that it uses some spatially systematic search behavior but without reference to individual resources. A crucial variable to be determined in future studies is the distance over which a group can detect previously unknown food platforms (or natural food patches); without a good estimate of this value, it is difficult to reject the hypothesis that observed movements between resources are due simply to sequential (re)discovery of food patches with no prior knowledge of their location. It is more difficult to explain away the significant effects of reward amount and time since the previous visit on the probability of visiting a given platform, although the effect of timing might happen to coincide with the natural rhythm of movement of the group through its home range. Further experiments are needed, varying the spatial density and time interval between food rewards to help clarify the extent to which the observed capuchin movements in this experiment reflect actual spatial and temporal knowledge of their food sources.

The Experimental Approach to Primate Social Ecology

I hope that the analyses given here, although incomplete, will inspire others to test additional such questions in their own study systems. Nevertheless, there are some pitfalls that potential experimentalists should be aware of. It is easy to design a bad experiment, especially when there is insufficient background information on the study species. The experimental design used here in the study of skewed feeding success made use of specific knowledge gained in a previous year-long descriptive study of the same capuchin population (e.g., group spreads, satiation amounts, typical resource dispersion) and of general knowledge of capuchin social interactions gained from studies in other areas (e.g., Janson, 1985). Without such data, it would have been easy to choose a range of parameter values that might have shown no effect even if the variable tested did have effects across a broader scale (e.g., in Fig. 3, there is relatively little effect of food amount on relative food intake within the ranges of 0–10 or 16–40 tangerines). It was crucial to separate the sites by twice the normal maximal group dispersion to ensure that the group did not use two or more platform sites at the same time. Additional problems that should be considered in the design of large-scale food manipulations are the potential impact of increased seed input of exotic species (if used for provisioning, as in this study), and the potential disruption of 'normal' patterns of social interaction and even increased disease transmission that may occur when provisioning is concentrated at a single site (see Goodall, 1986). Thus, I strongly urge any primate socio-ecologist to gain an observational understanding of the ecology and social behavior of their study species before trying experiments. Nevertheless, for some questions, even a small-scale experiment may produce more easily interpreted and reliable results than a large intensive observational study. I hope to see the day when researchers on other organisms begin routinely to cite primatological studies as exemplars of rigorous field studies, including experimental testing of established and novel hypotheses.

SUMMARY

I argue that the time is ripe for more extensive use of field experiments with wild primates to obtain more confident answers to long-standing questions in primate ecology and social behavior. I illustrate this proposition with three results of a large-scale food manipulation experiment on wild brown capuchin monkeys in subtropical Argentina, taking advantage of the fact that fruit production during the local winter is extremely low. First, I used artificial feeding platforms to vary systematically the abundance and spatial distribution of fruit within local feeding sites in an attempt to manipulate levels of aggression and feeding bias among members of a single group. Results show that different aspects of food abundance and distribution affect the food intake of different social 'cliques' within a group. Second, the large-scale spatial distribution of the feeding sites provided an opportunity to observe a group's spatial movements when the observer knew at least as much as the monkeys did about their food resources. The likelihood that a given site is the next one visited is affected significantly by its distance from the present site, its amonunt of food, and the interval since the previous visit to that site. Third, an analysis of the fraction of the fruits dropped from feeding platforms by different group members shows significant effects of social 'clique' and amount of food, consistent with the hypothesis that much of the apparent sloppiness of primate feeding on fruit is in fact extreme choosiness. Compared to observations on unmanipulated primates, the control available by the use of field experiments provides more confidence in confirming or rejuecting the effects of the variables examined on the behavior of interest. Compared to captive studies, field experiments are less well controlled, but offer the oppourtunity to test the importance of certain variables when embedded in the context of natural variation in other factors.

ACKNOWLEDGMENTS

This work would not have been possible without the stalwart assistance of numerous colleagues: Daniela Rode, Patricia Escobar-Páramo, Sandra Chediak, and especially Mario DiBitteti. I am grateful to Alejandro Brown for his encouragement to study the capuchins which his initial work made possible, and to the Argentine Administration of National Parks for permission to study in the Iguazu National Park and to use the facilities at the Centro de Investigaciones Ecológicas Subtropicales. Financial support for this research was provided by NSF grant BNS9009023. This is contribution #968 from the Graduate Program in Ecology and Evolution, State University of New York at Stony Brook.

REFERENCES

Altmann, J., and Muruthi, P., 1988, Differences in daily life between semiprovisioned and wild-feeding baboons, *Amer. J. Primatol.* 15:213–221.

Brown, A. D., and Zunino, G. E., 1990, Dietary variability in *Cebus apella* in extreme habitats: evidence for adaptability, *Folia. primatol.* 54: 187–195.

Brown, J. L., 1964, The evolution of diversity in avian territorial systems, *Wilson Bull.* 76: 160–169.

Cheney, D. L., and Seyfarth, R. M., 1990, *How Monkeys See the World*, Univ. of Chicago Press, Chicago IL. 377 pp.

Clutton-Brock, T. H., and Harvey, P. H., 1977, Species differences in feeding and ranging behavior in primates, *in Primate ecology: Studies of feeding and ranging behaviour in lemurs, monkeys, and apes* (Clutton-Brock, T. H., ed), Academic Press, London, pp. 557–584.

Coelho, A. M., Bramblett, C. A., Quick, L. B., and Bramblett, S. S., 1976, Resource availability and population density in primates: a sociobioenergetic analysis of the energy budgets of Guatemalan howler and spider monkeys, *Primates* 17: 63–80.

Crook, J. H., and Gartlan, J .S., 1966, Evolution of primate societies, *Nature* 210: 1200–1203.

DeVore, I., and Washburn, S. L., 1963, Baboon ecology and human evolution, *in African Ecology and Human Evolution* (Howell, F. C. and Bourliere, F., ed), Aldine, Chicago, pp. 335–367.

Fa, F. E., 1986, Use of time and resources by provisioned troops of monkeys: social behaviour, time, and energy in the Barbary Macaque (*Macaca sylvanus* L.) at Gibralter, *Contributions to Primatology* 23: 1–377.

Foster, S. A., and Janson, C. H., 1985, The relationship between seed size, gap dependence, and successional status of tropical rainforest woody species, *Ecology* 66: 773–780.

Garber, P. A., 1989, Role of spatial memory in primate foraging patterns: *Saguinus mystax* and *Saguinus fuscicollis, Amer. J. Primatol.* 19: 203–16.

Garber, P. A., and Hannon, B., 1993, Modeling monkeys: a comparison of computer-generated and naturally occurring foraging patterns in two species of neotropical primates, *Int. J. Primatol.* 14: 827–852.

Goodall, J., 1986, *The Chimpanzees of Gombe: Patterns of behavior*, Harvard Univ. Press, Cambridge MA. xii + 673 pp.

Grether, G. F., Palombit, R. A., and Rodman, P. S., 1992, Gibbon foraging decisions and the marginal value model, *International J. Primatology* 13: 1–17.

Harcourt, A. H., 1989, Environment, competition and reproductive performance of female monkeys, *Trends in Ecol. and Evol.* 4: 101–105.

Hauser, M. D. and Wrangham, R. W., 1987, Manipulation of food calls in captive chimpanzees, *Folia primatol.* 48: 207–210.

Howe, H. F., 1980, Monkey dispersal and waste of a neotropical fruit, *Ecology* 61: 944–959.

Janson, C. H., 1984, Female choice and mating system of the brown capuchin monkey *Cebus apella* (Primates: Cebidae), *Z. Tierpsych.* 65: 177–200.

Janson, C. H., 1985, Aggressive competition and individual food intake in wild brown capuchin monkeys, *Behav. Ecol. Sociobiol.* 18: 125–138.

Janson, C. H., 1987, Ecological correlates of aggression in brown capuchin monkeys, *International Journal of Primatology* 8: 431.

Janson, C. H., 1988, Food competition in brown capuchin monkeys (*Cebus apella*): Quantitative effects of group size and tree productivity. *Behaviour* 105:53–76.

Janson, C. H., 1990, Ecological consequences of individual spatial choice in foraging brown capuchin monkeys (*Cebus apella*), *Animal Behaviour* 38, 922–934.

Kinzey, W. G., 1974, Ceboid models for the evolution of human dentition, *J. Human Evol.* 3: 193–203.

Kinzey, W. G., 1977, Diet and feeding behaviour of *Callicebus torquatus*, in *Primate Behaviour: studies of feeding and ranging in lemurs, monkeys, and apes* (Clutton-Brock, T.H., ed.), Academic Press, London, pp. 127–151.

Kinzey, W. G. and Robinson, J. G., 1983, Intergroup loud calls, range size and spacing in *Callicebus torquatus*, *Amer. J. Physical Anthropol.* 60: 539–44.

Kortland, A., 1967, Experimentation with chimpanzees in the wild, in *Neue Ergebnisse der Primatologie — Progress in Primatology* (Starck, D., Schneider, R., and Kuhn, H. J., eds.), Fischer, Stuttgart, pp. 208–224.

Maynard Smith, J., 1982, *Evolution and the theory of games*, Cambridge Univ. Press, Cambridge. viii + 224 pp.

Mitani, J., 1987, Species discrimination of male song in gibbons, *Amer. J. Primatol.* 13: 413–23.

Robinson, J. G., 1988, Seasonal variation in the use of time and space by the wedge-capped capuchin monkey *Cebus olivaceus*: Implications for foraging theory, *Smithson. Contrib. Zool.* 431: 1–60.

Seyfarth, R. M., Cheney, D. L., and Marler, P., 1980, Vervet monkey alarm calls: semantic communication in a free-ranging primat, *Anim. Behav.* 28: 1070–94.

Sokal, R. R., and Rohlf, F. J., 1981, *Biometry*, 2nd Edition, W. H. Freeman, San Francisco.

Struhsaker, T. T., 1969, Correlates of ecology and social organization among African cercopithecines, *Folia primatol.* 11: 80–118.

Symington, M. M., 1988, Food competition and foraging party size in the black spider monkey (*Ateles paniscus chamek*), *Behaviour* 105: 117–134.

Terborgh, J. W., 1983, *Five New World primates: A study in comparative ecology*, Princeton Univ. Press, Princeton NJ, xiv + 260 pp.

Terborgh, J. W. and Janson, C. H., 1986, Socioecology of primates, *Annual Review of Ecology and Systematics* 17: 111–135.

van Schaik, C. P., 1983, Why are diurnal primates living in groups? *Behaviour* 87: 120–144.

van Schaik, C. P., 1989, The ecology of social relationships amongst female primates, in *Comparative Socioecology. The Behavioural Ecology of Humans and Other Mammals* (Standen, V. and Foley, R. A., eds), Blackwell, Oxford, pp. 195–218.

van Schaik, C. P., and van Noordwijk, M. A., 1988, Scramble and contest in feeding competition among female long-tailed macaques (*Macaca fascicularis*), *Behaviour* 105: 77–98.

van Schaik, C. P., and van Noordwijk, M. A., 1989, The special role of male *Cebus* monkeys in predation avoidance and its effect on group composition, *Behav. Ecol. Sociobiol.* 24: 265–276.

Waser, P. M., 1977, Feeding, ranging, and group size in the mangabey *Cercocebus albigena*, in: *Primate ecology: Studies of feeding and ranging behavior in lemurs, monkeys, and apes* (Clutton-Brock, T. H., ed), Academic Press, London, pp. 183–222.

Whitten, P. L., 1983, Diet and dominance among female vervet monkeys (*Cercopithecus aethiops*), *Amer. J. Primatol.* 5: 139–159.

Wrangham, R. W., 1979, On the evolution of ape social systems, *Soc. Sci. Inform.* 18: 335–68.

Wrangham, R. W., 1980, An ecological model of female-bonded primate groups, *Behaviour* 75: 262–300.

SECTION IV

New Perspectives on the Pitheciines

NEW PERSPECTIVES ON THE PITHECIINES

Alfred L. Rosenberger,[1] Marilyn A. Norconk,[2] and Paul A. Garber[3]

[1]Department of Zoological Research
National Zoological Park
Washington, D.C. 20008
[2]Department of Anthropology
Kent State University
Kent, Ohio 44242
[3]Department of Anthropology
University of Illinois
Urbana, Illinois 61801

If the callitrichines were the first major group of platyrrhines to benefit scientifically from the explosion of interest in platyrrhine biology in the last two decades, the sakis and uakaris are the surprise discovery. They are the evolutionary secret of the New World monkey radiation, hidden until now by the lack of a sound framework for platyrrhine systematics, the absence of any glimmerings of a fossil record and sheer ignorance of their behavior and ecology. Much the same situation existed for callitrichines. For nearly a hundred years, scientists have debated one way or the other - Are the callithrichines primitive or are they derived? No such uncertainties were ever associated with "pitheciines". Classifications dating to J.E. Gray and St. George Mivart in the middle 1800s show that taxonomists even then treated the three modern genera, *Pithecia*, *Chiropotes*, and *Cacajao* as a divergent, natural group. In modern terms, this implies they are monophyletic, related more closely to one another than any are to living non-pitheciine platyrrhines. Until recently, this legacy was the upshot of "pitheciine" biology: sakis and uakaris are behaviorally enigmatic and structurally bizarre, but they are an evolutionary cohesive group.

Although none would contest the notion that characteristic features of sakis and uakaris are derived, this observation addresses only one issue - monophyly. United by specialized craniodental anatomy that is quite divergent relative to other platyrrhines, we do not view these animals as marginal outliers, but as the survivors of a once diverse radiation whose origins can be traced through extant forms such as *Callicebus* and *Aotus* (Table 1). Obviously, our view hinges in part on a question of definition: What is a pitheciine? Here we break with tradition by including five living genera, not three - *Pithecia*, *Chiropotes*, *Cacajao* and *Callicebus* and *Aotus* (Table 1). This interpretation of the subfamily developed over the past two decades is based on studies of morphology (Kinzey, 1992; Rosenberger 1992), molecular genetics (Schneider, 1996; Schneider and Rosenberger, this

Table 1. Genus level classification of the Pitheciines

Family Atelidae
Subfamily Pitheciinae
Tribe Pitheciini
Subtribe Pitheciina
Pithecia - Sakis
Chiropotes - Bearded sakis
Cacajao -Uakaris
*Cebupithecia - Middle Miocene, Colombia
Subtribe Soriacebina
*Soriacebus - Early Miocene, Argentina
Tribe Homunculini
Subtribe Homunculina
*Homunculus - Early Miocene, Argentina
(*)*Aotus* - Owl monkeys; Middle Miocene, Colombia
*Tremacebus - Early Miocene, Argentina
Callicebus - Titi monkeys
Other pitheciines
*Carlocebus - Early Miocene, Argentina
*Lagonimico - Middle Miocene, Colombia
*Xenothrix - Pleistocene/Recent, Jamaica

*Extinct genus. (*) Living genus which includes an extinct species. See Schneider and Rosenberger (this volume) and Rosenberger (1994) for references and discussion. "Other pitheciines" include fossils whose relationships *within* Pitheciinae are uncertain.

volume) and feeding ecology (e.g. Ayres, 1989; Kinzey and Norconk, 1990; van Roosmalen et al., 1988). Schneider and Rosenberger (this volume) review the alternative phylogenetic interpretations. The point we wish to make here is that the Victorian-era pigeonhole of a three-genus subfamily - pitheciines - detracts from one's capacity to see the broader picture, such as the continuity linking the least derived "pitheciin" genus, *Pithecia*, with forms like *Callicebus* on the one hand, and *Chiropotes* and *Cacajao* on the other hand.

As Table 1 shows, there are more fossil genera classifiable as pitheciines by our criteria of monophyly than there are living pitheciines. Our tally of 12 genera, extinct and extant, means pitheciines are more abundant, generically, than any other platyrrhine subfamily. Moreover, they are morphologically diverse and geographically widespread. These points are profoundly important in considering the evolutionary history of pitheciines and their role within the platyrrhine radiation. The anatomical variety among these taxa provides not only the linkage that anchors sakis and uakaris to *Callicebus* and *Aotus*, but also the connection of this larger group to atelines (Schneider and Rosenberger, this volume).

The feeding ecology of the pitheciines is becoming well known. All of the long-term studies of the three larger pitheciins (saki-uakaris) have focused on feeding (Ayres, 1986, 1989; van Roosmalen et al, 1988; Kinzey and Norconk, 1990, 1993; Peres, 1993; Setz, 1994) and they agree that pitheciins occupy a predispersal seed predator niche in the Neotropics. As such, they can ingest fruit at early stages of maturity and may escape seasonal reduction in food resources during the dry season (Norconk, this volume). There are subtle differences in the diets of *Pithecia, Chiropotes,* and *Cacajao,* but they are similar in showing a preference for seeds of large-seeded fruit of the families Lecythidaceae and Sapotaceae (Ayres 1981, 1986; van Roosmalen et al., 1988; Kinzey and Norconk, 1990, 1993; Peres 1993). With evidence from long-term studies of feeding ecology, we are beginning to appreciate the tremendous value of seeds as dietary resources. It is significant that the feeding pattern of *Pithecia* fits well into this picture, for its dental specializations are far less extreme than the system shared by *Chiropotes* and *Cacajao*. As Kinzey (1992) and Rosenberger (1992) have discussed, the dentitions of *Callicebus* and *Aotus* are also best interpreted as part of the hard-fruit and seed-adapted continuum. In fact, *Callicebus*, and less so *Aotus*, are at the pole opposite that occupied by *Chiropotes* and *Cacajao*, with *Pithecia* nestled in the middle but closer to the latter. Morphologically, some of the fossils listed in Table 1 are also avatars of hypothetical morphotypes, filling in the anatomical gaps between, say, *Pithecia* and *Callicebus* (cf. *Soriacebus*).

Aotus may be the most generalized of the pitheciines ecologically. Wright (this volume) describes a diet of fruit supplemented with few leaves and abundant in insects. Ingestion of seeds was not mentioned at all by Wright for *Aotus* and they lack the narrow-tip, procumbent incisors that is characteristic of the other pitheciines. *Callicebus* and *Pithecia* spp. all ingest some leaves, but *Chiropotes* and *Cacajao* rarely take any. Ingestion of insects does not appear to indicate significant differences among the species - they all ingest insects. However, *Aotus* appears to be very general, lacking most of the important derived postcanine dental and feeding specialties shared by the larger pitheciines (although it does have an unusual, enlarged anterior dentition which is related to food harvesting; see Rosenberger, 1992).

There are two viable interpretations to the relatively generalized dentition and feeding behavior of *Aotus*. Either owl monkeys retain a more insectivorous postcanine dentition in concert with new "pitheciine" harvesting specializations of the anterior teeth; or the molars are derived as an insectivorous-folivorous adaptation associated with the shift to nocturnality. We favor the latter view, in part. Some of the fossil pitheciines, such as *Soriacebus* and *Lagonimico*, have postcanine teeth that do not resemble the flat-crown, crushing molars of saki-uakaris. This, too, is evidence for adaptive diversity and phylogenetic continuity within this broadly defined group.

While awareness of the dental specializations foreshadowed the demonstration of saki-uakaris as seed predators based on field observations, we have made little headway toward understanding their social behavior. Wright's (1989) comparative field study of the two smallest members of this subfamily confirmed that both *Callicebus moloch* and *Aotus* not only live in small family groups, but also that the adult members of the group exhibit behaviors that appear to reinforce long-term sociosexual bonds. In contrast, we can construct only superficial outlines of the social systems of the three larger members of this subfamily. *Chiropotes* and *Cacajao* form groups of 15 or more individuals that are not "family" oriented, in the traditional sense of monogamy. Small group sizes reported for *Pithecia* has led to the conclusion that they are monogamous (Buchanan et al., 1981, Robinson et al., 1987). However, preliminary observations of white-faced saki social behavior make it apparent that sakis challenge the traditional criteria we apply to primate monogamy.

First, in support of the interpretation that *Pithecia pithecia* is monogamous:

1. Based on the evidence of a few vocal playback experiments at Lake Guri, Venezuela, white-faced sakis responded as if they defend territories in a very traditional way. Adult males and females were attracted to playbacks of loud calls recording during an inter-troop encounter, by approaching the speaker and giving the same kind of call in response (Norconk and Araya, unpub). This represents the first suggestion of territorial behavior in *Pithecia* spp. Previously, *Pithecia* was considered non-territorial and group sizes larger than a family unit

were interpreted as aggregates of mated pair units (Happel, 1982; Oliveira et al., 1985; Soini, 1986).

2. Both males and females are forced to disperse from their natal group at about three years of age. We observed individuals of both sexes permanently segregated from, and interacting aggressively with, the core social group.

Second, confounding the view that *Pithecia pithecia* is monogamous:

1. More than one adult of each sex coexists in a social group. Our observations of adult males (Gleason and Norconk, 1995) suggest that sakis operate within a complex social context involving cooperation and competition among adult males and among adult females. Although our group of nine individuals may be unusually large due to the animals' inability to disperse from the island, similar observations of group size have been made in terra firme (Kinzey et al., 1988 for *P. pithecia*; Soini, 1986 for *P. hirsuta*).

2. There is no evidence of paternal care, although there is some very interesting data suggesting allomaternal care by full-sized daughters or other adult females (Ryan, 1995 for *P. pithecia*, Soini, 1986 for *P. hirsuta*). The behavior is not very complex, but consists of mother foraging or feeding 50 to 100 m away from the infant and "caretaker", within earshot of the infant who often gives a separation call.

3. It appears that both males and females jostle for reproductive position within a social group. After seven years of monitoring this group, we have never observed more than one female to give birth in the same year although our hormonal work (Scheideler and Norconk, unpubl.) shows that more than one female was reproductively active. We have also observed copulations by one male only, even when three adult males were resident.

In sum, *Pithecia pithecia* is not a "typical" monogamous primate. For the larger pitheciins, data on group size and social aggregates support a view that these sakis are organized on the "multiple male" theme; group sizes range from 15 to more than 30. Groups of *Chiropotes* and *Cacajao* divide up during feeding and coalesce during long distance travel although group fission may be more marked in *Cacajao calvus* than *Chiropotes satanas*. We are not yet sure of the relationship between the small feeding parties and social interactions. Ayres (1986) observed small groups of uakaris isolated for hours or days at a time, but as intriguing as it is, there is still little evidence to add support to the hypothesis that "large groups [of *Chiropotes*] might be relatively permanent aggregations of monogamous subunits" (Robinson et al., 1987:49). Nevertheless, the temporary unions which form for feeding and possibly reproductive reasons, is yet another point of continuity between the larger and smaller pitheciines; the big groups of *Chiropotes* and *Cacajao*, the intermediate-sized groups of *P. pithecia* and the small, pair-bonded units of *Callicebus* and *Aotus*. It suggests that the social organization of ancestral pitheciines may have been structured about the preference to form small parties to mitigate/benefit feeding and foraging strategies.

The growing body of data from field work, systematics and paleontology, when synthesized in an evolutionary perspective which recognizes the mosaic nature of change, provides evidence that the pitheciines are indeed an *adaptive radiation*. The fossil record amplifies this point as do the studies on extant species. Pitheciines as a whole are not radical and uninterpretable, but saki-uakaris are the rule breakers within the larger group that challenge and enlarge evolutionary models. We are coming to realize that pitheciins are a bizarre offshoot *within* a diversified ecological array pivoting on dental adaptations allowing hard-fruit harvesting.

REFERENCES

Ayres, J.M. 1981. Observações sobre a ecologia e o comportamento dos cuxiús (*Chiropotes albinasus* e *Chiropotes satanas*, Cebidae: Primates). Unpublished Master of Science Thesis, Instituto Nacional de Pesquisas da Amazônia e Fundação Universidade do Amazonas, Manaus, Brazil.

Ayres, J.M. 1986. Uakaris and Amazonian flooded forest. PhD thesis, University of Cambridge, Cambridge, UK.

Ayres, J.M. 1989. Comparative feeding ecology of the uakari and bearded saki, *Cacajao* and *Chiropotes*. *J. Hum. Evol.* 18:697–716.

Buchanan, D.B., Mittermeier, R.A., and van Roosmalen, M.G.M. 1981. The saki monkeys, genus *Pithecia* in:A.F. Coimbra-Filho, and R.A. Mittermeier (eds.), *The Ecology and Behavior of Neotropical Primates*, Vol. 1, pp. 371–417. Academia Brasiliera Ciências, Rio de Janeiro.

Gleason, T.M., and Norconk, M.A. 1995. Intragroup spacing and agonistic interactions in white-faced sakis. *Am. J. Primatol.* 36:125.

Happel, R.E., 1982. Ecology of *Pithecia hirsuta* in Peru. *J. Hum. Evol.* 11:581–590.

Kinzey, W.G. 1992. Dietary and dental adaptations in the Pitheciinae. *Am. J. Phys. Anth.* 88:499–514.

Kinzey, W.G., and Norconk, M.A. 1990. Hardness as a basis of food choice in two sympatric primates. *Am. J. Phys. Anthropol.* 81:5–15.

Kinzey, W.G., and Norconk, M.A. 1993. Physical and chemical properties of fruit and seeds eaten by *Pithecia* and *Chiropotes* in Surinam and Venezuela. *Int. J. Primatol.* 14(2): 207–227.

Kinzey, W.G., Norconk, M.A., and Alvarez-Cordero, E. 1988. Primate survey of eastern Bolívar, Venezuela. *Prim. Conserv.* 9:66–70.

Oliveira, J.M.S., Lima, M.G., Bonvicino, C., Ayres, J.M., and Fleagle, J.G. 1985. Preliminary notes on the ecology and behavior of the Guianan saki (*Pithecia pithecia*, Linnaeus 1766: Cebidae, Primates). *Acta Amazonica* 15:249–263.

Peres, C.A. 1993. Notes on the ecology of buffy saki monkeys (*Pithecia albicans*, Gray 1860): a canopy seed-predator. *Am. J. Primatol.* 31:129–140.

Robinson, J.G., Wright, P.C., and Kinzey, W.G. 1987. Monogamous cebids and their relatives: Intergroup calls and spacing. In B.B. Smuts, D.L. Cheney, R.M. Seyfarth, R.W. Wrangham, T.T. Struhsaker (eds.), *Primate Societies*, pp. 44–53. Chicago University Press, Chicago.

Rosenberger, A.L. 1992. Evolution of feeding niches in New World monkeys. *Amer. J. Phys. Anthropol.* 88(4):525–562.

Ryan, K. 1985. Preliminary report on the social structure and alloparental care in *Pithecia pithecia* on an island in Guri reservoir, Venezuela. *Am. J. Primatol.* 36:187.

Setz, E.Z.F. 1994. Ecologia alimentar de um grupo de parauacus (*Pithecia pithecia chrysocephala*) em um fragmento florestal na Amazonia Central. *Boletim da Sociedade Brasiliera de Mastozoologia* 28:5.

Soini, P. 1986. A synecological study of a primate community in Pacaya-Saimiria National Reserve, Peru. *Primate Conservation* 7:63–71.

van Roosmalen, M.G.M., Mittermeier, R.A., and Fleagle, J.G. 1988. Diet of the northern bearded saki (*Chiropotes satanas chiropotes*): a neotropical seed predator. *Am. J. Primatol.*14(1):11–35.

Wright, P.C., 1989. The nocturnal primate niche in the new world. *J. Hum. Evol.* 18:635–638.

17

THE EVOLUTION OF POSITIONAL BEHAVIOR IN THE SAKI-UAKARIS (*PITHECIA*, *CHIROPOTES*, AND *CACAJAO*)

S. E. Walker

Department of Anthropology
Humboldt State University
1 Harpst Street
Arcata, California 95521

INTRODUCTION

A major concern in the study of the postcranial skeleton and associated behaviors in primates has been distinguishing the effects of heritage from those of habitus. The skeleton represents many compromises between selective pressures and the effects of ancestry, and it is difficult to separate the influence of each. In order to interpret anatomical adaptations, knowledge of the positional behavior of animals in their natural habitat is of critical importance.

In order to hold heritage constant and examine the effects of habitus, I studied the positional behavior and habitat use of a monophyletic group of New World monkeys, the saki/uakaris, which comprise the tribe Pitheciini. One representative species from each of the three genera, *Pithecia*, *Chiropotes* and *Cacajao*, was observed. The evolutionary questions investigated deal with the selective advantage of certain positional behaviors; the first question concerns proximate causation, and the second, ultimate causation.

How are particular behavioral solutions advantageous to problems faced by pitheciin primates while engaged in positional behavior? Problems of locomotion and posture are typically imposed by the habitat and by an animal's anatomy, the latter of which limits the way a habitat can be used (Cant and Temerin, 1984). Primates of larger vs. smaller body size are each faced with different problems relating to balance, support use and the crossing of gaps in the canopy (Cartmill and Milton, 1977; Cant, 1992). The influence of habitat characteristics and phylogenetic features on the positional behavior of the saki/uakaris is investigated.

What are the primary selective pressures on the pitheciin positional repertoire; how did differences in positional behavior among the saki/uakaris arise? Data presented here contribute clues to the investigation of this challenging question. Of assistance in addressing the second question are the answers to the previous one. By demonstrating that some

Table 1. Body weights of pitheciin primates. From Ford and Davis (1992)

Species	Average weight (kg) - males	Average weight (kg) - females
Pithecia monachus -group*	2.6 (17)	2.0 (20)
P. pithecia	1.8 (32)	1.4 (19)
Chiropotesalbinasus	3.1 (20)	2.5 (18)
C. satanas	3.0 (73)	2.6 (72)
Cacajao calvus	3.8 (2)	3.2 (3)
C. melanocephalus	no data	2.7 (2)

*Includes *P. albinasus, P. irrorata, P. monachus* (excludes *P. aequatorialis* of Hershkovitz, 1987a)

positional behaviors appear to be more efficient for use in certain habitats, on particular supports, or to exploit certain food resources, the selective advantage for these behaviors can be inferred. Intraspecific comparisons can assist us as well, particularly for conspecifics occurring in different habitats. Information on the current distribution and ecology of the extant pitheciins are also considered in the investigation of this question.

BACKGROUND

The pitheciins are medium-sized platyrrhines, with some significant size differences among members of the tribe (Table 1). They occur, sometimes sympatrically, in a variety of habitat types in northern South America. *Chiropotes* and *Cacajao* are more restricted than is *Pithecia* in terms of their geographical range (Figure 1) and the habitat types in which they occur (Table 2).

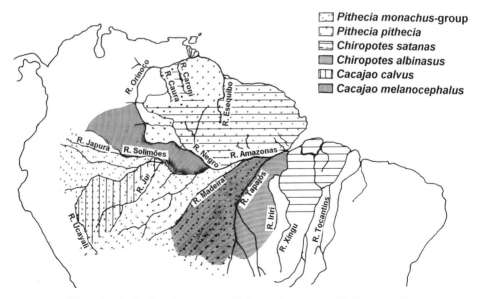

Figure 1. Distribution of the Pitheciini (adapted from Ayres, 1989; Emmons, 1990).

Table 2. Habitat preferences of the Pitheciini

Species	Primary Habitat	Additional Habitats	References
P. pithecia	terra firme		1,5,7
	primary and secondary forest		
	moist and dry forest		
	high and low forest		
P. monachus-group	terra firme	várzea	2,3,4,5,10
	primary forest	igapó	
	moist forest		
	high and low ground		
C. satanas	terra firme	igapó	1,5,6,8
	primary forest		
	moist forest		
	high forest		
C. albinasus	terra firme	igapó	2,8
	primary forest	várzea	
	high forest		
C. calvus	várzea		8
	high forest		
C. melanocephalus	igapó		2,8,9
	high forest		

Sources: 1) Mittermeier, 1977; 2) Mittermeier and Coimbra-Filho, 1977; 3) Freese et al., 1982; 4) Happel, 1982; 5) Wolfheim, 1983; 6) Johns and Ayres, 1987; 7) Kinzey et al., 1988; 8) Ayres, 1989; 9)G. Aymard, pers. comm.; 10) Peres, 1993.

Saki/uakaris are primarily seed predators, specialized to varying degrees for the harvesting, mastication and ingestion of seeds (Ayres, 1986). This dietary adaptation has been termed "sclerocarpivory" (Kinzey and Norconk, 1990). *Chiropotes* and *Cacajao* are more derived in this adaptation than is *Pithecia* (Ayres, 1989; Kinzey and Norconk, 1993; Rosenberger, 1988; Kinzey, 1992), which is the most eclectic in its dietary preferences. In addition to the dentition, cranial (Rosenberger, 1979) and postcranial features (Ford, 1986) as well unite *Chiropotes* and *Cacajao* into a lineage separate from *Pithecia*. *Cacajao* is unique in being the only New World monkey with a greatly reduced tail.

The Pitheciini constitute the last major group of platyrrhines to be examined in the wild, and long-term studies have been conducted on only one-half of the species of this group: *Pithecia pithecia* (Kinzey and Norconk, 1990, 1993; Walker, 1994a), *Pithecia albicans* (Peres, 1993), *Chiropotes satanas* (Frazão, 1991; Kinzey and Norconk, 1990, 1993; Walker, 1994a), *Chiropotes albinasus* and *Cacajao calvus* (Ayres, 1989). A current field project is underway for *Cacajao melanocephalus* (Boubli, pers. comm.). Several short-term field studies that have contributed significantly to the body of knowledge about saki/uakaris include Mittermeier and van Roosmalen (1981), Buchanan et al. (1981), Happel (1982), Oliveira et al. (1985), Johns (1986), and van Roosmalen et al. (1988). Museum-based studies have provided related information focused on either craniodental (Rosenberger, 1979; Kinzey, 1992) or postcranial (Fleagle and Meldrum, 1988; Meldrum and Lemelin, 1991; Ford, 1986) anatomy; these features and others have also been used for extensive taxonomic research on pitheciins (Hershkovitz 1979, 1985, 1987a, 1987b).

Research on the positional behavior and habitat use of wild pitheciins is limited; the most thorough previous study of pitheciin positional behavior (Fontaine, 1981) documented that of captive *Cacajao calvus rufus* living in a large, natural-type enclosure. Mittermeier (1977) and Fleagle and Mittermeier (1980) report on the positional behavior of *P.*

pithecia and *C. satanas* from Surinam; Oliveira et al. (1985) report on preliminary observations of *P. pithecia* as well in Brazil. However, the *Pithecia* individuals were unhabituated in these observations. Happel (1982) and Peres (1993) also report anecdotal observations of positional behavior on *Pithecia monachus* and *P. albicans*, respectively.

The representative species will be referred to by their generic names in this paper; however, this does not undermine the importance of possible intrageneric variation, which seems to be most extreme within the genus *Pithecia*. The reports by Happel (1982) and Peres (1993) of positional behavior, as well as preliminary evidence from the postcranial skeleton (Meldrum, pers. comm.; pers. obs.) are in accordance with other sources (e.g., Hershkovitz, 1987a) on the distinctiveness of *Pithecia pithecia* from members of the *P. monachus*-group (*P. monachus*, *P. aequatorialis*, *P. albicans* and *P. irrorata*).

STUDY SITES AND METHODS

I studied groups of the three pitheciin representatives in Brazil and Venezuela between 1988 and 1991. This involved short-term study of *Cacajao*, and extended study of *Pithecia* and *Chiropotes*.

The study of *Cacajao calvus calvus* was undertaken for six weeks, between July and September, 1988, at Teiú Lake, Brazil (3.22°S, 64.42°W), located in northwestern Amazonas between the Japurá River and Lake Mamirauá (Fig. 2). The forest on the shores of Lake Teiú is "várzea," seasonally inundated by the white water surrounding it (Ayres, 1986). The tallest trees reach 35 or 40 meters. My study group of *Cacajao*, a multimale group of 48–50 individuals, was the one observed by Ayres in 1983/84 after he established the site.

At Guri Lake, Venezuela, I studied habituated groups of *Pithecia pithecia* and *Chiropotes satanas* between 1989 and 1991, the former for fifteen months and the latter for

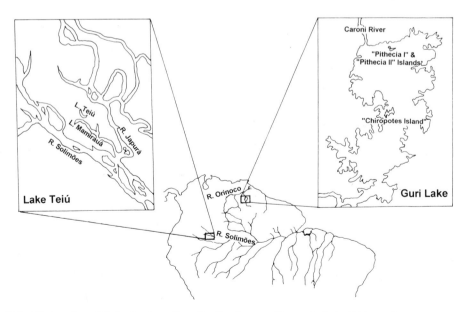

Figure 2. Location of study sites in northern Brazil and eastern Venezuela (Lake Teiú map from Ayres, 1989).

Table 3. Variables for habitat characteristics

Strip sampling variables. Collected within a 5 meter radius of trail marker.
Tree Density: number of trees with diameter of 2 cm or more
Liana Density: estimated density of lianas, based on 0-4 scale
Liana Form: single liana vs. vine tangle

Habitat plots and feeding trees. The variables listed below were recorded for habitat plots and feeding trees. In habitat plots, variables were recorded for each tree of 2 cm or greater. "*Pithecia* Island" I had 10 plots; "*Pithecia* Island" II, 20; and "*Chiropotes* Island," 50. For primary feeding trees, species are listed below, with sample size in ().
Diameter at Breast Height (dbh)
Tree Height: estimations of total tree height were practiced with the assistance of a clinometer until proficiency was achieved in estimating by eye.
Bole Height: height of tree trunk from ground to first main branch. This measurement was divided by tree height to provide a measure of *relative bole height*.
Crown Diameter: diameter (in meters) of crown at widest point. If crown was oblong, an average was calculated using the widest and narrowest dimensions.
Crown Overlap: visual estimate of percent of contact of tree (or near-contact, within 1 meter) with neighboring trees.
Branch Inclination: inclinations of most branches at:
  bottom 1/3 of crown
  middle 1/3 of crown
  top 1/3 of crown

Feeding tree species

"*Pithecia* Island II"	"*Chiropotes* Island"
Peltogyne floribunda (10)	*Pradosia caracasana* (8)
Connarus venezuelanus (10)	*Brosimum guianense* (7)
Licania discolor (10)	*Chrysophyllum lucentofolium* (7)
Erythroxylum steyermarkii (9)	*Melicoccus bijugatus* (8)
Coccoloba latifolia (5)	*Coccoloba fallax* (12)

ten. Guri Lake (mouth at 7.30°N, 63.00°W) covers 4300 km^2, and is located on the northern part of the Guayana shield (Schubert et al., 1986; Fig. 2). It was formed by the 1968 inundation of the Caroni River Basin for the Raul Leoni hydroelectric plant. Numerous islands in the lake support a wide variety of flora and fauna. *Pithecia pithecia* and *Chiropotes satanas* occur on separate islands in Guri Lake, although they are sympatric throughout much of their respective ranges. Before the inundation, these primates occurred in virtually the same habitat types as today (Pernía, 1985; L. Balbás, pers. comm.).

Pithecia occurs in the northern portion of the lake in the "Las Carolinas" zone, in tropical dry forest (Parolin, 1993). The two study groups of *Pithecia*, consisting of two and eight individuals, occurred on islands "*Pithecia* I" and "*Pithecia* II."

The *Chiropotes* study group fluctuated between fifteen and nineteen individuals over the course of the study. It is located 40 km to the south on "*Chiropotes* Island," in "transitional" moist tropical forest (Kinzey and Norconk, 1993); trees here reach 30 meters. For further description of habitats, see Parolin (1993) or Walker (1993).

Sampling Procedures

Forest structure was quantified for the habitats of *Pithecia* and *Chiropotes*. Support availability was determined in order to gauge the primates' preference for certain supports and to distinguish available forest pathways. I collected data through strip sampling (using already existing trail markers), 5 X 20 m habitat plots, and by sampling representatives of five primary feeding tree species for *Pithecia* and *Chiropotes*; variables are defined in Table 3. I focused on the connectedness of canopy (determined by tree density, liana density, liana form and crown overlap), support availability (in terms of branch inclination), and tree size.

Behavioral data were collected using focal animal instantaneous sampling at two minute intervals (after Garber, 1980; for a detailed description of methods see Walker, 1993). Adult males and females were observed for approximately the same amount of time, and results are pooled for the two sexes. In this paper, data are presented for positional behavior (defined in Table 4), tree portions used (Fig. 3), support number, inclination and diameter (Table 5), and the relative height of primates in the trees. The latter was obtained by dividing the height of the primate in the tree by the height of the tree itself. This measurement, referred to as tree level, is expressed as a percentage. Fewer variables and fewer samples were collected for *Cacajao* than for *Pithecia* and *Chiropotes*; results for *Cacajao* are presented elsewhere (Walker and Ayres, 1996). Since the goals of the three major activities of feed, travel and rest are distinct from one another, it is expected that the means to achieve them would also differ (Walker, 1994b). Therefore, results are presented and compared among the pitheciin species for each of these major

Table 4. Classification of positional behaviors

Each of the major categories below are those used in this quantitative analysis. However, each can be divided into subcategories based on the variation within. Likewise, the same general categories may be used for the positional behavior of all three pitheciin species, although they exhibit some qualitative differences, particularly in terms of leaping. These differences are primarily in terms of body orientation and limb segment positioning, and depend in large part upon characteristics of the supports used. For a more complete positional classification, and illustrations of each behavior, see Walker (1993).

Posture

Sit: Body generally upright; high degree of hindlimb flexion at both knee and hip, with forelimb adducted at shoulder and extended at elbow. Hand and foot positions vary depending upon support characteristics, and on orientation of anteroposterior body axis to support. One leg may dangle; haunches may be either on or off support. Forelimbs are often not involved in weight-bearing.

Stand: Long axis of body can be either parallel or perpendicular to the support; if perpendicular, a large degree of back flexion will result. Long axes of limbs are approximately 90° to the body, and are extended and adducted. Since hindlimbs are longer than forelimbs, they are slightly flexed at hip and knee (more so in *Pithecia* than in *Chiropotes* or *Cacajao*).

Vertical Cling: In this orthograde position, long axis of the body is approximately parallel to support, which is vertical or highly angled. Hindlimbs are abducted at the thigh and strongly flexed at all joints; sole of foot is appressed to the support, with digits encircling it and hallux opposed to them. Forelimbs are also flexed, but not as strongly as hindlimbs. Feet alone may be used for support while hands are involved in feeding. One hindlimb may release support and dangle for a short time.

Perch: Similar to sit, but haunches are not involved in either weight-bearing or balance; weight is transmitted and body balanced over feet alone, which grasp support. Body often hunched forward to aid in balance. Tail may be draped over a branch to provide additional support.

Bipedal Stand: Upper body is raised by hindlimb extension from quadrupedal stand, sit or perch, then a branch quickly grasped by at least one hand for additional balance. Torso orientation varies, but approaches the vertical.

Hindlimb suspension: Takes place underneath support, which is of small to medium diameter and usually occurs in or near the terminal branches.

Lie: Relaxed posture, whereby length of body rests along single support or across multiple supports. Most common is ventral lie, and a sprawling position is most common. Here, forelimbs are usually flexed, resting on support just anterior to head, or under chin. Lie on side is less common; here, body may be in a curled position, and multiple supports are often involved. Dorsal lie is infrequent.

Table 4. *(continued)*

Locomotion

Pronograde clamber: Follows Cant (1988), where movement is across multiple substrates in horizontal or diagonal direction. These multiple supports are often of various sizes, inclinations and orientations. Gait pattern is often irregular. Limbs are abducted at hip and shoulder; hip, knee and elbow are flexed, lowering center of gravity.

Pronograde clamber-run: Basically similar to pronograde clamber, but occurs at a faster speed of progression, approximating a trot.

Quadrupedal walk: Typical gait is symmetrical, with diagonal sequence/diagonal couplets (Hildebrand, 1967). Exhibited along single support. Overall body position similar to that of stand. The four limbs are used in propulsion and remain mostly extended. A slight degree of abduction occurs in swing phase, when each limb protracts and adducts to be placed upon the support. Feet are dorsiflexed and inverted, forearms are primarily pronated, and hands abducted at wrist to grasp supporting branch.

Quadrupedal run: Gait similar to that of quadrupedal walk in that diagonal sequence/diagonal couplets are used; however, opposite limbs strike support closer together in time, so that the gait approximates a trot, and speed of progression is increased. Bounding (a galloping gait) is much less often used than the trot; the two are pooled here.

Leap: This gap-crossing behavior involves rapid progression between discontinuous supports through propulsion by hindlimbs. Leaping is influenced by supports, position and speed of progression at take-off, which affect leap distance and body orientation.

Drop: Dropping is used to cross vertical gaps in the canopy; it provides rapid descent with little or no propulsive exertion by the limbs, and little horizontal displacement. Drop landings are usually onto flexible supports.

Hop: Both take-off and landing phases are hindlimb-dominated, and other leap phases are very short. The primary difference between hop and leap is that very little displacement occurs between take-off and landing.

Climb: Following Rose (1979) and Cartmill (1985): locomotion on supports with vertical or steeply sloping surfaces. Different forces act on hindlimbs than on forelimbs; for climb up, hindlimbs play a greater propulsive role, and forelimbs appear to act more as stabilizers, although they may exert tension as well. As support becomes more vertical, the more simultaneously opposite limbs strike substrate, causing lateral flexion in vertebral column. Climbing down is less common in the pitheciins.

Bridge: This gap-crossing behavior is accomplished by quadrupedally walking or pronograde clambering towards terminal end of support until it becomes too unstable (deformable). Forelimbs are then used to reach forward to grasp a support across the gap. Body is then pulled across while new support is being pulled towards body, until hindlimbs are close enough to grasp this second support. Continuous progression, usually pronograde clambering, follows.

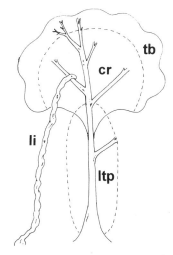

Figure 3. Tree portions. Cr, main crown; tb, terminal branches; ltp, lower tree portions; li, liana.

Table 5. Support variables

Support number
one
two
several (three or more)
Support inclination
horizontal (0-20°)
angled (20-70°)
vertical (70-90°)
deformable (support deforms under body weight)
mixed (combination of supports of various angles)
Support diameter
size category
1: < 2 cm
2: 2-5 cm
3: 6-10
4: 11-15
5: >15 cm
mix: combination of supports of various diameters

activities. Feeding involves the search for, preparation and ingestion of food, and can include movement within a feeding tree. Traveling is continuous movement, particularly between feeding trees or over long distances, but includes momentary postural pauses (<15 seconds) within a travel bout. Resting is defined as non-movement for at least fifteen seconds. Locomotion within an overall period of rest was scored as rest rather than travel if the individual moved for fewer than five seconds; typically for repositioning itself or moving to join another individual. Positional behavior during play is presented only for *Chiropotes*; although results are presented separately for play, these behaviors are exhibited primarily during the group's resting bouts. Qualitative information on differences in tail use among the species is also presented. Raw data used for graphs are presented in Appendices 1–5.

A Methodological Note

A primary tool by which to gauge the importance of various positional behaviors is how often they are performed by the animals. An indication of the importance of frequency data is demonstrated by studies (e.g., Fleagle, 1977; Ward and Sussman, 1979) that have established a functional correlation between certain often-used behaviors and particular anatomical features. Only recently have a rigorous set of statistical tests been applied to results of studies of positional behavior (e.g., Boinski, 1989; Fontaine, 1990; Doran, 1993; Dagosto, 1994; Gebo and Chapman, 1995; Remis, 1995). It has been noted that data on positional behavior are not particularly amenable to statistical testing, primarily because of problems of non-independence of the data (e.g., Mendel, 1976; Cant, 1988; Hunt, 1992). Dagosto (1994) discusses this problem, and suggests an approach involving randomization techniques and use of individuals rather than observations as the sample. This necessitates distinguishing between individuals during data collection.

Chi-square tests were used to gauge the strength of association between each species and its data set for positional behavior. The problem of non-independence of samples was dealt with in the field by noting at each sample as to whether or not the animal had moved

Table 6. Connectedness of canopy. Sample size in ( )

	"*Pithecia* I"	"*Pithecia* II"	"*Chiropotes*"
Crown overlap(avg %)	45	43	37
	(464)	(769)	(265)
Tree density (avg)	27.8	41.4	25.6
	(1444)	(9947)	(8589)
Liana density (avg)	1.7	0.6	2.9
	(1444)	(9947)	(8589)
Liana form(%)	(1444)	(9947)	(8589)
Single	61.4	89.1	9.9
Vine tangle	20.5	2.2	35.0
Both	18.2	8.7	55.1

at least one body length between samples; continuous samples could later be pooled for analysis. Differences between species were deemed significant if the associated Chi-square statistic had a probability value of 0.01 or less. SYSTAT (Wilkinson, 1988) was utilized to perform the statistical tests.

HABITAT CHARACTERISTICS

Connectedness of Canopy. Crown overlap is higher on the "*Pithecia*" islands (Table 6). *Tree density* is by far the highest on "*Pithecia* II." *Liana density* is highest on "*Chiropotes* Island" and lowest on "*Pithecia* II." *Liana form* also differs. Single lianas are by far most common on "*Pithecia* II," followed by "*Pithecia* I," while lianas at the sample points on "*Chiropotes* Island" are comprised of vine tangles, or occur both singly and in vine tangles. Single lianas frequently provide pathways between trees, while vine tangles are more often caught up in a single tree, or on/near the ground.

Support Inclination. Support inclination was recorded in the habitat plots of "*Pithecia* II" and "*Chiropotes* Island," as well as in five important feeding tree species on each of these islands (Fig. 4). In the habitat plots, horizontal supports are slightly more common on "*Chiropotes* Island," but angled supports more so on "*Pithecia* II." Vertical supports occur at similar frequencies on both islands. In feeding trees, horizontal supports are again more common on "*Chiropotes* Island," while both angled and vertical supports are more common on "*Pithecia* II."

Tree Size. The main differences among the trees of the habitat plots are the greater height and greater crown diameter of the trees of "*Chiropotes* Island" (Table 7). Bole height is greatest for "*Chiropotes* Island" trees and least for those of "*Pithecia* I." Relative bole height, however, is greater for the two "*Pithecia*" islands, indicating taller crowns for trees on "*Chiropotes* Island." Crown diameter is highest for trees on "*Chiropotes* Island." Tree size (in terms of dbh, height and crown diameter) is larger for feeding trees than those generally in the forest, as represented by the habitat plots. The mean dbh of the *Pithecia* feeding trees is less than one-half the size of those used by *Chiropotes*, and height and crown diameter are also somewhat larger for *Chiropotes* feeding trees.

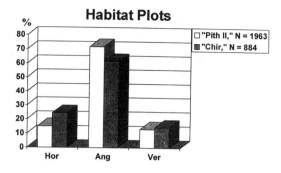

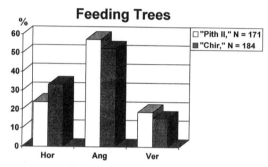

Figure 4. Availability of supports of various inclinations in habitat plots (top) and feeding trees (bottom). Hor, horizontal; ang, angled; ver, vertical. Neither *"Pithecia* II" vs. *"Chiropotes"* (habitat plots) nor *"Pithecia* II" vs. *"Chiropotes"* (feeding trees) demonstrate statistical significance

Table 7. Tree size

| Average | Habitat Plots | | |
	"Pithecia I"	*"Pithecia* II"	*"Chiropotes"*
dbh (cm)	7.4 (517)	6.6 (915)	8.4 (1328)
Height (m)	6.7 (508)	7.6 (868)	11.8 (270)*
Bole height (m)	3.2 (498)	4.0 (801)	5.1 (264)*
Relative bole height (%)	48 (498)	52 (801)	41 (264)*
Crown diameter (m)	3.8 (444)	4.1 (729)	7.7 (252)*

*Trees with dbh greater than 10 cm
(N) - different sample sizes are due to missing variables

| Average | Feeding Trees | |
	"Pithecia II" (44)	*"Chiropotes"* (42)
dbh (cm)	14.5	37.2
Height (m)	12.0	15.4
Bole height (m)	4.8	6.0
Relative bole height (%)	37	34
Crown diameter (m)	8.2	12.6

POSITIONAL BEHAVIOR AND HABITAT USE

Positional Behavior

During feeding, *Chiropotes* and especially *Cacajao* engage in more locomotor behaviors than does *Pithecia* (Fig. 5). For all species, pronograde clambering is the most frequent locomotor behavior, but is most used by *Cacajao*. Both *Chiropotes* and *Cacajao* walk quadrupedally more than does *Pithecia* during feeding. Sitting is the most frequent feeding postural behavior, although *Cacajao* sits less than the do the other species. Standing and perching are the second and third most frequent behaviors, respectively, with the exception of vertical clinging, exhibited only to a large degree by *Pithecia*. *Chiropotes* uses bipedal standing more than do the other species. Both *Chiropotes* and *Cacajao* were sampled using hindlimb suspension; *Cacajao* uses this behavior the most.

During travel, leaping frequency is high for all species, but most so for *Pithecia*. Pronograde clambering is used most by *Cacajao*, while quadrupedal walking and running,

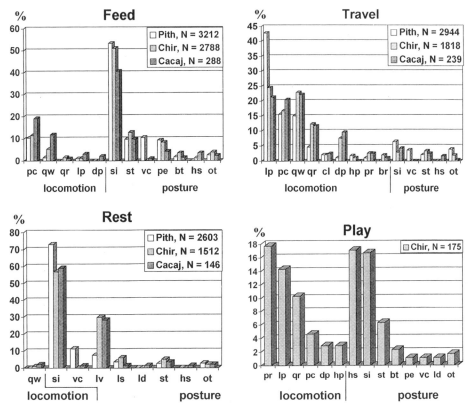

Figure 5. Frequencies of locomotor and postural behaviors used during Feed (upper left), Travel (upper right), Rest (lower left), and Play (lower right). Locomotor behaviors: pc, pronograde clamber; qw, quadrupedal walk; qr, quadrupedal run; lp, leap; dp, drop; cl, climb; hp, hop; pr, pronograde clamber-run; br, bridge. Postural behaviors: si, sit; st, stand; vc, vertical cling; pe, perch; bt, bipedal stand; hs, hindlimb suspension; lv, lie-ventrum; ls, lie-side; ld, lie-dorsum; ot, other. For all comparisons, P<0.001. Feed, *Pith* vs. *Chir*: χ^2=388.67; *Pith* vs. *Cacaj*: χ^2=300.89; *Chir* vs. *Cacaj*: χ^2= 118.82. Travel, *Pith* vs. *Chir*: χ^2 =553.01; *Pith* vs. *Cacaj*: χ^2=216; *Chir* vs. *Cacaj*: ns. Rest, *Pith* vs. *Chir*: χ^2=359.40; *Pith* vs. *Cacaj*: χ^2=170.51; *Chir* vs. *Cacaj*: χ^2=44.1.

and pronograde clamber-running are used more by both *Chiropotes* and *Cacajao*. Dropping and bridging are more often used by *Chiropotes* and *Cacajao* than by *Pithecia*. Momentary pauses (postural behavior) during travel bouts are most frequent for *Pithecia*; vertical clinging is uniquely used by this species, while all three use sitting and standing. *Cacajao* sometimes suspends (and swings) by its hindlimbs during these brief pauses.

For resting, sitting is the primary posture, and is most common for *Pithecia*. *Chiropotes* and *Cacajao* often lie on their ventrum, a much less common behavior for *Pithecia*. *Pithecia* often vertically clings during resting bouts.

Most play behaviors were recorded for juvenile and infant *Chiropotes*, but the adult male frequently participates as well. Play involves both locomotion and posture. The most common locomotor behavior, pronograde clamber-running, is approximately equal in frequency to hindlimb suspension and sitting, the two most common postural behaviors; these three behaviors comprise one-half of all play positions. Leaping is also important, as is quadrupedal running.

Tail Use in Positional Activities

In general, the tail does not play an active role in locomotion, but appears to be used as a counterbalance for *Pithecia* and *Chiropotes* in some locomotor activities, and, as in many monkey species (see Rose, 1974), is often draped over a nearby branch for added support in postural activities (van Roosmalen et al., 1981; pers. obs.). The tail of *Cacajao* is too short to be used in this manner. Use of the tail as an aid in locomotion is especially apparent in leaps of *Pithecia*, in which it may be used as a "rudder" to direct turning of the body in the mid-air phase. Tail carriage positions have been observed to vary among the species. *Pithecia*'s hangs straight down, while that of *Chiropotes* is often held up, and sometimes loops over the back (van Roosmalen et al., ibid.; Buchanan et al., 1981). Tail-wagging behavior is frequent for both *Chiropotes* and *Cacajao*, apparently used to signal excitement for a variety of stimuli (Fontaine, 1981; van Roosmalen et al., ibid.). In *Chiropotes*, it is frequently exhibited together with vocalizations at the end of a resting bout, immediately before travel begins.

Tree Portion Utilization

While feeding, the most apparent difference in use of tree portions is *Chiropotes*' frequent use of the main crown, and *Pithecia*'s frequent use of the lower and central tree portions (Fig. 6). While *Pithecia* occasionally feeds on the ground, *Chiropotes* has not been observed to do so. Data were not collected on *Cacajao*'s use of tree portions; however, inferring about tree portion use from support use indicates an overall similarity to *Chiropotes* while in the canopy. However, Ayres (1986) estimates that *Cacajao* feeds on the ground an estimated 30 to 40% of the time in the driest months at Lake Teiú to utilize seedling clumps, although ground use was not observed in this study.

During travel, *Chiropotes* uses the main crown and especially the terminal branches for travel more than does *Pithecia*. The lower and central tree portions are used much more frequently by *Pithecia* than by *Chiropotes*, and lianas somewhat more.

Tree portions used during rest differ greatly between *Pithecia* and *Chiropotes*, with the main crown used much more by *Chiropotes*, and the lower tree portions much more by *Pithecia*. Lianas are more often used by *Pithecia*.

For playing *Chiropotes*, the crown is the primary portion used, followed by the terminal branches. Lianas and other tree portions are much less frequently used; however, *Chiropotes* uses lianas more during play than during other activities.

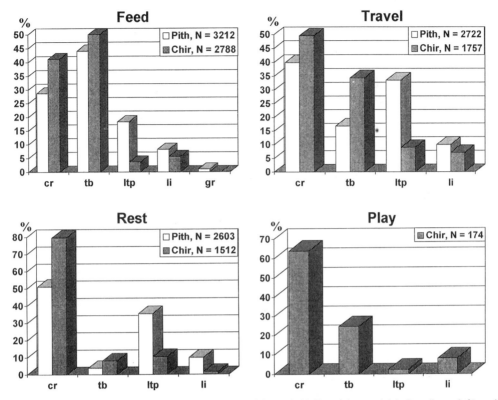

Figure 6. Frequencies of tree portions used during Feed (upper left), Travel (upper right), Rest (lower left), and Play (lower right). Abbreviations as in Figure 3; gr, ground. For all comparisons, P<0.001. *Pith* vs. *Chir*, Feed: χ^2=415.30; Travel: χ^2=553.01; Rest: χ^2=350.59.

Support Use

Pithecia and *Chiropotes* feed upon a single support in almost one-half of samples; *Cacajao* does so slightly less often (Figure 7). *Chiropotes* uses dual supports more frequently than do the other species. Several supports are used most by *Cacajao*, and least by *Chiropotes*. In terms of support inclination, *Pithecia* uses horizontal supports while feeding less than do *Chiropotes* or *Cacajao*, but uses angled and especially vertical supports more (*Chiropotes* doesn't use vertical supports at all). Deformable branches are used least by *Pithecia* and most by *Cacajao*, while a mix of supports is used most by *Pithecia* and least by *Cacajao*. Support diameter was recorded for *Pithecia* and *Chiropotes* only, and both species most often use supports of 2–5 cm for feeding. However, *Pithecia* is found more often on the smaller supports (<5 cm) than on the larger ones, while the larger supports are used slightly more by *Chiropotes*.

While traveling, *Pithecia* uses a single support more often than does either *Chiropotes* or *Cacajao*, and several supports less often (Figure 8). Horizontal supports were used least by *Pithecia*, and most by *Cacajao*; angled supports are used to a similar degree by *Pithecia* and *Chiropotes*, but less often by *Cacajao*. *Pithecia* is unique in its frequent use of vertical supports while traveling, while deformable supports are used about twice as often by *Chiropotes* and *Cacajao* than by *Pithecia*. *Cacajao* uses somewhat fewer mixed

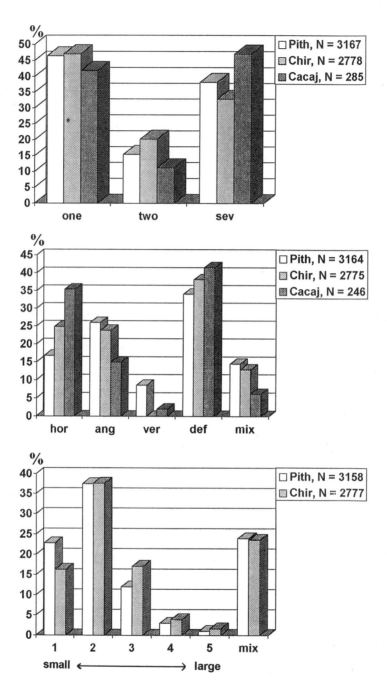

Figure 7. Frequencies of support use during Feed. Number of supports (top): sev, several. Inclination of supports (middle): hor, horizontal; ang, angled; ver, vertical; def, deformable; mix, mixed supports. Diameter of supports (bottom): size class 1, <2 cm; class 2, 2–5 cm; class 3, 6–10 cm; class 4, 11–15 cm; class 5, >15 cm. For all comparisons, P<0.001. Number, *Pith* vs. *Chir*: χ^2=37.6, *Pith* vs. *Cacaj*: ns; *Chir* vs. *Cacaj*: χ^2=26.42. Inclination, *Pith* vs. *Chir*: χ^2=253.31. *Pith* vs. *Cacaj*: χ^2=85.14. *Chir* vs. *Cacaj*: χ^2=33.37. Diameter, *Pith vs. Chir*: χ^2= 82.6.

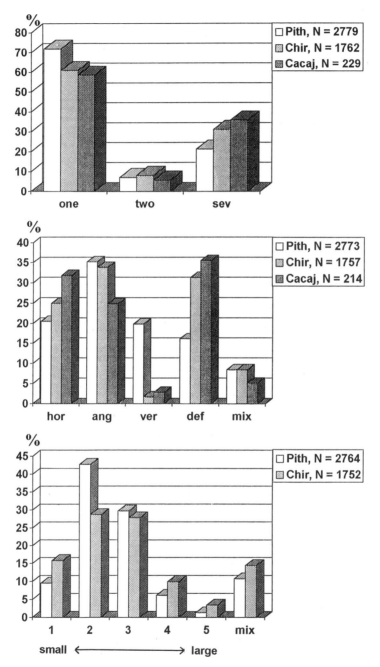

Figure 8. Frequencies of support use during Travel. Abbreviations same as for previous figure. For all comparisons, P<0.001, except where otherwise indicated. Number of supports (top), *Pith* vs. *Chir*: $\chi^2=63.58$; *Pith* vs. *Cacaj*: $\chi^2=26.23$; *Chir vs. Cacaj*: ns. Inclination of supports (middle), *Pith* vs. *Chir*: $\chi^2=399.51$; *Pith* vs. *Cacaj*: $\chi^2=94.14$; *Chir vs. Cacaj*: $\chi^2=13.29$, P=0.01. Diameter of supports (bottom), *Pith* vs. *Chir*: $\chi^2=147.50$.

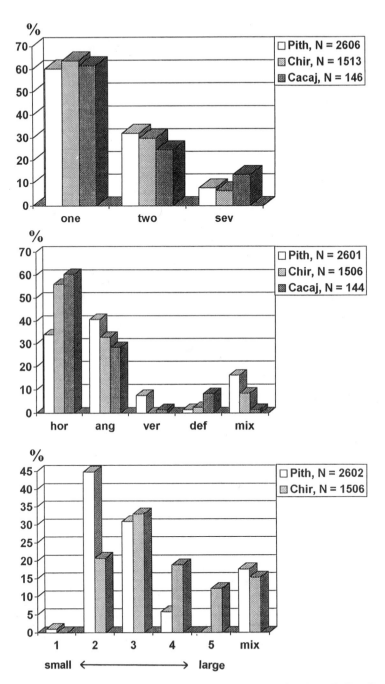

Figure 9. Frequencies of support use during Rest. Abbreviations same as for Figure 7. For all comparisons, P<0.001, except where otherwise indicated. Number of supports (top), *Pith* vs. *Chir*: ns; *Pith* vs. *Cacaj*: χ^2=9.88, P<0.01; *Chir* vs. *Cacaj*: χ^2=22.08. Inclination of supports (middle), *Pith* vs. *Chir*: χ^2=203.06; *Pith* vs. *Cacaj*: χ^2=85.05; *Chir vs. Cacaj:* χ^2=22.52. Diameter of supports (bottom), *Pith vs. Chir:* χ^2=410.94.

supports than do *Pithecia* and *Chiropotes*. The smallest support size (<2 cm) is used for traveling by *Chiropotes* considerably more than by *Pithecia*, while branches from the 2 to 5 cm category are used much more by *Pithecia*. *Chiropotes* travels on the largest supports (>10 cm) more than does *Pithecia*.

During rest, the number of supports used by the three species is quite similar, except that *Cacajao* more often uses several supports (Figure 9). Horizontal supports are used almost twice as frequently by resting *Chiropotes* and *Cacajao* as by *Pithecia*, which most often uses angled supports, and is virtually unique in its use of vertical supports. *Cacajao* uses deformable supports much more frequently than do *Pithecia* or *Chiropotes* for resting. Mixed supports are used by far the most by *Pithecia*, and the least by *Cacajao*. The smaller-sized support classes (<5 cm) are used much more often by *Pithecia* while resting, and the larger ones more by *Chiropotes*.

During play, *Chiropotes* most often uses a single support, but the use of several at a time is also common (Fig. 10). The inclination of most play supports is not distinguishable because they deform under body weight; however, angled and horizontal branches are also commonly used. The most important size classes of supports are those 2–5 cm in diameter, followed by the 6–10 cm size class, then a mix of supports.

Height of Primates in Trees

The absolute mean height at which each species is typically found is at 6.8 m for *Pithecia*, 13.5 m for *Chiropotes* 13.5, and 16.2 for *Cacajao*. Overall, *Pithecia* and *Chiropotes* follow a similar pattern of tree level use, but the frequency of use by each species differs at each level (Fig. 11). (Tree heights were not recorded for *Cacajao*'s trees.) During feeding, *Pithecia* uses the uppermost tree level more than does *Chiropotes*, but the next two lower tree levels are used more by *Chiropotes*. *Pithecia* is found more frequently than *Chiropotes* in the lower levels. While traveling, *Pithecia* is found less frequently at the two highest tree levels, and more frequently at those less than 80% of tree height. For resting, *Pithecia* uses the uppermost portion as well as the lowest two portions more than does *Chiropotes*, and the three middle levels less. *Chiropotes* tends to play relatively high in the trees; the levels between 70% and 90% of the tree's height make up over 80% of samples. *Chiropotes* seldom plays at less than 60% of a tree's height.

THE SELECTIVE ADVANTAGE OF POSITIONAL BEHAVIORS

The first of the two questions proposed earlier, which deals with problems and solutions of positional behavior, can be investigated by examining the relationships of positional behavior and habitat use with habitat type, body size and anatomical features. Strategies of each pitheciin differ depending upon these variables. The problems of positional behavior faced by primates, as well as their solutions, differ among the major activities of feed, travel and rest.

While the statistical tests demonstrate significant differences between most pairwise species comparisons, differences tend to be greater between *Pithecia* and each of the other two species. Certainly, in the "gestalt" of movement and in general patterns of habitat use, *Chiropotes* and *Cacajao* are much more similar to one another; this is also made evident by anatomical features (Fleagle and Meldrum, 1988). The much smaller sample size for *Cacajao* must be considered as well; it is predicted that an increased sample size would strengthen the association of the two larger taxa.

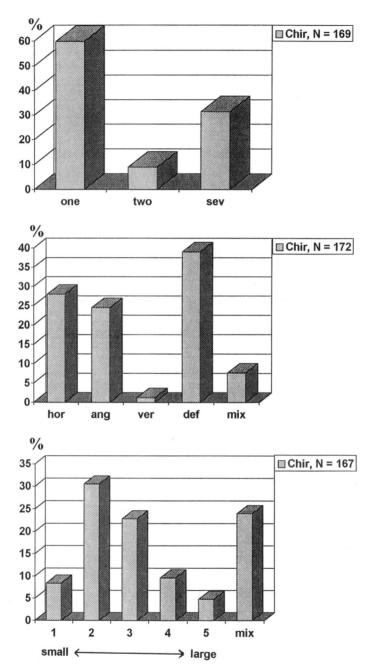

Figure 10. Frequencies of support use during play: number of supports (top), inclination of supports (middle), diameter of supports (bottom).

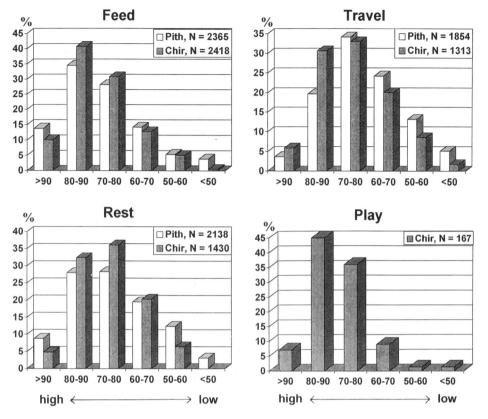

Figure 11. Frequencies of tree levels at which primates are found during Feed (upper left), Travel (upper right), Rest (lower left), and Play (lower right). >90 is the uppermost tree level, (at 90–100% of the tree's height) and <50% is the lowest. For all comparisons, P<0.001. *Pith* vs. *Chir*; Feed: χ^2=90.71; Travel: χ^2=84.95; Rest: χ^2=94.70.

Feed: Food Location as Ultimate Determinant of Positional Behavior

In feeding, the primary concern is access to food, which is often located on small terminal branches; food location and tree form determine the range of positional behaviors that can be used to obtain it. The majority of food for all three species is found in the terminal branches, and all species feed there more than in any other tree portion (Walker, 1995). However, the larger body size of *Chiropotes* and *Cacajao* introduces balance problems that are less important for *Pithecia*, which can more easily clamber out to and perch upon the small supports of the terminal branches. During feeding, the two main behavioral differences between *Chiropotes* and *Cacajao* on one hand, and *Pithecia* on the other, may relate to body size. These are the underbranch maneuvers, particularly hindlimb suspension, and the greater frequency of locomotion within the feeding trees by *Chiropotes* and *Cacajao*. It is easier for a large-bodied individual to hang underneath a branch than to balance atop it (Cartmill and Milton, 1977), and underbranch suspension may increase the feeding sphere (Grand, 1972). Van Roosmalen et al. (1981) as well have discussed hindlimb suspension in *Chiropotes* as a way to expand the feeding sphere. The higher frequency of movement by *Chiropotes* and *Cacajao* while feeding relates to the fact that these species more often carry foods to a solid support for ingestion (Fontaine, 1981; Ayres, 1986; Walker, 1994a). Greater hindlimb suspension and more frequent locomotion while feeding may comprise *Cacajao*'s responses to additional problems of balance imposed by the lack of a functional tail (Walker and Ayres, 1996). Although the actual frequency of hindlimb suspension in *Chiropotes* and *Cacajao* is not high, it is important because it is a unique, shared behavior, and is associated with anatomical features of the foot (Fleagle and Meldrum, 1988).

Pithecia is unique in its frequent use of vertical clinging. This behavior is related to its frequent use of the lower tree portions, usually comprised of vertical supports. *Pithecia* often clings to the trunk of one tree while feeding from the terminal branches of a neighboring, smaller tree. The use of specialized behaviors such as *Pithecia*'s vertical clinging, and the underbranch suspension of *Chiropotes* and *Cacajao*, allows primates access to foods in locations otherwise inaccessible to them.

A direct comparison between support availability and support use is difficult due to the presence of two categories in the behavioral sampling not used in habitat sampling: deformable and mixed supports (when support orientation could not be determined). However, if the samples from these categories are treated as missing variables, then a comparison is possible (Fig. 12). Both *Pithecia* and *Chiropotes* choose higher proportions of horizontal supports than their availability would predict, and *Chiropotes* avoids using vertical supports.

Ground use during feeding has been observed for both *Pithecia* and *Cacajao*. *Pithecia*'s descent to the ground is not surprising, since it often uses very low forest levels. However, *Cacajao* is a canopy primate, so descending to the ground constitutes a major shift in forest use. Ayres (1986) suggests that the dearth of terrestrial predators and competitors in its seasonally flooded habitat permits this behavior.

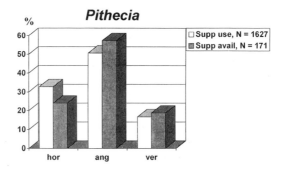

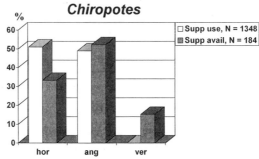

Figure 12. Use of supports of various inclinations during Feed by *Pithecia* (top) and *Chiropotes* (bottom) relative to their availability in feeding trees. *Pithecia*: ns; *Chiropotes*: χ^2=214.42, P<0.001.

For both *Pithecia* and *Chiropotes*, the average size of feeding trees was larger than most trees in the forest; trees in *Chiropotes*' habitat were larger than those of *Pithecia*. Tree size may also partly explain the greater frequency of locomotion by *Chiropotes* and *Cacajao* during feeding; larger trees with larger, spreading crowns require more movement within each tree.

Travel: How to Cross a Gap

In travel, positional problems relate primarily to the crossing of gaps in the canopy. Differences in habitat structure of the portions used and anatomical features are correlated with the solutions of each species.

Pithecia's habitat contains few emergent trees, and *Pithecia* typically moves through the forest at a consistent level, due partly to its ability to leap long distances as well as to its frequent use of low forest levels. *Pithecia* exhibits behavioral and anatomical specializations related to leaping (e.g., Fleagle and Meldrum, 1988; Walker, 1993) including long hindlimbs, the tendency to leap from a vertical position, maintenance of an orthograde body orientation throughout most leap phases, and use of solid supports for both take-off and landing. These behavioral characteristics are evident even for leaps which are initiated from a horizontal or low-angled support. This behavior appears to characterize *P. pithecia* alone in the genus, and are convergent upon those of some prosimian vertical clinger and leapers (e.g., Napier and Walker, 1967; Niemitz, 1983; Peters and Preuschoft, 1984). *Pithecia* can, in fact, be considered a vertical clinger and leaper.

In the habitats of *Chiropotes* and *Cacajao*, the various vegetation types diverge greatly in height, and emergent trees are often used for feeding. This necessitates frequent ascent and descent for traveling between trees, and is related to the more frequent dropping behavior of *Chiropotes* and *Cacajao*. Quadrupedal behaviors, particularly quadrupedal walk and run, pronograde clamber-run and bridge are used more frequently by *Chiropotes* and *Cacajao*.

Although canopy connectedness is higher on both "*Pithecia*" islands than on "*Chiropotes* Island," due to more crown overlap and more lianas in single form, *Pithecia* does not appear to take advantage of this overlap to cross gaps. Instead, it leaps between the portions of different trees not in contact with one another. Although *Pithecia* has a slight tendency to leap more at the lower levels, it is not statistically significant (Table 8).

Predictions have been proposed (e.g., 1977) and correlations established (Fleagle and Mittermeier, 1980) between small body size and frequent leaping (but see Gebo and Chapman, 1995), and larger size with bridging or climbing. Gaps would appear relatively larger to a smaller primate, which would also fare better in case of a fall than would a larger primate (Cartmill and Milton, ibid.). The positional behavior of pitheciins fits this prediction, with the smallest, *Pithecia*, taking its leaping behavior to an extreme.

During travel, *Pithecia* chooses more horizontal and vertical supports relative to their availability, and fewer angled supports, while *Chiropotes* chooses more horizontal and fewer vertical supports in relation to their availability (Fig. 13).

Rest: Balance and Avoidance of Predators

Balance while relaxing is the primary problem to be dealt with during resting bouts; the degree of predator pressure may also influence the use of positions which allow individuals to flee rapidly from a resting posture.

Table 8. Tree levels used in leaping vs. all travel bouts. Sample size
in ( ). Chi-square tests of all travel vs. leaping for each species
are not significant

Tree level	*Pithecia*		*Chiropotes*	
	All travel (1847)	Leaping (816)	All travel (1311)	Leaping (441)
90-100%	3.4	2.3	5.9	7.7
80-90	19.8	26.3	30.6	32.9
70-80	34.3	32.1	32.9	33.3
60-70	24.3	24.6	20.1	15.4
50-60	13.2	16.7	8.5	8.2
40-50	3.3	4.8	1.7	1.6
< 40	1.8	3.2	0.0	1.0

The major differences among the pitheciins in resting positions are that vertical
clinging is used almost exclusively by *Pithecia* as a resting posture, and that lying is more
frequent for *Chiropotes* and *Cacajao*. These relate to habitat structure, habitat use, and
possibly predator avoidance strategies. *Chiropotes* and *Cacajao* spend more resting time
on the large, horizontal branches in the main crown, which provide good lying supports.
Their trees tend to be larger-crowned than those of *Pithecia*, with more large, low-angled
supports. *Pithecia* uses lower tree portions and lianas more than do the other species; these
supports are less conducive to lying behavior. The high frequency of upright sitting rather

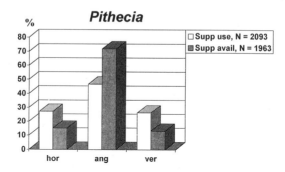

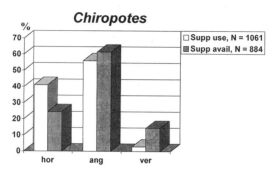

Figure 13. Use of supports of various inclinations during Travel by *Pithecia* (top) and *Chiropotes* (bottom) relative to their availability in trees of habitat plots. *Pithecia*: χ^2=532.52, P<0.001; *Chiropotes*: χ^2=103.52, P<0.001.

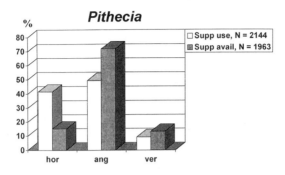

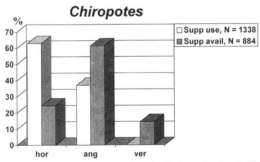

Figure 14. Use of supports of various inclinations during Rest by *Pithecia* (top) and *Chiropotes* (bottom) relative to their availability in trees of habitat plots. *Pithecia*: χ^2=483.42, P<0.001; *Chiropotes*: χ^2=277.10, P<0.001.

than lying may also relate to the fact that sitting can be used for leap take-off, particularly due to *Pithecia*'s maintenance of an orthograde leaping position, while reclining cannot.

The general pattern of *Pithecia*'s and *Chiropotes*' choice of supports is similar during rest: a higher proportion of horizontal supports, and a lower proportion of angled and vertical supports are chosen relative to their availability (Fig. 14). However, as in the other activities, *Pithecia* uses more vertical supports than *Chiropotes*, and *Chiropotes* more horizontal ones.

Play: Balance Again

The maintenance of balance while engaging in rough-and-tumble play presents a problem; an adult male *Chiropotes* was seen to fall from about fifteen meters, crashing through the vegetation to about two meters from the ground. Postural play often occurs above a mass of flexible branches, which are used to suspend above, and sometimes drop down onto. Hindlimb suspension is used much more during play than during feeding (Fontaine, 1981; Ayres, 1986), and may be even more frequent for playing *Cacajao* than *Chiropotes* (Walker and Ayres, 1996).

SELECTIVE FACTORS IN THE EVOLUTION OF POSITIONAL ADAPTATIONS

In attempting to explain the differences in positional behavior and habitat use among pitheciin taxa, the events leading to the evolutionary divergence of taxa must be considered, as well as the possible selective advantage of the various positional strategies. While *Chiropotes* and *Cacajao* share a similar positional pattern, *P. pithecia* diverges from it. Preliminary evidence from the *P. monachus* group demonstrates some similarities to the *Chiropotes/Cacajao* pattern, with the lack of the specialized vertical clinging and leaping adaptations of the smaller *P. pithecia*. Hershkovitz (1979) has suggested that *Pithecia* is the most primitive pitheciin genus, with *P. pithecia* derived within, a hypothesis that is supported here based on positional adaptations (keeping in mind the dearth of evidence from the *P. monachus*-group). Two alternative hypotheses are proposed to explain the differences among the studied taxa.

The different positional pattern observed between *P. pithecia* and *Chiropotes/Cacajao* may be causally related to historical patterns of geographical distribution and sympatry. The widespread sympatry between *Pithecia* species and *Chiropotes* or *Cacajao* (particularly between *P. pithecia* and *C. satanas*) may be quite ancient, and have influenced the evolution of the involved taxa.

Due to the principle of competitive exclusion (Gause, 1934), it would be expected that a niche such as seed predator cannot be shared by sympatric taxa. Given a sympatric situation, therefore, between an early *Pithecia* and *Chiropotes*, the latter, more derived in its adaptations to sclerocarpivory, would be expected to outcompete the less specialized *Pithecia* for fruit. *Pithecia*, in response, may have broadened its resource base, exploiting a wider variety of foods, including leaves, and shifting to a lower canopy level than that occupied by *Chiropotes*. This would permit sympatry between these two ecologically similar primates.

Milton (1984) and Peres (1993) have pointed out that the differing digestive strategy of *Pithecia* may reduce its dietary overlap with other pitheciins; Peres (ibid.) also discusses *Pithecia*'s adaptive divergence from the other pitheciins in use of forest levels and locomotor behavior. Lower food availability in adapting to this new strategy may also have selected for smaller body size (e.g., Case, 1978; Lawlor, 1982). This smaller size may have left *Pithecia* vulnerable to a wider variety of predators, and selected for a suite of predator avoidance strategies. *Pithecia* is the most cryptic of the pitheciins, and uses decoy behavior and freezing as part of its predator (and human observer) escape tactics (Oliveira et al., 1985; Walker, 1993). The use of solid supports for leaping allows for rapid, but relatively quiet, movement. Crypticity benefits as well from the small group size and relatively low activity level of *Pithecia*. The use of lower forest levels has often been associated with leaping behavior (e.g., Kinzey, 1976; Fleagle and Mittermeier, 1980; Davies, 1984; but see Gebo and Chapman, 1995), which is presumably the most efficient gap-crossing behavior for travel between discontinuous supports. In addition, the lower forest levels may afford *Pithecia* some degree of protection from raptors.

In contrast to *Pithecia*'s crypticity, *Chiropotes* and *Cacajao* are noisy travelers, due mostly to the loud rustling produced by the terminal branches used for leaping and dropping. Both of these taxa occur in large groups, typically in the middle to upper canopy, with numerous individuals to watch for predators.

An alternate hypothesis to explain the differences among pitheciin taxa is that of a forced dietary shift for *Pithecia*, caused either by a drastic lowering of food abundance in a particularly harsh season, or by the colonization of an unknown area. *Pithecia*, faced with such a situation, may have attempted to expand its resource base and use lower areas of the forest in its search for food. The capabilities for exploiting food in these areas of the forest could then be advantageous in allowing it to cope with other primates in subsequent, or existing, situations of sympatry.

The fact that the vertical clinging and leaping behavior of *Pithecia* can be biomechanically associated with numerous anatomical features indicative of leaping adaptations suggests that pressures imposed during travel have had a great influence on the evolution of its postcranial skeleton. *Pithecia*'s vertical clinging and leaping is here interpreted as a locomotor adaptation, evolved for rapid travel in lower forest levels. Now, the associated clinging capability is used during feeding, allowing *Pithecia* access to foods in the terminal branches of small trees.

In order to better understand the adaptive significance and evolution of saki/uakari features, the remaining taxa must be studied in their natural habitat. In addition, special attention should be paid to the distribution of resources in the canopy matrix as an important influence on positional behavior.

SUMMARY

The positional behavior and habitat use of a representative group of each of three genera of the tribe Pitheciini was observed in the wild in Venezuela and Brazil. Longterms studies were conducted of *Pithecia pithecia* and *Chiropotes satanas* at Guri Lake, Venezuela, and comparative data were collected over a shorter time period on *Cacajao calvus* at Teiú Lake, Brazil.

Phylogenetic constraints as well as habitat features affect pitheciin positional behavior during all activities. Location of food in trees influences the positional behaviors and tree portions used, and specific strategies, such as underbranch suspension, can be correlated with larger body size and anatomical specializations. Gap-crossing behaviors are influenced by divergence in vegetation height, as well as anatomical features. A great deal of variation exists within the Pitheciini in terms of positional behavior, to a large degree due to the influence of the vertical clinging and leaping adaptations on the entire positional repertoire of *Pithecia pithecia*. Vertical clinging behaviors are used in feeding, travel and rest, and are associated with this taxon's frequent use of the highly angled supports of lower tree portions. The primary locomotor behavior used by *Pithecia* is leaping, in which the body orientation remains orthograde, and solid, vertical supports often used for both take-off and landing.

Chiropotes and *Cacajao* tend to be found higher in the canopy, more frequently using the main crown and terminal branches for both feeding and travel. Their body size, larger than that of *Pithecia*, may constrain their food-reaching and balance capabilities in terminal branches; they compensate with hindlimb suspension to increase the feeding sphere, and moving to solid feeding supports. This problem may be exacerbated in *Cacajao* by its lack of a functional tail to aid in balance. In the more discontinuous forests of *Chiropotes* and *Cacajao*, there is greater need to ascend and descend between trees, the former accomplished by pronograde clambering and climbing, and the latter by leaping and dropping. Primarily quadrupedal, both of these species also exhibit substantial leaping. However, their leaps differ from the more specialized leaps of *Pithecia* in their pronograde body orientation, terminal branch landings and typical downward component.

The adaptive divergence of *Pithecia pithecia* from the other pitheciins is thought to have been due to the expansion of *Pithecia*'s resource base and use of new forest levels, with the impetus provided either by competition with an ecologically similar primate (of the *Chiropotes/Cacajao* lineage) or to a period of reduced food availability. *Pithecia*'s use of lower forest levels may have selected for leaping as an efficient, rapid locomotor mode, resulting in its vertical clinging and leaping specializations. This behavior, thought to have evolved as a locomotor adaptation, now allows *Pithecia* to exploit food in the terminal branches of small trees.

ACKNOWLEDGMENTS

I greatly appreciate the helpful comments of Drs. Alfred Rosenberger and Elizabeth Strasser, and two anonymous reviewers on an earlier version of the manuscript. I also thank Drs. Marilyn Norconk and Alfred Rosenberger for inviting me to participate in the Neotropical Primates Symposium. For research in Venezuela, I am deeply indebted to the members of EDELCA's Ecology Department in Guri; for my studies in Brazil, I appreciate the assistance of Dr. J. Márcio Ayres. This paper is dedicated to the memory of my mentor, Dr. Warren Kinzey, who made me think about the bigger picture. This research was supported by grants from the Exploration Fund of the Explorer's Club of New York and Wenner-Gren Foundation for Anthropological Research, National Science Foundation Dissertation Improvement Award (Grant BNS 89–13349), a fellowship from the Graduate Center of the City University of New York, and a National Science Foundation Grant to W.G. Kinzey (BNS 87–19800).

REFERENCES

Ayres, J.M., 1986, Uakaris and Amazonian Flooded Forest. Ph.D. Thesis, University of Cambridge.

Ayres, J.M., 1989, Comparative feeding ecology of the uakari and bearded saki, *Cacajao* and *Chiropotes*. *J. Hum. Evol.* 18:697–716.

Boinski, S., 1989, The positional behavior and substrate use of squirrel monkeys: ecological implications. *J. Hum. Evol.* 18(7):659–678.

Buchanan, D.B., Mittermeier, R.A., and van Roosmalen, M.G.M., 1981, The saki monkeys, *genus Pithecia*. In R.A. Mittermeier and A.F. Coimbro-Filho (eds.) Ecology and Behavior of Neotropical Primates, Academia Brasileira de Ciencias, pp. 391–418.

Cant, J.G.H., 1988, Positional behavior of long-tailed macaques *(Macaca fascicularis)* in northern Sumatra. *Am. J. Phys. Anthropol.* 76:29–37.

Cant, J.G.H., 1992, Positional behavior and body size of arboreal primates: a theoretical framework for field studies and an illustration of its application. *Am. J. Phys. Anthropol.* 88(3):273–284.

Cant, J.G.H. and Temerin, L.A., 1984, A conceptual approach to foraging adaptations in primates. In P.S. Rodman and J.G.H. Cant (eds.) Adaptations for Foraging in Nonhuman Primates. New York: Columbia University Press, pp. 304–342.

Cartmill, M., 1985, Climbing. In M. Hildebrand, D.M. Bramble, K.F. Liem, and D.B. Wake (eds.) Functional Vertebrate Morphology. Boston: Harvard University Press, pp. 73–88.

Cartmill, M., and Milton, K., 1977, The lorisiform wrist joint and the evolution of "brachiating" adaptations in the Hominoidea. *Am. J. Phys. Anthropol.* 47:249–272.

Case, T., 1978, A general explanation for insular body size trends in terrestrial vertebrates. *Ecol.* 59(1):1–18.

Dagosto, M., 1994, Testing positional behavior of Malagasy lemurs. A randomization approach. *Am. J. Phys. Anthropol.* 94(2):189–202.

Davies, A.G., 1984, An ecological study of the red leaf monkey (*Presbytis rubicunda*) in the dipterocarp forest of northern Borneo. Ph.D. Thesis, University of Cambridge.

Doran, D., 1993, The comparative locomotor behavior of chimpanzees and bonobos: the influence of morphology on locomotion. *Am. J. Phys. Anthropol.* 91:83–98.

Emmons, L.H., 1990, Neotropical Rainforest Mammals: A Field Guide. Chicago: University of Chicago Press.

Fleagle, J.G., 1977, Locomotor behavior and skeletal anatomy of sympatric Malaysian leaf-monkeys (*Presbytis obscura* and *Presbytis melalophos*). *Am. J. Phys. Anthropol.* 46:297–308.

Fleagle, J.G. and Meldrum, D.J., 1988, Locomotor behavior and skeletal morphology of two sympatric pitheciine monkeys, *Chiropotes satanas* and *Pithecia pithecia*. *Am. J. Primatol.* 16:227–249.

Fleagle, J.G. and Mittermeier, R,A., 1980, Locomotor behavior, body size, and comparative ecology of seven Surinam monkeys. *Am. J. Phys. Anthropol.* 52:301–314.

Fontaine, R., 1981, The uakari, genus *Cacajao*. In A.F .Coimbra-Filho and R.A. Mittermeier (eds.) Ecology and Behavior of Neotropical Primates. Rio de Janeiro: Academia Brasileira de Ciencias, pp. 443–494.

Fontaine, R., 1990, Positional behavior in *Saimiri boliviensis* and *Ateles geoffroyi*. *Amer. J. Phys. Anthropol.* 82:485–508.

Ford, S.M., 1986, Systematics of the New World Monkeys. In D.R. Swindler and J. Erwin (eds.) Comparative Primate Biology, Vol. 1: Systematics, Evolution and Anatomy. New York: Alan R. Liss, pp. 73–135.

Ford, S.M. and Davis, L.C., 1992, Systematics and body size: implications for feeding adaptations in New World monkeys. *Am. J. Phys. Anthropol.* 88:415–468.

Frazão, E., 1991, Insectivory in free-ranging bearded saki (*Chiropotes satanas chiropotes*) *Primates* 32(2):243–245.

Freese, C.H., Heltne, P.G., Castro, R., Whitesides, N.G., 1982, Patterns and determinants of monkey densities in Peru and Bolivia, with notes on distributions. *Intl. J. Primatol.* 3(1):53–90.

Garber, P.A., 1980, Locomotor Behavior and Feeding Ecology of the Panamanian Tamarin (*Saguinus oedipus geoffroyi*, Callitrichidae, Primates). Ph.D. Dissertation, Washington University.

Gause, G.F., 1934, The Struggle for Existence. Reprinted in 1964 by Hafner. Baltimore: Williams and Wilkins.

Gebo, D.L. and Chapman, C.A., 1995, Positional behavior in five sympatric Old World monkeys. *Amer. J. Phys. Anthropol.* 97(1):49–76.

Grand, T.I., 1972, A mechanical interpretation of terminal branch feeding. *J. Mammal.* 53(1):198–201.

Happel, R., 1982, Ecology of *Pithecia hirsuta* in Peru. *J. Hum. Evol.* 11:581–590.

Hershkovitz, P., 1979, The species of sakis, genus *Pithecia* (Cebidae, Primates), with notes on sexual dichromatism. *Folia Primatol.* 31:1–22.

Hershkovitz, P., 1985, A preliminary taxonomic review of the South American bearded saki monkeys, genus *Chiropotes* (Cebidae, Platyrrhini), with the description of a new subspecies. *Fieldiana (Zool.)* N.S. 27, 1363:1–46.

Hershkovitz, P., 1987a, The taxonomy of South American sakis, genus *Pithecia* (Cebidae, Platyrrhini): a preliminary report and critical review with the description of a new species and a new subspecies. *Am. J. Primatol.* 12:387–468.

Hershkovitz, P., 1987b, Uacaries, New World Monkeys of the genus *Cacajao* (Cebidae, Platyrrhini): a preliminary taxonomic review with the description of a new subspecies. *Am. J. Primatol.* 12:1–53.

Hildebrand, M., 1967, Symmetrical gaits of primates. *Am. J. Phys. Anthropol.* 26(2):119–130.

Hunt, K.D., 1992, Positional behavior of *Pan troglodytes* in the Mahale Mountains and Gombe Stream National Parks, Tanzania. *Am. J. Phys. Anthropol.* 87(1):83–106.

Johns, A.D., 1986, Notes on the ecology and current status of the buffy saki, *Pithecia albicans*. *Primate Conservation* 7:26–29.

Johns, A.D. and Ayres, J.M., 1987, Conservation of white uacaries in Amazonian várzea. *Oryx* 21(2):74–80.

Kinzey, W.G., 1976, Positional behavior and ecology in *Callicebus torquatus*. *Yrbk. Phys. Anthropol.*, Vol. 20:468–480.

Kinzey, W.G., 1992, Dietary and dental adaptations in the Pitheciinae. *Am. J. Phys. Anthropol.* 88:499–514.

Kinzey, W.G., Norconk, M.A., and Alvarez-Cordero, E., 1988, Primate survey of eastern Bolívar. *Primate Conservation* 9:66–70.

Kinzey, W.G. and Norconk, M.A., 1990, Hardness as a basis of fruit choice in two sympatric frugivorous primates. *Amer. J. Phys. Anthropol.* 81(1):5–15.

Kinzey, W.G. and Norconk, M.A., 1993, Physical and chemical properties of fruit and seeds eaten by *Pithecia* and *Chiropotes* in Surinam and Venezuela. *Intl. J. Primatol.* 14(2):207–227.

Lawlor, T., 1982, The evolution of body size in mammals: evidence from insular populations in Mexico. *Amer. Naturalist* 119(1):54–72.

Meldrum, D.J. and Lemelin, P., 1991, Axial skeleton of *Cebupithecia sarmientoi*. *Amer. J. Primatol.* 25(2):69–90.

Mendel, F., 1976, Postural and locomotor behavior of *Alouatta palliata* on various substrates. *Folia Primatol.* 26:36–53.

Milton, K., 1984, The role of food-processing factors in primate food choice. In P.S. Rodman and J.G.H. Cant (eds.) Adaptations for Foraging in Nonhuman Primates. New York: Columbia University Press, pp. 249–279.

Mittermeier, R.A., 1977, Distribution, synecology and conservation of Surinam monkeys. Ph.D. Thesis, Harvard University.

Mittermeier, R.A. and Coimbra-Filho, A.F., 1977, Primate conservation in Brazilian Amazonia. In Prince Rainier of Monaco and G. Bourne (eds.) Primate Conservation. New York: Academic Press, pp. 117–166.

Mittermeier, R.A. and van Roosmalen, M.G.M., 1981, Preliminary observations on habitat utilization and diet in eight Surinam monkeys. *Folia Primatol.* 36:1–39.

Napier, J.R. and Walker, A.C., 1967, Vertical clinging and leaping - a newly recognized category of locomotor behaviour of primates. *Folia Primatol.* 6:204–219.

Niemitz, C., 1983, New results on the locomotion of *Tarsius bancanus*, Horsfield, 1821. Annales des Sciences Naturelles. Zoologie, Paris. 13° Serie, Vol. 5, pp. 89–100.

Oliveira, J.M.S., Lima, M.G., Bonvincino, C., Ayres, J.M., and Fleagle, J.G., 1985, Preliminary notes on the ecology and behavior of the guianan saki (*Pithecia pithecia*, Linnaeus 1766: Cebidae, Primates). *Acta Amazonica* 15(1–2):249–263.

Parolin, P., 1993, Forest inventory in an island of Lake Guri, Venezuela. In W. Barthlott, C.M. Naumann, K. Schmidt-Loske, K.L. Schuchmann (eds.) Animal-Plant Interactions in Tropical Environments. Proceedings of German Society for Tropical Ecology, Bonn, pp. 139–147.

Peres, C.A., 1993, Notes on the ecology of buffy saki monkeys (*Pithecia albicans*), Gray 1860): a canopy seed-predator. *Amer. J. Primatol.* 31:129–140.

Pernía, J.E., 1985, Mapa de Fisiografia y Vegetacion del Area de Inundacion de la Tercera Etapa del Embalse el Guri - Estado Bolívar. Internal manuscript: CVG - Electrificacion del Caroní (EDELCA), Venezuela.

Peters, A. and Preuschoft, H., 1984, External biomechanics of leaping in *Tarsius* and its morphological and kinematic consequences. In C. Niemitz (ed.) Biology of Tarsiers. New York: Gustav Fischer Verlag, pp. 227–255.

Remis, M., 1995, Effects of body size and social context on the arboreal activities of lowland gorillas in the Central African Republic. *Am. J. Phys. Anthropol.* 97(4):413–434.

Rose, M.D., 1974, Postural adaptations in New and Old World monkeys. In F.A. Jenkins (ed.) Primate Locomotion. New York: Academic Press, pp. 201–222.

Rose, M.D., 1979, Positional behavior of natural populations: some quantitative results of a field study of *Colobus guereza* and *Cercopithecus aethiops*. In M.E. Morbeck, D. Preuschoft and N. Gomberg (eds.) Environment, Behavior and Morphology: Dynamic Interactions in Primates. New York: Gustav Fischer Verlag, pp. 75–94.

Rosenberger, A.L., 1979, Phylogeny, Evolution and Classification of New World Monkeys (Platyrrhini, Primates). PhD. Thesis, City University of New York.

Rosenberger, A.L., 1988, Pitheciinae. In I. Tattersall, E. Delson and J. van Couvering (eds.) Encyclopedia of Human Evolution and Prehistory. New York: Garland Publishing, pp. 454–455.

Schubert, C., Briceno, H.O., and Fritz, P., 1986, Paleoenvironmental aspects of the Caroni-Paragua River Basin (Southeastern Venezuela). *Interciencia* 11(6):278–289.

van Roosmalen, M.G.M., Mittermeier, R.A., and Milton, K., 1981, The bearded sakis, genus *Chiropotes*. In A.F. Coimbra-Filho and R.A. Mittermeier (eds.) Ecology and Behavior of Neotropical Primates, Vol. 1 Academia Brasileira de Ciências, pp. 419–441.

van Roosmalen, M.G.M., Mittermeier, R.A., and Fleagle, J.G., 1988, Diet of the northern bearded saki (*Chiropotes satanas chiropotes*): a neotropical seed predator. *Am. J. Primatol.* 14(1):11–35.

Walker, S.E., 1993, Positional Adaptations and Ecology of the Pitheciini. Ph.D. Thesis, City University of New York.

Walker, S.E., 1994a, Positional behavior and habitat use in *Chiropotes satanas* and *Pithecia pithecia*. In J.R. Anderson, N. Herrenschmidt, J.J. Roeder and B. Thierry (eds.), Proceedings of the XIV Congress of the Intl. Primatol. Soc., pp. 195–201.

Walker, S.E., 1994b, Habitat use by *Pithecia pithecia* and *Chiropotes satanas*. *Am. J. Phys. Anthropol.*, Suppl. 17:203.

Walker, S.E., 1995, Is there a relationship between diet and positional behavior? A case study of *Pithecia* and *Chiropotes*. *Am. J. Phys. Anthropol.*, Suppl. 20:217.

Walker, S.E. and Ayres, J.M., 1996, Positional behavior of the white uakari (*Cacajao calvus calvus*). *Am. J. Phys. Anthropol.*, 101: 161-172.

Ward, S.C. and Sussman, R.W., 1979, Correlates between locomotor anatomy and behavior in two sympatric species of *Lemur*. *Am. J. Phys. Anthropol.* 50:575–590.

Wilkinson, L., 1988, SYSTAT: The System for Statistics. Evanston, IL: SYSTAT, Inc.

Wolfheim, J., 1983, Primates of the World: Distribution, Abundance and Conservation. Seattle: University of Washington Press.

APPENDICES

Appendix 1. Availability of supports of various inclinations in habitat plots and feeding trees on "*Pithecia* II" Island and "*Chiropotes*" Island. Number in ( )

	Habitat Plots		Feeding Trees	
Inclination	"*Pithecia* II" (1963)	"*Chiropotes*" (884)	"*Pithecia* II" (171)	"*Chiropotes*" (184)
Horizontal	15.1	24.4	24.0	33.2
Angled	71.8	61.4	57.3	52.2
Vertical	13.1	14.4	18.7	15.2

Appendix 2. Frequencies of locomotor and postural behaviors expressed as percentages

Positional Behaviors	Feed			Travel			Rest			Play
	Pith	Chir	Cacaj	Pith	Chir	Cacaj	Pith	Chir	Cacaj	Chir
N	3212	2788	288	2944	1818	239	2603	1512	146	175
pron clmbr	10.4	11.5	19.1	15.6 %	16.7	20.5	0	0	0	4.6
quad walk	1.5	5.3	11.8	15.1	22.9	22.2	0	1	2.1	0
quad run	0	1.5	1.0	4.6	12.3	11.7	0	0	0	10.3
leap	1.0	1.2	3.1	39.7	24.5	21.3	0	0	0	14.3
drop	0	0	2.1	1.1	7.6	9.6	0	0	0	2.9
climb	0	0	0	2.1	2.1	2.5	0	0	0	0
hop	0	0	0	1.8	1.1	0	0	0	0	2.9
pron cl-run	0	0	0	1.0	2.6	2.5	0	0	0	17.7
bridge	0	0	0	0	2.0	1.0	0	0	0	0
sit	53.3	51.0	40.6	6.4	2.9	4.2	72.9	56.8	58.9	16.7
stand	9.8	12.8	9.7	2.1	3.3	2.5	2.5	5.0	3.4	6.3
vert. cl.	10.5	0	1.0	3.6	0	0	11.3	0	1.4	1.1
perch	9.3	8.3	4.2	0	0	0	0	0	0	1.1
bip. stand	1.7	3.5	1.4	0	0	0	0	0	0	2.3
hind. susp.	0	1.2	3.5	0	0	1.7	0	0	1.4	17.1
lie-ventrum	0	0	0	0	0	0	7.5	29.8	28.1	0
lie-side	0	0	0	0	0	0	3.6	6.1	1.4	0
lie-dorsum	0	0	0	0	0	0	0	0	1.4	1.1
other	2.5	3.8	2.3	2.9	1.9	0.4	2.7	2.0	2.0	1.7

Appendix 3. Frequencies of tree portions used by *Pithecia* and *Chiropotes*. Number in parenthesis

	Feed		Travel		Rest		
Tree portion	*Pith* (3212)	*Chir* (2788)	*Pith* (2722)	*Chir* (1757)	*Pith* (2603)	*Chir* (1512)	Play *Chir* (174)
Crown	28.7	41.1	39.9	49.5	51	79.5	63.8
Term br	44	49.9	16.8	34.3	3.8	8	24.7
Low, central portion	18.3	3.6	33.5	8.9	35.4	10.4	2.3
Liana	8.2	5.6	9.9	7	9.9	1.5	8
Ground	1	0	0	0	0	0	0

Appendix 4. Frequencies of support use expressed as percentages

Support	Feed Pith	Feed Chir	Feed Cacaj	Travel Pith	Travel Chir	Travel Cacaj	Rest Pith	Rest Chir	Rest Cacaj	Play Chir
N	3167	2778	285	2779	1762	229	2606	1513	146	169
one	46.4	47.0	41.8	71.8	60.8	58.5	60.3	63.7	61.6	59.8
two	15.4	20.2	11.2	7.1	8.1	5.7	31.9	29.8	24.7	8.9
several	38.2	32.8	47.0	21.2	31.1	35.8	7.8	6.5	13.7	31.4
N	3164	2775	246	2773	1757	214	2601	1506	144	172
horizont	16.8	24.8	35.4	20.5	24.9	31.8	34.1	56.0	60.4	27.9
angled	26.0	23.8	15.0	35.2	33.8	24.8	40.7	32.9	28.5	24.4
vertical	8.6	0	2.0	19.8	1.7	2.8	7.6	0	1.4	1.2
deform	34.1	38.1	41.5	16.1	31.3	35.5	1.4	2.4	8.3	39.0
mixed	14.5	12.9	6.1	8.4	8.4	5.1	16.2	8.5	1.4	7.6
N	3158	2777	no data	2764	1752	no data	2602	1506	no data	167
<2 cm	22.8	16.3	-	9.5	15.8	-	1.0	0	-	8.4
2-5 cm	37.5	37.6	-	42.7	28.7	-	44.8	20.7	-	30.5
6-10 cm	12.0	17.1	-	29.7	27.8	-	30.9	33.0	-	22.8
11-15 cm	3.0	3.9	-	6.1	9.9	-	5.8	18.8	-	9.6
>15 cm	1.0	1.6	-	1.3	3.4	-	0	12.2	-	4.8
mix	24.0	23.6	-	10.7	14.4	-	17.6	15.3	-	24.0

Appendix 5. Frequencies of the various tree levels at which *Pithecia* and *Chiropotes* are found expressed as percentages

Tree Level (%)	Feed Pith	Chir	Travel Pith	Chir	Rest Pith	Chir	Play Chir
N	2365	2418	1854	1313	2138	1430	167
>90	13.8	10.1	3.7	5.9	8.8	4.8	7.1
80-90	34.5	40.7	19.7	30.6	28.0	32.2	44.9
70-80	28.2	30.8	34.1	32.9	28.3	35.9	35.9
60-70	14.2	12.7	24.2	20.0	19.5	20.1	9.0
50-60	5.4	5.0	13.1	8.5	12.4	6.2	1.3
<50	3.9	0.7	5.1	1.7	3.1	0.8	1.3

18

THE NEOTROPICAL PRIMATE ADAPTATION TO NOCTURNALITY

Feeding in the Night (*Aotus nigriceps* and *A. azarae*)

Patricia C. Wright

Department of Anthropology
State University of New York
Stony Brook, New York 11794

INTRODUCTION

The only primate with a nocturnal lifestyle in the neotropics is *Aotus,* the owl monkey. The genus is found from Panama to northern Argentina, and is very diverse. Twelve karyotypically distinct forms have diploid chromosome counts ranging from 46–58 (Hershkovitz, 1983; Ma et al., 1985; Pieczarka et al., 1993) and recent morphological and molecular analyses support the taxonomic division into multiple species (Ford, 1994; Ashley and Vaughn, in press). The fossil record suggests that *Aotus* is an ancient taxon with a long period of adaptation to nocturnality (Setoguchi and Rosenberger, 1987). Morphologists have placed *Aotus* together with *Callicebus* (Ford, 1994; Hartwig, 1995), but some recent molecular and morphological analyses place *Aotus* within the pitheciine group (Rosenberger, 1995; Schneider and Rosenberger, this volume), with *Saimiri* and *Callicebus* (Kay, 1994), or with the callithrichid group. The difficulties of classifying *Aotus* within the Neotropical primates may be a result of its many primitive traits.

Few studies of *Aotus* have been conducted in the wild. Thorington and Vorek (1976), working on Barro Colorado Island, were the first to radiocollar and follow *Aotus* at night in the forest. Following this inspiration, Wright (1978; 1984; 1986; 1989) studied the behavior and ecology of *Aotus* groups in Peru and Paraguay. With a focus on harvesting *Aotus* for biomedical research and captive breeding, Aquino and Encarnación (1986a; 1986b; 1988; 1990; and 1994) have concentrated on surveys and sleeping tree site information of populations of *Aotus* from lowland forests of the Peruvian Amazon.

I found that the behavior and ecology of *Aotus* differed from a same-sized (1 kg) diurnal primate *(Callicebus moloch)* in the lowland rain forest in Manu National Park, Peru. First, although food eaten most of the year consisted of fruits, leaves and insects for both species, *Aotus* ate nectar and fruits in the season of scarce resources while *Callicebus* increased leaf-eating to nearly 50%. Second, at all times of year the small groups of *Callice-*

Adaptive Radiations of Neotropical Primates
edited by Norconk *et al*. Plenum Press, New York, 1996

bus were chased out of fruit trees by larger diurnal monkeys. Because other fruit trees were readily available most of the year, this changed the feeding choices of *Callicebus* only during the three months of fruit scarcity, but this harassment resulted in the *Callicebus* eating nearly 50% leaves. Third, population densities of *Aotus* were higher than population densities of *Callicebus*. Fourth, *Aotus* consistently slept in 5 trees over the annual cycle, while *Callicebus* used over 40 sleeping sites. Fifth, *Aotus* habitually exited sleep sites about 15 minutes after sunset and 15 minutes before sunrise during all times of year. The temporal range of exiting sleep sites varied seasonally for *Callicebus* with activity beginning as early as 5:20am and as late as 11:50am (Wright, 1985; 1989).

The third sympatric species of small monkey in Manu National Park, Peru, is the 1kg squirrel monkey. In many behaviors *Saimiri* differ from both *Callicebus* and *Aotus*. This monkey is noisy and active (in contrast to *Callicebus*), lives in large groups(25–45 individuals), and each group ranges over an area 20 times the size of areas used by *Callicebus* or *Aotus* groups (Terborgh, 1983; Mitchell, 1990; Mitchell et al., 1991).

Subsequent studies of *Aotus* in a drier habitat in Paraguay, without diurnal raptor predators or harassing diurnal monkeys, suggest that the owl monkey has behavioral flexibility in ranging patterns, sleep-tree site selection, and day time activity. On cold days, *Aotus* would forage in the daylight up to three hours, a behavior never seen in the rain forest of Manu (Wright, 1989). In five months the *Aotus* pair used 48 sleep sites in Paraguay, often sleeping in the open on a branch.

In this chapter I examine the nocturnal niche of *Aotus* in three dimensions. First, I review and refine the comparison between same-sized diurnal and nocturnal primates in Peru and Paraguay, concentrating on feeding behaviors. Second, I examine communities of diurnal and nocturnal arboreal mammals in a neotropical rain forest, comparing diet and body size. Third, I examine the neotropical nocturnal primate niche in the context of the paleotropics, concentrating on new data from Madagascar.

METHODS AND STUDY SITES

Peru

The study was conducted in the trail system of Cocha Cashu Biological Research Station in the Manu National Park of southeastern Peru (11° 51'S, 71° 19'W) (Fig. 1). The park, established in 1973, is one of the largest protected areas in the world, comprising 15,000 km² of rainforest on the western edge of the Amazon basin. Even before the establishment of the park, animal and plant populations were unaffected by hunting or logging because of the low human population density in the region and difficult accessibility. The study site is at 400m altitude. The annual rainfall is 2,000 mm with most of the rain falling from October to April. An extensive 28 km trail system covers a 5 to 7 km square area bordering the Manu River. The Manu National Park has twelve sympatric species of primates, eight of which have been thoroughly studied (Terborgh, 1983; Janson, 1984; 1986; Goldizen, 1987; Symington, 1988; Mitchell, 1990; Mitchell et al., 1991). The highest primate community biomass documented in Amazonia is found at this site (Janson and Emmons, 1990).

In Peru, nine groups of *Aotus nigriceps* (group size 2–5) and six groups of *Callicebus moloch bruneus* (group size 2–5) were surveyed and followed at monthly intervals over a 15 month period. One breeding pair formed the core of each group, with one offspring born each year, and offspring leaving the groups at three years of age. Two focal groups of *Callicebus*

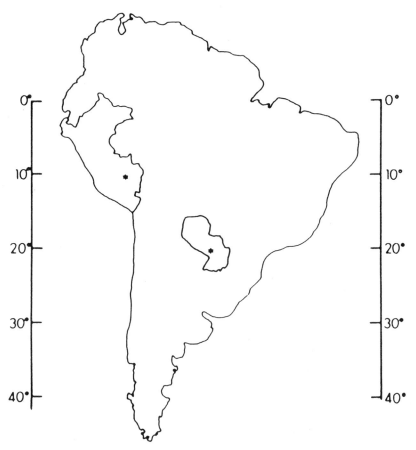

Figure 1. Map of South America with the study sites indicated by asterisks: Cocha Cashu Biological Research Station, Manu National Park, Madre de Dios, Peru and La Golondrina Ranch, Villa Hayes, Paraguay.

and one focal group of *Aotus* were followed throughout their activity cycle for five or six consecutive days or nights each month. Data on diet, social behavior, and ranging were taken on focal individuals at 5 minute intervals. Type of food (mature or new leaves, ripe or unripe fruit, insects or nectar), as well as feeding rates were documented. Plants were identified to the species level by the Missouri Botanical Garden. Nocturnal observations were aided by a Varo Noctron V image intensifier and an ITT image intensifier with an infra-red light source. Fecal samples of each species were taken opportunistically and examined for seeds, insect chitin, and fruit and leaf remains. To measure fruit abundance over an annual cycle, 40 plastic bags (fruit traps) were placed 25 m apart, and fruit contents were collected, identified and counted from each bag every two weeks (Terborgh, 1983). To quantify fluctuations in insect abundance over an annual cycle, black lights were set out all night bimonthly and insects were identified to family and counted.

Paraguay

This study site was located in the chaco of Paraguay (24° 25'S 58° 40'W) at La Golondrina Ranch in the department of Villa Hayes, 70 km north of Asunción on the Río

Confuso (Fig 1). This palm savanna area, grazed by herds of cattle, contained habitat islands of 5 to 20 ha of subtropical dry forest. These forests were separated by 500 m to 1km of savanna. Trails were established within the low spiny forest chosen as a study site. During the cold winter months, nighttime temperatures were as low as -5°C, and by September daytime temperatures were over 40°C. The study site was located at sea level. Annual rainfall for Asunción is 1,340 mm, but annual rainfall for the ranch was not available. The only other sympatric primate was *Alouatta caraya*. This study site was chosen because hunting had not been allowed on the ranch for over 50 years, and the forested area of the ranch was relatively undisturbed.

The same methodology was used in Paraguay as in Peru, except that one *Aotus azarae* group was followed for at least one five-day sequence each month over a four month period. Often, 24 hour samples were done with the help of a field assistant. The group consisted of a breeding pair.

RESULTS

Diet in the Presence of a Potential Competitor

Comparison of the diets of *Callicebus* and *Aotus* co-existing in the same territories in Peru showed that both species were frugivores that supplemented their diet with flowers, insects, nectar, and leaves. *Callicebus* fed more often (90% of the samples) in tree crowns of 10 m or less compared with *Aotus*. Sixty one percent of trees fed in by *Aotus* had crowns < 10 m in diameter) (Wright, 1986). More feeding minutes were spent in trees with crown diameters larger than 11 m in *Aotus* (85%; N = 10,620 min) compared with *Callicebus* (21%; N = 7,080min). *Callicebus* often fed consecutively in many small-crowned fruit trees of a single species, 50–100m apart. But *Aotus* preferred feeding in larger trees, resting, and then returning to feed in the same large-crowned tree or moving on to the next large-crowned tree (Wright, 1984; 1989).

Aotus foraged for insects at dawn and dusk and during moonlit nights. *Aotus* did not sit and scan for insects, like *Callicebus* did; rather, the night monkeys moved methodically along branches, grabbing insects out of the air or off the surface of branches. The insects eaten most often were large orthopterans, lepidopterans, coleopterans, and spiders. These preferences may reflect the nocturnal activity of these large insects. Diurnal *Callicebus* quietly sat and scanned for small cryptic insects including small homopterans, invertebrate cocoons, and spiders wrapped in webs. Although *Callicebus* spent 44 minutes a day (range = 29–74min) in all months of the year searching for insects, capture rate was low (only 6–10 small insects a day). Based on fecal samples of both monkey species, night monkeys ingested nearly five times more insects each day than the *Callicebus*. Insects were found in 14% of the *Callicebus* fecal samples compared with 64% of *Aotus* samples). In comparison, *Saimiri* were even more insectivorous with estimates of over 330 insects ingested eaten each day (Terborgh, 1983). Both *Saguinus* (Garber, 1992) and *Cebus* (Janson, 1985) consume more insects daily than either *Aotus* or *Callicebus*.

In Peru, *Callicebus* ingested more vegetation than *Aotus* during all months of the year. The diets of the two species were most similar during the wet season when fruits were abundant, and most different in the months of June, July, and August, when fruits and insects were scarce (Terborgh, 1983; Wright, 1985; Janson and Emmons, 1990). In the wet season, *Callicebus* groups spent about 25% of feeding time on vine leaves and bamboo leaf stems; in the dry season this leaf consumption increased to 40%. *Callicebus* ate leaves during all times of the day, but most leaf-eating bouts occurred in the late after-noon (between 15:00 and 17:30 hrs). *Aotus* ingested leaves less than 10% of feeding min-utes during all months of the year.

During periods of food scarcity, we saw more clearly the dynamics of niche separa-tion, and how competition for food affected these small-bodied monkeys. In the dry sea-son there were 25 times less fruit available in fruit traps, and insect sampling has shown that insects were very scarce during June, July, and August (Janson and Emmons, 1990). During these months, *Aotus* fed on nectar and figs. These were scarce foods that were also eaten by the diurnal monkeys. *Callicebus* did not eat these fruits from large crowned fruit trees, because of effective chasing (interference competition) by the larger diurnal species. *Saimiri*, another small-bodied diurnal monkey, was also chased from fruit trees by the large-bodied monkeys (*Ateles paniscus*, *Cebus apella* and *Cebus albifrons*), but the strategy was not effective. As some squirrel monkeys were chased from the tree, oth-ers continued to enter the fruit trees. The numbers of individuals in a *Saimiri* group can number up to 50, so that for each squirrel monkey chased, there were tens of them enter-ing a different part of the crown of the same tree. The large-bodied species ceased to chase the squirrel monkeys after about ten minutes, and it was assumed that the costs of the constant chasing was greater than the actual loss of fruit eaten by the squirrel mon-keys (Terborgh, 1983). *Saimiri* had a numerical advantage in feeding that *Callicebus* lacked.

Diet in the Absence of a Competitor

In the subtropical dry forests of Paraguay during June, July, and August (cold, dry season), fruits, figs or nectar were unavailable. *Aotus* ate leaves instead. These leaves in-cluded vine leaves, leaves of the bromeliad parasite, *Tillandsia* and new leaves of small trees. Although leaves were eaten at all times of the day and night, major leaf-eating bouts occurred in the late afternoon and at dusk (between 16:00 to18:30 hrs; Friedman two way analysis of variance by ranks: Chi Square = 9.87, df = 4, P<0.05) The least likely time for leaf eating was at dawn (Ganzhorn and Wright, 1995).

DISCUSSION

Aotus and *Callicebus*: Day and Night Niches

The problems that all nocturnal mammals must overcome to be night active include low light levels, colder temperatures, differences in nutritional value of plant foods and differences in activity levels of insects. The comparative approach to the behavior and ecology of the night monkey has revealed several important differences between the simi-lar-sized diurnal *Callicebus*, *Saimiri*, and the nocturnal *Aotus*, including the following: 1) In Peru, the ranging patterns of the night monkey were habitual, rigid, and often circular (Wright, 1989). These habitually traveled and scent-marked routes facilitated orientation in the dark. 2) Sleep sites are hidden in tree holes or dense vine tangles (Aquino and En-carnación, 1986b; 1994; Wright, 1989). 3) Night monkeys traveled more than twice as far during twilight and nights with a full moon, suggesting that low light levels may restrict activity levels. Further indication that light levels were important to *Aotus* behavior is the fact that social behaviors including intragroup calling, playing, and intergroup fighting oc-curred when the moon was bright and overhead (Wright, 1989; 1994). 4) Leaves were not preferred probably because they contained less protein and soluable carbohydrates during the night than during the day (Ganzhorn and Wright, 1995).

Advantages of being active at night in the Neotropics may include: 1) increased availability of active and calling insects, which is a good protein source (Janson and Emmons, 1990); 2) avoidance of the major avian predators, the day-active hawks and eagles (Rettig, 1978; Sherman, 1991); and 3) competition with the larger diurnal monkeys (Wright, 1989). Other small-bodied New World monkeys exhibited different ways of avoiding feeding interference. For example, larger group size for detection and confusion of predators is an effective deterrant used by *Saimiri* (Hamilton, 1971; Boinski and Mitchell, 1992). Active vigilance for detection of predators has been observed in *Saguinus* (Garber, 1986). Insect hunting style is changed during the day by 1) flushing sleeping and hidden insects (Terborgh, 1983); 2) searching for insects in tree cavities and hidden sites; and 3) peeling toxic caterpillars (Boinski and Fragaszy, 1989).

In Paraguay, changes in behavior of *Aotus* seemed to be reflective of changes in the threat of predation. Large hawks and eagles were absent, but the great horned owl was a predator. Habitual travel routes were not used and sleeping trees were randomly distributed in the forest (Wright, 1989). Although the monkeys were active six to nine hours during the night, they were also active from one to three hours during the day. This "release" from their strict nocturnal schedule, opened up new feeding opportunities. They fed on flower nectar at midday with the howler monkeys, and ate leaves in late afternoon. There is evidence from leaves collected hourly throughout the 24 hour day in Madagascar that leaves have both high protein and carbohydrate levels in late afternoon (Ganzhorn and Wright, 1994).

The differences in the composition of predators and competitors, and the ability of the monkeys to see at night may account for some of the differences between day and night niches. But food availability also changed across temporal niches with leaves higher in protein and carbohydrates during the day and insects more active and therefore more audible and visible in the night. This comparative research showed that some behavioral patterns of owl monkeys, including the monogamous social system, and long distance calling patterns, did not change when habitat changed. However, day time activity, sleep-site choice, and diet did change dramatically in a contrasting habitat.

Neotropical Rain Forest Arboreal Mammals: Day and Night Communities

Although *Aotus* is the only nocturnal primate in the Neotropics, two species of sloths, at least 7 species of marsupials, two species of anteaters, a porcupine, two cats, two raccoon-like carnivores, 54 species of bats, and five species of rodents are also active in the Neotropical rain forest trees at night (Fleming, 1979; Charles-Dominique et al., 1981; Charles-Dominique, 1983; Emmons, 1990)(Table 1).

(Bats, being volant, have a niche space more similar to birds, and are not considered in this analysis). When we compare the Neotropical nocturnal arboreal mammals with the diurnal arboreal community, there are about the same number of sympatric species, but the diurnal species average a kg heavier ($\bar{x}$ = 1.8kg - nocturnal; $\bar{x}$ = 2.8kg -diurnal) (Table 1, Fig. 2).

A comparison of broad diet types shows that most Neotropical arboreal mammals eat from a menu of fruit, nectar, and insects (Emmons, 1990) showing no significant difference in diet between day or night active species (Mann Whitney U = .36, P > .05). There are no nocturnal species larger than 6 kg, but in general there is a dearth of large ar-

Table 1. The South American nocturnal arboreal community: body mass and diet, ranked by body mass

Species	Mass (kg)	Diet
Choleoepus didactylus (southern two-toed sloth)	4.1-8.5	Leaves, fruit
Felis wiedii (margay)	3.0-9.0	Mammals, birds, reptiles
Tamandua tetradactyla (collared anteater)	3.6-8.4	Ants
Coendu prehensilis (brazilian porcupine)	3.2-5.3	Seeds, fruit
Bradypus variegatus (brown-throated 3-toed sloth)	2.3-5.5	Leaves
Potos flavus (kinkajou)	2.0-3.2	Fruit, insects
Bassaricyon gabbii (olingo)	1.1-1.4	Fruit, nectar, insects
Aotus spp. (owl monkey)	**0.78-1.25**	**Fruit, nectar, insects, leaves**
Didelphus marsupialis (common oppossum)	0.56-1.6	Fruit, mammals, reptiles, insects, nectar
Dactylomys dactylinus (amazon bamboo rat)	0.6-0.7	Bamboo leaves & stems
Caluromysiops irrupta (black-shouldered oppossum)	?	Fruit, nectar, insects
Caluromys lanatus (western woolly opossum)	0.31-0.41	Fruit, nectar
Philander opossum (common gray four-eyed opossum)	0.20-0.60	Reptiles, fruit, insects
Isothrix bistrata (yellow-crowned brush-tailed rat)	0.32-0.57	?
Echimys armatus (red-nosed tree rat)	0.34-0.41	Fruit, seeds
Cyclopes didactylus (pygmy anteater)	0.13-0.22	Fruit, insects
Mesomys hispidus (spiny tree rat)	0.08-0.15	Fruit, insects
Marmosops noctivagus (white-bellied slender mouse opossum)	0.06-0.09	Insects, fruit
Marmosa murina (murine mouse opossum)	0.04-0.06	Insects, fruit
Rhipidomys sp. (climbing rat)	0.04-0.17	Fruit, seeds, fungi, insects

boreal species in the Neotropics, with only a sloth and several monkeys in the 7 kg to 11 kg weight range. This pattern is also seen in Madagascar where no arboreal mammal is larger than 7.8kg (Emmons et al., 1983; Wright et al., in press).

   Aotus fits into the neotropical community as the largest of the night active "diet generalists" (eating an opportunistic mix of fruits, leaves, nectar and insects), while *Callice-*

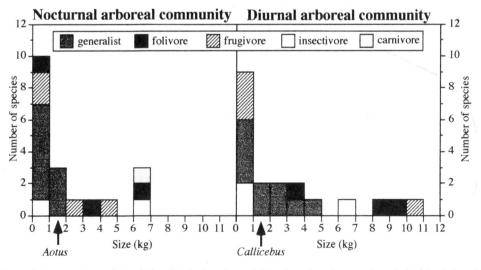

Figure 2. A comparison of the relative distributions (by weight and number of mammal species) of South American nocturnal versus diurnal dietary communities.

bus is in the middle of the size range of "diet generalists" eating during the day. This ranking affects success in interference competition (Wright, 1989). Furthermore, the potential for interspecific competition may be exacerbated in areas of high species richness (Thiollay, 1994). Perhaps as a result of this competition, the *Callicebus* diet appears less generalized than *Aotus*. Evidence from Amazon sites also confirms that *Aotus* is a "habitat generalist", using all habitat types from forest edge to interior, while *Callicebus* is a habitat specialist, confined to narrow habitat corridors (Kinzey, 1981; Emmons, 1984; Peres, 1993).

A comparison of social groupings shows that most neotropical arboreal nocturnal mammals are solitary, and *Aotus* has the largest foraging group size of terrestrial and arboreal mammals in the Neotropical night, not including bats (Emmons, 1990). I have hypothesized that predation pressure by hawks and eagles on day-active monkeys is greater in the South American rain forest than by owls at night (Wright, 1989; 1994). Emmons (1987) found diurnal monkey remains in felid feces, Peetz et al. (1992) have evidence of jaguar predation on *Alouatta*, and Rettig (1978) and Sherman (1991) confirmed that monkeys are a favorite prey of harpy eagles. Rain forest owls are small, probably not capable of taking prey larger than 500 g, but to my knowledge, South American owl prey items have not been studied (Brown et al, 1982). In Madagascar, owls ate small (ranging from 25 to 450gm) nocturnal primates voraciously (Goodman et al., 1993a,b).

The paleontological record shows that marsupials were in South America during the late Cretaceous (Simpson, 1967; 1980). Primates appeared in the fossil record in the Oligocene synchronously with, and later than, rodents (Ciochon and Chiarelli, 1980; Potts and Behrensmeyer, 1992; Meyers and Wright, 1993). The earliest fossil primates are dated from roughly 25 million years ago and are morphologically similar to African and modern South American forms (Fleagle, 1988; Rosenberger et al., 1991; Hartwig, 1995; Hartwig and Cartelle, 1996; Horowitz, 1995; Kay and Johnson, 1996). A primate fossil found in Colombia has been put in the genus *Aotus* and dated at 12 million years (Setoguchi and Rosenberger, 1987). This evidence suggests that *Aotus* evolved in a nocturnal environment occupied with marsupials and rodents. Procyonids, the other group of nocturnal competitors, arrived much later at the beginning of the Pliocene (Simpson, 1980; Potts and Behrensmeyer, 1992).

The biogeography and history of communities can provide information about the processes responsible for generating and maintaining diversity (Vermeij, 1992). Perhaps because marsupials radiated into many niches in South America first, their presence limited the radiation of nocturnal monkeys. An alternative suggestion is that constraints of evolving a nocturnal eye restricted adaptive radiation by nocturnal primates. Although the fossil record is very incomplete and it is difficult to know if other nocturnal monkeys existed and went extinct, the current evidence suggests that the genus *Aotus* has remained the same body weight, dental and limb morphology and nocturnal lifestyle for 12 million years (Setoguchi and Rosenberger, 1987).

What is the nocturnal primate niche?

There are different evolutionary solutions to solving the same problems. To best survive the problems of nighttime low temperatures, *Aotus* shares with the prosimians low basal metabolic rate (LeMaho et al., 1981; McNab and Wright, 1987) and dense fur. Perhaps to take advantage of better odor transmission in the night atmosphere, prosimians have olfactory bulbs and accessory olfactory bulbs that are much larger than the same structures in *Aotus* (Stephan et al., 1981). Most New World monkeys use scent marks and sexual signals to delineate territory, and the structure of the vomeronasal organ and nasopalatine ducts of *Aotus* are larger than other New World monkeys (Hunter et al., 1984), implying increased sensitivity to olfactory signals. Prosimians have moist rhinaria which increase olfactory abilities, but nocturnal haplorrhines like *Aotus* and *Tarsius* lack this feature.

To cope with the difficulties of perceiving substrates, foods, predators, and friends in the night, prosimians and *Aotus* have evolved special visual adaptations. Nocturnal prosimians have smaller orbits, but a reflective tapetum lucidum in the retina for keener night vision (Pariente, 1980). This layer, the tapetum lucidum, seems to reduce visual acuity, but enhances an animal's ability to see by "recycling" and amplifying all incoming light. In *Aotus* and other anthropoids, there is a different modification of the retina, called a fovea, a specialized central area where light-sensitive cells (cones) are packed closely together to allow for good visual acuity (Fleagle, 1988). The comparative research on lemur and anthropoid photo pigments reveal that the cone photopigment complements of the lemurs differ from the patterns of anthropoids (Jacobs and Deegan, 1993; Jacobs et al., 1994). New World monkeys and lemurs have two classes of cone photopigment, but the lemurs show no evidence of the widespread pigment polymorphism which is the hallmark of many platyrrhines (Jacob and Deegan, 1993). The spectral peak of sensitivity for lemurs with their two cone system in the retina is 543–545 nanometers, while *Aotus* tested in the same laboratory has a sensitivity of 700 nm. (Jacobs et al., 1994).

Experiments of spectral sensitivity of lemurs with and without tapeta lucidae suggest that the tapetum has little or no influence on photopic sensitivity. It is also important to note that lemurs may use signals from both rods and cones to discriminate between red and green (Jacobs and Deegan, 1993). In other words, *Aotus* and lemurs may not be much different in visual perception, even though they have evolved very different visual systems to do the same job.

Twenty percent of primate species and 28% of primate genera are nocturnal. Although early reports suggested nocturnal primates had small home ranges, were insectivorous, and weighed 1kg or less (Charles-Dominique, 1975; 1977), recent studies have shown that these primates of the night have a broader range of behaviors and sizes than previously known (e.g. Bearder, 1987). Night active primates range in size from 35g (Schmid and Kappeler, 1994) to 3.5kg (Erikson, 1994), and have diets that range from totally insectivorous *(Tarsius spectrum* - Gursky, 1994) to nearly totally folivorous (*Avahi* and *Lepilemur* - Ganzhorn et al., 1985; Harcourt, 1987). The home range size of nocturnal primates varies from very small (1 ha in *Tarsius spectrum, Avahi* and *Lepilemur* - MacKinnon and MacKinnon, 1980; Crompton and Andau, 1987; Gursky, 1994; Ganzhorn et al., 1985; Nash, 1995) to large (32–214 ha in *Daubentonia* - Sterling, 1993; Sterling and Richard, 1995).

When compared to other nocturnal primates, *Aotus* is medium-sized, eats a diet of fruits supplemented with insects and leaves, and has a medium sized territory (8 to10ha). A comparison with other nocturnal mammals in most of its range indicates that *Aotus* is half the body weight of kinkajous (2 kg), the same size as olingos, and larger than all but one sympatric opossum (7 species weighing from 55 g to 2,000 g) (Charles-Dominique et al., 1981; Emmons, 1990). *Aotus* in Peru can gain access to high quality resources at night. *Aotus* in Paraguay is forced to be more selective of the fewer or less nutritious resources and forced to expand its activity period to take advantage of leaves at this "most nutritious" time.

SUMMARY

In the rain forest of Peru, *Aotus* was primarily frugivorous. Data from my 15 month study in the Manu National Park comparing the behavioral ecology of the nocturnal monkey, *Aotus* and a sympatric same-sized diurnal monkey *(Callicebus)* suggested differences in diet items were most pronounced in the season of scarce resources (Wright, 1989). The diet of *Aotus* was next studied in the subtropical forests of the chaco of Paraguay. When the restrictions of the rain forest including diurnal predation and diurnal interference competition for food were absent in the chaco forests, *Aotus* diet changes accordingly. Leaf eating increased during the late afternoon, corresponding with the hours of diurnal primates' leaf eating (Ganzhorn and Wright, 1994). The reasons proposed for this late afternoon leaf eating in diurnal primates (Milton, 1979; Glander, 1982; Waterman, 1984; Chapman and Chapman, 1991) were not supported by the data on *Aotus* activity in the chaco. After the leaf-eating hours, *Aotus* did not rest, but remained active. *Aotus* has a lifestyle which allows flexibility of choice in the nocturnal monkey niche.

The niche of "nocturnal simian" seems to have remained conservative over 12–20 million years in the Neotropics since *Aotus dindensis*, a 12 million year old fossil from Colombia, appears to be postcranially, dentally and size similar to modern *Aotus* (Sarich and Cronin, 1980; Setoguchi and Rosenberger, 1987). Although our knowledge is limited by few studies of wild populations, the breadth of behavioral differences between *Aotus* in Panama to Argentina so far appear to be relatively narrow (Moynihan, 1964; Wright, 1989). We certainly know that there is little variation in body size or morphological measurements (Thorington and Vorek, 1976; Hershkovitz, 1983). The restriction of the adaptive radiation of nocturnal primates in the Neotropics to a single genus may be a result of historical and biogeographic reasons. Nocturnal opossums and bats occupied South America before the arrival of primates, and may add a biogeographic precedent that has restricted primate radiation (Croizat et al., 1974; Charles-Dominique, 1983; Hand, 1984).

ACKNOWLEDGMENTS

The Peruvian Ministry of Agriculture is acknowledged for permission to work in the Manu National Park. In Paraguay Diane and Anthony Espinoza graciously gave me permission to study *Aotus* on their ranch, provided logistic help and encouragement. David Sivertson and Peter Sherman assisted in field work in Peru. Patrick Daniels provided expert assistance in Paraguay, and made the 24h follows possible. John Terborgh, Charles Janson, Robin Foster, Louise Emmons, A. Goldizen, John Fleagle, provided encouragement and support in Peru. Robin Foster identified plants in Peru. J.J. Earhart and Dave Norman with the Peace Corps gave logistic support in Paraguay. John Oates gave expertise and criticisms on earlier versions of this research. Marilyn Norconk, Paul Garber, Amanda Wright and John Fleagle gave insightful comments on this manuscript. Jackie Stephans helped with typing and corrections. Jukka Jernvall is thanked for his assistance with the graphics. But none of this research would have happened without the guidance, encouragement, patience, support and good humor of Warren Kinzey. I will be always grateful for his passport into the world of "monkey business" and his support at every step and crisis.

REFERENCES

Aquino, R.; Encarnacion, F. (1986a). Population structure of *Aotus nancymae* (Cebidae: Primates) in Peruvian Amazon lowland forests. *American Journal of Primatology*, 11:1–7.

Aquino, R.; Encarnacion, F. (1986b). Characteristics and use of sleeping sites in *Aotus* (Cebidae: Primates) in the Amazon lowlands of Peru. *American Journal of Primatology,* 11:319–331.

Aquino, R.; Encarnacion, F. (1988). Population densities and geographic distribution of night monkeys (*Aotus nancymae* and *Aotus vociferans*) (Cebidae: Primates) in Northeastern Peru. *American Journal of Primatology*, 14:375–381.

Aquino, R.; Encarnacion, F. (1990). Supplemental notes on population parameters of northeastern Peruvian night monkeys, genus *Aotus* (Cebidae). *American Journal of Primatology*, 21:215–221.

Aquino, R.; Encarnacion, F. (1994). Owl monkey population in Latin America: Field work and conservation. In *Aotus: The Owl Monkey*, J.F. Baer, R.E. Weller, and I. Kakoma (eds.), pp. 59–96, Academic Press, New York.

Ashley, M.V.; Vaughn, J.L. (in press). Owl monkeys, *Aotus*, are highly divergent in mitochondrial cytochrome c. oxidase (COII) sequences. *International Journal of Primatology*.

Bearder, S.K. (1987). Lorises, bushbabies and tarsiers: Diverse societies in solitary foragers. In *Primate Societies*, B.B. Smuts, D.L. Cheney, R.M. Seyfarth, R.W. Wrangham, and T.T. Struhsaker (eds.), pp. 11–24, Chicago University Press, Chicago.

Boinski, S.; Fragaszy, D.M. (1989). The ontogeny of foraging in squirrel monkeys, *Saimiri oerstedi*. *Animal Behaviour*, 37:415–428.

Boinski, S.; Mitchell, C.L. (1992). Ecological and social factors affecting the vocal behavior of adult female squirrel monkeys. *Ethology*, 92:316–330.

Brown, L.H.; Urban, E.K.; Newman, K. (1982). *The Birds of Africa*, Academic Press, New York.

Chapman, C.A.; Chapman, L.J. (1991). The foraging itinerary of spider monkeys: When to eat leaves? *Folia Primatologica* 56:162–166.

Charles-Dominique, P. (1975). Nocturnal primates and diurnal primates: An ecological interpretation of these two modes of life by analysis of the higher vertebrate fauna in tropical forest ecosystems. In *Phylogeny of the Primates: A Multidisciplinary Approach*, W.P. Luckett and F.S. Szalay (eds.), pp. 69–88, Plenum Press, New York.

Charles-Dominique, P. (1977). *Ecology and Behavior of Nocturnal Primates: Prosimians of Equatorial West Africa*, Columbia University Press, New York.

Charles-Dominique, P. (1983). Ecology and social adaptions in didelphid marsupials: Comparison with eutherians of similar ecology. In *Advances in the Study of Mammalian Behavior*, J.F. Eisenberg (ed.), pp. 395–442, American Society of Mammalogists, Lawrence.

Charles-Dominique, P.; Atramentowicz, M.; Charles-Dominique, M.; Gerard, H.; Hladik, A.; Hladik, C.M.; Prevose, M.F. (1981). Nocturnal, arboreal, frugivorous mammals in a Guianean forest: Interrelations of plants and animals. *Revue D'Ecologie (Terra et la Vie)*, 35:341–435.

Ciochon, R.L.; Chiarelli, A.B. (1980). *Evolutionary Biology of the New World Monkeys and Continental Drift*, Plenum Press, New York.

Crompton, R.H.; Andau, P.M. (1987). Ranging, activity, rhythms and sociality in free-ranging *Tarsius bancanus*: A preliminary report. *International Journal of Primatology*, 8:43–71.

Croizat, L.; Nelson, G.; Rosen, D.E. (1974). Centers of origin and related concepts. *Systematic Zoology*, 23:265–287.

Emmons, L.H. (1984). Geographic variation in densities and diversities of non-flying mammals in Amazonia. *Biotropica*, 16:210–222.

Emmons, L.H. (1987). Comparative feeding ecology of felids in a Neotropical rain forest. *Behavioral Ecology and Sociobiology*, 20:271–281.

Emmons, L.H. (1990). *Neotropical Rainforest Mammals: A Field Guide*, University of Chicago Press, Chicago.

Emmons, L.H.; Gautier-Hion, A.; DuBost, G. (1983). Community structure of the frugivorous-folivorous forest mammals of Gabon. *J Zool Lond*, 199:209–222.

Erickson, C.J. (1994). Tap-scanning and extractive foraging in aye-ayes, *Daubentonia madagascariensis*. *American Journal of Primatology*, 335:235–240.

Fleagle, J.G. (1988). *Primate Adaption and Evolution*, Academic Press, Inc., New York.

Fleming, T.H. (1979). Neotropical mammalian diversity: Faunal origins, community composition, abundance and function. In *The abundance of animal in Malesian rain forest. Transactions of the 6th Aberdeen-Hull Symposium on Malesian Ecology.*, A.G. Marshall (ed.).

Ford, S.M. (1994). Taxonomy and distribution of the owl monkey. In *Aotus: The owl monkey*, J.F. Baer, R.E. Weller, and I. Kakoma (eds.), pp. 1–57, Alan R. Liss, New York.

Ganzhorn, J.U.; Abraham, J.P.; Razaqnanhoera-Rakotomalala, M. (1985). Some aspects of the natural history and food selection of *Avahi laniger*. *Primates*, 26:453–463.

Ganzhorn, J.U.; Wright, P.C. (1995). Temporal patterns in primate leaf eating: The possible role of leaf chemistry. *Folia Primatologica*, 63:203–208.

Garber, P.A. (1986). The ecology of seed dispersal in two species of callitrichid primate (*Saguinus mystax* and *Saguinus fuscicollis*). *American Journal of Primatology*, 10:155–170.

Garber, P.A. (1992). Vertical clinging, small body size and the evolution of feeding adaptations in the Callitrichinae. *American Journal of Physical Anthropology,* 88:469–482.

Glander, K.E. (1982). The impact of plant secondary compounds on primate feeding behavior. *Yearbook of Physical Anthropology,* 25:1–18.

Goldizen, A.W. (1987). Tamarins and marmosets: Communal care of offspring. In *Primate Societies,* B.B. Smuts, R.N. Seyfarth, R.W. Wrangham, and T.T. Struhsaker (eds.), pp. 34–43, Chicago University Press, Chicago.

Goodman, S.M.; Langrand, O.; Raxworthy, C.J. (1993a). The food habits of the barn owl, *Tyto alba* at three sites in Madagascar. *Ostrich,* 64:160–171.

Goodman, S.M.; Langrand, O.; Raxworthy, C.J. (1993b). Food habits of the Madagascar long-eared owl *Asio madagascariensis* in two habitats in southern Madagascar. *Ostrich,* 64:79–85.

Gursky, S.L. (1994). Infant care in the spectral tarsier (*Tarsius spectrum*) Sulawesi, Indonesia. *International Journal of Primatology,* 15(6):843–853.

Hamilton, W.D. (1971). Geometry for the selfish herd. *J. Theor. Biology,* 31:295–311.

Hand, S. (1984). Bat beginnings and biogeography: A southern perspective. In *Vertebrate Zoogeography and Evolution in Australasia,* M. Archer and G. Clayton (eds.), Hesperian Press, Carlisle.

Harcourt, C. (1987). Brief trap/retrap study of the brown mouse lemur (*Microcebus rufus*). *Folia Primatologica,* 49:209–211.

Hartwig, W.C. (1995). Allometry and the ancestral platyrrhine cranial morphotype. *American Journal of Physical Anthropology Supplement,* 20:107.

Hartwig, W.C.; Cartelle, C. (1996). *Protopithecus* and the evolution of ateline New World monkeys. *American Journal of Physical Anthropology Supplement,* 22:121.

Hershkovitz, P. (1983). Two new species of night monkeys, genus *Aotus* taxonomy. *American Journal of Primatology,* 4:209–243.

Horowitz, I. (1995). A phylogenetic analysis of the basicranial morphology of New World Monkeys. *American Journal of Physical Anthropology Supplement,* 20:113.

Hunter, A.J.; Fleming, D.; Dixson, A.F. (1984). The structure of the vomeronasal organ and nasopalatine ducts in *Aotus trivirgatus* and some other primate species. *J. Anat.,* 138(2):217–225.

Jacobs, G.H.; Deegan II, J.F. (1993). Photopigments underlying color vision in ringtail lemurs (*Lemur catta*) and brown lemurs (*Eulemur fulvus*). *American Journal of Primatology,* 30(3):243–256.

Jacobs, G.H.; Deegan II, J.F.; Neitz, J.; Crognale, M.A.; Neitz, M. (1994). Photopigments and color vision in the nocturnal monkey *Aotus*. *Vision Res.,* 33:1773–1783.

Janson, C.H. (1984). Female choice and mating system of the brown capuchin monkey *Cebus apella* (Primates: Cebidae). *Zeitschrift Tierpsychologie,* 65:177–200.

Janson, C.H. (1985). Aggressive competition and individual food consumption in wild brown capuchin monkeys (*Cebus apella*). *Behavioral Ecology and Sociobiology,* 18:125–138.

Janson, C.H. (1986). The mating system as a determinant of social evolution in capuchin monkeys (*Cebus*). In *Primate Ecology and Conservation,* J.C. Else and P.C. Lee (eds.), pp. 169–179, Cambridge University Press, Cambridge.

Janson, C.H.; Emmons, L.H. (1990). The ecological structuring of the nonflying mammal communities at Cocha Cashu Biological Station, Manu National Park, Peru. In *Four Neotropical Rainforests,* A.H. Gentry (ed.), pp. 314–338, Yale University Press, New Haven.

Kay, R.F. (1994). "Giant" tamarin from the Miocene of Colombia. *American Journal of Physical Anthropology,* 95:333–353.

Kay, R.F.; Johnson, D.D. (1996). New platyrrhines from the middle Miocene of Argentina. *American Journal of Physical Anthropology Supplement,* 22:136–137.

Kinzey, W.G. (1981). The titi monkeys, genus *Callicebus*. In *Ecology and Behavior in Neotropical Primates,* A.F. Coimbra-Filho and R.A. Mittermeier (eds.), pp. 241–276, Academia Brasileira de Ciencias, Rio de Janeiro.

LeMaho, Y.; Goffart, M.; Rochas, A.; Felbabel, H.; Chatonnet, J.C. (1981). Thermoregulation in the only nocturnal simian: The night monkey *Aotus trivirgatus*. *American Journal of Physiology,* 240:156–165.

Ma, N.S.F.; Aquino, R.; Collins, W.E. (1985). Two new karyotypes in the Peruvian owl monkey (*Aotus trivirgatus*). *American Journal of Primatology,* 9:333–341.

MacKinnon, J.R.; MacKinnon, K.S. (1980). The behavior of wild spectral tarsiers. *International Journal Primatology,* 1:361–380.

McNab, B.; Wright, P.C. (1987). Temperature regulation and oxygen consumption in the Philippine tarsier (*Tarsius syrichta*). *Physiological Zoology,* 60(5):596–600.

Meyers, D.M.; Wright, P.C. (1993). Resource tracking: Food availability and *Propithecus* seasonal reproduction. In *Lemur Social Systems and Their Ecological Basis,* P.M. Kappeler and J.U. Ganzhorn (eds.), pp. 179–192, Plenum Press, New York.

Milton, K. (1979). Factors influencing leaf choice by howler monkeys: A test of some hypotheses of food selection by generalist herbivores. *American Naturalist,* 114:362–378.

Mitchell, C.L. (1990).The ecological basis for female social dominance: A behavioral study of the squirrel monkey (*Saimiri sciureus*) in the wild. Ph.D. dissertation, Princeton University.

Mitchell, C.L.; Boinski, S.; van Schaik, C.P. (1991). Competitive regimes and female bonding in two species of squirrel monkeys (*Saimiri oerstedi* and *S. sciureus*). *Behavioral Ecology and Sociobiology*, 28:55–60.

Moynihan, M.A. (1964). Some behavior patterns of platyrrhine monkeys: The night monkey (*Aotus trivirgatus*). *Smith. Misc. Coll.,* 146:1–84.

Nash, L.T. (1995). Seasonal changes in time budgets and diet of *Lepilemur leucopus* from Southwestern Madagascar. In *International Conference on the Biology and Conservation of Prosimians*, 13–16 September, The North of England Zoological Society, Chester.

Pariente, G. (1980). Light available in the forest. In *Nocturnal Malagasy Primates: Ecology, Physiology and Behavior*, P. Charles-Dominique, H.M. Cooper, A. Hladik, C.M. Hladik, F. Pages, G. Pariente, J.J. Petter, A. Petter-Rousseaux, and A. Schilling (eds.), pp. 117–134, Academic Press, New York.

Peetz, A.; Norconk, M.A.; Kinzey, W.G. (1992). Predation by a jaguar on howler monkeys (*Alouatta seniculus*) in Venezuela. *American Journal of Primatology*, 28:223–228.

Peres, C.A. (1993). Structure and spatial organization of an Amazonian terra firme forest primate community. *Journal of Tropical Ecology*, 9:259–276.

Pieczarka, J.C.; Desouza Barros, R.M.; DeFana, F.M., Jr.; Nagamachi, C.V. (1993). *Aotus* from the Southwestern Amazon region in geographically and chromosomally intermediate between *A. azarae boliviensis* and A. *infulatus*. *Primates*, 34:197–204.

Potts, R.; Behrensmeyer, A.K. (1992). Late Cenozoic terrestrial ecosystems. In *Terrestrial Ecosystems Through Time: Evolutionary Paleoecology of Terrestrial Plants and Animals*, A.K. Behrensmeyer, J.D. Damuth, W.A. DiMichele, R. Potts, H. Sues, and S.L. Wing (eds.), pp. 418–519, Chicago University Press, Chicago.

Rettig, N.L. (1978). Breeding behavior of the harpy eagle (*Harpia harpyja*). *The Auk*, 95:629–643.

Rosenberger, A.L. (1995). The power of fossils, the pitfalls of parsimony-platyrrhine phylogeny. *American Journal of Physical Anthropology*, 20:184.

Rosenberger, A.L.; Hartwig, W.C.; Takai, M.T.; Setoguchi, T.; Shigehara, N. (1991). Dental variability in *Saimiri* and the taxonomic status of *Neosaimiri fieldsi*, an early squirrel monkey from La Venta, Colombia. *International Journal of Primatology*, 12:291–302.

Sarich, V.M.; Cronin, J.E. (1980). South American mammal molecular systems, evolutionary clocks, and continental drift. In *Evolutionary Biology of the New World Monkeys and Continental Drift*, R.L. Ciochon and A.B. Chiarelli (eds.), pp. 399–422, Plenum Press, New York.

Schmid, J.; Kappeler, P.M. (1994). Sympatric mouse lemurs (*Microcebus* spp.) in western Madagascar. *Folia Primatologica*, 63:162–170.

Setoguchi, T.; Rosenberger, A.I. (1987). A fossil owl monkey from La Venta, Columbia. *Nature*, 326(6114):692–694.

Sherman, P.T. (1991). Harpy eagle predation on a red howler monkey. *Folia Primatologica*, 56(1):53–56.

Simpson, G.G. (1967). The beginning of the age of mammals in South America Part 2. *Bulletin of the American Museum of Natural History*, 137:1–260.

Simpson, G.G. (1980). *Splendid Isolation, The Curious History of South American Mammals.* Yale University Press, New Haven.

Stephan, H.; Frahm, H.; Baron, G. (1981). New and revised data of volumes of brain structures in insectivores and primates. *Folia Primatologica*, 35:1–29.

Sterling, E.J. (1993). Patterns of range use and social organization in aye-ayes (*Daubentonia madagascariensis*) on Nosy Mangabe. In *Lemur Social Systems and Their Ecological Basis*, J. Ganzhorn and P. Kappeler (eds.), pp. 1–10, Plenum Press, New York.

Sterling, E.J.; Richard, A.F. (1995). Social organization of the aye-aye (*Daubentonia madagascariensis*) and the perceived distinctiveness of nocturnal primates. In *Creatures of the Dark: The Nocturnal Prosimians*, L. Alterman, G.A. Doyle, and M.K. Izard (eds.), pp. 439–452, Plenum Press, New York.

Symington, M.M. (1988). Demography, ranging patterns, and activity budgets of black spider monkeys (*Ateles paniscus chamek*) in the Manu National Park, Peru. *American Journal of Primatology*, 15:45–67.

Terborgh, J. (1983) *Five New World Primates: A Study in Comparative Ecology.* Princeton University Press, Princeton.

Thiollay, J.M. (1994). Structure, density and rarity in an Amazonian rainforest bird community. *Journal of Tropical Ecology*, 10:449–481.

Thorington, R.W., Jr.; Vorek, R.E. (1976). Observations on the geographic variation and skeletal development of *Aotus*. *Lab. Anim. Sci.,* 26(6):1006–1021.

Vermeij, G. (1992). Time of origin and biogeographic history of specialized relationships between northern marine plants and herbivorous molluscs. *Evolution*, 46:657–664.

Waterman, P.G. (1984). Food acquisition and processing as a function of plant chemistry. In *Food Acquisition and Processing in Primates*, D.J. Chivers, B.A. Wood, and A. Bilsborough (eds.), pp. 177–211, Plenum Press, New York.

Wright, P.C. (1978). Home range activity pattern and agonistic encounters of a group of night monkeys (*Aotus trivirgatus*) in Peru. *Folia Primatologica*, 29:43–55.

Wright, P.C. (1984). Biparental care in *Aotus trivirgatus* and *Callicebus moloch*. In *Female Primates: Studies by Women Primatologists*, M.E. Small (ed.), pp. 59–75, Alan R. Liss, Inc, New York.

Wright, P.C. (1985). The cost and benefits of nocturnality for *Aotus trivirgatus* (the night monkey). Ph.D. Dissertation, City University of New York.

Wright, P.C. (1986). Ecological correlates to monogamy. In *Primate Ecology and Conservation*, J.C. Else and P.C. Lee (eds.), pp. 159–168, Cambridge University Press, Cambridge.

Wright, P.C. (1989). The nocturnal primate niche in the new world. *Journal of Human Evolution*, 18:635–658.

Wright, P.C. (1994). The behavior and ecology of the owl monkey. In *Aotus: The Owl Monkey*, J.F. Baer, R.E. Weller, and I. Kakoma (eds.), pp. 97–111, Academic Press, Inc., New York.

Wright, P.C.; Heckscher, S.K.; Dunham, A. (in press). Predation on Milne-Edward's sifaka (*Propithecus diadema edwardsi*) by the fossa (*Cryptoprocta ferox*) in the rain forest of southeastern Madagascar. *Folia Primatologica*.

19

DIET AND FEEDING ECOLOGY OF MASKED TITIS (*Callicebus personatus*)

Klaus-Heinrich Müller

Deutsches Primatenzentrum
Kellnerweg 4
37077 Gёttingen, Germany

INTRODUCTION

With thirteen species (Hershkovitz, 1990), *Callicebus* is one of the most diverse but poorly studied genera of neotropical primates. Long-term field studies have been conducted on only two of the 13 recognized *Callicebus* species. Wright (1985) investigated dusky titis, *C. brunneus* (taxon replaces *C. moloch)* for 63 complete days during 11 months at the Cocha Cashu Biological Research Station in the Manu National Park of Peru. *Callicebus torquatus*, the yellow-handed titi has been the subject of a number of both short- and long-term studies by Freese (1975), Kinzey (1975, 1977, 1981), Kinzey & Gentry (1979), Izawa (1976) and Easley (1982). A short-term study which consisted of six complete days during October and November 1977 were made on the ecology of masked titis, *Callicebus personatus* (Kinzey & Becker, 1983).

Callicebus is restricted to South America; masked titis are endemic to the Atlantic rain forest of eastern Brazil. *C. personatus* are allopatric to all other *Callicebus* species and *C. torquatus* and *C. brunneus* have no overlapping ranges within the Amazonian rain forest (Hershkovitz, 1990).

Behavior of the members of genus *Callicebus* is characterized by three main features: (1) a nuclear family (or obligate monogamy) pattern of social organization (Defler, 1983; Easley, 1982; Kinzey, 1981; Mason, 1966; Welker et al., 1981); (2) the maintenance and defense of a territory (Kinzey and Robinson, 1983; Moynihan, 1976; Robinson, 1977; 1979a and b; 1981) and (3) a basically frugivorous feeding pattern (Easley, 1982; Kinzey, 1977; 1978; Kinzey & Becker, 1983; Wright, 1985). The frugivorous portion of the diets ranged from 54 % in *C. brunneus*, 74 % in *C. torquatus* to 81 % in *C. personatus* of the total diet. Differences occur in the non-frugivorous portion of the diet: *C. torquatus* supplements the frugivorous portion of its diet primarily with insects (Easley, 1982), *C. brunneus* with insects and leaves (Wright, 1985) and preliminary data suggested that *C. personatus* supplemented their diet with leaves only (Kinzey and Becker, 1983). These differences could be due to habitat type or to resource availability due to seasonal variation. It might be expected that seasonality has affected the diet of primates living so far from the equator like masked titis. The purpose of this study was to observe masked titi monkeys in a long-term-study to re-evaluate Kinzey & Becker`s (1983) results and to obtain information on seasonal changes of diet and feeding behavior.

This study provides the first detailed account of the diet of *C. personatus* recorded over a period of eleven months. I focused on plant species diversity in the diet and seasonal variation of the feeding behavior in order to compare my findings with the Amazonian *C. brunneus* and *C. torquatus*.

METHODS AND STUDY AREA

Study Area

The study site was a forest segment of the Estação Experimental Lemos Maia (ESMAI). ESMAI is an experimental field station of the Executive Cocoa Planting Commision (CEPLAC), located near Una (15°18`S, 39°06`W), Bahia, Brazil (Fig. 1). ESMAI consists of about 400 ha of protected Atlantic coastal rain forest, Mata Atlantica, at an elevation of less than 100 m. The study area of about 100 ha is surrounded on three sides by fields or dirt roads (Fig. 2). The fourth boundary is connected with other forest segments

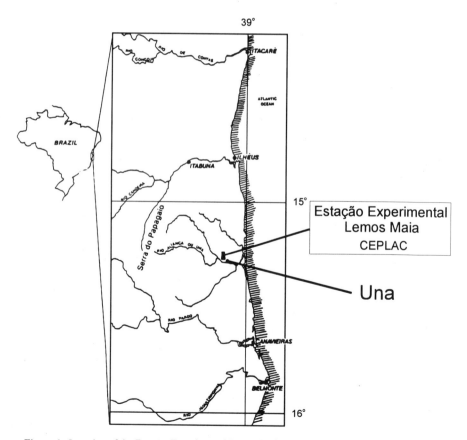

Figure 1. Location of the Estação Experimental Lemos Maia (ESMAI) in the South of Bahia, Brazil.

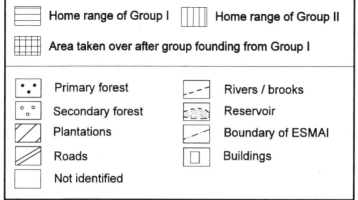

Figure 2. Study site and home ranges of the masked titi Group I and II at the ESMAI.

of ESMAI and privately owned forest. The terrain of the study site is gently undulating, with small streams, primary and secondary forest and sandy soil (Hohl, 1993). Habitat types within the study area have been classified by Rylands (1982) as (1) tall primary forest on sloping ground; (2) tall primary forest on low, wet ground; (3) tall primary forest on dry flat ground on hill tops; and (4) second growth in severely disturbed primary forest. I classified an additional forest type used fequently by the monkeys: (5) second growth after clearing (Müller, 1995a). The climate is characterized by a mean yearly temperature of 23.6° C and a high annual rainfall of 2,011 mm measured between 1968 and 1989 (Hohl, 1993). During this period average annual rainfall varied from 1,264 to 2,750 mm. Rainfall measured during the study totaled 1,522 mm between October 1992 and September 1993. Monthly rainfall is not markedly seasonal (Fig. 3). Average monthly temperature during study period was 24.0°C. The study period was divided into a warm season between November and May (average of monthly temperature ≥ 24°C) and a colder season between June and October (average of monthly temperature < 24°C).

Population density of 4 groups/km² was estimated using the range of group I and II. Using an average group size of 4 (an average of 3 known groups) I calculated a density of 17 titis /km². Other primates in the forest included *Leontopithecus chrysomelas* and *Callithrix kuhli*. *Cebus apella xanthosternos*, *Alouatta caraya* and *Brachyteles arachnoides* are locally extinct in this forest due to hunting (Mittermeier, 1987).

Capturing and Radio-Telemetry

Radio telemetry was essential to successful habituation of masked titis and all-day follows. Animals were darted using a carbon dioxide powered darting gun and reusable syringe darts (Telinject, Römerberg, Germany: Type Vario IV.3 1 NP). The monkeys were anesthetized using a mixture of Ketaset (30 - 45 mg Ketamine hydrochloride) and Rompun (6 - 9 mg Xylacine). A 40g radio transmitter pouch (K. Wagener, Köln, Germany)

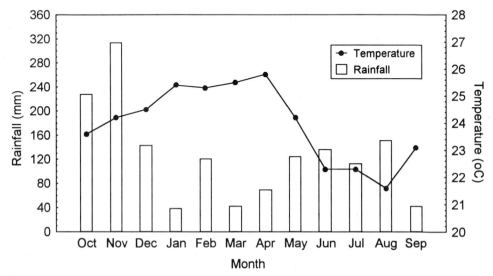

Figure 3. Monthly mean temperature (°C) and rainfall (mm) at the Estação Experimental Lemos Maia, Una, Bahia, Brazil during October 1992 and September 1993.

was fitted to the hip of the monkey and habituation, location and following the group were very successful using a receiver and H-antenna (K. Wagener, Kēln, Germany).

Study Groups

During preliminary stages of the study, Group I consisted of six animals: an adult pair, two subadults, a juvenile II and a juvenile I (for age classification, see Kinzey, 1981). Between the beginning of 1992 and August 1992 the adult male and a subadult disappeared and the juvenile II died. In October 1992 when I started systematic observations, Group I consisted of three animals: an adult female, an adult male (the previous subadult) and a juvenile II (the previous juvenile I). In December 1992, the adult male who was wearing a radio transmitter emigrated and founded a new group together with a new adult female and her infant. This group will be referred to as Group II (see Fig. 2). Details on emigration and group formation were described by Müller (1995b).

Group I was observed for 18 days in October 1992 and November 1992 for a total of 192 hours. After emigration of the mature offspring, Group II was observed from January through September 1993. The habituation of the adult female of Group II was accomplished in a two week period. Data on Group II were collected on 83 days for a total of 838 hours.

Data on behavior were collected using a scan sampling method as in previous studies of *Callicebus* (Kinzey et al., 1977; Easley, 1982). Samples were taken at exactly five minute intervals for each visible individual of the group.The following behaviors were recorded (as defined by Easley, 1982): locomoting, resting, playing and feeding. If feeding, additional data were collected on ingested food item (fruits, leaves, flowers or miscellaneous material as soil, insect etc.), ripeness of resource (immature, mature), and food source (tree, liana, epiphyte or parasite). All feeding trees *C. personatus* used were marked, identified and measured. Trees were identified to the level of species or genus by scientific name or by the Brazilian vernacular name with help from the Herbarium of CEPEX (Dr. Andre Carvalho and Ser Talmon). Lianas, ephiphytes and parasites could not be identified. The following measurements were taken on feeding trees: diameter at breast height (DBH), height of trees and crown diameter (CD). No phenological study or characterization of the vegetation structure of the study site was undertaken.

Two different feeding patterns were observed. Type I occurred when a tree had many ripe fruits or abundant young leaves. All members of the group entered the tree together, remained there feeding, and finally left the tree together, often leaving many fruits or leaves on the tree. The titis often rested after such a feeding session. Type II feeding pattern could also be characterized as foraging. The titi group moved slowly at a height of about 2 to 10 m and the individuals were spread over a large area. Thus the individuals of the group could be out of sight. Food sources were few and isolated, such as only one leaf or fruit of a liana species. I compared feeding samples of fruits and leaves during type I and type II. Masked titis were more likely to encounter fruits by the feeding pattern type I ("feeding") and leaves by the feeding pattern type II ("foraging"). If feeding lasted longer then five minutes, then the tree was called a "feeding site".

Data Analysis and Statistics

Data analysis was limited to adult and subadult titis (as described by Kinzey, 1977 and Easley 1982). Data of Group I and II were combined because group composition, home range size, and individuals of these two groups were similar. Percentages of behavior patterns were calculated from the scan samples. A total of 6683 feeding samples were analyzed: 1371 samples from Group I (adult female and subadult male) and 5312 samples from Group II (adult male and adult female). Frequencies of the monthly variation of feeding behavior and percentage of different food items used were calculated using the total number of feeding samples.

I defined a tree as a "feeding site", when one or more individuals were feeding during two or more consecutive time samples. Distance from one feeding site to another, day range length, home range size were calculated by superimposing a grid of 50 x 50 m quadrats (0.25 ha) on map of the study area. Parametric (c^2) and nonparametric (Spearman-rank-correlation test, Mann-Whitney U test) statistical tests, were performed with the help of STATISTICA/CSS-programs according to the methods outlined by Siegal (1956).

RESULTS

Daily Activity Patterns

The "alert period" (Chivers 1974) averaged 11h 12 min ± 54 min (range 8 hrs 40 min to 11 hrs 03 min). The monkeys rested during 39 % of the samples, were recorded as locomoting during 32 % of the samples, and feeding during 27 % of the samples. Data both on playing behavior or when the titis were out of sight were collected in about one percent of the total time samples. No difference in activity pattern could be found between Group I and II (c^2-test: p = 0.98). A significantly longer alert period (Mann-Whitney-U-

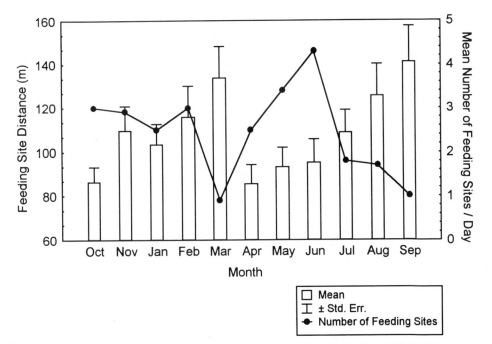

Figure 4. Mean distance traveled each month between feeding sites and mean number of daily feeding sites used during the study period.

test: p = 0.017) and a greater percentage of resting behavior (Mann-Whitney-U-test: p = 0.038) was observed during the warm season than during the cold season.

Travel and Feeding Patterns

Average daily path length during the study period was 1,007 m (SD ± 202 m) and the "straight- line distance between the two most remote points in a day's travel" (Mason, 1968:208–209) was 341 m (SD ± 45 m). Both of the two titi groups used an area of about 24 ha for their territory (Fig. 2). Masked titi monkeys traveled 109 m (SE = 19 m) between feeding sites. There was monthly variation in feeding site distance, which showed two peaks (Fig. 4). During March and September, the mean distance between feeding sites was greater than during October or April. The monthly average *number of feeding sites* used daily was negatively correlated with the *distance between feeding sites* (Spearman rank correlation: p = 0.028; r_s = - 0.657; N = 11). No correlation was found between *rainfall and feeding site distance* (p = 0.5; r_s = - 0.23) or *rainfall and number of feeding sites* (p = 0.09; r_s = - 0.52).

During the 11 month study the two masked titis groups spent an average of 76.6 ± 5.2 % of the feeding samples consuming fruits. No dietary differences were observed between Group I and II (c^2-test: p = 0.5). Pulp alone, seeds alone, pulp and seeds or the entire fruit were ingested (Table 1). Leaves were consumed in 17.2 ± 4.8 % of the feeding samples. About half of the leaves came from liana species (98% young, 2 % mature) while the other half were from trees species (≈50 % young; ≈50 % mature). In approximately 2% of the feeding samples, the titis consumed other food items such as a stem of tree, medulla, or soil (14 observations).

Composition of Flora and Dietary Diversity

The dietary sources of *C. personatus* included a minimum of 91 tree species (of which 88 were identified), belonging to 33 families of plants (Table 2). Myrtaceae (13 species) and Sapotaceae (10 species) provided the widest diversity of fruit species used as food by masked titis. Rubiaceae (4 species) and the Euphorbiaceae (4 species) provided the most diverse sources for leaves. Together, species in the families Myrtaceae, Sapo-

Table 1. Dietary composition of masked titi Group I
and II during 11 months

	Feeding samples	% of diet
Fruit		
Pulp	2820	42.2
Pulp and seeds	822	12.3
Entire fruit	13	0.2
Seeds	1464	21.9
Tree leaves		
Young	261	3.9
Mature	327	4.9
Liana leaves		
Young	555	8.3
Mature	7	0.1
Other	120	1.8
Not identified	294	4.4

Table 2. Diversity of the plant families in the diet of *C. personatus*. % Fruits = % of overall feeding samples while eating fruits; % Leaves = % of overall feeding time samples while eating leaves

No.	Plant family	Number of species	% Fruits	% Leaves	Relative% leaves
1	Myrtaceae	14	12.6	0.1	0.5
2	Sapotaceae	10	21.5	0.1	0.5
3	Rubiaceae	7	3.5	1.4	7.7
4	Euphorbiaceae	6	6.5	2.2	12.0
5	Caesalpiniaceae	5	1.9	0.0	0.0
6	Melastomataceae	5	0.7	0.1	0.5
7	Lauraceae	4	0.5	0.0	0.0
8	Annonaceae	3	1.5	0.0	0.0
9	Apocynaceae	3	0.2	0.0	0.0
10	Chrysobalanaceae	3	0.7	1.0	5.5
11	Moraceae	3	17.1	0.0	0.0
12	Clusiaceae-Guttifereae	2	0.0	<0.1	0.4
13	Humiriaceae	2	0.1	0.0	0.0
14	Sapindaceae	2	2.4	0.0	0.0
15	Anarcardiaceae	1	<0.1	0.0	0.0
16	Araliaceae	1	<0.1	0.0	0.0
17	Burseraceae	1	0.4	0.0	0.0
18	Caryocaraceae	1	1.8	0.0	0.0
19	Celastraceae	1	0.0	<0.1	0.4
20	Elaeocarpaceae	1	0.0	0.1	0.5
21	Fabaceae	1	0.0	0.1	0.5
22	Hippocrateaceae	1	<0.1	0.0	0.0
23	Lecythidaceae	1	2.9	0.0	0.0
24	Malpighiaceae	1	0.4	0.0	0.0
25	Mimosaceae	1	0.1	0.0	0.0
26	Myristicaceae	1	1.7	0.0	0.0
27	Nyctaginaceae	1	0.2	0.3	1.6
28	Papilionaceae	1	0.1	0.0	0.0
29	Rhizophoraceae	1	0.2	0.0	0.0
30	Simaroubaceae	1	0.0	0.2	1.1
31	Verbenaceae	1	<0.1	0.0	0.0
32	Violaceae	1	0.0	0.1	0.5
33	Vochysiaceae	1	0.0	0.7	3.8
	Unidentified	—	4.8	11.8	64.5
	Sum	88	81.7	18.3	100.0

taceae and Moraceae accounted for 51.4 % of the total feeding samples of the two groups. Rubiaceae and Euphorbiaceae accounted for 3.6 % of the total feeding samples and 19.7 % of leaf-feeding samples. Of the 91 species in the diet of Group I and II, 69 were used for fruits (including seeds), 30 for foliage, 1 for flower, and 9 species of which more than 1 part was eaten (Table 3).

Characteristics of Feeding Trees

Only 10 plant species accounted for about 50% of the masked titi diet (Table 4). The most important plant species were *Manilkara longifolia* and *Sprucella crassipedicellata*. These trees were characterized by a height of about 20 -25 m, with a crown diameter of

Table 3. Species, species characteristics and plant part consumed by masked titi monkey Group I and II from October 1992 to September 1993. Part = part eaten by masked titis: P = only pulp, S = only seed, P & S = pulp and seed, EF = entire fruit, YL = young leaf, ML = mature leaf, FLO = flower; N = Number of measured trees

No.	Scientific Nomenclature (when known)	Vernacular	Family	Part	N	DBH ± SD	HEIGHT ± SD	CD ± SD
1	Actinostemon spec.		Euphorbiaceae	ML	23	5.2 ±3.2	7.1 ±3.8	2 ±1.1
2	Allophylus spec.	Rhin verde	Sapindaceae	S	9	14.9 ±7.5	15.4 ±3.6	4.2 ±3.1
3	Andira spec.	Angelim	Papilionaceae	EF	3	9.5 ±2.4	12 ±1	4 ±1.7
4	Aniba cf. intermedia (Meissn.) Mez.	Louro–III	Lauraceae	P	1	6.4 ±0	7 ±0	2 ±0
5	Aparisthmium cordatum (A.Juss) Baill.	Lava-pratos-branco	Euphorbiaceae	S + ML	18	9.0 ±2	8.7 ±1.8	4.1 ±0.8
6	Apuleia leiocarpa (Vog.) Macbr.	Jitai-amarello	Caesalpiniaceae	ML	1	20.7 ±0	18 ±0	5.5 ±0
7	Arapatiella psilophylla (Harms.) Cowan.	Arapati	Caesalpiniaceae	S	8	28.1 ±9.7	18.3 ±3.3	6.7 ±2.8
8	Aspidosperma multiflorum DC.	Petia-amarello	Apocynaceae	S	2	24.8 ±7.6	20 ±2.8	6.5 ±0.7
9	Aspidosperma oblongum A.DC.	Pau-quina	Apocynaceae	S	-	-	-	-
10	Brosimum rubescens Taub.	Conduru-vermelho	Moraceae	S	40	20.9 ±6.3	19.2 ±3.1	6 ±2.5
11	Byrsonima sericea DC.	Murici-da-Mata	Malpighiaceae	P + ML	7	24.5 ±6.3	18.4 ±4.2	8.1 ±3.2
12	Caryocar edule	Petia-preta	Caryocaraceae	P	24	12.0 ±4.4	13.5 ±3.3	3.5 ±1.2
13	Copaifera langsdorfii Desf.	Pao-oleo-do-sertao	Caesalpiniaceae	ML	7	40.4 ±7.9	23 ±2.8	10.7 ±3.8
14	Couepia ovalifolia (schott.) Benth.	Oiti-Folha-pequen.	Chrysobalanaceae	ML	7	9.9 ±5.9	12.6 ±6.7	3.3 ±1.8
15	Dialium guiananse (Aubl.) Sandw.	Jitai-preto	Caesalpiniaceae	S	1	28.7 ±0	21 ±0	12 ±0
16	Didymopanax morototoni Done & Planch	Matatauba	Araliaceae	P	-	-	-	-
17	Ecclinusa ramiflora Mart.	Bapeba-de-nervura	Sapotaceae	P	5	16.8 ±6.9	14.8 ±4	4 ±1.5
18	Eschweilera ovata (Cambess.) Miers.	Biriba/Embiriba	Lecythidaceae	S	18	26.5 ±8.9	21.1 ±4.3	6.5 ±3.3
19	Eugenia spec. I	Araca I	Myrtaceae	P	5	18.0 ±6	20 ±5.1	7.9 ±4.9
20	Eugenia spec. II	Araca II	Myrtaceae	P	3	20.5 ±8.6	17.3 ±6.7	3.3 ±1.1
21	Eugenia spec. III	Araca branco I	Myrtaceae	P	4	27.5 ±12.5	19.5 0.6±	6.3 ±1.9
22	Eugenia spec. IX	Araca - laranja	Myrtaceae	P	2	11.3 ±2.9	14.5 ±0.7	3.8 ±0.4
23	Eugenia spec. V	Araca branco III	Myrtaceae	P	6	20.2 ±8.9	18.8 ±4.4	6.2 ±2.5
24	Eugenia spec. VI	Araca vermelho I	Myrtaceae	P	7	15.8 ±8.9	13.9 ±2.8	3.1 ±1.6
25	Eugenia spec. VII	Araca vermelho II	Myrtaceae	P	7	15.2 ±8.6	16.1 ±4.9	3.7 ±2.1
26	Eugenia spec. VIII	Araca vermelho III	Myrtaceae	P	13	17.7 ±10.6	16.9 ±4.3	4.8 ±3
27	Eugenia spec. X	Araca-casca amarella	Myrtaceae	P	3	20.8 ±3.5	19 ±1	4.3 ±1
28	Eugenia spec. XI	Araca-Germinosa	Myrtaceae	P	7	17.3 ±4.9	18.3 ±1.9	5 ±1.4
29	Eugenia spec. XII	Araca -casca furada	Myrtaceae	P	1	38.9 ±0	18 ±0	5 ±0
30	Eugenia spec.IV	Araca branco II	Myrtaceae	P	2	13.3 ±0.7	12.2 ±10.1	3.5 ±1.4

Table 3. (continued)

No.	Scientific Nomenclature (when known)	Vernacular	Family	Part	N	DBH ± SD	HEIGHT ± SD	CD ± SD
31	Euphorbia spec.		Euphorbiaceae	YL	7	13.0 ±1.7	16.1 ±1.1	3.2 ±1.3
32	Guatteria spec.	Pindaiba	Annonaceae	EF	4	14.4 ±5.9	14.8 ±5	5.2 ±1.2
33	Guettarda spec.	Arariba-II	Rubiaceae	EF	1	14.3 ±0	16 ±0	4 ±0
34	Helicostylis tomentosa (Poepig & Engl.) Rusby	Amora-preta	Moraceae	P	48	15.4 ±4.5	15.7 ±3.1	4.2 ±1.5
35	Henriettea succosa (Aubl.) DC.	Munduru-ferro	Melastomataceae	P	8	8.4 ±3.1	8.6 ±1.5	4.2 ±1.4
36	Humiria balsamifera (Aubl.) St.Hil.	Murtim	Humiriaceae	ML	1	25.5 ±0	17 ±0	2.5 ±0
37	Hyeronima alchorneoides Fr. Allem.	Cajueiro-vermelho	Euphorbiaceae	ML	2	23.4 ±10.1	14.5 ±5	5 ±0.7
38	Inga nuda	Inga-sabao	Mimosaceae	S	1	8.6 ±0	10 ±0	3.5 ±0
39	Licania hypoleuca Benth. var. hydera	Oiti-mirim	Chrysobalanaceae	P & S	1	70.1 ±0	28 ±0	12.5 ±0
40	Licania santosii Prance	Oiti-pintada	Chrysobalanaceae	P & S	5	24.3 ±8	21.6 ±3.1	8.2 ±2.5
41	Lucuma spec.	Mucuri	Sapotaceae	P	1	20.4 ±0	17 ±0	4.5 ±0
42	Manilkara longifolia	Paraju	Sapotaceae	P & S	101	51.3 ±17.1	22.9 ±3.4	11.9 ±3.5
43	Manilkara salzmannii	Macaranduba III	Sapotaceae	P & S	1	55.7 ±0	20 ±0	12 ±0
44	Manilkara spec.	Macaranduba II	Sapotaceae	P & S	1	29.3 ±0	24 ±0	8 ±0
45	Manilkara spec.	Macaranduba I	Sapotaceae	P & S	1	5.7 ±0	8 ±0	1.5 ±0
46	Maytenus spec.		Celastraceae	ML	1	13.1 ±0	9 ±0	3 ±0
47	Melanoxylon braunia Schott.	Brauna-preta	Caesalpiniaceae	P	1	41.4 ±0	11 ±0	3 ±0
48	Miconia spec. I		Melastomataceae	P	5	17.3 ±8.8	18 ±4.6	6.7 ±3.6
49	Miconia spec. II		Melastomataceae	YL	2	2.2 ±0.4	3 ±0	1.2 ±0.3
50	Miconia spec. III	Mundururu	Melastomataceae	P	12	8.4 ±3.7	9.8 ±4.7	3.6 ±1.7
51	Micropholis guyanensis (DC.) Pres.	Abiu-da-Mata	Sapotaceae	P	1	22.9 ±0	20 ±0	5 ±0
52	Micropholis spec.	Bapeba	Sapotaceae	P	9	18.2 ±4.1	14.8 ±3.8	4.7 ±2.2
53	Nees spec.	Farinha-seca	Nyctaginaceae	YL	7	13.7 ±9.6	12.6 ±7.5	3.9 ±2.7
54	Ocotea spec.	Oiti (b. B.)	Lauraceae	P & S	8	24.2 ±5.4	19.1 ±2.2	5.4 ±1.2
55	Ocotea spec.	Louro-prego	Lauraceae	P & S	4	22.2 ±9.4	17.5 ±1.3	5.5 ±1.3
56	Ocotea spec.	Louro	Lauraceae	P & S	6	17.0 ±13.9	14.3 ±5.6	4.2 ±2.6
57	Ophthalmoblapton pedunculare Muell. Arg.	Fruta-invermo	Euphorbiaceae	S + ML	37	8.9 ±4.8	10.3 ±3.3	3.6 ±1.3
58	Palicouria guianensis Aupl.	Arariba-I	Rubiaceae	ML	1	3.2 ±0	6 ±0	2 ±0
59	Pera glabrata (Schott.) Baillon	Pau-oleo-branco	Euphorbiaceae	P	4	16.2 ±10.1	14.8 ±5	3.6 ±0.5

Table 3. (*continued*)

No.	Scientific Nomenclature (when known)	Vernacular	Family	Part	N	DBH ± SD	HEIGHT ± SD	CD ± SD
60	Pera heteranthera (Schrank.) I.M. Jhtu.	Pau-de-cachimbo	Euphorbiaceae	P	2	10.7 ±2.1	12 ±0	3.5 ±2.1
61	Peschiera salzmanii		Apocynaceae	P + ML	4	2.4 ±1.4	2.6 ±0.9	2 ±1.4
62	Pourouma spec.	Tararanga	Moraceae	P	29	21.6 ±7.1	16.3 ±2.3	6.6 ±2.2
63	Pouteria spec.	Bapeba-vermelha	Sapotaceae	P + ML	18	22.9 ±6.5	19.2 ±2.5	5.4 ±1.8
64	Protium heptaphyllum (Aubl.) March.	Amescla-mirim	Burseraceae	P + ML	8	16.1 ±9.1	13.8 ±3.1	5.7 ±2.5
65	Psidium spec.	Batingucu	Myrtaceae	ML	3	18.4 ±2.4	17.3 ±1.5	4.3 ±0.6
66	Psychotria	Pau-cravo	Rubiaceae	P + ML	20	5.9 ±2.8	7.8 ±2.5	2.3 ±0.8
67	Pterocarpus violaceus Vog.	Pau-sangue	Fabaceae	ML	2	15.0 ±3.2	15 ±2.8	3 ±0
68	Rhizophora mangle L.	Mangue-vermelho	Rhizophoraceae	P + FLO	2	13.7 ±4	15 ±2.8	4 ±1.4
69	Rinorea guianensis Aublet.	Cinzeiro	Violaceae	ML	2	24.8 ±8.1	21.5 ±5	6 ±2.8
70	Rubiaceae I	Arvore pequena	Rubiaceae	YL	26	2.3 ±2.1	3.4 ±2.7	1.1 ±1.2
71	Rubiaceae II	Arvore maior	Rubiaceae	S + YL	6	15.1 ±7	14 ±5.6	5.8 ±3.1
72	Rubiaceae III		Rubiaceae	S	5	5.3 ±1.3	6.6 ±2.4	1.3 ±0.5
73	Schistostemon retusum (Ducke) Cuatr.	Bacore	Humiriaceae	P	4	23.2 ±2.6	18.3 ±2.7	5.3 ±1
74	Simaba glandulifera Gardn.	Falso-Pau-Paraiba	Simaroubaceae	YL	17	8.6 ±4.2	11.4 ±3.9	1.6 ±0.5
75	Sloanea obtusifolia Schum.	Gindiba	Elaeocarpaceae	ML	1	12.4 ±0	17 ±0	2.5 ±0
76	Sprucella crassipedicellata (Mart. & Eichl.)Pires	Bacumixa	Sapotaceae	P	72	50.6 ±14.3	20.5 ±2.7	11 ±3.2
77	Talisia esculenta Radlk.	Pitomba	Sapindaceae	P	7	10.5 ±6.8	12.9 ±7	3.1 ±2
78	Thrysodium schomburkianum Benth.	Manga-brava	Anarcardiaceae	P	1	11.1 ±0	15 ±0	3 ±0
79	Tibouchina francavillana Cogn. ex. char.	Quaresmeira	Melastomataceae	ML	-			-
80	Tontelea attenuata Miers.		Hippocrateaceae	EF	1	25.5 ±0	18 ±0	9 ±0
81	Tovomita guianensis Aubl.	Mangue-da-Mata	Clusiaceae	P	1	10.5 ±0	9 ±0	3 ±0
82	Virola officinalis (Mart.) Warb.	Bequiba branca	Myristicaceae	P	12	31.5 ±11	22.6 ±1.9	8.2 ±3.1
83	Vismia ferrugina H.B.K.	Copian-mirim	Clusiaceae-Guttifereae	ML	1	12.7 ±0	9 ±0	4.5 ±0
84	Vitex orinocensis H.B.K.	Carobucu	Verbenaceae	P	2	24.4 ±3.8	16 ±1.4	4 ±1.4
85	Vochysia cf. rudeliana Stalf.	Louro-da-agua	Vochysiaceae	ML	15	18.2 ±13.9	14 ±4.7	4.2 ±2.4
86	Xylopia spec.	Annonaceae	Annonaceae	S	2	9.2 ±1.8	9 ±0	3 ±1.4
87		Abacaxi da mata	Annonaceae	S	2	28.7 ±4.5	20.5 ±3.5	8.2 ±3.9
88		Genipapo-da-Mata	Rubiaceae	P & S	7	4.7 ±0.7	6.6 ±1.6	2.1 ±0.9
89		Imbinucu		ML	1	4.8 ±0	7 ±0	2 ±0
90	(all species pooled)	Murta	Myrtaceae	P	-		-	-
91		Virote-preto		ML	3	23.0 ±5.3	20 ±2.6	6.7 ±1.5

Table 4. The top 10 food items in the diet of *C. personatus* listed in order of prevalence in the diet. % Diet = % of overall feeding time samples; % Fruit = % of fruit feeding time samples; cumulative f. = cumulative frequencies of % of feeding time samples or fruit feeding time samples

Scientific Nomenclature	Family	% Diet	Cumulative f.	% Fruit	Cumulative f.
Manilkara longifolia	Sapotaceae	10.4	10.4	12.8	12.8
Sprucella crassipedicellata	Sapotaceae	7.4	17.8	9.1	21.9
Brosium rubescens	Moraceae	7.1	24.9	8.8	30.7
Helicostylis tomentosa	Moraceae	6.0	30.9	7.4	38.1
Aparisthium cordatum	Euphorbiaceae	4.8	35.7	5.9	44.0
Pouroma spec.	Moraceae	3.9	39.6	4.8	48.8
Eugenia spec. III	Myrtaceae	2.9	42.5	3.6	52.4
Eschweilera ovata	Lecythidaceae	2.9	45.4	3.6	56.0
Pouteria spec.	Sapotaceae	2.7	48.1	3.3	59.3
Miconia spec. I	Melastomataceae	2.4	50.5	3.0	62.3

about 11.0 - 11.9 m (see Table 2). Other tree species such as *Brosimum rubescens* and *Helicostylis tomentosa*, which have a height of 15 - 20 m and a crown diameter of 4.2 - 6 m, also played an important role in food for the titis. Leaf eating species were not among the top 10 tree species. The highest ranking plant species for leaf eating was *Euphorbia* sp. The mean crown diameter of this species was 3.2 m.

Masked titis expressed a preference for small feeding trees. Trees used by Group I and II in 60 % of feeding samples had a DBH ≤ 20 cm. Of these, half had a DBH of only 1 to 10 cm. In 26 % of feeding samples trees had a DBH between 21–40 cm and in 15 % between 41–100 cm. Crown diameters of feeding trees were likewise small. Only in 18.4 %

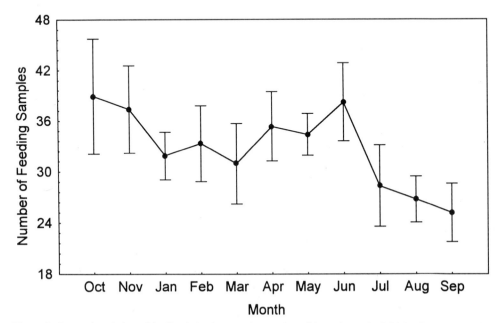

Figure 5. Seasonal variation of feeding behavior. Numbers and confidence intervals (95 %) of samples per day for each month.

of all feeding samples was a food item was eaten in a tree which had a crown diameter greater than 10 m.

Seasonal Variation of Feeding Behavior and Diet

Feeding activity varied from 17 feeding samples/day to 56 feeding samples/day. The months with lowest feeding activity were September and January-March (Fig. 5). Highest feeding activity was observed during June and October. No correlation of feeding activity to warm and cold season was found (Mann-Whitney-U-test: p = 0.46).

The variation of food items eaten by the titis during one year is shown in Fig. 6. The masked titis ate predominantly fruit during most months of the year. Between November and February, which were the hottest months in the year, the titis fed on fruits more than 70 % of feeding samples. Fruit feeding increased in September, where the titis used more then 80 % of feeding samples to eat fruits. If titis were feeding less on fruits, they compensated with leaf-eating or other food resources (Fig. 6: Other), as for example during April. Leaf-eating peaked in November (26 % of feeding samples) and was at its lowest point during September (11 % of feeding samples). Significant differences of fruit and leaf eating were observed during the warm and cold season. Masked titis ate more fruits and less leaves during the warm season, than during the cold season (Mann-Whitney-U-test: fruit eating, p = 0.0005; leaf eating, p = 0.002).

Annual variation in the use of different plant species for food was notable. The percentage of time samples the titis were feeding on different plant species changed almost completely within a year. Only 19 food species were eaten during October, the month which shows the lowest species diversity throughout the year; 39 plant species were used during March and May, the months of highest diversity. Annual variation was found in the frequency in plant species use. Some tree species were frequented by the monkeys throughout the year to eat fruit. A typical example was *Manikara longifolia* or *Sprucella crassipedicallata* (Fig. 7, A). Monthly feeding activity on these species varied from

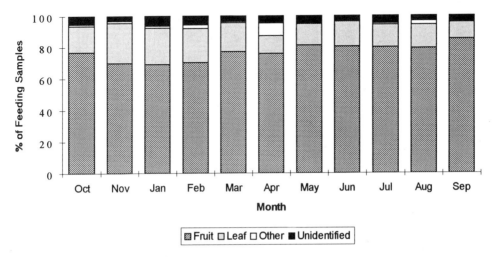

Figure 6. The seasonal variation of feeding samples of *C. personatus* spent eating different food items. Data were taken of Group I during October and November 1992 and of Group II between January and September 1993. The food category "Other" include flower,stem of tree, medulla, soils.

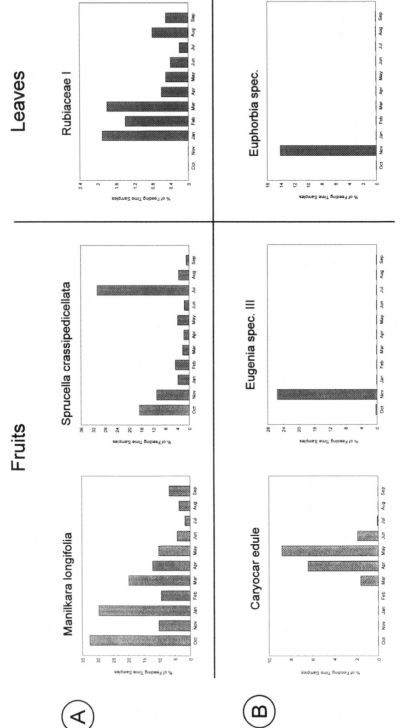

Figure 7. Examples of the annual variation in fruit-producing and leaf-producing trees used by *Callicebus personatus*.

Table 5. Comparison of the activity budget, daily path length, and home range size of
C. brunneus, C. torquatus, and *C. personatus*

	Activity budget			Daily path	Home range
	Resting (%)	Feeding (%)	Locomoting (%)	length (m)	size (ha)
C. brunneus[1]	34	38	22	670	6.9
C. torquatus[2]	63	16	17	820	18.0
C. personatus[3]	40	27	32	1007	24.0

[1]Wright, 1985a
[2]Easley, 1982
[3]Müller, 1985

1–33% of all feeding time samples. Other tree species were used for fruit eating during shorter periods, which could last only one month (e.g. *Caryocar edule* or *Eugenia* sp. III) (Fig.7, B). Similar patterns were found in tree species whose leaves were eaten by the titis (Fig. 7, A and B: *Rubiaceae* I and *Euphorbia* sp.).

DISCUSSION

The completion of three long-term studies (> 9 months) on activity patterns and feeding behavior now permits a preliminary comparison of *C. brunneus, C. torquatus* and *C. personatus*. Titis generally have an alert period of about 10 hrs., but there are interspecific differences in their time budgets (Table 5). *C. personatus* occupied the largest home range, showed a higher percentage of locomotion, and travelled a longer distances per day than either *C. torquatus* or *C. brunneus* (Easley, 1982; Wright, 1985). *C.personatus* was intermediate between *C. brunneus* and *C. torquatus* in time budgeted for feeding.

Differences in dietary preferences among the three titi species have also been observed. About 75% of the diet of *C. torquatus* consisted of fruits (Easley, 1982), but fruit only made up 54 % of the diet for *C. brunneus* (Wright, 1985). The portion of fruits ingested by *C. personatus* was similar to that of *C. torquatus*. Of the three species, the diet of *C. brunneus* was highest in leaves (28 % of the feeding samples, Wright, 1985), while *C. torquatus* or *C. personatus* fed less on leaves. The lowest daily feeding activity has been observed in *C. torquatus* who consumed leaves in only 9 % of feeding samples. The value for *C. personatus*, who ingested in around 17 % of feeding samples leaves, was under that for *C. brunneus*, who showed the highest feeding activity (see Table 5). Dietary differences might influence the foraging or feeding time. In contrast to these data for *Callicebus*, Clutton-Brock & Harvey (1977) assumed that "feeding or foraging time should also be negatively related to the proportion of foliage eaten". Data presented here showed, however, that *Callicebus* feeding time is positively related to the portion of leaves in the diet. For that reason I analyzed the feeding strategy of *C. personatus* and I will compare it with that of *C. brunneus* and *C. torquatus*. Feeding strategies were often described by parameters including energy intake versus energy cost or avoidance of plant toxins (Milton,1980; Lawes et al., 1990; Nakagawa, 1990; Ganzhorn, 1993). These studies did not describe how the different food items were exploited. The analysis of the duration of feeding on a resource showed a marked difference between fruit and leaf sources used by masked titis. Fruits were used for a longer duration during a feeding session than leaves (Müller, 1995a). Leaves, some of which were available throughout the year, were eaten only for a short duration, often opportunistically and very selectively from liana species or small trees from the understory. Masked titis used a few individual feeding sources and returned to them often prior to entering a sleeping tree. Similar results have been found by Wright (1985), who observed that *C. brunneus* "often fed on leaves opportunistically" and "the *Callicebus* strategy for leaf-eating is to browse through patches of vines or bamboo uniformly and to alternate these patches in such a way as to allow for renewal of resources". Similar results were found for *C. torquatus* (Easley, 1982). On that account it seems that leaf resources used by *Callicebus* tend to be more dispersed than fruit resources. Unfortunately no data on dispersion of leaf-eating sources are available yet.

I anticipated that latitudinal differences in the distribution of the three titi species would be reflected in dietary differences. *C. torquatus*, living near the equator, exhibited slight seasonal variation in food choice (Easley, 1982). *C. brunneus,* in southeastern Peru showed pronounced seasonal variation in time spent eating leaves. During the dry season when fruits were scarce, leaf consumption increased (Wright, 1985). I found a slight variation in food choice during cold and warm seasons, but *C. brunneus* showed greater seasonal variation in food choice than *C. personatus*. The recent reduction of other primates at the *C. personatus* site may have had a positive influence on the range of foods available since Wright (this volume) found that closely related *C. brunneus* probably suffers from feeding competition at Cocha Cashu where primate density is high.

The most obvious difference in food choice between the three *Callicebus* species was the predation on insects, which represents a food rich in protein. The amazonian *Callicebus* fed on insects, but *C. personatus* did not. Masked titis used an alternate food resource to supply the protein gap: young leaves and seeds. Eating young leaves and seeds presents a potential problem if these food items contain a high concentrations of digestion-inhibitors or toxins, for example tannins and alkaloids (Waterman, 1984). A tannin content of 5–13% dry weight has been shown for young leaves or seeds in other studies (Davies & Baillie, 1988; Wrangham et al., 1991). Only *C. personatus* was observed to practice geophagy (Müller, 1995a) that might be effective in absorbing these digestion-inhibitors.

The three species of *Callicebus* feed on a wide range of plant families and species. Studies of *C. torquatus* and *C. brunneus* documented plant species diversity ranging from 57 to 88 species and a plant family diversity of 24 families (Easley, 1982; Wright, 1985). In contrast, *C. personatus* used at least 91 identified plant species belonging to 33 plant families. I would estimate that the potential plant species used by titis vary from 100 to more than 150. In earlier studies, *Callicebus* has been described as a relatively specialized feeder (Kinzey, 1981); only 3 to 6 fruit species ingested comprised 60 % of fruit-feeding time. These results depended on season and duration of the study. Recent studies showed *Callicebus* as a more general feeder, because 8 fruit species (*C. torquatus*) and 10 fruit species (*C. personatus*) comprised > 60 % of fruit feeding time (Easley, 1982).

Tree characteristics used by the monkeys for alimentation have been analyzed by Wright (1985): *C. brunneus* fed primarily in trees with crown diameter less than 10 meters. I found similar results for *C. personatus*. Furthermore, Müller (1995a) analyzed the characteristics and use of tree species, which have a huge crown (> 10 meters). These trees could be characterized by (1) a great number of tree individuals used by the monkeys; (2) a availability of fruits throughout the year; and (3) a short duration of feeding during a feeding session. These trees therefore may play a similar role within the food of masked titis than the small crowned trees. These data indicate that masked titis use at least two different feeding strategies: feeding pattern type I: most of the fruits, which were available in quantity, were exploited with a longer duration (feeding), and 2) feeding pattern type II: leaves were harvested more opportunistically and for shorter duration than fruits (foraging). More comparative data among these species would be helpful in characterizing the importance of these differences. This interspecific variation may be influenced by habitat characteristics, diet availability and distribution, home range size or other parameters.

SUMMARY

Observations on feeding behavior of masked titi monkeys (*Callicebus personatus melanochir*) were carried out in the Atlantic coastal rain forest of eastern Brazil during 11 months with assistance of radio telemetry. Masked titis travelled an average of 1007 meters per day within a home range area of about 24 hectares. *C. personatus* is a mainly frugivorous primate, which consumed 77 % fruits and 17 % leaves of the overall diet. The dietary source includes at least 91 tree species belonging to 33 plant families. Seasonal variation in food choice was observed. During warm season *C. personatus* ingested more fruits than during cold season. *C. personatus* tended to use two different strategies to exploit a fruit or leaf source. Leaves were harvested more opportunistically than fruits. Comparison of the feeding behavior of *Callicebus* species showed, that in contrast to the Amazonian titis, *C. personatus* do not feed on insects. Furthermore, *C. personatus* living in a more seasonal habitat was not a more general feeder than *C. torquatus* or *C. brunneus*.

ACKNOWLEDGMENTS

This research was supported by Deutscher Akademischer Austauschdienst, Deutsche Forschungsgemeinschaft and Deutsches Primatenzentrum Gёttingen. Field work was authorized by Conselho Nacional de Pesquisa (CNPq) and Comissão Executiva do Plano Lavoura Cacaueira (CEPLAC). I am particulary grateful to Dr. Mauvis Gore for the stimulating discussion on the subject and for revision of English language and Dr. Marilyn Norconk for revision of earlier draft of this manuscript. I would like to thank Prof. Dr. Hans-Jürg Kuhn for supporting the Masked Titi Project. I thank Dr. Alcides Pissinatti (Centro de Primatologia do Rio de Janeiro, CPRJ) and Dr. José Inacio Lacerda (Estação Experimental Lemos Maia) for their helpful advise in Brazil.

Finally I am very grateful to my assistant, Cléa Serra-Müller, for her enormous help in the field.

REFERENCES

Chivers, D.J., 1974, The siamang in Malaya. *Contributions to Primatology* 4. Basel, S. Karger, 1974.

Clutton-Brock, T.H., 1977, Some aspects of intraspecific variation in feeding and ranging behaviour in primates. In: PRIMATE ECOLOGY: STUDIES OF FEEDING AND RANGING BEHAVIOUR IN LEMURS, MONKEYS AND APES. T.H. Clutton-Brock, ed., London, Academic Press, pp. 539–556.

Clutton-Brock, T.H., and Harvey, P.H., 1977, Species differences in feeding and ranging behaviour in primates. In: PRIMATE ECOLOGY: STUDIES OF FEEDING AND RANGING BEHAVIOUR IN LEMURS, MONKEYS AND APES. T.H. Clutton-Brock, ed., London, Academic Press, pp. 557–584.

Davies, A.G., and Baillie, I.C., 1988, Soil-eating by red leaf monkeys (*Presbytis rubicunda*) in Sabah, Northern Borneo. *Biotropica* 20(3): 252–258.

Defler, T.R., 1983, Some population characteristics of *Callicebus torquatus lugens* (Humbold, 1812) (Primates: Cebidae) in eastern Colombia. *Lozania* (Acta Zoologica Colombiana) 38:1–19.

Easley, S.P., 1982, The ecology and behavior of *Callicebus torquatus*, Cebidae, Primates. Unpublished PhD dissertation, Washington University, St. Louis.

Ganzhorn, J.U., 1993, Flexibility and constraints of Lepilemur ecology. In: LEMUR SOCIAL SYSTEMS AND THEIR ECOLOGICAL BASIS, P.M. Kappler and J.U. Ganzhorn eds., Plenum Press, New York, pp. 155–167.

Hershkovitz, P., 1990, Titis, New World Monkeys of the genus *Callicebus* (Cebidae, Platyrrhini): A preliminary taxonomic review. *Fieldiana Zoology* 55:1–109.

Hohl, A., 1993, Wiederherstellung der Artenvielfalt in den Agrarzonen tropischer Regenwälder. Unpublished PhD dissertation, Universität Hohenheim.

Kinzey, W.G., 1977, Diet and feeding behaviour of *Callicebus torquatus*. In: PRIMATE ECOLOGY: STUDIES OF FEEDING AND RANGING BEHAVIOUR IN LEMURS, MONKEYS AND APES. T.H. Clutton-Brock ed., London, Academic Press, pp 127–151.

Kinzey, W.G., 1978, Feeding behaviour and molar features in two species of titi monkey. In: RECENT ADVANCES IN PRIMATOLOGY, vol. 1, BEHAVIOUR: D.J. Chivers; J. Herbert eds., London, Academic Press, pp 375–385.

Kinzey, W.G., 1981, The titi monkey, genus *Callicebus*. In: ECOLOGY AND BEHAVIOR OF NEOTROPICAL PRIMATES. A.F. Coimbra-Filho and R.A. Mittermeier eds. Rio de Janeiro, Academia Brasileira de Ciencias, pp 241–276.

Kinzey, W.G., Rosenberger, A.L., Heisler, P.S., Prowse, D.,and Trilling, J., 1977. A preliminary field investigation of the yellow handed titi monkey, *Callicebus torquatus torquatus*, in Northern Peru. *Primates*, 18:159–181.

Kinzey, W.G. and Becker, M., 1983, Activity pattern of the masked titi monkey, *Callicebus personatus*. *Primates* 24(3):337–343.

Kinzey, W.G. und Robinson, J.G., 1983, Intergroup loud calls, range size, and spacing in *Callicebus torquatus*. *American Journal of Physical Anthropology* 60:539–44.

Lawes, M.J. Henzi, S.P., and Perrin, M.R., 1990, Diet and feeding behaviour of Samango monkeys (*Cercopithecus mitis labiatus*) in Ngoye Forest, South Africa. *Folia Primatologica* 54:57–69.

Mason, W.A., 1966, Social organization of the South American monkey *Callicebus moloch*: A preliminary report. *Tulane Studies in Zoology* 13:23–28.

Mason, W.A., 1968, Use of space by *Callicebus* groups. In: PRIMATES, STUDIES IN ADAPTION AND VARIABILITY. P.C. Jay ed., Holt, New York, pp. 200–216.

Milton, K., 1980, THE FORAGING STRATEGY OF HOWLER MONKEYS, A STUDY IN PRIMATE ECONOMICS, New York, Columbia University Press.

Mittermeier, R. A., 1987, Effects of hunting on rain forest primates. In: PRIMATE CONSERVATION IN THE TROPICAL RAIN FOREST, autor ed., Alan R. Liss, New York, pp. 109–146.

Moynihan, M., 1976, THE NEW WORLD PRIMATES. Princeton University Press.

Müller, K.-H., 1994, Capture and radio-telemetry of masked titi monkeys (*Callicebus personatus melanochir*, Cebidae, Primates) in tropical rain forest. *Neotropical Primates* 2(4):7–8.

Müller, K.-H., 1995a, Langzeitstudie zur Ökologie von schwarzkëpfigen Springaffen (*Callicebus personatus melanochir*, Cebidae, Primates) im atlantischen Küstenregenwald Ostbrasiliens. PhD dissertation Universität Berlin, Aachen, Shaker Verlag.

Müller, K.-H., 1995b, Individual separating from the group and establishment of a new territory in masked titi monkey (*Callicebus personatus*) in Brazil. *Neotropical Primates* 3(3): in press.

Nakagawa, N., 1990, Choice of food patches by Japanese monkeys (*Macaca fuscata*). *American Journal of Primatology* 21:17–29.

Robinson, J.G., 1977, Vocal regulation of spacing in the titi monkey *Callicebus moloch*. Unpublished PhD dissertation, University of North Carolina.

Robinson, J.G., 1979a, Vocal regulation of use of space by groups of titi monkeys *Callicebus moloch*. *Behavioral Ecology and Sociobiology* 5:1–15.

Robinson, J.G., 1979b, An analysis of the organization of vocal communication in the titi monkey *Callicebus moloch*. *Zeitschrift für Tierpsychologie* 49:381–405.

Robinson, J.G., 1981, Vocal regulation of inter- and intragroup spacing during boundary encounters in the titi monkey, *Callicebus moloch*. *Primates* 22(2):161–172.

Rylands, A.B., 1982, The behaviour and ecology of three species of marmosets and tamarins (Callitrichidae, Primates) in Brazil. Unpublished PhD dissertation, Cambridge.

Siegel, S., 1956, NONPARAMETRIC STATISTICS FOR THE BEHAVIORAL SCIENCES, New York, McGraw-Hill.

Waterman, P.G., 1984, Food acquisition and processing as a function of plant chemistry, In: FOOD ACQUISITION AND PROCESSING IN PRIMATES. D.J. Chivers; B.A. Wood; A. Bilsborough eds., New York, Plenum Press, pp. 177–211.

Welker, C., Rēber, J., Lührmann, B., 1981, Data on the carrying of young common marmosets *Callitrix jacchus*, cotton-head tamarins *Saguinus oedipus* and titi monkeys *Callicebus moloch* by other members of their family groups. *Zoologischer Anzeiger* 207(3–4):201–209.

Wrangham, R.W.; Conklin, N.L.; Chapman, C.A.; Hunt, K.D., 1991, The significance of fibrous foods of Kibale Forest Chimpanzes. *Philosophical transactions of the Royal Society of London, Series Biological sciences* 334: 171–178.

Wright, P.C., 1985, The cost and benefits of nocturnality for *Aotus trivirgatus* (The night Monkey). Unpublished PhD dissertation, New York, City University of New York.

20

SEASONAL VARIATION IN THE DIETS OF WHITE-FACED AND BEARDED SAKIS *(Pithecia pithecia* AND *Chiropotes satanas)* IN GURI LAKE, VENEZUELA

Marilyn A. Norconk

Department of Anthropology
Kent State University
Kent, Ohio 44242–0001

INTRODUCTION

Variation in rainfall provides the most readily quantifiable method to interpret seasonal influences on plant production and animal response to environmental change. But the effects of rainfall on an animal's behavior are complex, indirect, and less predictable, than the simple measurement of rainfall might convey. From the perspective of the primate consumer (or the human observer), periods of perceived stress or food shortage often correspond to periods of lower than average rainfall, but successful "response" depends on variables that may only be indirectly related to rainfall. Knowledge of an animal's diet, morphological or physiological adaptations of the digestive tract, variability in local plant production, duration of food shortage, body weight at the onset of the season, and reproductive or general health status all might help predict the degree of stress individuals might experience during seasonal changes in the resource base. Indeed, some examples of responses to seasonality by primates are well known, including apparent detrimental direct effects (e.g. weight loss presumably due to reduced food intake: Goldizen et al, 1988; Morland 1992) as well as indirect effects (e.g. timing of weaning: Pereira, 1992; Wright and Meyers, 1992). But there are also reports of an insignificant or lack of effect of seasonality on other population and environmental variables: infant mortality (Crockett & Rudran 1987), diet and range use (Chapman 1988), diversity of resource used in the diet (Garber 1993). Even when seasonality does appear to affect feeding behavior, the expectation of the *dry* season as the most stressful season is not always borne out (Cords, 1993).

The purpose of this paper is to examine seasonal effects on seed production and seed use by two pitheciins, white-faced sakis (*Pithecia pithecia*) and bearded sakis (*Chiropotes satanas*).

Adaptive Radiations of Neotropical Primates
edited by Norconk *et al*. Plenum Press, New York, 1996

We combined traditional measures estimating seasonal variation in food availability and diet breadth with new methods that characterized the changing physical and chemical composition of the fruit ingested over an annual cycle. We previously identified two critical features of saki diets that might limit access to specific foods (seed hardness and toxic properties of seed coats: Norconk & Kinzey 1994). These limitations are also examined within the context of seasonal variation.

The saki response to seasonal variation in rainfall is complex, but appears to be closely tied to fruiting phenophases of a few important resources. Sakis are seed predators of young seeds, primarily, and these seeds are most abundant during the dry season. The dry season may be a period of low fruit abundance or be characterized by a shift from fruit to leaves or insects for ripe fruit frugivores, but it is difficult to recognize this season as a period of scarcity or food stress for sakis. Both saki species are adapted to ingesting seeds (Kinzey and Norconk, 1993) and their ability to utilize fruit during more than one stage of maturity appears to dampen seasonal effects of food scarcity.

METHODS

Study Sites

Data were collected on two islands about 40 km apart in Guri Lake, eastern Bolívar State, Venezuela. Guri Lake is the catchment area for the Raúl Leoni dam and hydroelectric plant. It reached it's present size of 4,240 km^2 in 1986 exposing more than 100 islands of varying size (Alvarez et al, 1986; Kinzey et al, 1988). *Chiropotes satanas* occupy only southern islands and *Pithecia pithecia* only northern islands in the lake. Groups of *Cebus olivaceus* and *Alouatta seniculus* occupy islands throughout the lake.

We studied sakes on three islands in the lake, one island in the south and two islands in the north. Chiropotes Island is approximately 365 ha in size (7°21'N, 62°52'W) and had one group of *Chiropotes* and several groups each of *Cebus olivaceus* and *Alouatta seniculus*. Pithecia Island (also known as "Pithecia II" in Walker, this volume (Fig. 2) & Parolin, 1993) (7°45'N, 62°52'W) is about 15 ha in area and had one group of *Pithecia pithecia* and one group of *Alouatta seniculus*. Home range size in terra firme is not known exactly for either saki species, but the two study islands were chosen to roughly approximate the maximum area used by these species under free-ranging conditions. (Ayres (1981) estimated a home range for *Chiropotes albinasus* at 300 ha.) Base camp was located in the village of Guri at the site of the hydroelectric plant.

The study islands differed in vegetational composition as well as primate fauna. Pithecia Island was small, oval and completely forested. Trees rarely exceeded 18 m in height, averaged 11 cm in dbh, and undergrowth was composed primarily of saplings (see Parolin 1993). Although there was evidence of mixed flora from as far as the Caribbean areas in the north to the Gran Sabana in south, Aymard et al. (in press) found several examples of floristic endemism on northern islands (e.g. *Ouratea guianensis*, Ochnaceae; *Sloanea boliviensis*, Elaecarpaceae).

The southern islands in the vicinity of Chiropotes Island were about half the distance to the junction of Caroní and Paragua Rivers (the southern boundary of the lake) and tended to support taller, wetter forests on clay soil (e.g. *Ceiba pentandra*, Bombacaceae with dbh exceeding 200 cm and 28 m in height were not rare). Chiropotes Island was irregularly shaped, consisted of 4 peninsulas (up to 2 km long) running north-south away from a central strip of land that was as narrow as 200m in some places. The logistics of working on two islands each month resulted in a loss of one or two days at the beginning or end of the month. Thus, rainfall is presented as monthly averages in Fig. 1.

Vegetation Sampling

A total of 2,275 trees > 10 cm dbh were identified to species, measured (dbh, height, and canopy breadth) and mapped on Chiropotes Island. This sample represents both feeding trees of *Chiropotes* and strip samples collected by A. Peetz, I. Homburg, and S. Walker. In like manner, 3,570 trees were identified, measured and mapped on Pithecia Island. The Pithecia Island sample included smaller trees (> 5 cm dbh) measured in 16 ¼-ha quadrats (Parolin 1992) and feeding trees and lianas of *Pithecia*. Shrubs used by *Pithecia* were not measured, but they were identified.

Phenology trees were selected to focus on saki monkey feeding trees and were well-dispersed throughout the two islands. They were sampled monthly for leafing, flowering, and fruiting status by A. Peetz, I. Homburg, C. Wertis, and C. Butler. Leafing, flowering, and fruiting were given 0 to 4 scores on more than 300 trees monthly, where zero indicated lack of leaves, flowers, fruit and "4" indicated peak stage of leafing, flowering or fruiting. Data reported below are confined to fruiting activity from a subset of the sample, 11 of the top 25 species used by *Pithecia* and 12 of the top 25 species used by *Chiropotes*.

One hundred fruit traps constructed of nylon mesh draped over wire hoops (1 m in circumference) were suspended between saplings or stakes about 1.5 m off the ground and approximately 3 m on either side of main trails on both islands. Care was taken to position bags to avoid overhanging branches so that fruit fall was from the canopy. Contents of traps, collected once a month, were usually air-dried in the mesh bags unless rainfall was very recent. Fruit were divided into four categories: legumes, drupes/berries, nuts, and samaras, identified to species when possible, counted, and weighed. After Terborgh (1983:17), we estimated the volume of each trap to be 0.08 m^2 and the total weight the fruit in 100 traps was divided by 0.8 to provide a total weight measure (kg/ha) per month. Traps on Chiropotes Island were placed the entire length of a north-south peninsula (c. 2 km) and sampled both low-lying evergreen and deciduous slope or higher elevation habitats. Approximately 2/3 of the traps on Pithecia Island followed the perimeter of the island and the remainder were placed along a trail that crossed the crest of the island from north to south.

Study Animals and Collection of Feeding Samples

The term "ingestion" is used below to include the activities of both taking food into the mouth and mastication. Feeding data for five consecutive days each month (sleeping tree to sleeping tree) were collected on each island. Both *Pithecia* and *Chiropotes* were accustomed to our presence, but travel and feeding height, rapid movement, and relatively large group size of bearded sakis reduced our ability to collect focal samples on *Chiropotes*. Group size increased from 18 to 22 during the study (10 adult females, 2 to 3 adult males, 2 juveniles (matured into adults during the study) and the rest were infants less 2 years of age). Feeding samples reported for *Chiropotes* are derived from modified focal samples, using the entire group as the sampling unit (see Norconk & Kinzey 1993). Feeding duration was marked by the first animal into the feeding tree and the last out of the tree for a total of 30,692 feeding minutes. Data were collected opportunistically on insect foraging, to include "search" activities: for example, unrolling leaves and pulling up bark.

The *Pithecia* study group consisted of eight independently locomoting individuals 2 adult females, 3 adult males, 2 subadults (one male, one female) and 1 juvenile male. A female infant was born during the study. *Pithecia* were individually recognizable using distinctive facial patterns, and feeding activities (species of feeding tree & part of plant ingested) were recorded for all individuals that were visible at each 5 minute interval for a total of 44,776 feeding minutes. This rigorous sampling regime was possible because *Pithecia* have shorter total active days and visibility was better for *Pithecia* who move and feed in lower levels of the canopy than *Chiropotes*. The two sampling methods (group scores of *Chiropotes* vs. individual scores of *Pithecia*) confounded direct comparison of results (albeit both measured feeding duration) so monthly and seasonal averages were weighted by the total feeding time for each plant species.

Fruit and Seed Hardness Measurements and Biochemical Analysis

Methods used to estimate the dental force or resistance to puncturing pericarp and crushing seeds have been reported elsewhere (Kinzey & Norconk, 1990). All samples of whole or partial fruit and seeds were collected at the time the samples were dropped by the monkeys and measured on the same day.

Specimens of fruit and leaves were either collected while the monkeys were feeding, or we returned to the tree after feeding was completed and collected fruit and leaf samples that appeared to be representative of the stage of maturity at which feeding occurred. Samples were transported to base camp in Guri within 24 hours of sample collection and prepared for drying: seeds were separated from mesocarp, seed coat, pericarp and weighed ("wet weight"). Samples were dried at a uniform temperature (50°C) in an electric laboratory oven for up to two weeks until repeated weights ceased to change. Dried samples were transported to the U.S. and analyzed by Dr. N. Conklin-Brittain. Methods are described in Kinzey & Norconk (1993) and Norconk & Conklin (in prep). Only seasonal weighted averages for macronutrients (water soluble carbohydrates, crude protein, lipids), acid detergent fiber, and condensed tannins are presented below.

Analysis

Statistical comparisons between islands were made using nonparametric tests, Kendall's τ and Spearman rank correlation coefficient to measure association and Mann-Whitney test to examine differences in medians, with acceptance of significance level $\leq .05$.

RESULTS

Seasonal Variation in Rainfall and Fruit Availability

Guri Lake, at between 7° and 8° north latitude, has a single wet season and a single dry season, each lasting about 6 months annually. Annual rainfall averaged 1,258.6 mm (N = 4 years) on Chiropotes Island and 1,030.2 mm/yr (N = 3 years) on Pithecia Island. Wet and dry seasons were further subdivided into early and late periods so that each "season" represented three months. Annual rains began to fall in May and peaked in July or August. Average monthly rainfall was strongly correlated between the two islands (Spearman rank correlation coefficient, $r_s = .791$, $P < .01$, N = 12), but Chiropotes Island was

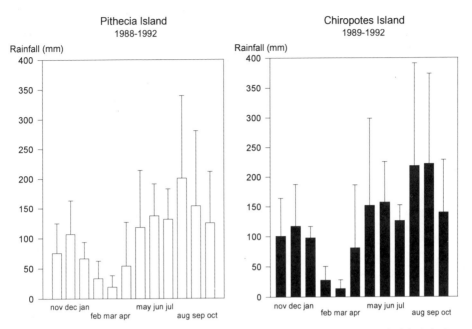

Figure 1. Rainfall data collected on Pithecia Island and Chiropotes Island (average & standard deviation) over a period of three to four years.

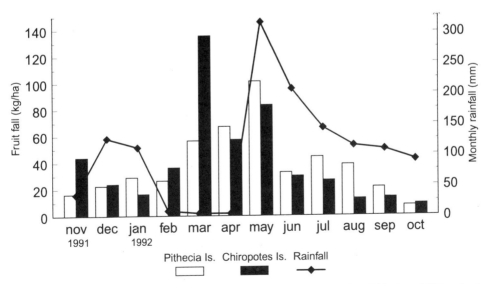

Figure 2. Monthly volume of fruit fall (kg/ha) collected from 100 fruit traps each on Pithecia and Chiropotes Islands compared with rainfall during the same period. Since the seasonal patterns of rainfall are the same between islands, rainfall from only Chiropotes Island was plotted.

significantly wetter (Mann Whitney test, $Z = 2.59$, $P < .01$) (Fig 1). Some months had similar rainfall totals between the years of the samples, but other months varied widely in rainfall from year to year. The mean annual deviation in monthly rainfall ($\sum |x - \bar{x}|/N$) averaged 39.6 mm (range 8.3 - 83.7 mm) for Pithecia Island and 58.6 mm (range 10.2 - 125.5 mm) for Chiropotes Island (Fig 1). Dry season months had less variation in rainfall than wet season months for both islands.

Fruit fall (kg/ha) was negatively correlated with rainfall on Chiropotes Island (τ = -.424, $P < .05$, N = 12 months), but rainfall was not correlated with fruit fall on Pithecia Island: $\tau = -.061$, ns) (Fig 2). Nevertheless, fruiting cycles showed a moderately strong correlation between islands ($\tau = .424$, $P < .05$). The wetter island, (Chiropotes Island) had significantly more fruit species in the monthly traps than Pithecia Island (Mann Whitney U: $Z = 2.94$, $P = .003$, N = 12 months), although there was no difference in average fruit weight between islands (Z: .51, ns) (Fig. 3). Because of the nature of fruit traps as "random" collections of canopy fruits, a few large, heavy fruits could cause marked fluctuations in monthly fruit weight estimates. Fruit from a range of fruiting stages were recovered from fruit traps: fruit that was very young and either aborted by the tree or dislodged by arboreal animals to legume pods that fell into traps long after the seeds had dispersed. Thus, it is difficult to obtain a monkey's view of fruit abundance from fruit traps.

About one-half of the top 25 plant species used by *Pithecia* and *Chiropotes* were sampled monthly for fruiting status. A phenological score (monthly score assessed on a scale of 0 to 4 divided by the number of sample trees for each species) provided information on the relative abundance of unripe and ripe fruit over a 12 month period (Tables 1 and 2). The most striking aspect of these results are the relatively low phenology scores. Although individual trees might produce abundant fruit crops, scores rarely exceeded 2.5 when averaged across all of the sample trees. Unripe fruit was available for a longer period than ripe fruit on both islands, although the difference was greater on the drier island. Pithecia Island fruit production averaged 6.6 months of unripe fruit and 4.7 months of ripe

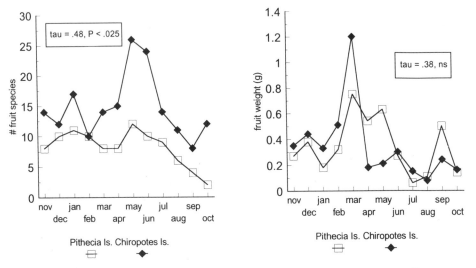

Figure 3. Two results from fruit trap data are compared from both study islands. There was a significant correlation between the two islands for diversity of fruit species found in the traps, but not for average weight of fruit. Note that fruit weight did not peak in the wet season with production of fleshy drupes. Samples were collected on both islands from November 1991 to October 1992.

Table 1. Phenology scores for 11 feeding tree species used by *Pithecia pithecia* on Pithecia Island (Guri Lake). Scores are averages and were calculated for unripe fruit/ripe fruit by dividing the sum of the total monthly score by the number of sample trees[1]. Trees were scored on a 0 to 4 scale for unripe and ripe fruit such that a score of 4.0/0 would indicate that all sample phenology trees were in full production of unripe fruit, but no trees had ripe fruit. Phenology sample was collected from November 1991 to October 1992

Plant species (N)[1]	Rank[2]	Nov	Dec	Jan	Feb	Mar	Apr	May	Jun	Jul	Aug	Sep	Oct
Connarus venezuelanus(10)	1	.9/0	.7/0	.7/.2	.4/.5	.3/.5	.3/.5	.4/.5	.1/.5	.1/.4	.2/.2	0/.1	0/0
Licania discolor (10)	2	.3/.9	.2/.9	.7/.2	.6/.2	.3/.6	.2/.4	0/.2	.4/.2	.2/.2	0/0	.6/0	.5/.7
Strychnos fendleri (10)	3	.1/0	0/0	0/0	0/0	0/0	0/0	1.2/0	2.8/0	2.8/0	.9/1.1	.4/1.0	0/0
Peltogyne floribunda (11)	4	.8/.5	.1/1.4	0/.9	0/.7	0/.4	0/.1	0/0	0/0	1.1/0	.5/.1	.5/.3	0/.1
Erythroxylum steyermarkii (14)	5	.1/0	.1/0	0/0	0/0	0/0	0/0	0/0	2.4/0	1.7/1.7	.5/.1	.1/.1	0/0
Strychnos mitscherlichii (7)	8	2.3/0	2.0/0	1.6/0	1.1/1.1	1.0/1.0	.7/1.0	.1/.1	0/0	0/0	0/0	0/0	0/0
Xylopia sericea (5)	9	1.6/.6	.3/.1	.8/.6	0/.1	0/0	0/0	1.6/0	1.4/0	1.4/0	.6/.6	.8/.8	0/0
Piptadenia leucoxylon (10)	10	1.8/0	1.8/0	1.7/0	.9/.6	0/.8	0/.5	0/0	0/0	1.8/0	0/0	0/0	0/0
Coccoloba striata (7)	12	0/0	0/0	0/0	0/0	0/0	0/0	1.7/0	1.3/0	.6/0	0/0	0/0	0/0
Alberita latifolia (5)	17	0/0	0/0	0/0	0/0	0/0	2.8/0	2.0/0	1.8/0	.4/0	0/0	0/0	0/0
Copaifera pubiflora (10)	25	1.3/0	1.4/0	1.4/0	1.2/.3	0.7	0/.1	0/0	0/0	0/0	0/0	0/0	0/0

[1] N = number of trees sampled monthly for phenological status

[2] Rank = feeding tree rank assessed after 18 months of feeding samples

Table 2. Summary of phenology scores for a sample of trees in the top 25 species of feeding trees used by *Chiropotes satanas* on Chiropotes Island. (Guri Lake). See Table 1 for note on how scores were calculated. The phenology sample was collected from November 1991 to October 1992, with the exception that the August sample is from 1991

Plant species (N)[1]	Rank[2]	Nov	Dec	Jan	Feb	Mar	Apr	May	Jun	Jul	Aug[4]	Sep	Oct
Pradosia caracasana (13)	1	3.4/0	2.7/0	2.2/0	1.8/0	1.8/0	.8/.3	0/.9	0/.6	0/.6	0/.1	.6/0	.3/0
Brosimum alicastrum (12)	3	1.9/0	2.3/0	1.8/0	1.3/0	1.3/0	1.3/0	1.5/0	1.2/.8	.1/.8	.4/.1	.1/.2	.2/0
Chrysophyllum lucentifolium (9)	4	2.9/0	2.6/0	2.6/0	2.1/0	2.1/0	2.1/0	.4/.2	.2/.2	.2/0	.1/0	0/0	0/0
Melicoccus bijugatus (8)	5	0/0	0/0	0/0	0/0	0/0	.5/0	.7/0	.8/0	0/.2	0/.2	0/.1	0/0
Sapium aubletianum (6)	7	0/0	0/0	0/0	0/0	0/0	0/0	.2/0	0/0	2.5/0	0/0	.8/1.0	0/0
Lepidocordia punctata (6)	9	0/0	0/0	0/0	0/0	0/0	0/0	0/0	0/0	0/0	2.0/0	0/0	0/0
Cecropia peltata (11)	13	0/.8	0/.3	0/.6	0/.5	0/.8	0/1.3	0/1.6	0/1.8	1.3/1.6	0/1.6	.3/.7	0/.1
Spondias mombin (12)	14	.3/.1	.3/.3	0/0	0/0	0/0	.3/0	1.8/0	2.0/0	1.8/.2	3.0/.1	.1/0	0/0
Hymenaea courbaril (4)	16	3.6/1.5	.8/3.5	0/2.8	0/1.8	0/1.5	0/.3	.3/.3	2.3/0	0/0	3.5/0	0/0	0/0
Maclura tinctoria (7)[3]	19	0/1.7	0/0	0/0	0/0	0/0	0/0	0/0	0/0	.4/0		0/0	0/0
Tabebuia serratifolia (8)	22	0/0	0/0	0/0	0/0	0/0	0/0	0/0	0/0	0/0	0/0	0/0	0/.1
Luehea speciosa (8)	26	0/0	0/0	1.0/0	1.5/0	1.3/0	.1/.6	0/.1	.2/0	.2/0	0/.6	0/.1	0/.1

[1] N = total number of trees of each species scored monthly for phenology.
[2] Rank = feeding tree rank based on group feeding durations over a period of 18 months.
[3] Phenology samples for *Maclura* began in September 1991.

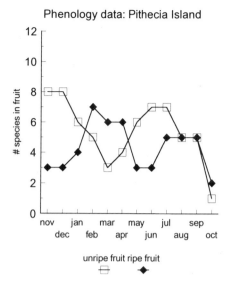
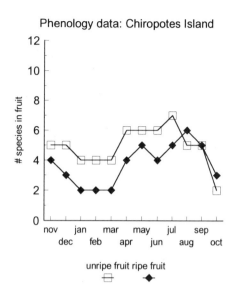

Figure 4. Relative proportion of feeding trees producing unripe and ripe fruit, documenting the number of plant species in fruit (not fruit abundance per species). Samples were obtained from phenology trees and the figures were derived from column totals of Tables 1 & 2. Chiropotes Island had a lower proportion of phenology species in fruit during the dry season and had more species producing unripe fruit than ripe fruit for all but two months in the wet season.

fruit compared with Chiropotes Island that averaged 4.9 months of unripe fruit and 3.75 months of ripe fruit.

The number of months that phenology tree species bore ripe and unripe fruit were compared between islands (Fig. 4). There were more trees species producing fruit (ripe or unripe) on Pithecia Island than Chiropotes Island, except during the late wet season when availability of fruit from high ranking species dropped precipitously. Unripe fruit was more abundant on high ranking food species throughout the dry season on Chiropotes Island, but trees on Pithecia Island showed a bimodal peak of unripe fruit production in early dry and early wet seasons. *Transitions* between seasons, particularly wet to dry, may be more difficult for the sakis than the dry season itself.

Fruiting periodicity detected by phenological methods and fruit traps is compared to the number of months the sakis ingested fruit of the same species in Tables 3 and 4. Fruiting cycles that characterized *high-ranking plant species* took two forms: 1) trees that fruited synchronously with very extended fruiting cycles (e.g. *Pradosia caracasana*, *Chrysophyllum lucentifolium* var. *pachycarpum*, and *Strychnos fendleri*) and 2) individual trees that fruited asynchronously so that at least some trees produced fruit in nearly every month of the year (e.g. *Connarus venezuelanus*, *Licania discolor*, and *Brosimum alicastrum* var. *bolivarense*). High ranking tree species achieved this rank because the sakis ingested fruit from the entire fruiting cycle. *Lower ranking feeding trees* were characterized by short, synchronous fruiting cycles where fruit matured quickly and monkeys ingested fruit as it became available (e.g. *Alibertia latifolia*, *Sapium aubletianum*, *Lepidocordia punctata*, *Erythroxylum steyermarkii*) or were tree species that had long cycles, but fruit were eaten only during one stage of maturity (e.g. *Peltogyne floribunda*, *Piptadenia leucoxylon* and *Spondias mombin*).

Table 3. Characteristics of fruiting cycles for 12 plant species in the diet of *Pithecia pithecia* and two methods for assessing fruiting cycles compared with number of months the fruit species was included in the diet. Phenology and fruit trap samples are reported for the same 15 month period from September 1991 to November 1992. Rank of feeding trees and "characteristics of fruiting cycle" were determined over a 19 month period. (Phenology data from Homburg, in prep.)

Plant species (N)[1]	Rank	Periodicity of fruiting[2]	Synchronicity of fruiting	Duration of fruiting[3]	Fruiting peak	Months in traps	Months in phenology sample	Months in feeding sample
Connarus venezuelanus (10)[4]	1	sub-annual	asynchronous	long	late dry-early wet	13	12	15[9]
Licania discolor (10)[4]	2	annual	asynchronous	short	early dry	3	13	13
Strychnos fendleri (10)[5]	3	annual	synchronous	long	early wet	8	10	9
Peltogyne floribunda (11)[6]	4	super-annual	asynchronous	long	early dry	13	13	6
Erythroxylum steyermarkii (14)[7]	5	annual	synchronous	short	late wet	5	7	5
Ouratea roraimae (1)[4]	7	-	-	-	-	3	2	3
Strychnos mitscherlichii (7)[5]	8	annual	synchronous	long	early dry	2	12	6
Xylopia sericea (5)[4]	9	annual	synchronous	long	late wet-early dry	0	13	4
Piptadenia leucoxylon (10)[8]	10	annual	synchronous	long	late wet-early dry	5	10	4
Coccoloba striata (7)[7]	12	annual[10]	asynchronous	short	early wet	4	3	2
Alibertia latifolia (5)[4]	17		synchronous	short	early wet	2	4	4
Copaifera pubiflora (10)[7]	25	super-annual	synchronous	short	early dry	6	7	1

[1] N = total number of trees (or lianas) of each species scored monthly for phenology (14 months).
[2] Periodicity: sub-annual = interval between fruiting is less than a year; super-annual = fruiting interval is greater than a year.
[3] Duration: short = fruiting less than 6 months; long = fruiting 6 months or more.
[4] Predate mature-sized seeds; color change of exocarp may not be complete.
[5] Ingest mesocarp; drop seeds below tree canopy.
[6] Predate young seeds; (usually) drop mature seeds.
[7] Predate mature seeds with mesocarp; do not ingest young fruit.
[8] Predate young seeds; mature seeds dispersed by wind (anemochorous).
[9] Months feeding on *Connarus* exceeded months in fruit trap and phenology samples because the monkeys went to the ground to ingest seeds from old, aborted fruit.
[10] Phenology sample collected for 12 months only.

Table 4. Characteristics of fruiting cycles for 12 plant species in the diet of *Chiropotes satanas* and two methods for assessing fruiting cycles compared with number of months the fruit species was included in the diet. Phenology and fruit trap samples are reported for the same 15 month period from September 1991 to November 1992. Rank of feeding trees and "characteristics of fruiting cycles" were determined over an 18 month period. (Phenology data from Peetz, in prep)

Plant species (N)[1]	Rank	Periodicity of fruiting[2]	Synchronicity of fruiting	Duration of fruiting[3]	Fruiting peak	Months in traps	Months in phenology sample	Months in feeding sample
Pradosia caracasana (13)[4]	1	annual	synchronous	long	early wet	2	14	15
Brosimum alicastrum (12)[4]	3	annual	asynchronous	long	early dry	4	13	12
Chrysophyllum lucentifolium (9)[4]	4	annual	synchronous	long	early dry	2	11	10
Melicoccus bijugatus (8)[4]	5	super-annual	asynchronous	short	early wet	0	5	11
Sapium aubletianum (6)[5]	6	annual	synchronous	short	early wet	4	4	3
Lepidocordia punctata (6)[6]	9	annual	synchronous	long	late wet	2	3	4
Cecropia peltata (11)[6]	13	annual	asynchronous	long	late wet	4	12	9
Spondias mombin (12)[7]	14	annual	synchronous	long	early wet	3	10	6
Hymenaea courbaril (4)[4]	16	annual	synchronous	long	late wet	1	6	7
Maclura tinctoria (7)[8]	19	super-annual	asynchronous?	short	early wet	0	2	2
Tabebuia serratifolia (8)[5]	22	super-annual	synchronous	short	early wet	1	2	2
Luehea speciosa (8)[5]	26	sub-annual	synchronous	short	late dry	1	12	6

[1] N = total number of trees of each species scored monthly for phenology (15 months)
[2] Periodicity: sub-annual = interval between fruiting is less than a year; super-annual: fruiting interval is greater than a year
[3] Duration: short = fruiting less than 6 months; long = fruiting six months or longer
[4] Predate young seeds; usually drop mature seeds and ingest mesocarp.
[5] Predate young seeds; mature seeds dispersed by wind (anemochorous).
[6] Predation of seeds with some dispersal (zoochorous) - (i.e. seeds found in feces).
[7] Ingests mesocarp of ripe fruit, drops seeds.
[8] Ingests whole fruit, including seed.

The total number of months in fruit (both immature and mature fruit) and number of months in which ingestion occurred were not significantly different for *Chiropotes* (Mann Whitney test: $Z = -.783$, ns), but time in the phenological sample was significantly longer than ingestion periods for *Pithecia* ($Z = -2.1$, $P = .036$). An exact overlap between fruiting phenophase and ingestion indicated that the sakis ingested parts of both ripe and unripe fruit, but masked an important detail of dietary selection mentioned above. Ingestion during the entire fruiting phase often meant that seeds were ingested from young fruit and mesocarp was ingested from mature fruit (e.g. *Pradosia caracasana*, *Brosimum alicastrum*, *Melicoccus bijugatus*, Table 4). Both sakis preferred young seeds over mature seeds, but only *Chiropotes* were found to use some plant species throughout fruit maturation.

The traps provided a better estimate of the period of fruit ingestion for *Pithecia* than *Chiropotes* (Tables 3 and 4). At 4% the size of Chiropotes Island, it is not surprising that 100 traps on Pithecia Island provided a closer fit to fruit availability than for the larger Chiropotes Island. However, some resources important to *Pithecia* were undersampled by traps because they were shrubs at the same height or lower than the traps (e.g. *Xylopia sericea*, *Morinda tenuiflora*, *Hirtella racemosa* var. *racemosa*).

Seasonal Variation in the Components of Saki Diets

Sakis are predominantly young seed predators and ingested seeds from a diversity of sources (Table 5). Seeds of nuts, capsules and drupes made up the dominant resources for both saki species and only became relatively rare in early and late wet seasons for *Chiropotes* and in the late wet season for *Pithecia*. In the *early wet season*, *Chiropotes* shifted from seeds to ingesting primarily ripe mesocarp of *Pradosia caracasana*, *Melicoccus bijugatus* and *Brosimum alicastrum* (Table 5). Seeds from these sources were mature and were usually dropped intact below the tree canopy. The *late wet season* was marked by a reduction in the ingestion of seeds by both sakis; *Pithecia* ingested mesocarp primarily and *Chiropotes* shifted to the soft, small drupes of the parasite *Oryctanthus alveolatus* (Loranthaceae).

Table 5. Seasonal differences in dietary composition of seeds and mesocarp ingested by *Pithecia* and *Chiropotes*. Sample is taken from the top 50 species ingested during an annual cycle (November 1991 to October 1992) and values are expressed as percent of fruit ingestion. The top two resoucess during each season are in bold-faced type

Plant Item	Early dry		Late dry		Early wet		Late wet	
	Pithecia	*Chiropotes*	*Pithecia*	*Chiropotes*	*Pithecia*	*Chiropotes*	*Pithecia*	*Chiropotes*
Seeds of nuts and drupes[1]	**62.47**	**58.25**	**62.78**	**61.41**	**76.09**	**44.47**	21.94	33.33
Winged seeds[2]	2.09	4.09	1.13	11.54	0.20	0.66	0	2.11
Legume seeds[3]	**33.76**	2.65	9.99	7.72	0.54	1.68	8.30	2.66
Parasite[4]	0.24	**30.31**	0.09	**16.97**	0	11.44	0.05	**42.71**
Whole fruit[5]	2.14	1.81	5.39	0.39	**21.54**	0.79	19.37	11.63
Mesocarp only[6]	0.29	2.90	**20.62**	1.96	1.64	**40.97**	**50.34**	8.55

Plant families represented in feeding samples include:
[1] Sapotaceae, Chrysobalanaceae, Connaraceae, Moraceae, Sapindaceae, Anacardiaceae, Rubiaceae.
[2] Bignoniaceae, Polygalaceae, Combretaceae, Tiliaceae.
[3] Caesalpiniaceae, Papilionaceae, Mimosaceae.
[4] Loranthaceae.
[5] Rubiaceae, Erythroxylaceae, Boraginaceae, Moraceae: monkeys ingested pericarp and seed.
[6] Primarily from Meliaceae, Loganiaceae, Sapotaceae.

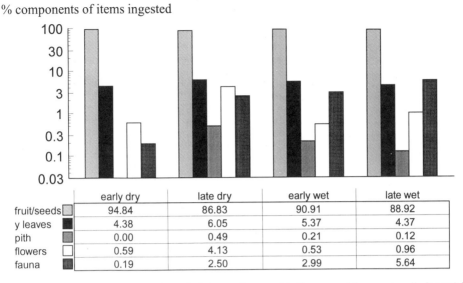

Figure 5. Seasonal variation in fruit and non-fruit (young leaves, pith, flowers, and fauna, primarily insects) in the diet of *Pithecia pithecia*. Percentage of the components are expressed as log values.

Many of the samaras ingested by the sakis were seeds from Bignoniaceae lianas that flowered in the late wet season and contributed to the "peak" of flower ingestion for *Chiropotes* (Figures 5 and 6). Winged seeds represented a larger proportion of the diet of *Chiropotes* than of *Pithecia*. Pod maturity and subsequent dehiscence occurred during windy weather conditions in the winter dry season and the ingestion of these seeds was reduced in the wet season diet. *Peltogyne floribunda* (Caesalpiniaceae) is the dominant tree species on Pithecia

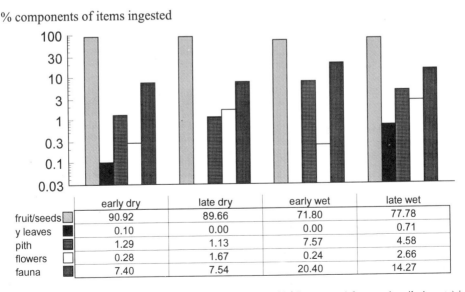

Figure 6. Seasonal variation in fruit and non-fruit (young leaves, pith, flowers, and fauna, primarily insects) in the diet of *Chiropotes satanas*. Percentage of the components are expressed as log values.

Island (Parolin, 1993) and young seeds from undehisced fruit of this species and *Piptadenia leucoxylon* (Mimosaceae) were both ranked in the top 10 plant species used by *Pithecia*. Fruiting and ingestion peaked in the early dry season on Pithecia Island (Table 5).

Three species of Loranthaceae have invaded the crowns of several species of trees in Guri Lake (e.g. *Protium tenuifolium* and *Chrysophyllum lucentifolium* var. *pachycarpum*), but are particularly notable on the highest ranked feeding tree of *Chiropotes*, *Pradosia caracasana*. Of the Loranthaceae species, *Oryctanthus alveolatus* is a major resource for *Chiropotes* and was ranked second in three of the four seasons (Table 5). Low ingestion rates on the northern island may be due to low abundance (and absence of principal host, *Pradosia caracasana*).

Mesocarp ingestion peaked during the wet season for both sakis. The annual peak of mesocarp ingestion by *Chiropotes* corresponded to the ripening of *Pradosia caracasana* and *Melicoccus bijugatus* in June, July, and August. Ingestion of mesocarp by *Pithecia* was tied closely to the fruiting phenophase of two species of *Strychnos*. *S. fendleri* peaked in availability in the early wet season; ingestion peaked once fruit reached full size in late wet season. Likewise, the fruiting peak for *S. mitscherlichii* occurred in the early dry season and fruit was ingested in the late dry season. The "whole fruit" category included both fleshy berries and arillate seeds like *Lepidocordia punctata* (Boraginaceae). Whole fruit appeared to be a more important resource for *Pithecia* than *Chiropotes*, representing close to 20% of *Pithecia* diet in the early and late wet seasons.

Non-fruit items were secondary resources for both sakis during all seasons (Figs 5 and 6) with one exception. Caterpillars were extremely abundant on *Pradosia caracasana* trees in the 1992 early wet season and represented 20% of the diet. Young leaves were relatively important to *Pithecia* and showed little annual variation by season. Pith was relatively important to *Chiropotes* and ingestion peaked in the early wet season. Flowers of six species were ingested by *Pithecia* in the late dry month of April and represented the flower ingestion increase in the late dry season. The only non-plant item of food intake was the "fauna" category. This included the odd egg and wasp nest taken by *Pithecia*, but was dominated by caterpillars in the wet season by *Chiropotes* and aseasonal ants, spiders, and caterpillars by *Pithecia*. Larger-bodied *Chiropotes* invested more time during all seasons in insect search and ingestion than did *Pithecia*.

Both saki species maintained a relatively diverse diet (plant species and plant parts used) year-round (Figure 7). The twelve plant species for which we have phenology data represented about half of the top 25 feeding species. In turn, these high ranking resources represented at most 40% of the monthly diet of *Pithecia* and at most 50% of the diet of *Chiropotes*. *Pithecia* were very uniform in the total number of plant species in each monthly diet. In contrast, *Chiropotes* appeared to increase diet breadth with a reduction in rainfall in the late dry season (Fig 7), although months of low rainfall did not always result in a more diverse diet (r_s = -.477, ns, N = 12 months). The sharp reduction of dietary diversity from April to May was due to the ripening of *Pradosia caracasana* (Sapotaceae) mesocarp on Chiropotes Island which was available for only about 8 weeks.

Using these data it is clear that the two saki species differed in their selection of food items by their degree of seed maturation. *P. pithecia* were more selective, ingesting fruit or seeds at specific stages (but not necessarily the same stage) of fruit development. For example, *Connarus venezuelanus* var. *orinocensis* seeds were ingested from mature (with reddish colored pericarp) fruit. *Peltogyne floribunda* and *Piptadenia leucoxylon* seeds were ingested only when they were immature. The mesocarp of *Strychnos* spp. was ingested more often at the mature than at the immature stage; seeds and exocarp were spat out. Berries of *Erythroxylum steyermarkii* and *Coccoloba striata* were ingested whole

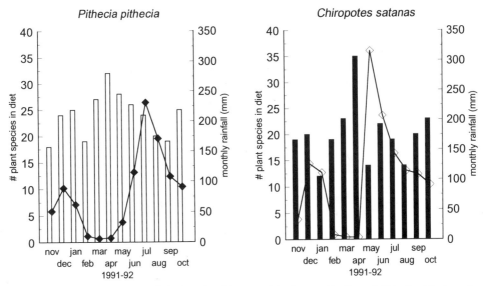

Figure 7. Comparison of the number of plant species (histograms) in the monthly diets of *Pithecia pithecia* and *Chiropotes satanas* from November 1991 to October 1992 compared to rainfall (line) during the same months.

only when ripe, but seeds of *Alibertia latifolia* were ingested throughout its relatively short fruiting cycle (see Table 3).

Seasonal Variation in the Mechanical and Chemical Composition of Fruit

There was little variation by season in the average hardness of fruit and seeds ingested by the sakis (Fig 8). Food items opened in the dry season were not harder than during the wet season, although there was some evidence of within season variability. Food items opened in the early wet season by *Chiropotes* were significantly harder than items in the late wet season (Wilcoxon $Z = 3.36$, $P < .01$). The high average and maximum puncture values for *Chiropotes* during the early wet season were due to the ingestion of both woody *Vitex compressa* (Verbenaceae) fruit and some mature *Pradosia caracasana* (Sapotaceae) seeds that were well-protected by hard, brittle seed testa.

Seasonal variation in *seed hardness* values was more marked than in *pericarp hardness* values (Fig. 8). *Pithecia* ingested (and masticated) harder seeds than *Chiropotes* during all seasons except the early dry season, when average values were comparable. Seeds were hardest in the late wet and early dry seasons for both *Pithecia* and *Chiropotes*. Although seed hardness and pericarp hardness cannot be compared directly since the methods of obtaining data and the measurements themselves are different, there was a weak inverse relationship between seed hardness and pericarp hardness particularly in the late wet season. Fruits of many species were mature at this time of year with a soft pericarp and mature seeds encased in a hard testa.

Weighted averages of macronutrients (water soluble carbohydrates (WSC), crude protein (CP), and lipids), fiber (ADF = cellulose + lignin), and condensed tannins are presented in Figure 9. These values represented 65% of the plant species used in 1991–1992 by *Pithecia* and 83% of annual diet of *Chiropotes*.

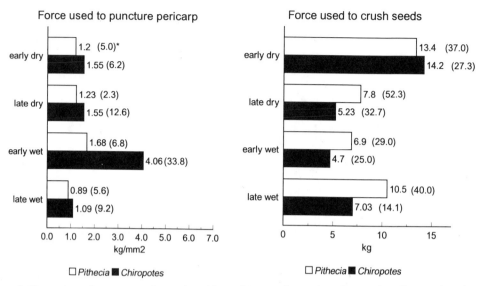

Figure 8. Comparison of average puncture and crushing resistance values and maximum values (in parentheses) of fruit and seeds ingested by *Pithecia* and *Chiropotes* in Guri Lake. Methods used to determine values were reported in Kinzey & Norconk, 1990.

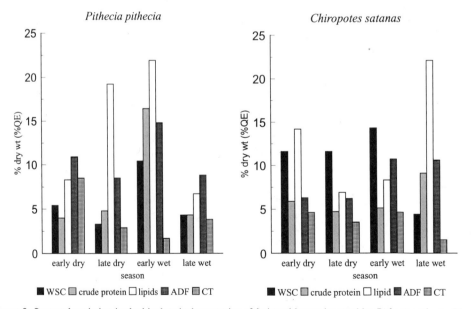

Figure 9. Seasonal variation in the biochemical properties of fruit and leaves ingested by *Pithecia pithecia* (N = 23 samples) and *Chiropotes satanas* (N = 19 samples). Values are averages of water soluble carbohydrates, crude protein, lipids, acid detergent fiber, and condensed tannins weighted by total feeding minutes.

The major resource for *Pithecia*, *Connarus venezuelanus* (Connaraceae) seeds, was found to be nearly 50% lipids/g dry weight and seeds were heavily ingested in late dry and early wet seasons. Lipid intake peaked in the early wet season with the addition of *Actinostemon schomburgkii* (Euphorbiaeae) seeds (24.2% lipids). There was no significant difference in an annual average of lipid intake for either primary species, although lipid intake peaked during opposite seasons, late wet and early dry, for *Chiropotes*. High lipid intake corresponded to peak ingestion of the parasite *Oryctanthus alveolatus* (Loranthaceae) at 38.3% lipids.

Chiropotes ingested food items that were significantly higher in water soluble carbohydrates (WSC) than *Pithecia*. The source of WSC for *Chiropotes* during the early and late dry periods and the early wet period appears to have been their major resource *Pradosia caracasana* whose seeds decreased from 34.0% to 14.6% as the dry season progressed and as the seeds matured. Protein intake peaked in the early wet season for *Pithecia*. A similar annual peak was not apparent for *Chiropotes*, but annual averages were not different (6.8% *Pithecia*; 6.0% *Chiropotes*). The source of the protein peak for *Pithecia* appears to have been *Actinostemon schomburgkii* and *Capparis muco* seeds (both > 22% CP by dry wt), but young leaves (e.g. *Capparis flexuosa*) also had high CP values (13.8%). The higher proportion of time invested in insect search and insect consumption may have off-set lower protein values of seeds and the negligible intake of leaves by *Chiropotes*.

Seasonal and annual ADF values were lower for *Chiropotes* than *Pithecia*. Higher fiber values not only reflected ingestion of more leaves by *Pithecia*, but some seeds ingested by *Pithecia* were also relatively high in lignin or cellulose (e.g. ADF values of seeds: *Licania discolor* 42.0%, *Amaioua corymbosa* 40.6%, *Angostura trifoliata* 40.0% exceeded values of young leaves: *Capparis flexuosa* 37.2% and *Coccoloba fallax* 36.6%). The ADF peak in the early wet season is not due to the ingestion of leaves, but rather to the ingestion of fibrous seeds.

Condensed tannin levels exhibited little seasonal variation and averaged less than 5% QE (quebracho units) annually for both sakis. The diversity of resources ingested may account for the low average since some high ranking resources (seeds) were relatively high in CT values (e.g. *Pradosia caracasana*: range 6.8% to 17.7% QE and *Licania discolor*: range 3.9% to 9.7% QE).

DISCUSSION

A diet that is largely seeds instead of fruit pulp may remove sakis from the seasonal effects of low rainfall. Sakis appear to monitor phenophases of individual plant species using two cues, change in seed or seed testa hardness and fluctuation in nutritional value. At the very least, the "dry season" cannot be considered to be a period of scarcity for sakis. They ingested many dry season resources in Guri Lake (winged seeds, legumes, nuts and immature drupes). Seed ingestion declined in the wet season with the ripening of fleshy fruits. The shift was more notable for bearded sakis with the ripening of their #1 resource (that was also ingested by capuchins, howlers, and macaws at this time). White-faced sakes did not have such an abundant resource available, although they too shifted to ripe berries and small drupes.

Resources that provided the sakis with food for long periods of time also tended to be high ranking resources by our measures. These were plant species that either produced relatively synchronous crops of fruit with prolonged periods of fruit development or were species that fruited asynchonously so that resources were effectively renewable, but on different trees.

The saki species, that are now separated by some 40 km in Guri Lake, were apparently allopatric and found on opposite banks of the Caroní River before flooding. While we found considerable complementarity in the categories of seeds ingested, some plant species are highly endemic in the north and there was little overlap in plant species used for food by the two saki species. Nevertheless, two findings emerged in this study to suggest differences in feeding adaptations: 1) *Chiropotes* appeared to maximize the use of some resources by ingesting seeds at the young stage of maturity and mesocarp at the mature stage of fruit maturity. A similar pattern is found in *Pithecia albicans*, intermediate in size between *Pithecia pithecia* and *Chiropotes* spp. (Peres, 1993). On the other hand, *P. pithecia* appeared to be more selective and ingested fruit only at one stage of maturity, early or late, seeds or mesocarp. 2) *P. pithecia* appeared to adjust to changes in fruit availability and seasonality by maintaining a nearly constant diversity of fruit species in their diet. *Chiropotes satanas* tended to increase diet breadth in the dry season and contract diet breadth in the wet season with maturation of their #1 ranked resource. Although seasonal variation was evident in the relative proportion of seeds, mesocarp, and whole fruit in their diets, it is difficult to find evidence of seasonal shortages.

The fact that *Pithecia* and *Chiropotes* are isolated on different islands contributes to differences in the plant species included in their diets, but patterns of use appear reflect true, albeit subtle, interspecific differences. Ayres (1981) found that terra firme, white-nosed bearded sakis (*Chiropotes albinasus*) ingested seeds and mesocarp from Sapotaceae and Moraceae and while these two categories made up most of their feeding activities, the sources of ripe fruit pulp were different than the sources of young seeds (Ayres 1981). The use of different resources to obtain different parts of fruit was more characteristic of *Pithecia* in our study and island-bound *Chiropotes satanas* in Guri appeared to have intensified their use of some resources (particularly *Pradosia caracasana* and *Brosimum alicastrum*). The seasonal shift to mesocarp was not only evident in Ayres's study of *Chiropotes albinasus* (1981) but also characterized closely related uakaris *Cacajao calvus* (Ayres 1986). This "sweet tooth" preference by the larger sakis of fruit high in water soluble carbohydrates was not evident in our *Pithecia* feeding samples. Indeed, relative proportion of WSC was one of the very few interspecific differences in the biochemical weighted averages between the Guri sakis.

Pitheciins Compared with Other Seed Predators

Ingestion and mastication of seeds as the major component of the diet is not limited to the pitheciins of the Neotropics, but is also found among some of the colobines (see Kool, 1993; Dasilva, 1994; and Maisels et al, 1994 for surveys of seed predation in colobines and Chivers, 1994 for a review of colobine gut anatomy), and some prosimians (*Propithecus diadema*: Wright, 1994; Yamashita, 1994). Both sakis and colobines prefer young seeds that provide high concentrations of valuable nutrients. Not only have some young seeds been found to have been comparable to young leaves in protein content (as did Dasilva, 1994 in her study of *Colobus polykomos*), but young seeds were also up to 20 times richer in lipids than were young leaves (Norconk, Kinzey & Conklin-Brittain, unpub; Maisels et al, 1994).

Lastly, the adaptation to seed predation appears to involve some mechanism to either penetrate hard seed coverings and thus guarantee access to mechanically protected young seeds or the ability to detoxify chemically protected seed coats or seed embryos (see Bell, 1978). Saki adaptations appear to be dental primarily, but preliminary work on saki digestion suggests that they may also have an enlarged cecum and a relatively slow transit time (Milton, 1984). Slowing the passage of food through the hindgut, for example, may increase their ability to extract nutrients from highly fibrous seeds. The presence of moderate to high levels of condensed tannins also provide a general indicator of food choice. Average CT values of seeds ingested by colobines were only slightly higher than those ingested by sakis. Sakis do ingest some seeds that are as high as 9% CT (Kinzey and Norconk, 1993), but seasonal averages were low suggesting that they might adjust the volume of intake or increase species diversity to minimize intake of high concentrations of tannins. For example, white-faced sakis were found to balance seed species diversity across all seasons. Colobines also appear to moderate tannin intake and shift back and forth between young leaves and seeds to optimize protein and lipid intake (Dasilva, 1994).

Seasonal variation in the diets of pitheciins is more accurately measured in plant species diversity, seed and pericarp hardness, and nutritional value, rather than abundance and rarity. Seeds are a reliable resource, available in Guri for extensive periods through the dry season (when fleshy fruits become rare) and into part of the wet season. Nutritionally, they are a valuable resource, often higher in lipids and protein than ripe fruit or young leaves.

SUMMARY

Seasonal variation in fruit availability and dietary diversity was compared between *Chiropotes satanas* and *Pithecia pithecia* studied over a period of 17 months in 1991 and 1992 in Guri Lake, eastern Venezuela. The study group of *Pithecia* inhabited a 15 ha island (dry tropical forest) in the northern part of the lake and the study group of *Chiropotes* occupied a 365 ha island (transitional wet forest) 40 km to the south in the southern area of the lake. We collected feeding data for five days each month on each saki group using modified focal animal samples that measured feeding durations on plant species. Fruit abundance and seasonal distribution was measured by fruit traps and phenology.

This region of eastern Venezuela is characterized by a wet season and dry season that are of nearly equal duration. Each season was sub-divided into early and late, wet and dry seasons. There was no difference in the monthly pattern of rainfall between islands, but the southern island was significantly wetter than the northern island.

Both primate species ingested seeds year-round and month to month overlap of plant species and part ingested was high (average of c. 45%). Ingestion of seeds was higher than 60% of the plant portion of the diet in the dry season and first half of the wet season for *Pithecia* and in the dry season for *Chiropotes*. Both sakis shifted to ingesting mesocarp as it became available (and as seeds matured) in the wet season. Non-fruit items were secondary resources during all seasons with the exception of early wet season ingestion of caterpillars by *Chiropotes*.

Three differences were found between the saki species: a) the tendency for *Chiropotes* to use fruit of the same plant species from young to mature seed and mesocarp; *Pithecia* ingested seeds, mesocarp and whole fruit from different species; b) *Pithecia* did not alter plant species diversity or diet breadth on a seasonal basis, but *Chiropotes* increased diet breadth in the dry season and reduced it in the wet season; c) *Chiropotes* ingested items high in water soluble carbohydrates, whereas lipid values were often higher in food items ingested by *Pithecia*.

Sakis are more likely to track phenological differences in resource availability than rainfall per se. Variation in fruit ingested on the same month in consecutive years ranged from a low of 9% (high predictability) to more 66%. But rainfall probably does provide some predictability. For example, a one month lag was seen in some species that was most likely related to a slight shift in rainfall synchronicity.

Pitheciins and colobines are the only two anthropoid subfamilies that maintain high levels of seeds predation. Seed predation may be related to anatomical specializations in both subfamilies: dental adaptations and perhaps hindgut enlargement in the Pitheciines and foregut adaptations in the digestive tract of Colobines.

ACKNOWLEDGMENTS

It was a privilege to work with Warren G. Kinzey on this project. He designed the project after identifying anatomical differences in the dental morphology of the sakis and procurred funding for the original project. We were assisted in the field by Willian Arteaga, Cynthia Butler, Ingo Homburg, Ric Lopez, Pia Parolin, Angela Peetz, Suzanne Walker and Catherine Wertis. Gerardo Aymard (UNELLEZ Herbarium, Guanare Venezuela and the Missouri Botanical Gardens) divided his time between field and lab to identify the plants listed herein. Logistical support was provided by Luis Balbás and the dedicated people of Estudios Basicos, EDELCA, Guri. The project was supported by NSF DBS 90–20614. I am grateful to Gerardo Aymard, John Oates, and Michael Power for offering critical comments on the chapter.

REFERENCES

Alvarez, E., Balbás, L., Massa, I. and Pacheco J. 1986 Aspectos ecológicos del Embalse Guri. *Interciencia* 11(6):325–333.

Aymard, G., Kinzey, W., and Norconk, M. (in press) Composición florística de bosques tropófitos macrotérmicos en islas en el embalse de Guri, bajo Río Caroní, Edo. Bolívar, Venezuela. *Biollania.*

Ayres, J.M. 1981 Observações sobre a ecologia e o comportamento dos cuxiús (*Chiropotes albinasus* e *Chiropotes satanas*, Cebidae: Primates). Unpublished Master of Science Thesis, Instituto Nacional de Pesquisas da Amazônia e Fundação Universidade do Amazonas, Manaus, Brazil.

Ayres, J.M. 1986 Uakaris and Amazonian flooded forest. PhD thesis, University of Cambridge, Cambridge, UK.

Bell, E.A. 1978 Toxins in seeds. In: BIOCHEMICAL ASPECTS OF PLANT AND ANIMAL COEVOLUTION. J.B. Harborne, ed., London, Academic Press, pp. 143–161.

Chapman, C. 1988 Patterns of foraging and range use by three species of neotropical primates. *Primates* 29(2):177–194.

Chivers, D. 1994 Functional anatomy of the digestive tract. In: COLOBINE MONKEYS:THEIR ECOLOGY, BEHAVIOUR, AND EVOLUTION. A.G. Davies & J.F. Oates, eds., Cambridge University Press, pp. 205–228.

Clutton-Brock, T.H. 1977 Some aspects of intraspecific variation in feeding and ranging behaviour in primates. In: PRIMATE ECOLOGY. T.H. Clutton-Brock, ed., London, Academic Press, pp. 539–556.

Cords, M. 1993 The behavior of adult female blue monkeys during a period of seasonal food scarcity. *Amer. J. Primatol.* 30(4):304–305.

Crockett, C.M. and Rudran, R. 1987 Red howler monkey birth data I: Seasonal variation. *Amer. J. Primatol.* 13(4):347–368.

Dasilva, G. 1994 Diet of *Colobus polykomos* on Tiwai Island: Selection of food in relation to its seasonal abundance and nutritional quality. *Int. J. Primatol.* 15(5): 655–680.

Galetti M. and Pedroni F. 1994 Seasonal diet of capuchin monkeys (*Cebus apella*) in a semideciduous forest in south-east Brazil. *J. Trop. Ecol.* 10–27–39.

Garber, P.A. 1993 Seasonal patterns of diet and ranging in two species of tamarin monkeys: Stability versus variability. *Int. J. Primatol.* 14(1):145–166.

Goldizen, A. W., Terborgh, J., Cornejo, F., Porras, D.T., and Evans, R. 1988 Seasonal food shortage, weight loss, and the timing of births in saddle-back tamarins (*Saguinus fuscicollis*). *J. Anim. Ecol.* 57(3):893–901.

Kinzey, W.G. 1992 Dietary and dental adaptations in the Pitheciinae. *Am. J. Phys. Anth.* 88:499–514.

Kinzey, W.G. and Norconk, M.A. 1990 Hardness as a basis of food choice in two sympatric primates. *Am. J. Phys. Anth.* 81:5–15.

Kinzey, W.G. and Norconk, M.A. 1993 Physical and chemical properties of fruit and seeds eaten by *Pithecia* and *Chiropotes* in Surinam and Venezuela. *Int. J. Primatol.* 14(2): 207–227.

Kinzey, W.G., Norconk, M.A., and Alvarez-Cordero, E. 1988 Primate survey of eastern Bolívar, Venezuela. *Prim. Conserv.* 9:66–70.

Kool, K. 1993 The diet and feeding behavior of the silver leaf monkey (*Trachypithecus auratus sondaicus*) in Indonesia. *Int. J. Primatol.* 14(5):667–700.

Maisels, F., Gautier-Hion, A., and Gautier, J-P. 1994 Diets of two sympatric colobines in Zaire: More evidence on seed-eating in forests on poor soils. *Int. J. Primatol.* 15(5):681–702.

Milton, K. 1984 The role of food processing factors in primate food choice. In: ADAPTATIONS FOR FORAGING IN NONHUMAN PRIMATES. P.S. Rodman, J.G.H. Cant , eds. New York, Columbia. pp 249–279.

Morland, H.S. 1992 Seasonal variation in the behavior of wild ruffed lemurs (*Varecia variegata variegata*). XIVTH CONGRESS OF THE INTERNATIONAL PRIMATOLOGICAL SOCIETY abstracts, p. 21. Strasbourg, IPS.

Norconk, M.A. and Kinzey, W.G. 1993 Challenge of neotropical frugivory: Travel patterns of spider monkeys and bearded sakis. *Amer. J. Primatol.* 34:171–183.

Parolin, P. 1993 Forest inventory in an island of Lake Guri, Venezuela. In: ANIMAL-PLANT INTERACTIONS IN TROPICAL ENVIRONMENTS. W. Barthlott, C.M. Naumann, K. Schmidt-Loske, and K.L. Schuchmann, eds. Bonn, pp. 139–147.

Pereira, M.E. 1992 Seasonal patterns in the growth and adult body weight of ringtailedlemurs. XIVTH CONGRESS OF THE INTERNATIONAL PRIMATOLOGICAL SOCIETY abstracts, p. 24. Strasbourg, IPS.

Peres, C.A. 1993 Notes on the ecology of buffy saki monkeys (*Pithecia albicans*, Gray 1860): a canopy seed-predator. *Am. J. Primatol.* 31:129–140.

Robinson, J.G. 1986 Seasonal variation in use of time and space by the wedge-capped capuchin monkey, *Cebus olivaceus*: Implications for foraging theory. *Smithsonian Contributions to Zoology*, no. 431. Washington D.C.: Smithsonian Press.

Struhsaker, T.T. 1975 THE RED COLOBUS MONKEY. Chicago: The University ofChicago Press.

Waterman P.G., Kool. K.M. 1994 Colobine food selection and plant chemistry. In: COLOBINE MONKEYS:THEIR ECOLOGY, BEHAVIOUR, AND EVOLUTION. A.G. Davies & J.F. Oates, eds., Cambridge University Press, pp. 251–284.

Wright, P.C. 1994 Ontogeny of tooth wear of *Propithecus diadema edwardsi* in relation toseed predation. XVTH CONGRESS OF THE INTERNATIONAL PRIMATOLOGICAL SOCIETY abstracts, p. 364.

Wright, P.C. and Meyers, D. 1992 Resource tracking: Food availability and *Propithecus* seasonal reproduction. XIVTH CONGRESS OF THE INTERNATIONAL PRIMATOLOGICAL SOCIETY abstracts, p. 21–22. Strasbourg, IPS.

Yamashita, N. 1994 Seed-eating among three sympatric lemur species in Madagascar. XVTH CONGRESS OF THE INTERNATIONAL PRIMATOLOGICAL SOCIETY abstracts, p. 365.

SECTION V

On Atelines

ON ATELINES

Walter Carl Hartwig,[1] Alfred L. Rosenberger,[2] Paul W. Garber,[3] and
Marilyn A. Norconk[4]

[1]Department of Anthropology
University of California
Berkeley, California 94720
[2]Department of Zoological Research
National Zoological Park
Washington, DC 20008
[3]Department of Anthropology
University of Illinois
Urbana, Illinois 61801
[4]Department of Anthropology
Kent State University
Kent, Ohio 44242

Looks can be deceiving. Although they are the largest and most easily recognizable New World monkeys, atelines are neither the most studied nor the best understood. Important aspects of their behavioral ecology and evolutionary history have yet to be researched in the field or collected from the fossil record, which increases the likelihood of making significant discoveries. For example, it has long been taken for granted that living atelines are the largest New World monkeys, but we have just found out that the surviving species are far from the largest. Their taxonomy has also been neglected; surprises may await, insight is inevitable. The most comprehensive systematic treatment of howler monkeys is over 60 years old (Lawrence, 1933); Froehlich's (1991) study of spider monkeys is the first assessment since the Kellogg and Goldman's revision of WWII vintage (1944); woolly monkeys haven't been looked at since the Kennedy administration (Fooden, 1963). The wisdom of our overall research strategies, which we usually associate with lengthy gestation if not maturation, is also suspect: the woolly spider monkey reached the very brink of extinction before rigorous field studies began (Aguirre, 1971; Valle et al., 1984; Milton, 1984; Strier, 1986).

Living atelines represent one of the obvious clades in primate systematics, characterized most prominently by a suite of derived postcranial adaptations to climbing locomotion. A growing number of molecular studies complement modern morphological interpretations. We argue strongly that the living atelines are a monophyletic group (Table 1; Schneider and Rosenberger, this volume), negating the once routine placement of

Table 1. Genus level classification of Atelines

Family Atelidae
Subfamily Atelinae
Tribe Alouattini
Alouatta - Howler monkeys
*Stirtonia - Middle Miocene, Colombia
*Protopithecus - ? Pleistocene/Recent, Brazil
*Paralouatta - Pleistocene, Cuba
Tribe Atelini
Subtribe Atelina
Ateles - Spider monkeys
*Caipora - Pleistocene, Brazil
Brachyteles - Woolly spider monkeys
Subtribe Lagotrichina
Lagothrix (*Lagothrix*) *lagotricha* - Woolly monkeys
?*Lagothrix (Oreonax) flavicauda* - Yellow-tailed woolly monkeys

• Extinct genus. See Schneider and Rosenberger (this volume) and Rosenberger (1988) for references and discussion. "Other atelines" include fossils whose relationships *within* Atelinae are uncertain.

Alouatta into a separate subfamily. The four extant genera, *Alouatta*, *Lagothrix*, *Ateles* and *Brachyteles*, neatly form a tight cluster in the panorama of New World monkey phylogeny, a quite distinct adaptive radiation within the platyrrhines (Rosenberger and Strier, 1989).

All atelines occupy a swathe of niche-space in the canopy of neotropical forests. Their basic diet shows relatively little variation among species, except in the proportion of fruits and leaves (see Strier, 1992). Locomotor behavior is dichotomous, ranging from a gradation of climbing and acrobatic suspension (*Ateles, Brachyteles, Lagothrix*) along one phylogenetic axis to climbing and deliberate quadrupedalism (*Alouatta*) along the other. Regarding social organization, the hallmark *Ateles* pattern - daily foraging parties that are flexible in size and composition - also characterizes *Brachyteles* and *Lagothrix* (e.g. Milton,1984; Strier, 1992; Peres, 1994, this volume; Defler and Defler, 1996). Group composition appears to be more stable on a day-to-day basis for *Alouatta* (Strier, 1992; Crockett, this volume). Thus atelines also exhibit differences in social behavior and locomotion despite basic similarities in habitat use and diet, and these correspond with the phylogenetic structure of the subfamily.

Phylogenetically, the interrelationships of the subfamily are still problematic. Cladograms based on morphology and molecules do not agree (Schneider and Rosenberger, this volume), and new fossil material adds another dimension. A difficulty in deciphering relationships exclusively from the living species is that the four genera are confounded by extreme deviations in the anatomy of howlers, coupled with primitive retentions in the limbs of non-howlers. Thus our sense of the evolutionary history of atelines is somewhat bipolar. However, it cannot be taken on faith that this taxonomic quartet is the truest foundation for ateline systematics. Lessons from higher phylogeny have had little reciprocal impact on issues of alpha and beta taxonomy, the impetus for which has been lost in the wake of Philip Hershkovitz's (see 1977, et seq.) aborted attempt to revise the entire platyrrhine infraorder. As with the genus *Callithrix*, which may include some species more closely related to the pygmy marmoset, *Cebuella pygmaea* (Barroso et al., in press), there may be a taxonomic problem with the genus *Lagothrix*, a point that Colin Groves brought to our attention.

Two species have been broadly accepted since Fooden's revision (1963), *L. lagotricha* and *L. flauvicauda*. The latter has received scant attention. It is a relict population, thought for a long time to have been extinct (Ruiz and Mittermeier, 1979), and it remains poorly represented in museum collections. Fooden, and Ruiz and Mittermeier, summarized what little was known of the species, but there has never been a detailed systematic analysis, certainly none that places *flauvicauda* in the context of atelin systematics. Cranial characters, usually the key referent for generic distinction, are suggestively different from *Lagothrix lagotricha*. The pattern of *L. flavicauda* includes a deeper, more inflated posterior mandible; more projecting and narrower snout; less inflated braincase; stronger postorbital constriction; more rounded nuchal plane; and, an assortment of basicranial features involving foramina, petrosal and ectotympanic shape, pterygoids, sphenoid, etc. Incisor proportions may differ and the molar cusp pattern is distinct; for example, the hypocone and talon region is less enlarged. In many respects, these characters resemble *Brachyteles* from eastern Brazil. Interestingly, Ruiz and Mittermeier (1979) note that the thumb of live, adult *flauvicauda* is markedly shorter than the second digit, which is a possible point of derived similarity shared with *Brachyteles* and *Ateles*.

We are left with no firm conclusions. Additional study, taking into account not only the other atelines but also *Alouatta* and the new Brazilian fossils discussed below is required. For the moment, we propose to keep open the questions: To whom is *flavicauda* most closely related, and how is it best classified? Thus we resurrect, with a query, the taxonomy presented when the animal first became well established in the literature (Thomas, 1927): ?*Lagothrix* (*Oreonax*) *flauvicauda*. The hypotheses we plan to test are that *Oreonax* is a "good" genus, perhaps more closely related to the *Ateles-Brachyteles* branch, perhaps a geographical relict of an Amazonian rather than Andean community, and perhaps with deeper implications for early ateline evolution than *L. lagotricha* and other living atelins.

Discoveries made more than 150 years apart are dramatic proof that fossil evidence - always painfully elusive - will be the ultimate arbiter in our rethinking of ateline evolution. The specimens are still rare (Rosenberger and Hartwig, in press) but they add richness to the artificially narrow frame of reference depicted by the surviving living forms. They show that atelines were even more widespread in the past than the present; that atelines are an old part of the platyrrhine fauna; that archaic forms may have persisted for long stretches of geological time; that our contemporary slice of the radiation, if not depauperate taxonomically, is certainly not archetypical in an adaptive sense. One fossil genus is known from the middle Miocene deposits of Colombia, *Stirtonia*. Another comes from the far flung Pleistocene or Quarternary of Cuba, *Paralouatta*. Both of these are howler-like forms, judging from craniodental parts, and they fit comfortably in the prevailing phylogenetic-adaptive evolutionary model of Rosenberger and Strier (1989). However, we now have evidence for atelines of types we could only barely imagine scientifically.

The third and fourth fossil genera are more revealing but also more difficult to interpret. Each is represented by a nearly complete skeleton from the Pleistocene, reported in 1993 by Dr. Castor Cartelle, a paleomammalogist working in cave deposits in eastern Brazil. One skeleton is a large juvenile (>20kg) that very closely resembles living spider monkeys cranially (Cartelle and Hartwig, 1996), *Caipora bambuiorum*. The other is an even larger adult (approximately 25 kg) that postcranially resembles the spider and woolly spider monkeys, but resembles howler monkeys cranially (Hartwig and Cartelle, 1996). The latter is referred to *Protopithecus brasiliensis*, first named by Peter Wilhelm Lund in 1838 for a partial femur and humerus he found in the Brazilian state of Minas Gerais (Hartwig, 1996).

With bodies complete, there is little doubt about the manifest mosaic: Heads looking like the two derived polar opposites of the ateline radiation (*Alouatta* vs. *Ateles*) but skeletons looking much alike, akin to the climbing-suspensory system of *Ateles*. Finding this combination in a howler relative, *Protopithecus*, was not predictable based on the four living genera. The new data make quite clear the need to document better the morphology of each genus in systematic detail, and reformulate the outlines of ateline evolution. It also reminds us there is limited resolution to studies which exclude fossil evidence and, by corollary, that informative fossils can provide a powerful test of detailed cladistic hypotheses.

Whereas the howler monkey axis was even recently considered so uniformly derived that it could shed little light on atelines broadly (see Rosenberger and Strier, 1989), these advances draw our attention to *Alouatta* as another focus in our rethinking of ateline evolution - keeping in mind that phylogenetic analysis of adaptive radiations is a process of historical triangulation. Clearly, *Protopithecus* is a closely related genus of relatively great size in which the signature cranial features of howlers - less the dental morphology - is combined with essential postcranial characters formerly associated only with the atelin lineage, which are related to climbing. Previously, we had inferred (Rosenberger and Strier, 1989) that the common ancestor of atelines was more of a quadruped than an acrobatic climber or brachiator, although climbing was indeed indicated by the behavioral repertoire and morphology of living howlers. We further argued that the deliberate quadrupedalism of howlers was somehow linked with a strategy to minimize energy expenditure, as befits a folivore (Strier, 1992). The evidence of *Protopithecus* suggests the following: 1) A confirmation that the alouattin lineage at its base was a climbing stock, with functional adaptations in fore- and hindlimb. 2) Adaptive anatomical specializations relating to vocal communication - not to diet - shaped the howler skull early on and was fundamental to the origins of the lineage. 3) Changes in stance and movement, which may be correlated with unique limb proportions, elbow joint morphology, shoulder and probably the rib cage anatomy - anatomical areas that are uniquely derived in *Alouatta* - may also be related primarily to the *production and delivery* of sound rather than foraging behavior or locomotor travel.

Primatology is not short of theoretical models for New World monkey evolution, in general (Rosenberger 1992) or ateline evolution, in particular (Rosenberger and Strier 1989). As we rethink atelines in light of a multidisciplinary front of advances, we must remember that a wealth of information has yet to be obtained from the animals themselves, whether from nature or the museum drawer.

REFERENCES

Aguirre, A. Continho 1971. *O Mono Brachyteles arachnoides (E. Geoffroy)*. Academia Brasileira de Ciências, Rio de Janeiro.

Barroso, C.M.L., Schneider, H., Schneider, M.P.C., Sampaio, M.L., Harada, M.L. Czelusniak, J., and M. Goodman In press. Update on the phylogenetic systematics of New World monkeys: further DNA evidence for placing the pygmy marmoset (*Cebuella*) within the marmoset genus *Callithrix. Int. J. Primatol.*

Cartelle C., and Hartwig W.C. 1996. A new extinct primate among the Pleistocene megafauna of Bahia, Brazil. *Proc. Nat. Acad. Sci.* 93:6405–6409.

Defler, T.R., and Defler, S.B. 1996. Diet of a group of *Lagothrix lagothricha lagothrica* in southeastern Colombia. *Int. J. Primatol.* 17(2):161–190.

Erickson G.E. 1963. Brachiation in New World monkeys and in anthropoid apes. *Symp. Zool. Soc, Lond.* 10:135–164.

Fooden J. 1963. A revision of the woolly monkeys (genus *Lagothrix*). *J. Mammal.* 44:213–247.

Froehlich, J.W., Supriatna, J., and Froehlich P.H. 1991. Morphometric analyses of *Ateles*: systematic and biogeographic implications. *Am. J. Primatol.* 25:1–22.

Hartwig W.C. 1995. *Protopithecus*: rediscovering the first fossil primate. *History and Philosophy of the Life Sciences* 17:447–460.

Hartwig W.C., and Cartelle C. 1996. A complete skeleton of the giant South American primate *Protopithecus. Nature* 381:307–311.

Hershkovitz, P. 1977. *Living New World Monkeys (Platyrrhini)*. Volume 1. University of Chicago Press, Chicago IL.

Kellogg R., and Goldman E.A. 1944. Review of the spider monkeys. *Proceedings of the United States National Museum* 96:1–45.

Lawrence, B. 1933. Howling monkeys of the *palliata* group. *Bull. Mus. Comp. Zool.* 75:315–354.

Milton K. 1980. *The Foraging Strategy of Howler Monkeys*. New York: Columbia University Press.

Milton, K. 1984. Habitat, diet, and activity patterns of free-ranging woolly spider monkeys (*Brachyteles arachnoides* E. Geoffroy 1806). *Int. J. Primatol.* 5:491–514.

Peres CA 1994. Diet and feeding ecology of gray woolly monkeys (*Lagothrix lagotricha cana*) in central Amazonia: comparisons with other atelines. *Int. J. Primatol.* 15:333–372.

Rosenberger A.L. 1988. Evolution of feeding niches in New World monkeys. *Am. J. Phys. Anthropol.* 88:525–562.

Rosenberger A.L., and Strier K.B. 1989. Adaptive radiation of the ateline primates. *J. Hum.Evol.* 18:717–750.

Ruiz, Hernando de Macedo, and Mittermeier, R.A. 1979 Redescubrimiento del primate Peruano *Lagothrix flavicauda* (Humboldt 1812) y primarias observaciones sobre su biología. *Revista de Ciencias* 71(1):78–92. Universidad National Mayor de San Marcos, Lima, Peru.

Schneider H., Schneider M..PC., Sampaio I., Harada M.L., Stanhope M., Czelusniak J., and Goodman M. 1993. Molecular phylogeny of the New World monkeys. *Molec. Phylogen. Evol.* 2:225–242.

Strier, K.B. 1986 *The Behavior and Ecology of the Woolly Spider Monkey, or Muriqui (Brachyteles arachnoides E. Goeffroy 1806)*. PhD dissertation, Harvard University, Cambridge MA.

Strier, K.B. 1992. Ateline adaptations: behavioral strategies and ecological constraints. *Am.J. Phys. Anthropol.* 88:515–524.

Thomas, O. 1927. A remarkable new monkey from Peru. *Ann. Mag. Nat Hist.* 19:156–157.

Valle, C.M.C., Santos, I.B., Alves, M.A., Pinto, C.A., and Mittermeier R.A. 1984. Algumas observacões preliminares sobre o comportamento do mono (*Brachyteles arachnoides*) em ambiente natural (Fazenda Montes Claros, municipio de Caratinga Minas Gerais, Brasil. In M. Thiago de Mello (ed.) *A Primatologia no Brasil* 1, pp. 271–283. Sociedad Brasileira de Primatologia, Brasilia, Brazil.

21

DENTAL MICROWEAR AND DIET IN A WILD POPULATION OF MANTLED HOWLING MONKEYS (*Alouatta palliata*)

Mark F. Teaford[1] and Kenneth E. Glander[2]

[1]Department of Cell Biology and Anatomy
Johns Hopkins University School of Medicine
725 N. Wolfe Street
Baltimore, Maryland 21205
[2]Duke University Primate Center
3705 Erwin Road
Durham, North Carolina 27705

INTRODUCTION

The mantled howling monkey (*Alouatta palliata*) was the subject of the first naturalistic study of nonhuman primate behavior - Carpenter's classic 1934 work in Panama. That study not only set the stage for future naturalistic behavioral-ecological work, it also set the tone for our perceptions of howling monkeys. For 40–50 years, howlers were essentially viewed as humid tropical forest leaf-eaters. It is only within the past 20–25 years that researchers have begun to appreciate the ecological variability and adaptability of howlers. Now, based on the work of Crockett, Glander, Milton, and others, researchers have gained a more realistic picture of howlers as folivorous frugivores which can inhabit a wide range of habitats including rain forests, swamp forests, and semideciduous forests (Crockett & Eisenberg, 1986). In fact, they are so adaptable that they are often the first neotropical primates to colonize new patches of secondary forest.

For more than 25 years, one of us (KG) has been studying mantled howlers at a site known as Hacienda La Pacifica in Guanacaste Province, Costa Rica. In contrast to public perceptions of Costa Rica, Guanacaste is *not* a tropical rain forest, but instead a markedly seasonal environment where significant changes in rainfall yield dramatic changes in resource availability (see figures 1 and 2). As a result, Glander's work has shown how howler diets vary from season to season, not only in terms of the percentages of basic food items, but also in terms of nutrients and secondary compounds (Glander, 1975, 1978b, 1981, 1992). With that work as a starting point, in 1989, we decided to start a collaborative study of diet and tooth use in the howlers at La Pacifica. Previous work had suggested that studies of microscopic wear patterns on teeth might shed new light on primate paleo-

Adaptive Radiations of Neotropical Primates
edited by Norconk *et al.* Plenum Press, New York, 1996

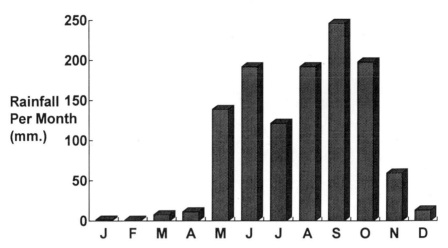

Figure 1. Monthly differences in rainfall at La Pacifica (note pronounced differences between dry season (November-April) and wet season (May-October)).

biology (e.g., Gordon, 1984, Grine 1986, Puech 1984, Teaford 1988, Teaford & Walker 1984, Walker et al. 1978). But, while many people couldn't resist the temptation to rush into analyses of fossil material, analyses of museum collections of modern primates (e.g., Teaford & Robinson 1989, Teaford & Runestad 1992) often raised more questions than they answered. At the same time, analyses of live wild-caught animals were proving extremely difficult. Thus, the initial goal of the collaborative work at La Pacifica was to demonstrate the feasibility of studies of live, wild-caught primates, so that we could begin to chart the limits of resolution of studies of tooth wear. How closely coupled were tooth wear and diet? What insights into diet and tooth use might be gained from analyses of tooth wear? Results of the 1989 pilot study were published in 1991 (Teaford and Glander 1991). This is a brief summary of what we've accomplished since then.

Hda. La Pacifica is a large ranch straddling the Inter-American highway near Canas, Costa Rica. It consists of a mix of dry tropical forest and pasture. The ranch was clear-cut 50 years ago, then protected from fires and allowed to partially regenerate. It is now maintained under a balance of two-thirds pasture and one-third dry forest (see figure 3). The only exceptions lie near the 3 rivers at the site: the Corobici, Tenorito, and Tenorio, where large primary trees still remain. These patches of riparian forest include a greater variety

Figure 2. Seasonal changes in foliage in Group 12's home range.

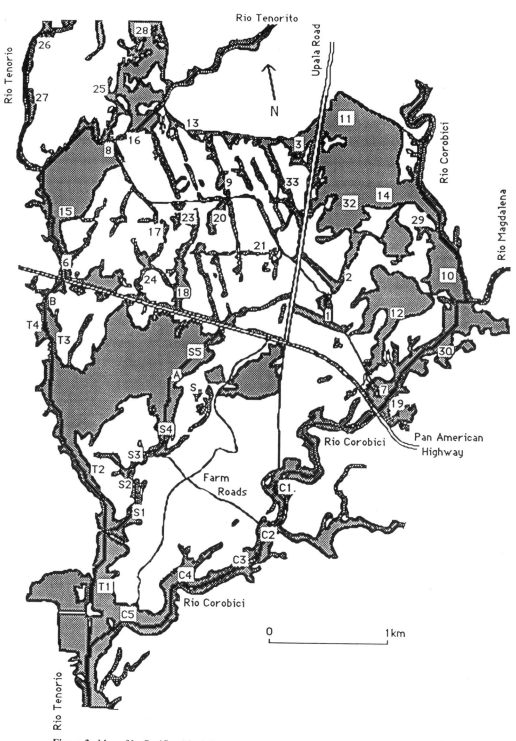

Figure 3. Map of La Pacifica (shaded areas = patches of forest, unshaded areas = pasture).

of tree species, resulting in significant differences in resource availability between river and non-river areas on the ranch.

Mantled howlers are generally the only nonhuman primates found at La Pacifica, except near the southwestern fringes of the ranch where *Cebus capucinus* occasionally is found. Given the seasonal, dry forest vegetation at the site, La Pacifica is an ideal location for a collaborative behavioral / morphological study because the animals are accessible and easily followed. They have also been studied extensively over the past 25 years (Clark et al., 1987; Clarke and Glander, 1981, 1984; Glander, 1975, 1978a, b, 1979, 1980, 1981, 1992; Glander et al., 1991; Moreno et al., 1991; Stuart et al., 1990). Thus, their diet is already known to have marked seasonal variations (Glander 1975, 1981); 42 howler social groups have been identified at the site (not to mention over 12,000 trees); and 315 of the estimated 572 howlers on the ranch have been permanently marked. As a result, individual animals can be recognized and followed throughout the day - facilitating observations of howler feeding behavior. In fact, the diet of *Alouatta* would seem to be particularly informative for studies of dental function, as all medium-to-large-sized primates eat varying amounts of leaves and fruit (Kay, 1984), and the seasonal changes in howler diets include changes in the proportions of young leaves, mature leaves, fruit, and flowers in the diet - i.e., just the sort of dietary subtleties that might prove useful in analyses of medium-to-large-sized fossil primates.

METHODS

Capture Techniques

One aim of the project was to gather data from large, cross-sectional samples of monkeys to document normal patterns of dental microwear and dental gross morphology in animals of known ages - with known seasonal and annual changes in resource availability. Thus, for this portion of the project, animals were captured and released from as many social groups on the ranch as possible. Another aim of the study was to gather short-term longitudinal data from the howlers to relate changes in rates of tooth wear to short-term changes in diet. With this in mind, 10–25 animals from 5 focal groups were recaptured (after a period of 3–4 days) during each season. Due to the inherent unpredictability of variations in the number of infants, number of mothers (with and without infants), and number of deaths in each social group, the individuals recaptured varied slightly from season to season and year to year.

As in previous work at La Pacifica, capture of the animals was accomplished using the methods of Glander et al. (1991). Thus a CO_2 powered gun was used to deliver a non-barbed dart ("Pneu-DartTM system", Williamsport, PA) loaded with "Telazol".[*] Extensive work by Glander has shown that "Telazol" is far more effective than other drugs (e.g., "Ketaset", "Sernylan") for capturing New World monkeys with prehensile tails (Glander et al., 1991). Its advantages include small volume doses which are easily and quickly administered by the capture darts, very rapid induction and immobilization, good muscle relaxation, a wide safety margin, and a rapid recovery. The dosage was 25 mg per Kg. This dosage has been successfully used during the past 7 years when over 800 individuals have been captured. Individuals were darted at distances of up to 20 meters. As very few trees at La Pacifica are taller than that, animals could be caught virtually anywhere at the site.

---

[*] (A.H. Robbins Co., Richmond, VA) "Telazol" is a Schedule IIIN drug, a nonnarcotic, nonbarbiturate, injectable anesthetic

Individuals weighing less than 2 kilos were not darted, although babies less than 1 kilo usually came down with their mothers if their mothers were darted.

The only negative effect of Telazol (or Ketaset or Sernylan) is that it occasionally interferes with the thermal-regulating ability of some individuals. To combat this, rectal temperature was closely monitored. Any rectal temperature of more than 39°C was treated by wetting the animal with water or immersing it in a bucket of cool water. As an additional safety factor, each animal was monitored via a Pulse Oximeter during the entire time it was anesthetized. This measures functional oxygen saturation of arterial hemoglobin, and pulse rate and, therefore, provides continuous information on the heart rate and breathing efficiency of the anesthetized animal.

Under the standard dosage, animals were generally immobilized within one minute, and fell from the trees within 3–5 minutes. Animals falling from the trees were caught in nylon mesh nets. When a darted animal did not fall, it generally hung from a branch by its tail. In such cases, the branch was either shaken or cut down with a saw attached to the end of an aluminum pole. The pole comes in six-foot sections which can be bolted together until it is long enough to reach the hanging animal.

Once captured, the monkeys were measured and marked (or re-marked if necessary) with a tattoo (on the inside left thigh) and with collars, anklets, or ear-tags. Markers of recaptured individuals were checked and replaced if necessary. Collars were used on adult female howlers, but the enlarged hyoid bone of male howlers necessitated the use of anklets in place of collars for adult male howlers. Ear-tags were used to mark immature individuals because collars and anklets do not expand as the animal grows. Data collected from these animals included dental impressions, body weights and measurements, footprints, and samples of blood, urine, feces, hair, and saliva. Data were collected at a field laboratory with access to electricity and running water. Animals recovering from the capture dosage before the procedures were completed were given injections of 3 mg / kg of Telazol, repeated as often as needed.

After completion of all procedures, the animals were placed in burlap bags until they recovered enough to walk or climb unaided. The bags were kept in the shade and are the best means of holding an animal until it recovers because, unlike standard plastic pet carriers, the bag reduces visual stimulus. After recovery, captured animals showed no significant changes in feeding or ranging behavior, and there were also no significant effects on non-captured group members.

Starting in 1992, there were two field seasons each year, one in the wet season and one in the dry season. Behavioral observations focused on the 5 focal groups out of 42 howler social groups on the ranch (groups 1, 7, 19, 30, and 33 in figure 3). For each focal group, as many animals as possible were caught in one day, and baseline dental impressions were taken before the animals were released back into their home range. For the next 3 days, 4–6 observers would watch 4–6 focal animals in the group, while the remainder of the research team would capture animals from additional social groups. At the end of 3 days, as many animals as possible were recaptured in the focal group and follow-up dental impressions were taken. In all, a total of 77 focal howlers were captured and recaptured during the 3 years of this project. 481 additional howlers were captured from other social groups on the ranch.

Behavioral Observations

Observations of focal animals insured that individuals within each focal group were observed in the wet and dry season each year, and it also insured that rates of tooth wear were obtained from animals for which we had good, current behavioral data. As in all previous work on this project, activities recorded were: move, rest, and feed. "Move" is defined as a change of location but not a change in position or posture. "Rest" is defined as the absence of either move or feed. "Feed" is defined as actively ingesting a food item. If feeding was the activity, the following additional data were gathered: (1) the part of the plant being ingested, i.e., mature leaves, new leaves, green or ripe fruits, flowers, twigs, petioles, pedicels, pulvinus; (2) which teeth were being used to harvest and process the selected items (i.e., incisors, canines, premolars, or molars); and (3) the length of time those teeth were in use. One of the nice things about La Pacifica is that the conditions of visibility there allow these kind of detailed feeding data to be gathered. The activity patterns are mutually exclusive and encompass all activities. Thus, this method of data collection provides durations of activities which can then be statistically analyzed using ANOVA (Glander, 1975, 1981). All data were either stored on portable Radio Shack computers, or written-down & transcribed later.

Dental Impression and Replication Techniques

Impression techniques were the same as those used previously by one of us (MT) on laboratory primates (Teaford and Oyen, 1989a) and in the pilot work for this project (Teaford and Glander, 1991). Thus, 10–15 minutes before impressions were taken, each animal was given a small dose (.05 mg per Kg) of atropine, to reduce salivation and to stabilize the heart rate. After careful cleaning and drying of the teeth (Teaford and Glander 1991, Teaford and Oyen 1989a), dental impressions were taken of the left mandibular tooth row (also including the right mandibular incisors) using a polysiloxane impression material ("President Jet, Regular Body", Coltene/Whaledent). The shrinkage of this material during polymerization is inconsequential for this work (see Teaford and Oyen 1989a for further discussions of the dimensional stability of dental replication materials). All impressions were stored in labeled zip-lock plastic bags and carried back to Baltimore where epoxy casts were poured approximately one month after the impressions were taken (using "Epotek" epoxy, Epoxy Technology, Billerica, MA). The epoxy casts were then used in scanning electron microscope (SEM) analyses and 3-D measurements of gross morphology.

Dental Measurements

SEM micrographs were taken at magnifications of 200X or 500X as outlined in previous publications (Teaford 1993, Teaford and Oyen, 1989b; Teaford and Robinson, 1989; Teaford and Runestad 1992; Teaford and Walker, 1984). Higher magnification micrographs were used to measure the incidence of microwear features on the enamel of LM_2.[†] To insure a good compromise between area surveyed in the micrograph and accurate measurement of small microwear features (Gordon, 1988; Grine, 1986), two micrographs were taken of the crushing/grinding areas of M_2. Thus, statistical analyses for this portion of the proposed project used mean values for the number of microwear features per micrograph for each individual.

Lower magnification micrographs were used to calculate rates of wear for LM_2 (Teaford and Glander, 1991; Teaford and Oyen, 1989b; Teaford and Tylenda, 1991). As in the

---

† More "traditional" microwear measurements (e.g., incidence of pits vs. scratches, and average size of pits) are still being computed for these samples.

high magnification work, micrographs were taken of 2 sites on the crushing/grinding facets on the second molar. Rates of wear were then determined as described in previous publications (e.g., Teaford & Tylenda 1991). Thus, the proportion of microscopic wear features created in 1 week was used in statistical analyses as an indicator of the rate of wear for each tooth.

More traditional measurements of macroscopic wear (size of the shearing facets along the buccal side of M_2) were also computed using casts from the pilot project and the present project. These measurements were computed using a 3-dimensional measuring microscope ("Reflex Microscope", Reflex Measurement Ltd.). Thus, 3-dimensional coordinates were computed for points of interest on the occlusal surface of the tooth, and the coordinates were then used to compute the measurements. Since standard measures of molar shearing crest length (e.g., Kay, 1975, 1978) were originally defined in terms of reference points on *un*worn teeth, 2 measurements of shearing facet size were used in this study (Teaford 1991): (1) the perimeter (and thus area) of each shearing facet, and (2) the endpoints of facets 1, 2, 3, and 4 (Kay, 1977) on each molar.

Dental Analyses

The incidence of microwear, and average rates of microscopic wear, were compared between individuals captured in different seasons or different microhabitats at La Pacifica using the nonparametric Mann-Whitney test. For individuals caught in both the wet and dry seasons, the nonparametric Wilcoxon paired-sample test was used to test for differences between rates of wear in the wet and dry season. Measures of gross morphology were compared using two different methods: (1) the Mann-Whitney test for comparisons of average *changes* in measurements (through time) (e.g., Teaford and Oyen, 1989c); and (2) Euclidean Distance Matrix Analyses (Lele 1993, Lele & Richtsmeier 1991) of 3-D coordinates marking the end-points of molar shearing facets.

RESULTS

Preliminary Results of Behavioral Studies

The activity of four social groups (1, 7, 19, 33) was recorded periodically from July 1989 to August 1993. Groups were chosen to represent the riverine (7 & 19) and non-river (1 & 33) microhabitats at the study site. A total of 6,521 hours of data were collected during 912 days for an average of 7.2 hours per dat. The study groups spent 72.5% of the observation time resting, 7.8% moving, and 18.1% feeding.

Analyses of the feeding data showed that, in the dry season, the howlers spent more time feeding than in the wet season (see figure 4). As might be expected, based on the dramatic changes in resource availability, the howlers spent significantly more time feeding on new leaves, flowers, and green fruit during the dry season, and significantly more time feeding on mature leaves during the wet season (see figure 5). Moreover, analyses also showed that there were differences in howler feeding time between microhabitats within the ranch, as howlers from riverine areas spent significantly less time feeding than did howlers from non-river areas (see figure 4).

Analysis of the howler diets based on phenophase also demonstrated intergroup differences in which plant parts were consumed. When tree phenophases were grouped into "hard" (leaves, new twigs, and vine-leaves), "soft" (new leaves, petioles, pulvinus, and

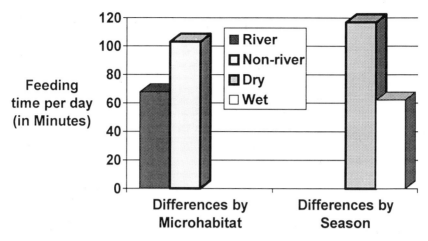

Figure 4. Seasonal and microhabitat differences in time spent feeding.

pedicels), and "fruit" (ripe and green), the main difference was between Group 7 and the other groups, with Group 7 spending much less time feeding on hard and fruit phenophases (see figure 6).

Preliminary Results of Dental Analyses

In the dry season, the howlers showed significantly more microwear on their molars than in the wet season (see figure 7) (Teaford et al., 1994). Likewise, the non-river howlers showed more microwear on their molars than did the river howlers (see figure 7) (Burnell et al., 1994). Were these microwear differences merely due to increased time spent feeding during the dry season and in the non-river microhabitats? We suspect that that would be a gross oversimplification, and we have begun to explore a number of possible explanations through preliminary analyses of howler feces and analyses of dust in the forest canopy (see discussion below).

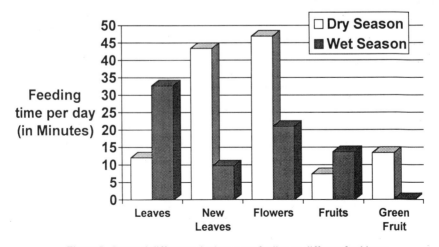

Figure 5. Seasonal differences in time spent feeding on different food items.

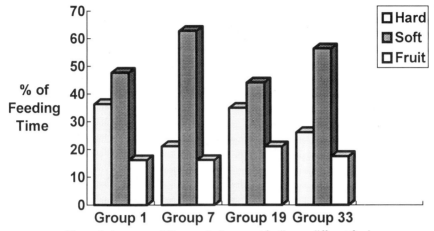

Figure 6. Intergroup differences in time spent feeding on different foods.

The rates of microscopic molar wear have also proven interesting and more compli-cated than we'd originally expected. If rates of microscopic wear were merely compared between wet and dry seasons, the dry season howlers showed a faster rate of molar wear (see figure 8) (Teaford et al., 1994). By contrast, there was no significant difference in rate of molar wear between animals caught in different microhabitats (see figure 8). How-ever, such summary-type figures also miss a great deal of information. In essence, individ-ual life events such as pregnancy and lactation seem to have dramatic effects on rates of wear (see figure 9, where the highest rates of wear (i.e., > 50%) are for individuals that are either pregnant, lactating, or under the stress of territorial disputes). In the process, they introduce significant amounts of variation into annual summary statistics thus complicat-ing interpretations. Probably the most reliable indicators of seasonal differences in rates of wear are provided by comparisons of the same individuals in different seasons (see fig-ure 10), where, again, dry season rates of molar wear are significantly greater than those

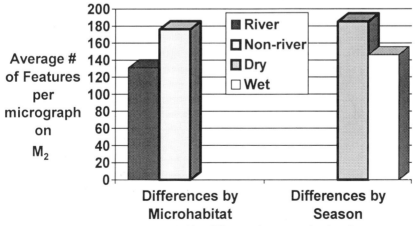

Figure 7. Seasonal and microhabitat differences in amount of molar microwear.

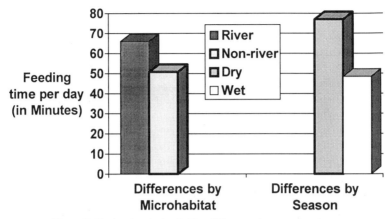

Figure 8. Seasonal and microhabitat differences in rates of molar wear.

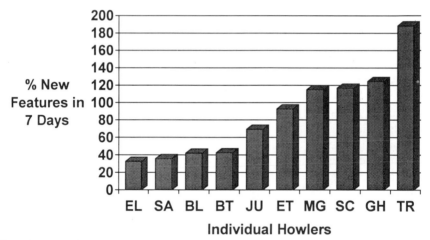

Figure 9. Rates of molar wear in individual howlers caught in the same field season (dry season of 1993).

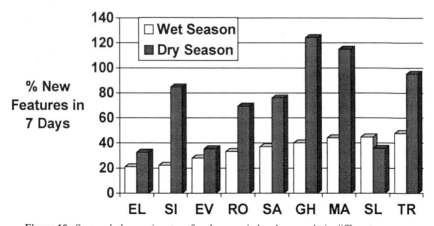

Figure 10. Seasonal *changes* in rates of molar wear in howlers caught in different seasons.

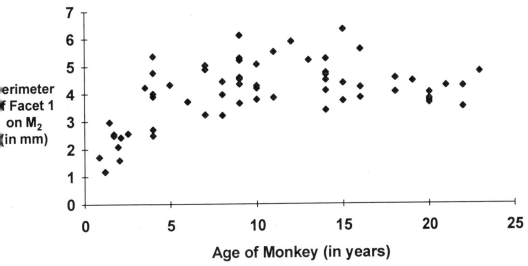

Figure 11. Molar shearing facet size in howlers of different ages.

from the wet season. Obviously, if we are to analyse the potentially interactive effects of variables such as age, sex, microhabitat, and season of collection, even larger samples will be needed.

To-date, analyses of macroscopic wear have focused on individuals caught during our 1989 season of pilot work, and during either 1992 or 1993. The reason for this is that the changes in tooth shape over only two years are still relatively subtle, whereas those over 3 or 4 years are more substantial and easier to measure. As in previous work (e.g., Teaford 1991), the endpoints of molar shearing facets, and molar shearing facet size, have been used in longitudinal and cross-sectional comparisons. The cross-sectional comparisons show that facet size increases with age in juveniles with mixed dentitions, but then levels off in adults (see figure 11). Longitudinal analyses of individuals caught in both 1989 and 1993 show that shearing facets on older individuals eventually begin to shrink in size (see figure 12). This finding is reflected in Euclidean Distance Matrix analyses (Lele 1993, Lele and Richtsmeier 1991) of 3-D coordinates marking the endpoints of the shearing facets on the buccal side of M_2 (see figure 13). In young adult howlers (those with minimal dentin exposure on M_2), molar shearing facets increase in size with wear. In older adult howlers (those with large dentin exposures on M_2), molar shearing facets begin decreasing with wear. This raises the possibility of interesting wear-related changes in dental function (see discussion below).

DISCUSSION

This study, if reported in the traditional manner, would merely give summary-type characterizations of howler diets and, in the process, mask a great deal of information. Instead, it is yielding the first hints of finer resolution differences in howler diets, such as seasonal, microhabitat, and intergroup differences in food and feeding, and maybe even differences in the physical properties of foods. Kinzey was well aware of the need for such work, as he and his students have emphasized it repeatedly in their publications (e.g., Rosenberger 1992, Rosen-

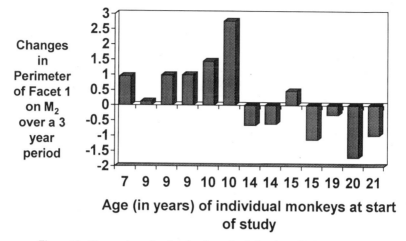

Figure 12. Changes in molar shearing facet size in howlers of different ages.

berger and Kinzey 1976, Rosenberger and Strier 1989, Kinzey 1978, 1992, Kinzey and Nor-conk 1990, 1993). Much of their work has focused on the implications of dietary variations on variations in tooth morphology. Until now, researchers have always had to make an intuitive leap of faith to make connections between the two. As Kinzey noted in discussing the dental adaptations of the pithecines, "This *assumes* that a relationship exists between morphology and ...behavior" (1992:500)(our italics). In other words, even if we could tell that "species A" was spending more time feeding on a certain food than was "species B", we could never tell for sure how, or where along the tooth row, that food was being processed. With the techniques used in this study, we can finally make a direct connection between the behavior and the morphology. We can truly monitor how the teeth are used - and with microscopic precison. We can close the gap between behavior and morphology and thus answer many questions that have heretofore been out of reach.

The current studies of the La Pacifica howlers are just beginning to answer exciting new questions, and raising still more in the process. For instance, pregnant and lactating

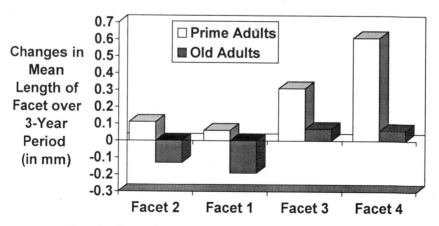

Figure 13. Changes in molar shearing facet size in young and old howlers.

females are showing a faster rate of tooth wear than are most other adult howlers at La Pacifica. This is the first time this sort of difference has ever been demonstrated. It certainly makes intuitive sense, because pregnant and lactating females may need to spend more time feeding to meet their increased nutritional needs. But it also raises another question: do pregnant and lactating females also change the composition of their diets to meet their nutritive needs, and could such changes lead to short-term increases in tooth wear? Clearly, more detailed studies of the feeding of these animals are needed to answer that qeustion.

As another example, during the dry season, the howlers show more microwear on their teeth together with a faster rate of microscopic wear. This is also exciting because it is the first time this phenomenon has been demonstrated in a wild-caught primate. However, it is also perplexing because, during the dry season, the howlers are feeding primarily on young leaves, buds, and green fruit - none of which is traditionally viewed as abrasive. What is the source of the extra abrasives? To begin to answer this question, we have begun pilot work in a number of areas. Since most primate foods are probably not hard enough to scratch tooth enamel (Lucas 1991), some of our pilot work has focused on other abrasives that might be in, or on, howler foods. Thus, during July 1993 and February 1994, we placed dust collectors (similar to those used by Ungar 1992) at various levels in the canopy at six sites on the ranch (Ungar et al., 1995). The preliminary results clearly show that there is a significant amount of dust in the canopy, even as high as 15 meters (see figure 14). They even suggest that there is more dust in the canopy at the non-river sites than at the river sites (Ungar et al., 1995). Unfortunately, the results of this work are still equivocal because the number of collecting sites is small. More sophisticated work is beginning, using air samplers routinely used to monitor air quality in industrial settings. Initial results (Pastor et al., 1995) reaffirm that there is more dust in the canopy at the non-river sites. However, even though many of the dust particles are silicates of volcanic origin, others are also sodium chloride and gypsum (which cannot scratch enamel) (Pastor et al, 1995). Clearly, more work, with larger samples is necessary.

Another obvious type of abrasive in howler diets would be phytoliths - the opaline abrasives that naturally grow in certain plant parts. However, even though phytoliths have

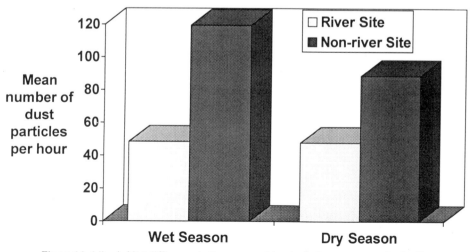

Figure 14. Microhabitat differences in the amount of dust in the forest canopy at La Pacifica.

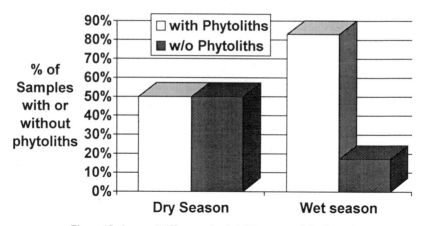

Figure 15. Seasonal differences in phytolith content of fecal samples.

long been thought to be a prime cause of dental microwear (Baker et al., 1959; Walker et al., 1978), we still know surprisingly little about the incidence of phytoliths in different plant parts, in plants of different species, and in plants of differing stages of maturity (Piperno 1988, 1989). Moreover, phytoliths are not the only siliceous abrasives known to occur in plants (Lucas and Teaford 1995). To begin to sort through these options, we've begun analyses of howler fecal samples, with the help of Dr. Dolores Piperno, of the Smithsonian Tropical Research Institute. Preliminary results indicate that most of the phytoliths are from dicot leaves, which is not surprising given the howlers' consumption of mature leaves during the wet season and new leaves during the dry season. However, significantly more of the *wet* season samples have phytoliths in them than do the dry season samples (see figure 15). Again, this would seem to run counter to the results of the tooth wear analyses. Are other siliceous abrasives routinely found in other howler foods? Are such abrasives found in plant parts other than leaves? Again, more work is necessary. It may well be that future analyses will shed light on a question raised by Rosenberger and Kinzey (1976) years ago: can critical food items (rather than those most frequently eaten) have a significant effect on teeth?

Finally, the analyses of wear-related changes in tooth shape suggest that, if facet size is indeed a useful monitor of tooth function (Teaford, 1991), juvenile and older howlers might not be processing their food as well as prime adults. The suggestion is intriguing, but must also be checked with further analyses of howler chewing efficiency, or digestive efficiency (as in Lanyon and Sanson's classic study (1986) of the koala).

CONCLUSIONS

Detailed behavioral analyses have demonstrated microhabitat and intergroup differences in the feeding and diet of the howlers at La Pacifica. Obviously, these findings open the way for a great deal of work, for example, on the feeding and ranging behavior of pregnant and lactating females, or dominant and subordinate males. But, just as Kinzey (1978) and Rosenberger and Kinzey (1976) showed that it is necessary to look past the primary dietary specializations to "critical food items", to completely understand dental adaptations, the work documented here shows that we must also look past traditional dietary categorizations of "frugivore", "insectivore", and "folivore", to truly understand dental microwear formation – as sex differences in time spent feeding, differences in phenophases consumed, and differences in the physical properties of food can also affect dental microwear. Of course, these complications are not without their paleobiological implications – from complexity can come insight. For instance, with a good background of data from modern animals, might we be able to document changes in food preferences (e.g., seasonal or geographic changes in diet) in prehistoric species? Male - female differences? Or even just interspecific differences in diet variability? Any of these would be an exciting addition to paleobiological interpretations.

SUMMARY

Analyses of microscopic wear patterns on teeth (or dental microwear analyses) have recently shown a great deal of promise in studies of diet and tooth use in modern and prehistoric mammals. However, in many cases, they've raised more questions than they've answered, thus emphasizing the need for better behavioral and dental data from living animals. The purpose of this study is to begin to fill that void.

Alouatta palliata, from a site known as Hacienda La Pacifica in Costa Rica, has been the subject of behavioral observation by KG for over 25 years. Since 1992, we have combined behavioral observation with analyses of tooth wear and morphology, Animals have periodically been captured and released so that dental impressions could be taken and observations could be made of known, marked animals. Scanning electron microscopy of high resolution epoxy casts was used to calculate the rates of tooth wear and to characterize patterns of microwear.

Preliminary results indicate that the howlers spend more time feeding in the dry season, when they consume more young leaves, flowers, and green fruit than in the wet season. In addition, howlers from riverine areas spend less time feeding than do howlers from non-river areas. These behavioral differences are coupled with significant differences in tooth wear, such that animals caught in the dry season show more microwear and faster tooth wear than do animals caught in the wet season. Likewise, animals caught in the riverine areas show less microwear than do those caught in the non-river areas.

Based on the above, it is tempting to say that feeding time is strictly correlated with the amount and rate of tooth microwear. However, this is probably an oversimplification. Individual life events, such as pregnancy, can also significantly affect rates of tooth wear and preliminary analyses of abrasives in the diet (e.g. plant phytoliths, dust in the forest canopy, etcs) have yielded conflicting results. Additional work should sort through these causal factors and open up new possibilities for paleobiological interpretations.

ACKNOWLEDGMENTS

As this is a long-term project, we cannot possibly thank everyone who has helped us with this work. However, we would like to give special thanks to Stephan Schmidheiny and the Board of Directors of Hacienda La Pacifica for their permission to work at La Pacifica and for their continued support and help. Thanks also to Lily and Werner Hagnauer, Tony and Vreni Leigh, Margaret Clarke, and Evan Zucker, for their many contributions to this project and for their friendship. Vivian Noble, Courtney Burnell, and Bob Pastor also deserve special thanks for their help with various aspects of this project. This research has been supported by the following grants - to MT: NSF DBC-9118876 (& R.E.U. supplements to that grant) and a Johns Hopkins University Provost's Award for Excellence in Undergraduate Research; to KG: NSF BNS-8819733, Duke University Research Council Grants, and PEW-COSEN grants.

REFERENCES

Baker, G., Jones, L.HP., and Wardrop, I.D., 1959, Cause of wear in sheep's teeth, *Nature.* 104:1583.

Burnell, C.L.B., Teaford, M.F., and Glander, K.E., 1994, Dental microwear differs by capture site in live-caught *Alouatta* from Costa Rica, *Am. J. Phys. Anthropol.* Suppl. 18:62.

Carpenter, C.R., 1934, A field study of the behavior and social relations of howling monkeys (*Alouatta palliata*), *Comp. Psych. Monogr.* 10:1–168.

Clark, S.B., Tercyak, A.M., and Glander K.E., 1987, Plasma lipo-proteins of free- ranging howling monkeys, *Comp. Biochem. Physiol.* 88B: 729–735.

Clarke, M.R., and Glander, K.E., 1981, Adoption of infant howling monkeys (*Alouatta palliata*), *Am. J. Primatol.* 1:469–472.

Clarke, M.R., and Glander, K.E., 1984, Female reproductive success in a group of free-ranging howling monkeys (*Alouatta palliata*) in Costa Rica. In: *Female Primates: Studies by Women Primatologists*, M. Small (ed.). New York: Alan R. Liss, pp.111–126.

Crockett, C.M., and Eisenberg, J.F., 1986, Howlers: Variations in group size and demography. In: *Primate Societies*. B.B. Smuts, R.M. Seyfarth, R.W. Wrangham, T.T. Struhsaker (eds.). Chicago: University of Chicago Press, pp. 54–68.

Glander, K.E., 1975, Habitat description and resource utilization: A preliminary report on mantled howling monkey ecology. In: *Socioecology and Psychology of Primates*. R.H. Tuttle (ed.). The Hague: Mouton, pp.37–57.

Glander, K.E., 1978a, Drinking from arboreal water sources by mantled howling monkeys (*Alouatta palliata* Gray), *Folia primatol.* 29:206–217.

Glander, K.E., 1978b, Howling monkey feeding behavior and plant secondary compounds: A study of strategies. In: *The Ecology of Arboreal Folivores*. G.G. Montgomery (ed.). Washington: Smithsonian Institution Press, pp.561–573.

Glander, K.E., 1979, Feeding associations between howling monkeys and basilisk lizards., *Biotropica* 11:235–236.

Glander, K.E., 1980, Reproduction and population growth in free-ranging howling monkeys, *Am. J. Phys. Anthropol.* 53:25–36.

Glander, K.E., 1981, Feeding patterns in mantled howling monkeys. In: *Foraging Behavior: Ecological, Ethological, and Psychological Approaches*. A. Kamil & T.D. Sargent. New York: Garland Press, pp.231–259.

Glander, K.E., 1992, Dispersal patterns in Costa Rican mantled howling monkeys, *Int. J. Primatol.* 13:415–436.

Glander, K.E., Fedigan, L.M., Fedigan, L. and Chapman, C., 1991, Capture techniques and measurements of three monkey species in Costa Rica, *Folia primatol.* 57:70–82.

Gordon, K.D., 1984, Hominoid dental microwear: Complications in the use of microwear analysis to detect diet, *J. Dent. Res.* 63:1043–1046.

Gordon, K.D., 1988, A review of methodology and quantification in dental microwear analysis, *Scanning Microsc.* 2:1139–1147.

Grine, F.E., 1986, Dental evidence for dietary differences in *Australopithecus* and *Paranthropus*: a quantitative analysis of permanent molar microwear, *J. Hum. Evol.* 15:783–822.

Kay, R.F., 1975, The functional adaptations of primate molar teeth, *Am. J. Phys. Anthropol.* 43:195–216.

Kay, R.F., 1977, The evolution of molar occlusion in the Cercopithecidae and early catarrhines, *Am. J. Phys. Anthropol.* 46:327–352.

Kay, R.F., 1978, Molar structure and diet in extant Cercopithecoidea. In: *Development, Function, and Evolution of Teeth*. P.M. Butler and K.A. Joysey (eds.). New York: Academic Press, pp.309–339.

Kay, R.F., 1984, On the use of anatomical features to infer foraging behavior in extinct primates. In: *Adaptations for Foraging in Nonhuman Primates*. P.S. Rodman and J.G.H. Cant (eds.). New York: Columbia University Press, 21–53.

Kinzey, W.G., 1978, Feeding behavior and molar features in two species of titi monkey. In: *Recent Advances in Primatology, Vol. 1, Behavior*. D.J. Chivers and J. Herbert (eds.). London: Academic Press, pp. 373–385.

Kinzey, W.G., 1992, Dietary and dental adpatations in the Pitheciinae, *Am. J. Phys. Anthropol.* 88:499–514.

Kinzey, W.G., and Norconk, M.A., 1990, Hardness as a basis of fruit choice in two sympatric primates, *Am. J. Phys. Anthropol.* 81:5–15.

Kinzey, W.G., and Norconk, M.A., 1993, Physical and chemical properties of fruit and seeds eaten by *Pithecia* and *Chiropotes* in Surinam and Venezuela, *Int. J. Primatol.* 14:207–228.

Lanyon, J.M., and Sanson, G.D., 1986, Koala (*Phascolarctos cinereus*) dentition and nutrition. II. Implications of tooth wear in nutrition, *J. Zool., Lond.* 209A:169–181.

Lele, S., 1993, Euclidean Distance Matrix Analysis (EDMA): Estimation of mean form and mean form difference, *Math. Geol.* 25(5):573–602.

Lele, S., and Richtsmeier, J.T., 1991, Euclidean Distance Matrix Analysis: a coordinate-free approach for comparing biological shapes using landmark data, *Am. J. Phys. Anthropol.* 86:415–427.

Lucas, P.W., 1991, Fundamental physical properties of fruits and seeds in primate diets. *Proceedings of the XIIIth Cong. of the Int. Primatol. Soc.*: 125–128.

Lucas, P.W., and Teaford, M.F., 1995, Significance of silica in leaves eaten by long-tailed macaques (*Macaca fascicularis*), *Folia Primatol.* 64:30–36.

Moreno, L.I., Salas, I.C. and Glander, K.E., 1991, Breech delivery and birth-related behaviors in wild mantled howling monkeys, *Am. J. Primatol.*, 23:197–199.

Pastor, R.F., Teaford, M.F., and Glander, K.E., 1995, Method for collecting and analyzing airborne abrasive particles from neotropical forests, *Am. J. Phys. Anthropol.* Suppl. 20:168.

Piperno, D.R., 1988, *Phytolith Analysis: An Archaeological and Geological Perspective.* New York: Academic Press.

Piperno, D.R. 1989, The occurence of phytoliths in the reproductive structures of selected tropical angiosperms and their significance in tropical paleoecology, paleoethnobotany, and systematics, *Rev. Paleobot. Palynol.* 61:147–173.

Puech, P-F, 1984, A la recherche du menu des premiers hommes, *Cah. Lig. Prehist. Protohist. N.S.* 1:45–53.

Rosenberger, A.L., 1992, Evolution of feeding niches in New World monkeys, *Am. J. Phys. Anthropol.* 88:525–562.

Rosenberger, A.L., and Kinzey, W.G., 1976, Functional patterns of molar occlusion in platyrrhine primates, *Am. J. Phys. Anthropol.* 45:281–298.

Rosenberger, A.L., and Strier, K.B., 1989, Adaptive radiation of the ateline primates, *J. Hum. Evol.* 18:717–750.

Stuart, M.D., Greenspan, L.L., Glander, K.E., and Clarke, M.R., 1990, A coprological survey of parasites of wild mantled howling monkeys, *Alouatta palliata, J. Wildl. Diseas.* 26:547–549.

Teaford, M.F., 1988, A review of dental microwear and diet in modern mammals, *Scanning Microsc.* 2:1149–1166.

Teaford, M. F., 1991, Measurements of teeth using the Reflex Microscope, *S.P.I.E. Biostereomet. Technol. Applic.* 1380:33–43.

Teaford, M.F., 1993, Dental microwear and diet in extant and extinct *Theropithecus*: Preliminary analyses. In: *Theropithecus: The Life and Death of a Primate Genus*, N.G. Jablonski (ed.). Cambridge: Cambridge University Press, pp. 331–349.

Teaford, M.F., and Glander, K.E., 1991, Dental microwear in live, wild-trapped *Alouatta* from Costa Rica, *Am. J. Phys. Anthropol.* 85:313–319.

Teaford, M.F., and Oyen, O.J., 1989a, Live primates and dental replication: new problems and new techniques, *Am. J. Phys. Anthropol.* 80:73–81.

Teaford, M.F., and Oyen, O.J., 1989b, *In vivo* and *in vitro* turnover in dental microwear, *Am. J. Phys. Anthropol.* 80:447–460.

Teaford, M.F., and Oyen, O.J., 1989c, Differences in the rate of molar wear between monkeys raised on different diets, *J. Dent. Res.* 68:1513–1518.

Teaford, M.F., Pastor, R.F., Glander, K.E., and Ungar, P.S., 1994, Dental microwear and diet: Costa Rican *Alouatta* revisited, *Am. J. Phys. Anthropol.* Suppl. 18:194.

Teaford, M.F., and Robinson, J.G., 1989, Seasonal or ecological differences in diet and molar microwear in *Cebus nigrivittatus, Am. J. Phys. Anthropol.* 80:391–401.

Teaford, M.F., and Runestad, J.A., 1992, Dental microwear and diet in Venezuelan primates, *Am. J. Phys. Anthropol.* 88:347–364.

Teaford, M.F., and Tylenda, C.A., 1991, A new approach to the study of tooth wear, *J. Dent. Res.* 70:204–207.

Teaford, M.F., and Walker, A., 1984, Quantitative differences in dental microwear between primate species with different diets and a comment on the presumed diet of *Sivapithecus, Am. J. Phys. Anthropol.* 64:191–200.

Ungar, P.S., 1992, *Incisor microwear and feeding behavior of four sumatran anthropoids.* Ph.D. Dissertation, State University of New York at Stony Brook.

Ungar, P.S., Teaford, M.F., Pastor, R.F., and Glander, K.E., 1995, Dust accumulation in the canopy: implications for the study of dental microwear in primates, *Am. J. Phys. Anthropol.* 97:93–99.

Walker, A.C., Hoeck, H.N., and Perez, L., 1978, Microwear of mammalian teeth as an indicator of diet, *Science,* 201:908–910.

22

SEASONAL DIFFERENCES IN FOOD CHOICE AND PATCH PREFERENCE OF LONG-HAIRED SPIDER MONKEYS (*Ateles belzebuth*)

Hernán G. Castellanos[1] and Paul Chanin[2]

[1]Universidad Experimental de Guayana, UNEG
Centro de Investigaciones Ecológicas de Guayana, CIEG
Av. Valmore Rodriguez, Upata, Estado Bolívar, Venezuela
[2]University of Exeter
Department of Biological Sciences
Hatherly Laboratories
Prince of Wales Rd.
Exeter EX4 4PS, Devon United Kingdom

INTRODUCTION

Studies concerned with seasonal variation in food choice by atelines have shown that dietary composition fluctuates seasonally and is influenced by phenological changes in food sources (van Roosmalen, 1985; Strier, 1987; Chapman and Chapman, 1991). Fruiting, flowering and leaf-flushing tree species are asynchronous in the tropics leading to complex changes in the type of food available at any one point in time as well as seasonal variation in the nutrients available to herbivores (Raemaekers et al., 1980; Foster, 1982; Leigh and Windsor, 1982; Gautier-Hion et al., 1985; Oates, 1987). These environmental variables are thought to influence food preference by primates (Hladik 1978a, Janzen 1978, McKey et al. 1981, Howe 1982, Wrangham and Waterman 1983).

Previous studies have shown that *Ateles* feeds mainly on fleshy fruits (Klein and Klein 1977, Hladik 1978a, Mittermier and van Roosmalen 1981, van Roosmalen 1985, Chapman 1987). Young leaves or shoots represent the principal source of protein in *Ateles* diet and are an important secondary components of the diet (Hladik 1978b). Fruits are generally low in protein and may be either lipid-poor or lipid-rich (Martinez del Rio and Restrepo 1993, Stiles 1993). Studies on preference for those types of fruit has been carried out in birds (Sorensen 1984, Johnson et al, 1985, Foster 1990), but preference for fruit quality in fruit-eating monkeys has received much less attention. One may ask whether monkeys prefer fruits according to either the nutritional value or the availability of fruits, or both.

Fruits eaten by *Ateles belzebuth* were either high in energy and lipids or low in energy and lipids and they fluctuated seasonally in abundance and availability (Castellanos

1995). Four of five palm tree species, eight tree species and one liana species used by *Ateles* produced fruits rich in energy. Although most of the palms fruited synchronously in time and space, some of them bore fruits throughout the year. The stem basal areas of palm trees were much smaller than many tree species and palms were lower in abundance (Castellanos, 1995). As palms have a very small crown, only a few monkeys can feed one at a time. However, fruits are very abundant in palms so that they can be ingested more rapidly. Thus, in high-energy patches, and particularly in palms, monkeys tended to feed in smaller groups and stayed for shorter periods than in patches with large stems and in those bearing fruits low in energy. One would predict, therefore, that monkeys should selectively feed on patches whose fruits are rich in energy at times of the year when rich patches are of smaller size and lower abundance. Thus, preference for patches rich in energy would be indicated by monkeys spending a disproportionate amount of time foraging in them. If there is no preference, then one would predict that monkeys would allocate their time in proportion to the availability of each patch type.

The aim of this chapter is to document seasonal differences in the nutrient content of food and availability by *Ateles belzebuth* inhabiting a hilly, tropical, moist forest in southern Venezuela. In addition, two environmental variables, patch abundance and patch availability, will be examined in order to determine whether or not monkeys fed selectively on particular foods in seasons when rich patches were of smaller size and lower abundance.

METHODS

Study Area

The study took place on the middle to lower reaches of the Tawadu River (06°21'49"N, 64°59'11"W), a tributary of the Nichare River. The study site is located in an undisturbed hilly evergreen forest within the El Caura Forest Reserve in southern Venezuela (Figure 1). The Reserve was established January of 1968 and is Venezuela's largest, comprising over 5,134,000 ha confined to the Caura River basin. Rainfall data were collected from April 1991 to June 1992.

Study Group and Sampling of Feeding Behavior

Data on the feeding behavior of one group of well-habituated spider monkeys were gathered by HGC from January 1991 to July 1992, except for a three-month gap in observations, from April to June 1991. The group consisted of 21 adult and subadult spider monkeys plus five juveniles and ten infants. Continuous observations of a focal sub-group (Altmann 1974) were made from 06:00 to 18:00, specifically recording the length of time spent by one sub-group within a single patch from the time they first entered it to the time they departed. The end of a feeding bout was recorded when individuals in a sub-group stopped feeding or changed to another activity incompatible with feeding, following Klein and Klein (1977). On the afternoon prior to each sampling day, individual monkeys were sought and followed to their sleeping trees. They were relocated in the morning at the sleeping sites, in the areas surrounding these sites, or by vocalizations. As far as possible, attempts were made to follow the same sub-group all day.

To determine the diet, each food type 1) fruit, 2) young leaves of trees, 3) young leaves of lianas, 4) flowers, and 5) others) handled and swallowed by monkeys was re-

Figure 1. Map of Venezuela showing the relative location of the study area in the Caura basin.

corded. Young leaves were recorded separately for trees and for lianas. The reason for this is that flushing tree leaves occurred seasonally, whereas liana leaves were replaced throughout the year. Other types of food regularly harvested by the monkeys were the young leaves or shoots of trees and high climbing lianas, and flowers or buds. Types of food eaten infrequently ("others") were the dry stems of lianas, leaf petioles, bromeliads. the bark of trees and lianas, and fungus.

Environmental Variables

To gather data on patch abundance and size, each fruiting patch visited by the monkeys was recorded and tagged with a pink numbered flag for subsequent identification and measurement. In this study, a patch was defined as one or more trees or lianas of the same species with overlapping fruit-bearing crowns. Patch abundance was obtained by counting the number of trees of the same species within the home range used by the monkeys and was estimated for each season (Castellanos, 1995). The patch size for trees was obtained by measuring the diameter at breast height (dbh) of every tree used within the patch. Patch size was obtained for lianas by measuring the dbh of one or more trees whose crowns were encompassed by the liana. This measure was used to account for the growth patterns and hence resource distribution peculiar to lianas.

Data Analysis

A total of 42 days of 72 were completed since some sampling days had to be suspended due to heavy rain in the wet season of 1991. To facilitate interannual comparisons, six complete days of 1992 were chosen at random in each season because only six days were completed from 06:00 to 18:00 in the wet season of 1991. Since the duration of feeding time does not fit the expectations of normally distributed data (Castellanos, 1995), it was transformed into arcsin squared values following Sokal and Rohlf (1981).

A General Linear Model (GLM) and two-factor analysis of variance (Minitab statistical software, release 8) was used to compare feeding time with food type between seasons and patch abundance between days and between seasons. A one-way analysis of variance was used to examine variation in feeding time spent in each patch type visited by monkeys.

Patch preference was estimated for each season using an "index of patch preference" for rich and poor patches and using group size and patch density as variables. "Rich patches" were defined as those containing fruits with a lipid content greater than 10% and Kcal value of greater than 5.5. All other patches were defined as "poor". The time individuals spent feeding in patches or patch residence time (PRT) (Stephens and Krebs, 1986) was not a consistent measure of time allocated to feeding on energy-rich or energy-poor patches. For example, sub-groups feeding in palms (rich patches) were consistently smaller in size and spent less time in them than in poor patches. Accordingly, monkeys should make decisions about which patch to visit based on availability, value: rich or poor, crown size: how many monkeys could feed in them, and abundance: how long they spent feeding. From these variables an "index of patch preference" is calculated. The Index

$$gs(d_i/prt_i)$$

where d is the proportion of fruiting trees per hectare of patch type i (rich and poor patches), gs proportion of whole subgroups staying in patches type i, prt = proportion of patch residence time spent by monkeys in patches type i. This index was calculated for rich and poor patches for each season and compared with the density of rich and poor patches in each season. If monkeys showed no preference for rich or for poor patches the ratio of rich patch preference to poor patch preference should be the same as the ratio of rich patch density to poor patch density. Thus the term 'preference' is referred as to choice of an particular patch type according to the intensity of use monkeys devoted to it.

Analysis of Fruits

Fruits were collected and processed following Hladik (1977). An average of 30 fruits of each species was collected, weighed and measured. Mesocarp or aril was separated from exocarp and seeds before drying, except for those fruits whose size were lesser than 5 mm. In these cases, whole fruit were dried and analyzed rather than just the seeds which may explain the high values of ash. Fresh samples were placed in Petri dishes and put into a dryer, maintained as far as possible at a constant temperature of 70 ºC. Subsequently, the dried samples were placed in glass containers and hermetically sealed. The dried fruit samples were analyzed at the Animal Nutrition Laboratory of the Faculty of Agriculture, Central University of Venezuela in Caracas.

RESULTS

Seasonality

The Tawadu basin is persistently cloudy throughout the year, with precipitation of 3,300 mm. Rainfall collected from April, 1991 to June,1992 shows a unimodal peak from June to October and short drought between February and March, 1992 (Fig 2). Additional climatic data were collected by EDELCA (Electrificación del Caroní) from Pie de Salto

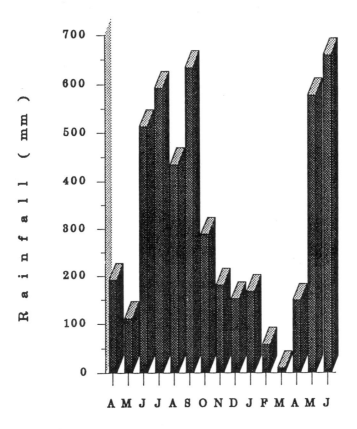

MONTHS (91-92)

Figure 2. Climatic records of the study area. Rainfall recorded with a gauge at Tawadu during fifteen months. The December record is under-estimated owing to the absence of personnel from the camp during the second half of that month.

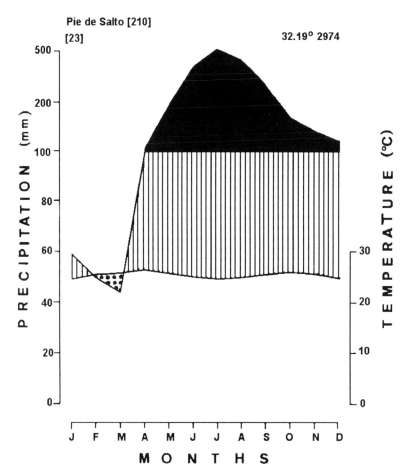

Figure 3. Climate diagram showing the temperature and precipitation recorded at the Pie de Salto Station, 210 m elevation. Data were collected for 23 years with an annual average temperature of 32.19 °C and annual average rainfall of 2,974 mm. Key to the climatic diagram: Abscissa represents the months of a year from January to December. Each division from 0 to 100 represents 20 mm rainfall on the left ordinate, and 10 °C on the right. Above 100 mm, each division represents 200 mm.

Meteorological Station located 51.25 km east of the study site. These data showed that the climate of the region is characterized by marked seasonal variation in rainfall throughout the year (Fig 3). A dry season of two months persists in February and March, followed by a variable wet season of 10 months. As shown in the climate diagram (method following to Walter and Medina, 1976), the latter season may be divided into three sub-seasons: 1) early-wet season (early rain between April and May); 2) wet season (heavy rain between June and September); and 3) late-wet season (late rain between October and January).

Diet Composition and Seasonal Variation

The diet of *Ateles* in Tawadu consisted chiefly of ripe, fleshy fruits. There is significant difference in feeding time between the five types of food (F=17.59, df=4, P<0.001) and between seasons (F=26.52, df=5, P<0.001). Figure 4 shows the percentage of time

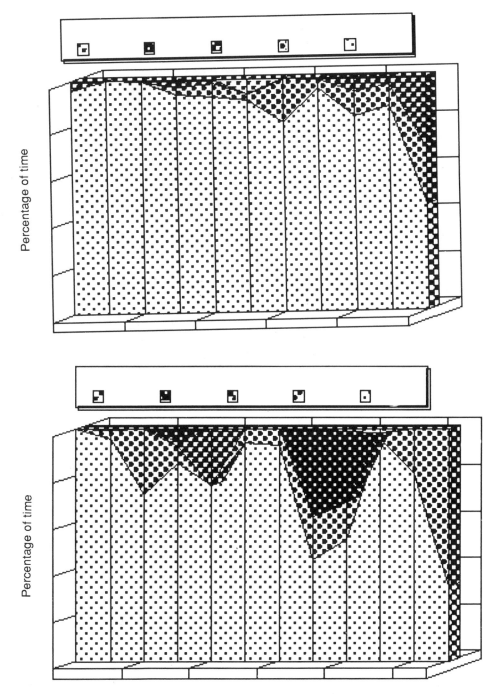

Figure 4. Percentage of hourly feeding time on five types of food: fruits (Fr), young leaves of lianas (Ll), young leaves of trees (Tl), flowers (Fl), and others (O) for each hour from 0600 hours to 1700: (top) the dry season of 1991, (bottom) the wet season of 1991.

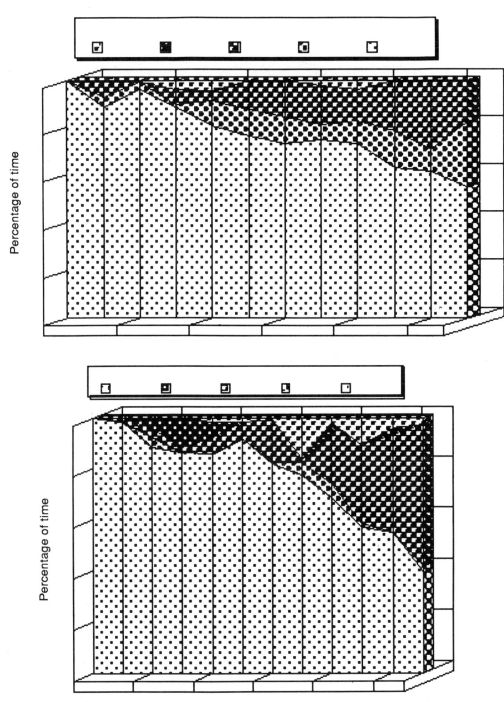

Figure 4. (*Continued.*) Percentage of hourly feeding time on five types of food: fruits (Fr), young leaves of lianas (Ll), young leaves of trees (Tl), flowers (Fl), and others (O) for each hour from 0600 hours to 1700: (top) the late-wet season of 1991, (bottom) the early-wet season of 1992.

spent feeding on each type of food versus hours of day for the three first seasonal periods of 1991. The late-wet season of 1991 and the early-wet season of 1992 show a similar trend in the consumption of fruits and the young leaves and shoots of trees, the latter being consumed more than of the young leaves of lianas.

Patch Abundance and Size

Spider monkeys in the Tawadu forest used a total of 815 patches (Table 1), grouped in 83 species of fruit. These species were grouped into lianas, palms and trees. Of these, the average patch size (dbh) of 14 liana species was 0.57±0.07 m (N=108) (see definition of liana patch above), of 5 palm species was 0.18±0.03 m (N=113), and of 65 tree species was 0.44±0.04 m (N=594) (Table 2).

A two-factor analysis of variance showed that there was significant difference in patch abundance *between seasons* (F=6.16, df=5, P<0.001), but no difference was found *between days* (F=1.93, df=5). A one-way analysis of variance showed that there was a significant difference between patch size and patch type (F=96.71, df=2, P<0.001). There was also significant difference between patch type and feeding time (F=33.48, df=2, P<0.001), as expected.

Nutrient Content in Fruits

A total of 44 fruit species, one fungus and two flowers were analyzed for nutritional content (Table 3). All fruits which had a high proportion of crude fat (>10%) were considered to be high in energy (Stiles, 1993). Eight species were rich in crude fat and 34 species rich in non-structural carbohydrates. Of all these species, fruits of *Dacroydes peruvianum*, *Gnetum urens*, *Licaria chrysophylla*, *Anomospermum reticulatum* and *Spondias mombin* had the highest protein values. Flowers of *Ampelocera edentula* and *Tabebuia serratifolia* also had high protein values.

Patch Preference

In order to explain differences in time allocated by monkeys in a patch type, a mean rate of ingestion of whole fruits was recorded for each patch type. For palms *Jessenia bacaba* and *Euterpe precatoria*, 13 and 19 fruits were ingested per minute, respectively and 4 fruit were ingested per minute for *Jessenia bataua* and *Attalea regia* (Castellanos, 1995). Seed sizes of *J. bataua* and *M. regia* were greater than 30 mm and number of fruit ingested per minute is much slower for these species than for *E. precatoria* and *J. bacaca* with fruits < 12 mm in size. At most, two or three monkeys visited a patch at a time since palm trees are branchless stems and have one or two bunches of fruits. The feeding time expended per subgroup in each palm was very short, 3 minutes on average. *J. bataua* and *M. regia* were relatively rare and the fruiting period of individual palms seem to be asynchronous, but restricted to the rainy season. Spider monkeys spent much of their time feeding on both species, even when only a single palm was bearing fruits. However, the use of them was interrupted by short resting periods caused probably by the bulk of fruits swallowed and the need to digest them. For trees and lianas, the mean rate of ingestion was highly variable and dependant upon fruit size (Castellanos, in press).

The Index of Preference showed that monkeys spent a disproportionate amount of time feeding in energy rich patches (Table 4). In the late wet season of 1991 and the wet season of 1992 monkeys used rich and poor patches in proportion to their density, but in

Table 1. Seasonal variation in patch abundance and basal area recorded from plant species used by spider monkeys that fruited between the dry season of 1991 and the wet season of 1992

	1991			1992			
	Dry	Wet	Late Wet	Dry	Early dry	Wet	Total
Abundance (N)							
Lianas	19	7	15	21	43	3	108
Palms	21	12	52	18	8	2	113
Trees	147	73	91	52	86	85	594
Total	**187**	**92**	**158**	**91**	**137**	**90**	**815**
Basal area (m2/ha)							
Lianas	5.80	5.80	2.40	9.22	8.69	5.31	37.22
Palms	.40	1.20	1.75	.46	.31	.03	4.15
Trees	28.50	17.54	21.51	42.06	24.18	16.32	150.11
Total	**34.70**	**24.54**	**25.66**	**51.74**	**33.18**	**21.66**	**191.48**

Table 2. Average patch size of fruit species mostly used by spider monkeys over a period of fourteen months measured as dbh in meters

Species	N	DBH mean ± SD (in meters)	Life form
Abuta imene	34	?	liana
Anacardium giganteum	11	0.83 ± 0.27	tree
Anomospermum reticulatum	7	?	liana
Catostema communne	14	0.50 ± 0.11	tree
Cecropia sciadophylla	24	0.37 ± 0.09	tree
Couma macrocarpa	37	0.60 ± 0.12	tree
Dialium guianensis	4	0.43 ± 0.17	tree
Doloicarpus dentatus	8	?	liana
Doloicarpus sp.	40	?	liana
Euterpes precatoria	36	0.15 ± 0.02	palm
Goupia glabra	16	0.61 ± 0.11	tree
Ficus sp.	12	0.31 ± 0.14	tree
Helicostylis tomentosa	13	0.26 ± 0.06	tree
Hyeronima alchoneoides	17	0.57 ± 0.18	tree
Iranthera laevis	29	0.29 ± 0.06	tree
Laetia procera	6	0.45 ± 0.11	tree
Maximiliana maripa	3	0.34 ± 0.01	tree
Micropholis cf. *eggensis*	28	0.35 ± 0.07	tree
Pouroma bicolor	32	0.30 ± 0.06	tree
Pouroma sp. 1	34	0.30 ± 0.06	tree
Pouroma sp. 2	14	0.26 ± 0.06	tree
Protium crenatum	17	0.42 ± 0.13	tree
Protium tenuifolium	11	0.24 ± 0.05	tree
Simarouba amara	11	0.63 ± 0.16	tree
Socratea exhorriza	18	0.13 ± 0.02	palm
Talisia sylvatica	9	0.29 ± 0.08	tree
Trichilia quadrijuga	6	0.25 ± 0.09	tree
Virola surinamensis	30	0.54 ± 0.11	tree

Table 3. Phytochemical analysis of lianas, palms, tree fruits, and other types of food eaten by spider monkeys. (fg = fungus; fr = fruit; fw = flower; st = stem; w = wood)

Species	% dry matter	% Ash	% Protein	% Fiber	% Lipid	% CHO	Kcal.	Food item
Lianas								
Cheiloclinium sp.	94.26	2.53	3.69	9.19	7.99	70.86	4.78	fr
Doliocarpus dentatus	–	–	7.26	–	–	–	6.65	fr
Doliocarpus sp.	83.16	4.02	3.16	3.65	3.65	68.27	4.10	fr
Gnetum urens	97.40	2.63	18.02	22.08	1.10	53.57	5.21	fr
Anomospermum reticulatum	82.47	9.17	13.39	10.10	1.52	48.29	3.95	fr
Uvero (*)	90.27	2.69	22.00	22.00	2.43	59.51	4.26	fr
Palms								
Euterpe precatoria	97.15	1.83	4.75	41.82	11.84	36.91	5.50	fr
Jessinia bacaba	97.16	1.43	7.03	25.81	28.30	34.59	6.04	fr
Jessenia bataua	95.38	2.96	4.96	21.39	34.80	31.38	6.47	fr
Maximiliana maripa	81.25	10.16	7.07	8.94	14.85	50.23	5.75	fr
Maximiliana maripa	95.75	7.23	7.19	29.06	1.17	51.10	4.52	w
Socratea exorrhiza	96.20	5.38	6.56	8.51	0.76	74.99	4.22	fr
Trees								
Ampelocera edentula	94.05	10.66	29.07	9.52	1.68	43.12	4.46	fw
Brosimum sp.	87.89	2.25	5.23	3.95	0.97	75.49	4.45	fr
Catostema commune	91.15	7.44	6.97	4.70	1.73	70.35	4.32	fr
Cecropia sciadophilla	94.52	5.26	7.18	30.14	10.28	41.66	4.37	fr
Clarisia racemosa	90.79	5.03	9.95	3.50	1.87	70.44	4.09	fr
Couma macrocarpa	83.33	2.18	2.18	1.94	9.41	54.32	5.11	fr
Coussapoa asperifolia	88.29	3.77	5.28	16.43	8.19	54.62	5.03	fr
Dacroides peruviana	84.76	4.61	13.68	39.54	1.75	34.18	4.57	fr
Dialium giuanensis	87.72	4.63	7.59	6.08	3.90	65.52	4.02	fr
Dipteryx odorata	93.82	16.46	4.87	12.93	5.19	54.37	4.86	fr
Ficus cf. guianensis	97.72	3.68	5.33	28.05	4.76	55.90	4.74	fr
Garcinia macrophylla	95.18	6.04	8.16	14.44	–	–	4.58	fr
Heliocostylis tomentosa	86.58	5.33	7.77	17.21	5.90	50.37	4.54	fr
Hyeronima alchoneoides	92.37	1.56	3.83	38.27	5.98	42.73	4.88	fr
Inga laurina	–	–	10.62	–	–	–	3.91	fr
Inga sp.	92.57	1.44	3.10	5.97	1.68	80.38	4.06	fr
Iranthera laevis	95.14	2.76	9.15	16.42	19.94	46.87	6.51	fr
Laetia procera	90.20	2.83	10.44	33.42	21.38	22.13	6.03	fr
Licaria chrysophylla	94.60	4.22	12.45	–	–	–	6.72	fr
Micranda minor	96.86	1.45	4.04	37.96	0.54	52.87	4.41	fr
Micranda sp.	96.72	1.72	5.38	45.80	1.45	42.37	4.15	fr
Micropholis cf. eggensis	94.66	5.84	6.99	6.81	0.97	74.15	4.07	fr
Micropholis melinomeana	87.00	6.23	6.33	6.79	6.11	61.54	4.69	fr
Ouratea castaneaefolia	94.56	2.70	5.75	–	38.57	–	6.38	fr
Protium crenatum	88.83	2.00	2.72	4.90	2.41	76.80	3.68	fr
Protium tenuifolium	87.01	4.37	8.51	2.72	1.96	69.45	4.48	fr
Sacoglottis guianensis	93.64	2.67	4.40	13.72	3.05	69.80	4.60	fr
Simarouba amara	90.06	3.48	5.87	2.15	0.20	78.36	4.18	fr
Spondias mombin	78.34	9.18	9.61	5.20	1.98	52.37	4.64	fr
Tabebuia sp.	94.05	5.83	22.42	12.79	5.76	47.71	4.89	fw
Trichilia quadrijuga	83.11	3.90	6.04	5.35	2.74	65.08	4.64	fr
Virola elongata	97.68	4.29	6.09	21.45	34.72	31.13	6.62	fr
Virola surinamensis	93.24	1.42	4.87	6.75	38.56	41.64	6.91	fr
Epiphytes								
Philodendron sp.	80.90	2.92	10.63	15.15	13.46	38.80	5.74	fr
Philodendron sp.	95.13	12.29	5.29	16.33	–	–	3.93	st
Kajuka (**)	92.34	1.52	8.73	30.91	2.50	48.68	4.23	fg

(*) = common name

(**) = Y'Kwana name

Table 4. Index of preference for energy-rich patches and ratios (rich:poor) of patch abundance, group residence time (GRT), and patch size (SBA) used by monkey comparing rich patches with poor patches. Rich patches had lipid values > 10% dry wt and Kcal > 5.5

Season	Index of preference	Index of preference rich: poor	Group Residence Time rich: poor	Stem Basal Area rich:poor
1991				
Dry	.49	1:6	1:37	1:12
Wet	4.17	1:5	1:4	1:26
Late wet	1.33	1:1	1:4	1:4.5
1992				
Dry	5.49	1:4	1:4.5	1:5.7
Early wet	.85	1:3	1:6	1:7
Wet	1.11	1:2.4	1:6	1:4

the wet season of 1991 they spent a disproportionate amount of time in energy poor trees. Group residence time was always longer in poor than rich patches and poor patches tended to be larger than rich patches (Table 4).

DISCUSSION

Diet Composition and Seasonal Variation

In all but one season, there was a consistent tendency for the spider monkeys to eat fewer fruits in the afternoon and shift to eating young leaves at that time. The exception was the wet season of 1991. During this season, the consumption of flowers in the afternoon increased considerably. During flowering periods, monkeys showed a strong preference for some flower species (*Ampelocera edentula*, *Micropholis melinoniana* and *M.* cf. *egensis*, *Tabebuia serratifolia*), and allocated much of their time to foraging for and feeding on flowers. As with flushing leaves, flowers are heavily depleted by spider monkeys, particularly *A. edentula* and *T. serratifolia*. Nutritional analysis of these flowers revealed a high protein concentration (O. Parra, pers. comm.) which suggests that monkeys also eat flowers to obtain protein. Flowering of *T. serratifolia* is of short duration and its flowers become depleted by monkeys as well as by wilting and falling. Flowering trees used by the monkeys flowered principally during dry seasons. Of particular importance is the fact that they ate flowers from the same tree species as fruit when it became available (*Micropholis melinoniana* and *M.* cf. *egensis*), except for *Tabebuia serratifolia* where the fruits were not eaten. The adaptive advantage for these tree species may lay in enhancing fruit production by removing excess flowers and thus assuring a good harvest of fruits. This might be explained by the reserve ovary model (Ehtlén 1991), which states that plants maintain a constant fruit:flower ratio despite high flower mortality or predation.

The consumption of young leaves was equally high in wet seasons of both years, alternating with the consumption of fruits throughout a day. Protein content of fleshy fruits tends to be lower than shoots and young leaves (Milton 1982). Moreover, high levels of essential micronutrients, such as magnesium, nitrogen, phosphorus, and potassium are also found in immature leaves (Waterman, 1984). In this study, young leaves were not analyzed to determine micronutrients and toxins since it was not possible to gather them. Nevertheless, shoots and young leaves have fewer toxins than mature leaves (McKey et al.

1981) and also have higher concentrations of phosphorus, nitrogen, gross energy, and total non-structural carbohydrates than mature leaves. Young leaves are also more digestible than mature leaves (Hladik 1978a, Waterman1984, McKey et al 1981), so a preference for young leaves over mature leaves by spider monkeys is to be expected (Hladik 1978b).

High climbing lianas used by monkeys were very common where the vegetation was associated with valley, glacis (habitats associated with slope of 5% or more), and riparian habitats (Castellanos 1995) and their leaves are eaten throughout the year. In contrast, flushing of tree leaves occurred early and late in the wet season, an event very short duration. The monkeys allocated most of their time to eating leaves at this time, depleting them rapidly. The spatial distribution of leaf flush in trees is very similar to that of fruiting trees (Castellanos 1995).

Tawadu spider monkeys also fed on bark, termite mud nests, and the dry stems of lianas, all of which might form a good source of micronutrients (Hladik 1978a). The allocation of time to feeding on these items occurred generally about mid-day or during the last feeding bouts of the day, by which time the monkeys have already ingested substantial quantities of foods rich in energy (fruits) and protein (young leaves). We found seasonal variation in the intake of these items in terms of proportion of feeding time, but at least two of these items formed part of the diet in every season. Of particular interest was an instance when the whole group fed on a dead palm trunk. Two monkeys at a time fed on the palm wood, while the rest of the group waited as if in a queue. Apart from obtaining micronutrients from such types of food, the consumption of these may aid digestion by absorbing secondary compounds and facilitating digestive action (Clutton-Brook 1977).

Patch Preference

Ateles belzebuth is a frugivore which spends most of its time feeding on fleshy, ripe fruits, suggesting that these provide the most significant part of its daily energetic requirements. The palms that were sampled, some tree species, and one liana species provided fruits with high caloric content, with a higher percentage of lipids than carbohydrates. This suggests that storing or accumulating fat by eating such fruits, when available, would provide an energetic reserve in long periods of either overnight feeding inactivity or scarcity.

Results of the Index of Preference showed that spider monkeys preferred rich-energy fruits even when those patches became scarce. However, discrepancies were found in the wet season of 1991 and between the dry seasons of 1991 and 1992. First, in the wet season of 1991 monkeys devoted much of their time to feeding in a palm bearing fruit rich in energy (*Attalea regia*) until the fruit were depleted. Second, the discrepancy in the dry seasons was due to the fact that the spider monkeys spent much of their time feeding in two fruiting tree species poor in energy: *Anacardium giganteum* and *Hyeronima alchoneoides*. The former is a common species but widespread, while the latter is rare and widespread (Castellanos, 1995). Furthermore, *H. alchoneoides*, as well as *Goupia glabra*, produced abundant small fruits (< 5.0mm). Consequently, monkeys preferred feeding on small fruits in these poor patches to feeding on larger fruits in rich patches. On the other hand, some rich patches in the dry seasons were still bearing fruits from the previous seasons. Third, most of the trees of *Micropholis melinoniana* fruited in the wet season of 1992. The monkeys preferred feeding in this poor patch to feeding in rich patches.

Palm fruits are rich in lipids and energy content. *Euterpe precatoria* and *Jessenia bacaba* are the most abundant palm species at Tawadu occurring in different habitats (Castellanos 1995). Fruits of *J. bacaba* and *J. bataua* are rich in lipids and high in gross energy, whereas the remaining palms have fruits with a high content of lipids but are lower in gross energy. Generally, there is a constant supply of palm fruits throughout the year, but the peak of heavy fruit production was in the late-wet season. The growth of palm trees is strongly influenced by light, and hence is gap-dependent, and their spatial distribution depends on the type of soil drainage (Kahn 1986, 1987). For example, *J. bacaba* occur preferentially on hilly habitats, *E. precatoria* and *J. bataua* in plain, valley, creek and edge habitats, but are most abundant in valleys. In contrast, *Attalea regia* is much scarcer than the other species, occurring mainly in edge forest and rarely in hilly and riparian forest. Yet its saplings are very common and widespread, particularly underneath resting and sleeping trees used by spider monkeys (Castellanos personal observations).

Apart from palms, there are some tree species that produce fruit rich in crude fat and energy content. They are *Iranthera laevis*, *Laetia procera*, *Licaria chrysophylla*, *Virola elongata* and *Virola surinamensis*, all of which were intensively used by monkeys. Information on nutrient and energetic contents of Neotropical fruits on the basis of fruit choice for primates is scarce. In Barro Colorado Island, Panamá, Howe (1982) found that the gross energy and lipid contents in fruits of *V. surinamensis* were 7.5 Kcal/g dry weight and 53.2%, respectively. These values are outside the values recorded in this study. Similarly, Foster and McDiarmid (1983) found high nutritional values in *Trichilia* (Meliaceae) arils in Costa Rica. Lipids have been shown to be an important factor in food preference for birds (Martinez del Rio and Restrepo 1993, Stiles 1993) and a similar preference for fruit with a high energy and lipid content may be also exhibited by monkeys.

CONCLUSIONS

The food choice by spider monkeys at Tawadu, Venezuela was markedly seasonal. It depended on food type, abundance, availability, and quality. Monkeys selected patches rich in energy content at times of the year when their abundance was low. Hence, one might infer that the preference for of rich patches is influenced seasonally on the basis of food abundance and availability of patches. It is concluded that the lower the high-energy patch availability, the higher its value (preference) relative to other available patches.

SUMMARY

A group of free-ranging spider monkeys (*Ateles belzebuth*) was studied from September 1990 to July 1992 in a moist tropical hill forest at the middle to lower Tawadu river in southern Venezuela. The main aim of this study was to document seasonal differences in food choice by one group of spider monkeys and to determine whether or not monkeys fed selectively and seasonally on particular food according to their quality. The diet of *Ateles* in Tawadu consisted chiefly of ripe, fleshy fruits (from 69% to 90% of the diet). Other types of food regularly harvested by monkeys were the young leaves of trees and high climbing lianas, and flowers or flower buds. It was found that there was a significant difference in feeding time between the types of food and between seasons. Spider monkeys in Tawadu forest used a total of 83 species of fruit tree found in patches of three types: lianas, palms and trees. Nutritional analyses of 44 species of fruits were carried out. Fruits could be separated into two categories, those which were high in gross energy (Kcal) and those with a high lipid content. An Index of Preference was used to determine whether monkeys selectively fed in patches where the fruit was high in energy or had a low energy content. It was found that monkeys preferred energy-rich patches in seasons when those patches were more abundant.

ACKNOWLEDGMENTS

This study was funded principally by Paignton Zoological and Botanical Garden; additional support was provided by British Airways; EcoNatura of Venezuela; Wildlife Conservation Society; and Consejo Nacional de Investigaciones Científicas y Tecnológicas (CONICIT) of Venezuela. Financial assistance and technical support from these organizations is gratefully acknowledged. We would like to thank Dr. Stuart Strahl, Mr. Peter Stevens and Dr. Ornella de Parra for their unfailing assistance through this study, and Simón, Aurora and Enrique Caura, Emilio and Germán Rodriguez for their field assistance and hospitality during fieldwork activities.

REFERENCES

Altmann, J. (1974). Observational study of behaviour: sampling methods. *Behaviour* 49: 227–267.
Castellanos, H.G. (1995). Feeding behaviour of *Ateles belzebuth* E. Geoffroy 1806 (Cebidae: Atelinae) in Tawadu Forest southern Venezuela. Ph.D. Dissertation (Unpublished). The University of Exeter, UK.
Castellanos, H.G. (1996). Ecología del comportamiento alimentario del marimonda (*Ateles belzebuth belzebuth*) en el Río Tawadu, Reserva Forestal El Caura. *Scientia Guianæ Monograph* (in press).
Chapman, C.A. (1987). Flexibility in diets of three species of Costa Rican primates. *Folia Primatologica* 49(**2**): 90–105.
Chapman (1988). Patch use and patch depletion by the spider monkey and howler monkey of Santa Rosa National Park Costa Rica. *Behaviour* 105(**1–2**): 99–116.
Chapman, C.A., and C.J. Chapman (1991). The foraging itinerary of spider monkeys - when to eat leaves. *Folia Primatologica* 56(**3**): 162–166.
Clutton-Brock, T.H. (1977). Some aspects of intraspecific variation in feeding and ranging behaviour in primates. In T.H. Clutton-Brock (editor). *Primate Ecology: Studies of feeding and ranging behaviour in lemurs, monkeys and apes*. pp: 539–556. Academic Press, London.
Ehtlén, J. (1991). Why do plants produce surplus flowers? A reserve-ovary model. *The American Naturalist* 138(**4**): 918–933.
Foster, B.R. (1982). The seasonal rhythm of fruitfall on Barro Colorado Island. In: E.G. Leigh, Jr., A.S. Rand, and D.M. Winsor. *The Ecology of a Tropical Forest*. pp: 151–172. Smithsonian Institution Press, Washington D.C.
Foster, M.S. (1990). Factors influencing bird foraging preference among conspecific fruit trees. *The Condor* 92: 844–854.
Foster, M.S., and R.W. McDiarmid (1983). Nutritional value of the aril of *Trichilia cuneata*, a bird-dispersed fruit. *Biotropica* 15(**1**): 26–31.
Gautier-Hion, A., J.M. Duplantier, L. Emmons, F. Feer, J.P. Decoux, G. Dubost, P. Hecketsweiler, A. Moungazi, R. Quris, and C. Sourd (1985). Coadaptation entre rythmes de fructification et frugivorie en foret tropicale humide du Gabon: mythe ou realite. *Rev. Ecol.* (Terre Vie) 40: 405–434.
Hladik, C.M. (1977). Field methods for processing food samples. In T. H. Clutton-Brock (editor). *Primate Ecology: Studies of feeding and ranging behaviour in lemurs, monkeys and apes*. pp: 595–601. Academic Press, London.
Hladik, A. (1978a). Phenology of leaf production in rain forest of Gabon: distribution and composition of food for folivores. In G.G. Montgomery (editor). *The Ecology of Arboreal Folivores*. pp: 51–71. The symposia of the National Zoological Park. Smithsonian Institution Press. Washington, D.C.
Hladik, A. (1978b). Adaptive strategies of primates in relation to leaf-eating. In G.G. Montgomery (editor). *The Ecology of Arboreal Folivores*. pp: 373–395. The symposia of the National Zoological park. Smithsonian Institution Press. Washington, D.C.
Howe, H.F. (1982). Fruit production and animal activity in two tropical trees. In: E.G. Leigh, Jr., A.S. Rand, and D.M. Winsor (editors). *The Ecology of a Tropical Forest*. pp: 83–94. Smithsonian Institution Press, Washington D.C.

Janzen, D.H. (1978). Complications in interpreting the chemical defenses of trees against tropical arboreal plant-eating vertebrates. In G.G. Montgomery, editor. *The Ecology of Arboreal Folivores*. pp: 73–84. The symposia of the National Zoological Park. Smithsonian Institution Press. Washington, D.C.

Johnson, R.A., M.F. Willson, J.N Thompson, and R.I. Bertin (1985). Nutritional values of wild fruits and consumption by migrant frugivorous birds. *Ecology* 66: 819–827.

Kahn, F. (1986). Life forms of Amazonian palms in relation to forest structure and dynamics. *Biotropica* 18(**3**): 214–218.

Kahn, F. (1987). The distribution of palms as a function of local topography in Amazonian terra-firme forest. *Experientia* 43: 251–259.

Klein, L.L. and D. Klein (1977). Feeding behaviour of the Colombian spider monkey. in T.H. Clutton-Brock (editor). *Primate Ecology: studies of feeding and ranging behaviour in lemurs, monkeys and apes*. pp: 153–181. Academic Press, London.

Leigh, E.G. and D.M. Windsor (1982). Forest production and regulation of primate consumers on Barro Colorado Island. In: E.G Leigh, Jr., A.S. Rand, and D.M. Winsor. *The Ecology of a Tropical Forest*. pp: 111–122. Smithonian Institution Press, Washington D.C.

Martinez del Rio, C., and C. Restrepo (1993). Ecological and behavioral consequences of digestion in frugivorous animals. In T.H. Fleming and A. Estrada (editors). *Frugivory and Seed Dispersal: Ecological and Evolutionary Aspects. Vegetatio* 107/108: 205–216.

McKey, D.B., J.S., Gartland, P.G. Waterman, and G.M. Choo (1981). Food Selection by black colobus monkeys (*Colobus satanas*) in relation to plant chemistry. *Biological Journal of the Linnean Society* 16: 115–146.

Milton, K. (1982). Dietary quality and population regulation in a howler monkey population. In: E.G Leigh, Jr., A.S. Rand, and D.M. Winsor (editors). *The Ecology of a Tropical Forest*. pp: 63–66. Smithsonian Institution Press, Washington D.C.

Mittermier, R.A. and M.G.M. van Roosmalen (1981). Preliminary observations on habitat utilization and diet in eight Surinam monkeys. *Folia Primatologica* 36:1–39.

Oates, J.F. (1987). Food distribution and foraging behavior. In: B.B. Smuts, D L Cheney, R.M. Seyfarth, R.W. Wrangham, and T.T. Struhsaker (editors). *Primate Societies*. pp: 197–209. Chicago, University of Chicago Press.

Raemaekers, J.J., F.P.G. Aldrich-Blake, and J.B. Payne (1980). The forest. In D.J. Chivers (editor). *Malayan Forest Primates. Ten years' study in tropical rain forest*. pp: 29–61. Plenum Press. New York and London.

Sokal, R.R. and F.J. Rohlf (1981). *Biometry*. 2nd edition. Freeman, San Francisco.

Sorensen, A.E. (1984). Nutrition, energy and passage time: experiments with fruit preference in European Blackbirds (*Turdus merula*). *Journal of Animal Ecology* 53: 545–557.

Stephens D.W., and J.R. Krebs (1986). *Foraging Theory*. Princeton University Press. Princeton, New Jersey.

Stiles E.W. (1993). The influence of pulp lipids on fruit preference by birds. In T.H. Fleming and A. Estrada (editors). *Frugivory and Seed Dispersal: Ecological and Evolutionary Aspects. Vegetatio* 107/108: 227–235.

Strier K.B. (1987). Activity budgets of woolly spider monkeys or muriquis *Brachiteles arachnoides*. *American Journal of Primatology* 13(**4**): 385–396.

Symington, M.M. (1988). Food competition and foraging group size in the black spider monkey (*Ateles paniscus chamek*). *Behaviour* 105(**1–2**): 117–134.

van Roosmalen, M.G.M. (1985). Habitat preferences, diet, feeding strategy and social organization of the black spider monkey (*Ateles paniscus paniscus* Linnaeus 1758) in Surinam. *Acta Amazonica* 15(**3/4**): supplement, 0–238.

Walter, H. and E. Medina (1976). Caracterizacion climatica de Venezuela sobre la base de climadiagramas de estaciones particulares. *Boletin de la Sociedad Venezolana de Ciencias Naturales* 29(**119–120**): 211–240.

Waterman P.G. (1984). Food acquisition and processing as a function of plant chemistry. In: Chivers, D.J., B.E. Wood, and A. Bilsborough (editors). *Food Acquisition and Processing in Primates*. pp: 177–211. Plenum Press, London.

Wrangham, R.W., and P.G. Waterman (1983). Condensed tanins in fruits eaten by chimpanzees. *Biotropica* 15(**3**): 217–222.

23

USE OF SPACE, SPATIAL GROUP STRUCTURE, AND FORAGING GROUP SIZE OF GRAY WOOLLY MONKEYS (*Lagothrix lagotricha cana*) AT URUCU, BRAZIL

A Review of the Atelinae

Carlos A. Peres[*]

Departamento de Ecologia
Universidade de São Paulo
Caixa Postal 11.461
São Paulo, S.P. 05422–970, Brazil

INTRODUCTION

Until recently, most published references on wild populations of the two recognized species of woolly monkeys (Fooden 1963) — the common (or Humboldt's, 'smokey' or 'lowland') woolly monkey, *Lagothrix lagotricha*, and the yellow-tailed (or Hendee's) woolly monkey, *L. flavicauda* — had come from short-term observations (Durham 1975, Kavanagh and Dresdale 1975, Ramirez 1980, 1988, Freese et al. 1982, Johns 1986, Peres 1990, 1991a). Little ecological data are available on *L. flavicauda* — a montane species endemic to the cloud forests of an isolated Andean mountain range in northern Peru — for this species is yet to be studied systematically (but see naturalistic accounts by Graves and O'Neill 1980, Parker and Barkley 1981, Leo Luna 1984, 1987). In comparison, several long-term socioecological studies have addressed other ateline genera (e.g. Klein and Klein 1975, van Roosmalen 1985, Symington 1987, Strier 1987, Chapman 1990, see Strier 1992 for a review), which have inevitably overrepresented most of what is known of the biology of the largest-bodied platyrrhine subfamily (ateline taxonomy follows Napier and Napier 1967, including only *Ateles*, *Lagothrix*, and *Brachyteles*).

Woolly monkeys are thus the least known member of the ateline clade. Yet speculations as to how they should be aligned within several gradients of primate socioecology

[*] Present Address: School of Environmental Sciences, University of East Anglia, Norwich NR4 7TJ, and England. e-mail: C. Peres@uea. ac. uk.

have often been cautiously inferred from short-term studies, or extrapolated from closely related genera (e.g. Robinson and Janson 1987, Rosenberger and Strier 1989). The indeterminate position of *Lagothrix* along the folivory-frugivory continuum typical of all large-bodied primates, for instance, had until recently obscured unclouded inferences on socioecological features closely related to diet (but see Peres 1994a). In particular, the question of whether *Lagothrix* groups do or do not exhibit some form of fission-fusion sociality (*sensu* Kummer 1971) has remained largely unanswered, provoking much speculation. Yet the social organization of *Lagothrix* has been placed with that of spider monkeys, *Ateles* (Strier 1992, but see Strier 1994), which in broad geographic terms may co-occur with woolly monkeys, but often partitioning adjacent forest types (Rylands 1987, Peres 1993, unpubl. data). The fission-fusion social organization exhibited by spider monkeys (Klein 1972) converges with that of chimpanzees, *Pan* (Wrangham 1979) in that both of these genera live in fairly discrete groups or "communities", a mutually exclusive social network of animals which usually interact peacefully, but characterized by small subgroups, often changing in size and composition, that move independently of one another (Symington 1990, Chapman et al. 1995). Groups of *Ateles* and *Pan*, however, tend to be separated from one another by agonistic interactions, such as vocal confrontations involving charges by adult males from each group (*A. belzebuth*: Klein 1972; *A. geoffroyi*: Cant 1977; *A. paniscus paniscus*: van Roosmalen 1985; *A. p. chamek*: Symington 1987; *Pan troglodytes*: Goodall 1986; *Pan paniscus*: White 1992). A similar grouping pattern has also been described for muriquis (*Brachyteles arachnoides*: Milton 1985, Strier et al. 1993, but see Strier 1989), a species more easily studied in small forest patches than in remaining continuous forests of southeastern Brazil, where they appear to occur at lower densities (Pinto et al. 1993).

More recently, however, the obvious gap in the quantity and quality of field data available for *Lagothrix* and other atelines has been largely abridged by studies of *L. lagotricha poeppigii* in Peru (Soini 1986a, 1986b), *L. lagotricha lagotricha* (Defler 1989a, 1989b, Nishimura 1990) and *L. l. lugens* in Colombia (Stevenson et al. 1994, Nishimura 1994), and *L. lagotricha cana* in Brazil (Peres 1991a, 1993, 1994a, 1994b). In this paper I present data on home range use, spatial behavior, group elasticity, and foraging group size of a group of gray woolly monkeys (*Lagothrix lagotricha cana* E. Geoffroy, 1812) at an entirely undisturbed, unflooded (hereafter, terra firme) forest of central-western Amazonia. I describe and review observed patterns of overall group cohesion and elasticity, and examine how these central features of *Lagothrix* social organization diverges from that of other atelines. Finally, I attempt to explain the unusually diffuse pattern of spatial group structure found in this genus in light of underlying environmental variables, such as resource quality, dispersion, and availability, quantified concurrently at this site.

METHODS AND STUDY SITE

This study was carried out in a 900-ha plot of undisturbed, terra firme forest, located 4 km or more from the south bank of the upper Urucu river (4°50'55"S, 65°16'05"W), 181 km directly south of the nearest town, Tefé, Amazonas, Brazil (Fig. 1). Gray woolly monkeys were the largest-bodied of the 12 primate species co-occurring in this area, for black spider monkeys (*Ateles paniscus chamek*) in this watershed were largely restricted to seasonally flooded (igapó) forest along the Urucu river (Peres 1993). Field data at this site were obtained continuously between February 1988 and September 1989, after a line-transect primate census in April–May 1987.

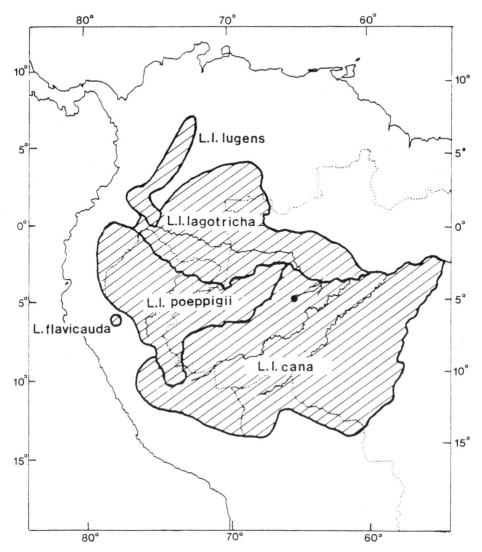

Figure 1. Geographic distribution of the four recognized subspecies of *Lagothrix lagotricha* (Fooden 1963), and the monotypic *L. flavicauda*. Solid dot indicates the location of the Urucu forest site.

The Urucu study plot consists largely of high terra firme forest (93%) on undulating terrains (52–71 m above sea level), which have been mapped at a contour resolution of 5 m. This was a welcome legacy from a brief incursion, prior to this study, of a team of topographers hired by an oil company. Two low-lying forest types — creekside forest and palm swamps — located in water-logged soils along a black-water perennial forest stream accounted for a remainder 6% of the study plot. Annual rainfall at Urucu, as measured on a daily basis for over 6 full years, has averaged 3028 mm (range = 2650–3610 mm/yr, 1988–1993). On average, however, only 11% of this total (range = 6–13%) falls between July and September, which comprise the driest months of the year (hereafter, dry season). More detailed descriptions of the forest structure and composition, patterns of plant phenology, and the primate community occurring in the upper Urucu are presented elsewhere (Peres 1991b, 1993, 1994a, 1994c, 1994d).

Lagothrix data from the Urucu forest are supplemented by those obtained during a series of line-transect censuses of central-western Amazonian primate communities in which woolly monkey numbers have not been seriously reduced by selective overharvest (Peres 1990, 1991a, unpubl. data). These data, albeit restricted to more cursory observations of unhabituated groups and estimates of population density, can serve to ascertain some of the generalities derived from the focal study of gray woollies at Urucu.

Behavioral Sampling and Definitions

With the help of two field assistants, one group of woolly monkeys—consisting of 39–41 independently locomoting individuals (i.e. 7 adult males, 12–14 adult females carrying 5–8 infants, 2 subadult males, and 15–18 subadult females and juveniles)—was followed from dawn-to-dusk during monthly periods of 4–6 days. Sampling was discontinued for nearly 4 months (April–July 1989), when this group had vacated the 900-ha study plot, after moving beyond the southernmost limits of our extensive trail grid. A total of 594 h of observations were obtained during 53 days in the field evenly distributed over 11 months. More ephemeral, less systematic observations were also conducted on another 32 days since February 1988.

Data on the feeding ecology of gray woollies presented here were quantified in the form of feeding bouts, defined as a single visit by one or more animals feeding in a discrete food patch (e.g. a feeding tree, epiphyte, woody liana, or vine) simultaneously. During each feeding bout, food tree size was determined by logger-tape measurements of its Diameter at Breast Height (DBH), and the height of trees and lianas were visually estimated with an error of 4.5% (n=35). Feeding party size is defined as the maximum number of independent animals (unweaned infants excluded) observed feeding simultaneously on plant materials of a single patch, often consisting of a single plant. Because the study group was almost always widely dispersed, only a few individuals could be sampled by a single observer at any one time. Complementary spatial and feeding data were thus obtained by a second (R. Nonato), and less frequently a third independent observer (L. Lopes) who followed different parts of the same group. Because observers were rarely near one another during group follows, follows were coordinated with the help of a pair of two-way walkie-talkie radios with an active reception range of *ca.* 350 m. Otherwise, simultaneously recorded field notes obtained by observers farther apart were pooled together and compared at the end of the day in order to establish patterns of group dimension and cohesion beyond the capability of a single observer. Because of sampling heterogeneity, however, only one group location (usually obtained by the primary observer) is considered for each 10 min interval for the purposes of quadrat occupancy analysis. This resulted in a total of 2,052 group locations, which were made possible within our study plot by an accurate 120 x 120 m reference grid of linear transects, which had been laid out by professional trail cutters using both a theodolite and Global Positioning System fixes at all major intersections.

In this paper, I define group spread as the farthest mapped distance between geometrically opposite flanks of the entire group at any one time, as estimated from simultaneous group locations by at least two independent observers. Group spread thus usually amounted to the major axis of an approximate ellipse, for the spatial configuration of large groups of woollies rarely approached a circular shape. Given that the *Lagothrix* group as a whole was typically spread out, estimates of home range size, in the interest of comparability, are given for two quadrat sizes: one smaller (0.92 ha: 96 x 96 m) and one fourfold larger (3.7 ha: 192 x 192 m). Home range lacunae (i.e. quadrats with no occupancy records) observed here are mostly an artefact of limitations in our joint observation capability, and scale discrepancies between group dimension and quadrat size. After testing against observed patterns of group cohesion, however, these two quadrat sizes were considered most appropriate given that quadrats smaller than 0.92 ha would account for only a small fraction of the large area used by the study group at any one time, which almost always straddled the boundary of these smaller quadrats. This problem of scale was partly corrected for by estimating range size based on 3.7 ha quadrats, thus capturing a much greater proportion, if not all, of the area used by unobserved parties of woollies located within an observer's quadrat. On the other hand, increasingly unrealistic overestimates of range size are derived from quadrats larger than 3.7 ha which tend to incorporate areas which may not have been used by the group.

Ecological Sampling

Tree DBH, height, and species identity were determined for a random set of 3612 stems [3]10 cm DBH contained by 14 floristic plots encompassing a 5-ha area (twelve of 0.25 ha, and two of 1 ha) within the home range of the study group. The same measurements were obtained from an additional set of 1304 trees selected by a point-quadrant method along 4.9 km of transects bisecting the group's home range. A 14-month phenological survey involving the monthly inspection of presence and abundance (ranked on a 0–4 scale) of young leaves, flowers, and unripe and ripe fruits was undertaken with a set of 996 random trees located along transects, which were number-tagged with aluminium plates (see Peres 1994c). For each tree, the crown height was estimated with a range finder, and the crown diameter and distance to the nearest random point along the transect were measured. Plotless tree density estimates (D) could then be calculated as $D=1/d^2$ (Pielou 1959) where 'd' is the average distance between each tree and its random point. Crown diameter and height were then combined to calculate a measure of crown volume (m^3) using eccentricity ratios for each tree in order to generate a hypothetical ellipsoid volume. Considering the diversity of geometric shapes of the wide array of tree species sampled, this assumption seems unavoidable in face of the more accurate but time-consuming alternative of determining multiple crown dimensions according to the geometric configuration of each tree. At Urucu, DBH was a reasonably robust predictor of a tree size, explaining 51% (r= 0.71, p<0.001) of the variation in the area projected by the crown, 54% of crown volume (r= 0.73, p<0.001), and 65% of tree height (r= 0.81, p<0.001) for this set of 996 trees (Peres 1991b). Short of exhaustive counts of entire fruit crops, DBH has also been considered an accurate estimator of fruit production by tropical forest trees elsewhere (Leighton and Leighton 1982, Chapman et al. 1992).

Calculations of ateline group biomass are based on age-, sex-, and subspecies-specific body weight data from wild-caught museum specimens (Fooden 1963), chemically restrained individuals from wild populations (Glander et al. 1991, Lemos de Sá and Glander 1993), or fresh carcasses obtained by subsistence hunters (Peres 1994e, unpublished data). These were then weighed across different age-sex classes whenever data on group composition were available (see references in Table 2).

RESULTS

Home Range Use

Considering a 0.92-ha quadrat resolution, members of the study group were known to use a minimum home range of 935 ha (= 1,015 quadrats of 96 x 96 m), even though our sampling protocol—based on 10-min group locations—yielded only 656 quadrats or 605 ha (Fig. 2). This resulted in a somewhat spurious home range configuration in which 359

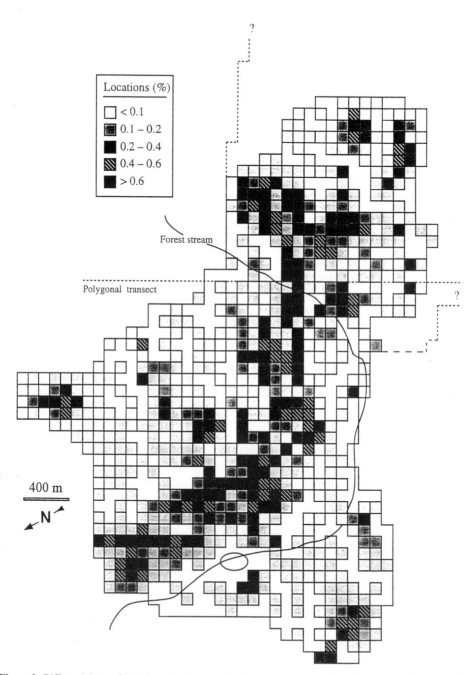

Figure 2. Differential use of 0.92-ha (96x96 m) quadrats by one group of *L. lagotricha cana* in the Urucu forest. Each quadrat is assigned to one of five classes of occupation intensity, defined as the number of 10-min group locations (n = 2052) obtained throughout the study. Interspersed lacunae (in white) represent quadrats containing no group locations, but which may have been used by members of the group. Solid line represents a perennial forest stream, which flowed along creekside forest and palm-swamps. Dashed lines represent the estimated boundaries of an uncharted portion of the group's home range.

quadrats containing suitable habitat had no group locations assigned to (see Methods). Considering 3.7 ha quadrats and 10-min locations, however, a home range of 962 ha (= 261 quadrats) was obtained, an area similar to that derived from 0.92-ha quadrats used by the group at all times. Finally, a home range size of 1,021 ha was estimated using the larger quadrat size, irrespectively of the 10 min sampling interval.

We were unable to find the main study group between April and July 1989, despite our relentless search effort over the entire study area, which consisted of three observers moving independently on different transects during 20–25 days per month. The study group thus became conspicuously absent from the 900-ha study area during this 4 month period, after moving beyond the southern limits of our trail system. This conclusion is based primarily on the fact that our large group of woollies was very conspicuous when travelling, feeding, and vocalizing, and thus could not have remained unnoticed by trained observers during at least part of any given search, day after day. Home range size figures presented here are, therefore, minimum estimates, for a portion of what appears to be the annual home range of the study group—representing at least 30% of the area documented here—is yet to be charted. A plausible estimate of the total home range size of our study group would, therefore, be in the order of 1200–1350 ha.

Woolly monkeys preferred high terra firme forest to low-lying forest types below 53 m a.s.l., which remained water-logged for most months of the year. This is based on a comparison between the availability of quadrats in different habitat types and the number of group locations allocated to them (χ^2=10.8, df=2, p<0.001). The study group spent 97.4% of its time within high forest quadrats, whereas this habitat accounted for 93% of its home range. In contrast, woollies spent only 2.6% of their time within the strip of creekside forest, and interspersed small patches of palm swamps, which ran along a perennial stream and occupied 6% of the study plot (Peres 1994d).

Spatial Group Structure of *Lagothrix l. cana*

Woolly monkeys at Urucu were typically found in large, highly uncohesive social units which almost always occupied several to many hectares of forest at any one time. Although grouping behavior was variable both within and between different seasons of the year, woollies clearly did not exhibit a typical fission-fusion social system (*sensu* Kummer 1971). Rather, they were found either (i) as a single, albeit uncohesive, scatter of isolated individuals and more tightly-grouped feeding parties (ca. 60% of the time), or (ii) as two or three subgroups which by and large *did not* move independently of one another (ca. 40% of the time). The very concept of a "subgroup" as applied to *Lagothrix* thus needs to be redefined here because previous use of this term, particularly with reference to an ateline primate, is deeply entrenched with connotations to the subgrouping pattern typically found in spider monkeys and chimpanzees (Symington 1990, Chapman et al. 1993, 1995). Subgroups in *L. l. cana* were relatively loose and less discrete units of 11–25 individuals, which often included a few adult males, several adult females, and several nonadults. For example, a typical "subgroup" consisted of 3 adult males, 3–4 lactating females with 1 infant each, 2 subadult males, 5–6 adult and subadult females carrying no dependent infants, and 2–3 juveniles. No all-male or all-female subgroups were ever observed, and solitary animals (i.e. >100 m from nearest conspecifics) were almost never found. Unlike *Ateles*, members of a subgroup did not retain priority of access to a relatively discrete core area within the group's home range to the partial or entire exclusion of other subgroups. Rather, members of all subgroups used the group's home range in its entirety, which they usually shared in a concurrent fashion through relatively coordinated movements.

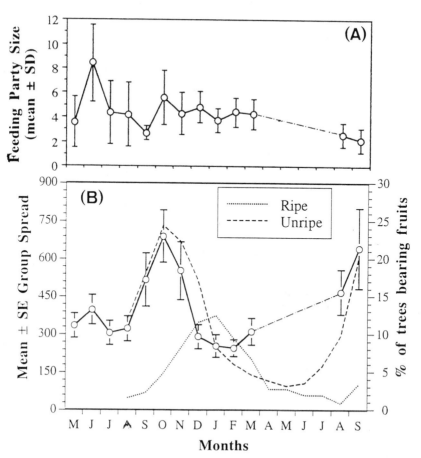

Figure 3. Seasonal variation in overall group cohesion, as defined by (A) feeding party size, and (B) group spread during different months of the year. Dotted and dashed lines (B) represent the variation in the proportion of 996 trees [3]10 cm DBH, inspected on a monthly basis, which were observed bearing ripe and unripe fruits during a 14-month phenological study.

Indeed the key feature of the spatial group structure of gray woolly monkeys is not a clearly discernible subgroup unit (which by and large did not take place) but the overall group dimension, which in turn affected the likelihood of fission into scattered parties. The overall spatial cohesion of the study group as a whole was typically low to extremely low. Group members were usually spread out over a major group axis greater than 300 m, but which sometimes exceeded 1000 m. Group spread during the entire study averaged 431 m (sd = 414 m, range = 60–1920 m, n = 76 mapped simultaneous locations), and varied from being most cohesive during the mid-wet season (December–February) to least cohesive during the late dry season (Fig. 3).

The overall integrity of the group was largely maintained by a varied repertoire of medium- and long-distance vocal signals exchanged between members of the same or different subgroups. These were often echoed sequentially by individuals in neighboring parties, reverberating onwards in an intriguingly staggered fashion. Although individuals on opposite flanks of the group were often outside one another's hearing range, their calls were relayed by spatially intermediate group members. Typical contact calls emitted by both males and females included frequent "clucks", "chucks", "long-chucks", and the infrequent "loud-neighs" produced mostly by adult males. Under contexts of threat, adult males usually approached alarm calls emitted by other males, including repeated "clucks", whimpering "chirps", puppy-like "yapps", and seal-like "barks", which were often used in communal mobbing, branch-shaking, and branch-crashing displays directed at a specific stimulus. It is therefore presumed that the unusual spatial coordination of such a large uncohesive group, spanning considerable group dimensions, was greatly assisted by the sequential propagation of such calls. Moreover, the transmission efficiency of long-distance signals appears to be greater in the forest canopy—where woollies spend a vast proportion of their time (see Peres 1993)—for the hearing range of, for instance, an adult male's "loud neigh" could elicit responses from distances exceeding 400 m.

Food Patch Structure and Size of Feeding Parties

Woolly monkeys at Urucu fed primarily within large- to very large-crowned trees and lianas, a pattern expected of the largest-bodied primate occurring in the study plot. Food trees used by the study group averaged 47.7 ± 24.0 cm in DBH (range = 6–158 cm, n = 597) and 33.0 ± 8.8 m in crown height (range = 9–60 m, n = 610; Fig. 4). Mean heights of feeding parties during feeding bouts (30.3 ± 6.8 m, range = 11–55 m, n = 610) were only marginally lower than that of feeding trees, indicating that animals tended to use the upper levels of the crown. That increasingly fewer feeding bouts took place in the largest tree size classes is largely explained by the fact that these large trees were rare (see Fig. 5 for relative size-class abundance of 4,884 trees sampled at this site). Gray woollies thus clearly *selected* (as defined by use/availability ratios) large food patches (see also Peres 1994b), regardless of feeding party size. This trend would become even more obvious should one consider the amount of time allocated to different-sized trees (rather than individual feeding bouts), for larger patches also allowed longer feeding bouts (Peres, unpubl. data).

Feeding parties in woolly monkeys can be viewed as yet another level of grouping and spatial organization—beyond the coarser scales of groups and subgroups as defined here—for, unlike *Ateles*, subgroup size usually far outnumbered feeding party size. In fact, single animals feeding alone within a tree or liana crown accounted for the most frequently observed (36%, N=511) feeding party size (Fig. 4), although feeding bouts engaged by solos were far more likely to occur in sources of foliage (62 of 103 bouts) than those of seeds, mesocarps, and other fruit parts (101 of 369 bouts; G=36.8, df=1, p<0.001, Fig. 6). This reflects the more dispersed and less synchronous foraging pattern when animals fed on foliage. In general, size of feeding parties averaged 3.48 ± 3.21 individuals (range = 1–22 ind., N=511; Fig. 4). The most number of animals ever observed feeding together (22 ind.) was thus only slightly outnumbered by the largest observed subgroups (25 ind.), perhaps imposing an upper threshold in subgroup size. Whole subgroups, therefore, only rarely fed together within a given food crown. Rather, different feeding parties belonging to the same subgroup usually had access to a large fruiting tree in a sequential manner, while other subgroup members rested in neighboring trees. Order of access to major feeding trees thus apparently followed an implicit (non-overt) male dominance hierarchy (but see Nishimura 1994 for priority of access to resources in *Lagothrix* under different contexts).

There was considerable variation in the size of feeding parties accommodated by food patches bearing different categories of food items (Table 1). The consequences of food patch size on the size of feeding parties was considerably greater for reproductive

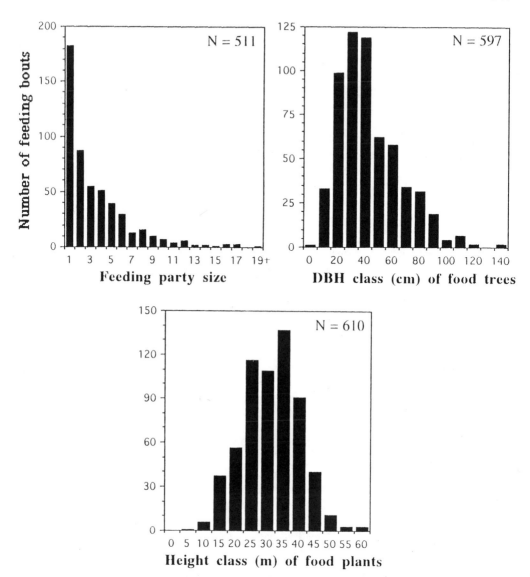

Figure 4. Frequency distribution of woolly monkey (*Lagothrix lagotricha cana*) feeding party size (top left), and DBH (top right) and height (bottom) of food patches during independent feeding bouts.

plant parts (e.g. flowers, floral nectar, seeds, and ripe and unripe mesocarps) than for young leaves and leaf petioles (Fig. 6). Indeed, differences in the size of fruit and flower patches explained 43% the variation in woolly monkey feeding party size, or nearly four times that explained by patches of foliage (11%). This is not surprising since the spatiotemporal distribution of vegetative matter is inherently more diffuse than that of fruits both within and between patches, and woollies resorted to foliage primarily during periods of acute fruit scarcity (Peres 1994a), when group cohesiveness was necessarily low. Numbers of woollies sharing preferred, low-density patches, such as those of fruits and flowers, thus appeared to more closely reflect the limits in feeding party size imposed by within-patch feeding competition.

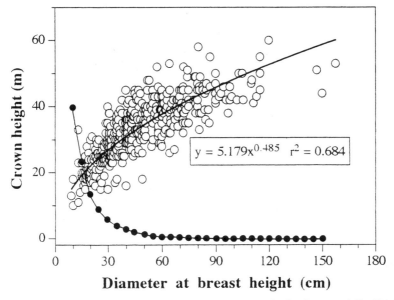

Figure 5. Size of food trees used by the study group during independent feeding bouts, as defined by the relationship between tree DBH and crown height. Line dotted by solid circles represents the availability of different DBH classes for a random set of 4,884 trees measured within the home range of the study group.

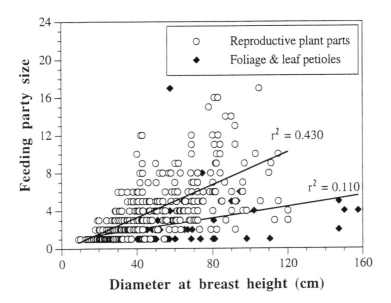

Figure 6. Relationship between the diameter (DBH) of food trees and the number of woolly monkeys observed feeding simultaneously during a given feeding bout for fruit parts and flowers (open circles), and vegetative matter (solid diamonds).

Table 1. Size of food patches (mean ± SD), and Pearson correlations (and coefficients of determination) between one estimate of food patch size (DBH) and feeding party size for different major food categories

| Major food types | Size of Food Patches | | Feeding | | | | |
	DBH (cm)	Height (m)	Party Size	r	r^2	p	N
Foliage	55.3±29.2	37.5±7.5	2.3±2.6	0.34	0.11	*	111
Flowers	34.5±20.8	29.0±10.8	1.7±1.3	0.83	0.69	*	29
Unripe seeds	51.2±23.4	33.4±9.1	4.9±4.2	0.69	0.48	*	106
Ripe fruits (excl. figs)	43.3±19.1	31.5±8.2	3.3±2.4	0.65	0.42	*	304
Ripe figs[**]	– – – –	28.3±7.1	6.2±5.5	0.04	0.02	ns	26
Parkia pod exudates	77.8±24.4	40.0±9.8	6.6±4.6	0.25	0.06	ns	29
Unripe pericarps	42.1±22.9	27.8±8.2	2.6±1.8	0.79	0.62	*	17
All fruit parts (excl. figs)	46.6±22.2	32.2±8.7	4.0±3.3	0.56	0.31	*	457

* $p < 0.001$.

** Because all fig species (*Ficus* spp.) used by *L. l. cana* at Urucu, with the exception of one strangler, were hemi-epiphytic, their DBH were very poor predictors of crown volume.

Food Patch Density and Patterns of Group Cohesion

Mean feeding party size during 10 months of study matching concurrently obtained data on food resource phenology was not positively correlated with the density of young leaf patches ($r_s = 0.32$, p=0.34, N=10), flower patches ($r_s = -0.53$, p=0.114, N=10), and unripe fruit patches ($r_s = 0.13$, p=0.703, N=10; see Peres 1994c, 1994d for patterns of plant phenology at this site). Feeding party size was, however, moderately greater during periods of higher density of ripe fruit patches ($r_s = +0.52$, p=0.122, N=10). This positive correlation became only mildly stronger when I incorporated measures of patch size (crown volume) and patch productivity (rank of ripe fruit production) for each phenology tree into a composite Index of Standing Fruit Crop Size, SFCS ($r_s = +0.54$, p=0.106), for all else being equal, larger patches containing a higher density of fruits were expected to accommodate increasingly larger feeding parties.

The pronounced seasonal variation in our measure of group spread was more clearly correlated with density of mature fruit patches within the group's home range. In contrast, the study group was found to be most spread out during the peak of unripe fruit availability (Fig. 3), which suggests that large patches of very young or immature fruits were either mostly ignored, or failed to attract commensurately larger feeding parties. Group spread increased with the density of unripe fruit patches ($r_s = 0.84$, p=0.012, N=10), but decreased as the species richness of ripe fruits ($r_s = -0.56$, p=0.091, N=10) and absolute density of ripe fruit patches increased ($r_s = -0.54$, p=0.106, N=10). Group spread was lowest (or group cohesion highest) during the three mid-wet season months, a period of greatest ripe fruit availability (Fig. 3). Group spread was highest (group cohesion lowest) during the mid-late dry season, after the large crops of exudate-rich pods of an emergent legume (*Parkia nitida*)—a key resource to *L. l. cana* at this time of year (Peres 1994a)—began to dwindle. Relatively large feeding parties were often found clustered around large fruiting *Parkia* trees, which partly explains the unexpectedly high group cohesion between May and August 1988, a period of otherwise extreme scarcity of fleshy fruits. In contrast, group spread appeared to be unaffected by the temporal variation in density of overstory trees producing young foliage ($r_s = 0.15$, p=0.649, N=10), but was positively correlated with that of flowering trees ($r_s = 0.73$, p=0.028, N=10), probably because the availability of flowers at Urucu followed an inverse pattern to that of mature fruits ($r_s = -0.75$, p=0.007, N=14).

Greater group cohesion thus appeared to result from higher ripe fruit availability as a function of, firstly, (i) increased density of ripe fruit patches allowing shorter nearest-neighbor distances between feeding parties, and to a lesser extent (ii) increased patch size, for larger fruiting crowns were able to accommodate increasingly larger parties. Food patch size explained only a limited amount of the variation in feeding party size, since larger patches did not necessarily result in commensurately greater feeding parties. This is because there is an obvious physical limit in the upper size of discrete food patches to which woollies could converge, and even the largest trees failed to contain but a fraction of the entire group. For example, the most number of noninfants observed feeding in a single tree (22 ind.) was just under half the entire group size. Moreover, large-crowed food plants capable of accommodating a larger feeding became increasingly rare at the Urucu forest (Peres 1991b).

DISCUSSION

Group Size and Group Biomass

The prodigious size of lowland woolly monkey groups has been remarked upon since the earliest forays of European naturalists into western Amazonia. Unhunted groups of *L. l. cana* in the vicities of Tefé (formerly, Ega), Brazil, were found to be "very numerous" by H. Bates (1892), and large groups of this subspecies were observed by L. Miller during the Roosevelt Expedition to Brazil in the upper Solimões river (Allen 1916). More recently, a group of 70 animals was reported by Nishimura and Izawa (1975), but later this was thought to be the sum of two groups travelling in tandem (Nishimura 1990). I, for one, have counted a group of 62 or more *L. l. cana* (excluding infants) at a terra firme site 30 km northwest of the study plot (Oleoduto do Rio Tefé). There are also reliable reports by rubber-tappers of *L. l. cana* groups (or coalescences of neighboring groups) of as many as 130 independently locomoting individuals using dense stands of fruiting *Couma macrocarpa* (Apocynaceae) and *Theobroma* sp. (Sterculiaceae), in an inland terra firme forest site 80 km north of the Urucu site.

From data on group sizes obtained to date (Table 2), it thus appears that the combined weight of *Lagothrix* groups can be greater than that of either *Ateles* or *Brachyteles*: at unhunted sites of western Brazilian Amazonia, for example, the combined weight of entire woolly monkey groups can exceed 320 kg (unpublished data). Group mass is not sig-

Table 2. Group size, home range size, group biomass, and home range area per unit of group biomass in ateline populations studied to date

Taxon	Study site, Country	Group size	Home range (ha)	Group[1] mass (kg)	Spatial[2] requirement	Length of study (mo)	Sources
Lagothrix							
L. l. cana	Rio Urucu, Amazonas, Brazil	44-49	1021+	270	≥3.8	11	This study
L. l. poeppigii	Cahuana Island, Río Pacaya, Peru	17-20	350	112	3.1	12	Soini 1986
L. l. poeppigii	Paquitza, Río Manu, Peru	14	400+	105	≥3.8	1	Ramirez 1980
L. l. poeppigii	Paquitza, Río Manu, Peru	10	250+	75	≥3.3	1	Ramirez 1980
L. l. poeppigii	Cocha Otorongo, Río Manu, Peru	15	150+	84	≥1.8	1	P. Herrera, in litt.
L. l. lugens	Tinigua N.P., La Macarena, Colombia	19	169	96	1.8	12	Stevenson et al. 1994
L. l. lagotricha	Río Apaporis, Vaupés, Colombia	20-24	732	123	6.0	60	Defler 1989
L. l. lagotricha	Río Peneya (Pto. Japón), Colombia	42-43	1100	260	4.2	9	Izawa 1976
L. l. lagotricha	Río Peneya (Pto. Tokio), Colombia	13	350	86	4.1	9	Nishimura 1990
L. l. lagotricha	Río Peneya (Pto. Tokio), Colombia	45	450	245	1.8	9	Nishimura 1990
Ateles							
A. geoffroyi	Barro Colorado Island, Panama	15	115	77	1.5	12	Dare 1974
A. geoffroyi	Tikal, Petén, Guatemala	?	280	?	?	9	Cant 1977
A. geoffroyi	Santa Rosa N.P., Costa Rica	42	170	214	0.8	24	Chapman 1990
A. geoffroyi	La Selva, Costa Rica	30+	114	153+	≥0.7	16	Campbell and Sussman 1993
A. b. belzebuth	La Macarena N.P., Colombia	18	260-390	100	2.6-3.9	15	Klein and Klein 1976
A. p. paniscus	Voltzberg Reserve, Suriname	18	220	110	2.0	26	van Roosmalen 1985
A. p. chamek	Cocha Cashu, Río Manu, Peru	37	231	232	1.0	24	Symington 1988
A. p. chamek	Cocha Cashu, Río Manu, Peru	40	153	248	0.6	24	Symington 1988
Brachyteles							
B. arachnoides	Barreiro Rico, São Paulo, Brazil	7	70	52	1.3	8	Milton 1984
B. arachnoides	Montes Claros, Minas Gerais, Brazil (1984)	23-26	168	184	0.9	14	Strier 1987
B. arachnoides	Montes Claros, Minas Gerais, Brazil (1990)	43	300[3]	248	1.2	30	Strier et al. 1993
B. arachnoides	F. Esmeralda, Minas Gerais, Brazil	15-18	40	120	0.3	7	Lemos de Sá 1988

[1]Estimates of group biomass take into account group composition and sex- and age-related differences in body weight whenever those could be obtained from the studies or best weight data available in the literature (e.g. Fooden 1963, Glander et al. 1991, Ford and Davis 1992, Lemos de Sá and Glander 1993, Peres 1994e).

[2]Given in terms of home range area (ha) per unit of group biomass (kg)

[3]Estimated from Strier et al. (1993): as of July 1990, this range size "had nearly doubled" from the 168-ha figure for the same (but smaller) group as of 1984.

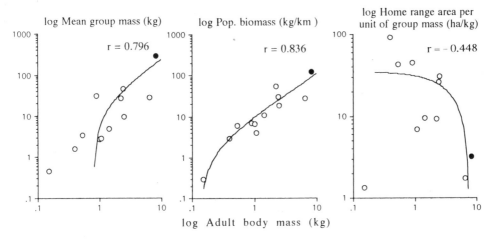

Figure 7. Synecological position of *Lagothrix lagotricha cana* (solid dot) — in terms of group biomass (left), crude population biomass (center), and ranging requirements (right) — relative to all other primate species co-occurring at the Urucu study plot (data from Peres 1993). The extreme outlier in the left scatterplot is represented by pygmy marmosets (*Cebuella pygmaea*) which required exceptionally small home ranges per unit of group mass.

nificantly different between the three genera, however, when one examines data from systematic ateline studies (Kruskal-Wallis anova, H=1.2, df=2, p=0.56, Table 2). At Urucu, the mean group mass for *L. l. cana*, based on accurate counts of 2 groups and body weights of different age/sex classes (Peres 1994a), was 291 kg (range = 270–312 kg). This is almost twice that of the combined group mass of all other 11 primate species sympatric at this site (158 kg; Fig. 7). Such group biomass can exceed that of any other living platyrrhines, being surpassed mainly by considerably larger-bodied primates released from the contraints of arboreality (see Grant et al. 1992 for a review). *Lagothrix* populations in Amazonian terra firme forests can, therefore, reach the highest primate group biomass in the neotropics, and rival that of extant arboreal species in the paleotropics. In unfragmented habitats of pre-Colombian days, however, large groups of muriquis (*B. arachnoides*) or even the 20% larger "mega-*Brachyteles*" (*B. brasiliensis*) that lived to end of the Pleistocene (Cartelle 1993), may have weighed considerably more than those of *Lagothrix* today. In the foothills of the State of São Paulo Paranapiacaba mountain range, for instance, groups of "50 or more" *B. arachnoides* were sighted as late as 1928 (Lane 1990). Such exceedingly large group biomass are clearly attractive from a hunter's standpoint: whole Indian tribes and expeditions have subsisted almost entirely on *Lagothrix* (e.g. 562 Waorani kills over 275 days: Youst and Kelly 1983) and *Brachyteles* meat (Lane 1990), respectively. Selective overkills of these game species are however best described as an ephemeral bonanza, for all atelines can easily be driven to local extinction by even moderate hunting pressure (Peres 1990, 1991a).

Use of Space and Ranging Requirements

Of all strictly arboreal primates, *Lagothrix* groups studied to date can use some of the largest home ranges, which in at least two Amazonian sites have exceeded 1000 ha (Table 2). Such a large area may contain over 850 individuals and at least 225 food species of trees, lianas, epiphytes, and hemi-epiphytes, including at least 169 species of fruit which are consumed when ripe (Peres 1994a). Preliminary data indicate that *L. lagotricha* returned to and monitored a sufficient number of relatively ephemeral sources of fruits spaced over its large home range throughout the annual cycle. This supports the often suggested, but rarely tested possibility that adult males and females of this relatively long-lived species had a mental map (or at least good spatial cognition) of a large proportion of the large fruit patches they exploited [the longevity of *Lagothrix* in the wild remains unknown, but an adult female at the Woolly Monkey Sanctuary, Looe, England, lived to be 38 years of age (R. Havesi, pers. comm.)]. Selective pressure for greater cognitive ability may partly explain the relatively large, convoluted brain and the high neocortical indices found in *Lagothrix* (Stephan et al. 1981). At the Urucu, home range size and relative volume of the neocortex (which is key to the analysis and storage of spatial information: O'Keefe and Nadel 1978) in *Lagothrix* were comparable only to that of *Cebus* and *Saimiri*. For example, the relative neocortical volume (NV) accounts for 75% of the variation in a log-transformed average of home range size (HR) for nine genera of Urucu primates (HR = 23.61 * NV − 13.18). In this case, home ranges of different species overlapped one another within the same study plot, thus controlling for between-site differences in forest productivity that are usually disregarded in allometric relationships involving neocortex size (e.g. Dunbar 1992, Sawagushi 1992).

There is preliminary evidence to suggest that the home range size, year-round residency status, and overall group cohesion of woolly monkeys in lowland Amazonia is affected by forest type. Once factors such as selective hunting are controlled for, it appears that terra firme forests near nutrient-rich white-water floodplains and flooded forest (várzea) mosaics are able to support higher group densities of *Lagothrix* within smaller home ranges than are terra firme forests in remote interfluvial regions such as the Urucu (unpublished data). Moreover, year-round ranging behavior in eutrophic forest sites need not exhibit the pronounced seasonal shifts in home range position observed in at least certain terra firme populations (Johns 1986, Rylands 1987, Peres 1993). Home range sizes in areas under the direct influence of major rivers, on the other hand, appear to be smaller (Table 2). During a single dry season month, two troops of *L. l. poeppigii* used a forest area of 250 and 400 ha at the entrance of Manu National Park (Paquitza), southern Peru (Ramirez 1980). A "clan" of 20 animals in another Peruvian population of this subspecies— confined to a flooded forest within a fluvial island of the Pacaya river—used a home range of 350 ha over several years (Soini 1986a), and a group of 19 *L. l. lugens* studied in the western lowlands of the Colombian La Macarena Mountains used an area of 169 ha (Stevenson et al. 1994). Home range sizes more similar to that of *L. l. cana* of the Urucu terra firme forest, however, include *L. l. lagotricha* at two eastern Colombian sites near the Peneya river (Nishimura and Izawa 1975, Izawa 1976, Nishimura 1990), and one site near the Apoporis river (Defler 1989a, 1989b).

Within the Atelinae, home range sizes are on average largest in *Lagothrix*, intermediate in *Ateles*, and smallest in *Brachyteles* (Table 2 and 3). However, only *Lagothrix* have significantly larger ranges than either one of the two other genera (Kruskal-Wallis anova, H=8.6, p=0.014). The relative amount of space used by a primate group, whether one considers consumer pressure in terms of total group biomass (ha/kg) or the total number of individuals (ha/ind.), is also significantly greater in *Lagothrix* than in the other ateline genera (ha/kg: K-W anova, H=12.4, p=0.002; ha/ind.: H=12.6, p=0.002). The amount of area required to sustain a group mass of 1 kg is 1.8–6.0 ha for *Lagothrix* and 0.6–3.9 ha for *Ateles*. This occurs despite the fact that *Ateles* is equally, if not more, frugivorous than *Lagothrix* (see Peres 1994a for a review), degree of frugivory in primary consumers being positively correlated with relative range size. This ratio is lowest in the ateline genus ex-

Table 3. Mean ± SD (and range) home range size, group mass, and spatial requirements
in ateline primate study groups

Genus	Home range size (ha)	Group mass (kg)	Home range requirement		No. of studies[1]
			(ha/kg)	(ha/ind.)	
Lagothrix	497±339	146±79	3.37±1.32	20.9±8.7	10
	(150-1100)	(75-270)	(1.76-5.95)	(8.9-33.3)	
Ateles	201±77	162±69	1.41±0.95	8.0±5.4	8
	(114-325)	(77-248)	(0.62-3.25)	(3.8-18.1)	
Brachyteles	145±117	151±84	0.95±0.45	4.9±2.2	4 [2]
	(40-300)	(52-248)	(0.33-1.35)	(2.2-7.0)	

[1]See references in Table 2
[2]Because of a dramatic change in group size, data from Fazenda Montes Claros, Minas Gerais (Strier 1987, Strier et al. 1993) are considered as part of two separate studies.

hibiting the highest degree of folivory—*Brachyteles*—less than 1 ha of seasonal semideciduous forest of southeastern Brazil being required to sustain a group mass of 1 kg (Table 2). This ratio, however, is clearly affected by the extent to which home ranges of neighboring groups overlap, for the biomass density of a population is a better measure of overall resource use than that of a group.

Lagothrix studies undertaken so far attest to the fact that woolly monkey groups (i) do not defend a home range, (ii) retain few, if any, exclusive home range areas, and (iii) the combined range overlap with all neighboring groups is usually complete or approach 100% (Kavanagh and Dresdale 1975, Defler 1989a, Nishimura 1990, this study). A similar situation also occurs in *Brachyteles*: two groups contained in a 800-ha forest fragment overlapped 46–88% of their ranges with one another (Strier 1987). In marked contrast, *Ateles* occupies fairly discrete home ranges, only 10–15% of which overlaps with those of neighboring groups (van Roosmalen 1985, Symington 1988).

Spatial Group Structure in *Lagothrix*

The economics of sociality as determined by environmental correlates can be best examined in primate species such as *L. lagotricha*, which exhibit considerable elasticity in grouping behavior over a wide range of group sizes. Individual tendencies along the group size and cohesion gradients could then be viewed as responses to changes in abundance and distribution of resources. In primates, a small number of species—including geladas and hamadryas baboons (Kummer 1971), pygmy and common chimpanzees (Kuroda 1979, Goodall 1986, White 1992), and spider monkeys (Klein and Klein 1975, van Roosmalen 1985, Symington 1987b, Chapman 1990)—come closest to this theoretical fluid state. These species, despite considerable variability in grouping patterns (e.g. Chapman et al. in press), generally share a fission-fusion social system in which a large "community" of associated individuals often breaks up into smaller subunits of flexible, and often unpredictable, composition.

Lagothrix at Urucu and at several upper Amazonian sites do not fit this pattern. Unlike *Ateles*, the study group did not segregate into spatially independent feeding parties occupying different core areas within a larger home range. Groups of gray woollies were usually found as a *single*, highly elastic and uncohesive social unit, but which also broke up into ephemeral subunits that after a few to several hours coalesced again for extended periods (cf. Durham 1975, Nishimura and Izawa 1975, Izawa 1976, Defler 1989b, Nishimura 1990, 1994). This rather amorphous scatter of individuals — even when separated into different subgroups — could easily have been mistaken by short-term observations for a single large, uncohesive group. In contrast, a spider monkey community is almost always fragmented into highly clustered, discrete subgroups, which despite their apparent recognition and tolerance of other subgroups, will gather together in one place for only a few weeks of the year (van Roosmalen 1985, Symington 1987a, pers. obs.).

All studies of *Lagothrix* elsewhere are generally consistent with the grouping pattern described here, despite some variation along the gradients of physical group dimension and fluidity. Kavanagh and Dresdale (1975), after their February (dry season) observations in northern Colombia, were "unable to state whether or not woolly monkeys form fairly constant groups...but may have been members of a more complex or fluid society with various levels of association". Observations on two groups of 10 and 14 *L. l. poeppigii* by Ramirez (1988) indicate that they "split into temporary sub-groups", which "remained separated from the rest of the group for up to two hours". Elsewhere in Peru, the social organization of this subspecies was reported by Soini (1986a, 1986b) to consist of 2–3 "not entirely independent troops" (subgroups) which formed "a larger social unit or clan" (group). Defler (1989b) also states that his group of 20–24 *L. l. lagotricha* "often breaks up into two partially independent subunits...the members of each interact among themselves more than with members of other subgroups". Nishimura (1990) suggests that this subspecies of *L. lagotricha* elsewhere in eastern Colombia "lives in cohesive groups" but that "the distance between group members is not short". A group of 42–43 woollies similar in size to that described here exhibited only "temporary divisions into clusters, which rarely stayed far apart to make visual or vocal contact" (Nishimura and Izawa 1975). In the case of groups of *L. l. lugens*, Stevenson et al. (1994) also recognized their "higher cohesiveness in social structure...compared to the closely related spider monkeys". Finally, a more recent study by Nishimura (1994) on *L. l. lugens* confirmed that "group members of *Lagothrix* monkeys move in a single foraging unit like many other nonhuman primates".

The overall group structure of *L. l. cana* was, therefore, generally in accordance to that of previous studies, except for one important caveat. While this study confirms that at any one time (i) subgroup clusters in *Lagothrix* are not necessarily clearly discrete from one another, and (ii) the spatial continuity of individuals within a group is far greater than, for instance, that of *Ateles*, it does not necessarily follow that *Lagothrix* groups are *more* cohesive. In fact, our measurements of group dimensions showed that the study group was often extremely uncohesive even when discrete subgroups could not be distinguished, much more so than it could be physically possible for segregated parties of *Ateles* confined to the relatively small home ranges reported for this genus (Klein 1972, van Roosmalen 1985, Symington 1988, Fedigan et al. 1988, Chapman 1990). Perhaps a less ambiguous form of quantifying subgrouping cohesion or dispersion in atelines should, therefore, involve measures of *both within- and between-subgroup spread* (or a ratio between the two). This can be illustrated by the fact that, for instance, single subgroups of *Ateles* tend to be *more* cohesive than those of *Lagothrix*, but *less* cohesive at the level of entire groups. The ambiguity created by vague references to the concepts of grouping "cohesion", "fluidity", and "flexibility" could be resolved once it is accepted that spatial aggregation parameters (quantified via either overall dimensions or nearest-neighbor distances) at the level of subgroups need not correspond to that of whole groups. In part, this stems from a methodological field problem in that tracking whole groups (or several subgroups) necessarily involve multiple observers operating simultaneously. The previous overemphasis in the primate literature on the spatial behavior of single subgroups, therefore, merely reflects the absence of secondary observers as it has been attempted in this study.

The grouping pattern of woolly monkeys thus represents a radical departure from that of most Neotropical primates, which tend to live in either (i) relatively small, cohesive groups where visual contact among its members is possible throughout the day, or (ii) fission-fusion subgroups, the spatial behavior of which is largely oblivious of one another. However, two cases of other New World primate genera presenting a similar grouping behavior have been documented. Chapman (1988) describes a group of 40 howlers (*Alouatta palliata*) which "frequently fragmented into 2 or more subgroups that were spatially separated by up to 2 km, for periods that lasted as long as 4 weeks". A group of 23–26 woolly spider monkeys (*Brachyteles arachnoides*) has also been observed to "modify their grouping associations in response to patchy food resources and yet remain within calling proximity while resting and foraging" (Strier 1989).

To a certain extent grouping patterns within the Atelinae may reflect between-genera differences in levels of aggression, which in turn appear to be correlated with diet. Agonistic displacements over food sources are rare in the more folivorous ateline genera (*Brachyteles*: Milton 1984, Strier 1989; *Lagothrix*: Ramirez 1988, Defler 1989b, Nishimura 1990, 1994, Peres unpubl. data; see also Macleod 1993 for a study in captivity), but regular in the more frugivorous *Ateles* (Symington 1987a).

The grouping behavior of gray woollies can be seen as extremely elastic, for they often failed to split into discrete subgroups despite considerable group dimensions. At Urucu, this grouping pattern was even more diffuse than that of other *Lagothrix* populations in upper Amazonia. It appears that group elasticity is closely tied to the strongly demarcated seasonality in ripe fruit availability at Urucu, compared, for instance, with the much more aseasonal Apoporis site studied by T. Defler (1989a) in eastern Colombia, where groups as a whole appear to be more cohesive. Ripe fruit patch density at Urucu declined from 95 to 6 patches per hectare towards the single annual trough of nearly absolute fruit scarcity (Peres 1994c). At the Apoporis site, mean monthly rainfall never falls below 250 mm and the staggered fruiting peaks of different forest types provide a nearly constant source of ripe fruits for *Lagothrix* (as quantified by a 3-year fruit trapping program: Defler 1989a). Unfortunately, there are almost no other concurrent studies on plant phenology and overall group dimension for other atelines (but see Symington 1987b for correlation between plant phenology and subgroup size) to allow a comparative analysis. More parallel studies of resource patch density and structure, and the spatial behavior of both whole groups (mediated by multiple observers) and isolated subgroups, are thus required to refine our understanding of ateline grouping ecology.

ACKNOWLEDGMENTS

This study was funded by the World Wildlife Fund-US (Project 6199), the Wildlife Conservation Society, and the Brazilian Science and Research Council (CNPq). Thanks are due to the Brazilian Oil Company (Petrobrás S.A.), and its senior geologists, for kindly providing me with critical access to and logistical support in the upper Urucu river. I am immensely indebted to the monkey watchers of Urucu, particularly Raimundo Nonato and Luis Pereira Lopes, for their dedicated and untiring assistance during fieldwork. I wish to thank Marc van Roosmalen and Meg Symington for our early discussions on ateline ecology, and Colin Chapman and Karen Strier for comments on the manuscript.

REFERENCES

Allen, J.A., 1916, Mammals collected on the Roosevelt Brazilian Expedition, with field notes by Leo E. Miller, *Bull. Am. Mus. Nat. Hist.* 35:559–610.

Bates, H.W. 1892. The Naturalist on the River Amazons. John Murray, London.

Campbell, A.F., and Sussman, R.W. 1993. The value of radio tracking in the study of neotropical rain forest monkeys. *Am. J. Primatol.* 32:291–301.

Cant, J.G.H. 1977. Ecology, locomotion, and social organization of spider monkeys (*Ateles geoffroyi*). Unpubl. PhD. diss., University of California, Davies.

Cartelle, C. 1993. Achado de *Brachyteles* do Pleistoceno final. *Neotropical Primates* 1(1):8.

Chapman, C.A., 1988, Patch use and patch depletion by spider and howling monkeys of Santa Rosa National Park, Costa Rica. *Behaviour* 105:99–116.

Chapman, C.A., 1990, Association patterns of spider monkeys: the influenece of ecology and sex on social organization. *Behav. Ecol. Sociobiol.* 26:409–414.

Chapman, C.A., Chapman, L.J., Wrangham, R.W., Hunt, K., Gebo, D., and Gardner, L. 1992. Estimators of fruit abundance of tropical trees. *Biotropica* 24:417–421.

Chapman, C.A., White, F.J., and Wrangham, R.W. 1993. Defining subgroup size in fission-fusion societies. *Folia Primatol.* 61:31–34.

Chapman, C.A., Wrangham, R.W, and Chapman, L.J. 1995. Ecological constraint on group size: an analysis of spider monkey and chimpanzee subgroups. *Behav. Ecol. Sociobiol.* 36:59–70.

Chapman, C.A., White, F.J., and Wrangham, R.W. in press, Party size in chimpanzees and bonobos: a reevaluation of theory on two similarly forested sites. In Behavioural diversity of chimpanzees.

Dare, R. 1974. The social behavior and ecology of spider monkeys, Ateles geoffroy, on Barro Colorado Island. PhD Thesis, University of Oregon, Eugene.

Defler, T.R., 1989a, Recorrido y uso del espacio en un grupo de *Lagothrix lagothricha* (Primates: Cebidae) mono lanudo churuco en la Amazonia Colombiana. *Trianea* 3:183–105.

Defler, T.R., 1989b, Wild and woolly. *Animal Kingdom* 92(5):36–43.

Dunbar, R.I.M. 1992. Neocortex size as a constraint on group size in primates. *J. Human Evol.* 20:469–493.

Durham, N.M. 1975. Some ecological, distributional, and group behavioral patterns of Atelinae in southern Peru, with comments on interspecific relations. In Socioecology and psychology of primates, ed. R.H. Tuttle, Mouton Publ., The Hague., pp. 87–101

Fedigan, L.M., Fedigan, L., Chapman, C., and Glander, K.E. 1988. Spider monkey home ranges: a comparison of radio telemetry and direct observation. *Am. J. Primatol.* 16:19–29.

Fooden, J., 1963, A revision of the wooly monkeys (genus *Lagothrix*). *J. Mammal.* 44:213–217.

Ford, S.M. and Davis, L.S. 1992. Systematics and body size: implications for feeding adaptations in New World monkeys. *Am. J. Phys. Anthropol.* 88:415–468.

Freese, C.H., Heltne, P.G., Castro R., and Whitesides, G. 1982. Patterns and determinants of monkey densities in Peru and Bolivia, with notes on distributions. *Int. J. Primatol.* 1:53–90.

Glander, K.E., Fedigan, L.M., Fedigan, L. and Chapman, C. 1991. Field methods for capture and measurement of three monkey species in Costa Rica. *Folia Primatol.* 57:70–82.

Goodall, J. 1986. The chimpanzees of Gombe: patterns of behaviour. Belknap Press, Cambridge, MA.

Grant, J.W.A., Chapman, C.A., and Richardson, K.S. 1992. Defended versus undefended home ranges of carnivores, ungulates and primates. *Behav. Ecol. Sociobiol.* 31:149–161.

Graves, G.R., and O'Neill, J.P., 1980, Notes on the yellow-tailed woolly monkey (*Lagothrix flavicauda*) of Peru, *J. Mammal.* 61:345–347.

Izawa, K., 1976, Group sizes and compositions of monkeys in the upper Amazon basin, *Primates* 17:367–399.

Johns, A.D. 1986. Effects of habitat disturbance on rainforest wildlife in Brazilian Amazonia. Unpubl. Report to WWF-US, Washington, DC.

Kavanagh, M. and L. Dresdale. 1975. Observations on the woolly monkey (*Lagothrix lagotricha*) in Northern Colombia. *Primates* 16:285–294.

Klein, L.L. 1972. The ecology and social organization of the spider monkey, *Ateles belzebuth*. Unpubl. Ph.D. diss., University of California, Berkeley.

Klein, L.L. and D.J. Klein. 1975. Social and ecological contrasts between four taxa of Neotropical primates. in: Socioecology and psychology of primates (R. Tuttle, ed). Mouton, The Hague. pp. 59–85

Kummer, H. 1971. Primate societies. Aldine-Atherton, Chicago.

Kuroda, S. 1979. Grouping of pygmy chimpanzees. *Primates* 20:161–183.

Lane, F. 1990. A hunt for "monos" (*Brachyteles arachnoides*) in the foothills of the Serra da Paranapiacaba, São Paulo, Brazil. *Primate Conservation* 11:23–25.

Leighton, M. and D.R. Leighton. 1982. The relationship of feeding aggregate to size of food patch: howler monkeys (Alouatta palliata) feeding in Trichilia cipo fruit trees on Barro Colorado Island. Biotropica 14:81–90.

Lemos de Sá, R.M. 1988. Situação de uma população de Mono Carvoeiro, *Brachyteles arachnoides*, em um fragmento de Mata Atlântica (M.G.) e implicações para sua conservação. M.Sc. Thesis, Universidade de Brasília, Brasília, D.F.

Lemos de Sá, R.M., and Glander, K.E. 1993. Capture techniques and morphometrics for the woolly spider monkey, or muriqui (*Brachyteles arachnoides*, E. Geoffroy 1806). *Am. J. Primatol.* 29, 145–153.

Leo Luna, M. 1984. The effects of hunting, selective logging and clear-cutting on the conservation of the yellow-tailed woolly monkey (*Lagothrix flavicauda*). M.A. thesis, University of Florida, Gainesville.

Leo Luna, M. 1987. Primate conservation in Peru: a case study of the yellow-tailed woolly monkey. *Primate Conservation*, 8:122–123

Macleod, M.C. 1993. Another patrilineal primate? A behavioural study of a captive group of woolly monkeys, *Lagothrix lagotricha*. MSc. thesis, University College London, London.

Milton, K. 1984. Habitat, diet, and activity patterns of free-ranging woolly spider monkeys (*Brachyteles arachnoides* E. Geoffroy 1808). *Int. J. Primatol.* 5:491–514.

Milton, K. 1985. Mating patterns of woolly spider monkeys, *Brachyteles arachnoides*: implications for female choice. *Behav. Ecol. Sociobiol.* 17:53–59.

Napier, J.R., and Napier, P.H. 1967. A Handbook of Living Primates. Academic Press, New York.

Nishimura, A. 1990. A sociological and behavioral study of woolly monkeys, *Lagothrix lagotricha*, in the upper Amazon. *The Science and Engineering Review of Doshisha University* 31(2):1–121.

Nishimura, A. 1994. Social interaction patterns of woolly monkeys (*Lagothrix lagotricha*): a comparison among the atelines. *The Science and Engineering Review of Doshisha University* 35(2):91–110.

Nishimura, A., and Izawa, K. 1975. The group characteristics of wolly monkeys (*Lagothrix lagotricha*) in the upper Amazonian basin. in: Contemporary Primatology (S. Kondo, M. Kawai. A. Ehara, and S. Kawamura, eds.), pp. 351–357. Karger, Basel.

O'Keefe, J. and Nadel, L. 1978. The hippocampus as a cognitive map. Clarendon Press, Oxford.

Parker, T.A. III and L.J. Barkley. 1981. New locality for the yellow-tailed woolly monkey. *Oryx* 16:71–72.

Peres, C.A. 1990. Effects of hunting on western Amazonian primate communities. *Biol. Conserv.* 54:47–59.

Peres, C.A. 1991a. Humboldt's woolly monkeys decimated by hunting in Amazonia. *Oryx* 25:89–95.

Peres, CA. 1991b. Ecology of mixed-species groups of tamarins in Amazonian terra firme forests. Ph.D. Thesis, University of Cambridge, Cambridge.

Peres, C.A. 1993. Structure and spatial organization of an Amazonian terra firme primate community. *J. Trop. Ecol.* 9:259–276.

Peres, C.A. 1994a. Diet and feeding ecology of gray woolly monkeys (*Lagothrix lagotricha cana*) in central Amazonia: comparisons with other atelines. *Int. J. Primatol.* 15(3):333–372.

Peres, C.A. 1994b. Plant resource use and partitioning in two tamarin species and woolly monkeys in an Amazonian terra firme forest. In Current Primatology, Vol. I: Ecology and Evolution (B. Thierry, J.R. Anderson, J.J. Roeder, and N. Herrenschmidt, eds.), pp. 57–66. Université Louis Pasteur, Strasbourg, France.

Peres, C.A. 1994c. Primate responses to phenological changes in an Amazonian terra firme forest. *Biotropica* 26:98–112

Peres, C.A. 1994d. Composition, density, and fruiting phenology of arborescent palms in an Amazonian terra firme forest. *Biotropica* 26:285–294.

Peres, C.A. 1994e. Which are the largest New World monkeys? *J. Human Evol.* 26:245–249.

Pielou, E.C. 1959. The use of point-to-plant distances in the study of the pattern of plant populations. *J. Ecol.* 47:607–613.

Pinto, L.P.S., Costa, C.M.R., Strier, K.B., and Fonseca, G.A.B. 1993. Habitat, density and group size of primates in a Brazilian tropical forest. *Folia Primatol.* 61:135–143.

Ramirez, M., 1980, Grouping patterns of the woolly monkey, *Lagothrix lagotricha*, at the Manu National Park, Peru, *Am. J. Phys. Anthropol.* 52:269.

Ramirez, M., 1988, The woolly monkey, genus *Lagothrix*. in Ecology and behavior of Neotropical primates, Vol. 2 (R.A. Mittermeier, A.B. Rylands, A.F. Coimbra-Filho, and G.A.B. Fonseca, eds.). World Wildlife Fund, Washington, D.C. pp 539–575.

Robinson, J.G. and C.H. Janson. 1987. Capuchins, squirrel monkeys, and Atelines: socioecological convergence with old world primates. In: Primate Societies (B.B. Smuts, D.L. Cheney, R.M. Seyfarth, R.W. Wrangham, and T.T. Struhsaker, eds.), University of Chicago Press, Chicago.

van Roosmalen, M.G.M. van., 1985, Habitat preferences, diet, feeding strategy and social organization of the black spider monkey (*Ateles paniscus paniscus* Linnaeus 1758) in Surinam. *Acta Amazonica* 15(Suppl):1–238.

Rosenberger, A.L., and Strier, K.B. 1989. Adaptive radiation of the ateline primates. *J. Human Evol.* 18:717–750.

Rylands, A.B., 1987, Primate communities in Amazonian forests: their habitats and food resources. *Experientia* 43:265–279.

Sawaguchi, T. 1992. The size of the neocortex in relation to ecology and social structure in monkeys and apes. *Folia Primatol.* 58:131–145.

Soini, P. 1986a. A synecological study of a primate community in the Pacaya-Samiria National Reserve, Peru. *Primate Conservation* 7:63–71.

Soini, P. 1986b. Informe preliminar de la ecologia y dinamica poblacional del 'choro', *Lagothrix lagotricha* (Primates). Unpubl. Report, Informe de Pacaya No. 20.

Stephan, H., Frahm, H., and Baron, G. 1981. New and revised data on volumes of brain structures in insectivores and primates. *Folia Primatol.* 35:1–29.

Stevenson, P.R., Quiñones, M.J., and Ahumada, J.A., 1994, Ecological strategies of woolly monkeys (*Lagothrix lagotricha*) at Tinigua National Park, Colombia. *Am. J. Primatol.* 32:123–140.

Strier, K.B., 1987, Ranging behavior of woolly spider monkeys, or muriquis, *Brachyteles arachnoides*. *Int. J. Primatol.* 8:575–591.

Strier, K.B., 1989, Effects of patch size on feeding associations in muriquis (*Brachyteles arachnoides*), *Folia Primatol.* 52:70–77.

Strier, K.B., 1992, Ateline adaptations: behavioral strategies and ecological constraints. *Am. J. Phys. Anthropol.* 88:515–524.

Strier, K.B. 1994. Brotherhoods among atelins: kinship, affiliation, and competition. *Behaviour* 130:151–167.

Strier, K.B., Mendes, F.D.C., Rímoli, J., and Rímoli, A.O. 1993. Demography and social structure of one group of muriquis (*Brachyteles arachnoides*). *Int. J. Primatol.* 14:513–526

Symington, M.M., 1987a, Ecological correlates of party size in the black spider monkey, *Ateles paniscus chamek*. Unpubl. Ph.D. Diss., Princeton University, Princeton.

Symington, M.M., 1987b, Food compatition and foraging party size in the black spider monkey (*Ateles paniscus chamek*). *Behaviour* 105:117–134.

Symington, M.M., 1988, Demography, ranging patterns, and activity budgets of black spider monkeys (*Ateles paniscus chamek*) in the Manu National Park, Peru, *Am. J. Primatol.* 15:45–67.

Symington, M.M., 1990, Fission-fusion social organization in *Ateles* and *Pan. Int. J. Primatol.* 11:47–61

White, F.J. 1992 Pygmy chimpanzee social organization: variation with party size and between study sites. *Am. J. Primatol.* 26:203–214.

Wrangham, R.W., 1979. On the evolution of ape social systems. *Soc. Sci. Inform.* 18:335–368.

24

THE RELATION BETWEEN RED HOWLER MONKEY (*Alouatta seniculus*) TROOP SIZE AND POPULATION GROWTH IN TWO HABITATS

Carolyn M. Crockett

Regional Primate Research Center Box 357330
University of Washington
Seattle, Washington 98195–7330
and National Zoological Park
Smithsonian Institution
Washington, D.C. 20008

INTRODUCTION

No consensus yet exists to explain the diversity of primate social organization. Although ecological and social factors are both involved in the evolution and expression of social organization, they do not necessarily act together and may differ for the two sexes (Wrangham, 1987). Differing social factors usually are rooted in a species mating system—the degree of monopolization of mates by each sex—resulting from the effects of local ecology on the dispersion of the sex investing more in the production of offspring (Emlen and Oring, 1977; Vehrencamp and Bradbury, 1984). However, some features of primate social organization are phylogenetically conservative and resist change in response to varying ecological situations (Di Fiore and Rendall, 1994). At any given point in time, the particular social organization exhibited by a population also reflects recent demographic events (Altmann and Altmann, 1979; Dunbar, 1979). Among these life history variables are birth rates, which may vary directly in response to environmental variation, and sex ratios, which may vary randomly at birth but are affected thereafter by differential mortality (Dunbar, 1987). Social organization, then, is the result of a complex interplay of behavioral responses to ecological conditions, tempered by recent demographic events and constrained by phylogeny (Strier, 1994).

One aspect of a species' social organization is group size. For solitary primate species, "group" size is usually one, except when dependent offspring associate with the nurturing parent; for monogamous primate groups consisting of the mated pair and their dependent offspring, average group size is small, rarely exceeding five; average sizes of polygynous groups, however, vary considerably across species as well as within species, with means ranging from less than 5 to more than 80 (Smuts et al., 1987). This degree of

Adaptive Radiations of Neotropical Primates
edited by Norconk *et al.* Plenum Press, New York, 1996

variation is perplexing and is not simply explained by variation in habitat and diet (Strier, 1994; Wrangham, 1987). Factors which favor primate grouping and which constrain the size of social groups have been the focus of investigation, discussion, and controversy (Isbell, 1991; Janson and Goldsmith, in press; van Schaik, 1983; Wrangham, 1987; Wrangham, 1980). It is not clear whether group size is more appropriately viewed as a component or a consequence of social organization. For example, a recent phylogenetic analysis included the relative size of foraging units to reproductive units as a social organization trait but omitted total group size (Di Fiore and Rendall, 1994).

Longitudinal data from red howler monkeys (*Alouatta seniculus*) in two Venezuelan habitats help clarify the relative importance of ecological, demographic, and social factors affecting group size in this species. Most howler species, including red howlers, live in relatively small troops (Crockett and Eisenberg, 1987). I use the term "troop" to refer to bisexual social groupings of howlers that have produced offspring, much as the term "pride" is used for social groups of lions (Pusey and Packer, 1987). This is to distinguish a permanent breeding unit from a temporary group of dispersers or a subgroup that might form during foraging.

After the first two years of my study, red howler troop size averaged 9.4 individuals, including 1.4 adult males and 2.7 adult females (Crockett, 1984). In this species, both sexes disperse from their natal troops (Crockett, 1984; Crockett and Pope, 1993). Contrary to the "myth of the typical primate," more primate species exhibit bisexual dispersal than female philopatry (Strier, 1994). Female philopatry appears to be a phylogenetically derived trait most characteristic of Old World monkeys, the most ecologically diverse taxa (Di Fiore and Rendall, 1994). Here, I focus on the relation between troop size and population growth, and discuss the role of female dispersal in maintaining small troop size. Some of these patterns may apply to other species with bisexual dispersal.

METHODS

Study Site

In 1969, Hato Masaguaral, Venezuela (8° 35' N, 67° 35' W), a cattle ranch owned by the Tomás Blohm family, became the site of the first detailed study of red howler monkeys (Neville, 1972). The owner maintains the cattle at a density which does little damage to howler habitats. With the exception of rattlesnakes (which occasionally kill cattle), the ranch's numerous native wildlife species have been protected since 1944. Potential predators of howlers known to inhabit the area include puma (*Felis concolor*), ocelot (*F. pardalis*), jaguar (*Panthera onca*), spectacled caiman (*Caiman crocodilus*), and the snake *Boa constrictor*. The harpy eagle (*Harpia harpyja*), the best documented predator of howler monkeys in rain forest (Eason, 1989; Rettig, 1978; Sherman, 1991), is absent from this tropical dry forest zone (Thomas, 1979).

Much of the habitat is naturally open, being a type of tropical savanna known locally as *llanos* (Sarmiento, 1984; Troth, 1979). The red howler populations were monitored in two habitat types with study areas of similar size, a woodland habitat of 3.45 km² and a gallery forest of 3.9 km². The woodland comprises patches of grassland and scattered groves of trees up to ~ 14 m (Troth, 1979, "discrete mata bajío" and "shrub woodland bajío"). The gallery forest, about 4 km to the east, is more continuously forested with a more closed canopy seldom exceeding 20 m. In both habitats, many trees are deciduous, losing their leaves in the November to April dry season, and the lowest areas flood during the May to October wet season (Robinson, 1986; Troth, 1979). Annual rainfall during 1976–1983 averaged 1693 mm (Crockett and Rudran, 1987b).

Red howler monkeys were the only primates in the woodland, but they shared the gallery forest with wedge-capped capuchins (*Cebus olivaceus*). There was some dispersal of howlers between the habitats, which are separated by several kilometers of grassland and scattered shrubs (Crockett, 1985). Over time, since the first studies by Neville (1972), the habitat available to howlers improved because of fire control and changes in drainage patterns, resulting in an increase in woody vegetation. These changes were documented in aerial photographs of the area.

Census Methods

In 1975, Eisenberg (1979; Crockett and Eisenberg, 1987) censused the woodland troops first studied by Neville (1972), and laid the groundwork for Rudran's study which began in 1976 (1979). I first visited Hato Masaguaral in October 1978 for initial training and subsequently spent two years there (March 1979 - February 1981). Four follow-up census periods occurred through 1987. My research objectives were to extend comparative studies of red howlers to the gallery forest and to continue censuses of woodland troops initiated by Rudran. Detailed observations on behavioral ecology of one troop in each habitat type were supplemented with approximately monthly censuses of as many troops as I could locate in both habitats (Crockett, 1985; Crockett, 1987, and unpublished). By February 1981, I was regularly censusing 26 woodland and 17 gallery troops (Crockett, 1984).

The overall goal of my censuses was to locate all red howlers in troops, noting births, immigrations, disappearances and emigrations, as well as identifying possible extratroop (dispersing) howlers. Censuses began early in the morning to make use of dawn howls (Sekulic, 1982b) in locating troops. Censuses ceased at mid-day when howlers are usually inactive (Crockett, 1987; Sekulic, 1982b). Late afternoon censuses, beginning ~1530 h, occurred in the woodland but not in the gallery, which had to be reached by jeep. Approximately one-third of each study area could be searched systematically on foot in one morning (5–6 h). Transect trails established by Robinson (1986) were followed in the gallery forest, but the woodland census consisted of checking trees known to be used for sleeping or resting by various troops. The open nature of the woodland habitat permitted visual location of troops hundreds of meters away. In both habitats, particular trees usually were used by only one to three different troops, and most individuals were so habituated that their ongoing behavior was not interrupted. Observations were made with Leitz 10 x 40 binoculars which provide an exceptionally clear view of detail.

Once a troop was located, its identity was ascertained from troop identification sheets on which each troop member was described in drawings, words, and—when present—ear tag position. In 1981, 153 howlers were ear-tagged, including 26 gallery individuals, 110 previously unmarked woodland individuals, and 17 of 35 woodland howlers marked in 1978 (Crockett and Pope, 1988; Thorington et al., 1979). No gallery howlers had ear tags prior to January 1981. Continuity of identification of woodland individuals was based on descriptions by previous researchers (Mack, 1978 unpub. report; Rudran, pers. comm.). All troop adults and most younger animals were recognized as individuals during the monthly censuses from March 1979 through February 1981. Red howlers differ in characteristics such as brow shape and depigmentation spots on nipples, anogenital region, and soles of the feet. In addition, nearly 40% of howlers had injuries, many of which were permanent features such as torn ears, scarred lips, and damaged digits (Crockett,

1984; Crockett and Pope, 1988). All troop members were described in the troop identification sheets, but younger individuals had less detailed descriptions (e.g., fewer scars). Infants were distinguished from one another primarily by size and sex, but this was not a serious problem because there were rarely more than three infants (≤ 1 year old) per troop. Births occur during all months, but are less frequent in May-July (Crockett and Rudran, 1987a).

Each monkey was given an ID number which was used to register its presence in a census notebook. The first two digits indicated the individual's troop (where first identified), the third coded sex (odd=male, even=female) and initial age class (e.g., 1=adult male, 3=subadult male, 5=juvenile male, 7=infant male), and the fourth coded the specific individual; for juveniles and infants, the fourth digit assigned was the same as the mother's when possible. Infants born after 2/81 were assigned an additional decimal digit (e.g., .1 if born in 1981). If individual identity was uncertain, the monkey was listed by age-size/sex class along with any identifying characteristics. The troop's location and any notable behaviors (e.g., copulation, intertroop interactions, alloparenting, aggression) were recorded. If any individuals present during the previous census were missing, a thorough search was made of the area. New infants were sexed if possible (this could be achieved within a few days of birth, if provided a clear view), and maternal identity noted. Only some extratroop animals were recognized individually, and the rest were noted as to age/sex class and identifying features so that they might be distinguished if encountered again. Extratroop monkeys' location and any interaction with troop members were recorded.

Troop sizes reported for 1979–1981 are for the month of February. Follow-up census data were collected during 3–4 week periods in 12/81 (plotted as 1982 on graphs), 2/83, 2/84, and 4/87. Visibility was much better in these dry season months because the dominant, deciduous tree species dropped their leaves. In most cases, each troop was contacted at least twice during these follow-up censuses. Because I was unable to census the woodland in 1987, troop counts from Rumiz' dissertation (1992) for May were used for 1987 woodland data points. Some census counts of particular troops were contributed by R. Sekulic and T. Pope in conjunction with their dissertation research (Pope, 1989; Sekulic, 1981). Earlier data analyzed come from the following sources: woodland 1969 (Neville, 1972), 1975 (Eisenberg, 1979, and pers. comm.), 1978 (Rudran, 1979); gallery forest 1978 (Robinson, 1979, and pers. comm.).

Density Calculation

Population density was calculated as ecological density, the number of individuals divided by area of suitable habitat, and as raw or absolute density, the number of individuals divided by total area censused (Eisenberg, 1979). Area censused was determined by aerial photographs and hand-drawn maps derived from dry-season surveys. A contiguous area of forest with a fairly discrete boundary of open savanna on three sides and a small stream on the east defined the gallery forest (3.9 km^2) (Crockett, 1985). Although it was somewhat variable in microhabitat, all areas of the gallery forest were regarded as potentially suitable howler habitat (i.e., ecological density = absolute density). The woodland area censused was 3.45 km^2, but approximately 25% of the area was not deemed suitable howler habitat, consisting of open grassland on sandy rises, or low areas which were the last to dry out in the dry season and which supported sparse vegetation and scattered palms. These areas were sometimes traversed by dispersing howlers or those traveling from one part of their home range to another, but were not used for foraging. Thus, to estimate woodland ecological density, the number of individuals was divided by 2.6 km², whereas absolute density calculation used 3.45 km². In order to be consistent with the method used here, ecological density estimates for 1969 and 1978 were recalculated from published information (Neville, 1972; Rudran, 1979), as those sources estimated density from a subset of the area. The total population size upon which my 1987 woodland density estimate was based includes the May 1987 counts for the 30 troops studied by Rumiz, plus my last count of four other troops which I presume were still in the area (34 troops were said to be regularly censused (Rumiz, 1992)). I also added 6% to the total, previously having calculated this percentage of the population to be extratroop in 1983 (Crockett, 1985).

Because troops in the gallery forest were located gradually, I could not directly calculate population density there in 1979. Also, not all of the woodland troops in the circumscribed census area had been included in Rudran's (1979) regular census, and I added several more to the list over time. Based on troops censused during the two years preceding 1981, the gallery population grew at an annual rate of 21% while the woodland population grew more slowly, at about 6% per year. Thus, I was able to calculate the population densities for 1979 and 1980 using these per annum growth rates and 1981 densities.

Ecological Monitoring

The gallery forest is floristically more diverse than the woodland (Troth, 1979). The gallery study troop consumed food items from 62 different plant species compared with only 39 eaten by the woodland troop (Crockett, 1987). I collected phenology data monthly from August 1979 to December 1980 (Crockett and Rudran, 1987a). Twenty-four howler food species were sampled in each habitat within the home ranges of the two study troops for a total of 109 individuals of 31 different species. Plant abundance for all individuals of howler food species (> 2.5 cm in diameter, distinguishing individuals ≥ 4 m in height and those > 18 cm in diameter at breast height) was determined from 5-m-wide east-west transects every 50 m in the two home ranges (woodland: ~ 6 ha; gallery: ~ 20 ha) (Crockett, unpublished). Robinson (1986) presents data on phenology and abundance of *Cebus olivaceus* food plant species, 59% of which are also eaten by gallery forest howlers; 61% of gallery howler plant food species are also eaten by *Cebus* (Crockett, unpublished).

Statistical Analysis

Statistical tests reported here were performed with Data Desk 4.1 (Velleman, 1993). The dependent variables, troop size per year, were tested for normality and homogeneity of variance and did not require transformation. For tests involving comparisons between habitats, because the number of troops was not constant, each year was tested separately or in pairwise comparisons with selected years. The specific tests are described in the results.

RESULTS

Population Density: Habitat Differences and Growth over Time

By 1981, after two years of censusing and behavioral studies, a general pattern had emerged (Crockett, 1984; Crockett, 1985). Howlers in the woodland habitat lived at high

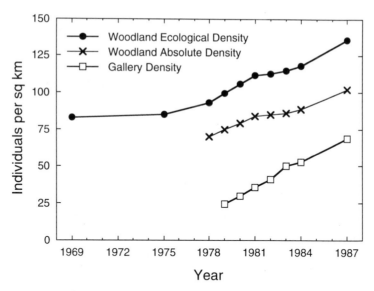

Figure 1. Population density in two habitats over time.

densities, in small home ranges, and in troops including 2 or 3 adult males more often than just one. Gallery forest howlers were more often found in troops with a single adult male, mean troop size was smaller and population density was less than half that in the woodland. In 1979, ecological density was ~ 25 individuals per square kilometer in the gallery vs. ~ 100/km² in the woodland. Follow-up censuses showed that the populations were not stable and in fact both were growing (Fig. 1). Population density increased in both habitats, but growth slowed in the woodland between 1981 and 1984. My 1987 density estimate for the woodland is based on several assumptions (see methods) and may be slightly overestimated, but the population definitely continued to expand. Figure 1 plots both ecological and absolute density for the woodland because the habitat was improving over time in ways that were not measured.

In the gallery, steady growth occurred since at least 1977, when the howler population was low and troop size averaged less than 6 (Robinson, 1979). In contrast, in 1970 Neville (1972) was able to census five troops in the gallery forest which averaged 9.6 individuals. The much lower gallery density at the onset of my study was attributed to a population crash in the mid-1970's during a period of drought and the associated crop failure of an important howler fruit, *Spondias mombin*, found only in the gallery forest (Crockett, 1985; Robinson, pers. comm.). *Spondias* fruit was one of the most frequently consumed items eaten by gallery howlers during the wet season (Crockett, unpubl.), but was rarely eaten by *Cebus* (Robinson, 1986). Disease seems unlikely to account for a crash in just one habitat but is a possibility; disease in *A. palliata* appeared to be significantly higher in riparian than deciduous habitats of a Costa Rican ranch (Jones, 1994). There are no records of yellow fever epidemics in Venezuela in the time frame of interest (Collias and Southwick, 1962, attributed severe decline in Barro Colorado *A. palliata* to yellow fever). Whatever the cause, all the evidence supports a severe decline in the gallery population in the mid-1970s.

In 1987, gallery population density was still below the ecological density estimated for the woodland in the 1970s. However, absolute density in the woodland in 1978 is about the same as gallery density in 1987. Persistent differences in ecological density suggest that there are differences in the carrying capacity between the two habitats related to the abundance and distribution of howler foods. For example, the single most important food source year around, *Ficus pertusa*, is much more abundant in the woodland habitat than in gallery forest (31.6 individuals > 4 m tall per ha vs. 5.1/ha; Crockett, unpubl.). The two habitats had similar seasonal patterns in leaf flush and, to a lesser extent, flowering, but differed in fruiting pattern: the woodland appears to have a more uniform year-around fruit availability. Also, phenology scores (percentage presence of a phenophase–young leaves, flowers, or fruits) were significantly higher for the woodland species sampled (Crockett and Rudran, 1987a).

The howlers and cebus feed on some of the same foods, so the potential carrying capacity in the gallery forest may be lessened by the presence of an interspecific competitor. *Ficus pertusa* is a major fruit source for both species (Robinson, 1986, and Crockett, unpubl.). However, on a number of occasions I observed both species feeding together in the same tree with little interspecific interference. The two monkeys primarily compete for the fruits of plant species that both eat, as howlers eat many more leaves and do not eat the insects and small vertebrates consumed by *Cebus*. Furthermore, the *Cebus* population was increasing between 1977 and 1986 at an annual rate of ~ 8% (Robinson, 1988a). Thus, the impact of *Cebus* on the howler carrying capacity (and vice-versa) seems small.

Changes in Numbers and Average Size of Troops

The most interesting finding from long-term monitoring was that differences in mean troop size between habitats diminished over time and were largely accounted for by different proportions of newly formed troops in the population. In this study, newly formed troops were defined as those that first produced infants in 1979 or later, whereas established ones were those that first produced infants prior to 1979. The new troops were formed from dispersing individuals, not from group fissioning (Crockett, 1984; Crockett and Pope, 1993). Some new troop members were known to have emigrated from other troops in the same habitat, but the origins of many were unknown and some undoubtedly came from outside the two study populations.

The woodland was a more saturated habitat, with less room for new troops to form. Initially, only a few new troops formed in the woodland (Fig. 2). Although more did form over time, most experienced poor reproductive success and some failed (Pope, 1992; Rumiz, 1992). For example, 12 woodland troops that formed in 1984–1985 had a 50% failure rate, measured by troop dissolution or 100% infant mortality (Pope, 1992). By contrast, of 14 new gallery troops that first produced offspring between 1980 and 1984, 13 still existed in 1987. In the initially low-density gallery forest, more than half of the troops in the area in 1987 had formed since 1979.

An important factor influencing the average troop size per habitat is the proportion of newly formed troops. Because these new troops are smaller than longer-established troops, they reduce the overall mean troop size (Crockett, 1985; Crockett and Eisenberg, 1987; Rumiz, 1992). When troops that were newly formed were distinguished from established ones, habitat differences in the mean size of the established troops diminished over time and converged at 10–11 individuals (compare Figs. 3 and 4). The mean troop size of recently formed troops remained smaller. Woodland troops were significantly larger than gallery troops from 1978 through 1981 (one-way ANOVAs, $p < .01$; Table 1). By 1982, there were enough newly formed troops to perform 2-way ANOVAs of troop size, with factors Habitat and Established vs. New (Est.v.New). In 1982, both Habitat ($p = .003$) and

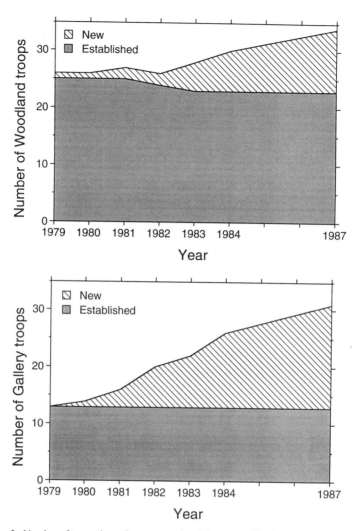

Figure 2. Number of troops in study area over time (above: woodland; below: gallery forest).

Est.v.New (p ≤ .0001) were significant, with woodland troops and established troops (habitats combined) being larger. In 1983, 1984, and 1987, only factor Est.v.New was significant (p ≤ .0001 each year). Furthermore, mean group size of the subset of newly formed troops increased significantly between 1981 (n=4) or 1983 (n=14) and 1987 (Table 1).

Troop Size Related to Population Density

A previous analysis found that mean troop size in howlers increased significantly with population density, when data from *A. seniculus* and *A. palliata* were considered separately (Crockett and Eisenberg, 1987). However, this analysis combined data points from different locations with those from the same location at different times. In a new analysis, I examined mean troop size (established and new troops combined) as a function

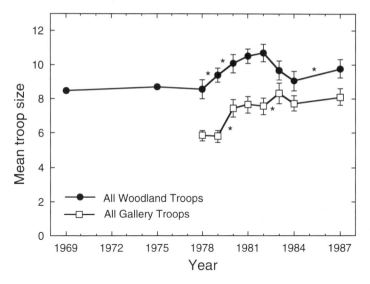

Figure 3. Mean troop size (± SE) over time in two habitats. * p < .05, paired t-test, comparing adjacent years within habitats.

of population density at Hato Masaguaral. Figure 5 plots mean troop size vs. population density; both absolute and ecological densities are plotted for woodland troop size. Because both populations grew from year to year, the data points within habitats are in temporal order. It is notable that when population density of the gallery (in 1987) converged with woodland absolute density (in 1978), mean troop size in the two habitats was very similar.

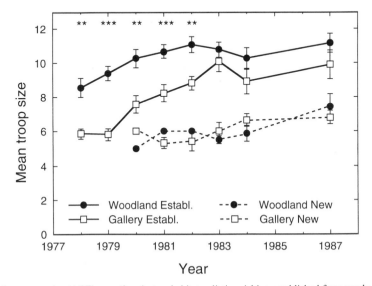

Figure 4. Mean troop size (± SE) over time in two habitats, distinguishing established from newly formed troops. ** p < .01 and *** p < .001, within-year ANOVAs comparing habitats (established and new combined). See text and Table 1.

Table 1. Summary of statistical tests

Fig.	Troop size	Habitat	Est.v.New	Test	Factor	Statistic	df	p =	Signif
3	1979-1978	Gallery	combined	paired t-test (t)	N/A	1.16	7	.28	NS
3	1980-1979	"	"	"		5.68	10	.0002	***
3	1981-1980	"	"	"		1.34	12	.21	NS
3	1982-1981	"	"	"		1.66	14	.12	NS
3	1983-1982	"	"	"		3.54	18	.002	***
3	1984-1983	"	"	"		-0.95	19	.35	NS
3	1987-1984	"	"	"		1.24	24	.23	NS
3	1979-1978	Woodland	"	"		2.63	21	.02	*
3	1980-1979	"	"	"		3.23	24	.004	**
3	1981-1980	"	"	"		1.44	25	.16	NS
3	1982-1981	"	"	"		1.20	24	.24	NS
3	1983-1982	"	"	"		-1.11	24	.28	NS
3	1984-1983	"	"	"		-1.41	25	.17	NS
3	1987-1984	"	"	"		2.15	22	.04	*
4	1987-1981	combined	New	"		3.46	3	.04	*
4	1987-1983	"	"	"		3.55	13	.004	***
4	1978	Factor	combined	ANOVA (F)	Habitat	8.07	1,28	.008	**
4	1979	"	"	"	"	26.95	1,34	≤.0001	***
4	1980	"	"	"	"	9.41	1,37	.004	**
4	1981	"	"	"	"	17.54	1,39	.0002	***
4	1982	Factor A	Factor B	ANOVA (F)	Habitat	9.63	1,42	.003	**
					Est.v.New	24.57	1,42	≤.0001	***
4	1983	Factor A	Factor B	ANOVA (F)	Habitat	0.32	1,47	.57	NS
					Est.v.New	58.11	1,47	≤.0001	***
4	1984	Factor A	Factor B	ANOVA (F)	Habitat	0.64	1,51	.43	NS
					Est.v.New	25.20	1,51	≤.0001	***
4	1987	Factor A	Factor B	ANOVA (F)	Habitat	2.32	1,58	.13	NS
					Est.v.New	29.57	1,58	≤.0001	***
5	1978-87	combined	combined	ANCOVA (F)	Habitat	0.02	1,12	.88	NS
					Density	5.33	1,12	.04	*
5	x = mean troop size	combined	combined	regression (F)	y = density	50.50	1,13	≤.0001	***
5	"	Woodland	"	"	"	0.81	1,6	.40	NS
5	"	Gallery	"	"	"	5.84	1,5	.06	trend
6	x = troop size 1980	combined	only 2 new	regression (F)	y = size 1987	0.22	1,31	.64	NS
6	"	Woodland	only 1 new	"	"	0.12	1,18	.73	NS
6	"	Gallery	only 1 new	"	"	0.002	1,11	.97	NS

When troop size is analyzed with ANCOVA, with habitat as a factor and density as a covariate, density is significant (p = .04) but habitat is not (p = .88). This implies that troop size is explained by density. In fact, when troop size is regressed on density across habitats and years, adjusted R^2 = 78% (p = ≤ .0001, df = 13). However, within the woodland habitat, troop size is not explained by density (adjusted R^2 = -2.8%, p = .40, df = 6). This is because woodland troop size declined in 1983 and 1984 (Figs. 3 and 4). There is a trend for gallery troop size to increase with density (R^2 = 45%, p = .06, df = 5). When woodland absolute density is substituted in the statistical tests, the p-values and R^2 do not change meaningfully.

These analyses suggest that if the gallery population reached the densities observed in the woodland, mean gallery troop size would be indistinguishable from the woodland.

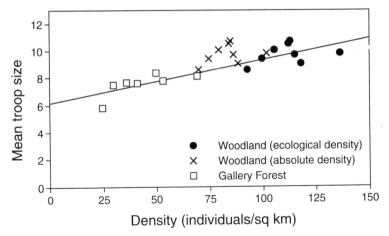

Figure 5. Mean troop size as a function of population density. Regression line is linear fit of troop size on ecological density. Plots of troop size versus woodland absolute density also shown.

However, the ecological monitoring data hint that the gallery carrying capacity might never support the highest population densities of the woodland. Furthermore, the relative proportion of newly formed troops, reflecting stage of population growth, would influence the overall mean troop size.

Predicting Future Troop Size from Initial Troop Size

If troop size is directly related to home range quality, variations in home range quality lead to the prediction that the relative size of troops should remain constant over the years, i.e., small groups should stay smaller while large troops stay relatively larger. For red howlers, whose folivore-frugivore diet places energetic constraints on travel, variations in home range quality could translate into variations in birth rate and survival. Red howler troops, with small home ranges and some overlap with neighboring troops' ranges (Sekulic, 1982b), demonstrate strong site fidelity, and occupy approximately the same home ranges for years. The winner in intertroop encounters depends on location rather than troop size (Sekulic, 1982c). In heterogeneous habitats, such as both woodland and gallery, home ranges may differ markedly from one another in food availability and distribution, and in structural characteristics that could expose individuals to predation (Peetz et al., 1992). My ecological data on habitat quality is restricted to that of the two study troops rather than across the entire census areas. However, Robinson (1986) presents relative density and spatial distribution for several tree species across the entire gallery forest area that I censused. There is considerable variation, for example, in *Pterocarpus acapulcensis*, the dominant canopy tree in my gallery study troop's range. Other features, such as proportion of "edge" habitat also vary markedly from one troop's range to another.

Other theories also predict that, under conditions in which being small is disadvantageous, small troops would tend to stay smaller and larger troops would remain larger. This would hold for species and ecological circumstances where larger groups are more successful in displacing smaller groups from clumped resources and translate better nutrition into higher birth rate and lower mortality (Robinson, 1988b; Wrangham, 1980). A similar prediction might follow where smaller groups are more vulnerable to predation and,

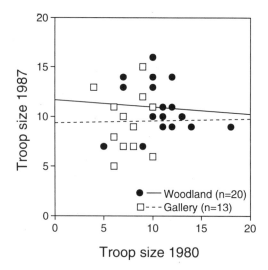

Figure 6. Troop size in 1987 as a function of size in 1980, with regression lines for woodland troops (n=20) and gallery forest troops (n=13) (both NS).

hence, stay small (van Schaik, 1983). On the other hand, relative size might not be constant over time if larger size were disadvantageous. Then troop size might grow until reversed by some demographic process.

To examine whether large troops stayed large, and small troops, small, troop size in 1987 was regressed on its size in 1980 for all 33 troops for which both counts were available. Changes in home range size over this period, while pertinent, are unknown. The sample consisted of only one new troop per habitat in 1980 (as mentioned above, troops that were recently formed did grow significantly between 1981 or 1982 and 1987). The results indicated that the size of troops in 1980 was not significantly related to their size in 1987 (woodland and gallery habitats tested separately or combined) (Fig. 6; Table 1).

DISCUSSION

Troop Size Related to Stage of Population Growth

These results suggest that troop size in red howler monkeys at Hato Masaguaral is more a function of troop age, stage of overall population growth, and demographic events such as immigrations and emigrations rather than specific features of the home range such as food availability. At Masaguaral, mean troop size increased from 1978 through 1982 in the woodland and from 1979 through 1983 in the gallery, somewhat paralleling increases in population density. Although density continued to increase, troop size gives the impression of leveling off (Figs. 3 and 5). Growth in troop size occurred through births, retention of maturing individuals, and subsequent births to females breeding natally, as well as from immigrations, usually by males. Reductions in troop size were the result of emigrations by both sexes, deaths, and disappearance.

The difference in average troop size between habitats over the years was due primarily to a greater proportion of newly formed troops in the gallery, which was in a different stage of population growth. The convergence in mean troop size over time was due in part to the "aging" and growth, primarily through births, of the newly formed troops in the gallery. Whether the slight increase in overall troop size between the 1970s and 1980s in both habitats is related to increasing population density (e.g., by reducing options for dispersal and new troop formation) or improved habitat providing more food in a given home range, or both, is uncertain. Unfortunately, no quantitative long-term habitat monitoring was done. However, the lack of relationship between the size of individual troops in 1980 and 1987 (hence, reflecting their range quality), tends to rule out the latter alternative.

If the apparent difference in carrying capacity between the two habitats is real, population densities should remain higher in woodland than gallery forest for the foreseeable future (barring catastrophic crash of only one habitat, an event that does appear to have precedent). Growth between 1981 and 1987 seemed remarkably parallel in the two habitats, with woodland ecological density being ~65 animals per square kilometer more than the gallery each year. At the same time, differences in average group size diminished, and were due mostly to the gallery forest including a greater proportion of smaller new troops. It appears that these new troops were able to form more easily in the gallery because of the much lower density. General growth in the population may be related to a combination of improvement in habitat quality as well as displacement of animals from outside the area. It is suspected that howlers were being displaced from habitats beyond ranch boundaries in areas being destroyed by agricultural development (Robinson, pers. comm.). Clarification of the long-term pattern may be forthcoming from the results of ongoing studies (Agoramoorthy, 1994).

Average troop size is related to population density in that they generally increase together. However, population density is limited by carrying capacity and in howlers varies much more across habitats than does troop size (Crockett and Eisenberg, 1987).

Comparisons with Other Howler Species

Changes in troop size and numbers of troops related to stages of population growth have been reported for other howler species. At a high density phase, mean troop size of *A. caraya* was larger than during a low density stage, and the distribution of sizes was bimodal, suggesting the inclusion of smaller, recently formed troops (Rumiz, 1990). Mean troop size in Belizean *A. pigra* was very small, 4.4, after a devastating hurricane, and most troops appeared to be monogamous (Bolin, 1981). Several years later, troop size had increased to 6.2 and most troops had two or three adult females (Horwich and Gebhard, 1983). Correlated reductions in average troop size and population density were inferred when a population crash in 1951 attributed to yellow fever was associated with record low average troop size among the mantled howlers (*A. palliata*) of semi-evergreen Barro Colorado Island (BCI), Panama (Collias and Southwick, 1962; Milton, 1982). Historically, mantled howler troop size on BCI has averaged ~18–19 individuals (Carpenter, 1934; Milton, 1982). Carpenter had predicted that the normal average troop size would continue to be 18 indefinitely (Carpenter, 1965). During the BCI population crash of 1951, troop size averaged only 8 individuals, and more troops were of unimale composition (Collias and Southwick, 1962). By 1959, the average had climbed to 18.5 (Carpenter, 1965). In 1967, the average declined somewhat to ~14 (Chivers, 1969). According to Chivers, as population density increased, so, initially, did troop size; as the density increase continued, larger troops were reduced in size and there were more, smaller troops. How the reductions in size and proliferation of new troops occurred was not known, but Chivers suspected intragroup tension and fragmentation of larger troops. Carpenter (1965) also speculated that new troop formation was through fissioning. In the 1970s, average troop size of BCI howlers increased again and stabilized at ~19 as the island became saturated at ~90 individuals/km^2 (Milton, 1982). The Masaguaral woodland red howler troops also showed a pattern of troop size increasing (1978–1982) and then declining (1982–1984) while density continued to increase, followed by an increase in mean troop size (1984–1987) (Figs. 3 and 5).

Similar stability in average troop size alternating with periods of lower mean size has been recorded for Costa Rican mantled howlers at La Pacífica ranch, tropical dry forest habitat in Guanacaste Province. The connection to population density is not as clear, and the typical mean size is somewhat smaller. Howlers at La Pacífica did not change significantly in population density, number of troops, or mean troop size between the mid-1970s and 1984 (Clarke et al., 1986). In 1991, total numbers of animals had not changed notably, but there were 30% more troops, their mean size was significantly smaller, and there was an increase in the number of troops with a single male (Clarke and Zucker, 1994). The decrease in mean troop size was due to fewer large troops (≥ 15 individuals) as well as a substantial increase in the number of small ones. Part of the reduction in overall mean troop size was due to the inclusion of new troops, some of which were likely smaller. Several cases of new troop formation were documented in this population (Glander, 1992). Some troops have gotten smaller because older animals that disappeared (presumably died) were not replaced by new immigrants (M. Clarke, pers. comm.). It is unclear whether and how changes in irrigation practices, deforestation and regeneration at La Pacífica are contributing to the increase in numbers of troops and reduction in their average size (Clarke and Zucker, 1994).

From the mid-1970s through 1984, mean troop size at La Pacífica was 16 individuals; however, in 1991, the mean had dropped to 12.6 (Clarke and Zucker, 1994). Population density was high, ~100 monkeys/km^2. The 1991 troop mean is remarkably similar to the overall average of 11.9 monkeys per troop during the years 1967–1970 at La Pacífica. Taboga, a nearby location, had similar mean troop size of 11.5 during 1966–1971, and both populations were judged "depressed," based on low ratios of infants to adult females (Heltne et al., 1976). In Santa Rosa National Park, also located in Costa Rica's Guanacaste Province, mean group size of mantled howlers in 1984 was 13.6; their population density was very low (~3–4 howlers/km^2) and seemed to be stable or expanding only very slowly (Fedigan, 1986). In the tropical wet forest of La Selva in northeastern Costa Rica, mantled howlers living at low density (~7–15 individuals/km^2) were found in troops averaging 11 monkeys (Stoner, 1994). In evergreen forests of Veracruz, Mexico, mantled howlers living at a density of ~23/km^2 averaged only 9 individuals (Estrada, 1982).

The mean size of red howler troops at Masaguaral (~9) is thus similar to the lowest means for mantled howlers, yet is at the high end for red howlers and the other five howler species (*A. belzebul, A. caraya, A. coibensis, A. fusca, A. pigra*) (Bonvicino, 1989; Crockett and Eisenberg, 1987; Horwich and Gebhard, 1983; Mendes, 1989; Milton and Mittermeier, 1977; Neville et al., 1988; Rumiz, 1990; Schaller, 1983; Schwarzkopf and Rylands, 1989). Mantled howlers have more adult females, up to 14, and more adult males, up to 6, than the other species, which usually have 2–3 adult females and 1–2 adult males. Mean troop size and average numbers of adult males and adult females are significantly larger in mantled howlers versus red howlers, but the species do not differ significantly with respect to the population densities at which they are found (Crockett and Eisenberg, 1987).

Home Range Size, Population Density, and Troop Size

Understanding the relation between home range size, home range overlap, and population density is pertinent to understanding the relation between troop size and population density. It is unfortunate that detailed home range data are not available for many red howler troops at Masaguaral, especially in the continuous canopy gallery forest, to see how range size or overlap changed as density increased. There is indirect evidence that howler home range size decreases and overlap increases as population density increases (Baldwin and Baldwin, 1972; Chivers, 1969; Crockett and Eisenberg, 1987). The general locations of the troops I censused in both Venezuelan habitats did not change over the years, and new troops sometimes formed in areas used by existing troops. Some increase in home range overlap must follow, but I did not get the impression of home range shrinkage. Investigating shrinkage would require detailed ranging data from a number of troops, and home range reduction might occur over a period of years. The inverse relation between home range size and population density across howler species and study areas (Crockett and Eisenberg, 1987), could result from differences between habitats, as found between the woodland and gallery forest, and not necessarily reflect what would happen if population density increased within one location. Home ranges might be constricted only when ranging activities begin to be limited by increasing population density or forest fragmentation. Howler home ranges can be as small as 1 to 4 ha (Chiarello, 1993; Neves and Rylands, 1991), but in some areas with extremely low population density can exceed 100 ha (Schlichte, 1978).

Correlations between demographic and ecological variables in eight contiguous *A. palliata* troops on Barro Colorado Island suggested that troop size was more strongly related to nonecological social factors, such as male invasions, than to food availability (Froehlich and Thorington, 1982). However, other studies of BCI howlers found foraging group size to be positively correlated with the abundance of certain fruiting trees (Gaulin et al., 1980; Leighton and Leighton, 1982). Possibly mantled howlers are better able to translate increased food availability into larger troop size than other howlers, and this may explain why "typical" group size varies between BCI (~18–19) and drier Guanacaste Province (~13–16). The foraging strategy of all howler species seems remarkably similar, but the tendency of larger mantled howler troops to form smaller foraging parties (Chapman, 1990) may allow total troop size to be larger in optimal habitats.

Female Emigration and Troop Size

Mantled howlers also differ from red howlers in their emigration patterns, and this contrast largely explains species differences in troop size and typical female composition. Details of these provocative differences can be found elsewhere (Crockett and Pope, 1993; Glander, 1992; Moore, 1992). Briefly, howlers (probably all species) have bisexual dispersal from natal troops, yet some individuals of both sexes are philopatric and breed in their birth troop (in red howlers, ~20–30% of females born breed natally versus 2% of males). Mantled howlers differ from red howlers in that males disperse younger than females, more females than males disperse, emigrant females typically immigrate into established troops, and new troop formation is less common. In both species, dispersers may spend months extratroop in small parties or solitary. What strongly differentiates red howlers (and probably the other howlers) from mantled howlers is that red howler females rarely succeed in entering established troops, unless deaths have reduced female membership to one, so they must form new troops with dispersing males in order to achieve reproductive success. Female-female aggression appears to be the proximate factor stimulating most female emigrations as well as preventing immigrations (Crockett, 1984; Crockett and Pope, 1988; Crockett and Pope, 1993; Sekulic, 1982a). Female-female aggression also is involved in mantled howler emigration and immigration, and females must fight their way up the hierarchy to gain troop membership (Glander, 1992; Jones, 1980). However, in red howlers, the resistance to female immigration, as well as the strong negative correlation between number of troop females and likelihood of natal female recruitment (Crockett and Pope, 1993), results in a sharply limited number of females. I have never observed a red howler troop to have more than four reproductive females, and few have even four (> 75 *different* troops censused). Long-term studies of other howler species (excepting mantled) have also reported maximum female group sizes of four (Crockett and Eisenberg, 1987; Rumiz, 1990).

Thus, the numbers of breeding females in red howler troops seems to be truncated at four. In red howlers, this pattern is explained on a proximate level by the emigration of females. On an ultimate level, the dispersal patterns of female red howlers require further explanation, given that it is more costly for a female to emigrate than to breed natally. Compared with natal females, dispersing females (1) range in much larger, sometimes unfamiliar, areas; (2) have deficient diets, probably related to lack of familiarity with the area or its marginal quality; (3) first breed at significantly older ages; and (4) if they succeed in breeding, it is usually in newly-formed troops, which, in the woodland, have a rather high rate of male invasion, infanticide, and dissolution (Crockett and Pope, 1993). Emigrating females, by ranging in unfamiliar areas, are assumed to have a higher mortality rate than females who stay put (Isbell et al., 1990, present evidence for *Cercopithecus aethiops*).

These costs suggest that females should rarely choose to emigrate. There are probably insufficient numbers of marked individuals with known kinship and reproductive success to compare inclusive fitness payoffs for dispersal versus nondispersal, as recently calculated for dwarf mongooses (*Helogale parvula*) (Creel and Waser, 1994). Although all major evolutionary explanations for dispersal, including seeking better or unexploited resources, intrasexual competition, seeking better access to mates, or avoiding inbreeding (Pusey, 1992), could be argued for red howler females, intrasexual competition seems the most important. The main factor determining whether or not a female stays is the number of reproductive females already resident in her natal troop (Crockett and Pope, 1993; Rumiz, 1992). Thus, assuming that emigration is highly correlated with low reproductive success, a female's fate is determined to a large extent by demography. (When population density is very low relative to carrying capacity, it is possible that females might emigrate to seek unexploited resources through new troop formation, and they might thereby avoid inbreeding if low density is correlated with long tenures of breeding males.)

The ultimate question raised, however, is why variation in red howler troop size is so restricted, and why more than four reproductive females have never been found in a red howler troop? Red howler troop size, excluding adult females, increases significantly as a function of the number of females, not surprising in that considerable variation in troop size is accounted for by the immature offspring of those females (Crockett, 1984). Previously I argued that the number of reproductive positions in red howler troops is limited by the females themselves, because of the relation between female group size and troop size (Crockett, 1984; Crockett and Eisenberg, 1987). This argument was based on sociobiological theory which predicts that the sex that invests more energy in offspring should show behaviors that maximize food intake required for the production of those offspring (Silk, 1993). In mammals it is invariably the females that invest more since they carry fetuses during pregnancy and then nurse the young after birth. Thus, it is in the reproductive interest of females to limit group size and thereby increase food intake relative to foraging effort. The most direct and feasible way that red howler females can affect troop size is to prohibit the recruitment of new reproductive females through aggression (Crockett, 1984).

Recently a more plausible factor, infanticide, has emerged to explain small troop size in red howlers, and this factor is directly related to female group size rather than its indirect effects on troop size via offspring produced (Crockett and Janson, 1993, Crockett and Janson, in preparation). Janson and I suggest that, whereas food may limit troop size theoretically, red howler troops rarely grow to the size limited by food because larger troops with more adult females are more likely to attract invading males that commit infanticide. Thus, ultimately, excess females are excluded by other females because their presence increases vulnerability to takeover.

Infanticide is a potent factor in red howler demography. Infant mortality averages only ~ 20%, but infanticide is suspected to account for at least 44% of it (Crockett and Rudran, 1987b). At least 14 infanticides have been observed at Masaguaral between 1978 and 1994, and many more have been inferred from infant disappearances and bite wounds (Agoramoorthy, 1994; Crockett and Sekulic, 1984; Rudran, 1979; Sekulic, 1983). Infanticide in red howlers associated with male invasions also has been well documented in Colombia, where 67% of infant births resulted in observed or suspected infanticide (Izawa and Lozano M., 1991; Izawa and Lozano M., 1994). Infanticide under similar circumstances also has been reported for *A. palliata*, *A. caraya* and *A. fusca* (Clarke and Glander, 1984; Galetti, 1994; Rumiz, 1990; Zunino et al., 1986). The evidence is building that infanticide in association with male invasions or status changes is a regular feature of howler monkey societies. Red howlers are among several primate species in which infanticide accounts for more deaths annually than predation (Cheney and Wrangham, 1987). Thus, we should not be surprised if it emerges as a potent factor in shaping social organization.

Conclusions

Troop size in two habitats which differed in population density by ~65 howlers per square kilometer converged on a similar mean (~9) after a period of population growth. One habitat apparently suffered an extreme population decline a few years before the study began, and mean troop size was very small at the onset of censusing. New troops formed in both habitats, but considerably more did so in the lower density gallery forest. These new troops were smaller than established troops and depressed the overall mean of the gallery forest compared with the woodland population, which had not experienced a population crash and had fewer new troops form. Most of the variation in troop size between habitats was related to stage of population growth.

What causes troop size to stop growing is a more vexing problem. "Typical" group sizes have been described for many primate species, and they sometimes vary as a function of habitat, as suggested by small but persistent differences in troop size over time between BCI and Guanacaste mantled howlers. More difficult to explain is the difference in troop size between mantled versus red howlers. Red howler troops are smaller primarily because they have fewer reproductive females. Female membership seems to be limited by female-female aggression which promotes female emigration and prevents female immigration. It is easier for dispersing female red howlers to form new troops with dispersing males than to enter an established troop. This explains why old troops do not grow indefinitely, why new troops form, and why new troops are small. However, it does not explain why female membership is limited to four to begin with.

My previous inference, that troop size was limited by food competition, seems less tenable as more information is gathered. It is difficult to explain such a truncated female composition, especially since mantled howlers, a species with a similar diet, ranging pattern, and habitat occupancy, does not show a dramatic limitation of female numbers. Female mantled howlers also emigrate, but they typically enter established troops even though they meet resistance from other females. There is some evidence that mantled howler group size is limited by food competition, or at least may vary as a function of it. Red howler troops, however, rarely seem to reach sizes where food competition limits them. An emerging hypothesis, developed elsewhere, suggests that the number of females in red howler troops is limited by infanticide, because the rate of infanticide deaths increases with the number of adult females. The proposed mechanism is that more females (i.e., potential mates) attract males to invade and takeover the troop.

If this hypothesis were to be extended to mantled howlers, it would predict that dispersing females would preferentially enter troops with fewer females than their natal troop. However, it does not explain why mantled howler females seldom form new, smaller troops, presumably unattractive to male invaders.

SUMMARY

Long-term census data from red howler monkeys (*Alouatta seniculus*) in the Venezuelan *llanos* help clarify the relative importance of ecological, demographic, and social factors affecting troop size in this species. Troop size in gallery forest (3.9 km^2) and woodland (3.45 km^2) habitats, which differed in ecological density by ~65 howlers/km^2 over a decade, converged on a similar mean (~9) after a period of population growth. Ecological monitoring suggested that the gallery forest has a lower carrying capacity for howlers than the woodland habitat. Most of the variation in troop size between habitats was related to stage of population growth. The gallery habitat apparently suffered an extreme population decline a few years before the study began, and mean troop size was very small at the onset of censusing. New troops formed in both habitats, but considerably more did so in the lower density gallery forest. These new troops were smaller than established troops and depressed the overall mean of the gallery forest compared with the woodland population, which had not experienced a population crash and had fewer new troops form. Troop size was positively correlated with population density when data for habitats and years were combined, but not when habitats were tested separately. Although density continued to increase year by year, mean troop size decreased after several years of successive increases. Red howler troops remain small primarily because reproductive female membership seems to be limited to ≤ 4 by female aggression promoting female emigration and preventing immigration. An hypothesis to account for female group size limitation, developed elsewhere, suggests that the rate of infanticide deaths increases with the number of adult females because more potential mates attract males to invade and takeover the troop.

ACKNOWLEDGMENTS

This research was funded by the Smithsonian Institution International Sciences Program, Friends of the National Zoo, National Geographic Society, and Harry Frank Guggenheim Foundation. Valuable bibliographic information was provided by the Primate Information Center, University of Washington Regional Primate Research Center (RPRC). The RPRC also provided incidental support. I thank the T. Blohm family for providing Hato Masaguaral as a research site over the years. At the onset of my research, I had the benefit of prior information on the population collected by Neville, Eisenberg, and Rudran in the woodland, and by Robinson who studied *Cebus olivaceus* in the gallery. D. Mack provided useful descriptions of woodland individuals (Mack, 1978 unpub. report). Three excellent dissertations contributed substantially to the information on the Masaguaral howlers (Pope, 1989; Rumiz, 1992; Sekulic, 1981). I especially thank my husband Bob Brooks for his invaluable support in the field and to this day. Warren Kinzey, to whom this volume is dedicated, also played an important role in my research by forcing me to think about the implications of howler studies for understanding the evolution of human behavior (Crockett, 1987). Thanks go to Paul Garber and two other reviewers for comments on the manuscript, and to Marilyn Norconk for inviting me to contribute to this volume.

REFERENCES

Agoramoorthy, G. 1994. An update on the long-term field research on red howler monkeys, *Alouatta seniculus*, at Hato Masaguaral, Venezuela. *Neotrop. Primates* 2: 7–9.

Altmann, S. A., and Altmann, J. 1979. Demographic constraints on behavior and social organization. In I. S. Bernstein and E. O. Smith (Eds.), *Primate Ecology and Human Origins*, pp. 47–64. Garland STPM Press, New York.

Baldwin, J. D., and Baldwin, J. I. 1972. Population density of howling monkeys (*Alouatta villosa*) in southwestern Panama. *Primates* 13: 371–379.

Bolin, I. 1981. Male parental behavior in black howler monkeys (*Alouatta palliata pigra*) in Belize and Guatemala. *Primates* 22: 349–360.

Bonvicino, C. R. 1989. Ecologia e comportamento de *Alouatta belzebul* (Primates: Cebidae) na mata Atlântica. *Rev. Nordestina Biol.* 6: 149–179.

Carpenter, C. R. 1934. A field study of the behavior and social relations of howling monkeys (*Alouatta palliata*). *Comparative Psychology Monographs* 10: 1–168.

Carpenter, C. R. 1965. The howlers of Barro Colorado Island. In I. DeVore (Ed.), *Primate Behavior*, pp. 250–291. Holt, Rinehart and Winston, New York.

Chapman, C. A. 1990. Ecological constraints on group size in three species of neotropical primates. *Folia Primatol.* 55: 1–9.

Cheney, D. L., and Wrangham, R. W. 1987. Predation. In B. B. Smuts, D. L. Cheney, R. M. Seyfarth, R. W. Wrangham, and T. T. Struhsaker (Eds.), *Primate Societies*, pp. 227–239. University of Chicago Press, Chicago.

Chiarello, A. G. 1993. Home range of the brown howler monkey, *Alouatta fusca*, in a forest fragment of southeastern Brazil. *Folia Primatol.* 60: 173–175.

Chivers, D. J. 1969. On the daily behavior and spacing of howling monkey groups. *Folia Primatol.* 10: 48–102.

Clarke, M. R., and Glander, K. E. 1984. Female reproductive success in a group of free-ranging howling monkeys (*Alouatta palliata*) in Costa Rica. In M. F. Small (Ed.), *Female Primates: Studies by Women Primatologists*, pp. 111–126. Alan R. Liss, New York.

Clarke, M. R., and Zucker, E. L. 1994. Survey of the howling monkey population at La Pacifica: a seven-year follow up. *Int. J. Primatol.* 15: 61–73.

Clarke, M. R., Zucker, E. L., and Scott, N. J. J. 1986. Population trends of the mantled howler groups of La Pacifica, Guanacaste, Costa Rica. *Am. J. Primatol.* 11: 79–88.

Collias, N., and Southwick, C. 1962. A field study of population density and social organization in howling monkeys. *Proc. Am. Philos. Soc.* 96: 143–156.

Creel, S. R., and Waser, P. M. 1994. Inclusive fitness and reproductive strategies in dwarf mongooses. *Behav. Ecol.* 5: 339–348.

Crockett, C. M. 1984. Emigration by female red howler monkeys and the case for female competition. In M. F. Small (Ed.), *Female Primates: Studies by Women Primatologists*, pp. 159–173. Alan R. Liss, New York.

Crockett, C. M. 1985. Population studies of red howler monkeys (*Alouatta seniculus*). *Nat. Geogr. Res.* 1: 264–273.

Crockett, C. M. 1987. Diet, dimorphism and demography: Perspectives from howlers to hominids. In W. G. Kinzey (Ed.), *The Evolution of Human Behavior: Primate Models*, pp. 115–135. SUNY Press, New York.

Crockett, C. M., and Eisenberg, J. F. 1987. Howlers: Variations in group size and demography. In B. B. Smuts, D. L. Cheney, R. M. Seyfarth, R. W. Wrangham, and T. T. Struhsaker (Eds.), *Primate Societies*, pp. 54–68. University of Chicago Press, Chicago.

Crockett, C. M., and Janson, C. H. 1993. The costs of sociality in red howler monkeys: infanticide or food competition? *Am. J. Primatol.*

Crockett, C. M., and Pope, T. 1988. Inferring patterns of aggression from red howler monkey injuries. *Am. J. Primatol.* 15: 289–308.

Crockett, C. M., and Pope, T. R. 1993. Consequences of sex differences in dispersal for juvenile red howler monkeys. In M. E. Pereira and L. A. Fairbanks (Eds.), *Juvenile Primates: Life History, Development and Behavior*, pp. 104–118. Oxford University Press, New York.

Crockett, C. M., and Rudran, R. 1987a. Red howler monkey birth data I: Seasonal variation. *Am. J. Primatol.* 13: 347–368.

Crockett, C. M., and Rudran, R. 1987b. Red howler monkey birth data II: Interannual, habitat, and sex comparisons. *Am. J. Primatol.* 13: 369–384.

Crockett, C. M., and Sekulic, R. 1984. Infanticide in red howler monkeys (*Alouatta seniculus*). In G. Hausfater and S. B. Hrdy (Eds.), *Infanticide: Comparative and Evolutionary Perspectives*, pp. 173–191. Aldine, New York.

Di Fiore, A., and Rendall, D. 1994. Evolution of social organization: A reappraisal for primates by using phylogenetic methods. *Proc. Natl. Acad. Sci. USA* 91: 9941–9945.

Dunbar, R. I. M. 1979. Population demography, social organization, and mating strategies. In I. S. Bernstein and E. O. Smith (Eds.), *Primate Ecology and Human Origins*, pp. 67–88. Garland STPM Press, New York.

Dunbar, R. I. M. 1987. Demography and reproduction. In B. B. Smuts, D. L. Cheney, R. M. Seyfarth, R. W. Wrangham, and T. T. Struhsaker (Eds.), *Primate Societies*, pp. 240–249. University of Chicago Press, Chicago.

Eason, P. 1989. Harpy eagle attempts predation on adult howler monkey. *The Condor* 91: 469–470.

Eisenberg, J. F. 1979. Habitat, economy, and society: Some correlations and hypotheses for the Neotropical primates. In I. S. Bernstein and E. O. Smith (Eds.), *Primate Ecology and Human Origins*, pp. 215–262. Garland STPM, New York.

Emlen, S. T., and Oring, L. W. 1977. Ecology, sexual selection, and the evolution of mating systems. *Science* 197: 215–223.

Estrada, A. 1982. Survey and census of howler monkeys (*Alouatta palliata*) in the rain forest of "Los Tuxtlas," Veracruz, Mexico. *Am. J. Primatol.* 2: 363–372.

Fedigan, L. M. 1986. Demographic trends in the *Alouatta palliata* and *Cebus capucinus* populations of Santa Rosa National Park, Costa Rica. In J. G. Else and P. C. Lee (Eds.), *Primate Ecology and Conservation*, pp. 285–293. Cambridge University Press, Cambridge.

Froehlich, J. W., and Thorington, R. W. 1982. The genetic structure and socioecology of howler monkeys (*Alouatta palliata*) on Barro Colorado Island. In J. E. G. Leigh, A. S. Rand, and D. M. Windsor (Eds.), *The Ecology of a Tropical Forest: Seasonal Rhythms and Long-term Changes*, pp. 291–305. Smithsonian, Washington, D. C.

Galetti, M. 1994. Infanticide in the brown howler monkey, *Alouatta fusca*. *Neotrop. Primates* 2: 6–7.

Gaulin, S. J. C., Knight, D. H., and Gaulin, C. K. 1980. Local variance in *Alouatta* group size and food availability on Barro Colorado Island. *Biotrop.* 12: 137–143.

Glander, K. E. 1992. Dispersal patterns in Costa Rican mantled howling monkeys. *Int. J. Primatol.* 13: 415–436.

Heltne, P. G., Turner, D. C., and Scott, N. J. 1976. Comparison of census data on *Alouatta palliata* from Costa Rica and Panama. In R. W. Thorington and P. G. Heltne (Eds.), *Neotropical Primates: Field Studies and Conservation*, pp. 10–19. National Academy of Sciences, Washington, D. C.

Horwich, R. H., and Gebhard, K. 1983. Roaring rhythms in black howler monkeys (*Alouatta pigra*) of Belize. *Primates* 24: 290–296.

Isbell, L. A. 1991. Contest and scramble competition: patterns of female aggression and ranging behavior among primates. *Behav. Ecol.* 2: 143–155.

Isbell, L. A., Cheney, D. L., and Seyfarth, R. M. 1990. Costs and benefits of home range shifts among vervet monkeys (*Cercopithecus aethiops*) in Amboseli National Park, Kenya. *Behav. Ecol. Sociobiol.* 27: 351–358.

Izawa, K., and Lozano M., H. 1991. Social changes within a group of red howler monkeys (*Alouatta seniculus*). III. *Field Studies of New World Monkeys, La Macarena, Colombia* 5: 1–16.

Izawa, K., and Lozano M., H. 1994. Social changes within a group of red howler monkeys (*Alouatta seniculus*). V. *Field Studies of New World Monkeys, La Macarena, Colombia* 9: 33–39.

Janson, C. H., and Goldsmith, M. in press. Predicting group size in primates: foraging costs and predation risks. *Behav. Ecol.*

Jones, C. B. 1980. The functions of status in the mantled howler monkey, *Alouatta palliata* GRAY: Intraspecific competition for group membership in a folivorous Neotropical primate. *Primates* 21: 389–405.

Jones, C. B. 1994. Injury and disease of the mantled howler monkey in fragmented habitats. *Neotrop. Primates* 2: 4–5.

Leighton, M., and Leighton, D. R. 1982. The relationship of size and feeding aggregate to size of food patch: howler monkeys (*Alouatta palliata*) feeding in *Trichilia cipo* fruit trees on Barro Colorado Island. *Biotrop.* 14: 81–90.

Mendes, S. L. 1989. Estudo ecológico de *Alouatta fusca* (Primates: Cebidae) na Estação Biológica de Caratinga, MG. *Rev. Nordestina Biol.* 6: 71–104.

Milton, K. 1982. Dietary quality and demographic regulation in a howler monkey population. In J. E. G. Leigh, A. S. Rand, and D. M. Windsor (Eds.), *The Ecology of a Tropical Forest: Seasonal Rhythms and Long-term Changes*, pp. 273–289. Smithsonian, Washington, D. C.

Milton, K., and Mittermeier, R. A. 1977. A brief survey of the primates of Coiba Island, Panama. *Primates* 18: 931–936.

Moore, J. 1992. Dispersal, nepotism, and primate social behavior. *Int. J. Primatol.* 13: 361–378.

Neves, A. M. S., and Rylands, A. B. 1991. Diet of a group of howling monkeys, *Alouatta seniculus*, in an isolated forest patch in central Amazonia. *A Primatologia no Brasil* 3: 263–274.

Neville, M. K. 1972. The population structure of red howler monkeys (*Alouatta seniculus*) in Trinidad and Venezuela. *Folia Primatol.* 18: 56–86.

Neville, M. K., Glander, K. E., Braza, F., and Rylands, A. B. 1988. The howling monkeys, genus *Alouatta*. In R. A. Mittermeier, A. B. Rylands, A. F. Coimbra-Filho, and F. A. B. da Fonseca (Eds.), *Ecology and Behavior of Neotropical Primates, Volume 2*, pp. 349–453. World Wildlife Fund, Washington, D. C.

Peetz, A., Norconk, M. A., and Kinzey, W. G. 1992. Predation by jaguar on howler monkeys (*Alouatta seniculus*) in Venezuela. *Am. J. Primatol.* 28: 223–228.

Pope, T. R. 1989. *The influence of mating system and dispersal patterns on the genetic structure of red howler monkey populations*. Unpublished Ph. D. Dissertation, University of Florida.

Pope, T. R. 1992. The influence of dispersal patterns and mating system on genetic differentiation within and between populations of the red howler monkey (*Alouatta seniculus*). *Evolution* 46: 1112–1128.

Pusey, A. E. 1992. The primate perspective on dispersal. In N. C. Stenseth and W. Z. Lidicker (Eds.), *Dispersal: small mammals as a model*, pp. 243–259. Chapman & Hall, .

Pusey, A. E., and Packer, C. 1987. The evolution of sex-biased dispersal in lions. *Behaviour* 101: 275–310.

Rettig, H. L. 1978. Breeding behavior of the harpy eagle (*Harpia harpyja*). *The Auk* 95: 629–643.

Robinson, J. G. 1979. *Progress report 2* : Smithsonian Institution.

Robinson, J. G. 1986. *Seasonal variation in use of time and space by the wedge-capped capuchin monkey Cebus olivaceus: Implications for foraging theory*. Smithsonian, Washington, D. C.

Robinson, J. G. 1988a. Demography and group structure in wedge-capped capuchin monkeys, *Cebus olivaceus*. *Behaviour* 104: 202–232.

Robinson, J. G. 1988b. Group size in wedge-capped capuchin monkeys *Cebus olivaceus* and the reproductive success of males and females. *Behav. Ecol. Sociobiol.* 23: 187–197.

Robinson, J. G. pers. comm. .

Rudran, R. 1979. The demography and social mobility of a red howler (*Alouatta seniculus*) population in Venezuela. In J. F. Eisenberg (Ed.), *Vertebrate Ecology in the Northern Neotropics*, pp. 107–126. Smithsonian, Washington, D. C.

Rumiz, D. I. 1990. *Alouatta caraya*: Population density and demography in northern Argentina. *Am. J. Primatol.* 21: 279–294.

Rumiz, D. I. 1992. *Effects of Demography, Kinship, and Ecology on the Behavior of the Red Howler Monkey, Alouatta seniculus*. Unpublished Ph. D. Dissertation, University of Florida.

Sarmiento, G. 1984. The Ecology of Neotropical Savannas (O. Solbrig, Trans.). Harvard University Press, Cambridge.

Schaller, G. B. 1983. Mammals and their biomass on a Brazilian ranch. *Arquivos de Zoologia* 31: 1–36.

Schlichte, H.-J. 1978. A preliminary report on the habitat utilization of a group of howler monkeys (*Alouatta villosa pigra*) in the National Park of Tikal, Guatemala. In G. G. Montgomery (Ed.), *The Ecology of Arboreal Folivores*, pp. 551–559. Smithsonian Institution Press, Washington, D. C.

Schwarzkopf, L., and Rylands, A. B. 1989. Primate species richness in relation to habitat structure in Amazonian rainforest fragments. *Biol. Conserv.* 48: 1–12.

Sekulic, R. 1981. *The significance of howling in the red howler monkey Alouatta seniculus*. Unpublished Ph. D. Dissertation, University of Maryland.

Sekulic, R. 1982a. Behavior and ranging patterns of a solitary female red howler (*Alouatta seniculus*). *Folia Primatol.* 38: 217–232.

Sekulic, R. 1982b. Daily and seasonal patterns of roaring and spacing in four red howler *Alouatta seniculus* troops. *Folia Primatol.* 39: 22–48.

Sekulic, R. 1982c. The function of howling in red howler monkeys (*Alouatta seniculus*). *Behaviour* 81: 38–54.

Sekulic, R. 1983. Male relationships and infant deaths in red howler monkeys (*Alouatta seniculus*). *Zeitschrift für Tierpsychologie* 61: 185–202.

Sherman, P. T. 1991. Harpy eagle predation on a red howler monkey. *Folia Primatol.* 56: 53–56.

Silk, J. B. 1993. The evolution of social conflict among female primates. In W. A. Mason and S. P. Mendoza (Eds.), *Primate Social Conflict*, pp. 49–83. State University of New York Press, Albany.

Smuts, B. B., Cheney, D. L., Seyfarth, R. M., Wrangham, R. W., and Struhsaker, T. T. (eds.). 1987. *Primate Societies*. University of Chicago Press, Chicago.

Stoner, K. E. 1994. Population density of the mantled howler monkey (*Alouatta palliata*) at La Selva Biological Reserve, Costa Rica: A new technique to analyze census data. *Biotrop.* 26: 332–340.

Strier, K. B. 1994. Myth of the typical primate. *Yrbk. Phys. Anthropol.* 37: 233–271.

Thomas, B. T. 1979. The birds of a ranch in the Venezuelan llanos. In J. F. Eisenberg (Ed.), *Vertebrate Ecology in the Northern Neotropics*, pp. 213–232. Smithsonian, Washington, D. C.

Thorington, R. W. J., Rudran, R., and Mack, D. 1979. Sexual dimorphism of *Alouatta seniculus* and observations on capture techniques. In J. F. Eisenberg (Ed.), *Vertebrate Ecology in the Northern Neotropics*, pp. 97–106. Smithsonian, Washington, D. C.

Troth, R. G. 1979. Vegetational types on a ranch in the central llanos of Venezuela. In J. F. Eisenberg (Ed.), *Vertebrate Ecology in the Northern Neotropics*, pp. 17–30. Smithsonian, Washington, D. C.

van Schaik, C. P. 1983. Why are diurnal primates living in groups? *Behaviour* 87: 120–144.

Vehrencamp, S. L., and Bradbury, J. W. 1984. Mating systems and ecology. In J. R. Krebs and N. B. Davies (Eds.), *Behavioural Ecology*, 2nd ed., pp. 251–278. Sinauer, Sunderland, Mass.

Velleman, P. F. 1993. *Data Desk: The New Power of Statistical Vision*. Data Description Inc., Ithaca, NY.

Wrangham. 1987. Evolution of social structure. In B. B. Smuts, D. L. Cheney, R. M. Seyfarth, R. W. Wrangham, and T. T. Struhsaker (Eds.), *Primate Societies*, pp. 282–296. University of Chicago Press, Chicago.

Wrangham, R. W. 1980. An ecological model of female-bonded primate groups. *Behaviour* 75: 262–299.

Zunino, G. E., Chalukian, S. C., and Rumiz, D. I. 1986. Infanticidio e desaparición de infantes asociados al reemplazo de macho en grupos de *Alouatta caraya*. *A Primatol. Brasil* 2: 185–190.

25

REPRODUCTIVE ECOLOGY OF FEMALE MURIQUIS (*Brachyteles arachnoides*)

Karen B. Strier

Department of Anthropology
University of Wisconsin-Madison
1180 Observatory Drive
Madison, Wisconsin 53706

INTRODUCTION

Reproductive seasonality, or the clustering of reproductive events, is well-documented among New World primates, and has been associated, in part, with the effects of seasonal rainfall on food availability (Lindberg, 1987). Food availability may influence the optimal timing of conceptions and births depending on the energetic and nutritional requirements of mothers and infants during the critical periods of gestation, lactation, and weaning (Crockett and Rudran, 1987; Bercovitch and Harding, 1993). It has been difficult, nonetheless, to identify consistent relationships between food availability and reproduction because ecological conditions and nutritional requirements may vary across species, different populations of the same species, and different years in the same population. For example, although some populations of both *Ateles* and *Alouatta* exhibit birth peaks, in *Ateles* births are concentrated during the early wet season when preferred fruits required by lactating mothers are most abundant (Chapman and Chapman, 1990). In *Alouatta*, by contrast, the avoidance of early wet season births has been attributed to the importance of food availability during the weaning period (Crockett and Rudran, 1987).

In some species, the benefits of reproductive synchrony may be more significant than the precise timing of reproductive events (Rowell and Richards, 1979). Social factors that may select for reproductive synchrony independent of ecological conditions include the availability of potential allomothers, particularly among the Old World female-bonded primates (Nicolson, 1987) and the opportunitiy to form nursery groups in which some aspects of infant care such as predator detection can be shared as has been proposed for the highly synchronized *Saimiri* (Boinski, 1987). Early socialization experiences with cohorts that ultimately become allies when infants mature in their natal groups (Walters, 1987) or transfer to other groups could also be important (Pusey and Packer, 1987).

The ability of females to synchronize their reproduction in response to either ecological or social stimuli may be confounded by female life history variables such as age and length of association with one another, or by differential investment strategies in sons and daughters. Irregular ovulatory cycles in young nulliparous females may result in departures from the optimal timing of conceptions and consequently births (reviewed in Clarke, et al., 1992). Among the many New World primates in which adolescent females migrate between groups (Strier, 1990, 1994a), the absence of strong or continuous associations between young immigrant females and older, long-term female residents may limit social contact and thus interfere with the transmission of pheromonal cues that stimulate ovulatory synchrony (McClintock, 1983).

Similarly, differences in when sons and daughters are weaned may affect when mothers resume cycling and conceive again, thereby altering their subsequent reproductive patterns. If offspring sex is related to female rank, as has been proposed for *Ateles* (Symington, 1987; but see Chapman and Chapman, 1990) and some Old World monkeys (e.g., Altmann, et al., 1978; Nicolson, 1987), then it is possible that reproductive patterns of dominant and subordinate females reflect differences in their respective investments in offspring.

Efforts to distinguish between the interacting effects of ecological, social, and physiological variables on reproduction in wild New World primates have been limited by the absence of visual cues that signal female reproductive condition (Dixson, 1983). Recent advances in the development of non-invasive fecal steroid assays have made it possible to begin to evaluate extrapolations based on observations of sexual behavior and subsequent births, and to correlate endocrinological condition with ecology and behavior (e.g., Strier and Ziegler, 1994).

This paper examines the reproductive ecology of one group of muriquis (*Brachyteles arachnoides*), the largest and one of the most endangered New World primates. Muriquis are restricted to what remains of the Atlantic forest of southeastern Brazil, an ecosystem characterized by pronounced seasonality in climate, rainfall, and forest phenology (Milton, 1984; Torres de Assumpção, 1983; Strier, 1986; Lemos de Sá, 1988; Ferrari and Strier, 1992). Many aspects of muriqui behavior at the study site have been associated with their seasonal ecology. For example, although muriquis devote equivalent proportions of their time to resting, feeding, and traveling activities throughout the year, they shift their diurnal activity patterns to coincide with the warmest hours during the dry winter months and with the coolest hours during the hot summer months (Strier, 1987a). Muriquis also travel shorter distances during the dry winter months when mature leaves comprise a greater proportion of their diet, and cover longer distances at faster rates during the wet summer months as fruits and flowers become abundant and new leaves increase in availability (Strier, 1987b, 1991a).

Muriqui reproductive activities also exhibit seasonal patterns. Sexual interactions tend to be concentrated during the early-mid rainy months (Strier, 1987c, 1994b) and births tend to be clustered during the peak dry months (Strier, 1991b; Strier and Ziegler, 1994). Nonetheless, substantial variation in the timing of reproduction exists from year-to-year and from female to female in any particular year. Relationships between the timing of reproduction and annual variability in rainfall, differences in female life histories, including reproductive history, age, and length of residency in their reproductive group, and infant sex are explored here. The potential applications of non-invasive fecal steroid monitoring to elucidate these relationships are discussed.

STUDY SUBJECTS AND METHODS

Data on female muriqui reproductive ecology were obtained during an ongoing field study of one of the muriqui groups inhabiting the 800 ha forest at the Estação Biologica de

Table 1. Group size and demographic events, 1982–1994

	N	Males		Females		
		Adults	Immatures	Adults	Immatures	Unknown
June 1982	22	6	4	8	4	
December 1994	52	7	13	17	15	
Births	42		15		26	1
Immigrations	9				9	
Emigrations	12				12	
Deaths/disappearances	9	3	2	1	2	1

Caratinga, located on Fazenda Montes Claros in Minas Gerais. The forest and longterm project have been described in detail elsewhere (e.g., Strier, 1987b, 1991a,b, 1992a; Strier, et al., 1993). All individuals in the study group have been identified by their natural markings and monitored since the onset of the study in June 1982. The group has increased in size from its original 22 members to 52 members as of December 1994 due to births, deaths, and migrations (Table 1).

At the onset of the study period in 1982, six of the 8 adults females present in the group were carrying dependent infants. Although the prior reproductive histories of these females were not known, their subsequent reproductive careers have been followed. Each of these females has reproduced at least once since then, justifying their classification as multiparous females in the following analyses. Both of the other two original females were visibly nulliparous at the onset of the study period. One of these females has copulated routinely over the years without reproducing; the other gave birth to her first infant in July 1983 as a primiparous female and has subsequently reproduced as a multiparous female. Complete reproductive histories have been recorded for this female, all females which have immigrated into the study group, and the sole natal female known to have reproduced in the study group (Table 2). In addition to primiparous and multiparous classifications based on documented reproductive histories, female residency histories were distinguished as immigrant mothers (all of the females which immigrated since 1982), the sole natal mother, and longterm resident mothers (the reproductively successful females which were present in the group in 1982).

All but one of the 42 infants born in the group since June 1982 were single infants. The one set of twins is treated here as one of the total 41 reproductive events. Infant sex was determined for all infants born during the study period except the one twin which disappeared and is presumed to have died at 13 months of age.

Dates of birth were documented by trained observers for 37 of the 41 reproductive events. Two infants were born from August 1984-early September 1985 and two were born from late October 1985-April 1986, the only periods since July 1983 when no observers were accompanying the group. These infants are included only in analyses involving female reproductive histories and infant sex.

Like all New World primates, muriquis lack any obvious visible signs of pregnancy (Dixson, 1983), and consequently, documented births provided the only evidence that successful conceptions had occurred. Extrapolations from the last observed copulations to subsequent births had yielded an estimated gestation length of 6.5–8 months (Strier, 1987c). More recently, the development of fecal steroid assay techniques has made it possible to obtain endocrinological data from wild muriquis without invasive procedures, and to correlate ovarian cycling with sexual behavior (Strier and Ziegler, 1994). Preliminary

Table 2. Muriqui interbirth intervals, 1983–1994 (n = 19), mean = 36.4 ± 4.3 mos; median = 36.0 mos.

Females	Interval (mo)				No. of Births	No. sons	No. daughters
	I	II	III	IV			
BS	??	26	32	36	5*	1	4
NY	??	38	35	35	5*	3	2
DD	??	??	44		5*@	3	2
RO	??	??	38	36	5*	1	4
AR	??	??	39		4*	3	1
MO	34	38	36		4+	0	4
CH	??	36	35		4	1	3
BL	37				2	1	1
SY	??				2*	0	2
BR	47				2	0	2
FE	34				2	0	2
JU	35				2	1	1
HE	—				1	0	1
PL	—				1	0	1
TZ	—				1	1	0
KA	—				1	1	0
IZ	—				1	1	0
LS	0				0*		

??One or both birth dates in interval are unknown.
—Only 1 infant to date.
*Present in study group in 1982.
+Present as nulliparous female in 1982, birth of one set of twins treated here as single event.
@One of sons died within a week of birth.

analyses indicate a direct association between sexual activity and ovarian cycling in muriquis (Strier and Ziegler, 1994). Furthermore, gestation length determined from fecal steroid analyses may be more closely estimated at 7 months (Strier and Ziegler, in prep.). The timing of successful conceptions is therefore extrapolated here by subtracting the estimated 7 month gestation from the known birth month.

Rainfall data presented here were collected on a daily basis by trained assistants from July 1986-June 1994 with a Taylor Rain Gauge situated in an open site near the edge of the forest. Rainfall during this 8-year period was consistent with precipitation patterns recorded in the same location from June 1983-July 1984, and is presumed to correspond to similar patterns in the availability and abundance of fruit, flower, and new leaf resources that have been identified previously at this site (Strier, 1986, 1991a; Ferrari and Strier, 1992).

SEASONALITY

Births

The 37 known birthdates were nonrandomly distributed across the year (Figure 1a). The occurrence of at least one birth in all but three months reflects the absence of a discrete birth season. Nonetheless, births exhibited a distinct peak, with nearly two-thirds (65%) occurring between June and August and 86% occurring between April and September. Rainfall was also strongly seasonal during this study period (Figure 1b). Average annual rainfall over the 8 consecutive years monitored was 1,119 mm (sd=189 mm), with

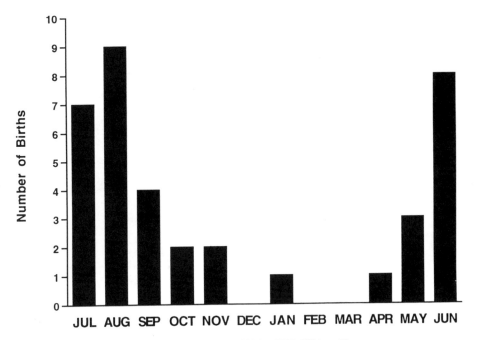

Figure 1a. Distribution of births, 1983–1994, n=37.

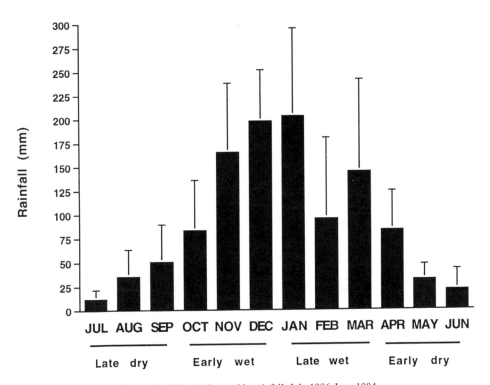

Figure 1b. Mean (sd) monthly rainfall, July 1986-June 1994.

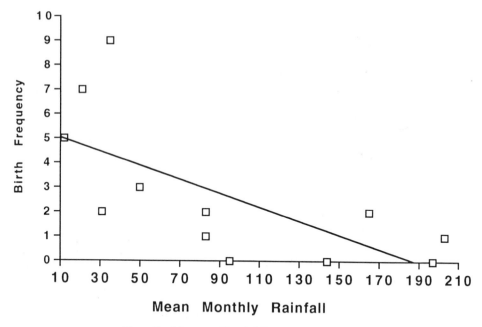

Figure 2a. Mean monthly rainfall and birth frequencies.

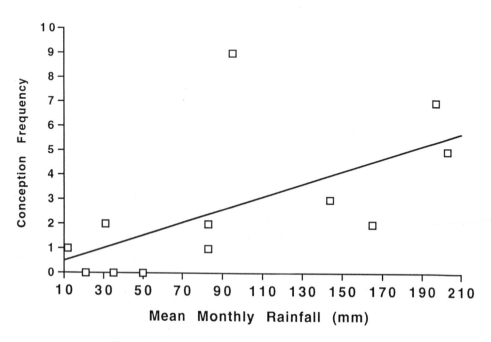

Figure 2b. Mean monthly rainfall and conception frequencies.

nearly 80% of annual precipitation falling during October through March. Monthly rainfall was negatively associated with the timing of births and positively associated with the timing of estimated conceptions (r=0.671, n=12, p<0.01; Figures 2a and b).

Variable rainfall patterns may be responsible for some of the year-to-year variation observed in the timing of conceptions and births (Figure 3). Both years in which early wet season births occurred were also exceptionally wet , with cumulative rainfall exceeding the overall average for the other 6 years by roughly 25% in 1990–91 and nearly 50% in 1991–92 (Figure 3).

Conceptions

Higher than average accumulated rainfall characterized all three of the early dry season conceptions, but only 7 of the 18 wet season conception months (Figure 4). Thus, although high rainfall accumulations appear to be responsible for extending conceptions into the dry season and births into the wet season, they do not predict the precise timing of conceptions during the wet season months in any year.

Positive trends in rainfall levels between a conception month and the preceding month, characterized by an increase of at least 100 mm in the conception month, or roughly 10% of mean annual rainfall, were more common in the early wet months (6/7) than in late wet (5/11) or early dry months (0/3). Of the 21 conception months, 9 were characterized by negative rainfall trends, 7 by positive rainfall trends, and 5 by comparable month-to-month rainfall. However, for each of the 8 years for which rainfall data are available, at least one conception occurred in a month in which rainfall declined by at least 100 mm from the previous month's rainfall. By contrast, conceptions were associated with positive monthly rainfall trends in only 3 of the 8 years.

Resumption of Sexual Activity

The average interval between parturition and the first post-partum copulation observed was 23.5 months (sd=5.5, median=25, n=15 intervals for 12 different females). Post-partum copulations spanned an average of 5.64 months (sd=4.41, median=4, n=11) but females rarely remained sexually active into the dry season (Figure 5). Consequently, females that failed to conceive during the wet season in which they resumed copulating experienced a full year delay in reproduction unless unusually wet conditions, such as those in 1990–91 and 1991–92, prevailed.

Females who conceived during the early wet months and gave birth during the early dry months tended to resume sexual activity sooner (mean =22.25 mos, sd=4.99, median=21.5, n=4) than females which had conceived during the late wet months (mean=23.88, sd=6.88, median=26, n=8) or early dry months (mean=24.33, sd=1.16, median=25, n=3), but these differences were not significant (Kruskall-Wallis test=0.23, df=2, p=0.89). In fact, independent of when their prior parturitions had occurred, most females resumed copulating in the early wet season and conceived in the late wet season (Figures 6a and b). Females resumed copulating when rainfall was either increasing (8/11 females) or equivalent (3/11) to rainfall in the prior months, suggesting that rainfall trends and their effects on food availability and female condition may be the proximate stimuli for the onset or resumption of ovarian cycling. However, the fact that females did not resume copulating while they were still lactating during the rainy seasons following parturition indicates that rainfall and food availability, or other seasonal ecological variables such as photoperiod, are not the only cues that trigger ovulation in muriquis.

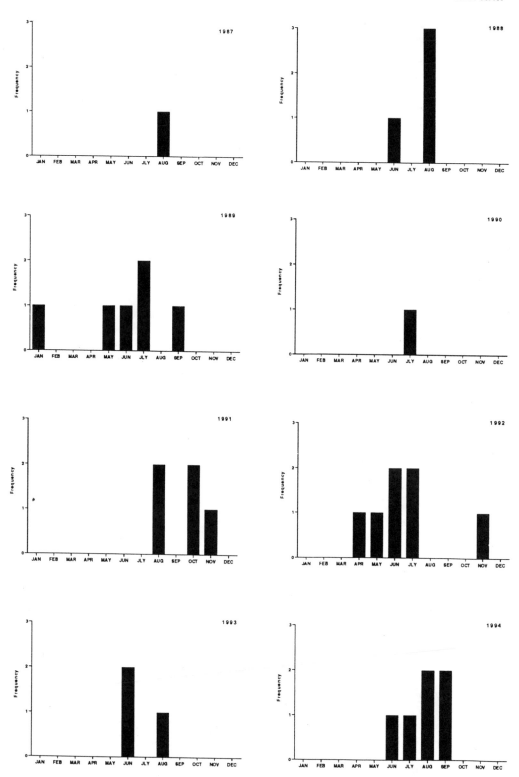

Figure 3. Annual variation in timing of births.

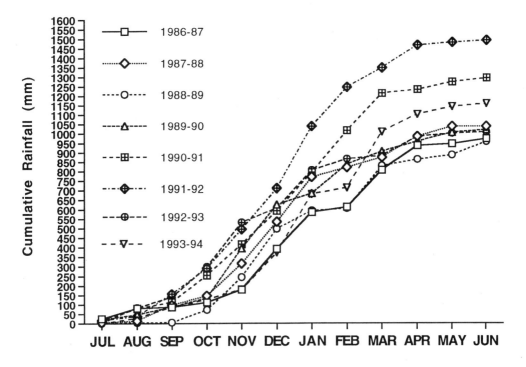

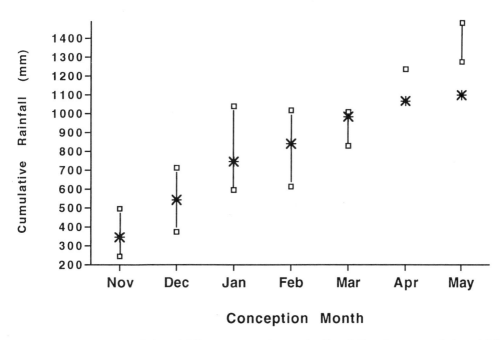

Figure 4. Variation in cumulative rainfall across conception months. Stars indicated mean cumulative rainfall each month (1986–1994); boxes indicate minimum and maximum cumulative rainfall for conceptions in each month.

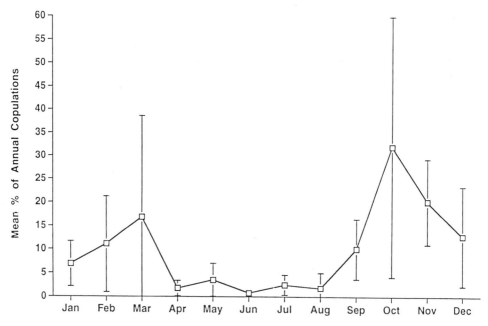

Figure 5. Distribution of copulations, March 1990-March 1994.

Interbirth Intervals

Interbirth intervals, calculated from 19 successive pairs of births involving 11 different females, averaged 36.37 months (sd=4.3, median=36, range=26–47). This was consistent with previous reports based on subsets of these data (Strier, 1991b; Strier and Ziegler, 1994). Interbirth intervals were only weakly correlated with the post-partum resumption of sexual behavior (r=0.57, n=11, p=0.065), and were unrelated to when prior births had occurred (Figure 7a) or to when females resumed post-partum sexual activity (Figure 7b).

The timing of successive births was only partially synchronized among females which reproduced in the same year. For example, in 1988, three multiparous females gave birth in the same month, but only one had resumed sexual activity and ovarian cycling by October 1990 (Strier and Ziegler, 1994). This female, together with a second female which had given birth 2 months earlier than her, reproduced again within weeks of one another in August 1991, while the two other females from the 1988 cohort gave birth 3 and 8 months later. Similarly, of two immigrant females which gave birth to their first infants in the same early wet month in 1991, one resumed sexual activity 2 months earlier, and gave birth to her second infant one month earlier, than the other.

FEMALE SOCIAL AND LIFE HISTORIES

Female Dispersal

Muriqui females routinely disperse from their natal groups as adolescents, prior to any evidence of sexual interest or activity (Strier, 1987c, 1991b). Three of the four female infants that were present in the group when the study began in June 1982 and all nine females born in the group between then and August 1988 had emigrated by November 1994.

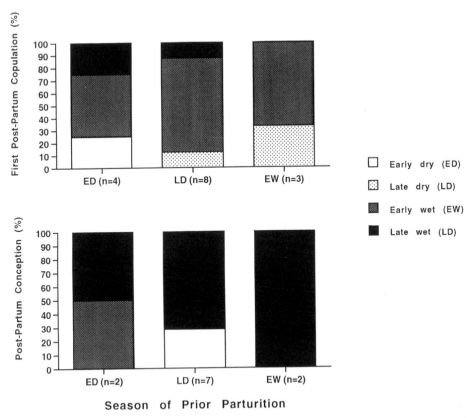

Figure 6. a) Timing of resumption of post-partum sexual activity. **b)** Timing of conceptions, as a percentage of females whose prior parturitions occurred in different seasons.

The mean and median age at emigration for the 7 emigrants whose birthdates were known was 72 months (sd=5.5, range=64–78 months). No other natal females had reached this age or emigrated as of December 1994.

Nine females immigrated from an adjacent group and subsequently reproduced in the study group between June 1982 and December 1994. Although the ages of these immigrants are not known, their resemblance to emigrant natal females in size, overall physical appearance, and visibly nulliparous condition suggests that their ages were similar.

Intergroup encounters appeared to provide the stimulus for both emigrations and immigrations. Emigrant females sometimes returned to their natal group temporarily before subsequently, and apparently permanently, migrating. Females born in the same year often emigrated within a few weeks of one another, and several immigrant females, presumably from the same natal cohort, have joined the study group together or in close succession. Both intergroup encounters and female migrations have occurred throughout the year, but recent observations suggest that there may be a peak in these events during the wet season (Strier, in prep.).

Migration appears to be stressful for adolescent females. All of the females which ultimately immigrated did so only after repeated threats and chases initiated by resident females (Strier, 1991b, 1992a, 1993). Even when immigrant females were tolerated by members of the study group, they tended to remain peripheral for several months, interact-

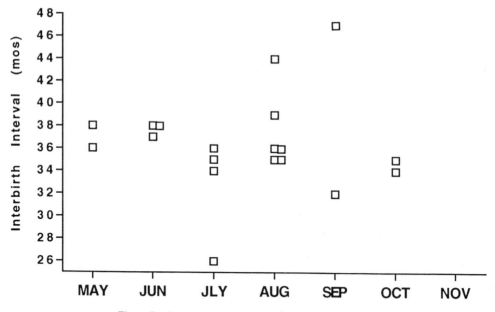

Figure 7a. Interbirth intervals from month of prior parturition.

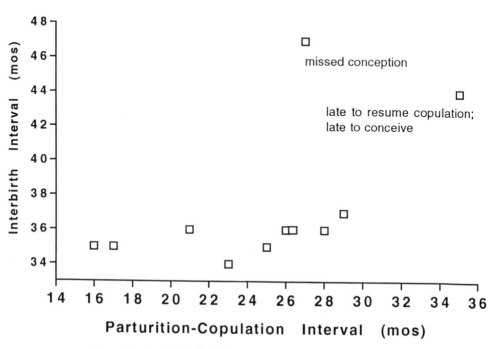

Figure 7b. Interbirth intervals from month when sexual activity resumed.

ing primarily with juveniles and adolescent males rather than adults of either sex. Several of the natal females which subsequently emigrated made repeated returns to their natal group prior to their ultimate transfers. The period when adolescent females are moving between their natal and future reproductive groups may result in restricted access to preferred foods due either to difficulties in locating food in an unfamiliar area or to active exclusion by hostile group residents. Migrating females may also suffer energetic costs associated with greater travel demands and antagonistic or demanding social encounters.

Age at First Reproduction

The only natal female that reproduced in the group did so at roughly 7.5 years of age (Strier, 1991b). This female provides the sole reference for age at first reproduction in wild muriquis. Immigrant females exhibited variable latencies before becoming sexually active, and the interval between immigration and first reproduction ranged from approximately 1–8 years (Table 3). Assuming these females immigrated at the same average age that natal females emigrated, they would have been older (mean=9 years, sd=2, median=8.5, n=9) than the natal female at first reproduction.

It is difficult, however, to distinguish between the effects of age and the costs of migration on female reproduction. It is possible that energetic or social stresses during the transitional migration period result in delays in reproduction. However, the natal female experienced an unusually long delay between the births of her first and second offspring (47 mos) compared to immigrant females giving birth to their second infant (mean and median=35 months, sd=1.53, n=3). Therefore, although migration may delay the age at first reproduction, an important life history parameter affecting fitness, it may have minimal, if any, effects on a female's subsequent reproductive career if migrant females can compensate for this delay through shorter interbirth intervals.

Effects of Residency Histories

The indication that immigrant females were older at first reproduction than the one natal female which reproduced in the group (see above) is consistent with the hypothesis that migration involves energetic and social stresses with at least short-term reproductive consequences. It is difficult, nonetheless, to distinguish between the effects of age and length of residency on the timing of reproduction. Seven of the eight immigrant females whose dates of first reproductions were known gave birth within 1 month of the multiparous mothers reproducing in the same years, but only one reproduced in the same month. Two of the four early dry month births by primiparous females involved immigrants with the longest periods of residencies in the group, but these females were also likely to have been older at their first reproductions than other primiparous females.

Young female muriquis may experience irregular ovarian cycles like other young primates. Alternatively, pheromonal cues known to synchronize female ovarian cycling in other species (McClintock, 1983) may require closer or more frequent associations than those that occur between resident and newly-immigrant female muriquis. For example, the one natal female that reproduced in the group did so in September instead of May-July like the three older females who gave birth the same year. The fact that she had associated with two of these females throughout her life and with one of these females, which immigrated in 1984, since she was 2 years old, suggests that longterm associations may not be sufficient to promote reproductive synchrony.

Table 3. Reproductive histories of primiparous females

Female	History	Years in group prior to parturition	1st Parturition: Estimated age	Date	Same mo	w/in 1 mo
MO	Present 6/82, nullip	?	?	July 83	0	0
CH	Immigrated 7/83-12/83, nullip	1-2	7-8	8/84-9/85	?	?
BL	Immigrated 1/84-6/84, nullip	4.5-5	10.5-11	6/89	--	+M
KA	Immigrated 7/86-12/86, nullip	7.5-8	13.5-14	6/94	--	+M
BR	Natal infant in 6/82	7.5	7.5	9/89	--	--
FE	Immigrated 10/89-1/90, nullip	1.7-2	7.7-8	10/91	+P	+M
JU	Immigrated 10/89-1/90, nullip	1.7-2	7.7-8	10/91	+P	+M
HE	Immigrated 10/89-1/90, nullip	2.5	8.5	6/92	+P	+M
PA	Immigrated 10/89-1/90, nullip	2.5	8.5	6/92	+P	+M
TZ	Immigrated 10/89-1/90, nullip	3	9	11/92	--	--
IZ	Immigrated 1/92, nullip	2.75	8.75	9/94	+M	+M

+P indicates parturition in same month as a primiparous female.
+M indicates parturition in same month or within 1 month as a multiparous female.

Indeed, the fact that the first reproductions of immigrant females were also not synchronized with multiparous females implies that association patterns may override the effects of ecological variables on reproduction. Of the 8 months in which at least two births occurred in the same month, 5 involved multiparous females and 2 involved primiparous females. Only one of the 8 cohort months involved females with mixed reproductive histories, suggesting that reproductive synchrony tends to be segregated by female age and social histories.

Reproductive Experience

Different ecological pressures appeared to be operating on females with different reproductive histories. Multiparous females tended to give birth more often during the late dry months (63%), perhaps because of the need to replenish their energy stores after the long period of infant dependency. By contrast, births by primiparous females which had not yet been subjected to the energetic costs of reproduction were evenly distributed across the early dry, late dry, and early wet months (Figure 8).

Experienced mothers may be more efficient at weaning their offspring than first time mothers, but the timing of weaning did not appear to affect subsequent reproductive opportunities. Although multiparous females resumed post-partum sexual activity slightly sooner (median=23 months, range=16–35, n=10) than primiparous females (median=25 months, range=22–29, n=5), the differences were not significant (Mann Whitney U=17.5, p>0.05, 2-tail). Similarly, interbirth intervals for primiparous females giving birth to their second infant (median=35.0 months, range=34–47, n=5) were similar to those of more experienced multiparous females (median=36 months, range=26–44, n=14; Mann Whitney U=30.5, p>0.05 2-tail). Multiparous females which resumed post-partum cycling sooner tended to cycle over a more extended period of time before conceiving than primiparous females.

INFANT SEX

Daughters (63.4%) were more common than sons (36.6%) among the 41 infants whose sex was confirmed. This sex ratio is comparable to that among the 6 infants which

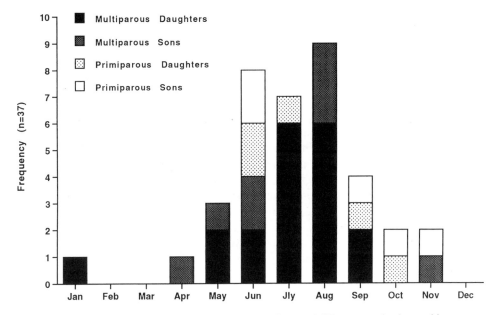

Figure 8. Distribution of son and daughter births by femaes of different reproductive condtions.

were present in the group when the study began (66.7% females). Despite such a marked female bias, infant sex did not appear to be related to the timing of reproductive events in muriquis. There were no differences in the frequency of sons or daughters born in the early versus late dry months (Fisher exact 1-tail test=0.08) or in the dry versus early wet months (Fisher exact 1-tail test=0.12) over the years (Figure 8), and the conceptions of 12 males and 18 females were equally divided between months with positive and negative trends in rainfall.

There were no sex differences among the 17 infants born in the same months or the 15 infants born in different months during years in which at least two births occurred (Fisher exact 2-tailed test=0.71). However, multiparous females produced twice as many daughters (n=20) as sons (n=10), whereas primiparous females produced similar numbers of daughters (n=6) and sons (n=5). Although the sample size is too small to reach statistical significance (Fisher exact 2-tailed test=0.72), female reproductive histories may be related to the sex of infants produced. Infant sex has been associated with female age and reproductive condition in other primates (e.g., Altmann, et al., 1978; Paul and Thommen, 1984), although the relationship may be different in muriquis because of other differences in their social organization (see below).

Behavioral observations on the same study subjects over a 9-month period in 1989–90 indicate that weaning in muriquis tends to occur 2–4 months earlier for sons than daughters (Rímoli, 1992). However, the resumption of post-partum sexual activity was not significantly different for mothers of sons (median=25.0 monhs, range=17–29, n=5) or daughters (median=24.5 months, range=16–35, n=10; Mann-Whitney U=24.5, p>0.05 1-tail). Similarly, interbirth intervals following the births of sons (median=36 months, range 35–39, n=7) and daughters (median=36.0 months, range=26–47, n=12) did not differ (Mann-Whitney U=37, p>0.05 1-tail). The resumption of sexual activity and the timing of subsequent conceptions and births therefore appear to be influenced more strongly by seasonal factors affecting mothers than by sex differences in weaning.

DISCUSSION

Ecological Correlates of Reproduction

The distribution of muriqui conceptions and births can be attributed to general patterns of rainfall and their effects on the availability of preferred foods and presumably female nutritional condition. During a year of typical rainfall at this site (June 1983-July 1984), monthly rainfall was significantly correlated with both the availability of new leaves and the proportion of feeding time devoted to new leaves (Strier, 1991a), suggesting that these food items may be particularly important for muriqui reproduction. In the present study, rainfall, and by inference, the availability and consumption of new leaves, was positively associated with annual conceptions and negatively associated with births. Unusually heavy and prolonged rainfall in two consecutive years extended the conception and birth peaks in those years.

Females that failed to resume copulating or conceive during the first post-weaning rainy season tended to cease sexual activity and thus conception opportunities until the following year. Nonetheless, females that resumed post-partum copulating sooner did not necessarily benefit by shorter interbirth intervals.

The fact that all but one conception occurred after successive months of increasing rainfall independent of when experienced mothers resumed post-partum sexual activity suggests that females may need to attain a minimum threshold of energy or nutritional reserves before conception is possible. At the Centro de Primatologia de Rio de Janeiro, the only captive breeding facility for muriquis, mothers resume copulating and may conceive more than a full year sooner than the wild population described here (Coimbra-Filho, et al., 1993). The nutritious diet, including vitamin supplements, provided to the captive muriquis, together with the reduced energetic expenditures inherent in any captive setting, may permit more rapid recovery of energy reserves.

The dry season birth peak observed in the present study contrasts with the early wet season birth peak described in *Ateles,* a closely-related taxa which resembles muriquis in their age of first reproduction (7–8 years) and their long interbirth intervals (32–36 months; Symington, 1987; Chapman and Chapman, 1990). The fact that muriquis are more similar to *Alouatta* in their avoidance of dry season conceptions may be related to the more folivorous diet and cohesive groups that distinguish both muriquis and *Alouatta* from *Ateles* (Strier, 1992b).

Muriquis may benefit by giving birth during the dry season for a number of reasons (Strier, 1991b). First, because the dry season is not typically a period of high fruit productivity (Strier, 1986, 1991a), a dietary emphasis on localized leaves at this time results in significantly shorter day ranges (Strier, 1987b). Both mothers and newborn infants may benefit by the reduced travel demands during the first awkward months post-partum when mothers must help to secure their uncoordinated infants. Second, muriqui infants are highly dependent on their mothers for food and transport during the first 6 months of life. Young infants rarely travel more than a few meters on their own, and are in physical contact with their mothers at least 50% of the time (Rímoli, 1992). Thus, the energetic costs of carrying growing infants while also lactating are likely to increase for females during the early post-partum months (see Altmann, 1980). Giving birth during the dry winter months insures that mothers enter their most energetically stressful periods of infant care, when they are lactating and still transporting their infants, during the early part of the rainy summer months when high quality foods, including new leaves, are most abundant (Strier, 1991a), and when large patches of energy-rich fruit and flower sources permit muriquis to camp-out for days at a time (Strier, 1987b). Furthermore, because weaning does not begin until roughly 18 months post-partum (Strier, 1991b; Rimoli, 1992), infants born in the dry season one year would reach weaning age during the following year's rainy season when high quality preferred foods, which may also be important weaning foods, are most readily available.

The fact that the only two infant mortalities (other than the twin) involved the only infants born prior to May in any year provides indirect support for the benefits of giving birth in the dry season. However, both of these infant mortalities may have been exceptional for different reasons. The January 1989 birth involved a visibly aged mother who had not been associated with an infant since the daughter she was carrying in 1982 emigrated. This mother spent much of her time after her January parturition alone with her daughter, separated from the rest of the group, and has not reproduced since her daughter disappeared and was presumed to have died at 2.8 years of age.

The April 1992 birth involved an experienced mother whose prior mating records suggest that she may have given birth at least one month prematurely. She abandoned her visibly dead son by the fifth day post-partum, and resumed copulating two months later despite the low frequency with which sexual activity has been observed during the dry season (Strier, 1987c, 1994b). Although this female copulated over several consecutive months, she did not give birth again until the following June. The timing of conception thus appears to be strongly influenced by seasonal ecological variables rather than female ovarian cycling *per se* (see below).

By contrast to the late wet season births, all four of the infants born in the early wet months of October 1991 and November 1991 and 1992 have survived to their present ages of 3 1/2 and 4 1/2 years of age, respectively. Both females which gave birth in October 1991 subsequently reproduced again in August and September 1994. The two females which gave birth in November of 1991 and 1992 were still copulating, apparently without conceiving, through December 1994.

Despite the potential benefits that a dry season birth peak may confer on muriqui mothers and infants, it is difficult to assess the degree to which rainfall at the time of conception provides an accurate indication of rainfall and food conditions at the time of birth 7 months later, during the first 4–6 months of lactation, or at weaning the following year without continuous data on forest phenology and food availability. Nonetheless, the low incidence of sexual activity and cycling during the dry season months suggests that muriqui behavior and physiology, like that of *Alouatta*, may be sensitive to decreasing trends in rainfall that trigger the arrival of the dry season (Crockett and Rudran, 1987).

Several possible mechanisms could be responsible for the apparent shut-down in female sexual activity, and presumably ovulatory cycling (see Strier and Ziegler, 1994) at these times. These mechanisms include shorter day lengths during the dry winter months, declining female nutritional condition, or the inclusion or exclusion of specific dietary components that enhance or inhibit fertility. The one female which conceived in June and gave birth in January was unusual in her solitary behavior and low reproductive success during the study period (see above). Nonetheless, both she and one other female, who lost her infant within a week post-partum and resumed copulating but did not conceive during the dry season, provide indirect evidence that day length or diet *per se* do not inhibit sexual activity entirely.

Dietary changes in the dry season may cause females to lose weight that they have acquired during the prior rainy season, leading to a cessation in ovulatory activity until food availability improves during the following rainy season. Although no weight data are available for muriquis at this site, weights obtained from muriquis at a nearby forest during the late dry season were lower than those previously described for this species (Lemos de Sá and Glander, 1993). Further comparative data are needed to determine whether these differences reflect inter-population variability or real patterns of seasonal weight loss that may ultimately affect female reproduction. Detailed analyses of female diets and behavior throughout their reproductive cycles will provide greater insights into the dietary correlates of muriqui reproduction (Nogueira, in prep.).

The potential fertility-regulating foods in muriqui diets should not be discounted. Muriquis exploit many species belonging to the Leguminosae, the dominant plant family at this forest, (Hatton, et al., 1983; Strier, 1991a). Legumes are well-known for their phytosteroids and effects on mammalian reproduction, but whether they are responsible for regulating muriqui fertility will require results from ongoing biochemical analyses of these and other muriqui foods year round.

Social and Demographic Correlates of Reproduction

Female Histories and Timing of Reproduction. While rainfall and its effect on food availability and muriqui diets can explain some of the annual reproductive patterns described here, female histories may account for some of the variation in the timing of conceptions and births within any particular year. Prime-aged multiparous and older primiparous females consistently conceived in the wet season and gave birth in the dry season, whereas both young primiparous females and older multiparous females tended to conceive and reproduce at the extreme tails of the yearly conception and birth peaks.

The segregation of birth cohorts by female reproductive experience suggests that close associations between individual females may promote reproductive synchrony among them. Multiparous females with longterm associations in the group, as well as young primiparous females which immigrated together and maintained close associations (Nogueira, in prep.), tended to reproduce as cohorts. However, although both older multiparous females and the one natal female which reproduced in the group had similar opportunities to associate, they tended to reproduce asynchronously from other females in those years. This tendency, together with the reproductive patterns of other young immigrants, suggests that age-related irregularities in female ovarian cycling may confound the effects of social histories on the timing of reproduction. Analyses of fecal steroid data from females with different ages and reproductive histories will provide greater insights into the age-dependent correlates of muriqui reproduction.

Social Variables. The distribution of reproductive events within any particular year implies that social variables favoring reproductive synchrony in muriquis may be weak. Evidence of allomothering was rare throughout most of the study period described here, although recently the incidence of infant carrying and even suckling by nonmothers has been observed with increasing frequency (Rímoli, in prep.). Similarly, the paucity of muriqui predators at this site makes it unlikely that females would benefit by shared vigilance, and is consistent with absence of tight reproductive synchrony. Furthermore, although infants may benefit by opportunities to establish social relationships with future allies (see below), there is currently no evidence that infants born in the same month develop stronger associations than infants born in the same year.

Infant Sex. The strongly female-biased infant sex ratio may have swamped any relationship between the timing of son versus daughter conceptions and births. The fact that mothers tend to wean sons 2–4 months earlier than daughters (Rímoli, 1992) may account for the faster resumption of post-partum sexual activity after the births of sons. However, the absence of a relationship between infant sex and subsequent interbirth intervals implies that ecological factors override any effects that differential investment in offspring may have on the timing of subsequent conceptions.

Muriquis appear to differ markedly from their close *Ateles* relatives, where significantly longer interbirth intervals following the birth of sons have been reported (Symington, 1987; but see Chapman and Chapman, 1990). However, such differential investment in sons and daughters may be an artifact of the inverse relationship Symington (1987) observed between mothers' ranks and the production of sons in *Ateles*. If female rank is age-dependent in *Ateles*, as it is in *Alouatta* (Glander, 1980; Jones, 1980), then it is possible that infant sex and investment strategies could reflect differences in mother age, and presumably experience.

Opportunities to evaluate the effects of female rank on infant sex and subsequent reproductive patterns in muriquis are limited because no evidence of hierarchical relationships among female muriquis have yet been detected (Strier, 1990). In the present study, multiparous females produced roughly twice as many daughters as sons, whereas primiparous females gave birth equally to sons and daughters. While this finding may be a consequence of the limited sample size available, it is consonant, nonetheless, with predictions based on muriqui social organization. Immigrant females that produce a son early in their reproductive careers may benefit by the continued (and apparently life-long) presence of a related ally in the group. By about 4 years of age, natal males begin to associate regularly with other adult males, and by the time they reach adulthood at 7 years of age they are fully integrated into male society (Strier, 1993). Adult sons frequently embrace their mothers, but only one instance of a known mother-son copulation has been observed over the years (Strier, 1994b). Producing too many sons might limit the mating options available to females and, if produced in close succession, could interfer with their reproductive careers.

Although muriqui infants of both sexes play together, they begin to exhibit sex differences in behavior by at least 4 years of age (Strier, 1993). Young males increase their associations with one another and with older males, while young females become increasingly more solitary until they disperse as adolescents. Females born in the same year that disperse together or in close succession may benefit by having familiar associates in their new groups to ease the stress of immigration. Males appear to remain together in their natal groups for life, but we do not yet know whether males born in the same year maintain stronger associations with one another into adulthood.

Conclusion

Muruiqui reproduction is evidently regulated by multiple interacting ecological, social, and physiological variables. Yet, despite more than 12 years of data on muriqui reproductive ecology, many outstanding questions about the interacting effects of these variables persist. Delayed maturation and long interbirth intervals in muriquis compared to most other New World primates result in relatively small sample sizes that prohibit definitive statistical analyses and conclusive assertions. Furthermore, many questions, including those pertaining to the relationship between seasonality in sexual behavior and ovarian cycling and to distinctions between successful and unsuccessful conceptions and pregnancies in inexperienced and experienced mothers, will require the corresponding endocrinological data that non-invasive fecal steroid assays can provide. Continued monitoring of these variables and their effects on muriquis will extend our understanding of comparative primate reproductive ecology, and provide much needed insights into the conservation of muriquis and other endangered species.

SUMMARY

Reproductive patterns examined in one population of muriqui monkeys (*Brachyteles arachnoides*) at the Estação Biologica de Caratinga, Fazenda Montes Claros, Minas Gerais, Brazil, were related to seasonal rainfall cycles and individual differences in female life histories. Monthly rainfall was negatively related to the timing of documented births and positively related to the timing of estimated conceptions. Independent of when births occurred, most females resumed post-partum sexual activity during the early wet months and conceived during the late wet months. Multiparous females resumed post-partum sexual activity slightly earlier than primiparous females, but subsequent interbirth intervals did not vary with female reproductive histories. Multiparous females gave birth to a greater proportion of daughters than sons, while infant sex ratios of primiparous females were not female-biased. Infant sex was not, however, related to subsequent interbirth intervals. Mother age and length of associations with other females accounted for some of the annual variation in the timing of reproductive events. In particular, reproduction tended to be synchronized among both multiparous and primiparous females with long-term associations with one another.

ACKNOWLEDGMENTS

Permission to conduct research in Brazil was provided by CNPq, with sponsorship by Professors Celio Valle, César Ades, and Gustavo de Fonseca. The research was supported by NSF grants BNS 8305322, BNS 8619442, and BNS 8959298, the Fulbright Foundation, grant no. 213 from the Joseph Henry Fund of NAS, Sigma Xi, the L.S.B. Leakey Foundation, the World Wildlife Fund, the Seacon Fund of the Chicago Zoological Society, the Liz Claiborne and Art Ortenberg Foundation, and the Graduate School of the University of Wisconsin-Madison. E.Veado, F. Mendes, J. Rímoli, A.O. Rímoli, F. Neri, P. Coutinho, A. Carvalho, L. Oliveira, C. Nogueira, S. Neto, W. Teixeira, R. Printes, and M.A. Maciel contributed to the long-term demographic data records reported here. T.E. Ziegler, P. Garber, and an anonymous reviewer provided valuable comments on an earlier version of this manuscript. I am grateful to M. Norconk and A. Rosenberger for organizing the symposium in honor of Warren Kinzey at which a version of this paper was presented, and to the late Warren Kinzey for his pioneering contributions to the study of New World primates.

REFERENCES

Altmann, J. 1980, *Baboon Mothers and Infants*. Cambridge: Harvard University Press.

Altmann, J., Altmann, S.A., and Hausfater, G. 1978, Primate infant's effects on mother's future reproduction. *Science* 201:1028–1030.

Bercovitch, F.B. and Harding, R.S.O. 1993, Annual birth patterns of savanna baboons (*Papio cynocephalus anubis*) over a ten-year period at Gilgil, Kenya. *Folia Primatol.* 61:115–122.

Boinski, S. 1987, Birth synchrony in squirrel monkeys (*Saimiri oerstedi*). *Behav. Ecol. Sociobiol.* 21:393–400.

Chapman, C.A. and Chapman, L.J. 1990, Reproductive biology of captive and free-ranging spider monkeys. *Zoo Biology* 9:1–9.

Clark, A.S., Harvey, N.C., and Lindburg, D.G. 1992, Reproductive coordination in a nonseasonally breeding primate species, *Macaca silenus. Ethology* 91:46–58.

Coimbra-Filho, A.F., Pissinatti, A. and Rylands, A.B. 1993, Breeding muriquis, *Brachyteles arachnoides*, in captivity: the experience of the Rio de Janeiro Primate Centre (CPRJ-FEEMA). *Dudo, J. Wildl. Preserv. Trusts* 29:66–77.

Crockett, C.M. and Rudran, R. 1987, Red howler monkey birth data I: seasonal variation. *Am. J. Primatol.* 13:347–368.

Dixson, A.F. 1983, Observations on the evolution and behavioral significance of "sexual skin" in female primates. *Advances in the Study of Behaviour* 13:63–106.

Ferrari, S.F. and Strier, K.B. 1992, Exploitation of *Mabea fistulifera* nectar by marmosets (*Callithrix flaviceps*) and muriquis (*Brachyteles arachnoides*). *J. Trop. Ecol.* 8:225–239.

Glander, K.E. 1980, Reproduction and population growth in free-ranging mantled howling monkeys. *Am. J. Phys. Anthropol.* 53:25–36.

Hatton, J.C., Smart, N.O.E., and Thomson, K. 1983, An Ecological Study of the Fazenda Montes Claros Forest, Minas Gerais, Brazil, Interim report, Department of Botany and Microbiology, University College London.

Jones, C.B. 1980, The functions of status in the mantled howler monkey, *Alouatta palliata* Gray: Intraspecific competition for group membership in a folivorous neotropical primate. *Primates* 21:389–405.

Lemos de Sá, R.M. and Glander, K.E. 1993, Capture techniques and morphometrics for the woolly spider monkey, or muriqui (*Brachyteles arachnoides*, E. Geoffroy 1806). *Am. J. Primatol.* 29:145–153.

Lindburg, D.G. 1987, Seasonality of reproduction in primates. In: *Comparative Primate Biology, Volume 2B: Behavior, Cognition, and Motivation*, G. Mitchell and J. Erwin (eds.). New York: Alan R. Liss, pp. 167–218.

McClintock, M.K. 1983, Pheromonal regulation of the ovarian cycle: enhancement, suppression, and synchrony. In: *Pheromones and Reproduction in Mammals*, J.G. Vandenbergh (ed.). New York: Academic Press, pp. 113–149.

Milton, K. 1984, Habitat, diet, and activity patterns of free-ranging woolly spider monkeys (*Brachyteles arachnoides* E. Geoffroy 1806). *Int. J. Primatol.* 5:491- 514.

Nicolson, N.A. 1987, Infants, mothers, and other females. In: *Primate Societies*, B.B. Smuts, D.L. Cheney, R.M. Seyfarth, R.W. Wrangham, and T.T. Struhsaker (eds.). Chicago: University of Chicago Press, pp. 330–342.

Paul, A. and Thommen, D. 1984, Timing of birth, female reproductive success and infant sex ratio in semitree-ranging Barbary macaques (*Macaca sylvanus*). *Folia Primatol.* 42:2–16.

Pusey, A.E. and Packer, C. 1987, Dispersal and philopatry. In: *Primate Societies*, B.B. Smuts, D.L. Cheney, R.M. Seyfarth, R.W. Wrangham, and T.T. Struhsaker (eds.). Chicago: University of Chicago Press, pp. 250–266.

Rímoli, A.O. 1992, *O Filhote Muriqui (Brachyteles arachnoides): mm Estudo do Desenvolvimento da Independência*. Unpublished MA Thesis, Universidade de São Paulo, São Paulo, Brazil.

Rowell, T.E. and Richards, S.M. 1979, Reproductive strategies of some African monkeys. *J. of Mammal.* 60:58–69.

Strier, K.B. 1986, *The Behavior and Ecology of the Woolly Spider Monkey, or Muriqui (Brachyteles arachnoides, E. Geoffroy 1806)*. Unpublished Ph.D. Thesis, Harvard University, Cambridge, MA.

Strier, K.B. 1987a, Activity budgets of woolly spider monkeys, or muriquis (*Brachyteles arachnoides*). *Am. J. Primatol.* 13: 385–395.

Strier, K.B., 1987b, Ranging behavior of woolly spider monkeys. *Int. J. Primatol.* 8: 575–591.

Strier, K.B. 1987c, Reprodução de *Brachyteles arachnoides*. In: *A Primatologia no Brasil*, M. Thiago de Mello (ed.). Brasilia: Sociedade Brasileira de Primatologia, pp. 163–175.

Strier, K.B. 1990, New World primates, new frontiers: Insights from the wooly spider monkey, or muriqui (*Brachyteles arachnnoides*). *Int. J. Primatol.* 11: 7–19.

Strier, K.B. 1991a, Diet in one group of woolly spider monkeys, or muriquis (*Brachyteles arachnoides*). *Am. J. Primatol.* 23: 113–126.

Strier, K.B. 1991b, Demography and conservation in an endangered primate, *Brachyteles arachnoides. Conservation Biology* 5: 214–218.

Strier, K.B. 1992a, *Faces in the Forest: the Endangered Muriqui Monkeys of Brazil*. New York: Oxford University Press.

Strier, K.B. 1992b, Atelinae adaptations: behavioral strategies and ecological constraints. *Am. J. Phys. Anthropol.* 88: 515–524.

Strier, K.B. 1993, Growing up in a patrifocal society: sex differences in the spatial relations of immature muriquis (*Brachyteles arachnoides*). In: *Juveniles: Comparative Socioecology*, M.E. Pereira and L.A. Fairbanks (eds.). New York: Oxford University Press, pp. 138–147.

Strier, K.B. 1994a, Myth of the typical primate. *Yrbk Phys. Anthropol.* 37:233–271.

Strier, K.B. 1994b, Female mate choice in muriqui monkeys. *Am. J. Primatol.* 33:242.

Strier, K.B. and Ziegler, T.E. 1994, Insights into ovarian function in wild muriqui monkeys (*Brachyteles arachnoides*). *Am. J. Primatol.* 32:31–40.

Strier, K.B., Mendes, F.D.C., Rimoli, J., and Rimoli, A.O. 1993, Demography and social structure in one group of muriquis (*Brachyteles arachnoides*). *Int. J. Primatol.* 14:513–526.

Symington, M.M. 1987, Sex ratio and maternal rank in wild spider monkeys: when daughters disperse. *Behav. Ecol. Sociobiol.* 20:333–335.

Torres de Assumpção, C. 1983, Ecological and behavioral information on *Brachyteles arachnoides*. *Primates* 24:584–593.

Walters, J.R. 1987, Transition to adulthood. In: *Primate Societies*, B.B. Smuts, D.L. Cheney, R.M. Seyfarth, R.W. Wrangham, and T.T. Struhsaker (eds.). Chicago: University of Chicago Press, pp. 358–369.

26

TRANSLATIONS OF CHAPTER SUMMARIES

Laura Cancino and Anthony B. Rylands

CLASSIFICAÇÃO E FILOGENIA DOS PLATIRRÍNEOS: CONTRIBUÇÃO A UMA SÍNTESE DE MOLÉCULAS E MORFOLOGIA

H. Schneider e A.L. Rosenberger

Análise cladística das seqüências dos genes Epsilon-globina e IRBP fornece informação complementar importante para um esboço das principais linhas da filogenia dos macacos do Novo Mundo. As abordagens morfológicas e de genética molecular são razoavelmente consistentes com as evidências disponíveis através do registro fóssil, significando que as formas modernas fornecem uma boa base para o desenvolvimento de uma classificação dos platirríneos, e que o entendimento das relações entre fósseis podem ser facilitados com a inclusão dos gêneros viventes nas análises. Os estudos moleculares e morfológicas fortalecem a idéia de três grandes grupos modernos, possivelmente divergindo num intervalo de tempo relativamente curto. Considerando as discordâncias nos estudos da sistemática de platirríneos nas últimas décadas - a correta localização filogenética de *Cebus*, *Saimiri*, *Aotus* e *Callicebus* - a combinação das evidências colocam *Callicebus* definitivamente como parente dos pitecíneos. Elas reforçam também a ligação entre *Saimiri* e os calitriquíneos, o elo entre *Cebus* e *Saimiri*, e sua associação com calitriquíneos como uma linhagem monofilética do grupo dos "cebídeos". Os dados de DNA divergem, porém, com a colocação de *Aotus* como uma linhagem basal desse agrupamento, um achado inconsistente com as evidências morfológicas. A análise de DNA também aponta à necessidade de uma reconsideração da taxonomia do gênero *Callitrhix*, que talvez não seja monofilética. Os dados confirmam parcialmente o padrão de ramificação do clade dos atelídeos, posicionando *Alouatta* como a linhagem mais velha. Problemas que permanecem dentro dos calitriquíneos e atelíneos incluem: 1)os afinidades precisas entre os atelíneos, *Lagothrix*, *Ateles* e *Brachyteles*; e 2) a seqüência de ramificação dentre os calitriquíneos, i.e., *Callithrix/Cebuella*, *Leontopithecus*, *Saguinus* e *Callimico*.

PRIMATAS DA MATA ATLÂNTICA: ORIGEM, DISTRIBUIÇÕES, ENDEMISMO E COMUNIDADES

Anthony B. Rylands, Gustavo A.B. da Fonseca, Yuri L.R. Leite e Russell A. Mittermeier

Vinte e quatro espécies e subespécies de primatas ocorrem na Mata Atlântica brasileira. Vinte dessas espécies e subespécies são endêmicas. Sequindo uma breve descrição das formações vegetais da região, descrevemos as distribuições dessas taxa e discutimos a composição e estrutura das comunidades de primatas e a evidência disponível para explicações sobre suas origens durante or Terciário. Enquanto *Cebus apella* mostra uma diferenciação somente á nível de subespécie, *Callithrix*, *Callicebus* e *Alouatta* são representados por espécies distinctas, e *Leontopithecus* e *Brachyteles* são gêneros endêmicos. Razões para isto são discutidas, e na parte final abodamos a situação atual em termos da conservação dos primatas da Mata Atlântica, 18 sendo considerados ameaçados de extinção.

POPULACIONAIS DE PRIMATES DO LESTE DA AMAZÔNIA

Stephen Ferrari e Maria Aparecida Lopes

Apesar de englobar pelo menos treze taxa subespecíficos, as comunidades de primatas das florestas de terra firme localizadas ao leste do rio Xingú são entre as menos diversas da Amazônia. Isto relete o gradiente leste-oeste de diversidade comum a muitos outros grupos de fauna e flora. Densidades populacionais e biomassa também são relativamente baixas em comparação com as comunidades da Amazônia ocidental, que pode ser devido, pelo menos parcialmente, a diferenças na qualidade de habitat. A diversidade também cai em habitats marginais como as formações de cerrado, manquezais e especialmente florestas perturbadas ou secundárias, mesmo que a densidade populacional de algumas espécies possa aumentar. Quatro espécies diúrnas predominam na maioria da região: *Alouatta belzebul*, *Cebus apella*, *Chiropotes satanas* and *Saguinus midas*. A distribuiçao geográfica e densidade populacional das outras espécies, em particular *Callithrix argentata* e *Cebus kaapori*, parecem ser limitadas por fatores como a qualidade do habitat e a competição interespecífica. Juntos, o desmatamento extensivo e descontrolado, a extração de madeira e a caça constituem um problema de conservação significativo em toda a região, mas especialmente ao leste do rio Tocantins. Este problema é exacerbado pela atual falta de dados sobre a ecologia dos primatas da região.

PRIMATES DEL ESCUDO DE LA GUAYANA: VENEZUELA Y LAS GUIANAS

Marilyn A. Norconk, Robert Sussman y Jane Phillips-Conroy

La distribución de primates modernos y el raro endemismo de primates en los Neotrópicos norteños sugiere que un amplio intercambio de poblaciones ha sido posible desde el Brasil amazónico a pesar de la extensión del Escudo de la Guayana. Aparentemente, dos corredores han permitido a los primates el colonizar la región: en dirección al Norte desde la Amazonía Este hacia las Guianas, y en dirección al Norte desde la Amazonía Oeste (Brasil y Colombia) hacia el Suroeste de Venezuela (Eisenberg 1989). La distribución contínua de ocho especies de primates en las Guianas es interrumpida en el Oeste de la Guyana por varias posibles barreras: el río Essequibo, hábitats de Sabanas, cadenas de montañas en la frontera Brasil-Guyana y el delta del Orinoco en la frontera norte entre Guyana y Venezuela. La barrera más obvia para la dispersión de primates en el Este de Venezuela es la Gran Sabana, una vasta sabana salpicada con mesetas antiguas. Los bosques son discontínuos y están limitados a las laderas de los tepuis y a los bosques de galería a lo largo de arroyos perennes.

Los unicos primates endémicos de la región, *Aotus lemurinus* y *Ateles belzebuth hyridusque* ingresaron al Noroeste de Venezuela a través de un corredor andino, tienen una distribución muy limitada y son muy vulnerables ya que sus hábitats continúan siendo amenazados. Las poblaciones de primates probablemente han sido protegidas por las bajas densidades poblacionales humanas, pero la caza, la minería y las explotación maderera en Venezuela, así como en Guyana y Surinam y la expansión de poblaciones humanas en el interior de las Guianas aparentan constituir ahora un amenaza seria, particularmente para los atelinos y pitheciinos.

LA OTRA FACETA DE LA ALIMENTACIÓN GOMÍVORA EN LOS CALLITRÍQUIDOS: DIGESTIBILIDAD Y VALOR NUTRICIONÁL

Michael L. Power

La alimentación a base de goma es una rara adaptación alimentaria en los mamíferos. Entre los primates sin embargo, es un comportamiento relativamente común. Entre los primates del nuevo mundo, los miembros de Callitrichinae son los más *gomívoros*. Las marmosetas (los géneros *Callithrix* y *Cebuella*) aparentan ser más gomívoros que los otros miembros de la subfamilia, sin embargo, dentro del género *Callithrix* existe una amplia variación en el grado de alimentación con goma. Las marmosetas tienen una serie de adaptaciones morfológicas que facilitan la alimentación con goma. Estas caracteristicas incluyen modificaciones en la dentición inferior-anterior (Coimbra-Filho y Mittermeier, 1977; Rosember, 1978) y un alargamiento del caecum y colon en relación al intestino delgado (Coimbra-Filho et al., 1980; Power, 1991; Ferrari y Martins, 1992; Ferrari et al., 1993; Power y Oftedal, 1996).

A pesar de la falta de éstas u otras adaptaciones claramente relacionadas con la alimentación con goma en los géneros *Saguinus* y *Leontopithecus*, este tipo de alimentación ha sido sugerido como un factor importante en la radiación de los callitríquidos (Garber, 1980; Sussman y Kinzey, 1984). Para explorar este punto, llevé a cabo una investigación sobre digestión y respuestas digestivas a una dieta de goma en cinco especies (*Cebuella pygmaea, Callithrix jacchus, Saguinus oedipus, S. fuscicollis, y Leontopithecus rosalia*). A los animales se les proveyó con una dieta idéntica y homogénea de un sólo ítem, con y sin el añadido de polvo de goma arábiga. El tiempo de tránsito y los coeficientes de digestibilidad aparente de materia seca y energía fueron evaluados para cada animal en cada dieta.

Para todas las especies con excepción de las marmosetas pigmeas, estos parámetros digestivos mostraon una allometría positiva bajo la dieta sin goma. Las marmosetas pigmeas se diferenciaron al tener los tiempos de tránsito absolutos más largos y los coeficientes de digestibilidad aparente iguales a los de los tamarinos, que eran cuatro veces más grandes. En contraste, las marmosetas comunes no se diferenciaron de los tamarinos y tamarinos leones. Esto implica que un pasaje rápido de digesta es probablemente una condición ancestral y que las marmosetas pigmeas son las más derivadas en términos de parámetros digestivos entre estas especies.

La adición de goma a la dieta no tuvo efecto en la digestión en ninguna especie de marmoseta, pero decreció significativamente la digestión de la dieta de las otras tres especies. Los resultados de estas pruebas de digestión ofrecen evidencia adicional de que las marmosetas tienen adaptaciones derivadas para la alimentación con goma que otros callitríquidos carecen. Propongo que el callitríquido ancestral era muy probablemente un frugívoro-faunívoro. A pesar de que el tamaño pequeño de los callitríquidos probablemente los predispone a la alimentación con goma, solamente en el linaje de las marmosetas se han presentado algunas adaptaciones significativas para una estrategia de alimentación con goma.

LOCOMOÇÃO USO DO SUBSTRATO PELO MICO-LEÃODOURADO (*Leontopithecus rosalia*) NA NATUREZA

Brian J. Stafford, Alfred L. Rosenberger, Andrew J. Baker, Benjamin B. Beck, James M. Dietz, e Devra G. Kleiman

Nós abordamos neste artigo o comportamento locomotor de três grupos de mico-leão-dourado (*Leontopithecus rosalia*); um em cativeiro, outro em semi-liberdade e o terceiro na natureza. *L. rosalia* tem uma forma de locomoção que é mais quadrúpede do que os outros calitriquíneos em cativeiro, e utiliza uma forma quadrúpede de progressão singular, provavelmente relationada com a mao alongada da espécie. Propomos que essa forma de locomoção em *L. rosalia* resulta da incorporação de técnicas especializades de forrageamento no sistema locomotor, não sendo uma adaptação sensu stricto. Encontramos que a locomoção diferiu significativamente entre os três grupos, e que existe uma relação íntima entre os tipos específicos de substrato e certos comportamentos locomotores. Propomos que os ambientes diferentes resultam em diferenças nos comportamentos locomotores. As diferenças na locomoção foram mais notáveis entre os grupos de cativeiro e em liberdade. Existiram poucas diferenças entres os grupos em liberdade, apesas do fato de ficarem aparentes diferenças substanciais nos substratos utilizados. Propomos que as limitações morfológicas explicam esta constáncia relativa no comportamento locomotor entre os grupos em liberdade.

IMPLICANCIAS FUNCIONALES Y FILOGENÉTICAS DE LA MORFOLOGÍA DEL TOBILLO EN LOS MONOS DE GOLDEI

Lesa Davis

Debido a su intrigante variedad de características platirrinas callitríquidas y no callitríquidas, el *Callimico goeldii* ha figurado prominentemente en los controversiales asuntos de la filogenia y sistemática de los platirrinos. Estudios de campo de la ecología y comportamiento de los *Callimico* han documentado un número de patrones ecológicos y conductuales específicos que distinguen aún más a este mono. Aunque nuestra comprensión de la ecología y comportamiento de esta especie es aún incompleta, varios estudios de campo han notado la propensión al uso de soportes verticales como plataformas para el salto y suspensión vertical en estos monos, a pesar de no consumir goma.

Pocos estudios de su anatomía esquelética han sido realizados. El presente estudio examina el significado funcional y las afinidades filogenéticas de las características esqueléticas de la región del tobillo en el *Callimico* en comparación con otras 14 especies de platirrinos.

Para una apreciación más precisa de la compleja relación entre morfología y comportamiento, el análisis fue llevado a cabo al nivel de especies. La muestra comparativa comprende *Cebuella pygmea, Callithrix jaccus, C. penicillata, Saguinus fuscicollis, S. geoffroyi, S. labiatus, S. midas, S. mystax, S. nigricollis, Leonthopitecus rosalia, Saimiri sciureus, Callicebus torquatus, Pithecia pithecia,* y *Ateles geoffroyi.* Un total de 53 características cuantitativas y 10 cualitativas de la tibia distal, astragalus y elementos del pie fueron examinados. Análisis multivariados y bivariados fueron llevados a cabo.

En primer lugar, análisis de los componentes principales en las 53 características cuantitativas se realizaron para explorar en grandes rasgos la posición del *Callimico* en los platirrinos. Para facilitar las comparaciones entre especies con tamaños corporales distintos todas las medidas cuantitativas fueron corregidas según el tamaño. Pruebas de ANOVA y Turkey-Kramer fueron usadas para identificar diferencias morfológicas significativas entre las especies y para identificar cúales especies eran distintas al *Callimico.* La discusión se centró en aquellas características en las que el *Callimico* era diferente al resto de los callitrícinos. La información acerca del comportamiento locomotor y de postura del *Callimico* fue tomada de la literatura.

El análisis de la mayoría de las características indica que el *Callimico* está cercanamente alineado con los callitricinos. Estos resultados concuerdan con los estudios previos que documentan un espacio distinto para las características postcraniales que está compartido por los callitrícinos y el *Callimico.* Sin embargo, a pesar de sus afinidades con los callitrícinos, y su comportamiento posicional y tamaño corporal parecido al de los callitrícinos, el *Callimico* comparte varias características con los platirrinos no-callitrícinos. El *Callimico* tiene una distintiva area malleolar tibial grande y una correspondiente area medial astragalar en facetas. Estas dos características, unidas con una faceta tibial malleolar articular distintivamente bulbosa, forman un complejo morfológico que es exclusivamente compartido con los *Pithecia pithecia.* En especies relativamente grandes, este complejo anatómico facilita una mayor estabilidad del tobillo en una posición completamente dorsofleccionada, "close-packed" del tobillo, y es consistente con la propensión de esta especie para la suspensión vertical. Estas características compartidas pueden ayudar a elucidar las adaptaciones posicionales de un platirrino fósil, *Cebupithecia sarmientoi.*

Se ha encontrado que el *Callimico* tiene también la mayor longitud media troclear-astragalar de todos los callitrícinos, y se sitúa en el rango para los no-callitrícinos en la muestra. Análisis preliminares sugieren que la troclea alargada en los *Callimico* representa una retención primitiva. Análisis de dos características astragalares, el índice de "trochlear wedging" y la presencia de una interrupción distal de la tibia con respecto a la troclea, falla en confirmar sugerencias previas con respecto a una relación forma/función exclusiva. El índice de "trochlear wedging", que ha sido predecido en estudios previos como bajo para los saltadores, fue significativamente alto en *Callimico* y *P. pithecia,* ambos frecuentes saltadores. Ademas, el índice de "trochlear wedging" en *Callimico* y *P. pithecia* no fue significativamente diferente al de *Ateles geoffroyi,* un trepador frecuente. El significado, si acaso hay alguno del "trochlear wedging" permanece poco claro hasta ahora. La presencia de una interrupción distal de la tibia ("tibial stop") con respecto a la troclea astragalar ha sido previammente sugerida como una correlación con la postura de suspensión vertical. Sin embargo, en el presente estudio, esta característica es demasiado variable entre la mayoría de especies de los platirrinos como para representar un correlato anatómico consistente con la suspensión vertical.

Los resultados de este estudio enfatizan que las contribuciones relativas y las correlaciones de tamaño corporal, comportamiento posicional y herencia filogenética deben tomarse en cuenta para el análisis funcional de los elementos esqueléticos. Las correlaciones forma/función que son adecuadas para un grupo no pueden ser asumidas como tales para otro grupo taxonómico. En corcondancia con ésto, los análisis a nivel de especies proveen una figura más adecuada de la compleja relación entre morfología y comportamiento. Aún cuando el *Callimico* comparte varias características morfológicas con los platirrinos no-callitrícidos, en total, se encontró una consistencia morfológica remarcable entre el *Callimico* y los callitrícidos. Estos resultados añaden mayor soporte a los sistemas de clasificación de los platirrinos que ubican al *Callimico* en la familia Callitrichidae.

ECOLOGIA DOS SAGÜIS DO SUDESTE DO BRASIL (Callithrix aurita e Callithrix flaviceps): QUÃO DISTINTOS E QUÃO SIMILARES?

Stephen F. Ferrari, H. Kátia M. Corrêa e Paulo E.G. Coutinho

Endêmicos das serras do sudeste brasileiro, os saguis "meridionais", *Callithrix aurita* e *Callithrix flaviceps*, enfrentam o que seriam provavelmente as condições climáticas mais extremas encontradas por qualquer calitriquíneo na natureza. Os resultados de dois estudos de longo prazo, apresentados aqui, indicam que as características gerais da ecologia das duas espécies são bastante parecidas, e que a maioria dos contrastes se devem, aparentemente, a diferenças na composição do habitat, ás condições e o tamanho do grupo nos dois sítios, ao invez de diferenças inter-específicas de adaptações para o forrageio. Apesar disto, existe um contraste fundamental nos padrões de acasalamento nos dois grupos estudados apenas uma única fêmea foi ativa reprodutivamente em dado momento no grupo de *C. flaviceps*, mas duas fêmeas de *C. aurita* reproduziram ao longo de todo o período do estudo. Como o grupo de *C. flaviceps* continha até seis fêmeas adultas, as razões para tal diferença são obscuras. De um modo geral, este e outros padrões observados nos dois estudos fazem mais para reenfatizar a flexibilidade ecológica dos saguis do que esclarecer os caracteres ou adaptações específicos a uma dada espécie.

ATIVIDADE E PADRÕES DE USO DA ÁREA DOMICILIAR DO SAGÜI-COMUM (*Callithrix jacchus*): IMPLICAÇÕES PARA ESTRATÉGIAS REPRODUTIVAS

Leslie Digby e Claudio E. Barreto

Apresentamos aqui dados sobre os padrões de actividade e uso de espaço em três groupos selvagens do sagüi comum (*Callithrix jacchus*). Durante épocas quando não haviam infantes no grupo, os adultos passaram aproximadamente 43% do seu tempo forrageando ou se alimentando, 30% em decanso, 14% se locomovendo, e 14% em atividades sociais. No entanto, durante os primeiros dois meses após um nascimento, indivíduos car-

regando filhotes aumentaram significantivamente seu tempo em descanso (59%) e diminuíram o tempo forrangeando e comendo (12%). Os padrões de uso de espaço nesses
grupos se mostrou comparável aos relatados em outros estudos dessa espécie (área de uso:
3,9 a 5,2 ha; caminhos percorridos durante o dia: 912 a 1.243 ha). Tanto a área utilizada
quanto as estimativas de distâncias percorridas a cada dia diminuíram em adultos carre-
gando filhotes quando comparadas as áreas e distâncias percorridas pelo grupo sem filhotes jovens. Surgerimos que as mudanças nas atividades e no uso de espaço indicam que
indivíduos buscam minimizar seus gastos energéticos quando carrengando filhotes.
Sugerimos que também que a capacidade de compensar e minimizar os custos de cuidados
parentais se relaciona, provavelmente, com tais fatores como área de uso, tamanho do
grupo, e o desenvovimento do filhotes. Cada um desses fatores precisa ser levado em
conta na tentativa to decifrar as causas da variação, tanto intra- quanto interespecífica, na
organização social na callitrichini.

PATRONES DE CUIDADO PATERNAL Y VIGILANCIA EN TITÍ BLANCO EN LIBERTAD

Anne Savage, Charles T. Snowdon, y Humberto Giraldo

Los patrones de reproducción de los titíes blancos en cautiverio y en libertad son
diferentes. Los titíes en libertad tienen crías una vez al año, mientras que los titíes en cautiverio se reproducen más frecuentemente.

Los infantes de los titíes en libertad exhiben patrones de desarrollo similares a los de
los infantes nacidos en colonias en cautiverio, y tienen una alta probabilidad de sobrevivir
hasta el primer año (82%).

Un factor importante que parece influenciar la supervivencia de los infantes de titíes
en libertad es el tamaño del grupo. Los grupos grades tuvieron más éxito al criar infantes
que los grupos más pequeños. Los grupos grandes no solamente tenían más individuos
para ayudar a cargar y proteger a los infantes, sino que tenían mayores probabilidades de
incrementar su eficiencia al recolectar alimentos y detectar a los depredadores, dis-
tribuyendo así la carga energética del cuidado de los infantes, la eficiencia de la
recolección del grupo y la detección de depredadores entre varios individuos del grupo.

EVALUANDO PARADIGMAS DE APRENDIZAJE EN EL CAMPO: EVIDENCIA PARA EL USO DE INFORMACIÓN ESPACIAL Y PERCEPTUAL Y RECOLECCIÓN BASADA EN REGLAS EN PICHICOS BIGOTUDOS (*Saguinus mystax*) EN LIBERTAD

Paul A. Garber y Francine L. Dolins

Basada en una serie de experimentos naturales, hay evidencia de que los Pichicos
bigotudos peruanos salvajes *(Saguinus mystax)* aprendieron a distinguir posiciones en el
espacio de 16 plataformas individuales de alimentación que contenían carnada, y a usar
información temporal de visitas anteriores para predecir la posición presente de las recompensas (comida). Estos monos fueron sensibles aún a pequeñas alteraciones en información ambiental y aprendieron rápidamente las reglas de recolección de alimentos,

aplicándolas para resolver problemas de recolección nuevos. Los pichicos exhibieron un nivel alto de flexibilidad para el aprendizaje y adoptaron un conjunto de reglas de recolección tales como ganar-retornar, perder-cambiar, perder-retornar, ganar-cambiar en respuesta a diferencias en la disponibilidad espacial y temporal de recursos alimenticios.

En el primer conjunto de experimentos, la posición de ocho plataformas conteniendo plátanos verdaderos y la de ocho plataformas conteniendo plátanos falsos fue constante en el tiempo. Los pichicos aprendieron rápidamente las posiciones en el espacio que les daban recompensas y las que no se las daban, y hacia el día número cuatro más del 80% de las plataformas visitadas contenían plátanos reales. La introducción de dos señales locales (banderas rojas) en cada plataforma que contenía plátanos reales (1B) resultó en un incremento en la precisión de la búsqueda en los pichicos al 100%.

Dadas algunas limitaciones del diseño de esta condición experimental, el grado en el que solamente las señales rojas fueron responsables por este incremento en el desempeño de los pichicos permanece poco claro. En la condición 1C, la posición de los plátanos falsos y verdaderos fue rotada 90 grados en sentido horario. A pesar de la disponibilidad de señales rojas y de señales olfatorias, el primer día de rotación los monos retornaron a las plataformas que previamente habían contenido los plátanos como premio. Estos resultados sugieren fuertemente que los pichicos usan información espacial más que señales rojas y olfatorias como factor primario para localizar sitios de alimentación.

En una segunda serie de experimentos, un esquema de renovación de recursos interdiario fue introducido en el diseño de la investigación de tal forma que: a) las plataformas que contenían plátanos verdaderos en la mañana contenían plátanos de plástico en la tarde, y b) las plataformas que contenían plátanos de plástico en la mañana contenían plátanos reales en la tarde. Los pichicos exhibieron un alto grado de flexibilidad conductual y aprendieron rápidamente a recordar y asociar información temporal con información respecto a la localización de actividades de recolección exitosas y no exitosas previamente en el día. Estos primates adoptaron una estrategia de recolección ganar-cambiar/perder-retornar para predecir la localización de los sitios de alimentación que contenían carnada en la tarde.

Otros días sin embargo, el horario de renovación de recursos fué realizado al azar y los pichicos aplicaron un conjunto diferente de reglas de recolección. Los resultados de estos experimentos sugieren que los pichicos no usaron señales olfatorias para navegar hacia los lugares de alimentación.

En conclusión, estudios controlados en un ambiente de campo proveyeron un método excelente para examinar las formas en las que los primates no-humanos usan información ambiental cuando se dirigen y seleccionan lugares de alimentación. Los experimentos de campo proporcionan la oportunidad de explorar hipótesis acerca de las diferencias de aprendizaje espacial para cada especie, el desarrollo de reglas de recolección y la jerarquía de reglas perceptuales usadas por los primates en libertad cuando toman decisiones para la recolección de alimentos.

DEFINICIÓN Y DIFERENCIACIÓN DE ESPECIES DE ACUERDO CON EL ESQUELETO POST-CRANIAL DE LOS CAPUCHINOS (*Cebus spp*)

Susan Ford y David M. Hobbs

Un aspecto crítico de cualquier estudio de procesos evolutivos es la comprensión de la naturaleza y la distinción de las especies. Aún cuando las especies son generalmente reconocidas como entidades reales, la identificación de los miembros de una especie sigue siendo problemática, particularmente para los miembros de una paleoespecie. La mayoría de los investigadores busca analogías en las diferencias que definen una especie moderna, pero poco de este trabajo se ha enfocado en el esqueleto postcranial. Estudios recientes han demostrado que la mayoría de diferencias entre especies se manifiestan en los tejidos blandos, pero varios estudios han documentado algunas diferencias en tejidos duros entre especies cercanamente relacionadas. Este estudio examina las diferencias en numerosas regiones del esqueleto apendicular de especies del género *Cebus*, que son generalmente caracterizadas como cautos cuadrúpedos arborícolas. Mucho aspectos del esqueleto post-cranial no demuestran diferencias significativas entre las especies de *Cebus*. Sin embargo, varias diferencias significativas fueron halladas. *C. apella* tiene extremidades anteriores cortas, segmentos distales cortos en las extremidades y superficies articulares más pequeñas. Además difiere marcadamente de las otras especies en dieta y locomoción, utilizando más objetos duros incluyendo semillas de palma y siendo más "deliberadamente" cuadrúpedo con técnicas de recolección de alimentos más poderosas. En contraste, *C. capucinus* tiene extremidades largas y grandes superficies de articulación en el codo y la rodilla. *C. albifrons* se parece más a *C. capucinus* pero es más pequeño, mientras que *C. olivaceous*, que carece de copete, muestra algunas similitudes con *C. apella*, que lo tiene.

La capacidad de estas diferencias para diagnosticar o apoyar la identidad de las especies difiere según los distintos conceptos de especie, y estos conceptos distintos llevan a interpretaciones alternativas de las implicancias de las diferencias postcraniales en la especiación.

COORDINACIÓN VOCAL DEL MOVIMIENTO DE TROPAS EN MONOS ARDILLA (*Saimiri sciureus* y *Saimiri oerstedi*) y EN CAPUCHINOS DE CARA BLANCA (*Cebus capucinus*)

Sue Boinski

Aunque muchos grupos sociales parecen recordar cómo está distribuída la comida porque se mueven entre areas de recolección de alimentos de forma bastante directa y eficiente, cómo es que se logra el movimiento coordinado en los grupos es aún oscuro.

Tres estudios recientes sobre el comportamiento vocal de monos en los Neotrópicos documentan que al menos dos especies (capuchinos de cara blanca y los monos ardilla de Costa Rica), pero no todas (monos ardilla peruanos), se basan en llamadas especializadas para iniciar y dirigir el movimiento de las tropas. En algunos casos, las vocalizaciones que coordinan el movimiento de tropas pueden ser consideradas bajo el rubro general de "llamadas de contacto", ya que finalmente actúan manteniendo la cohesión del grupo. Sin embargo, estas vocalizaciones se distinguen del concepto general de llamada de contacto, porque el transmisor no solo transmite información sino que manipula a los otros para que sigan sus decisiones acerca de los movimientos.

El uso de vocalizaciones de coordinación para los viajes en capuchinos de cara blanca y monos ardilla costarricenses está probablemente ligada de manera cercana a un conocimiento de la distribución espacial, tácticas de recolección de alimentos y posición social de cada miembro del grupo. Como resultado, este comportamiento vocal especializado provee puntos ventajosos sobre los cuales considerar las habilidades cognoscitivas relativas y las complejidades sociales de los individuos y las especies.

Las tropas de monos ardilla peruanos, aunque cruzan fácilmente áreas mucho mayores que las otras dos especies, no dieron evidencia de coordinación vocal para el movimiento de tropas en sus repertorios vocales. En vez de esto, los monos ardilla peruanos forman tropas de especies mixtas con especies simpátricas de capuchinos, para beneficiarse con una recolección de fruta más eficiente. Las tropas de capuchinos, a su vez, parecen determinar la ruta a seguir para el grupo de especies mixto.

ECOLOGÍA COMPORTAMENTÁL DE LOS CAPUCHINOS (*Cebus olivaceus*)

Lynne Miller

Las hembras adultas en el grupo de estudio pequeño experimentaron desventajas en la recolección en relación con aquellas en el grupo grande, manifestadas en fluctuaciones estacionales en la ingestión de alimentos.

Las tasas de ingestión para las hembras del grupo pequeño estuvieron, a su vez, relacionadas positivamente con la disponibilidad de recursos. Durante la estación húmeda, cuando el alimento era abundante, el volúmen diario de comida para las hembras del grupo pequeño era alto, significativamente más alto que el de las hembras del grupo grande. Durante la estación seca, en cambio, cuando la comida era escasa, el consumo fue significativamente menor, tanto en comparación con los niveles de la estación húmeda como en relación con el consumo de las hembras del grupo grande.

La restricción en el consumo de alimentos durante los meses normales de gestación puede deprimir la reproducción de las hembras del grupo pequeño y así la membrecía en un grupo pequeño puede ser costosa en términos de adaptación individual. La variación anual en el consumo de alimentos de las hembras en ambos grupos estuvo relacionada con la variación en los patrones de actividad.

Los patrones de las hembras del grupo pequeño, sin embargo, variaron con las estaciones. Cuando la disponibilidad y el consumo de alimentos fueron altos, los individuos dedicaron un tiempo considerable a la recolección y viaje y poco tiempo al descanso, maximizando así el consumo de energía. Cuando la comida era escasa y el consumo bajo, las hembras redujeron significativamente el tiempo dedicado al viaje e incrementaron el tiempo de descanso, conservando así energía.

Comparaciones de patrones de actividad entre las tropas indican que el bajo consumo de alimentos de las hembras del grupo pequeño durante la estación seca no fue el resultado de un esfuerzo de recolección menor, sino de una incapacidad aparente para localizar y/o mantener el acceso a recursos alimentarios, posiblemente debido al desplazamiento periódico que sufrieron por parte de grupos grandes.

La existencia contínua de grupos pequeños a pesar de las aparentes desventajas en la recolección que enfrentan sus miembros; podría ser facilitada por las estrategias conductuales que los individuos emplean para mitigar los costos de adaptación. Al reducir su gasto de energía durante los tiempos de escasez, las hembras de los grupos pequeños podrían reducir el alcanze que las deficiencias nutricionales tienen en su suceso reproductivo. El incremento de su consumo de comida durante los períodos de abundancia de alimentos les permitiría acumular grasa y así mejorar aún más sus probabilidades de manejar sus bajos niveles de ingestión durante los meses secos.

Un patrón así representa la adaptación conductual a los aspectos del ambiente físico del individuo tales como las fluctuaciones en la disponibilidad de recursos y al ambiente social, tales como la incapacidad para competir por recursos alimenticios en encuentros entre grupos.

Los individuos en el grupo de estudio grande experimentaron costos más altos de competencia intragrupal que se manifestaron en las tasas de interacciones agresivas. El miembro promedio del grupo grande participó en peleas significativamente con mayor frecuencia que su contraparte en el grupo más pequeño. Aún más, la disparidad entre los dos grupos de estudio se incrementó durante la estación seca. Mientras las tasas de agresión dentro del grupo pequeño permanecieron virtualmente constantes durante todo el año, las peleas en el grupo grande se incrementaron significativamente cuando la comida era escasa. Ya que las peleas continuas probablemente reduzcan la adaptación individual, la membrecía a un grupo grande es costosa, especialmente durante las épocas de escasez de recursos.

Las tasas relativas de agresión para los grupos estudiados podrían, hipotéticamente, ser explicadas en términos de espaciamiento; esto es, si los miembros del grupo grande están más próximos entre sí, pelearán con mayor frecuencia. Sin embargo, lo contrario es cierto. Cuando la comida se volvió escasa y las tasas de agresión se elevaron en el grupo grande, los individuos en realidad incrementaron su espacio individual. Se esparcieron significativamente más de lo que lo habían hecho durante la estación húmeda y más que los del grupo pequeño. Para aquellos en el grupo pequeño, en contraste, la frecuencia de las peleas y el espacio individual fueron constantes a través de las diferentes estaciones. Si los miembros del grupo grande hubieran mantenido la cercanía en la estación seca, las tasas de agresión podrían haber sido aún más altas que las observadas. El cambio en el espaciamiento en el grupo grande podría haber sido un medio para controlar el crecimiento en las tasas de agresión, minimizando así los costos de pertenencia a un grupo grande.

Los resultados de esta investigación sugieren que no solamente existen costos de pertenencia a grupos grandes y pequeños, pero también que el ambiente físico, especialmente las fluctuaciones en la abundancia de recursos, pueden influir en la magnitud de estos costos. Los miembros de grupos pequeños pueden sufrir una falta de habilidad competitiva en encuentros intergrupales, particularmente en períodos de baja disponibilidad de recursos. Aquellos en grupos grandes pueden sufrir altas tasas de agresión, especialmente cuando los recursos son escasos. Sin embargo, la selección natural debería alentar a los individuos a desarrollar estrategias para reducir los costos de membrecía a un grupo. Los datos aquí presentados apoyan esta hipótesis al elucidar los cambios en patrones de comportamiento correlacionados con los incrementos en los costos. Estos resultados indican que los primates pueden, y de hecho desarrollan, estrategias adaptativas en respuesta a la presión selectiva de sus ambientes físicos y sociales.

MIRA COMO CRECEN: RASTREANDO POBLACIONES DE MONOS CAPUCHINOS DE CARA BLANCA (*Cebus capucinus*) EN UN BOSQUE SECO REGENERADO EN COSTA RICA

Linda M. Fedigan, Lisa M. Rose, y Rodrigo Morera Avila

La expansión de los refugios de hábitats a través de la restauración de terrenos y la subsiguiente regeneración del bosque tropical es un instrumento crítico en la conservación biológica. Describimos el crecimiento poblacional de los capuchinos de cara blanca (*Ce-*

bus capucinus) en un bosque tropical seco bajo regeneración en el Parque Santa Rosa, Costa Rica. Los datos provienen de censos anuales de todo el parque, complementados por observaciones actuales de tres grupos sociales bajo estudio intensivo.

Entre 1984 y 1992, la población aumentó de 393 a 526 individuos. Principalmente el aumento se debió a un crecimiento en el tamaño de los grupos sociales existentes y no a la formación de nuevos grupos. El porcentaje de machos en la población y la proporción de machos adultos:hembras aumentó significativamente entre 1983 y 1992. Esto podría deberse a una inmigración de machos desde afuera del parque y/o una tasa de nacimientos posiblemente favorable a los machos. Una muestra de 44 nacimientos mostró una inclinación hacia los machos de casi 3:1. Los datos de nuestros grupos en estudio demuestran una tasa promedio de nacimientos de 0.47, con 39% de sobrevivencia hasta los cinco años de edad. Las tasa de nacimientos no fueron afectadas por el tamaño del grupo y ni la tasa de nacimientos, el tamaño del grupo o el aumento en el tamaño del grupo fueron afectados por el tipo de hábitat. Sin embargo, las tasas de nacimientos se correlacionaron positivamente con la cantidad de precipitación de las primeras lluvias del año anterior y los nacimientos fueron significativamente sesgados hacia la estación seca.

Concluimos que la necesidad de frutos y agua para beber en la estación seca evita que los capuchinos de cara blanca vivan exclusivamente en el bosque recientemente regenerado y además evita la formación de nuevos grupos como respuesta a la expansión del hábitat. A la vez sugerimos que por lo menos durante las primeras etapas de restauración del bosque los grupos existentes crecen en tamaño y amplían su distribución en un hábitat nuevo disponible, manteniendo el acceso a la fruta esencial y a los recursos acuáticos en áreas del bosque más viejo. La proporción creciente de machos en la población sugiere que el sexo en dispersión puede beneficiarse más por la protección y expansión de hábitats.

HACÍA UNA SOCIOECOLOGÍA EXPERIMENTAL DE LOS PRIMATES: EJÉMPLOS DE MONOS CAPUCHINOS MARRONES ARGENTINOS (*Cebus apella*)

Charles Janson

Propongo que éste es el momento adecuado para un uso más extenso de experimentos de campo con primates en libertad para obtener respuestas más confiables a las eternas preguntas sobre ecología y comportamiento social de primates. Ejemplifico esta proposición con tres resultados de experimentos a gran escala de manipulación de alimentos en monos capuchinos marrones en libertad en la Argentina subtropical, tomando ventaja del hecho que la producción de fruta durante el invierno local es extremadamente baja. Primero usé plataformas de alimentación artificial para variar sistemáticamente la abundancia y distribución espacial de fruta dentro de los sitios de alimentación local en un intento por manipular los niveles de agresión y sesgo alimentario entre los miembros de un solo grupo. Los resultados demuestran que diferentes aspectos de la distribución y abundancia de alimentos afectan el consumo de alimento de diferentes "pandillas" sociales dentro de un grupo. En segundo lugar, la distribución espacial a gran escala de los sitios de alimentación proveyó la oportunidad para observar los movimientos espaciales del grupo cuando el observador sabía por lo menos tanto como los monos acerca de sus recursos alimenticios. La probabilidad de que un sitio sea el siguiente en ser visitado es afectada significativamente por su distancia con respecto al sitio en el que se encuentran los monos en el presente momento, la cantidad de comida y el intervalo transcurrido desde la visita previa a ese sitio. En tercer lugar, un análisis de la fracción de fruta dejada caer desde la plataforma de alimentación por diferentes miembros del grupo muestra efectos significativos con respecto a la "pandilla" social y la cantidad de comida consistentes con la hipótesis que mucha de la aparente chapucería de los primates cuando se alimentan con fruta es en realidad signo de una extremada selectividad. En comparación con las observaciones de primates no manipulados, el control disponible con el uso de experimentos de campo provee de mayor confiabilidad al confirmar o rechazar los efectos de las variables examinadas en los comportamientos de interés. En comparación con los estudios en cautiverio, los experimentos de campo son menos controlados, pero ofrecen la oportunidad de evaluar la importancia de ciertas variables cuando se hallan inmersas en el contexto de la variación natural con otros factores.

EVOLUCIÓN DEL COMPORTAMIENTO POSICIONAL EN LOS SAKI-UAKARIS (*Pithecia, Chiropotes, y Cacajao*)

Suzanne E. Walker

El comportamiento posicional y uso de hábitats de un grupo representativo de cada uno de los tres géneros de la tribu Pitheciini fueron observados en libertad en Venzuela y Brasil. Estudios de largo plazo fueron llevados a cabo con *Pithecia pithecia* y *Chiropotes satanas* en el lago Guri, Venezuela y datos comparativos sobre *Cacajao calvus* se recolectaron durante un período de tiempo más corto en Lago Teiú, Brasil.

Las restricciones filogenéticas asi como características del hábitat afectan el comportamiento posicional de los pithecinos durante todas sus actividades. La localización de la comida en los árboles influencia los comportamientos posicionales y las porciones de los árboles que son utilizadas, y las estrategias especificas utilizadas tales como suspensión bajo las ramas pueden ser relacionadas con un mayor tamaño corporal y especializaciones anatómicas. Los comportamientos de cruce de claros son influenciados por la divergencia en la altura de la vegetación, asi como por características anatómicas. Una gran variación existe entre los pithecinos en términos de comportamiento posicional, en gran parte debido a la influencia del colgamiento vertical y adaptaciones para el salto en el repertorio posicional entero de *Pithecia pithecia*. Las conductas de colgamiento vertical son usadas en la alimentación, movilización y descanso y están asociadas con el frecuente uso que este taxon hace de los soportes fuertemente angulares de las porciones inferiores de los árboles. El prinicpal comportamiento locomotor de *Pithecia* es el salto, en el cual la posición del cuerpo permanece vertical y soportes sólidos y verticales son frecuentemente usados para el inicio del salto y el aterrizaje.

Chiropotes y Cacajao tienden a ser encontrados más arriba en la bóveda, usando más frecuentemente la corona principal y las ramas terminales para la alimentación y movilización.

Sus tamaños corporales, más grandes que los de *Pithecia*, podrian restringir sus habilidades para alcanzar los alimentos y mantener el balance en ramas terminales; y compensan con la suspensión con las extremidades inferiores para incrementar la esfera de alimentación y también al desplazarse hacia soportes de alimentación sólidos. Este problema puede estar exacerbado en *Cacajao* ya que carece de una cola functional para ayudarlo en el balance. En los bosques más discontinuos de *Chiropotes y Cacajao*, se presenta una mayor necesidad de ascender y descender de los árboles, lo primero es realizado a través del trepado, lo ultimo saltando y dejándose caer.

Principalmente cuadrúpedas, ambas especies exhiben también un grado sustancial de salto. Sin embargo, sus saltos difieren de los saltos más especializados de *Pithecia* en su orientación corporal pronogrado, aterrizaje en ramas terminales y un componente típico de orientación hacia abajo.

Se cree que la divergencia adaptativa de *Pithecia pithecia* con respecto a los otros pitheciinos se ha debido a la expansión de la base de recursos de *Pithecia* y el uso de nuevos niveles en el bosque, ya sea por el ímpetu proveído por la competencia con un primate ecológicamente similar (del linaje de *Chiropotes/Cacajao*), o por un período de escasa disponibilidad de alimentos. El uso de los niveles más bajos del bosque por parte de *Pithecia* podría haber seleccionado el salto como una forma de locomoción rápida y eficiente, dando como resultado el colgamiento vertical y las especializaciones en el salto. Este comportamiento, que se considera ha evolucionado como una adaptación locomotriz, permite ahora a *Pithecia* el explotar los alimentos encontrados en las ramas terminales de árboles pequeños.

ADAPTACIÓN DE LOS PRIMATES DE LOS NEOTROPICOS A LA NOCTURNALIDAD: ALIMENTACIÓN EN LA NOCHE (*Aotus nigriceps* y *Aotus azurae*)

Patricia C. Wright

En el bosque tropical del Perú, *Aotus* era predominantemente frugívoro. Datos provenientes de mi estudio de 15 meses en el Parque Nacional del Manu comparando la ecología conductual del mono nocturno *Aoutus* y un mono diurno del mismo tamaño que habita el mismo territorio, *Callicebus*, sugirió que las diferencias en las dietas eran más pronunciadas en las estaciones en las que los recursos eran escasos (Wright, 1989). La dieta del *Aotus* fue luego estudiada en el bosque subtropical en el Chaco del Paraguay.

Cuando las restricciones del bosque tropical, incluyendo depredación diurna y competencia e interferencia diurna por alimentos, fueron eliminadas en los bosques del Chaco, los cambios en la dieta revelaron un resultado interesante. El consumo de hojas se incrementó y los horarios de consumo correspondieron con las horas en las que se sabe que los primates diurnos comen hojas (Ganzhorn y Wright, 1955). Las razones propuestas para la ingestión de hojas en las últimas horas de la tarde en los primates diurnos (Milton, 1979; Glander, 1982; Waterman, 1984; Chapman y Chapman, 1991) no fueron confirmadas por los datos obtenidos sobre la actividad del *Aotus* en el Chaco.

Aotus tiene una dieta generalizada que le permite flexibilidad para escoger en el nicho de los monos nocturnos. El nicho para el "simio nocturno" parece haberse conservado por 12 a 20 millones de años en los Neotrópicos, ya que *Aotus dindensis*, un fósil de 12 millones de años en Colombia parece ser postcranialmente, en dentadura y en tamaño, similar al *Aotus* moderno (Sarich y Cronin, 1980; Setoguchi y Rosenberger, 1987).

Aunque nuestro conocimiento está limitado a unos pocos estudios de poblaciones salvajes, la amplitud de las diferencias en el comportamiento entre los actuales taxa de Panamá a la Argentina hasta ahora parecen ser bastante estrechas (Moynihan, 1964; Wright, 1989). Con seguridad sabemos que hay poca variación en el tamaño del cuerpo o medidas morfológicas (Thorington y Vorek, 1976; Hershkovitz, 1983).

La restricción de la radiación adaptativa de los primates nocturnos en los Neotrópicos a un solo género puede ser el resultado de razones históricas y biogeográficas. Zarigüeyas y murciélagos nocturnos ocuparon América del Sur antes de la llegada de los primates y pueden añadir un precedente biogeográfico que haya restringido la radiación de los primates (Croizat, Nelson y Rosen, 1974; Charles-Dominique, 1983; Hand, 1984).

ECOLOGIA ALIMENTAR DO SAUÁ (*Callicebus personatus*)

Klaus-Heinrich Müller

Observações sobre comportamento alimentar em sauás (*Callicebus personatus melanochir*) foram realizadas na floresta Atlântica do leste do Brasil durante 11 meses, utilizando-se equipamento de rádio-telemetria. Os sauás percorreram uma média de 1.007 m por dia numa área de uso de aproximadamente 24 ha. A dieta de *C. personatus* foi composta principalmente de frutos. Foi observado comendo frutos em aproximadamente 77% do registros (*scans*), enquanto folhas compuseram aproximadamente 17%. A ingestão de polpa de fruto e sementes juntos representaram cerca de 64% da dieta. Folhas jovens foram ingeridas em cerca de 8% das amostras, enquanto folha maduras foram comidas muito raramente, < 1% das amostras. Os principais frutos se originaram de três famílias: Sapotaceae, Moraceae e Myrtaceae, embora um total de 91 espécies de plantas (árvores) de 33 famílias fizerram parte da dieta. Houve pouca variação sazonal na dieta, com frutos compondo mais do que 70% das amostras durante a maioria dos meses. *C. personatus* mostrou uma tendência de utilizar estratégias diferentes na exploração de fontes de frutos e folhas: folhas foram colhidas numa maneira mais oportunista do que frutos. Na latitude de 15° S, esperava-se uma influência de sazonalidade no comportamento alimentar de *C. personatus*. Contudo, *C. brunneus* mostrou mais variação sazonal na ingestão de folhas do que *C. personatus*. Em contraste com o que tem sido registrado em sauás da Amazônia (*C. torquatus* ou *C. brunneus*), *C. personatus* não foi observado comendo insetos.

VARIACIÓN ESTACIONAL EN LAS DIETAS DE SAKIS DE CARA BLANCA Y CHIROPOTES (*Pithecia pithecia* y *Chiropotes satanas*) EN EL LAGO GURI, VENEZUELA

Marilyn A. Norconk

La variación estacional en la disponibilidad de fruta y diversidad en la dieta fueron comparadas entre *Chiropotes satanas* y *Pithecia pithecia*, estudiados en un período de 17 meses en 1991 y 1992 en el Lago Guri, al Este de Venezuela. El grupo estudiado de *Pithecia* habitaba un isla (bosque tropical seco) de 15 ha. en la parte Norte del lago y el grupo estudiado de *Chiropotes* ocupaba una isla de 365 ha. (bosque húmedo transicional) ubicada a 40 kms. al Sur de la anterior, en el area Sur del lago. Recolectamos datos referentes a la alimentación por cinco días cada mes para cada grupo de sakis utilizando una forma modificada de muestreo de animal focal que medía la duración de las sesiones de alimentación en cada especie de planta. La abundancia de frutas y distribución estacional fue medida utilizando trampas de frutas y fenología.

Esta región del Este de Venezuela está caracterizada porque la estación húmeda y la estación seca tienen casi la misma duración. Cada estación fue subdividida en temprana y tardía, húmeda y seca. No se encontraron diferencias en el patrón mensual de lluvias entre las islas, pero la isla en el Sur era significativamente más húmeda que la isla en el Norte.

Ambas especies de primates consumieron semillas durante todo el año y la superposición mensual en las especies y partes de plantas ingeridas fue alta (promedio cerca a 45%). La ingestion de semillas fue mayor que el 60% de la porción vegetal de la dieta en la estación seca y la primera mitad de la estación húmeda para los *Pithecia* y en la estación seca para los *Chiropotes*. Ambos sakis empezaron a ingerir mesocarpo cuando éste estuvo disponible (y mientras las semillas maduraban) en la estación húmeda. Alimentos no-frutales fueron recursos secundarios en todas las estaciones, con la excepción de la ingestión de orugas en la estación húmeda temprana por parte de los *Chiropotes*.

Se encontraron tres diferencias entre las especies de sakis: a) la tendencia de los *Chiropotes* a usar la misma especie de planta para semillas tiernas y maduras y mesocarpo; los *Pithecia* ingerían semillas, mesocarpo y frutas enteras de diferentes especies de plantas; b) los *Pithecia* no alteraron la diversidad de las especies de plantas o la amplitud de sus dietas en una base estacional, pero los *Chiropotes* incrementaron la amplitud de sus dietas durante la estación seca y la redujeron durante la estación húmeda; c) los *Chiropotes* ingirieron ítems altos en carbohidratos solubles en agua, mientras que los valores en lípidos fueron altos para los alimentos ingeridos por los *Pithecia*.

Los sakis probablemente rastreen diferencias fenomenológicas en la disponibilidad de recursos, más que la cantidad de lluvia en sí. La variación en la cantidad de fruta ingerida en el mismo mes en años consecutivos fluctuó entre un bajo 9% (alta predictabilidad) y más de 66%. Sin embargo, la precipitación de lluvias probablemente sí provee alguna predictabilidad. Por ejemplo, cuando un retraso de un mes fue visto en algunas especies de plantas, éste muy probablemente estuvo relacionado con una pequeña variación en la sincronización de las lluvias.

Los Pitheciinos y Colobinos son las únicas dos familias de antropoides que mantienen niveles altos de depredación de semillas. El consumo de semillas puede estar relacionado con especializaciones anatómicas en ambas subfamilias: adaptaciones dentales en los Pitheciinos y adaptaciones del intestino en el aparato digestivo de los Colobinos.

MICRODESGASTE DENTAL Y DIETA EN UNA POBLACIÓN SALVAJE DE MONOS CONGOS (*Alouatta palliata*)

Mark Teaford y Kenneth Glander

El análisis de los patrones microscópicos de desgaste en los dientes (o análisis de microdesgaste dental) han demostrado recientemente ser muy prometedores en los estudios de dieta y uso de dientes en mamíferos modernos y prehistóricos. Sin embargo, en muchos casos han creado más preguntas de las que han respondido, enfatizando así la necesidad de mejores datos conductuales y dentales provenientes de animales actualmente existentes. El propósito de este estudio es el empezar a llenar ese vacio.

Alouatta palliata de un sitio conocido como Hacienda La Pacifica en Costa Rica ha sido el objeto de observación conductal por parte de KG por más de 25 años. Desde 1992 hemos combinado observaciones conductuales con análisis de desgaste y morfologia dentales. Los animales han sido capturados periódicamente y luego liberados para tomar impresiones dentales y para poder hacer observaciones de animales conocidos y marcados. Su usó la técnica de microscopía electrónica con barrido de alta resolución en moldes de epoxia para calcular las tasas de desgaste y para caracterizar patrones de microdesgaste.

Los resultados preliminares indican que los aulladores pasan mayor tiempo alimentándose en la estación seca, cuando consumen más hojas tiernas, flores y fruta verde que en la estación lluviosa. Además, los aulladores de áreas ribereñas pasan menos tiempo alimentándose que los aulladores de áreas no ribereñas. Estas diferencias conductuales van emparejadas con diferencias significativas en desgaste dental, tanto así que los animales atrapados durante la estación seca presentan mayor microdesgaste y desgaste dental más acelerado que los animales atrapados en la estación lluviosa. Similarmente, animales atrapados en áreas ribereñas presentan menor microdesgaste que aquellos atrapodos en áreas no ribereñas.

Basados en lo anterior, es tentador decir que el tiempo de alimentación correlacionaba estrictamente con la cantidad y tasas de microdesgaste dental. Sin embargo, ésto es probablemente una sobresimplificación. Eventos individuales en la vida de los animales, tales como la preñez, también pueden afectar significativamente las tasas de desgaste dental; y análisis preliminares de abrasivos en la dieta (ej: fitolitos en las plantas, polvo en la bóveda del bosque, etc.) han dado resultados conflictivos. Los trabajos adicionales deben distinguir entre estos factores causales y abrir nuevas posibilidades para las interpretaciones paleobiológicas.

DIFERENCIAS ESTACIONALES EN PREFERENCIAS DE COMIDAS Y DE LUGARES DE ALIMENTACIÓN EN MARIMONAS (*Ateles belzebuth*)

Hernán Castellanos y Paul Chanin

Un grupo de marimonas (*Ateles belzebuth*) fue estudiado en condiciones naturales, desde Septiembre de 1990 hasta Julio de 1992, en un bosque húmedo en el bajo medio río Tawadu al sur de Venezuela. El objetivo principal del presente estudio fue evidenciar diferencias estacionales en la escogencia del alimento por un grupo de marimonas y determinar si o no ellos se alimentaron selectiva y estacionalmente de algún alimento particular, según su calidad. La dieta de ellos en Tawadu consistió de frutas maduras, comprendiendo entre 69% y 90% de la dieta. Otros tipos de alimentos regularmente cosechados por ellos fueron: hojas tiernas de árboles y de plantas trepadoras, flores o capullos. Entre éstos y los períodos estacionales fue encontrado que hubo diferencias significativas en el tiempo de alimentación. Los marimonas en el bosque de Tawadu usaron un total de 83 especies frutales arbóreas, encontrados en parches de tres tipos: árboles, bejucos y palmas. Un total de 44 especies frutales fueron analizadas. Las frutas fueron separadas en dos categorias, aquellas que presentaron valores altos en energía y aquellas con un alto contenido de lipidos. Un indice de selectividad fue empleado en determinar si los marimonas se alimentan selectivamente en parches donde frutas poseen alto o bajo valor en energía. Como resultado de este indice, fue encontrado que los marimonas prefirieron parches ricos en energía en aquellos períodos estacionales donde la disponibilidad de tales parches fue más alta.

USO DO ESPACIO E ESTRUTURA DE GRUPOS DE FORRAGEIO DE UM GRUPO DE MACACAO BARRIGUDOS CINZAS (*Lagothrix lagotricha cana*) DA URUCU, BRASIL

Carlos Peres

Um grupo de macacos barrigudos cinzas (*Lagothrix lagotricha cana*), que variou entre 44 e 49 indivíduos, foi estudado durante um período de 17 meses por observadores múltiplos operando simultâneamente numa floresta de terra firme inalterada da Amazônia centro-ocidental, Urucu, Amazonas, Brasil. Neste sítio de estudo, macacos barrigudos detém o maior porte entre 12 espécies simpátricas de primates, vivem em áreas de uso superiores a 1000 ha, e formam bandos numerosos e pouco coesos que atingem a maior biomassa de grupo já registrada para quaisquer platirríneos. Este trabalho descreve os padrões de usodo espaço, a estrutura de grupos de forrageio, e a elasticidade de grupo em *L.l. cana*, e examina como essas características importantes da organização social de *Lagothrix* diverge da de outros atelíneos. Ao longo do ciclo anual, o diâmetro de grupo apresentou-se inversamente correlacionado à disponibilidade de frutos maduros no dossel, o grupo tornando-se o mais coêso durante períodos de maior abundância de frutos. O tamanho de grupos durante atividades de forrageio foi parcialmente determinado pela estrutura de agregados (manchas) de recursos, mas não necessariamente pela disponibilidade de recursos, já que somente uma pequena fração do grupo pôde ser fisicamente contida pela copa individual de árvoles e cipós alimentíceos. A estrutura espacial de grupo altamente elástica de *Lagothrix* diverge claramente daquela apresentada por macacos-aranha (*Ateles*), mas muito menos da de muriquís (*Brachyteles*), já que subgrupos espacialmente discretos (como definido até hoje para a subfamília Atelinae) foram raramente discernidos. Essas diferenças são discuitidas sob a ótica de padrõs observados de qualidade, dispersão, e disponibilidade de recursos.

RELACIÓN ENTRE TAMAÑO DE TROPA Y CRECIMIENTO POBLACIONÁL EN DOS HABITATS DE ARAGUATOS (*Alouatta seniculus*)

Carolyn Crockett

Datos provenientes de censos de larga duración en monos araguatos rojos (*Alouatta seniculus*) en los llanos venezolanos, ayudan a clarificar la importancia relativa de los factores ecológicos, demográficos y sociales que afectan el tamaño de las tropas de esta especie. Tamaños de tropas en bosques de galería (3.9 km2) y en hábitats boscosos (3.45 km2), que diferían en densidad ecológica en aprox. 65 araguatos/km2 durante una década, convergieron en una media similar (aprox. 9) luego de un período de crecimiento poblacional. El monitoreo ecológico sugería que el bosque de galería tiene una capacidad de carga menor para los araguatos que el hábitat boscoso.

La mayor parte de la variación en el tamaño de las tropas entre los hábitats fue atribuída a una fase en el crecimiento poblacional. El hábitat de la galería aparentemente sufrió una disminución extrema de la población unos pocos años antes del inicio del estudio, y el tamaño promedio de tropa era muy pequeño al comenzar el censo. Se formaron nuevas tropas en ambos hábitats, pero considerablemente más en el bosque de galería de baja densidad. Estas tropas nuevas eran más pequeñas que las ya establecidas y redujeron la media total para el bosque de galería en comparación con la población del hábitat boscoso, que no había experimentado una caída rápida de la población y tenía menos tropas recién formadas.

El tamaño de la tropa está positivamente correlacionado con la densidad poblacional cuando los datos para los hábitats y años fueron combinados, pero no cuando los hábitats fueron evaluados separadamente. A pesar de que la densidad continuó aumentado año tras año, el tamaño promedio de tropa decreció luego de varios años de sucesivos incrementos. Las tropas de monos araguatos rojos permanecen pequeñas básicamente porque la membrecía de hembras reproductoras parece estar limitada a ≤ 4 ya que la agresión femenina promueve la emigración de las hembras y previene la inmigración.

Una hipótesis para explicar la limitación del tamaño del grupo de hembras, desarrollada en otra publicación sugiere que la tasa de muertes por infanticio se incrementa con el número de hembras adultas porque más parejas potenciales atraen machos a invadir y tomar las tropas.

ECOLOGIA REPRODUTIVA DE FÊMEAS DE MURIQUI
(*Brachyteles arachnoides*)

Karen Strier

Padrões reprodutivos investigados numa população de muriquis (*Brachyteles arachnoides*) na Estação Biológica de Caratinga, Fazenda Montes Claros, Minas Gerais, Brasil, foram relacionados a ciclos sazonais de chuva e diferenças individuais em parâmetros reprodutivas vitais. Precipitação mensal mostrou uma correlação negativa com as datas dos nascimentos registrados, e uma correlação positiva com as datas estimadas de concepção. A maioria das fêmeas reiniciaram atividade sexual pós-parto durante o início da época de chuva, independentemente da data dos nascimentos, e conceberam novamente durante os últimos meses da época de chuva. Fêmeas multíparas anteciparam um pouco as fêmeas primiparas no reinicio dessa atividade sexual pós-parto, embora intervalos entre nascimentos subsequentes não variaram com as histórias dos parâmetros vitais individuais. Fêmeas multíparas pariram um maior numero de fêmeas do que machos, enquanto a razão sexual para fêmeas primíparas não mostrou desvio. O sexo do filhote não foi, porem, relacionado aos intervalos subsequentes entre nascimentos. A idade da mãe e a duração de associações com outras fêmeas explicou parcialmente a variação anual na temporalidade de eventos reprodutivos. Em especial, reprodução mostrou um tendência de sincronia em fêmeas, tanto multíparas quanto primíparas, com associações a longo prazo entre elas. Monitoramento não-invasivo de esteroides fecais de muriquis selvagens tem demonstrado potencial na obtenção de informação detalhada sobre a biologia da reprodução neste e em outros primatas ameaçados.

INDEX

Activity budget, 176–181, 275, 278–281, 347, 365,
    397, 438–439, 463, 512
Allopatric populations, 221
Aotus dindensis, 376, 378

Biomass, 62–63, 471, 79–483
Birth rate, 294–295, 298
Body weight, 33, 89–92, 102–103, 118, 135, 202, 232,
    244, 336
Body size, 89, 134, 150, 222, 377
Bootstrap analysis, 8–13
Brazilian Shield, 55, 57

Caipora, 428–429
Capture methods, 436–438
Carlocebus, 330
Cebupithecia, 15, 149–150, 330
Chilecebus, 15, 220
Cladistics, 4–7, 7–15, 428, 430
Coalitions, 223
Conservation
    in Brazil, 44–46, 64–65
    in Costa Rica, 290, 299
    in Venezuela, 80–81
Cranial anatomy, 220, 429

Dental measurements, 39
Dental morphology, 331, 439, 443
Dental wear, 438–447
Diameter at breast height (dbh), 59–61, 339, 344,
    394–395, 459–460, 470–471, 477–478
Derived traits, 11, 89
Digestibillity (ADDM), 100–104
Dispersal of males and females, 160–161, 174, 191,
    194, 223–224, 297–298, 301–304, 370, 387,
    520–523, 529
Dolichocebus, 15, 91, 220
Dominance, 161, 314–317, 320–322
Durophagy, 34

Endemism, 23, 41, 69, 157
Energy (ADE), 101–104
Energy budget, 173–174, 179–183

Energy expenditure, 188, 198, 272,
Epsilon-globin gene, 8–16, 220
Evolution
    of Atelinae, 9–11, 41–44, 91, 427–430
    of Callitrichinae, 12–16, 42–44, 87–93, 107–108
    of Cebinae, 12, 42–44, 91, 376
    of Pitheciinae, 12, 41–43, 329–333, 358
Extractive foraging, 122
Extramutation, 10

Faunivory, 225
Fecal steroids, 513–514
Fecundity, 283
Feeding competition, 263, 272, 284–285, 311,
    314–317, 320–321, 373–374, 376
Feeding platforms, 203
Fetal loss, 196
Fitness, 272, 283, 310, 315
Folivory, *see* leaf eating
Food patch, 311
Food preference, 317, 321
Foraging rules, 201–205, 208–214
Fruit eating, 34, 41, 90, 97, 162–164, 170–171, 174, 203,
    206–211, 231, 274–286, 311–324, 330–331, 337,
    353–355, 369–378, 383, 389, 395–397, 403–422,
    439, 456–465, 475–476, 482, 495
Fruit hardness, 406, 417
Fungivory, 163

Group size
    of Atelinae, 452, 480, 489, 494–500
    of Callitrichinae, 192–197
    of Cebinae, 271, 292, 298–300, 311, 374
    of Pitheciinae, 331–332, 369–370, 405–406
Guayana Shield, 69–81
Gum arabic, 101–104, 107–108
Gum feeding, 32, 36–40, 63, 90, 92, 97–108, 137, 142,
    162–167, 176, 181, 203

Habitat preference, 137, 232
Habitat type
    brejos, 22
    caatinga, 26, 40, 58, 173